Encyclopedia of
Natural
Resources

Volume II—Water & Air

Encyclopedias from the Taylor & Francis Group

Encyclopedia of
Natural
Resources

Volume II—Water & Air

Edited by
Yeqiao Wang

CRC Press
Taylor & Francis Group
Boca Raton London New York

CRC Press is an imprint of the
Taylor & Francis Group, an **informa** business

CRC Press
Taylor & Francis Group
6000 Broken Sound Parkway NW, Suite 300
Boca Raton, FL 33487-2742

© 2014 by Taylor & Francis Group, LLC
CRC Press is an imprint of Taylor & Francis Group, an Informa business

No claim to original U.S. Government works

Printed on acid-free paper
Version Date: 20140501

International Standard Book Number-13: 978-1-4398-5258-3 (Hardback)

Visit the Taylor & Francis Web site at
http://www.taylorandfrancis.com

and the CRC Press Web site at
http://www.crcpress.com

Land

Water

Air

Editorial Advisory Board

Contributors

Jazuri Abdullah / *Department of Civil and Environmental Engineering, Engineering Research Center, Colorado State University, Fort Collins, Colorado, U.S.A.*

Gayatri Acharya / *Environmental Economist, World Bank Institute (WBI), Washington, District of Columbia, U.S.A.*

Vicenç Acuña / *Catalan Institute for Water Research, Girona, Spain*

Gordon N. Ajonina / *CWCS Coastal Forests and Mangrove Programme, Cameroon Wildlife Conservation Society, and Institute of Fisheries and Aquatic Sciences, University of Douala, Yabassi, Douala, Cameroon*

Erhan Akça / *Adiyaman University, Adiyaman, Turkey*

Jose A. Amador / *Laboratory of Soil Ecology and Microbiology, University of Rhode Island, Kingston, Rhode Island, U.S.A.*

Kathryn M. Anderson / *Department of Zoology, University of British Columbia, British Columbia, Vancouver, Canada*

Kim A. Anderson / *Environmental and Molecular Toxicology, Oregon State University, Corvallis, Oregon, U.S.A.*

Konstantinos M. Andreadis / *Jet Propulsion Laboratory, California Institute of Technology, Pasadena, California, U.S.A.*

Frank Asche / *Department of Industrial Economics, University of Stavanger, Stavanger, Norway*

Richard Aspinall / *Honorary Research Fellow, James Hutton Institute, Aberdeen, U.K.*

Peter V. August / *Department of Natural Resources Science, University of Rhode Island, Kingston, Rhode Island, U.S.A.*

James L. Baker / *Department of Agricultural and Biosystems Engineering, Iowa State University, Ames, Iowa, U.S.A.*

Andrew H. Baldwin / *Department of Environmental Science and Technology, University of Maryland, College Park, Maryland, U.S.A.*

Robert D. Ballard / *Graduate School of Oceanography, University of Rhode Island, Narragansett, Rhode Island, U.S.A.*

Roger G. Barry / *Cooperative Institute for Research in Environmental Sciences, National Snow and Ice Data Center, Boulder, Colorado, U.S.A.*

Austin Becker / *Emmett Interdisciplinary Program in Environment and Resources (E-IPER), Stanford University, Stanford, California, U.S.A.*

Jürgen Bender / *Federal Research Institute for Rural Areas, Forestry and Fisheries, Thünen Institute of Biodiversity, Braunschweig, Germany*

Jagtar S. Bhatti / *Canadian Forest Service, Northern Forestry Centre, Edmonton, Alberta, Canada*

John Boardman / *Department of Geographical and Environmental Science, University of Cape Town, Cape Town, South Africa and Environmental Change Institute, University of Oxford, Oxford, U.K.*

Derek B. Booth / *Center for Water and Watershed Studies, University of Washington, Seattle, Washington, U.S.A.*

John Borrelli / *Department of Civil Engineering, Texas Tech University, Lubbock, Texas, U.S.A.*

Virginie Bouchard / *School of Natural Resources, The Ohio State University, Columbus, Ohio, U.S.A.*

Thomas Boving / *Department of Geosciences/Department of Civil and Environmental Engineering, University of Rhode Island, Kingston, Rhode Island, U.S.A.*

James Boyd / *Resources for the Future, Washington, District of Columbia, U.S.A.*

John Van Brahana / *Division of Geology, Department of Geosciences, University of Arkansas, Fayetteville, Arkansas, U.S.A.*

Helen Bramley / *The UWA Institute of Agriculture, University of Western Australia, Crawley, Western Australia, Australia*

Michael L. Brennan / *Center for Ocean Exploration, University of Rhode Island, Narragansett, Rhode Island, U.S.A.*

Sylvie M. Brouder / *Department of Agronomy, Purdue University, West Lafayette, Indiana, U.S.A.*

Jaclyn N. Brown / *Wealth from Oceans National Research Flagship, CSIRO Marine and Atmospheric Research, Hobart, Tasmania, Australia*

Bill Buffum / *Department of Natural Resources Science, University of Rhode Island, Kingston, Rhode Island, U.S.A.*

Thomas J. Burbey / *Department of Geosciences, Virginia Tech University, Blacksburg, Virginia, U.S.A.*

Mary T. Burke / *UC Davis Arboreturm, University of California, Davis, California, U.S.A.*

Robert H. Burris / *Department of Biochemistry, College of Agriculture and Life Sciences, University of Wisconsin, Madison, Wisconsin, U.S.A.*

Richard Burroughs / *Department of Marine Affairs, University of Rhode Island, Kingston, Rhode Island, U.S.A.*

Tim P. Burt / *Department of Geography, University of Durham, Durham, U.K.*

Thomas J. Butler / *Cary Institute of Ecosystem Studies, Millbrook, and Cornell University, Ithaca, New York, U.S.A.*

Frederick H. Buttel / *Department of Rural Sociology, University of Wisconsin, Madison, Madison, Wisconsin, U.S.A.*

Carrie J. Byron / *Gulf of Maine Research Institute, Portland, Maine, U.S.A.*

Steven X. Cadrin / *Department of Fisheries Oceanography, University of Massachusetts, Fairhaven, Massachusetts, U.S.A.*

Rebecca L. Caldwell / *Department of Geological Sciences, Indiana University, Bloomington, Indiana, U.S.A.*

Daniel E. Campbell / *Atlantic Ecology Division (AED), National Health and Environmental Effects Research Laboratory (NHEERL), U.S. Environmental Protection Agency (EPA), Narragansett, Rhode Island, U.S.A.*

Lucila Candela / *Department of Geotechnical Engineering and Geo-Sciences, Technical University of Catalonia (UPC), Barcelona, Spain*

Edward Capone / *Northeast River Forecast Center, National Oceanic and Atmospheric Administration (NOAA), Taunton, Massachusetts, U.S.A.*

Jennifer Caselle / *Marine Science Institute, University of California, Santa Barbara, California, U.S.A.*

Don P. Chambers / *College of Marine Science, University of South Florida, St. Petersburg, Florida, U.S.A.*

Thomas N. Chase / *Department of Civil, Environmental and Architectural Engineering and Cooperative Institute for Research in the Environmental Sciences (CIRES), University of Colorado at Boulder, Boulder, Colorado, U.S.A.*

Gargi Chaudhuri / *Department of Geography and Earth Science, University of Wisconsin, La Crosse, Wisconsin, U.S.A.*

Long S. Chiu / *Department of Atmospheric, Oceanic and Earth Sciences, George Mason University, Fairfax, Virginia, U.S.A.*

L. M. Chu / *School of Life Sciences, Chinese University of Hong Kong, Hong Kong, China*

Daniel L. Civco / *Department of Natural Resources and the Environment, University of Connecticut, Storrs, Connecticut, U.S.A.*

John C. Clausen / *Department of Natural Resources and the Environment, University of Connecticut, Storrs, Connecticut, U.S.A.*

Sharon A. Clay / *Plant Science Department, South Dakota State University, Brookings, South Dakota, U.S.A.*

Michael T. Clegg / *Department of Ecology and Evolutionary Biology, University of California, Irvine, Irvine, California, U.S.A.*

Gopalasamy Reuben Clements / *School of Marine and Tropical Biology, James Cook University, Cairns, Queensland, Australia*

David A. Cleveland / *Environmental Studies Program, University of California, Santa Barbara, California, U.S.A.*

Janet Coit / *Rhode Island Department of Environmental Management, Providence, Rhode Island, U.S.A.*

Jill S. M. Coleman / *Department of Geography, Ball State University, Muncie, Indiana, U.S.A.*

Sean D. Connell / *Southern Seas Ecology Laboratories, School of Earth and Environmental Science, University of Adelaide, Adelaide, South Australia, Australia*

William H. Conner / *Baruch Institute of Coastal Ecology and Forest Science, Clemson University, Georgetown, South Carolina, U.S.A.*

Jeffrey D. Corbin / *Department of Biological Sciences, Union College, Schenectady, New York, U.S.A.*

Richard T. Corlett / *Xishuangbanna Tropical Botanical Garden, Chinese Academy of Sciences, Yunnan, China*

Ray Correll / *Commonwealth Scientific and Industrial Research Organisation (CSIRO), Adelaide, South Australia, Australia*

Robert Costanza / *Institute for Sustainable Solutions, Portland State University, Portland, Oregon, U.S.A.*

Roland C. de Gouvenain / *Department of Biology, Rhode Island College, Providence, Rhode Island, U.S.A.*

Marinés de la Peña-Domene / *Department of Biological Sciences, University of Illinois at Chicago, Chicago, Illinois, U.S.A.*

Eddy De Pauw / *International Center for Agricultural Research in the Dry Areas (ICARDA), Aleppo, Syria*

Sherri DeFauw / *New England Plant, Soil and Water Laboratory, Agricultural Research Service, U.S. Department of Agriculture (USDA-ARS), University of Maine, Orono, Maine, U.S.A.*

Ahmed M. Degu / *Department of Civil and Environmental Engineering, Tennessee Technological University, Cookeville, Tennessee, U.S.A.*

Heidi M. Dierssen / *Department of Biology, Norwegian University of Science and Technology, Trondheim, Norway, and Department of Marine Sciences/Geography, University of Connecticut, Groton, Connecticut, U.S.A.*

Michael E. Dietz / *Center for Land Use Education and Research (CLEAR), University of Connecticut, Storrs, Connecticut, U.S.A.*

Peter Dillon / *Commonwealth Scientific and Industrial Research Organisation (CSIRO), Adelaide, South Australia, Australia*

Endre Dobos / *Department of Physical Geography and Environmental Sciences, University of Miskolc, Miskolc-Egyetemváros, Hungary*

Lindsay M. Dreiss / *Department of Natural Resources and the Environment, University of Connecticut, Storrs, Connecticut, U.S.A.*

Douglas A. Edmonds / *Department of Geological Sciences, Indiana University, Bloomington, Indiana, U.S.A.*

David Ehrenfeld / *Department of Ecology, Evolution, and Natural Resources, Rutgers University, New Brunswick, New Jersey, U.S.A.*

Florent Engelmann / *Institute of Research for Development (IRD), Montpellier, France*

Hari Eswaran / *National Resources Conservation Service, U.S. Department of Agriculture (USDA-NRCS), Washington, District of Columbia, U.S.A.*

M. Tomedi Eyango / *Institute of Fisheries and Aquatic Sciences, University of Douala, Yabassi, Douala, Cameroon*

Daniel B. Fagre / *Northern Rocky Mountain Science Center, U.S. Geological Survey (USGS), West Glacier, Montana, U.S.A.*

Souleymane Fall / *College of Agriculture Environment and Nutrition Science, and College of Engineering, Tuskegee University, Tuskegee, Alabama, U.S.A.*

Lynn Fandrich / *Oregon State University, Corvallis, Oregon, U.S.A.*

Thomas E. Fenton / *Soil Morphology and Genesis, Agronomy Department, Iowas State University, Ames, Iowa, U.S.A.*

Charles W. Fetter, Jr. / *C. W. Fetter, Jr. Associates, Oshkosh, Wisconsin, U.S.A.*

Joseph Fiksel / *Center for Resilience, The Ohio State University, Columbus, and Office of Research and Development, U.S. Environmental Protection Agency (EPA), Cincinnati, Ohio, U.S.A.*

Joshua B. Fisher / *Jet Propulsion Laboratory, California Institute of Technology, Pasadena, California, U.S.A.*

Patrick J. Fitzpatrick / *Mississippi State University, Stennis Space Center, Mississippi, U.S.A.*

Graham E. Forrester / *Department of Natural Resources Science, University of Rhode Island, Kingston, Rhode Island, U.S.A.*

Neil W. Foster / *Canadian Forest Service, Ontario, Sault Ste. Marie, Ontario, Canada*

Thomas G. Franti / *Department of Biological Systems Engineering, University of Nebraska, Lincoln, Nebraska, U.S.A.*

Alan J. Franzluebbers / *Agricultural Research Service, U.S. Department of Agriculture (USDA-ARS), Watkinsville, Georgia, U.S.A.*

Lisa Freudenberger / *Center for Development Research (ZEF), University of Bonn, Bonn, Germany*

Congbin Fu / *Institute of Atmospheric Physics, Chinese Academy of Sciences, Beijing, China*

Scott Glenn / *Coastal Ocean Observation Laboratory, Institute of Marine and Coastal Sciences, School of Environmental and Biological Sciences, Rutgers University, New Brunswick, New Jersey, U.S.A.*

Brij Gopal / *Centre for Inland Waters in South Asia, Jaipur, India*

L. R. Gopinath / *M.S. Swaminathan Research Foundation, Chennai, India*

Josef H. Görres / *Department of Plant and Soil Science, University of Vermont, Burlington, Vermont, U.S.A.*

Felicity S. Graham / *Wealth from Oceans National Research Flagship, CSIRO Marine and Atmospheric Research, and Institute for Marine and Antarctic Studies, University of Tasmania, Hobart, Tasmania, Australia*

David Gregg / *Rhode Island Natural History Survey, Kingston, Rhode Island, U.S.A.*

Burak Güneralp / *Department of Geography, Texas A&M University, College Station, Texas, U.S.A.*

Sally D. Hacker / *Department of Integrative Biology, Oregon State University, Corvallis, Oregon, U.S.A.*

C. Michael Hall / *Department of Management, University of Canterbury, Christchurch, New Zealand, Centre for Tourism, University of Eastern Finland, Savonlinna, and Department of Geography, University of Oulu, Oulu, Finland*

Pertti Hari / *Department of Forestry, University of Helsinki, Helsinki, Finland*

Christopher D. G. Harley / *Department of Zoology, University of British Columbia, British Columbia, Vancouver, Canada*

Mark E. Harmon / *Oregon State University, Corvallis, Oregon, U.S.A.*

John L. Havlin / *Department of Soil Science, North Carolina State University, Raleigh, North Carolina, U.S.A.*

Michael J. Hayes / *National Drought Mitigation Center, Lincoln, Nebraska, U.S.A.*

Hong S. He / *School of Natural Resources, University of Missouri, Columbia, Missouri, U.S.A.*

Richard W. Healy / *U.S. Geological Survey (USGS), Lakewood, Colorado, U.S.A.*

Alan D. Hecht / *Office of Research and Development, U.S. Environmental Protection Agency (EPA), Washington, District of Columbia, U.S.A.*

Robert W. Hill / *Biological and Irrigation Engineering Department, Utah State University, Logan, Utah, U.S.A.*

Curtis H. Hinman / *School of Agricultural, Human and Natural Resource Sciences, Washington State University, Tacoma, Washington, U.S.A.*

Kyle D. Hoagland / *School of Natural Resources, University of Nebraska, Lincoln, Nebraska, U.S.A.*

Faisal Hossain / *Department of Civil and Environmental Engineering, Tennessee Technological University, Cookeville, Tennessee, U.S.A.*

Jinrong Hu / *Yuen Yuen Research Centre for Satellite Remote Sensing, Institute of Space and Earth Information Science, Chinese University of Hong Kong, Hong Kong, and Laboratory of Coastal Zone Studies, Shenzhen Research Institute, Shenzhen, China*

T. G. Huntington / *U.S. Geological Survey (USGS), Augusta, Maine, U.S.A.*

Thomas P. Husband / *Department of Natural Resources Science, University of Rhode Island, Kingston, Rhode Island, U.S.A.*

Stephen Hutchinson / *Baruch Institute of Coastal Ecology and Forest Science, Clemson University, Georgetown, South Carolina, U.S.A.*

Kristen C. Hychka / *Atlantic Ecology Division, Office of Research and Development, U.S. Environmental Protection Agency (EPA), Narragansett, Rhode Island, U.S.A.*

C. Rhett Jackson / *Daniel B. Warnell School of Forest Resources, University of Georgia, Athens, Georgia, U.S.A.*

Jennifer P. Jorve / *Department of Zoology, University of British Columbia, British Columbia, Vancouver, Canada*

Pierre Y. Julien / *Department of Civil and Environmental Engineering, Engineering Research Center, Colorado State University, Fort Collins, Colorado, U.S.A.*

Manjit S. Kang / *Department of Plant Pathology, Kansas State University, Manhattan, Kansas, U.S.A.*

Selim Kapur / *University of Çukurova, Adana, Turkey*

Darryl J. Keith / *Atlantic Ecology Division, U.S. Environmental Protection Agency (EPA), Narragansett, Rhode Island, U.S.A.*

Olivia Kellner / *Department of Earth, Atmospheric, and Planetary Sciences, Purdue University, West Lafayette, Indiana, U.S.A.*

D. Q. Kellogg / *Department of Natural Resources Science, University of Rhode Island, Kingston, Rhode Island, U.S.A.*

Jaehoon Kim / *Department of Civil and Environmental Engineering, Engineering Research Center, Colorado State University, Fort Collins, Colorado, U.S.A.*

Jinwoo Kim / *The Ohio State University, Columbus, Ohio, U.S.A.*

Marcos Kogan / *Integrated Plant Protection Center, Oregon State University, Corvallis, Oregon, U.S.A.*

Josh Kohut / *Coastal Ocean Observation Laboratory, Institute of Marine and Coastal Sciences, School of Environmental and Biological Sciences, Rutgers University, New Brunswick, New Jersey, U.S.A.*

Rai Kookana / *Commonwealth Scientific and Industrial Research Organisation (CSIRO), Adelaide, South Australia, Australia*

Rebecca L. Kordas / *Department of Zoology, University of British Columbia, British Columbia, Vancouver, Canada*

Ken W. Krauss / *National Wetlands Research Center, U.S. Geological Survey (USGS), Lafayette, Louisiana, U.S.A.*

Betty J. Kreakie / *Atlantic Ecology Division, Office of Research and Development, U.S. Environmental Protection Agency (EPA), Narragansett, Rhode Island, U.S.A.*

Jürgen Kreuzwieser / *Institute of Forest Botany and Tree Physiology, University of Freiburg, Freiburg, Germany*

Anne Kuhn / *Atlantic Ecology Division, U.S. Environmental Protection Agency (EPA), Narragansett, Rhode Island, U.S.A.*

William P. Kustas / *Hydrology and Remote Sensing Lab, Agricultural Research Service, U.S. Department of Agriculture (USDA-ARS), Beltsville, Maryland, U.S.A.*

R. Lal / *Carbon Management and Sequestration Center, The Ohio State University, Columbus, Ohio, U.S.A.*

Matthias Langensiepen / *Department of Modeling Plant Systems, Humboldt University of Berlin, Berlin, Germany*

Jean-Claude Lefeuvre / *Laboratory of the Evolution of Natural and Modified Systems, University of Rennes, Rennes, France*

Gene E. Likens / *Cary Institute of Ecosystem Studies, Millbrook, New York, and University of Connecticut, Storrs, Connecticut, U.S.A.*

Yuling Liu / *Cooperative Institute for Climate and Satellites, University of Maryland, College Park, Maryland, U.S.A.*

Zhong Lu / *Cascades Volcano Observatory, U.S. Geological Survey (USGS), Vancouver, Washington, U.S.A.*

Ariel E. Lugo / *International Institute of Tropical Forestry, Forest Service, U.S. Department of Agriculture (USDA-FS), Río Piedras, Puerto Rico*

Alison MacNeil / *Northeast River Forecast Center, National Oceanic and Atmospheric Administration (NOAA), Taunton, Massachusetts, U.S.A.*

Carol Mallory-Smith / *Oregon State University, Corvallis, Oregon, U.S.A.*

Jeffrey A. Markert / *Biology Department, Providence College, Providence, Rhode Island, U.S.A.*

Toshihisa Matsui / *NASA Goddard Space Flight Center, National Aeronautics and Space Administration (NASA), Greenbelt, and ESSIC, University of Maryland, College Park, Maryland, U.S.A.*

Thomas Joseph McGreevy, Jr. / *Biology Department, Boston University, Boston, Massachusetts, U.S.A.*

Gregory McIsaac / *Natural Resources and Environmental Sciences, University of Illinois, Urbana, Illinois, U.S.A.*

Ernesto Medina / *International Institute of Tropical Forestry, Forest Service, U.S. Department of Agriculture (USDA-FS), Río Piedras, Puerto Rico, and Center for Ecology, Venezuelan Institute for Scientific Research, Caracas, Venezuela*

Miguel A. Medina, Jr. / *Department of Civil and Environmental Engineering, Duke University, Durham, North Carolina, U.S.A.*

Mallavarapu Megharaj / *Commonwealth Scientific and Industrial Research Organisation (CSIRO), Adelaide, South Australia, Australia*

Cevza Melek Kazezyılmaz-Alhan / *Department of Civil Engineering, Istanbul University, Istanbul, Turkey*

Forrest M. Mims III / *Geronimo Creek Observatory, Seguin, Texas, U.S.A.*

Emily S. Minor / *Department of Biological Sciences and Institute for Environmental Science and Policy, University of Illinois at Chicago, Chicago, Illinois, U.S.A.*

Niti B. Mishra / *University of Texas at Austin, Austin, Texas, U.S.A.*

Vasubandhu Misra / *Department of Earth, Ocean and Atmospheric Science and Center for Ocean-Atmospheric Prediction Studies, Florida State University, Tallahassee, Florida, U.S.A.*

David M. Mocko / *NASA Goddard Space Flight Center, National Aeronautics and Space Administration (NASA), Greenbelt, and SAIC, Beltsville, Maryland, U.S.A.*

Adam H. Monahan / *School of Earth and Ocean Sciences, University of Victoria, Victoria, British Columbia, Canada*

Miranda Y. Mortlock / *School of Agriculture and Food Sciences, University of Queensland, Brisbane, Queensland, Australia*

Ravendra Naidu / *Commonwealth Scientific and Industrial Research Organisation (CSIRO), Adelaide, South Australia, Australia*

V. Arivudai Nambi / *M.S. Swaminathan Research Foundation, Chennai, India*

Jocelyn C. Nelson / *Department of Zoology, University of British Columbia, British Columbia, Vancouver, Canada*

John Nimmo / *U.S. Geological Survey (USGS), Menlo Park, California, U.S.A.*

Dev Niyogi / *Department of Agronomy and Department of Earth, Atmospheric, and Planetary Sciences, Purdue University, West Lafayette, Indiana, U.S.A.*

Barry R. Noon / *Department of Fish, Wildlife, and Conservation Biology, Colorado State University, Fort Collins, Colorado, U.S.A.*

David Noone / *Department of Civil, Environmental and Architectural Engineering and Cooperative Institute for Research in the Environmental Sciences (CIRES), University of Colorado at Boulder, Boulder, Colorado, U.S.A.*

Jesse Norris / *Centre for Atmospheric Science, School of Earth, Atmospheric and Environmental Sciences, University of Manchester, Manchester, U.K.*

Alyssa Novak / *University of New Hampshire, Durham, New Hampshire, U.S.A.*

Brittany L. Oakes / *Department of Biological Sciences, Union College, Schenectady, New York, U.S.A.*

Micheal D. K. Owen / *Department of Agronomy, Iowa State University, Ames, Iowa, U.S.A.*

Antônio R. Panizzi / *Embrapa Trigo, Passo Fundo, Brazil*

Peter W. C. Paton / *Department of Natural Resources Science, University of Rhode Island, Kingston, Rhode Island, U.S.A.*

Jan Pergl / *Department of Invasion Ecology, Institute of Botany, Academy of Sciences of the Czech Republic (CAS), Průhonice, Czech Republic*

Debra P. C. Peters / *Jornada Experimental Range, Agricultural Research Service, U.S. Department of Agriculture (USDA-ARS), Las Cruces, New Mexico, U.S.A.*

Manon Picard / *Department of Zoology, University of British Columbia, British Columbia, Vancouver, Canada*

Roger A. Pielke, Sr. / *Cooperative Institute for Research in Environmental Sciences (CIRES), University of Colorado at Boulder, Boulder, Colorado, U.S.A.*

Finn C. Pillsbury / *Jornada Experimental Range, Agricultural Research Service, U.S. Department of Agriculture (USDA-ARS), Las Cruces, New Mexico, U.S.A.*

David Pimentel / *Department of Entomology, Cornell University, Ithaca, New York, U.S.A.*

Håkan Pleijel / *Applied Environmental Science, Göteborg University, Göteborg, Sweden*

Tony Prato / *Department of Agricultural and Applied Economics, University of Missouri, Columbia, Missouri, U.S.A.*

Stephen A. Prior / *National Soil Dynamics Laboratory, Agricultural Research Service, U.S. Department of Agriculture (USDA-ARS), Auburn, Alabama, U.S.A.*

Seth G. Pritchard / *Department of Biology, College of Charleston, Charleston, South Carolina, U.S.A.*

Jeffrey P. Privette / *National Climate Data Center, National Environmental Satellite Data Information Service, National Oceanic and Atmospheric Administration (NOAA), Asheville, North Carolina, U.S.A.*

Petr Pyšek / *Department of Invasion Ecology, Institute of Botany, Academy of Sciences of the Czech Republic (CAS), Průhonice, Czech Republic*

Nageswararao C. Rachaputi / *Queensland Department of Primary Industries, Kingaroy, Queensland, Australia*

Deeksha Rastogi / *Department of Atmospheric Sciences, University of Illinois, Urbana, Illinois, U.S.A.*

John P. Reganold / *Department of Crop and Soil Sciences, Washington State University, Pullman, Washington, U.S.A.*

Paul Reich / *National Resources Conservation Service, U.S. Department of Agriculture (USDA-NRCS), Washington, District of Columbia, U.S.A.*

Heinz Rennenberg / *Intitute of Forest Botany and Tree Physiology, University of Freiburg, Freiburg, Germany*

John S. Roberts / *New Technology Department, Research and Development, Rich Products Corporation, Buffalo, New York, U.S.A.*

Gilbert L. Rochon / *Tuskegee University, Tuskegee, Alabama, U.S.A.*

Cathy A. Roheim / *Department of Agricultural Economics and Rural Sociology, University of Idaho, Moscow, Idaho, U.S.A.*

Somnath Baidya Roy / *Department of Atmospheric Sciences, University of Illinois, Urbana, Illinois, U.S.A.*

G. Brett Runion / *National Soil Dynamics Laboratory, Agricultural Research Service, U.S. Department of Agriculture (USDA-ARS), Auburn, Alabama, U.S.A.*

Bayden D. Russell / *Southern Seas Ecology Laboratories, School of Earth and Environmental Science, University of Adelaide, Adelaide, South Australia, Australia*

Grace Saba / *Coastal Ocean Observation Laboratory, Institute of Marine and Coastal Sciences, School of Environmental and Biological Sciences, Rutgers University, New Brunswick, New Jersey, U.S.A.*

Sergi Sabater / *Catalan Institute for Water Research, Girona, Spain*

Uriel N. Safriel / *Center of Environmental Conventions, Jacob Blaustein Institutes for Desert Research, and Department of Ecology, Evolution and Behavior, Hebrew University of Jerusalem, Jerusalem, Israel*

William Saunders / *Northeast River Forecast Center, National Oceanic and Atmospheric Administration (NOAA), Taunton, Massachusetts, U.S.A.*

Mary C. Savin / *Department of Crop, Soil, and Environmental Sciences, University of Arkansas, Fayetteville, Arkansas, U.S.A.*

Jennifer L. Schaeffer / *Food Safety and Environmental Stewardship Program, Oregon State University, Corvallis, Oregon, U.S.A.*

Neil T. Schock / *Institute for Great Lakes Research, CMU Biological Station, and Department of Biology, Central Michigan University, Mt. Pleasant, Michigan, U.S.A.*

Oscar Schofield / *Coastal Ocean Observation Laboratory, Institute of Marine and Coastal Sciences, School of Environmental and Biological Sciences, Rutgers University, New Brunswick, New Jersey, U.S.A.*

David M. Schultz / *Centre for Atmospheric Science, School of Earth, Atmospheric and Environmental Sciences, University of Manchester, Manchester, U.K.*

Yongwei Sheng / *Department of Geography, University of California, Los Angeles, Los Angeles, California, U.S.A.*

Xun Shi / *Department of Geography, Dartmouth College, Hanover, New Hampshire, U.S.A.*

Frederick T. Short / *University of New Hampshire, Durham, New Hampshire, U.S.A.*

Paul Short / *Canadian Sphagnum Peat Moss Association, St. Albert, Alberta, Canada*

C. K. Shum / *Division of Geodetic Science, School of Earth Sciences, The Ohio State University, Columbus, Ohio, U.S.A.*

Kadambot H. M. Siddique / *The UWA Institute of Agriculture, University of Western Australia, Crawley, Western Australia, Australia*

Thomas R. Sinclair / *Crop Genetics and Environmental Research Unit, U.S. Department of Agriculture (USDA), University of Florida, Gainesville, Florida, U.S.A.*

Michael J. Singer / *Department of Land, Air, and Water Resources, University of California, Davis, California, U.S.A.*

Ajay Singh / *Department of Biology, University of Waterloo, Waterloo, Ontario, Canada*

Martin D. Smith / *Nicholas School of the Environment and Department of Economics, Duke University, Durham, North Carolina, U.S.A.*

Daniela Soleri / *Geography Department, University of California, Santa Barbara, California, U.S.A.*

Bretton Somers / *Department of Geography and Anthropology, Louisiana State University, Baton Rouge, Louisiana, U.S.A.*

Jan Hanning Sommer / *Center for Development Research (ZEF), University of Bonn, Bonn, Germany*

Roy F. Spalding / *Water Science Laboratory, University of Nebraska, Lincoln, Nebraska, U.S.A.*

Jean L. Steiner / *Agricultural Research Service, U.S. Department of Agriculture (USDA-ARS), El Reno, Oklahoma, U.S.A.*

Deborah Stinner / *Ohio Agriculture Research and Development Center, The Ohio State University, Wooster, Ohio, U.S.A.*

Mark H. Stolt / *Department of Natural Resources Science, University of Rhode Island, Kingston, Rhode Island, U.S.A.*

David A. Stonestrom / *U.S. Geological Survey (USGS), Menlo Park, California, U.S.A.*

Nigel E. Stork / *Environment Futures Centre, Griffith School of Environment, Griffith University, Brisbane, Queensland, Australia*

Matthew L. Stutz / *Department of Chemistry, Physics, and Geoscience, Meredith College, Raleigh, North Carolina, U.S.A.*

Donglian Sun / *Department of Geography and Geoinformation Science, George Mason University, Fairfax, Virginia, U.S.A.*

Philip Sura / *Department of Earth, Ocean and Atmospheric Science and Center for Ocean-Atmospheric Prediction Studies, Florida State University, Tallahassee, Florida, U.S.A.*

Takehiko Takano / *Tohoku Gakuin University, Sendai, Japan*

Albert E. Theberge, Jr. / *Central Library, National Oceanic and Atmospheric Administration (NOAA), Silver Spring, Maryland, U.S.A.*

Ralph W. Tiner / *National Wetlands Inventory (NWI), U.S. Fish and Wildlife Service, Hadley, Massachusetts, U.S.A.*

Ronald F. Turco / *Department of Agronomy, Purdue University, West Lafayette, Indiana, U.S.A.*

Donald G. Uzarski / *Institute for Great Lakes Research, CMU Biological Station, and Department of Biology, Central Michigan University, Mt. Pleasant, Michigan, U.S.A.*

Fernando Valladares / *Centro de Ciencias Medioambientales, Madrid, Spain*

Arnold G. van der Valk / *Ecology, Evolution and Organismal Biology, Iowa State University, Ames, Iowa, U.S.A.*

George F. Vance / *Department of Ecosystem Sciences and Management, University of Wyoming, Laramie, Wyoming, U.S.A.*

Kathleen J. Vigness-Raposa / *Marine Acoustics, Inc., Middletown, Rhode Island, U.S.A.*

Dale H. Vitt / *Department of Plant Biology and Center for Ecology, Southern Illinois University, Carbondale, Illinois, U.S.A.*

John C. Volin / *Department of Natural Resources and the Environment, University of Connecticut, Storrs, Connecticut, U.S.A.*

Daniel von Schiller / *Catalan Institute for Water Research, Girona, Spain*

James A. Voogt / *Department of Geography, University of Western Ontario, London, Ontario, Canada*

Jonathan M. Wachter / *Department of Crop and Soil Sciences, Washington State University, Pullman, Washington, U.S.A.*

H. Jesse Walker / *Department of Geography and Anthropology, Louisiana State University, Baton Rouge, Louisiana, U.S.A.*

Ivan A. Walter / *Ivan's Engineering, Inc., Denver, Colorado, U.S.A.*

Pao K. Wang / *Atmospheric and Oceanic Sciences, University of Wisconsin, Madison, Wisconsin, U.S.A.*

Yeqiao Wang / *Department of Natural Resources Science, University of Rhode Island, Kingston, Rhode Island, U.S.A.*

Zhuo Wang / *Department of Atmospheric Sciences, University of Illinois at Urbana-Champaign, Urbana, Illinois, U.S.A.*

Owen P. Ward / *Department of Biology, University of Waterloo, Waterloo, Ontario, Canada*

Elizabeth R. Waters / *Department of Biology, San Diego State University, San Diego, California, U.S.A.*

Hans-Joachim Weigel / *Federal Research Institute for Rural Areas, Forestry and Fisheries, Thünen Institute of Biodiversity, Braunschweig, Germany*

W. W. Wenzel / *Institute of Soil Research, University of Natural Resources and Life Sciences, Vienna, Austria*

French Wetmore / *French & Associates, Ltd., Park Forest, Illinois, U.S.A.*

Ryan L. Wheeler / *Institute for Great Lakes Research, CMU Biological Station, and Department of Biology, Central Michigan University, Mt. Pleasant, Michigan, U.S.A.*

Penny E. Widdison / *Department of Geography, University of Durham, Durham, U.K.*

Robert L. Wilby / *Department of Geography, University of Loughborough, Loughborough, U.K.*

Donald A. Wilhite / *National Drought Mitigation Center, Lincoln, Nebraska, U.S.A.*

John Wilkin / *Coastal Ocean Observation Laboratory, Institute of Marine and Coastal Sciences, School of Environmental and Biological Sciences, Rutgers University, New Brunswick, New Jersey, U.S.A.*

Victoria A. Wojcik / *Pollinator Partnership, San Francisco, California, U.S.A.*

Graeme C. Wright / *Department of Primary Industries, Queensland Department of Primary Industries, Kingaroy, Queensland, Australia*

Marcia Glaze Wyatt / *Department of Geology, University of Colorado at Boulder, Boulder, Colorado, U.S.A.*

Jason Yang / *Department of Geography, Ball State University, Muncie, Indiana, U.S.A.*

Jian Yang / *State Key Laboratory of Forest and Soil Ecology, Institute of Applied Ecology, Chinese Academy of Sciences, Shenyang, China*

Qingsheng Yang / *Department of Resources and Environment, Guangdong University of Business Studies, Guangzhou, China*

Xiaojun Yang / *Department of Geography, Florida State University, Tallahassee, Florida, U.S.A.*

Xu Yi / *Coastal Ocean Observation Laboratory, Institute of Marine and Coastal Sciences, School of Environmental and Biological Sciences, Rutgers University, New Brunswick, New Jersey, U.S.A.*

Kenneth R. Young / *University of Texas at Austin, Austin, Texas, U.S.A.*

Yunyue Yu / *Center for Satellite Applications and Research, National Environmental Satellite Data Information Service, National Oceanic and Atmospheric Administration (NOAA), College Park, Maryland, U.S.A.*

Baiping Zhang / *State Key Lab for Resources and Environment Information System, Institute of Geographic Sciences and Natural Resources Research, Chinese Academy of Sciences, Beijing, China*

Yuanzhi Zhang / *Yuen Yuen Research Centre for Satellite Remote Sensing, Institute of Space and Earth Information Science, Chinese University of Hong Kong, Hong Kong, and Laboratory of Coastal Zone Studies, Shenzhen Research Institute, Shenzhen, China*

Fang Zhao / *State Key Lab for Resources and Environment Information System, Institute of Geographic Sciences and Natural Resources Research, Chinese Academy of Sciences, Beijing, China*

Jinping Zhao / *Ocean University of China, Qingdao, China*

Yuyu Zhou / *Joint Global Change Research Institute, Pacific Northwest National Laboratory, College Park, Maryland, U.S.A.*

Carl L. Zimmerman / *Department of Urban and Environmental Policy and Planning, Tufts University, Medford, Massachusetts, U.S.A.*

Michael A. Zoebisch / *German Agency for International Cooperation, Tashkent, Uzbekistan*

Contents

Land

Land (cont'd.)

Water

Water (cont'd.)

Air

Topical Table of Contents

Land

Biodiversity

Coastal Issues

Land (cont'd.)

Land (cont'd.)

Watershed

Land (cont'd.)

Water

Water (cont'd.)

Marine Issues (cont'd.)

Coastal Environment (cont'd.)

Marine Resources

Air

Atmosphere

Climate

Meteorology

Preface

With unprecedented attention on global change, one of the focuses of debate is about the availability and sustainability of natural resources and about how to balance the equilibrium between what society demands from natural environments and what the natural resource base can provide. It is critical to gain a full understanding about the consequences of the changing resource bases to the degradation of ecological integrity and the sustainability of life. Natural resources represent such a broad scope of complex and challenging topics; in this field, a reference book must cover a vast number of topics in order to be titled an encyclopedia.

A total of 275 scholars from 25 countries including Australia, Austria, Brazil, China, Cameroon, Canada, Czech Republic, Finland, France, Germany, Hungary, India, Israel, Japan, New Zealand, Norway, Puerto Rico, Spain, Sweden, Syria, Turkey, United Kingdom, United States, Uzbekistan, and Venezuela contributed 188 entries to the two volumes of this printed set. Volume I (*Land*) includes 98 entries that cover the topical areas of renewable and nonrenewable natural resources such as forest and vegetative; soil; terrestrial coastal and inland wetlands; landscape structure and function and change; biological diversity; ecosystem services, protected areas and management; natural resource economics; and resource security and sustainability. In Volume II, Water includes 59 entries and Air includes 31 entries. The *Water* entries cover topical areas such as fresh water, groundwater, water quality and watersheds, ice and snow, coastal environments, and marine resources and economics. The *Air* entries cover topical areas such as atmosphere, meteorology and weather, climate resources, and effects of climate change.

This *ENR* is a reference series that is under continuous development. In addition to the entries included in this printed set, continuously developed entries will be published through the online version of the *ENR* and will be included in the future edition of the printed copy. I look forward to working with a broader scope of contributors and reviewers for the further development of this series of publication.

Yeqiao Wang
University of Rhode Island

Acknowledgments

I am honored to have had this opportunity and privilege to work on this important publication. Many people who helped tremendously during the process deserve special acknowledgment. First and foremost, I thank the 275 contributors from 25 countries and all the reviewers from different countries around the world. Their expertise, insights, dedication, hardwork, and professionalism ensure the quality of this encyclopedia.

I appreciate the guidance and contribution of the members of the Editorial Advisory Board: Drs. Peter V. August of University of Rhode Island; Kate Moran of the Ocean Networks Canada at the University of Victoria; Roger A. Pielke, Sr. of Colorado State University and University of Colorado at Boulder; Russell G. Congalton of University of New Hampshire; Giles Foody of the University of Nottingham; Jefferson M. Fox of the East-West Center; Xiaowen Li of the Chinese Academy of Science; Hui Lin of the Chinese University of Hong Kong; Jing Ming Chen of University of Toronto; and Darryl J. Keith of the U.S. Environmental Protection Agency.

I wish to express my gratitude to my respected colleagues and scholars, in particular Drs. Peter V. August, Peter W.C. Paton, Thomas R. Husband, Jose A. Amador, Mark H. Stolt, Graham E. Forrester, Cathy A. Roheim, Thomas Boving, Richard Burroughs, Bill Buffum, Even Preisser, Serena M. Moseman-Valtierra, Nancy Karraker, Arthur J. Gold, Scott McWilliam, Laura Meyerson, David Abedon, Daniel L. Civco, John C. Volin, John C. Clausen, Jiansheng Yang, and Yuyu Zhou, for their encouragement, support, contribution, and service during the preparation of this publication.

The preparation for the development of this publication was started in spring 2010. I appreciate the initial contact from Irma Shagla, Editor for Environmental Sciences and Engineering of the Taylor & Francis Group/CRC Press, for the conceptual development of this publication. Claire Miller and Molly Pohlig from the Encyclopedia Program of the Taylor & Francis Group managed the preparation of this publication with leadership, insights, and the highest standard of professionalism. A special tribute goes to Susan Lee, a former editor of the encyclopedia program of the Taylor & Francis Group. Susan was dedicated to the development of this encyclopedia for 3.5 years from the very beginning of the project till the very last days of her very young life. It would have been impossible to achieve what we have accomplished today without Susan's diligence, gentle diplomacy, and superior organizational skills. I am so sorry that Susan could not be with us to see the publication of this encyclopedia.

The inspiration for working on this reference series came from my 30 years of research and teaching experiences in different stages of my professional career. I am grateful for the opportunity of working with many top-notch scholars, staff members, administrators, and enthusiastic students throughout the time.

As always, the most special appreciation is due to my wife and daughters for their love, patience, understanding, and encouragement during the preparation of this publication. I wish my parents and previous professors of soil ecology and of climatology from the School of Geography of the Northeast Normal University could see this set of publications.

Aims and Scope

Land, water, and air are the most precious natural resources that sustain life and civilization. Maintenance of clean air and water and preservation of land resources and native biological diversity are among the challenges that we are facing for the sustainability the well-being of all on the planet Earth. Natural and anthropogenic forces have affected constantly land, water, and air resources through interactive processes such as shifting climate patterns, disturbing hydrological regimes and alternating landscape configurations and compositions. Improvements in understanding of the complexity of land, water, and air systems and their interactions with human activities represent priorities in scientific research, technology development, education programs, and administrative actions for conservation and management of natural resources.

The *Encyclopedia of Natural Resources* (ENR) consists of two volumes (Volume I: *Land* and Volume II: *Water* and *Air*). The *ENR* is designed to provide an authoritative and organized informative reference covering a broad spectrum of topics such as the forcing factors and habitats of life, their histories, current status, future trends, and their societal connections, economic values, and management. The topical entries are authored by world-class scientists and scholars. The contents represent state-of-the-art science and technology development and perspectives of resource management. Efforts were made for each topical entries to be written at a level that allows a broad scope of audience to understand. Public and private libraries, educational and research institutions, scientists, scholars, and students will be the primary audience of this encyclopedia.

About the Editor-in-Chief

 Dr. Yeqiao Wang is a professor at the Department of Natural Resources Science, College of the Environment and Life Sciences, University of Rhode Island. He earned his BS from the Northeast Normal University in 1982 and his MS degree in remote sensing and mapping from the Chinese Academy of Sciences in 1987. He earned an MS and a PhD in natural resources management & engineering from the University of Connecticut in 1992 and 1995, respectively. From 1995 to 1999, he held the position of assistant professor in the Department of Geography and Department of Anthropology, University of Illinois at Chicago. He has been on the faculty of the University of Rhode Island since 1999. In addition to his tenured position, he held an adjunct research associate position at the Field Museum of Natural History in Chicago. He has also served as a guest professor and an adjunct professor at universities in the United States and in China. Among his awards and recognitions, Dr. Wang was a recipient of the prestigious Presidential Early Career Award for Scientists and Engineers (PECASE) by former U.S. president William J. Clinton in 2000.

Dr. Wang's specialties are in terrestrial remote sensing and applications of geoinformatics in natural resources analysis and mapping. His research projects have been funded by different agencies that supported his scientific studies in various regions of the United States, in East and West Africa, and in regions in China. Besides peer-reviewed journal publications, Dr. Wang edited *Remote Sensing of Coastal Environments* and *Remote Sensing of Protected Lands* published by the CRC Press in 2009 and 2010, respectively. He has also authored and edited over 10 science books in Chinese.

Encyclopedia of
Natural Resources

Water and Air

Volume II
Aquaculture–Wind
Pages 589–1088

Aquaculture—
Bathymetry

Coastal—
Evaporation

Field—Low

Marine—
Pollution

Riparian—
Wetlands

Acid—
Atmospheric

Circulation—
Crops

Dew—Wind

Aquaculture

Carrie J. Byron
Gulf of Maine Research Institute, Portland, Maine, U.S.A.

Abstract
Aquaculture is the process of cultivating aquatic organisms in a controlled setting and may include some or all of the processes of breeding, rearing, and harvesting. Aquaculture is increasing on a global scale with disproportionate growth in Asia. Rearing of some species, particularly salmon and shrimp, over time, has received criticism for their negative influence on environmental health. New technologies and innovative approaches to aquaculture, such as integrated multitrophic aquaculture and utilizing carrying capacity for an ecosystem approach to aquaculture, provide a sustainable path forward for the management of aquaculture, particularly in coastal waters.

INTRODUCTION

Aquaculture is the process of cultivating aquatic organisms in a controlled setting and may include some or all of the processes of breeding, rearing, and harvesting. Aquaculture can be done on land in tanks, freshwater habitats such as ponds, brackish water habitats such as estuaries, or marine habitats such as in the coastal zone and open ocean. Mariculture is often used to describe brackish or marine aquaculture. Sea ranching describes the process of rearing early life stages in controlled systems and releasing them back to the ocean often for the purpose of restocking a population.[1] Tuna ranching is a relatively recent development whereby young wild fish are captured and reared in net pens to marketable size.[2,3]

GLOBAL PERSPECTIVE

Aquaculture is increasing on a global scale. However, the increase in aquaculture is not uniform across the globe.[4] In 2010, global aquaculture production equaled 63.6 million tonnes.[4] Asia comprised 89% (53,301,157 million tonnes) of the world aquaculture production by volume, with China as the leading nation comprising 61.4% (36,734,215 million tonnes) of global production. The remaining global aquaculture production came from the Americas (4.3%; 2,576,428 million tonnes), Europe (4.2%; 2,523,179 million tonnes), Africa (2.2%; 1,288,320 million tonnes), and Oceana (0.3%; 183,516 tonnes).[4]

The National Aquaculture Act of 1980 states that it is "in the national interest, and it is the national policy, to encourage aquaculture development in the U.S." Despite this national policy, the United States is not one of the top 10 leading nations in aquaculture production. Aquaculture production has grown most strongly in developing countries, particularly in Asia.[4]

Freshwater is the dominant habitat type for aquaculture production comprising 62% of global production and 58.1% of global value of products produced.[4] Of all the animal species produced in freshwater, 91% (33.9 million tonnnes) of the production are freshwater finfishes such as carps, tilapia, and catfish.[4] Marine aquaculture comprised 30.1% of global production and 29.2% of global value.[4] Of all the animal species produced in marine water, three-quarters are molluscs (75.5%; 13.9 million tonnes) such as clams, cockles, oysters, and mussels.[4] Brackish water habitats comprised only 7.9% of global aquaculture production and 12.8% of its value.[4] Of all the animal species produced in brackish water, more than half are crustaceans (57.7%; 2.7 million tonnes) such as shrimp.[4]

Most aquaculture species are produced for human consumption. Some ornamental fish, such as clownfish, are produced for aquarium trade.[5,6] Some algae and small finfish species are produced for consumption by other farmed animals including shellfish, fish, chickens, pigs, and other mammals. Of the animal species being produced through aquaculture for human consumption, there has been a proportionally large increase in freshwater finfish and molluscs over the past four decades.[4] More than half (56.6%; 33.7 million tonnes) of global aquaculture production is freshwater finfish and nearly one-quarter (23.6%; 14.2 million tonnes) is molluscs.[4] Crustaceans (9.6%; 5.7 million tonnes), diadromous fish (6.0%; 3.6 million tonnes), marine fish (3.1%; 1.8 million tonnes), and other assorted animals (1.4%; 814,3000 tonnes) such as sea cucumbers and softshell turtles comprise the rest of the aquaculture production by species type.[4]

Encyclopedia of Natural Resources DOI: 10.1081/E-ENRW-120047572

SPECIES AND HABITATS

Finfish are vertebrate fish species and include freshwater, diadromous, and marine fish. Diadromous fish such as salmon live part of their life in freshwater and another part in salt water. Most freshwater finfish are grown on land in tanks or raceways (long narrow troughs). U.S. freshwater aquaculture is dominated by channel catfish (*Ictalurus punctatus*) but is shadowed by the dramatic increase in the production of pangas catfish (*Pangasius* spp.) in Vietnam in recent years.[4,7] Trout, tilapia, and some bass species are also grown in land-based facilities often using a recirculating aquaculture system designed to reduce and reuse water. Land-based aquaculture facilities require high amounts of electricity to maintain operations, and efficient systems that cause little pollution are generally required to maintain these facilities, which have a little impact on the environment.[8] Recirculating aquaculture systems clean and reuse effluent water within the facility instead of discharging dirty water into the environment.

Although some marine species can also be raised on land, most are raised in the ocean in net pens or cages (Fig. 1). Atlantic salmon represents many of the environmental and social debates surrounding the efficacy of finfish farming. Salmon is farmed around the world with the two leading producers of farmed Atlantic salmon being Norway and Chile.[4] Criticisms for finfish aquaculture concern environmental and human health. Escaped farm salmon may breed with wild salmon, thereby polluting "wild" genetic stocks with genes from domesticated stocks, which may not be derived from local populations.[9] The mixed-gene offspring have lower survival in natural habitats than their wild counterparts, thereby further hindering the persistence of an already threatened and endangered species.[10] Another environmental concern is the fecal waste from farmed fish as well as the uneaten fish food on farms, which can pollute surrounding habitats.[11,12] Additionally, like any dense monoculture of species, disease also becomes more prevalent. Sea lice are obligate ectoparasites that attach to and graze on the skin of the fish, thereby causing shallow lesions that can permit bacterial infections or lead to death, which devalue the seafood product.[13,14]

The commercially formulated pellet feed that many commercially aquacultured fish are fed is composed of proteins, lipids, and carbohydrates that vary in proportion with each farm-raised species.[15,16] The protein meal is derived from a variety of sources including fish caught specifically to produce fish meal and terrestrial-based crops such as soya and maize. Little is known regarding the impact of terrestrial-derived nutrients in marine food webs.[17]

Advances in technology and scientific knowledge are minimizing the impacts of finfish culture in the coastal environment. The risk of escapees from pens has also declined over the years with increased net-pen technology. Technologies for feeding the fish are becoming more advanced where very little food goes uneaten, thereby reducing waste to the environment. Some countries, such as the United States, require pens to remain fallow for a period of time between stocking cycles to further reduce the loading of fecal waste to an area and to mitigate the spread of diseases and pests, such as sea lice.[18,19] Furthermore, innovative approaches to aquaculture such as integrated multitrophic aquaculture and utilizing carrying capacity for an ecosystem approach to aquaculture hold promise for further reductions on environmental impact in the future.[20–23]

Crustacean species that are most commonly grown using aquaculture are primarily Penaeid marine shrimp and freshwater prawns.[4] In 2010, the fastest growing crustacean production was the whiteleg shrimp (*Penaeus vannamei*). Thailand, China, and Indonesia are the major exporting countries of shrimp, whereas the United States and Japan are the primary importers of shrimp.[4] Like salmon, there are several environmental and social issues surrounding shrimp farming including the destruction of mangrove habitats for the construction of rearing ponds, salinization of groundwater, depletion of wild fish stocks for formulated feeds, depletion of wild broodstocks, and disease outbreaks[24] (http://www.asc-aqua.org; http://www.wildlife.org). Shrimp farming is highly unregulated in the global market and poses several environmental threats. Recent technologies hold promise for a more sustainable method of culturing shrimp in the future. Pilot studies of native shrimp species grown in Aquapod™ net pens in the open ocean on natural productivity, which serves as a food source without herbicides, pesticides, commercial fish oil, or processed fish meal, have shown that there is minimal impact on the environment and that it alleviates pressure on fish stocks and wild broodstocks[5,7] (http://www.olazul.org). Other finfish species can also be grown in various designs of large pens in the open ocean, which present additional engineering and economic challenges but would alleviate environmental impacts in crowded and heavily used coastal areas[25–29] (http://www.amac.unh.edu).

Fig. 1 Floating salmon net pen. Notice the netting to keep out predators, the pneumatic feeder in the center of the cage, and the floating feed-supply house to the left and behind the cage.

Bivalve molluscs such as clams, cockles, oysters, and mussels are commonly grown worldwide (Figs. 2, 3).[4] Unlike finfish farming, bivalves grown in coastal habitats filter feed directly on ambient plankton and do not require additional additive feeds. Bivalve aquaculture in coastal environments has the potential to alter the community structure of the ecosystem, nutrient dynamics in the water column, and deposition of sediment.[30–39] Nitrogen is the primary nutrient of concern in coastal habitats where bivalves live naturally and are grown. In extreme cases, excessive nitrogen can lead to large algae blooms and subsequently anoxia in the water column, which can kill finfish and shellfish.[8] As bivalves consume algae and other particulate organic matter in the water, nitrogen gets assimilated into their hard shell, thereby removing it from water and making it unavailable to other organisms, which is a method of cleansing the water in high nutrient systems.[40–42] Bioextraction describes the process of using bivalves to clean the water by extracting excessive nutrients. Some economists are exploring ways to trade nitrogen credits in a way that mirrors the carbon credit trading system[43] (http://www.motu.org.nz/research/detail/nutrient_trading).

Globally, cultivation of algae is dominated by marine macroalgae or seaweeds.[4] Seaweeds are algae and can include both small microscopic species such as *Isochrysis* sp. primarily used to feed aquaculture-grown bivalve shellfish and macroalgae, commonly called kelp. In 2010, algae production equaled 19 million tonnes and came primarily from eight countries: China, Indonesia, the Philippines, the Republic of Korea, Democratic People's Republic of Korea, Japan, Malaysia, and the United Republic of Tanzania.[4] Aquatic algae production by volume increased at an average annual rate of 9.5% in the 1990s, which is comparable to rates for farmed aquatic animals.[4] The most-cultivated alage is the seaweed called Japanese kelp (*Saccharina japonica*) consisting of 98.9% of global production of algae cultured.[4] One of the many methods for cultivation involves growing kelp spores into seedlings in land-based facilities before being transplanted to long ropes or other rack-and-line structures in coastal waters. Seaweeds are used in several food products, fertilizers, and animal feeds.

Fig. 2 Oyster aquaculture. This photo demonstrates one of many techniques for raising oysters in coastal waters. Notice the rack-and-bag system. The rack, or cage, sits on the bottom under the water and is attached to a rope and surface buoy. The oyster farm is not visible from the surface of the water, other than a few buoys marking the boundaries of the lease area.

INTEGRATED MULTITROPHIC AQUACULTURE

Integrated multitrophic aquaculture (IMTA) is an extension of polyculture, whereby the culture of fed organisms are combined with the culture of organisms that extract dissolved inorganic nutrients (seaweeds) or particulate organic matter (shellfish) or both, thereby linking trophic

A

B

Fig. 3 Cultured oysters (**A**) almost grown to market size in bags; (**B**) juvenile oyster spat, sometimes called oyster seed.
Source: Figure 3B by Trisha Towanda, 2008.

levels.[44] The term trophic level describes the position an organism occupies in the food chain. One of the goals of IMTA is to reduce organic pollution to the environment from fish culture. Another goal is to improve the growth of all the IMTA species due to a rich environment in terms of food. Species from lower trophic levels such as mussels and algae ingest particulate and dissolved nutrients from the fish farm, thereby "cleansing" the water[25] while getting additive nutrition. Mussels may also be capable of the infectious pressure of the bacteria *Vibrio anguillarum*.[45] Sea cucumbers (Echinodermata: Holothuroidea) can also be integrated into IMTA and have a growing market in aquaculture, independent of IMTA.[46] Coupled aquaculture–agriculture systems have been practiced throughout time primarily on a small scale in developing countries and have undoubtedly led to the initial conceptualization of IMTA.[40,47]

MANAGEMENT

Marine Spatial Planning

The coastal environment, where much aquaculture activity takes place, is a busy place that is used by several industries and activities (e.g., shipping, fishing, recreational boating, renewable energy, protected habitat areas, etc.). Conflict between aquaculture and other coastal activities presents a hurdle for the aquaculture industry, often making it difficult to obtain permits for operation. Effective spatial planning of coastal habitats is a way to mitigate these types of user conflicts. Marine spatial planning allows for optimizing the use of the coastal zone. For example, by coupling IMTA aquaculture farms with wind turbine farms, the unfishable and unnavigable spaces between turbines can be efficiently used for food production.[15, 48–51]

Ecosystem Approach to Aquaculture

The ecosystem approach to aquaculture is defined as "a strategic approach to the development and management of the sector aiming to integrate aquaculture within the wider ecosystem such that it promotes sustainability of interlinked social-ecological systems" and is guided by three main principles:[52] i) "Aquaculture development and management should take account of the full range of ecosystem functions and services, and should not threaten the sustained delivery of these to society;" ii) "Aquaculture should improve human well-being and equity for all relevant stakeholders;" and iii) "Aquaculture should be developed in the context of other sectors, policies and goals." The principles of the ecosystem approach to aquaculture follow closely from the principles of ecosystem-based management, which emphasize the interdependencies of ecological and social goals within the same system (http://www.ebmtools.org/about_ebm.html).

Carrying Capacity

There are four types of carrying capacity that have been adapted to describe coastal aquaculture where a farm is located in an ecosystem. The first type is physical carrying capacity, which is the total area of marine farms that can be accommodated in the available physical space.[53] The definition of physical carrying capacity is limited in that it only considers available space and not the function or impacts of the farm in the ecosystem. The second type is production carrying capacity, which is the stocking density at which harvests are maximized.[53] Every successful farmer knows the production carrying capacity for the farm because it is directly linked to sales and profits. The definition of production carrying capacity is limited to the farm site and does not consider the larger ecosystem in which the farm is located. To extend this definition, consider ecological carrying capacity, which is the stocking or farm density above which would cause unacceptable ecological impacts.[53] And similarly, social carrying capacity is the level of farm development that causes unacceptable social impacts.[53] Both ecological and social carrying capacity recognize that the farm is simply part of the larger ecosystem, which is important when taking an ecosystem approach to aquaculture. As useful as these two definitions may be, they are also somewhat vague. How does one define what is unacceptable in an ecosystem? McKindsey et al.[54] suggest that ecological and social carrying capacity must be considered together. Society, represented by a group of stakeholders, needs to decide what is acceptable.[20] Once acceptability limits are set, carrying capacity can be calculated.[21,22] Ecological carrying capacity can be calculated using mass–balance ecosystem modeling as described by Byron et al.[20–22] Social carrying capacity can be calculated using economic modeling, spatial modeling, and/or resource valuation. These techniques are still in development and are not yet well established. Ultimately, ecosystem carrying capacity and social carrying capacity can be used as techniques for implementing an ecosystem approach to aquaculture.

The aquaculture industry is growing, with most of that growth occurring in Asia. The United States National Policy on Aquaculture encourages further sustainable growth in U.S. coastal waters but currently contributes only a very small fraction to global production. Now is an appropriate time to set measures in place to ensure sustainable growth within the capacity of the ecosystem and society.

REFERENCES

1. Schraga, T.S.; Cloern, J.E.; Dunlavey, E.G. Primary production and carrying capacity of former salt ponds after reconnection to San Francisco Bay. Wetlands (Wilmington, NC) **2008**, *28* (3), 841–851.

2. De Stefano, V.; Van Der Heiden, P.G.M. Bluefin tuna fishing and ranching: a difficult management problem. New Medit. **2007**, *2*, 59–64.

3. Longo, S.B.; Clark, B. The commodification of bluefin tuna: the historical transformation of the mediterranean fishery. J. Agrarian Change **2012**, *12* (2–3), 204–226.

4. FAO. *The State of the World Fisheries and Aquaculture 2012*; FAO: Rome, Italy 2012; 209.

5. Bren. Spatial planning and bio-economic analysis for sustainable offshore shrimp aquaculture. Donald Bren School of Environmental Science and Management & University of California, Santa Barbara. 2010; 33. https://www.box.com/s/3aa1bb89b09cbd7329cb.

6. Molina, L.; Segade, A. Aquaculture as a potential support of marine aquarium fish trade sustainability. In *Management of Narual Resources, Sustainable Development and Ecological Hazards III*. Brebbia, C., Zubir, S. WIT Press: Southhampton, UK, 2012.

7. Olivares, A.E.V. *Design of a Cage Culture System for Farming in Mexico*; The United Nations University, Fisheries Training Programme; Reykjavik, Iceland, 2003; 47 pp.

8. Howarth, R.; Anderson, D.; Cloern, J.; Elfring, C.; Hopkinson, C.; Lapointe, B.; Malone, T.; Marcus, N.; McGathery, K.; Sharpley, A.; Walker, D. Nutrient pollution of coastal rivers, bays, and seas. Issues in Ecology. 2000. http://www.esa.org/science_resources/issues/TextIssues/issue7.php.

9. Houde, A.L.S.; Fraser, D.J.; Hutchings, J.A. Fitness-related consequences of competitive interactions between farmed and wild Atlantic salmon at different proportional representations of wild–farmed hybrids. ICES J. Mar. Sci. **2010a**, *67* (4), 657–667.

10. Houde, A.L.S.; Fraser, D.J.; Hutchings, J.A. Reduced antipredator responses in multi-generational hybrids of farmed and wild Atlantic salmon (*Salmo salar* L.). Conserv. Genet. **2010b**, *11* (3), 785–794.

11. Hargrave, B.T. Empirical relationships describing benthic impacts of salmon aquaculture. Aquaculture Environ. Interact. **2010**, *1* (1), 33–46.

12. Piedecausa, M.A.; Aguado-Gimenez, F.; Valverde, J.C.; Llorente, M.D.H.; Garcia-Garcia, B. Influence of fish food and faecal pellets on short-term oxygen uptake, ammonium flux and acid volatile sulphide accumulation in sediments impacted by fish farming and non-impacted sediments. Aquaculture Res. **2012**, *43* (1), 66–74.

13. Krkosek, M. Host density thresholds and disease control for fisheries and aquaculture. Aquaculture Environ. Interact. **2010**, *1* (1), 21–32.

14. Revie, C.; Dill, L.; Finstad, B.; Todd, C.D. Salmon aquaculture dialog working group report on sea lice. commissiond by the salmon aquaculture dialogue. 2009. http://www.nina.no/archive/nina/PppBasePdf/temahefte/039.pdf

15. Buck, B.H.; Krause, G.; Rosenthal, H. Extensive open ocean aquaculture development within wind farms in Germany: the prospect of offshore co-management and legal constraints. Ocean Coastal Manag. **2004**, *47* (3–4), 95–122.

16. Southgate, P.C. Foods and Feeding. In *Aquaculture. Farming Aquatic Animals and Plants*. Lucas, J.S., Southgate, P.C. Wiley-Blackwell: Oxford, UK, 2012; 188–213.

17. Spivak, A.C.; Canuel, E.A.; Duffy, J.E.; Richardson, J.P. Nutrient enrichment and food web composition affect ecosystem metabolism in an experimental seagrass habitat. PLoS ONE **2009**, *4* (10), e7473.

18. Brauner, C.J.; Sackville, M.; Gallagher, Z.; Tang, S.; Nendick, T.L.; Farrell, A.P. Physiological consequences of the salmon louse (*Lepeopthteirus salmonis*) on juvenlie pink salmon (*Onchorhynchus gorbuscha*): implications for wild salmon ecology and management, and for salmon aquaculture. Philos. Trans. R. Soc. **2012**, *367* 1770–1779.

19. USASAC. *Annual Report of the U.S. Atlantic Salmon Assessment Committee 22*. Prepared for U.S. Section to NASCO: Portland, ME, 2010.

20. Byron, C.; Bengtson, D.; Costa-Pierce, B.; Calanni, J. Integrating science into management: ecological carrying capacity of bivalve shellfish aquaculture. Mar. Policy **2011a**, *35* (3), 363–370.

21. Byron, C.; Link, J.; Bengtson, D.; Costa-Pierce, B. Calculating carrying capacity of shellfish aquaculture using mass-balance modeling: Narragansett Bay, Rhode Island. Ecol. Model. **2011b**, *222*, 1743–1755.

22. Byron, C.; Link, J.; Costa-Pierce, B.; Bengtson, D. Modeling ecological carrying capacity of shellfish aquaculture in highly flushed temperate lagoons. Aquaculture **2011c**, *314* (1–4), 87–99.

23. Neori, A.; Chopin, T.; Troell, M.; Buschmann, A.H.; Kraemer, G.P.; Halling, C.; Shpigel, M.; Yarish, C. Integrated aquaculture: rationale, evolution and state of the art emphasizing seaweed biofiltration in modern mariculture. Aquaculture **2004**, *231* (1–4), 361–391.

24. Hopkins, J.S.; Devoe, M.R.; Holland, A.F. Environmental impacts of shrimp farming with special reference to the situation in the continental United States. Estuaries **1995**, *18* (1A), 25–42.

25. Chambers, M.D.; Howell, W.H. Preliminary information on cod and haddock production in submerged cages off the coast of New Hampshire, USA. ICES J. Mar. Sci. **2006**, *63* (2), 385–392.

26. Fredriksson, D.W.; DeCew, J.; Swift, M.R.; Tsukrov, I.; Chambers, M.D.; Celikkol, B. The design and analysis of a four-cage grid mooring for open ocean aquaculture. Aquacultural Eng. **2004**, *32* (1), 77–94.

27. Fredriksson, D.W.; Muller, E.; Baldwin, K.; Robinson Swift, M.; Celikkol, B. Open ocean aquaculture engineering: system design and physical modeling. Mar. Technol. Soc. J. **2000**, *34* (1), 41–52.

28. Fredriksson, D.W.; Swift, M.R.; Eroshkin, O.; Tsukrov, I.; Irish, J.D.; Celikkol, B. Moored fish cage dynamics in waves and currents. J. Ocean Eng. **2005**, *30* (1), 28–36.

29. Howell, W.H.; Chambers, M.D. Growth performance and survival of Atlantic halibut (*Hippoglossus hippoglossus*) grown in submerged net pens. Bulletin of the Aquaculture Assoc. Canada **2005**, *9*, 35–37.

30. Callier, M.D.; Richard, M.; McKindsey, C.W.; Archambault, P.; Desrosiers, G. Responses of benthic macrofauna and biogeochemical fluxes to various levels of mussel biodeposition: an in situ "benthocosm" experiment. Mar. Pollut. Bull. **2009**, *58* (10), 1544–1553.

31. Forrest, B.M.; Keeley, N.B.; Hopkins, G.A.; Webb, S.C.; Clement, D.M. Bivalve aquaculture in estuaries: Review and synthesis of oyster cultivation effects. Aquaculture **2009**, *298* (1–2), 1–15.

32. McKindsey, C.; Landry, T.; O'Beirn, FX.; Davies, IM. Bivalve aquaculture and exotic species: a review of ecological considerations and management issues. J. Shellfish Res. **2007**, *26* (2), 281–294.

33. McKindsey, C.W.; Lecuona, M.; Huot, M.; Weise, A.M. Biodeposit production and benthic loading by farmed mussels and associated tunicate epifauna in Prince Edward Island. Aquaculture **2009**, *295* (1–2), 44–51.

34. Nakamura, Y.; Kerciku, F. Effects of filter-feeding bivalves on the distribution of water quality and nutrient cycling in a eutrophic coastal lagoon. J. Mar. Syst. **2000**, *26* (2), 209–221.

35. Nugues, M.M. Benthic community changes associated with intertidal oyster cultivation. Aquaculture Res. **1996**, *27*, 913–924.

36. Sara, G. Aquaculture effects on some physical and chemical properties of the water column: a meta-analysis. Chem. Ecol. **2007a**, *23* (3), 251–262.

37. Sara, G. Ecological effects of aquaculture on living and non-living suspended fractions of the water column: a meta-analysis. Water Res. **2007b**, *41* (15), 3187–3200.

38. Sara, G. A meta-analysis on the ecological effects of aquaculture on the water column: dissolved nutrients. Mar. Environ. Res. **2007c**, *63* (4), 390–408.

39. Smaal, A.; van Stralen, M.; Schuiling, E. The interaction between shellfish culture and ecosystem processes. Canad. J. Fisheries Aquatic Sci. **2001**, *58*, 991–1002.

40. Costa-Pierce, B.A. *Ecological Aquaculture: The Evolution of the Blue Revolution*. John Wiley & Sons Blackwell Science, Oxford, 2008.

41. Stadmark, J.; Conley, D.J. Mussel farming as a nutrient reduction measure in the Baltic sea: consideration of nutrient biogeochemical cycles. Mar. Pollut. Bull. **2011**, *62* (7), 1385–1388.

42. White, C.; Halpern, B.S.; Kappel, C.V. Ecosystem service tradeoff analysis reveals the value of marine spatial planning for multiple ocean uses. Proc. Natl. Acad. Sci. **2012**, *109* (12), 4696–4701.

43. Kerr, S.; Lauder, G.; Fairman, D. Towards design for a nutrient trading programme to improve water quality in lake Rotorua 07-03. Motu Work. Paper **2007**.

44. Chopin, T. Integrated multi-trophic aquaculture. What it is, and why you should care... and don't confuse it with polyculture. Northern Aquaculture **2006**, *4*.

45. Pietrak, M.R.; Molloy, S.D.; Bouchard, D.A.; Singer, J.T.; Bricknell, I. Potential role of *Mytilus edulis* in modulating the infectious pressure of *Vibrio anguillarum* 02β on an integrated multi-trophic aquaculture farm. Aquaculture **2012**, *326–329*, 36–39.

46. FAO. *Sea Cucumbers. A Global Review of Fisheries and Trade*. FAO Fisheries and Aquaculture Technical Paper: Rome, Italy, 2008; 516.

47. Murshed-E-Jahan, K.; Pemsl, D.E. The impact of integrated aquaculture-agriculture on small-scale farm sustainability and farmers livelihoods: Experience from Bangladesh. Agricultural Syst. **2011**, *104*, 392–402.

48. Buck, B.H.; Ebeling, M.;Michler-Cieluch, T. Mussel cultivation as a co-use in offshore wind farms: potentials and economic feasibility. Aquaculture Econ. Manag. **2010**, *14* (4), 1365–7305.

49. Buck, B.H.; Krause, G.; Michler-Cieluch, T.; Buchholz, B.M.; Busch, C.M.; Fisch, J.A.; Geisen, R.; Zielinski, M. Meeting the quest for spatial efficiency: progress and prospects of extensive aquaculture within offshore wind farms. Helgoland Mar. Res. **2008**, *62*, 269–281.

50. Michler-Ceiluch, T.; Krause, G.; Buck, B. Marine aquaculture within offshore wind farms: Social aspects of multiple use planning. GAIA Ecol. Perspect. Sci. Soc. **2009a**, *18* (2), 158–162.

51. Michler-Ceiluch, T.; Krause, G.; Buck, B. Reflections on integrating operation and maintanence activities of offshore wind farms and mariculture. Open Coastal Manag. **2009b**, *52* (1), 57–68.

52. Soto, D.; Manjarez, J.A., Eds.; *Building an Ecosystem Approach to Aquaculture. FAO/Universitat de les Illes Balears Expert Workshop. 7–11 May 2007*; FAO Fisheries and Aquaculture Proceedings: No. 14. Rome, FAO., Palma de Mallorca, Spain, **2008**.

53. Inglis, G.J.; Hayden, B.J.; Ross, A.H. *An Overview of Factors Affecting the Carrying Capacity of Coastal Embayments for Mussel Culture*. Client Report CHC00/69: NIWA, Christchurch, **2002**.

54. McKindsey, C.W.; Thetmeyer, H.; Landry, T.; Silvert, W. Review of recent carrying capacity models for bivalve culture and recommendations for research and management. Aquaculture **2006**, *261* (2), 451–462.

Aquifers: Groundwater Storage

Thomas J. Burbey
Department of Geosciences, Virginia Tech, Blacksburg, Virginia, U.S.A.

Abstract

Aquifer storage describes the volume of water released from or added to a unit area of aquifer resulting from a unit change in hydraulic head. The source of water released from storage in confined aquifers is matrix compression and water expansion with the storage coefficient ranging from 0.005 to 0.00005. The source of water released from unconfined aquifers is largely due to dewatering of the pore space and is called specific yield with values ranging from 0.01 to 0.30. The volume of water released from confining units and horizontal strain can yield significant quantities of water to aquifers during long-term pumping events.

INTRODUCTION

All rocks or sediments that contain pore space have the capacity to store water. However, the concept of aquifer storage has more to do with how water is released from the rock or sediments than the fact that water is simply contained in these geologic materials. The idea that water can be "squeezed" out of a rock or sediment infers the concept that aquifers can store and release water. This idea was perhaps first formally described by O.E Meinzer,[1] who noted the occurrence of abrupt water-level fluctuations due to the passing of trains over a railroad near a well. Meinzer concluded that "aquifers are apparently all more or less compressible and elastic though they differ widely in the degree and relative importance of these properties." Shortly thereafter, Theis[2] formulated one of the most commonly used expressions today that describe the cylindrical drawdown response of a pumping well as being dependent on the aquifer transmissivity and storativity (storage coefficient) without formulating a rigorous description of the storage coefficient. Low-storage materials tend to have broader and deeper drawdown cones than materials with larger storage values. Theis[3] recognized that water released to wells from storage was a result of aquifer compression in artesian (confined) aquifers or due to the physical drainage of water in unconfined aquifers. Jacob[4] derived a mathematical formulation of the storage coefficient by assuming that the compressibility of the aquifer material under stress from a reduction in hydraulic head occurs solely in the vertical direction and that storage in artesian aquifers is dependent on both skeletal (aquifer) and fluid compressibility. Thus, to provide a complete description of aquifer storage, the concept of compressibility must first be developed.

COMPRESSIBILITY

Compressibility is simply defined as the ratio of strain over stress and describes how a material or fluid deforms as a result of an applied external stress. Strain refers to how a material deforms relative to an original or previous state. When Jacob[4] defined strain, he made a convenient mathematical decision to assume that all the strain induced by an imposed stress occurs in the vertical direction only. For now, we will also make this assumption and then later qualitatively address this issue in three dimensions. Stress represents the weight per unit area acting at any given location within an aquifer system. Consider an arbitrary vertical plane through a saturated porous media as depicted in Fig. 1. According to Terzaghi's principle of effective stress,[5] the total stress acting on the plane is balanced by the pore pressure and effective stress as

$$\sigma_T = p + \sigma_e. \qquad (1)$$

That is, the total weight per unit area acting on the plane from above, from both the porous material and fluid, is supported by the buoyant fluid stress and the grain-to-grain stress of the matrix. The differential of each term in Eq. 1 describes how the stress state may change in time. In confined aquifer systems, the total stress can be considered to be constant; therefore, $d\sigma_e = -dp$. Under such a condition, a decline in the pore pressure (say, from pumping) will result in an equal and opposite increase in effective stress. That increase in effective stress is what causes the matrix to compress. The pore pressure can be converted to hydraulic head as $p = \rho_w g(h - z)$ where ρ_w is the fluid density, g is the gravitational constant, h is the hydraulic head, and z is the elevation head. If z is constant (negligible vertical compaction at the plane of interest relative to the change in

Encyclopedia of Natural Resources DOI: 10.1081/E-ENRW-120047545

Total stress

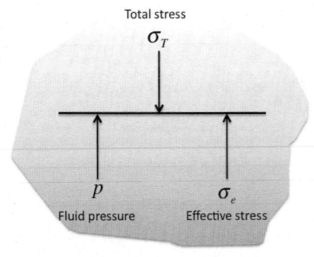

Fig. 1 Depiction of Terzaghi's principle of effective stress on an arbitrary plane within a saturated aquifer.

hydraulic head), then the change in pressure is directly related to a change in hydraulic head as

$$dp = \rho_w g dh. \tag{2}$$

Now that the components of compressibility are defined, one can surmise that compressibility represents a material property that can be applied to the pore fluid (water), the granular matrix or aquifer material, and the individual grains that make up the aquifer system framework. This latter type of compressibility is typically ignored in groundwater systems because its contribution is typically considered to be negligible in most aquifer environments. Now a reduction in hydraulic head can both cause the fluid to expand and the matrix to contract or compress. Thus, both matrix or porous media compressibility and fluid compressibility need to be defined before the storage contributions from these constituents can be properly defined.

Compressibility of the porous media is defined as

$$\alpha = \frac{dV_T / V_T}{d\sigma_e}. \tag{3}$$

where dV_T is the change in the total volume of the medium (aquifer material) from its initial state V_T. Again, for convenience, here we assume that the volume change occurs only in the vertical direction. The total volume is the sum of the volume of solids, V_s and the volume of voids, V_e. That is, $V_T = V_s + V_v$. If we examine the change in volume resulting from a change in effective stress, then $dV_T = dV_s + dV_v$ but because the change in the volume of solids is considered negligible (that is, the compressibility of individual grains), $dV_s = 0$. Therefore, $dV_T = dV_v$.

Table 1 The range of compressibility of some common geologic materials and pure water

Material	Compressibility, α (m^2/N)
Clay	10^{-6} to 10^{-8}
Sand	10^{-7} to 10^{-9}
Gravel	10^{-8} to 10^{-10}
Fractured rock	10^{-8} to 10^{-10}
Unfractured rock	10^{-9} to 10^{-11}
Water (β)	4.4×10^{-10}

Source: Modified from Freeze and Cherry.[17]

Water compressibility under isothermal conditions is defined as

$$\beta = \frac{-dV_w / V_w}{dp} \tag{4}$$

where dV_w is the change in the volume of water from an initial state V_w. The negative sign indicates that when the pressure decreases, there is an increase in fluid volume. In other words, fluid expands when the pressure declines. This is analogous to an expanding helium balloon that ascends into the atmosphere as a result of the declining atmospheric pressure with increasing altitude. Table 1 lists the range of compressibility for common geologic materials and pure water.

CONFINED AQUIFER STORAGE

The storage coefficient in confined aquifers is defined[6] as the volume of water released from storage (dV_w) over a unit cross-sectional area of aquifer (A) per unit decline in hydraulic head ($-dh$) and is expressed as

$$S = \frac{-dV_w}{A dh}. \tag{5}$$

The specific storage, S_s, is related to the storage coefficient through the aquifer thickness, $S = S_s b$, and is defined as the volume of water released by a unit volume of aquifer due to a unit decline in hydraulic head

$$S_s = \frac{-dV_w}{V_T dh}. \tag{6}$$

It should be stressed that the confined aquifer is assumed to remain saturated; that is, the head, even after the water-level decline, remains higher than the top of the aquifer as depicted in Fig. 2.

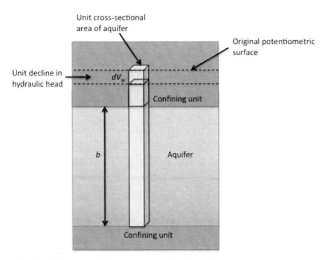

Fig. 2 Diagram illustrating confined aquifer storage.

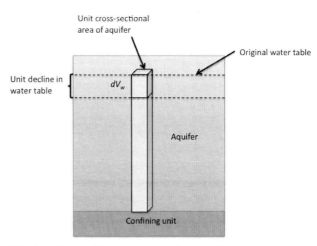

Fig. 3 Diagram illustrating unconfined aquifer storage.

The question then is what is the source of the volume of water, dV, that is, either released from storage when the water level is lowered or added to storage if the water level is increased?

We saw from Terzaghi's principle of effective stress[5] that the volume of water comes from the sum of the volumes of water released from expansion of water and compression of the aquifer matrix. Confined aquifers, by definition, are saturated so that $dV_T = dV_v = dV_w$. Substituting this relation into Eq. 3 and rearranging to solve for dV_w yields

$$dV_w = aV_T d\sigma_e = -aV_T \rho_w g dh. \tag{7}$$

Equation 4 can also be rearranged to solve for dV_w and when combined with Eq. 2, we obtain

$$dV_w = -\beta V_w \rho_w g dh. \tag{8}$$

Now summing Eqs. 7 and 8 and recognizing that $V_w = nV_T$, yields

$$dV_w = -\rho_w g dh V_T (a + n\beta).$$

Substituting the above expression into Eq. 5 yields the expression for specific storage (1/L) used in the groundwater flow equation:

$$S_s = \rho_w g (a + n\beta). \tag{9}$$

The related storage coefficient (dimensionless) is

$$S = \rho_w g b (a + n\beta). \tag{10}$$

The range of storage coefficient values for confined aquifers is typically from 0.005 to 0.00005.

UNCONFINED AQUIFER STORAGE

The source of water released from (or added to) storage in unconfined aquifers is largely the result of draining or dewatering of the pore space due to a water table decline and is referred to as specific yield. The specific yield is defined as the volume of water released from storage per unit surface area of aquifer per unit decline in the water table:[6]

$$S_y = \frac{dV_w}{A dh}. \tag{11}$$

Unconfined storage is depicted in Fig. 3. Because the specific yield represents a draining of the pore space, it is directly related to the porosity by the simple expression $n = S_y + S_r$, where S_r is the specific retention and represents the portion of the water retained in the matrix after gravity drainage. The specific yield can never be larger than the porosity.

The total storativity (equivalent to storage coefficient) of unconfined aquifers is accurately represented as

$$S = S_y + hS_s \tag{12}$$

where the S_s here represents the sources of storage from aquifer compression and water expansion. Because the quantity hS_s is usually 2–4 orders of magnitude less than S_y, the unconfined storativity is typically considered to be the specific yield. The typical range of values for specific yield is 0.01–0.30.

Even though S_y is much larger than typical values of S_s, the total storativity of confined aquifers is often much larger than that of unconfined aquifers because confined aquifers tend to be much thicker and laterally far more extensive than unconfined aquifers.

CONFINING UNITS

Confining units or aquitards often bound aquifers above and below and do not affect the storativity of the aquifer itself, but they can potentially greatly influence the quantity of water available from storage in an aquifer that is being pumped. Confining units, because they are generally composed largely of clays, have a very high compressibility and porosity (Table 1) and therefore a high specific storage relative to the aquifer. However, the confining units also have a very low hydraulic conductivity, and therefore the head change across the confining unit can greatly lag behind that of the aquifer. The one-dimensional diffusion equation can be used to describe how the head change propagates through the confining layer over time. As this occurs, the confining unit releases water to the aquifer in which the decline in head has occurred, thus providing additional storage to the aquifer that can be significant, especially when pumping occurs over long time periods—typically years.

THREE-DIMENSIONAL STORAGE

In the previous analysis, the compressibility, a, is represented as a one-dimensional quantity where strain (numerator in Eq. 3) is assumed to occur in the vertical direction only. The assumption of purely vertical strain proposed by Jacob[4] was made on the basis of mathematical expediency and intuitional reason. However, Biot[7,8] and later Rice and Cleary[9] formulated expressions for the coupling of fluid flow with three-dimensional matrix deformation. These formulations result from the fact that the strain term in a must now be written as the sum of the directional components of strain, $\varepsilon_x + \varepsilon_y + \varepsilon_z$, whose sum represents the volume strain. These three terms along with fluid pressure or hydraulic head result in a system of four equations and four unknowns. The formulation of these coupled expressions leads to a new definition of specific storage that contains two elastic parameters for characterizing volume compressibility of the aquifer matrix:[10]

$$S_s = \rho_w g \left(\frac{3(1-2v)}{2G(1+v)} + n\beta \right) \tag{13}$$

where G is the shear modulus and v is Poisson's ratio. The shear modulus is the ratio of shear stress to shear strain while the Poisson's ratio is the negative ratio of transverse to axial strain. It should be pointed out that this formulation is not compatible with the standard groundwater flow equation. The derivations of the equations leading to Eq. 13 can be found in Burbey and Helm[11] and Burbey,[10] with Verruijt[12] providing the mathematical foundation from first principles.

The importance of horizontal strain became apparent from the deformation that resulted from oil withdrawal in the Wilmington Oil Field beginning in Long Beach,

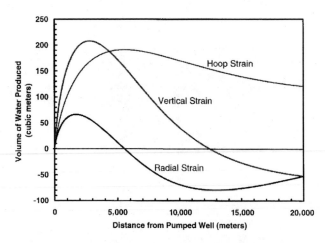

Fig. 4 Simulation showing relative volumes of water removed from storage (positive) or added to storage (negative) from the three strain components as a function of distance from the pumped well for a purely homogeneous isotropic aquifer.
Source: Adapted from Burbey.[16]

California, in the 1930s.[13] Wolff[14] was perhaps the first to conduct a controlled aquifer test to explicitly measure horizontal strain and found the horizontal strains to be significant and form a radial pattern as a function of distance from the pumped well. Burbey[15] measured the horizontal strains from a pumping test and used them to characterize aquifer properties in the absence of water-level data.

Figure 4 illustrates the potential storage contribution of each strain component (vertical, radial, and hoop or tangential strain) from a numerical analysis of a 100-day pumping test of an infinite homogeneous and isotropic confined aquifer where the overlying and underlying confining units do not influence the simulation results.[16] The results show that radial strain provides a sizable contribution of the water released from storage near the pumping well. Furthermore, as the distance from the pumping well increases, hoop strain becomes the primary source of water released from storage. The importance of horizontal strain can vary depending on the aquifer system being evaluated.

CONCLUSION

Storage is one of the most important aquifer properties that influences the distribution and shape of drawdown and directly affects the quantity of water that can be extracted from pumping wells. Classical formulations of storage only consider vertical strain in the formulation of aquifer storage. However, more recent investigations show that the volume of water released from storage in confined aquifers during pumping is the result of both vertical and horizontal aquifer compression and water expansion in the vicinity of the pumping well. Unconfined storage is largely due to dewatering of the aquifer by gravity drainage during a decline in the water table.

REFERENCES

1. Meinzer, O.E. Compressibility and elasticity of artesian aquifers. Econ. Geol. **1928**, *23*, 263–291.
2. Theis, C.V. The relation between the lowering of the piezometric surface and the rate and duration of discharge of a well using ground-water storage. Trans. Am. Geophys. Union **1935**, *16*, 519–524.
3. Theis, C.V. The significance and nature of the cone of depression in ground-water bodies. Econ. Geol. **1938**, *33*, 889–902.
4. Jacob, C.E. On the flow of water in an elastic artesian aquifer, EOS Trans. Am. Geophysi. Union **1940**, *21*, 574–586.
5. Terzaghi, K. Principles of soil mechanics: IV. settlement and consolidation of clay. Eng. News-Rec. **1925**, *95*, 874–878.
6. Hantush, M.S. Drawdown around a partially penetrating well. Am. Soc. Civil Eng. Trans. **1962**, *27* (part 1), 268–283.
7. Biot, M.A. General theory of three-dimensional consolidation. J. Appl. Phys. **1941**, *12* (2), 155–164.
8. Biot, M.A. Theory of elasticity and consolidation for a porous anisotropic solid. J. Appl. Phys. **1955**, *26* (2), 182–185.
9. Rice, J.R.; Cleary, M.P. Some basic stress diffusion solutions for fluid-saturated elastic porous media with compressible constituents. Rev. Geophys. Space Phys. **1976**, *14*, 227–241.
10. Burbey, T.J. Effects of horizontal strain in estimating specific storage and compaction in confined and leaky aquifer systems. Hydrogeol. J. **1999**, *7* (6), 521–532.
11. Burbey, T.J.; Helm, D.C. Modeling three-dimensional deformation in response to pumping of unconsolidated aquifers. Environ. Eng. Geosci. **1999**, *5* (2), 199–212.
12. Verruijt, A. Elastic storage of aquifers. In *Flow Through Porous Media*; DeWiest, R.J.M., Ed.; Academic Press: New York, 1969; 331–376.
13. Yerkes, R.F.; Castle, R.O. Surface deformation associated with oil and gas field operations in the United States. In *Land Subsidence*; Tison L.J., Ed., Vol 1. International Association. Sci Hydrol Pub. **1969**, *88*, 55–64.
14. Wolff, R.G. Relationship between horizontal strain near a well and reverse water level fluctuation. Water Resour. Res. **1970**, *6*, 1721–1728.
15. Burbey, T.J. Three-dimensional deformation and strain induced by municipal pumping, part 1: Analysis of field data. J. Hydrol. **2006**, *319* (1–4), 123–142.
16. Burbey, T.J. Storage coefficient revisited: Is purely vertical strain a good assumption? Ground Water **2001**, *30* (3), 458–464.
17. Freeze, R.A.; Cherry, J.A. *Groundwater*; Prentice-Hall, Inc.: Englewood Cliffs, New Jersey, 1979; 604 p.

BIBLIOGRAPHY

1. Green, D.H.; Wang, H.F. Specific storage as a poroelastic coefficient. Water Resour. Res. **1990**, *26*, 1631–1637.
2. Lohman, S.W. Ground-water hydraulics. In *U.S. Geological Survey Professional Paper*; 1972; Vol. 708, 70 p.
3. Wang, H.F. *Theory of Linear Poroelasticity*; Princeton University Press: Princeton, New Jersey, 2000; 287 p.

Aquifers: Recharge

John Nimmo
David A. Stonestrom
U.S. Geological Survey (USGS), Menlo Park, California, U.S.A.

Richard W. Healy
U.S. Geological Survey (USGS), Lakewood, Colorado, U.S.A.

Abstract
Aquifer recharge is water that moves from the unsaturated zone into the saturated zone. It might be more precisely called "aquifer-system" or "saturated-zone" recharge. Recharge varies considerably with time and location. Quantitative estimation of recharge rate contributes to the understanding of large-scale hydrologic processes.

INTRODUCTION

Aquifer recharge was defined by Meinzer[1] and Heath[2] as water that moves from the land surface or the unsaturated zone into the saturated zone. This definition excludes saturated flow between aquifers, which avoids double-accounting in large-scale studies, so it might be more precisely called "aquifer-system" or "saturated-zone" recharge. *Recharge rate* designates either a flux [L^3/T] into a specified portion of aquifer, or a flux density [L/T] into an aquifer at a point. Sources of water for recharge include precipitation that infiltrates, permanent or ephemeral surface water, irrigation, and artificial recharge ponds. Recharge may reach the aquifer directly from portions of rivers, canals, or lakes,[3] though usually it first travels by various means through the unsaturated zone.

Recharge varies considerably with time and location. Temporal variation occurs, for example, with seasonal or short-term variations in precipitation and evapotranspiration (ET). This variability is especially evident in thin unsaturated zones, where recharge may occur within a short time of infiltration. In deep unsaturated zones, recharge may be homogenized over several years so that it may occur with essentially constant flux even though fluxes at shallow depths are erratic. Spatial variation occurs with climate, topography, soils, geology, and vegetation. For example, a decrease of slope or increase of soil permeability may lead to greater infiltration and greater recharge. Many applications use a concept of recharge that is time-averaged or areally averaged.

Both the amount of infiltration and the fraction of it that becomes recharge tend to be greater with more abundant water, so the recharge process is most efficient if infiltration is concentrated in space and time. Because ET may extract most or all of the water that infiltrates, water is more likely to become recharge if it moves rapidly below the root zone. Temporal concentration occurs during storms, floods, and snowmelt, when ongoing processes such as ET are overwhelmed. Spatial concentration typically occurs in depressions and channels, where higher water contents promote rapid movement by increasing the hydraulic conductivity (K), the amount of preferential flow, and the downward driving force at a wetting front. Quantitative estimation of recharge rate contributes to the understanding of large-scale hydrologic processes. It is important for evaluating the sustainability of ground water supplies, though it does not equate with a sustainable rate of extraction.[4] Because it represents a first approximation to the rate of solute transport to the aquifer, the recharge rate is also important to estimate contaminant fluxes and travel times from sources near the land surface. Methods for obtaining a quantitative estimate of recharge mostly require a combination of various types of data which themselves may be hard to estimate, so in general it is wise to apply multiple methods and compare their results.

WATER BUDGET METHODS

The water balance for a basin can be stated as

$$P + Q_{on}^{sw} + Q_{on}^{gw} = ET^{sw} + ET^{uz} + ET^{gw} + Q_{off}^{sw} \\ + Q_{off}^{gw} + Q_{bf} + \Delta S^{snow} \\ + \Delta S^{sw} + \Delta S^{uz} + \Delta S^{gw} \quad (1)$$

where P is precipitation and irrigation; Q_{on} and Q_{off} are water flow on and off of the site, respectively; Q_{off}^{sw} is runoff; Q_{bf} is baseflow (ground water discharge to streams or springs); and ΔS is change in water storage. Superscripts refer to surface water, ground water, unsaturated zone, or snow, and all parameters are in units of L/T (or volume per

Encyclopedia of Natural Resources DOI: 10.1081/E-ENRW-120010040

unit surface area per unit time). For the saturated zone only, a water balance can be written for a defined area as

$$R = \Delta Q^{gw} + Q_{bf} + ET^{gw} + \Delta S^{gw} \tag{2}$$

where R is recharge and ΔQ^{gw} is the difference between ground water flow off of and onto the basin. This equation implies that water arriving at the water table: 1) flows out of the basin as ground water flow; 2) discharges to the surface; 3) is evapotranspired; or 4) goes into storage. Substitution in Eq. (1) produces a simpler form of the water balance:

$$R = P + Q_{on}^{sw} - Q_{off}^{sw} - ET^{sw} - ET^{uz} - \Delta S^{snow} \\ - \Delta S^{sw} - \Delta S^{uz} \tag{3}$$

Water budget methods include all techniques based, in one form or another, on one of these water balance equations.

The most common water budget method is the "residual" approach: all other components in the water budget are measured or estimated and R is set equal to the residual. Water budget methods can be applied over a wide range of space and time scales. The major limitation of the residual approach is that the accuracy of the recharge estimate depends on the accuracy with which other components can be measured. This limitation can become significant when the magnitude of R is small relative to other variables. The time scale for applying water budget methods is important, with more frequent tabulations likely to improve accuracy. If the water budget is calculated daily, P can greatly exceed ET on a single day, even in arid settings. Averaging over longer time periods tends to dampen out extreme precipitation events and hence underestimate recharge. Annual recharge estimated with water budgets range from 23 mm in a region of India[5] to 400 mm at a site in the eastern United States.[6]

Watershed, surface water flow, and ground water flow models constitute an important class of water budget methods that have been used to estimate of recharge.[7] An attractive feature of models is their predictive capability. They can be used to gauge the effects of future climate or land-use changes on recharge rates.

METHODS BASED ON SURFACE WATER OR GROUND WATER DATA

Fluctuations in ground water levels can be used to estimate recharge to unconfined aquifers according to

$$R = S_y \, dh/dt = S_y \, \Delta h/\Delta t \tag{4}$$

where S_y is specific yield, h is water table height, and t is time. The method is best applied over short time periods in regions with shallow water tables that display sharp water-level rises and declines. Analysis of water-level fluctuations can also, however, be useful for determining the magnitude of long-term change in recharge rates caused by climate or land-use change. The method is only appropriate for estimating recharge for transient events; recharge occurring under steady flow conditions cannot be estimated.

Difficulties lie in determining S_y and ensuring that fluctuations are due to recharge, not to changes in pumping rates or atmospheric pressure or other phenomena. Recharge rates estimated by this technique range from 11 mm over a 26-month period in Saudi Arabia[8] to 541 mm yr^{-1} over 1 yr for a small basin in the United States.[9]

Ground water levels can also be used to estimate flow, Q, through a cross-section of an aquifer that is aligned with an equipotential line. Multiplying K by the hydraulic gradient normal to the section times the area of the section calculates Q. Recharge is determined by dividing Q by the surface area of the aquifer upgradient from the section.

Methods of estimating recharge based on surface-water data include the Channel Water Balance Method (CWBM) and determination of baseflow by hydrograph separation. The CWBM involves measuring discharge at two gauges on a stream; the difference in discharge between the upstream and downstream gauges is the transmission loss. This loss may become recharge, ET, or bank storage. Hydrograph separation involves identifying what portion of gauged stream flow is derived from ground water discharge. Rutledge and Daniel[10] developed an automated technique for this purpose and applied the method to estimate recharge at 15 sites. Drainage areas for the sites ranged from less than 52 km^2 to more than 5200 km^2; estimated annual recharge was between about 13 cm and 64 cm.

DARCIAN METHODS

Applied in the unsaturated zone, Darcy's law gives a flux density equal to K times the driving force, which equals the recharge rate if certain conditions apply. Matric-pressure gradients must be measured or demonstrated to be negligible. Some types of preferential flow are inherently non-darcian and if important would need to be determined separately. Accurate measurements are necessary to know K adequately under field conditions at the point of interest. For purposes requiring areal rather than point estimates, additional interpretation and calculation are necessary.

In the simplest cases, in a region of constant downward flow in a deep unsaturated zone, gravity alone drives the flow. With a core sample from this zone, laboratory K measurements at the original field water content directly indicate the long-term average recharge rate.[11]

In the general case, transient water contents and matric pressures must be measured in addition to K.[12] Transient recharge computed with Darcy's law can relate to storms or other short-term events, or provide data for integration into temporal averages.

TRACER METHODS

Increasing availability and precision of physical and chemical analytical techniques have led to a proliferation in applications of tracer methods for recharge estimation. Isotopic and chemical tracers include tritium, deuterium, oxygen-18, bromide, chloride, chlorine-36, carbon-14, agricultural chemicals, dyes, chlorofluorocarbons, and noble gases. In practice, concentrations measured in pore water are related to recharge by applying chemical mass-balance equations, by matching patterns inherited from infiltrating water, or by determining the age of the water. Tracer methods provide point and areal estimates of recharge. Multiple tracers used together can test assumptions and constrain estimates.

The most common tracer for estimating recharge is chloride. Chloride continually arrives at the land surface in precipitation and dust. Chloride is conservative in many environments and is non-volatile. Under suitable conditions,

$$R = P[Cl_p]/[Cl_r] \qquad (5)$$

where P is precipitation and $[Cl_p]$ and $[Cl_r]$ are chloride concentrations in precipitation and pore water, respectively. Chloride mass-balance methods can be applied to unsaturated profiles[13] and entire basins.[14]

Isotopic composition of water provides a useful tracer of the hydrologic cycle. The isotopic makeup of precipitation varies with altitude, season, storm track, and other factors. Recharge estimates using isotopic varieties of water usually employ temporal or geographic trends in infiltrating water.

Non-conservative tracers can indicate the length of time that water is isolated from the atmosphere, that is, its "age." Recharge rates can be inferred from water ages if mixing is small. If ages are known along a flow line,

$$R = \theta L / (A_2 - A_1) \qquad (6)$$

where θ is volumetric water content, A_1 and A_2 are ages at two points, and L is separation length. One point is often located at the water table. Preindustrial water can be dated by decay of predominately cosmogenic radioisotopes, including carbon-14 and chlorine-36. The abundance of tritium and other radioisotopes increased greatly during atmospheric weapons testing, labeling recent precipitation. Additional compounds for dating modern recharge include chlorofluorocarbons, krypton-85, and agricultural chemicals.[15,16]

Heat is yet another tracer of ground water recharge. Daily, seasonal, and other temperature fluctuations at the land surface produce thermal signals that can be traced through shallow profiles.[17,18] Water moving through deeper profiles alters geothermal gradients, which can be used in inverse modeling to obtain recharge rates.[19]

OTHER METHODS

Additional geophysical techniques provide recharge estimates based on the water-content dependence of gravitational, seismic, and electromagnetic properties of earth materials. Repeated high-precision gravity surveys can indicate changes in the quantity of subsurface water from recharge events.[20] Similarly, repeated surveys using seismic or ground-penetrating-radar equipment can resolve significant changes in water-table elevation associated with transient recharge.[21] In addition to surface-based techniques, cross-bore tomographic imaging can provide detailed three-dimensional reconstructions of water distribution and movement during periods of recharge.[22]

REFERENCES

1. Meinzer, O.E. *The Occurrence of Ground Water in the United States, with a Discussion of Principles*; U.S. Geological Survey Water-Supply Paper 489: 1923; 321 pp.
2. Heath, R.C. *Ground-Water Hydrology*; U.S. Geological Survey Water-Supply Paper 2220: 1983; 84 pp.
3. Winter, T.C.; Harvey, J.W.; Franke, L.O.; Alley, W.M. *Ground Water and Surface Water a Single Resource*; U.S. Geological Survey Circular 1139: 1998; 79 pp.
4. Bredehoeft, J.D.; Papadopoulos, S.S.; Cooper, H.H., Jr. *Groundwater: The Water-Budget Myth, Scientific Basis of Water-Resource Management: Studies in Geophysics*; National Academy Press: Washington, DC, 1982; 51–57.
5. Narayanpethkar, A.B.; Rao, V.V.S.G.; Mallick, K. Estimation of groundwater recharge in a basaltic aquifer. Hydrol. Proc. **1994**, *8*, 211–220.
6. Steenhuis, T.S.; Jackson, C.D.; Kung, S.K.; Brutsaert, W. Measurement of groundwater recharge in eastern long Island, New York, U.S.A. J. Hydrol. **1985**, *79*, 145–169.
7. Bauer, H.H.; Mastin, M.C. *Recharge from Precipitation in Three Small Glacial-Till-Mantled Catchments in the Puget Sound Lowland*, U.S. Geological Survey Water-Resources Inv. Rep. 96-4219, Washington, 1997; 119 pp.
8. Abdulrazzak, M.J.; Sorman, A.U.; Alhames, A.S. Water balance approach under extreme arid conditions—aquifers: Recharge 57 case study of Tabalah Basin, Saudi Arabia. Hydrol. Proc. **1989**, *3*, 107–122.
9. Rasmussen, W.C.; Andreasen, G.E. *Hydrologic Budget of the Beaverdam Creek Basin*; U.S. Geol. Survey. Water-Supply Paper 1472, Maryland, 1959; 106 pp.
10. Rutledge, A.T.; Daniel, C.C. Testing an automated method to estimate ground-water recharge from streamflow records. Ground Water **1994**, *32*(2), 180–189.
11. Nimmo, J.R.; Stonestrom, D.A.; Akstin, K.C. The feasibility of recharge rate determinations using the steady state centrifuge method. Soil Sci. Soc. Am. J. **1994**, *58*, 49–56.
12. Freeze, A.R.; Banner, J. The mechanism of natural ground-water recharge and discharge 2. Laboratory column experiments and field measurements. Water Resour. Res. **1970**, *6* (1), 138–155.
13. Bromley, J.; Edmunds, W.M.; Fellman, E.; Brouwer, J.; Gaze, S.R.; Sudlow, J.; Taupin, J.-D. Estimation of rainfall inputs and direct recharge to the deep unsaturated zone of southern Niger using the chloride profile method. J. Hydrol. **1997**, *188–189* (1–4), 139–154.
14. Anderholm, S.K. *Mountain-Front Recharge along the Eastern Side of the Middle Rio-Grande Basin;* U.S. Geological Survey Water-Resources Investigations Report 00-4010, Central New Mexico, 2000; 36 pp.
15. Ekwurzel, B.S.P.; Smethie, W.M., Jr.; Plummer, L.N.; Busenberg, E.; Michel, R.L.; Weppernig, R.; Stute,

M. Dating of shallow groundwater: comparison of the transient tracers 3H/3He, chlorofluorocarbons, and 85Kr. Water Resour. Res. **1994**, *30*(6), 1693–1708.

16. Davisson, M.L.; Criss, R.E. Stable isotope and groundwater flow dynamics of agricultural irrigation recharge into groundwater resources of the central valley, California. In *Proc. Isotopes in Water Resources Management*, Vienna (Austria), Mar 20–24, 1995; International Atomic Energy Agency: Vienna, 1996; 405–418.

17. Lapham, W.W. *Use of Temperature Profiles Beneath Streams to Determine Rates of Vertical Ground-Water Flow and Vertical Hydraulic Conductivity*; U.S. Geological Survey Water-Supply Paper 2337: 1989; 34 pp.

18. Constantz, J.; Thomas, C.L.; Zellweger, G. Influence of diurnal variations in stream temperature on streamflow loss and groundwater recharge. Water Resour. Res. **1994**, *30*, 3253–3264.

19. Rousseau, J.P.; Kwicklis, E.M.; Gillies, D.C.; Eds. *Hydrogeology of the Unsaturated Zone, North Ramp Area of the Exploratory Studies Facility, Yucca Mountain*, U.S. Geological Survey Water Resources Investigations Report 98-4050, Nevada, 1999.

20. Pool, D.R.; Schmidt, W. *Measurement of Ground-Water Storage Change and Specific Yield Using the Temporal-Gravity Method Near Rillito Creek, Tucson*; U.S. Geological Survey Water-Resources Investigations Report 97-4125, Arizona, 1997; 30 pp.

21. Haeni, F.P. *Application of Seismic-Refraction Techniques to Hydrologic Studies*; U.S. Geological Survey Open File Report 84-746, 1986; 144 pp.

22. Daily, W.; Ramirez, A.; LaBrecque, D.; Nitao, J. Electrical resistivity tomography of vadose water movement. Water Resour. Res. **1992**, *28*(5), 1429–1442.

Archaeological Oceanography

Michael L. Brennan
Center for Ocean Exploration, University of Rhode Island, Narragansett, Rhode Island, U.S.A.

Robert D. Ballard
Graduate School of Oceanography, University of Rhode Island, Narragansett, Rhode Island, U.S.A.

Abstract

Archaeological oceanography is a multidisciplinary field combining oceanography, archaeology, and ocean engineering to access and explore cultural sites in deep water. This subdiscipline is focused on the location, documentation, characterization, and in situ preservation of archaeological sites underwater. Additionally, these sites are utilized as platforms from which environmental studies can be conducted of the surrounding ocean. These data thereby aid interpretations of historical sites through the lens of the modern landscape that has come into equilibrium with the environment since the time of a site's abandonment. This field emphasizes the use of oceanography and science in the investigations of archaeological sites to maximize the potential for hard-to-reach sites at depth.

INTRODUCTION

Modern deep submergence technology has advanced in recent decades, increasing our ability to find archaeological sites in the deep ocean.[1,2] As the depths of exploration have expanded below those accessible by human divers, so too has our reliance on oceanographic methods and data for the investigation and interpretation of these sites. Archaeological oceanography is a multidisciplinary field that combines archaeology, oceanography, and ocean engineering to develop ways to locate and document archaeological sites in deep water. The methods with which we approach, investigate, and conserve these sites require an understanding of the environments in which these wrecks lie, and how these alien materials have impacted and have been impacted by the marine environment. Over the past few decades, archaeology has been increasingly reliant on the application of scientific techniques for the analysis of cultural materials. As the capabilities of archaeological science continue to expand, so too does our responsibility to survey, accurately document, and conserve these sites. The objective of archaeological oceanography as a multidisciplinary field is the location, documentation, characterization, and finally in situ preservation of underwater cultural heritage. These sites are then used as platforms from which long-term monitoring can be conducted to better understand the surrounding environment and how each site has come into equilibrium with it as part of the modern landscape.[3]

Why the title "Archaeological Oceanography"? George Bass, one of the founders of underwater archaeology, wrote that it is easier to teach an archaeologist to work underwater than it is to teach archaeology to an accomplished diver.[4] Similarly, Ballard writes, "An archaeological oceanographer is an archaeologist working in the ocean."[1] However, the compounding factor for conducting archaeology underwater is depth. The development of maritime archaeology included the adaptation of terrestrial archaeological methods to underwater sites: grids were set up over a wreck, the site was drawn, and then excavated by divers in its entirety.[4] This was when technological access was limited to shallow coastal zones. The discovery of ancient shipwrecks in deep water has typically indicated a better degree of preservation than shallow water–wreck sites due to their deposition in low-energy environments, where they are removed from dynamic coastal processes and human activity, and where sedimentation rates are low.[1] However, the cost, time, and technology required to work at sites at these depths vastly limit any attempt to fully excavate and recover a shipwreck site. Therefore, in the case of deepwater sites, the goal must shift from full-scale excavation to high-resolution imaging, environmental characterization, and in situ preservation. The inclusion of oceanography and ocean engineering then becomes essential, not only for the technology and methodology required to access deepwater sites, but also to develop and refine ways to approach and document them.

LOCATION

The exploration of our deep oceans was greatly furthered by President Clinton's Panel on Ocean Exploration, which recommended that the United States develop a

Encyclopedia of Natural Resources DOI: 10.1081/E-ENRW-120047573

national program in ocean exploration that includes both natural and social sciences.[5] The first objective listed in the panel's report is "Mapping the physical, geological, biological, chemical, and archaeological aspects of our ocean, such that the U.S. knowledge base is capable of supporting the large demand for this information from policy makers, regulators, commercial ventures, researchers, and educators."[5] Archaeological oceanography parallels this exploration initiative by aiming to document the locations of previously unknown archaeological sites in deep water.

Archaeological survey techniques in deep water require both acoustic and visual imaging including side-scan sonar, sub-bottom profiling, and video and still-camera photography from remote or autonomous vehicle platforms.[2,6] Locating archaeological sites involves not only historical parameters such as trade routes, but also oceanographic parameters such as currents, sedimentation rates, seafloor morphology, and modern trawling patterns.[7] Systematic surveys are important for documenting the submarine landscapes surrounding archaeological sites to interpret processes that have affected the formation of the modern site. These processes can include the ship's impact on the seabed, such as the landslide created by the German battleship, *Bismarck*;[7,8] chemical and physical impacts of shipwrecks on ocean ecology;[9,10] damage to shipwreck sites by bottom-trawling activities;[11,12] and environmental parameters affecting a site's preservation on the seabed.[13] Surveys should also include areas where archaeological sites are not expected to be found, as well as where they are; an absence of sites helps support interpretation of the archaeology of a region.[6]

DOCUMENTATION

Excavation of shipwrecks in deep water is difficult because of the cost and limitations of accessing the sites. Therefore, the focus of archaeological oceanography shifts to high-resolution imaging and documentation of the wrecks. Imaging in deep water faces unique challenges that include underwater navigation, light attenuation, and the physics of mapping three-dimensional, non-planar sites.[14] Such challenges, along with the need for precise mapping for archaeological interpretations of sites, have helped drive the development of deepwater imaging systems.[2] Current remotely operated vehicle (ROV)-mounted mapping systems are able to produce subcentimeter-level precision bathymetric maps of archaeological sites over hundreds of square meters, which meet the standards of maps drawn by divers at shallow sites. A combination of high-frequency multibeam sonar, stereo cameras, and structured light laser profile imaging have created some of the highest-resolution images of shipwreck sites to date (Fig. 1).[15,16]

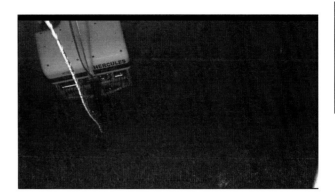

Fig. 1 High-definition video capture of remotely operated vehicle (ROV) *Hercules* surveying an ancient shipwreck with structured light laser profile-imaging system.
Source: Copyright the Institute for Exploration (IFE)/the Ocean Exploration Trust (OET).

Photomosaic and bathymetric maps of shipwrecks are important products for the investigation of deepwater sites because they can provide high-resolution images of a site's plan, orientation, and artifact provenience that excavations at these depths cannot (Figs. 2 and 3). These surveys with a single ROV take mere hours, and are therefore practical, given the challenges of working in deep water. Such imagery of shipwrecks can also be used for oceanographic studies, including the analysis of the changing conditions of wreck sites to document damage to the seabed by bottom trawling.[16] These high-resolution imaging systems have broad applications beyond archaeological sites, such as the detailed bathymetric imaging of hydrothermal vents and other geological features.[16,17] However, the mapping of shipwreck sites has been important for the development of imaging techniques because artifacts such as amphoras are of well-known shapes and sizes, which greatly aid the fine-tuning of the imaging calibrations for use on unknown objects, such as those of a geological nature.[17]

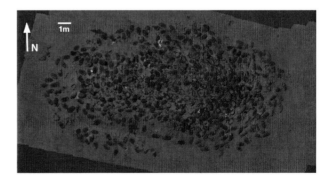

Fig. 2 Photomosaic of *Byzantine C* shipwreck site, Knidos C, located south of Datcha, Turkey.
Source: Image by Chris Roman.

Fig. 3 Multibeam bathymetric map of *Byzantine C* shipwreck site, Knidos C, located south of Datcha, Turkey.
Source: Image by Chris Roman.

CHARACTERIZATION

Following the inception of underwater archaeology, George Bass wrote[4] "No archaeologist specializes in the environment in which he works." Over the past two decades, most archaeologists have realized the importance of collaborating with scientists to gain a more rounded picture of the site and the processes that went into forming it. The birth of the field of geoarchaeology, for example, is testament to the need to understand a site's complete history. Likewise, in marine archaeology, there is much that can be learned about a shipwreck by studying the modern site with respect to the environment in which it is deposited.[7] Archaeological oceanography not only investigates cultural sites, but also the history of those sites from the time of their deposition underwater to the present day, documenting the modern landscapes as part of the environment.[13] Understanding the history of a shipwreck site from its sinking to the present day is as important as learning about the history of the ship from its construction to its sinking. Characterizing the physical and chemical actions that have continuously affected a site is necessary to define ongoing site formation processes. This is a principle objective behind archaeological oceanography.

A prime example of this application can be found in the Black Sea. The stratification of this body of water into three layers—oxic, suboxic, and anoxic—provides an environment with a high preservation potential for cultural materials.[11,13] Additionally, investigations of archaeological sites in these waters, particularly those located close to the interface between these layers, can help provide a greater understanding of the physical and chemical dynamics of the Black Sea.[11,13] Over the past decade, expeditions by the Institute for Exploration and the Ocean Exploration Trust have located over a dozen shipwrecks off the southern Black Sea coast, near Sinop, Turkey, all in various states of preservation. These wrecks, located between 100 and 300 m depth, span the shelf break and the oxic/anoxic

interface. In addition to documenting the wreck sites with high-resolution imagery, sediment and water samples were collected in the area of these wrecks to begin to characterize the oceanographic environment in which they sit, including microbiological, geochemical, meiofaunal, and mineralogical analyses.[11] Most of these wrecks lie between 100 and 115 m depth, which is likely due to a combination of physical parameters rather than historical. Heavy bottom trawling along the shallower, coastal shelf areas, and slumping along the steep slopes at the shelf break may have destroyed or concealed some shipwreck sites.[11] At the same time, anoxic waters have likely moved up along the shelf during episodes of internal wave action, which has been hypothesized to have contributed to shipwreck preservation off Crimea, Ukraine,[20] helping to preserve wrecks in this specific depth range. This research in the Black Sea illustrates the importance of including oceanographic research in deepwater archaeological investigations, and the need for environmental specialists during such research.

IN SITU PRESERVATION

The scientific excavation of underwater cultural heritage sites versus their salvage by commercial companies has been a topic of rigorous legal and ethical debates.[3,21–23] The 2001 UNESCO Convention on the Protection of the Underwater Cultural Heritage is the result of an international response to the illegal looting of shipwreck sites. The recent technological ability to access sites in deep water, in both academic and private sectors, opens new discussions about the responsibility of investigating and preserving these sites.[23] The UNESCO Convention calls for the in situ preservation of underwater archaeological sites as a first option, followed by the careful recovery of materials for either scientific or protective purposes.[24] Recent research has shown that the expansion and better enforcement of marine protected areas can have a positive impact on the protection of sites from bottom trawlers, and that such steps toward their in situ preservation may be more effective than their salvage.[15]

Technological developments in underwater imaging and remote sensing, both surficial bathymetry and sub-bottom profiles, require us to consider seriously when excavation in deep water is in fact appropriate because sites can be documented in great detail remotely. The UNESCO Convention also states, "Responsible non-intrusive access to observe or document in situ underwater cultural heritage shall be encouraged to create public awareness, appreciation, and protection of the heritage except where such access is incompatible with its protection and management."[24] Such goals of the Convention are in line with the objectives of archaeological oceanography, where the non-intrusive documentation and characterization of sites are the primary goals, as discussed earlier. Additionally, recent expeditions by the *E/V Nautilus* broadcast live video from the seafloor for educational outreach, as public access to the deep sea is severely limited, helping to increase knowledge and excitement of ocean exploration, as well as awareness of the processes, human and environmental, that threaten both marine environments and underwater cultural sites.[25]

CONCLUSION

Archaeological oceanography does not only apply to the study of sites in deep water, but can be applied to any submerged cultural site. Underwater archaeology has been a successful discipline for decades; the application of archaeological oceanography emphasizes the inclusion of oceanographic methods and technology as a collaborative study between these disciplines. Although the study of archaeological sites in deep water requires oceanographic technology and resources for access and interpretation, the investigation of these sites can aid ocean exploration as well. Once shipwreck sites are dated, based on artifact type or by other means, they can serve as important references for sediment processes, such as for detecting differences in sedimentation observed at two wrecks off Knidos, Turkey from the Archaic Greek and Byzantine periods.[15] Additionally, the fixed locations of shipwreck sites can be used as platforms for long-term monitoring of oceanographic environments, such as in the Black Sea, where these sites are helping to pinpoint the depths and locations of internal wave activity and the dynamics of the Black Sea water column.[11,13]

Many archaeologists working underwater today utilize a variety of scientific methods to great effect. Archaeological oceanography does not modify the methodology of marine archaeology, but rather reinforces the importance of oceanographic research principles and practices in such studies. Beyond the technology required to access cultural sites in deep water, oceanography should also be employed to gather simultaneous, related data sets from the shipwrecks in their in situ environments. These data can broaden our understanding of both the ocean environment and the condition of the wreck site. These sites have come into equilibrium with the ocean over the course of hundreds to thousands of years underwater, and much can be learned about both the ships and the environment through the investigations of the modern sites. The application of oceanography to archaeological studies aims to maximize the potential information that can be gathered from the investigation of these sites and provides critical insights into site dynamics and management decisions. Given the time and cost associated with accessing sites in deep water, multidisciplinary collaboration is essential for the documentation and preservation of underwater cultural heritage.

ACKNOWLEDGMENTS

The authors wish to thank Katy Croff Bell, Alexis Catsambis, Dwight Coleman, Dan Davis, Jim Delgado, Carter DuVal, Gabrielle Inglis, Chris Roman, Art Trembanis, and Sandra Witten for their help in preparing this manuscript.

REFERENCES

1. Ballard, R.D. Ed. Introduction. In *Archaeological Oceanography*; Princeton University Press: New Jersey, 2008; ix-x pp.
2. Mindell, D.A.; Croff, K.L. Deep water, archaeology and technology development. MTS J. **2002**, *36* (3), 13–20.
3. Brennan, M.L. Quantification of trawl damage to premodern shipwreck sites: Case studies in the Aegean and Black Sea. UNESCO Scientific Colloquium on Factors Impacting the Underwater Cultural Heritage. 2011. http://www.unesco.org/new/fileadmin/MULTIMEDIA/HQ/CLT/pdf/UCH_S2_Brennan.pdf.
4. George B. F. *Archaeology under Water*; Praeger, New York, 1966.
5. U.S. Dept of Commerce/NOAA. Discovering Earth's Final Frontier: A U.S. Strategy for Ocean Exploration. The Report of the President's Panel on Ocean Exploration. U.S. Department of Commerce/National Oceanic and Atmospheric Administration. 2000. explore.noaa.gov/media/http/pubs/pres_panel_rpt.pdf.
6. Coleman, D.F.; Ballard, R.D. Oceanographic methods for underwater archaeological surveys. In *Archaeological Oceanography*; Ballard, R.D., Ed.; Princeton University Press: New Jersey, 2008; 3–14.
7. Ballard, R.D. The search for contemporary shipwrecks in the deep sea: lessons learned. In *Archaeological Oceanography*; Ballard, R.D., Ed.; Princeton University Press: New Jersey, 2008; 95–127.
8. Ballard, R.D. *The Discovery of the Bismarck*; Scholastic/Madison Press: New York, 1990.
9. Delgado, J.P. Recovering the past of the *USS Arizona*: symbolism, myth, and reality. His. Archaeol. **1992**, *26* (4), 69–80.
10. Kelly, L.W.; Barott, K.L.; Dinsdale, E.; Friedlander, A.M.; Nosrat, B.; Obura, D.; Sala, E.; Sandin, S.A.; Smith, J.E.; Vermeij, M.J.; Williams, G.J.; Willner, D.; Rohwer, F. Black reefs: iron-induced phase shifts on coral reefs. Int. Soc. Microbial. Ecol. **2011**, *6*, 638–649.
11. Brennan, M.L.; Davis, D.; Roman, C.; Buynevich, I.; Catsambis, A.; Kofahl, M.; Ürkmez, D.; Vaughn, J.I.; Merrigan, M.; Duman, M. Ocean dynamics and anthropogenic impacts along the southern Black Sea shelf examined by the preservation of pre-modern shipwrecks. *Continental Shelf Res.* **2013**, *53*, 89–101.
12. Foley, B. Archaeology in deep water: Impact of fishing on shipwrecks. Woods Hole Oceanographic Institution. 2008. http://www.whoi.edu/sbl/liteSite.do?litesiteid=2740&articleId=4965. (accessed December 2009).
13. Brennan, M.L.; Ballard, R.D.; Croff Bell, K.L.; Piechota, D. Archaeological oceanography and environmental

characterization of shipwrecks in the Black Sea. In *Geology and Geoarchaeology of the Black Sea Region: Beyond the Flood Hypothesis*; Buynevich, I., Yanko-Hombach, V., Gilbert, A., Martin, R.E., Eds.; Geological Society of America Special Paper, **2011**; *473*, 179–188.

14. Singh, H.; Roman, C.; Pizarro, O.; Foley, B.; Eustice, R.; Can, A. High-resolution optical imaging for deep-water archaeology. In *Archaeological Oceanography;* Ballard, R.D., Ed.; Princeton University Press: New Jersey, 2008; 30–40.

15. Brennan, M.L.; Ballard, R.D.; Roman, C.;. K.L.C.; Buxton, B.; Coleman, D.F.; Inglis, G.; Koyagasioglu, O.; Turanli, T. Evaluation of the modern submarine landscape off southwestern Turkey through the documentation of ancient shipwreck sites. *Continental Shelf Res.* **2012**, *43*, 55–70.

16. Roman, C.; Inglis, G.; Rutter, J. *Application of Structured Light Imaging for High Resolution Mapping of Underwater Archaeological Sites.* IEEE OCEANS: Sydney, 2010.

17. Roman, C.N.; Inglis, G.; Vaughn, J.I.; Williams, S.; Pizarro, O.; Friedman, A.; Steinberg, D. Development of high-resolution underwater mapping techniques. In *New Frontiers in Ocean Exploration*; Bell, K.L.C., Fuller, S.A., Eds.; The *E/V Nautilus* 2010 Field Season. Oceanography **2011**, *24* (1), supplement, 14–17.

18. Murray, J.W.; Stewart, K.; Kassakian, S.; Krynytzky, M.; DiJulio, D. Oxic, suboxic, and anoxic conditions in the Black Sea. In *The Black Sea Flood Question: Changes in Coastline, Climate and Human Settlement*; Yanko-Hombach, V., Gilbert, A.S., Panin, N., Dolukhanov, P.M., Eds.; *Dordrecht*; Springer: 2007; 1–22.

19. Ward, C.; Ballard, R. Black Sea shipwreck survey 2000. Int. J. Nautical Archaeol. **2004**, *33*, 2–13.

20. Trembanis, A.; Skarke, A.; Nebel, S.; Coleman, D.F.; Ballard, R.D.; Fuller, S.A.; Buynevich, I.V.; Voronov, S. Bedforms, hydrodynamics, and scour process observations from the continental shelf of the northern Black Sea. In *Geology and Geoarchaeology of the Black Sea Region: Beyond the Flood Hypothesis*; Buynevich, I., Yanko-Hombach, V., Gilbert, A., Martin, R.E., Eds.; Geolo. Soc. Am. Special Paper, **2011**; *473*, 165–178.

21. Aznar-Gomez, M.J. Treasure hunters, sunken state vessels and the 2001 UNESCO Convention on the Protection of the Underwater Cultural Heritage. Int. J. Marine Coastal Law **2010**, *25*, 209–236.

22. Delgado, J.P. Underwater archaeology at the dawn of the 21st century. His. Archaeol. **2000**, *34* (4), 9–31.

23. Greene, E.S.; Leidwanger, J.; Leventhal, R.M.; Daniels, B.I. Mare nostrum? Ethics and archaeology in Mediterranean waters. Am. J. Archaeol. **2011**, *115*, 311–319.

24. UNESCO. Convention on the Protection of the Underwater Cultural Heritage 2001. *UNESCO.* 2001. http://portal. unesco.org/en/ev.php-URL_ID=13520&URL_DO=DO_ TOPIC&URL_SECTION=201.html.

25. Witten, A; O'Neal, A. Education and outreach activities enabled by telepresence technology. In *New Frontiers in Ocean Exploration: The E/V Nautilus 2010 Field Season*; Bell, K.L.C, Fuller, S.A., Eds.; Oceanography **2011**, *24* (1), supplement, 12–13.

Arctic Hydrology

Bretton Somers
H. Jesse Walker
Department of Geography and Anthropology, Louisiana State University, Baton Rouge, Louisiana, U.S.A.

Abstract

Arctic hydrology considers the interconnected environmental processes and physical phases of the water present in the Arctic, which is a high latitude region where most water is in a frozen state for nine or more months a year. Of major importance is the vast distribution of snow, ice, and permafrost. Snow, because of its high albedo, has a major affect on local as well as regional climate. In late spring, snow melt leads to a short period of runoff which contributes about 10% of the Earth's river discharge into the Arctic Ocean. Changes in river discharge along with decreasing sea-ice cover, permafrost degradation, and glacial melting are an indication of the importance of the Arctic in global change.

INTRODUCTION

After a discussion about the boundaries of the Arctic, its hydrological elements are analyzed. It is noted that, in the Arctic, the solid phase of water dominates the landscape during most of the year. In the solid phase, water occurs as snow, river and lake ice, sea ice, glacial ice, and ground ice. Ground ice is integral to permafrost, which is nearly continuous. Although all of the hydrologic elements can be considered separately as though each is in storage, they are in fact interconnected and should be considered as an integral part of a cycle that progresses through all three phases among the atmosphere, lithosphere, and cryosphere. Recent research has shown that many of the Arctic's hydrologic elements are changing rapidly.

THE ARCTIC

Although no one denies the existence of the region known as the Arctic, few actually agree on its southern boundary. Astronomers usually use the Arctic Circle, oceanographers use the distribution of sea ice, biologists use the tree line, and cryosphere specialists use a permafrost boundary.[1] In one way or another, all of these boundaries are of significance to hydrologists, because they must contend with distributions that often spread well beyond the Arctic. For example, many of the rivers that drain into the Arctic Ocean originate in temperate latitudes (Fig. 1), the Arctic Ocean is impacted by flow from both the Atlantic and Pacific Oceans, and the atmosphere can be affected by non-arctic weather patterns. Nonetheless, the main hydrologic factor that distinguishes the Arctic from other climatic zones is the long period of time during which snow and ice dominate the landscape. From a regional standpoint, the Arctic may best be considered a moderately sized ocean almost completely surrounded by a fringe of land, i.e., a configuration that is the opposite of its counterpart in the Southern Hemisphere, the Antarctic.

THE OCEAN, RIVERS, LAKES, AND GLACIERS

The major features of the Arctic are the Arctic Ocean with its numerous bordering seas, the rivers that drain into it, its lakes and ponds, and the numerous islands (many with glaciers) that occur north of the continents (Fig. 1). Some scientists contend that the Arctic Ocean should be classified as a Mediterranean sea rather than an ocean because in addition to being nearly surrounded by land, it occupies less than 4% of the Earth's ocean area and only about 1% of the Earth's ocean volume.

The Arctic Ocean is also unique in that it is only 40% as large as the continental area that drains into it. The rivers that enter the ocean vary greatly in size and discharge. Included among them are four of the Earth's 12 longest—the Yenisey (5870 km), the Ob (5400 km), the Lena (4400 km), and the Mackenzie (4180 km). They contribute nearly 60% of the fresh water that drains from the continents.[2] This total equals some 11% of the Earth's runoff from the continents. Because of its unique relationship between land and sea, the Arctic Ocean is impacted more by water draining into it than any other ocean.[3]

In addition to its seas, rivers, and islands, the Arctic also possesses numerous lakes and ponds. Most Arctic lakes are small. One of the most studied lake types is the oriented

Encyclopedia of Natural Resources DOI: 10.1081/E-ENRW-120042337

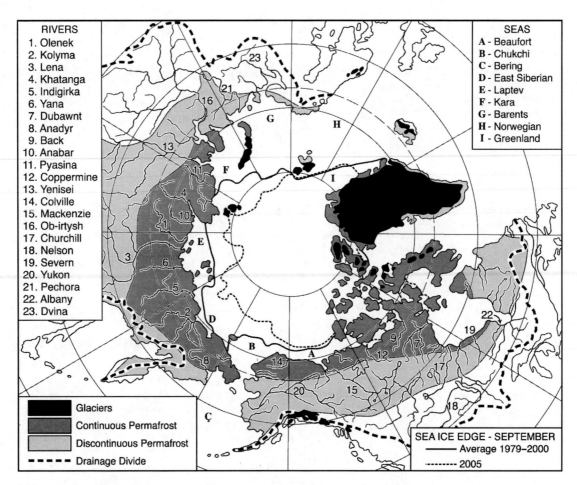

Fig. 1 Map of the Arctic showing the major rivers, permafrost distribution, September sea-ice locations, and bordering seas in the Arctic Ocean. **Source:** Adapted from Walker & Hudson.[9]

lake. Generally elliptical in shape, oriented lakes originate as thaw lakes and range in length up to several kilometers. However, by far, the most common bodies of water dotting the land surface are small ponds whose existence is partly the result of the poor drainage and limited infiltration that characterizes permafrost environments.

Glaciers in the Arctic, except for the Greenland ice cap, are generally limited to the Canadian and Russian arctic islands. Arctic glaciers are major contributors of water and ice bergs to the ocean.

SNOW, ICE, AND PERMAFROST

One of the unique characteristics of water is that it is the only chemical compound that occurs on Earth in three phases, i.e., as a gas (water vapor), a liquid, and a solid. In its solid phase, water appears as snow, surface ice (river, lake, and sea ice), and ground ice.

Snow

Although snow is not limited to high latitudes, it is there that it is nearly ubiquitous. Over most of the Arctic, it covers

the surface for nine or more months. Snow fall is related to the volume of water vapor in the air, a volume that is temperature dependent. Therefore, the actual amount of snow that falls is limited in quantity. Once it falls, it tends to retain its solid form until melt season. However, because most of the Arctic has low-lying vegetation (tundra), snow drifting is extensive. On flat surfaces, such as lake ice, the snowpack is thin or even missing. Uneven surfaces, such as river banks, trap snow into sizable drifts, some of which may last through most of the summer.

Snow is important in several major ways in the Arctic. Because it is highly reflective, much of the sun's energy that reaches it is returned to space. The reflected energy, or albedo, of fresh snow is more than 75%, whereas that from water is usually less than 30% and from tundra less than 20%.[4] Further, snow, because of its low thermal conductivity, is a good insulator—affecting the occurrence and maintenance of permafrost and in helping keep vegetation and fauna from freezing. Snow, when it melts, provides the bulk of the water that feeds the rivers and streams that drain into the Arctic Ocean. Because the melt season is relatively short and the melt rate is rapid, most rivers in the Arctic have peaks of discharge that correlate closely with river breakup (Fig. 2).

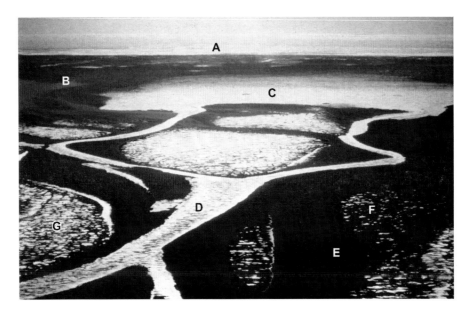

Fig. 2 The mouth of the Colville River, Alaska showing the distribution of snow, water, river ice and sea ice during breakup. (**A**) sea ice, (**B**) floodwater on top of sea ice, (**C**) sea ice floating on top of river floodwater, (**D**) river ice floating on top of floodwater, (**E**) flooded mudflats, (**F**) flooded island with remnant snow patches, (**G**) snow and ice covering ice-wedge polygons. Photograph by Donald Nemeth.

River and Lake Ice

In the Arctic, the temperature regime is such that surface ice, whether on rivers, lakes, or the sea, is present for most of the year. On fresh-water bodies, the ice reaches thicknesses of about 2 meters. Because most of the ponds and many of the streams have depths less than that, water freezes to the bottom. In deeper lakes and rivers, water will remain in the liquid state beneath the ice. In some rivers, especially those originating in lower latitudes (Fig. 1), flow continues throughout the winter beneath the ice. Ice in the deeper lakes lasts longer than ice on rivers, where breakup follows after snow-melt water enters the river channels.

Sea Ice

Some authors maintain that sea ice is the most dramatic feature of the Arctic. Formed by the freezing of sea water, sea ice covers nearly all of the Arctic Ocean during winter when it is attached to much of the continental coastline of Siberia and North America. Except for a narrow band of fast ice, the bulk of the sea ice is in motion, steered by ocean currents and winds. Its predominant direction of flow is clockwise. Because its motion is erratic, pressure ridges many meters high and open bodies of water, known as leads, develop. First year sea ice averages about 2 meters in thickness whereas ice that survives through the summer becomes thicker during the following winter. The minimum extent of sea ice in the Arctic is in September (Fig. 1). Sea ice suppresses the energy exchange between the ocean and the atmosphere, affects the circulation of ocean currents, and dampens wave action.[5]

Permafrost and Ground Ice

Permafrost is defined as earth material in which the temperature has been below 0°C for two or more years. Because it is defined only by its temperature, water is not necessary for its existence. However, most permafrost, which underlies more than 20% of the Earth's land area, does contain ice in various amounts and forms. Ground ice occurs in the pores of the soil as lenses or veins and in large forms such as ice wedges.[6] By volume, pore ice is the largest, although ice wedges are more conspicuous. Where ice wedges are well developed, they may occupy as much as 30% of the upper 2 or 3 meters of the land surface. Their surface expression is distinctive and takes the form of ice-wedge polygons. In the Arctic, permafrost is continuous (Fig. 1) except beneath those water bodies that are more than 2 meters deep and do not freeze to the bottom during winter. Permafrost is also present beneath near-shore waters off Siberia and North America.

Associated with permafrost is the active layer—the portion of terrain that thaws and freezes seasonally. The active layer can vary in thickness from a few centimeters to several meters, depending mainly on vegetation cover and soil texture.

THE HYDROLOGIC CYCLE IN THE ARCTIC

In the above discussion, each hydrologic element was considered individually, as if it was a pool of water in storage as a gas, a liquid, or a solid. However, in reality, the hydrologic elements are interconnected and mobile, moving from one phase to another through time. The examination of water

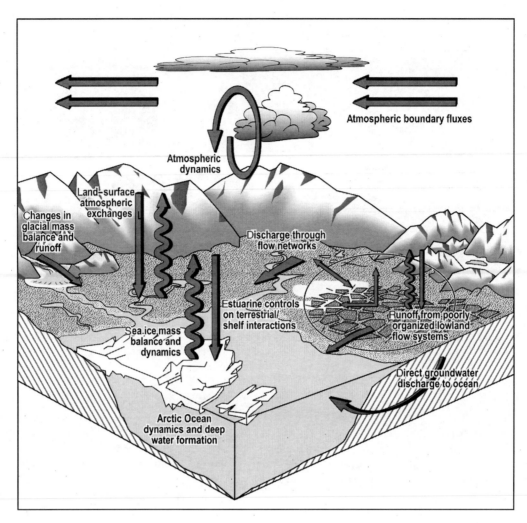

Fig. 3   Conceptual model of the arctic hydrologic cycle.
Source: Adapted from Vörösmarty, et al.[3]

from this standpoint is best done through the concept of the hydrologic or water cycle, a cycle that is considered the most fundamental principle of hydrology.[7] The conceptual model (Fig. 3) illustrates hydrologic links among the atmosphere, the land, and the ocean.[3] As a concept, it can be applied to small units within the hydrosphere or large units such as the Arctic. In the Arctic, the movement of water from one phase or location is, to a large extent, dependent on freeze-thaw cycles of snow, ice, and permafrost.

ARCTIC HYDROLOGY AND THE FUTURE

In recent years, research on climate change has increased dramatically. Much of it has been conducted in and about the Arctic and especially about arctic hydrology. Numerous changes in the hydrologic elements of the Arctic have been recently documented. Included are a shortening of the snow-cover season, later freeze-up and earlier breakup

of river, lake, and sea ice, increased fresh-water runoff, melting glaciers, the degradation of permafrost and an increase in active-layer thickness, increased groundwater flow, decreased sea-ice extent (Fig. 1), and decreased albedo, among others. In 2005, Dan Endres of the National Oceanic and Atmospheric Administration, in a discussion on climate change, stated, "Whatever is going to happen is going to happen first in the Arctic and at the fastest rate."[8] This is a prophecy that is especially applicable to virtually every hydrologic and ecological element in the Arctic.

REFERENCES

1. Walker, H.J. Global change and the arctic. Curr. Top. Wetland Biogeochem. **1998**, *3*, 44–60.
2. http://www.awi-bremerhaven.dc/GEO/APARD/Backgr. html (accessed November 2005).

3. Vörösmarty, C.J.; Hinzman, L.D.; Peterson, P.J. et al. *The Hydrologic Cycle and Its Role in Arctic and Global Environmental Change: A Rationale and Strategy for Synthesis Study*; Arctic Research Consortium of the U.S.: Fairbanks, Alaska, 2001.

4. Collier, C.G. Snow. In *Encyclopedia of Hydrology and Water Resources*; Herschy, R.W., Fairbridge, R.W., Eds.; Kluwer Academic Publishers: Dordrecht, 1998; 621–624.

5. Untersteiner, N. Structure and dynamics of the Arctic ocean ice cover. In *The Arctic Region*; Grantz, A., Johnson, L., Sweeney, J.F., Eds.; The Geological Society of America, Inc.: Boulder, Colorado, 1990; 37–51.

6. Carter, L.D.; Heginbottom, J.A.; Woo, M.-K. Arctic lowlands. In *Geomorphic Systems of North America*; Graf, W.L., Ed.; Geological Society of America: Boulder, CO, 1987; Centennial Special; Vol. 2, 583–628.

7. Maidment, D.R.; Ed. *Handbook of Hydrology*; McGraw-Hill, Inc.: New York, 1992.

8. Glick, D. Degrees of change. Nat. Conservancy **2005**, *55*, 40–50.

9. Walker, H.J.; Hudson, P.F. Hydrologic and geomorphic processes in the Colville River Delta, Alaska. Geomorphology **2002**, *56*, 291–303.

Arctic Oscillation

Jinping Zhao
Ocean University of China, Qingdao, China

Abstract

The Arctic Oscillation (AO) is a seesaw pattern in which sea level pressure (SLP) at the polar and middle latitudes in the Northern Hemisphere fluctuates between positive and negative phases. This pattern is obtained by the first mode of the Empirical Orthogonal Function (EOF) of the SLP north of 20°N. The standardized time coefficient of EOF first mode is defined as the Arctic Oscillation Index (AOI), which indicates the polarity of the AO and reflects the atmospheric circulation variation of the lower atmosphere. AO can be easily confused with the North Atlantic Oscillation (NAO), as both indices correlate very well. Some scientists believe that AO and NAO are the same process described in different ways. A coupled signal of AO and a strong fluctuation at the 50 hPa level was identified, which indicates that AO is the surface signature of the Northern Hemisphere Annular Mode. AO not only influences the climate, sea ice, and ocean change in the Arctic but also produces global impacts on surface air pressure and temperature, precipitation, and the frequency and track of storms. The spatial variation of AO was revealed to reflect the variation of spatial area AO dominated. The spatial variation index of AO indicates the three stages of long-term variation of the AO.

INTRODUCTION

The Arctic Oscillation (AO) is a seesaw pattern in which sea level pressure (SLP) at the polar and middle latitudes in the Northern Hemisphere fluctuates between positive and negative phases (Fig. 1). This pattern is obtained by the first mode of Empirical Orthogonal Function (EOF) of the SLP north of 20°N.[1] It is characterized by a pattern of one sign in the polar region with the opposite sign anomaly in the middle latitude centered about 37–45°N.[2]

The standardized leading principal component (PC) time series is defined as the Arctic Oscillation Index (AOI, Fig. 2). The monthly mean SLP anomaly fields are regressed upon the AOI. Based on this definition, the AO index is dimensionless and the regression maps have the unit of the SLP anomaly field. The amplitudes shown in the regression maps therefore correspond to anomaly values.

The polarity of the AO is defined by the AOI. If the AOI is in the positive phase, the AO in that time is called positive AO, otherwise, the AO is negative. The AO was found to exhibit high polarity or positive AOI in the early 1990s,[5] but has switched to a near-neutral or negative phase since 1996.[6] The year-to-year persistence of positive or negative values and the rapid transition from one to the other is often referred to as "regime-like." The AO presents a "positive phase" with relatively lower pressure over the polar region and higher pressure at mid-latitudes, and a "negative phase" with relatively higher pressure over the polar region and lower pressure at mid-latitudes.

AO AND ATMOSPHERIC CIRCULATION

The AOI in fact reflects the dominant mode of variation in SLP and zonal wind. Also, the time series of 50 hpa is strongly correlated with AOI. AO extends through the depth of the troposphere. During the "active season," January through March, it extends upward into the stratosphere where it modulates in the strength of the westerly vortex that encircles the Arctic polar cap region.[7] So AOI is an index to reflect the main characteristics of the overall atmospheric circulation in hemispheric scale.

In the negative phase of AO, the polar region is dominated by higher pressure, and the zonal wind becomes weaker (Figs. 3 and 4). A very persistent, strong ridge of high pressure, or "blocking system," near Greenland allows polar cold air to easily extend to the south,[9] which causes a frigid winter in the eastern part of the U.S.A. and the Mediterranean. However, when the AO index is positive, surface pressure is low in the polar region (Fig. 4). This helps the middle latitude jet stream to blow strongly and consistently from west to east, thus keeping cold Arctic air locked in the polar region.

During the twentieth century, the AO mostly fluctuated between its positive and negative phases until the 1970s. Since then, the oscillation has stayed mostly in the positive phase for about 30 years, especially from 1988 to 1996. During this period, the U.S.A., China, and Europe kept at a lower air pressure and higher temperature. Since this century, the phase of the AO has tended to a neutral state. Obvious negative AO appeared in the winters of 2010/2011 and 2011/2012, which is attributed to the causal factor of

Encyclopedia of Natural Resources DOI: 10.1081/E-ENRW-120048079

AO for the extremely cold winter in the U.S.A., Europe, and Eastern Asia.

In February 2010, the AO reached its most negative monthly mean value, −4.266 (http://www.cpc.ncep.noaa. gov/products/precip/CWlink/daily_ao_index/monthly.ao. index.b50.current.ascii). The colder winter appeared in eastern U.S.A., such as New York, Washington D.C., and Baltimore. That month was characterized by three separate historic snowstorms that occurred in the mid-Atlantic region of the U.S.A.[8]

AO exerts strong influences on wintertime climate in the Northern Hemisphere.[10] However, the high correlation of negative AO and an excessive colder winter does not necessarily mean that extreme weather will surely occur. For a certain location, the climate is influenced not only by AO, but also by some other processes related with tropical or regional factors.

GLOBAL EFFECTS OF AO

As the AO reflects the variation of the atmospheric system, it will also embody a global effect. In the case with positive AO, four main phenomena are notable:

1. Surface air pressure. Positive AO produces a lower-than-normal pressure over the polar region. The westerly moves to the north and cold air is restricted in the polar region.
2. Air temperature. In the positive phase, frigid winter air does not extend as far into the middle of North America as it would during the negative phase of the oscillation. This keeps much of the U.S.A. east of the Rocky

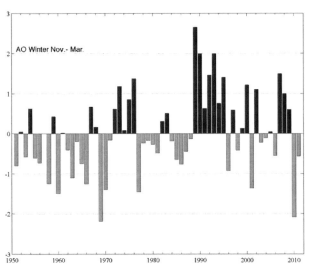

Fig. 1 The Arctic Oscillation Index in winter (November to March). **Source:** Data from NCEP.[3]

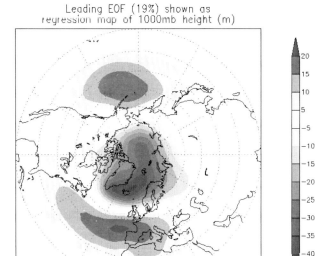

Fig. 2 The seesaw-like pattern of the Arctic Oscillation. **Source:** Data adapted from http://www.cpc.ncep.noaa.gov/products/precip/CWlink/daily_ao_index/ao.shtml.[4]

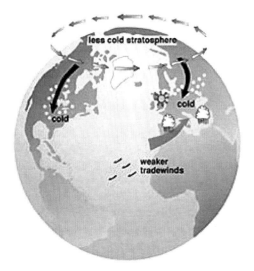

Fig. 3 Positive and negative Arctic Oscillation. **Source:** Courtesy of J. Wallace, University of Washington.

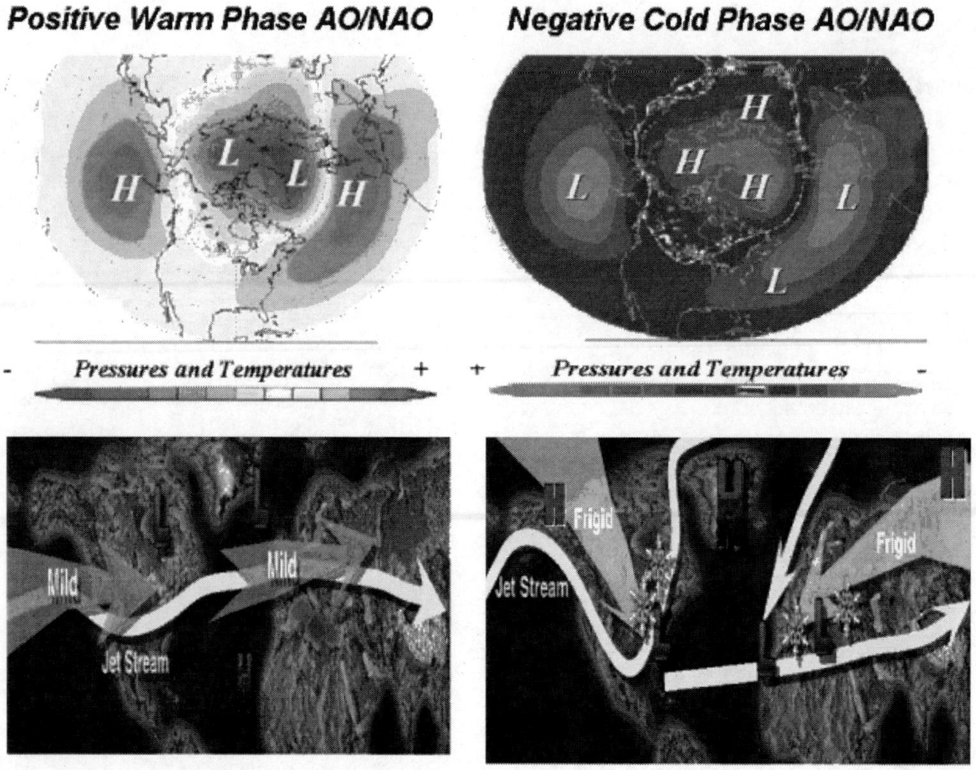

Fig. 4 Atmospheric circulations in positive or negative AO.
Source: Data adapted from http://www.newx-forecasts.com/ao_2.html.[8]

Mountains warmer than normal, but leaves Greenland and Newfoundland colder than usual (Fig. 5A).

3. Flood and drought. With the positive AO, the atmospheric circulation brings wetter weather to Alaska, Scotland, and Scandinavia, as well as drier conditions to the western part of the U.S.A. and the Mediterranean, Spain, and the Middle East (Fig. 5B).

4. Ocean storms. With the positive AO, higher pressure at mid-latitudes drives ocean storms farther north.

Weather patterns in the negative phase are in general "opposite" to those of the positive phase.

AO AND NORTHERN HEMISPHERE ANNULAR MODE

In addition to defining AO, Thompson and Wallace[1] also pointed out a coupled signal of AO and a strong fluctuation at the 50 hPa level on the intraseasonal, interannual, and interdecadal time scales. The leading modes of variability of the extratropical circulation in both hemispheres are characterized by deep, zonally symmetric, or "annular" structures, with geopotential height perturbations of opposing signs in the polar cap region and in the surrounding zonal ring centered near 45° latitude.[5] Compared with the pattern of the annular circulation mode of the lower stratosphere in the Southern Hemisphere, AO is recognized in essence as the Northern Hemisphere annular mode (NAM)[4] and is

interpreted as the surface signature of modulations in the strength of the polar vortex aloft (Fig. 6).[4,10]

The structures of the Northern Hemisphere and Southern Hemisphere annular modes are shown to be remarkably similar, not only in the zonally averaged geopotential height and zonal wind fields, but in the mean meridional circulations as well. Both exist year-round in the troposphere, but they amplify with height upward into the stratosphere during midwinter in the Northern Hemisphere.[5]

The NAM is shown to exert a strong influence on wintertime climate, not only over the Euro-Atlantic half of the hemisphere, but over the Pacific half as well. It affects not only the mean conditions but also the day-to-day variability, modulating the intensity of mid-latitude storms and the frequency of occurrence of high-latitude blocking and cold air outbreaks throughout the hemisphere.[11]

AO AND NORTH ATLANTIC OSCILLATION

North Atlantic Oscillation (NAO) is a seesaw-like dipole oscillation along the Atlantic sector and defined by the difference of air pressure between Iceland and the Azores.[12,13] NAO was discovered in the 1920s by Walker[14] and has been studied extensively.[15] Since the NAO index correlates well with the AOI,[16] some studies focused on the physical reality of the AO and the connection between AO and NAO.

A

DJF Temperature Anomaly (°C) by AO PHASE

B

ASO Precipitation Anomaly (mm/day) by AO PHASE

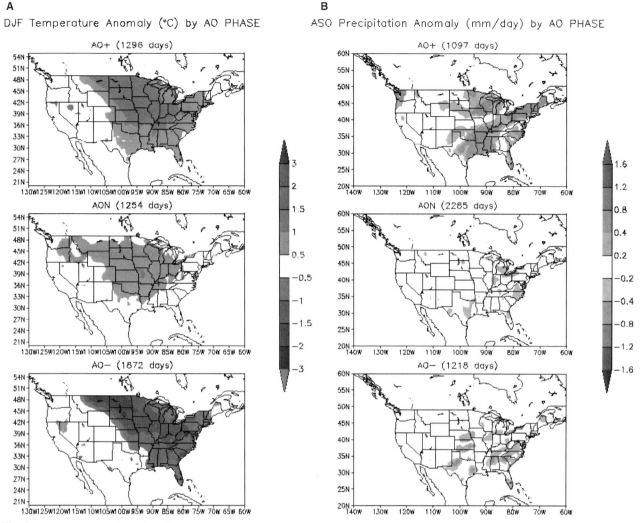

Fig. 5   Temperature and precipitation anomalies by AO phase.
Source: Data adapted from http://www.cpc.ncep.noaa.gov/research_papers/ncep_cpc_atlas/8/figures/temp/, http://www.cpc.ncep.noaa.gov/research_papers/ncep_cpc_atlas/8/figures/precip/.[8]

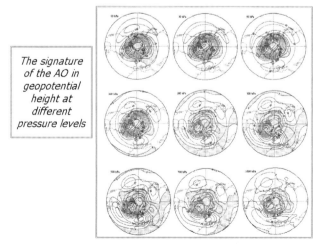

The signature of the AO in geopotential height at different pressure levels

Fig. 6   The signature of the AO in geopotential height at different pressure levels.
Source: Data adapted from http://www.newx-forecasts.com/ao_2.html.[8]

Although the definitions of AO and NAO were reached by quite different methods, the interesting thing is that both are closely correlated to each other with a correlation coefficient higher than 0.89. Some scientists believe that AO and NAO are the same process obtained from a different visual angle. They ignore the little differences between them and write it as AO/NAO to avoid the possible misunderstanding.

However, there is an argument about AO and NAO in regard to their scientific mechanism, whether one or the other is more fundamentally representative of the atmospheric dynamics. Some investigators argue that the NAO paradigm may be more physically relevant to and robust in the Northern Hemisphere than the AO paradigm.[17] Others proposed, based on the nonlinear principal component analysis (PCA), that the variability of the atmospheric system in the Northern Hemisphere is better characterized by an Arctic–Eurasian oscillation state. In the North Atlantic, such a state is replaced occasionally by a split-flow configuration state; AO represents a compromise between the two states.[18]

Deser[19] and Ambaum et al.[18] pointed out the inconsistency to demonstrate that the NAO is the more robust paradigm. Itoh[20] tried to explain the inconsistency by the concepts of true AO and apparent AO. However, AO is more strongly coupled to surface air temperature fluctuations over the Eurasian continent than the NAO. AO resembles the NAO in many respects, but its primary center of action covers more of the Arctic, giving it a more zonally symmetric appearance.[1] According to the National Snow and Ice Data Center, the AO and NAO are virtually "different ways of describing the same phenomenon."[9] Both are impossible to distinguish with purely statistical methods, as the two models are mathematically identical.[21]

AO AND ARCTIC WARMING

The Arctic region is experiencing obvious climatological warming at a rate that is nearly twice the global average during the past decades.[22] Most climatological indicators show a linear trend. The notable ones include the decline of the perennial ice coverage at a rate of 9% per decade,[23] and the thinning of the ice draft of more than a meter from two to four decades ago to the 1990s.[24] However, AO, which is believed to associate with the variation of atmospheric circulation in the Northern Hemisphere, behaves differently from such regularities. In AO, instead, a seesaw-like oscillation is still dominant.

In recent years, many studies have revealed the relationship between the AO/NAO and Arctic sea-ice condition. For example, when the AO/NAO turns to a high-index polarity, the coherent changes include the following: 1) a spin down of the Beaufort Gyre; 2) a slight increase in the ice advection out of the Arctic Basin through the Fram Strait; 3) an increase in ice import from the Barents/Kara Seas and ice advection away from the coast of the East Siberian and Laptev; 4) a decrease in ice advection from the western Arctic into the eastern Arctic; 5) a weakening of the Transpolar Drift Stream; and 6) rapid advection of ice out of the western Arctic with increased melting and inhibited accumulation of thicker ice.[25–30]

SPATIAL VARIATION OF AO

Based on the introduction section, AO was derived by the EOF method, which is a popular method to analyze spatio-temporal variation[31] and gives a space-stationary and time-fluctuating characteristic of a variable.[18] However, as the AO is a spatially variable phenomenon, only the space-stationary leading mode of EOF is not sufficient[32] to reflect the spatial variation of the AO.

By calculating the running correlation coefficients (RCCs) of AO index and gridded SLPs, a special region was identified by Zhao et al. called the Arctic Oscillation Core Region (AOCR),[33] in which SLP variation has never been influenced by a non-AO process as shown in Fig. 7A. The normalized average SLP of AOCR is always consistent with the negative AOI since 1950 with a correlation coefficient of −0.949, both are nearly interchangeable each other (Fig. 7B).[33] The AOCR includes the GIN Seas (Greenland Sea, Iceland Sea, and Norwegian Sea), part of the Barents Sea, and the area north of the Fram Strait. The high correlation of normalized averaged SLP and the AOI assigns the AOI a new physical significance, namely, the negative averaged SLP of the AOCR.

Based on the concept of AOCR, the high correlation of gridded SLP with AOI means the high negative correlation with the average SLP of AOCR. By calculating the RCC of gridded SLP with AOI, the ensemble of grids with high RCC in a certain time window could be identified, which is called the "AO-dominant region" as in this time span the gridded SLPs vary similar with AOI. By moving the time window with time, the AO-dominant region spatially varies by shape changing, area shrinking and extending, and boundary swinging, which is used to expresses the spatial variation of AO. The spatial variation of AO was then defined as "the temporal variation of the ensemble of grids with SLP varying consistent with the AO index in certain time span."[34] The difference of positive and negative relative area of the AO-dominant region is defined as a spatial variation index of AO (Fig. 8). With this index, the three stages of long-term spatial variation of the AO were clearly identified. The positive SLP anomaly area dominated

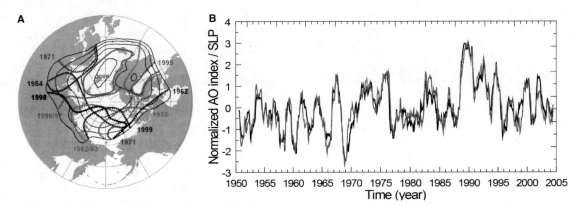

Fig. 7 (**A**) AOCR and (**B**) the normalized AO and negative average SLP of AOCR.
Source: Data adapted from Zhao et al.[33]

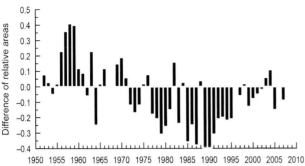

Fig. 8 The AO spatial variation index of January of every year. **Source:** Data adapted from Zhao et al.[34]

before 1970, showing the state before global warming. The negative SLP anomaly area dominated during 1971–1995, indicating the effect of global warming before the Arctic warming became apparent. Since 1996, both positive and negative SLP anomaly areas have been all small, possibly caused by the sea-ice retreat during the Arctic warming.

REFERENCES

1. Thompson, D.W.J.; Wallace, J.M. The Arctic Oscillation signature in the wintertime geopotential height and temperature fields. Geophys. Res. Lett. **1998**, *25* (9), 1297–1300.

2. Joint Institute for the study of the Atmosphere and Ocean, Arctic Oscillation (AO) time series, 1899 - June 2002, http://jisao.washington.edu/data/aots/

3. NOAA, Climate Indicators - Arctic Oscillation. http://www.arctic.noaa.gov/detect/climate-ao.shtml

4. NOAA Climate Prediction Center, Arctic Oscillation (AO). http://www.cpc.ncep.noaa.gov/products/precip/CWlink/daily_ao_index/ao.shtml

5. Thompson, D.W.J.; Wallace, J.M. Annular modes in the extratropical circulation, Part I: Month-to-month variability. J. Climat. **2000**, *13* (5), 1000–1016.

6. Overland, J.E.; Wang, M. The Arctic climate paradox: The recent decrease of the Arctic Oscillation. Geophys. Res. Lett. **2005**, *32* (6), L06701. doi: 10.1029/ 2004GL021752.

7. National Snow and Ice Data Center, Arctic Oscillation, http://nsidc.org/arcticmet/glossary/

8. NEWxSFC, Arctic Oscillation, http://www.cpc.ncep.noaa.gov/research_papers/ncep_cpc_atlas/8/figures/temp/, http://www.cpc.ncep.noaa.gov/research_papers/ncep_cpc_atlas/8/figures/precip/

9. Department of Environment and Natural Resources, Natural Climate Fluctuations, http://www.enr.gov.nt.ca/_live/pages/wpPages/soe_natural_fluctuations.aspx

10. Thompson, D.W.J.; Wallace, J.M.; Hegerl, G.C. Annular modes in the extratropical circulation. Part II: Trends J. Climat. **2000**, *13* (5), 1018–1036.

11. Thompson, D.W.; Wallace, J.M. Regional climate impacts of the Northern Hemisphere annular mode. Science **2001**, *293* (5527), 85–89.

12. Barnston, A.G.; Livezey, R.E. Classification, seasonality and persistence of low-frequency atmospheric circulation patterns. Mon. Weather Rev. **1987**, *115* (6), 1083–1126.

13. Hurrell, J.W. Decadal trends in the North Atlantic Oscillation region temperatures and precipitation. Science **1995**, *269* (5224), 676–679.

14. Walker, G.T.; Bliss, E.W. World weather V. Mem. Roy. Meteorol. Soc. **1932**, *4* (36), 53–84.

15. Stephenson, D.B.; Wanner, H.; Brönnimann, S.; Leterbacher, J. The history of scientific research on the North Atlantic oscillation. In *The North Atlantic Oscillation: Climate Significance and Environmental Impact*; Hurrell, J.W., Ed.; American Geophysical Union: Washington, DC, 2003; 37–50.

16. Hurrell, J.W.; Kushnir, Y.; Ottersen, G.; Visbeck, M. An overview of the North Atlantic oscillation. In *The North Atlantic Oscillation: Climate Significance and Environmental Impact*; Hurrell, J.W., Ed.; American Geophysical Union: Washington, DC, 2003; 1–35.

17. Ambaum, M.H.P.; Hoskins, B.J.; Stephenson, D.B. Arctic oscillation or North Atlantic oscillation? J. Climat. **2001**, *14* (16), 3495–3507.

18. Monahan, A.H.; Fyfe, J.C.; Flato, G.M. A regime view of Northern Hemisphere atmospheric variability and change under global warming. Geophys. Res. Lett. **2000**, *27* (8), 1139–1142.

19. Deser, C. On the teleconnectivity of the Arctic oscillation. Geophys. Res. Lett. **2000**, *27* (6), 779–782.

20. Itoh, H. True versus apparent arctic oscillation. Geophys. Res. Lett. **2002**, *29* (8), 109.1–109.4. doi: 10.1029/2001GL013978.

21. Christiansen, B. Comment on "True versus apparent arctic oscillation." Geophys. Res. Lett. **2002**, *29* (24), 2150. doi: 10.1029/2002GL016051.

22. Hassol, S.J. *Impacts of a Warming Arctic*, Arctic Climate Impact Assessment (ACIA) Overview Report; Cambridge University Press: Cambridge, UK, 2004; p 8.

23. Comiso, J.C. A rapidly declining perennial sea ice cover in the Arctic. Geophys. Res. Lett. **2002**, *29* (20), 1956. doi: 10.1029/2002GL015650.

24. Rothrock, D.A.; Zhang, J.; Yu, Y. The Arctic ice thickness anomaly of the 1990s: a consistent view from observations and models. J. Geophys. Res. **2003**, *108* (C3), 3083. doi: 10.1029/2001JC 001208.

25. Arfeuille, G.; Mysak, L.A.; Tremblay, L.B. Simulation of the inter-annual variability of the wind-driven Arctic sea-ice cover during 1958–1998. Clim. Dyn. **2000**, *16* (2–3), 107–121.

26. Kwok, R.; Cunningham, G.F.; Wensnahan, M.; Rigor, I.; Zwally, H.J.; Yi, D. Thinning and volume loss of the Arctic Ocean sea ice cover: 2003–2008. J. Geophys. Res. **2009**, *114* (C7), C07005. doi: 10.1029/2009JC005312.

27. Tucker, W.B.; Weatherly, J.W.; Eppler, D.T.; Farmer, L.D.; Bentley, D.L. Evidence for rapid thinning of sea ice in the western Arctic Ocean at the end of the 1980s. Geophys. Res. Lett. **2001**, *28* (14), 2851–2854.

28. Rigor, I.G.; Wallace, J.M.; Colony, R.L. On the response of sea ice to the Arctic Oscillation. J. Clim. **2002**, *15* (18), 2648–2668.

29. Rigor, I.G.; Wallace, J.M. Variations in the age of Arctic sea-ice and summer sea-ice extent. Geophys. Res. Lett. **2004**, *31*, L09401. doi: 10.1029/2004GL019492.

30. Rampal, P.; Weiss, J.; Marsan, D.; Bourgoin, M. Arctic sea ice velocity field: general circulation and turbulent like

fluctuations. J. Geophys. Res. **2009**, *114* (C10), C10014. doi: 10.1029/2008JC005227.

31. Hannachi, A.; Joliffe, I.; Stephenson, D. Empirical orthogonal functions and related techniques in atmospheric science: a review. Int. J. Climatol. **2007**, *27* (9), 1119–1152. doi: 10.1002/ joc.1499.

32. Dommenget, D.; Latif, M. A cautionary note on the interpretation of EOFs. J. Clim. **2002**, *15* (2), 216–225.

33. Zhao, J.; Cao, Y.; Shi, J. Core region of Arctic oscillation and the main atmospheric events impact on the Arctic. Geophys. Res. Lett. **2006**, *33* (22), L22708. doi: 10.1029/ 2006GL027590.

34. Zhao, J.; Cao, Y.; Shi, J. Spatial variation of the Arctic Oscillation and its long-term change. Tellus **2010**, *62A* (5), 661–672.

Artificial Reservoirs: Land Cover Change on Local Climate

Faisal Hossain
Ahmed M. Degu
Department of Civil and Environmental Engineering, Tennessee Technological University, Cookeville, Tennessee, U.S.A.

Abstract

This entry presents a review of what is currently known regarding the potential impact of large artificial reservoirs and dams on the local climate from the paradigm of land use and land cover (LULC) change. The review is based on the premise of systematic change in climate-sensitive LULC that most dams initiate after their construction and operation. To better understand the implication of change in local climate near dams and artificial reservoirs, it is therefore important to have a broader view of the changes to the landscape in the post-dam era. However, in the future, other factors, such as urbanization, atmospheric aerosols and global warming, need to be explored as well.

INTRODUCTION

Understanding the terrestrial water cycle is important for any study that concerns future water availability. The dynamic nature of the Earth's water cycle means that the water that is available naturally in various forms (precipitation, stream flow, ground water, snow etc.) is highly variable in both space and time. Consequently, this means that the supply of water that is available for harnessing directly from nature does not always meet the rising demand to sustain our human civilization (for energy, food, water supply needs). Historically, therefore, one of the common engineering solutions to guarantee a more steady water supply against such rising demands has been to construct surface water impoundments on rivers. Such large-scale infrastructures are commonly known as dams and artificial reservoirs. These structures trap a sufficiently large amount of water from the local water cycle to make up for a shortfall when demand exceeds the variable supply from nature.

Recently published research on sustainable development stresses the need to improve our understanding of how the extremes of climate and water availability are changing. Of the many important factors, land use and land cover (LULC) change represents a major human-induced activity critical to extremes of water cycle and fresh water availability.[1–3] One example of human-induced LULC change is the construction of engineering facilities for irrigation, hydroelectric power generation, industrial and domestic water supply once a large dam is built nearby. In particular, irrigation is one of the more pervasive drivers of change in the water cycle. During the last century,

irrigable land increased from 40 million hectares (Mha) to 215 Mha.[4] About 40% of the current irrigable land is supplied with surface water that is impounded by large artificial reservoirs and dams built on rivers.[5] Hereafter the term "dam" will be used interchangeably with "artificial reservoir."

The world has around 845,000 dams.[6] The water stored in these large dams amounts to about 10% (or more) of the annual river flow and covers one-third of the Earth's natural lake areas.[6] Though no accurate data are available on the volume of water impounded behind dams, estimates show that up to 10,800 cubic kilometers (km³) of water may have been impounded. This volume is equal to a volume of the world's ocean water created to a depth of 30 mm.[7] To get an idea of the distribution of dams around the world, one can refer to the most comprehensive compilation of large dams archived by the Global Water Project and known as the GRanD database.[8] Figure 1 shows the global distribution of large dams. According to this GRanD database, about 34% of these large dams are engaged in irrigation.

In the United States, statistics may suggest that dam building is a thing of the past and not a current focus for the civil engineering profession (Fig. 2).[10,11] However, for vast regions of the underdeveloped or developing world, large dam construction projects will continue in increasing numbers for tackling the rising water deficit for emerging economies. Examples of such large dam projects are the Southeast Anatolia Project or GAP (Turkish acronym) project in Turkey comprising 22 dams on the Tigris and Euphrates rivers,[12] the Three Gorges Dam (TGD) in China,[13] Itaipu Dam in Brazil,[14] and the proposed

Encyclopedia of Natural Resources DOI: 10.1081/E-ENRW-120047593

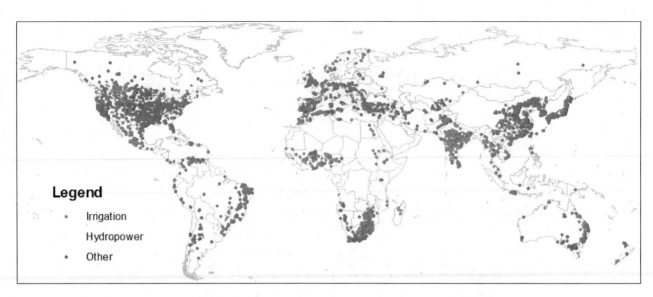

Fig. 1 The global distribution of large dams and their main purpose as archived in the GRanD database.

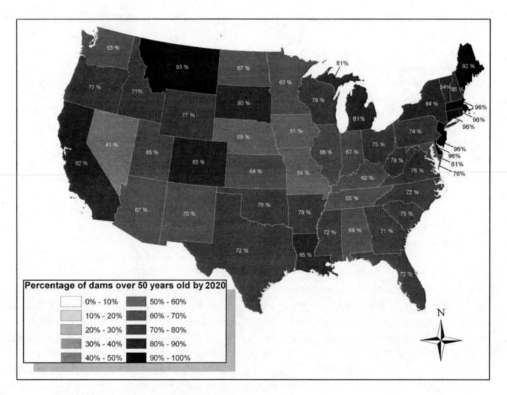

Fig. 2 The percentage of dams per state that will be over 50 years old in 2020.
Source: Reproduced from U.S. Army Corps of Engineers report.[9]

Indian River Linking Project.[15] Thus, an understanding of the impact of dams on local climate and water cycle is important for sustainable development in the developing world.

In current literature, the impact of global climate change (which is usually perceived as a large-scale phenomenon based on a globally rising temperature trend) on dams and water supply has been studied at regional scales for some time.[16,17] This is also evident from a comprehensive literature synthesis published by the U.S. Bureau of Reclamation (USBR) on climate change adaptation.[18] The converse (impact of reservoirs on local climate), which is the focus of this entry, has not been explored in depth yet.[19,20]

A broader view of the change a dam typically triggers needs to be considered if we are to understand the changes a dam can initiate on the local climate. At a minimum, a dam changes the pre-existing landscape to an open body of water, which then leads to a change in surface albedo, surface roughness, and sensible and latent heat fluxes. A flood control or a hydropower dam can also accelerate the pace of urbanization in the downstream and flood-safe valley regions. Irrigation dams intensify agricultural production in the vicinity of the reservoir.

As mentioned earlier, of the various LULC changes due to dams, irrigation is the most widespread. Global data and simulation analysis by Biemans et al.[21] report that artificial reservoirs contribute significantly to irrigation water supply in many regions around the world. The additional contribution of reservoirs to irrigation has increased spectacularly from 5% (around 1900 A.D.) to almost 40% in the 21st century.[21] When such changes to the land cover and land use (LCLU) are assessed in relation to existing knowledge on their impact on climate,[22–24] it becomes apparent that a typical dam-reservoir system can, in principle, change the local climate through a gradual change in the landscape. Examples of the effect of local LULC on precipitation include Gero et al., Lei et al., Shepherd et al., and Niyogi et al.[25–28] Recently, there have been studies that report that there is global scale impact from irrigation,[29] of which a significant fraction of this irrigation is the result of water made available for vegetation from artificial reservoirs.

The change to local climate near dams is expected to be particularly noticeable on temperature and precipitation patterns during the growing season. For example, Mahmood et al.[30] showed that irrigation in the Great Plains during growing season (May through September) might have caused regional cooling by 1.41°C of growing season mean maximum temperatures at irrigated locations during the post-1945 period. A few other studies support the notion that atmospheric moisture added by irrigation can also increase rainfall, provided that the mesoscale conditions are appropriate.[31,32] DeAngelis et al.[33] have reported an increase of 10–30% of rainfall downwind of the Ogallala aquifer that has been used for ground water-based irrigation in the Great Plains. A statistical analysis of observed rainfall frequency downwind of irrigated lands in South Spain (upper and lower Vegas and lower Guadalquivir) found a significant increase in the number of months with minimum precipitation after irrigation when compared with pre-irrigation records.[34] In summary, dams can, therefore, clearly cause a change of climate in their vicinity through systematic change in the land use and land cover.

While the climate effects of LULC change attributable to large dams have been explored, the hydrology impact is not understood as well yet and therefore needs to be explored. Qualitative investigation of databases archived by the National Performance of Dams Program (NPDP; http://npdp.stanford.edu) reveals that historically, the number of reported incidences of embankment overtopping far exceeds that of structural failures. According to our NPDP survey, a total of 1133 dams have overtopped hydrologically between 1889 and 2006. Of these, 625 experienced a complete "hydrologic performance failure," most likely due to increased reservoir inflow from upstream. Herein, it is important to note that the land use changes can amplify the surface runoff generation mechanism in two ways: (1) through modification of precipitation rates leading to increased infiltration–excess runoff and (2) through enhancement of rainfall partitioning as runoff due to increased imperviousness. The former cause is akin to a "strategic" cause that occurs through gradual change in the local climate in the post-dam era, while the latter is a more "tactical" (instantaneous) cause (of increasing imperviousness). Both causes may be equally important and may work together to magnify the intensification of upstream runoff, as the dam ages.

IS THE IMPACT OF DAMS DETECTABLE FROM OBSERVATIONAL RECORDS?

The previous section dwelt on a "can"-type question on how a dam–reservoir system is physically capable of changing the pre-dam climate. This section addresses the following question: *If dams can really change local climate in principle, can we detect it from observational records?* Wu et al.[35] examined the effect of the TGD of China, built in 1998, on the regional precipitation around its vicinity by analyzing satellite precipitation data. Based on their 8-year data analysis (1998–2006), they concluded that filling up of the TGD reservoir was found to correlate with an increase in precipitation in the nearby region. However, at other regions, precipitation was found to experience a decrease.

For a more in-depth and comprehensive analysis of observational data (comprising several dams and multidecadal data), a recent effort by Hossain et al.[20] is recounted here as is hereafter with kind permission granted by the American Society of Civil Engineers (ASCE). Ninety-two dams located in the United States and defined as large, according to the standards set by the International Commission on Large Dams, were recently studied in a manner similar to a recent study by Degu et al.[19] The database on dams was available from a series of world dam registers published by the Global Water Systems Project Digital Water Atlas.[36] These dams were spread across eight distinct Köppen climate zones (Fig. 3). For the observational record of climate, reanalysis data from the National Center for Environmental Prediction, North American Regional Reanalysis (NARR),[37,38] were used. NARR provides good quality and finer-resolution climate dataset for North America. Daily NARR fields span a period from 1979 to 2009 (30 years) and focus on the daily average of the surface Convective Available Potential Energy (CAPE; J/kg) as a proxy signature of dams.[39] Among the many important ingredients required for rainfall, CAPE can be considered an important atmospheric

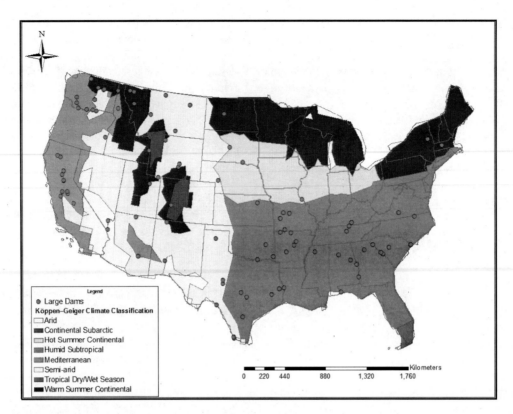

Fig. 3 Location of the 92 large dams according to the Köppen climate map.
Source: Reprinted with kind permission from American Society of Civil Engineers.

indicator of the presence of heavy rainfall process. The objective was to identify detectable changes in the CAPE climatology (available at 32 km spatial scales) in the vicinity of the dams for a given climate zone.

Because the NARR record of CAPE does not date back before the construction of most large dams in the United States, the ergodic assumption was invoked to answer the "detection" and "have"-type question posed in this section. The spatial average of CAPE over a region far away from dams (referred to as "no-dam" region) was considered as a substitute for the temporal average of CAPE during the pre-dam era. In other words, the average CAPE over the "no-dam" region was used to represent the "pre-dam" climatology of convective atmospheric instability. Vice versa, the average CAPE over the "dam" region was used as a proxy for the post-dam climatology. Herein, the "no-dam" region was defined as the annular region 100 km outside the dam's spillway having the same area as the "dam" region (Fig. 4). To attribute the unique and local impact due to the dam alone, a "dam" region was defined as a circular area having a 50 km radius around the dam location (Fig. 4).

For a statistical generalization of the 92 dams, Table 1 shows the relative ranking of the dams in terms of the % change (increase or decrease in the climatologic average of CAPE) between the "dam" and "no-dam" region. Here, the % change in CAPE is defined as

$$\text{Percent increase} = \frac{\left(\text{mean of "Dam"} - \text{mean of "No dam"}\right)}{\text{mean of No dam}} \times 100\%$$

Herein, the "mean" refers to the 30-year climatologic average for the period 1979–2009. It is very clear from Table 1 that dams in the Mediterranean climate (Koppen class name is Dry Summer Subtropical) consistently experience the highest alteration of CAPE near the reservoir ("dam") region. Subsequently, this change is matched by dams located in hot or cold arid climates. Dams in humid or arctic climates exhibit spatially insignificant increases (<5%) in CAPE near the reservoir. An important aspect to note is the likely absence of irrigation in humid and arctic climates that may be one of the contributing factors for spatial uniformity of CAPE near and away from the reservoir. As an example, Fig. 5 shows the time series of 30-year CAPE climatology for two dams in contrasting climates (Walter Dam in Alabama and Coolidge Dam in Arizona). Two possible reasons may be attributed for the intensification of CAPE by dams in arid and Mediterranean climates: (1) more widespread irrigation; and (2) stronger spatial gradients in humidity and surface evaporation around the reservoir–shoreline interface.

Table 1 Ranking of dams studied according to % change in CAPE climatology between dam and no-dam regions. Only a % change of up to 1% is reported in the table

Köppen–Geiger Climate	Dam	State	% increment
Dry Summer Subtropical	New Bullards Bar	California	105.01
Dry Summer Subtropical	Oroville	California	99.00
Dry Summer Subtropical	New Exchequer	California	80.91
Dry Summer Subtropical	Shasta	California	66.89
Dry Summer Subtropical	Folsom	California	56.86
Dry Summer Subtropical	Don Pedro	California	56.06
Dry Summer Subtropical	Trinity	California	42.41
Dry Summer Subtropical	Monticello	California	33.79
Dry Summer Subtropical	Pine Flat Lake	California	31.31
Dry Summer Subtropical	Swift	Washington	27.22
Warm Summer Continental	Conklingville	New York	26.36
Dry Summer Subtropical	Dworshak	Idaho	25.73
Cold Semiarid	Kingsley	Nebraska	21.27
Cold Semiarid	Yellowtail	Montana	20.61
Humid Subtropical	Hartwell	Georgia	20.42
Warm Summer Continental	Albeni Falls	Idaho	18.01
Warm Summer Continental	Hungry Horse	Montana	16.66
Warm Summer Continental	Kerr	Montana	15.35
Cold Arid	Hoover	Nevada	14.16
Humid Subtropical	Smith Mountain	Virginia	13.45
Humid Subtropical	Pensacola	Florida	13.09
Warm Summer Continental	Seminoe	Wyoming	12.13
Warm Summer Continental	Palisades	Idaho	11.19
Hot Arid	Davis	Arizona	11.17
Humid Subtropical	Center Hill	Tennessee	10.87
Hot Semiarid	Robert Lee	Texas	7.86
Humid Subtropical	Little River	South Carolina	7.49
Continental Subarctic	Blue Mesa	Colorado	7.26
Dry Summer Subtropical	Mossyrock	Washington	6.13
Humid Subtropical	Sardis	Mississippi	5.97
Cold Semiarid	Owyhee	Oregon	5.47
Cold Semiarid	Pathfinder	Wyoming	4.78
Dry Summer Subtropical	New Melones	California	4.24
Humid Subtropical	Richard B. Russell	Georgia	4.02
Humid Subtropical	Wolf Creek	Kentucky	3.97
Humid Subtropical	Eufaula	Oklahoma	3.75
Humid Subtropical	Dale Hollow	Tennessee	3.38
Continental Subarctic	Oahe	South Dakota	3.35
Cold Semiarid	Coolidge	Arizona	3.28
Cold Semiarid	Buffalo Bill	Wyoming	3.03
Warm Summer Continental	Libby	Montana	2.98
Humid Subtropical	Stockton	Missouri	2.79
Cold Semiarid	Tiber	Montana	2.23
Hot Semiarid	Twin Buttes	Texas	2.19
Cold Semiarid	Fort Peck	Montana	1.86

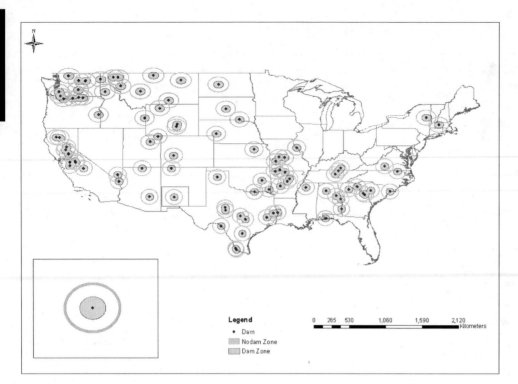

Fig. 4 The 50 km buffer zone showing the "dam" and the 100 km away annular region (outer encircle) as "no-dam region."
Source: Reprinted with kind permission from American Society of Civil Engineers.

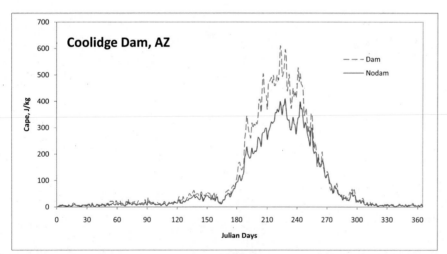

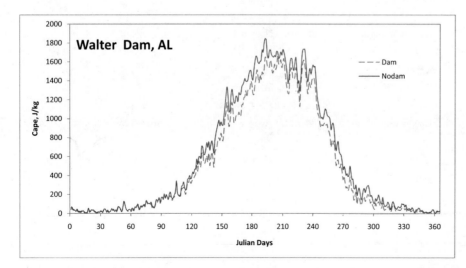

Fig. 5 Comparison of Convective Available Potential Energy (CAPE) climatology (30 year) for dam- and no-dam region for Coolidge (in arid climate, Arizona) and Walter (in humid climate, Alabama) dams.
Source: Reprinted with kind permission from American Society of Civil Engineers.

CONCLUSION

This entry presented a review of what is currently known about the potential impact of large artificial reservoirs and dams on the local climate based on the premise of systematic change in climate-sensitive LCLU that most dams initiate. To better understand the implication of local climate changes, it is, therefore, important to have a broader view of the changes to the landscape in the post-dam era. However, in the future, other factors, such as urbanization,[40] atmospheric aerosols,[28] and global warming,[18] need to be explored as well. These factors are well known to initiate changes in local climate, particularly to precipitation. Systematic changes in precipitation patterns can also gradually alter the hydrology of the impounded river basin, which is currently not well understood. For example, more intensified storms near the reservoir shorelines or overly dense irrigated landscapes may result in more runoff, erosion, and water-logging issues. The groundwater table may rise and gradually change infiltration–excess runoff areas to a more excess saturation type. Overall, it is intuitive to expect that the hydrological impact due to changing rainfall patterns will differ upstream and downstream of the reservoir due to the hydraulic head differences and seepage issues. For better water resources management, it is now timely to investigate the changing hydrology of aging dams, which undergo a shift in local rainfall dynamics.

ACKNOWLEDGMENTS

This entry was reproduced from an earlier forum article that appeared in ASCE Journal of Hydrologic Engineering titled "Climate Feedback-based Considerations to Dam Design, Operations and Water Management in the 21st Century" (vol. 17(8), pp. 837–850, doi:10.1061/(ASCE) HE.1943-5584.0000541).

REFERENCES

1. Chase, T.N.; Pielke Sr., R.A.; Kittel, G.F.; Nemani, R.R.; Running, S. W. Simulated impacts of historical land cover changes on global climate in northern winter. Clim. Dyn. **2000**, *16*, 93–105.
2. Vörösmarty, C.J.; Sahagian, D. Anthropogenic disturbance of the terrestrial water cycle. Bioscience **2000**, *50* (9), 753.
3. Hossain, F.; Niyogi, D.; Adegoke, J.; Kallos, G.; Pielke Sr., R. Making sense of the water resources that will be available in future use, EOS Forum **2011**, *92* (17).
4. Freydank, K.; Siebert, S. Towards mapping the extent of irrigation in the last century: a time series of irrigated area per country. Germany: Frankfurt Hydrology Paper 08, Institute of Physical Geography, University of Frankfurt (Main), 2008.
5. Lempérière, F. The role of dams in XXI century: Achieving a sustainable development target. International Journal of Hydropower and Dams, http://www.hydropower-dams.com/Issue-Three.php?c_id=60 (accessed May 2011).
6. Jacquot, J. Dams, from Hoover to Three Gorges to the Crumbling Ones, Discover Magazine. March 2009, http://discovermagazine.com/2009/mar/08-dams-hoover-three-gorges-crumbling-ones (accessed May 2011).
7. Chao, B.F.; Wu, Y.H.; Li, Y.S. Impact of artificial reservoir water impoundment on global sea level. Science **2008**, *320*, 212–214.
8. Lehner, B.; Döll, P. Development and validation of a global database of lakes, reservoirs and wetlands. J. Hydrol. **2004**, *296*, 1–22.
9. Hossain, F.; Jeyachandran, I.; Pielke, R., Sr. Have Large Dams altered extreme precipitation patterns?, EOS-AGU **2009**, *90* (48), 453–454.
10. Graf, W.L.; Wohl, E.; Sinha, T.; Sabo, J.L. Sedimentation and sustainability of western American reservoirs. Water Resour. Res. **2010**, *46*, W12535, (doi:10.1029/2009WR008836).
11. Graf, W.L. Downstream hydrologic and geomorphic effects of large dams on American rivers. Geomorphology **2006**, *79* (3–4), 336–360.
12. Unver, O.H.I. South-eastern Anatolia integrated development project (GAP), Turkey: an overview of issues of sustainability Int. J. Water Resour. Develop. **1977**, *13* (2), 187–208.
13. Shen, G.; Xie, Z. Three Gorges project: chance and challenge, Science **2004**, *681* (doi:10.1126/science.304.5671.681b).
14. Pierce, F. The biggest dam in the world New Sci. **1995**, *145* (1962), 25–29.
15. Misra, A.K.; Saxena, A.; Yaduvanshi, M.; Mishra, A.; Bhauduriya, Y.; Takur. A. Proposed river linking project of India: Boon or bane to nature? Env. Geol. **2007**, *51* (8), 1361–1376. (doi: 10.1007/s00254-006-0434-7).
16. Hamlet, A.F.; Lettenmaier, D.P. Effects of climate change on hydrology and water resources in the Columbia River basin J. Amer. Water Res. Assoc. **1999**, *35* (6), 1597–1623.
17. Christensen, N.S.; Wood, A.W.; Voisin, N.; Lettenmaier, D.P.; Palmer, R.N. The effects of climate change on the hydrology and water resources of the Colorado river basin. Clim. Change **2004**, *62* (1–3), 337–363 pp.
18. USBR. Literature synthesis on climate change implications for reclamation's water resources, USBR Technical Services Report (Technical Memorandum 86-68210-091), 2nd Edition, 2009, available online at http://www.usbr.gov/research/docs/climatechangelitsynthesis.pdf (last accessed May 2011).
19. Degu, A.M.; Hossain, F.; Niyogi, D.; Pielke, R., Sr.; Shepherd, J.M.; Voisin, N.; Chronis, T. The influence of large dams on surrounding climate and precipitation patterns. Geophys. Res. Lett. **2011**, *38* (4), L04405, doi:10.1029/2010GL046482.
20. Hossain, F.A.M. Degu; Yigzaw, W.; Niyogi, D.; Burian, S.; Shepherd, J.M.; Pielke, R., Sr. Climate feedback-based considerations to dam design, operations and water management in the 21st century. ASCE J. Hydrol. Eng. **2011**, *17* (8), 837–850.
21. Biemans, H.; Haddeland, I.; Kabat, P.; Ludwig, F.; Hutjes, R.W.A.; Heinke, J.; von Bloh, W.; Gerten, D. Impact of reservoirs on river discharge and irrigation water supply during

the 20th century, Water Resour. Res. **2011**, *47*, W03509 (doi:10.1029/2009WR008929).

22. Pielke Sr., R.A. Land use and climate change. Science **2005**, *310*, 1625–1626.

23. Feddema, J.J.; Oleson, K.W.; Bonan, B.; Mearns, L.O.; Buja, L.E.; Meehl, G.A.; Washington, W.M. The importance of land cover change in simulating future climates. Science **2005**, *310*, 1674–1678.

24. Pielke, Sr., R.A.; Adegoke, Sr., J.; Beltran-Przekurat, A.; Hiemstra, C.A.; Lin, J.; Nair, U.S.; Niyogi, D.; Nobis, T.E. An overview of regional land use and land cover impacts on rainfall. Tellus B **2007**, *59*, 587–601.

25. Gero, A.F.; Pitman, A.J.; Narisma, G.T.; Jacobson, C.; Pielke, R.A., Sr. The impact of land cover change on storms in the Sydney Basin. Global Planet. Change **2006**, *54*, 57–78.

26. Lei, M.; Niyogi, D.; Kishtawal, C.; Pielke , R., Sr.; Beltrán-Przekurat, A.; Nobis, T.; Vaidya, S. Effect of explicit urban land surface representation on the simulation of the 26 July 2005 heavy rain event over Mumbai, India, Atmos. Chem. Phys. **2008**, *8*, 5975–5995.

27. Shepherd, J.M.; Carter, W.M.; Manyin, M.; Messen, D.; Burian, S. The impact of urbanization on current and future coastal convection: a case study for Houston. Env. Plan. **2010**, *37*, 284–304.

28. Niyogi, P. Pyle, Lei, M.; Arya, S.; Kishtawal, C.; Shepherd, M.; Chen, F.; Wolfe, B. Urban modification of thunderstorms - an observational storm climatology and model case study for the Indianapolis urban region. J. Appl. Meteor. Climatol **2011**, *50*, 1129–1144.

29. Puma, M. J.; Cook, B.I. Effects of irrigation on global climate during the 20th century, J. Geophys. Res. **2011**, *115*, D16120 (doi:10.1029/2010JD014122).

30. Mahmood, R.; Hubbard, K.G.; Carlson, C. Modification of growing-season surface temperature records in the Northern Great Plains due to land use transformation: Verification of modeling results and implications for global climate change. Int J Climatol. **2004**, *24*, 311–327.

31. Lohar, D.; Pal, B. The effect of irrigation on pre-monsoon season precipitation over Southwest Bengal, India. J. Clim. **1995**, *8* (10), 2567–2570.

32. Barnston, A.G.; Schickedanz, P.T.; The effect of irrigation on warm season precipitation in the Southern Great Plains. J. Appl. Meteorol. Clim. **1984**, *23* (6).

33. DeAngelis, A.; Dominguez, F.D.; Fan, Y.; Robock, A.; Kustu, M.D.; Robinson. D. Evidence of enhanced precipitation due to irrigation over the Great Plains of the United States. J. Geophys. Res. **2010**, *115*, D15115 (doi:10.1029/2010JD013892).

34. Jodar, J.; Carrera, J.; Cruz, A. Irrigation enhances precipitation at the mountains downwind. Hydrol. Earth Syst. Sci. Discussion, **2010**, *7*, 3109–3127.

35. Wu, L.G.; Zhang, Q.; Jiang, Z.H. Three Gorges Dam affects regional precipitation. Geophys. Res. Lett. **2006**, 33L13806 10.1029/2006gl026780.

36. GWSP, Digital Water Atlas, Map 41: Dams and capacity of artificial reservoirs (V1.0), Bonn, Germany, 2008 (available at http://atlas.gwsp.org/)

37. Mesinger, F. North American Regional Reanalysis, Bull. Am. Meteorol. Soc. **2006**, *87* (3), 343–360.

38. Rutledge, G.K.; Alpert, J.; Ebuisaki, W. NOMADS: A Climate and Weather Model Archive at the National Oceanic and Atmospheric Administration. Bull. Am. Meteorol. Soc. **2006**, *87*, 327–341.

39. Pielke Sr., R.A. Influence of the spatial distribution of vegetation and soils on the prediction of cumulus convective rainfall. Rev. Geophys. **2001**, *39*, 151–177.

40. Shepherd, J.M.; Pierce, H.; Negri, A.J. On rainfall modification by major urban areas: Observations from spaceborne radar on TRMM. J. Appl. Meteorol. **2002**, *41*, 689–701.

Bathymetry: Assessment

Heidi M. Dierssen
*Department of Biology, Norwegian University of Science and Technology, Trondheim, Norway, and
Department of Marine Sciences/Geography, University of Connecticut, Groton, Connecticut, U.S.A.*

Albert E. Theberge, Jr.
Central Library, National Oceanic and Atmospheric Administration (NOAA), Silver Spring, Maryland, U.S.A.

Abstract
Covering over 70% of the Earth's surface, the undersea landscape or bathymetry has been revealed only within the past century due to advances in technology for measuring the depth of the ocean. Current depth-measuring techniques employ vehicles ranging from near-bottom remotely operated vehicles to ships on the sea surface to satellites high above the Earth. Depending on vehicle and mission, these systems utilize acoustics, optics, or radar altimetry to either directly measure or infer bathymetry. Each method provides different spatial resolution and can probe depths ranging from shallow coastlines to the deepest trenches. Coastal estuaries and bays must also be evaluated in terms of tidal currents that can change bathymetry on hourly time scales. Improvements are continually made to the types of data and analysis methods used for estimating bathymetry across the global ocean.

INTRODUCTION

The term "bathymetry" most simply refers to the depth of the seafloor relative to sea level, but the concepts involved in measuring bathymetry are far from commonplace. The mountains, shelves, canyons, and trenches of the seafloor have been mapped with varying degrees of accuracy since the mid-nineteenth century. Today, the depth of the seafloor can be measured from kilometer- to centimeter-scale using techniques as diverse as multibeam sonar from ships, optical remote sensing from aircraft and satellite, and satellite radar altimetry. Various types of bathymetry data have been compiled and modeled into gridded matrices spanning the seafloor at 1 arc-minute spatial resolution per pixel (<2 km) or even finer resolution for portions of the globe (e.g., <100 m resolution for U.S. coastal waters).[1] Even so, errors are prevalent in these modeled datasets and remote regions of the global ocean have yet to be accurately mapped. Indeed, the topography of Mars and Venus may be better known than our own seafloor.

Bathymetry is measured using "remote sensing" methods that investigate the seafloor indirectly without making physical contact. The majority of methods for estimating are based on the concept of using time to infer distance. Specifically, sensors emit a beam of sound, light, or radio waves and measure the round-trip travel time it takes for the beam to be reflected from a surface and return back to the sensor. The time elapsed is then related to the distance the beam traveled and used to infer bathymetry. The longer the elapsed time for the beam to return, the greater the distance traveled. However, limitations are inherent with all of these methods and no single technique is ideal for measuring the diversity and complexity of the underwater landscape and coastline. In addition to collecting additional datasets across the global ocean, new techniques are required to measure bathymetry and to interpret and process the datasets that have already been collected.

METHODS FOR ASSESSING BATHYMETRY

This section outlines the primary modern techniques for estimating bathymetry from acoustics to the use of both the visible and radio wave portions of the electromagnetic spectrum. Measurements are made through the water medium in the case of acoustics, through the water and air for visible light measurements, and through the air medium alone for radar altimetry remote sensing to obtain an estimate of the seafloor depth. As will be outlined further, each method has its own advantages and disadvantages in terms of scale of feature observed, accuracy, and water depth that can be sensed (Fig. 1). Acoustics is applicable for determining centimeter- to kilometer-scale features and is the prevailing method used to validate optical and radar bathymetry. Because of limits to penetration of visible light in a water medium, measurements using active and passive

Encyclopedia of Natural Resources DOI: 10.1081/E-ENRW-120048588

light are only applicable in shallow water (generally <60 m) and are the primary means for delineating coastlines. Radar altimetry measurements from satellites provide large-scale bathymetry across the deep ocean and are particularly valuable in remote regions of the world ocean with little ship traffic. Marine radars from ships and shore have also been used to map bathymetry of shallow seas. The platforms for measuring bathymetry are also variable and include satellites orbiting high above the Earth, suborbital aircraft, ships on the sea surface, and remotely operated vehicles hovering along the seafloor.

Bathymetry measurements are not static. The rise and fall of the tides can change bathymetry up to several meters in height, depending on when and where measurements are taken. For bathymetric maps seaward of the continental shelf, usually no tidal corrections are made to ship soundings. Most international navigational charts used over shallow areas such as the Great Bahama Bank commonly adjust soundings to the lowest astronomical tide while mean lower low water is the datum for United States charts. However, tidal models have not yet been made perfect for shallow water mapping and tidal forecasting is an active area of research.[2] Having accurate tidal models are required for delineating coastlines, planning for storm surges and flooding,[3] and assessing bathymetry from satellite altimetry.

Acoustics

Acoustic bathymetry or echo sounding came into play in the 1920s and was instrumental in determining the configuration of the seafloor as we know it today.[4] The initial technique was to emit a single pulse of sound and measure the elapsed time for it to travel to the seafloor and be reflected and be detected by a shipboard hydrophone. One-half of the round-trip travel time multiplied by the velocity

of sound in sea water equals the depth at a given point. As noted by Leonardo da Vinci in 1490 and later by Benjamin Franklin in 1762, sound travels far in water with little attenuation relative to air.[5] Sound also travels significantly faster in water than in air and great ocean depths can be probed acoustically without substantial degradation of the signal. The velocity of sound in seawater is ~1500 m/sec, but the precise velocity depends on ocean temperature, salinity, and pressure. In addition to the velocity of sound, the character of the seabed, vegetative cover, biota, and other particles in the water column, can affect the accuracy of the measured depth.

Today, high-resolution measurements of ocean depth and bottom reflectivity are produced by multibeam sonar. Each ping of a multibeam sonar emits a single wide swath of sound (up to 153°) that reflects off the seafloor (Fig. 2). The return echo is received by an array of transducers and then separated electronically into a number of individual beams for each of which a depth is calculated. Very high resolution is attainable in shallow water, but the swath width is decreased. Conversely, the efficiency of ship operation is increased in deep water as the swath width expands geometrically but resolution is decreased. A swath of the seafloor is acoustically imaged with each pass of a survey ship as it follows a pattern similar to "mowing the grass." A series of overlapping swaths produce a bathymetric map of the area being surveyed.

Various sound frequencies (e.g., 12–400 kHz) are employed for different depth ranges; the lower the frequency, the deeper the depth measurement possible, while higher frequency is associated with higher resolution but shallower depth. Bathymetry is now being estimated at levels of resolution and accuracy that were previously unattainable. For example, shallow-water multibeam systems can measure bathymetry roughly at a 10 cm scale in 10 m

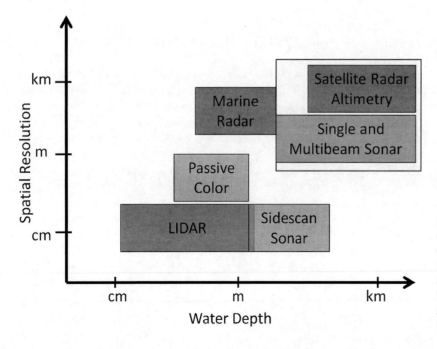

Fig. 1 Schematic showing the applicability of different techniques for estimating bathymetry in terms of spatial resolution of measurement and the range in water depths that can be sampled. The yellow box indicates the datasets that are integrated and interpolated to form the global gridded bathymetry datasets at 1 arc-minute resolution.

of water and have been used to effectively map coastal waters of the United States.[6,7] Concentrated mapping programs also incorporate "sidescan" sonar for a qualitative view of seafloor reflectivity characteristics.[7] Acoustic sensors can also be placed on tethered remotely operated vehicles or autonomous underwater vehicles, such as gliders, that can maintain a constant position relative to the seafloor and provide high-resolution bathymetry.[8]

The primary disadvantages of acoustic measurements are the time and cost associated with making measurements from a ship in deep waters or a small vessel in shallow waters. In order to build up coherent images at a high resolution, many survey lines with overlapping tracks must be run. Because the swath width decreases in shallow water, many more ship or glider tracks are required in coastal estuaries and bays with shallower water. Hence, detailed surveys in coastal regimes require considerable time and effort to cover relatively small portions of the sea bed. In general, acoustic methods can be used throughout all oceanic depths from shallow estuaries to the deepest trenches. However, ship time is costly even in deep water, and because of increasing time and effort to operate in shallow waters, acoustic systems are not ideal for such tasks as monitoring bathymetric changes and shoreline configuration changes caused by such phenomena as tidal currents, storm surge, and sea-level change.

Optics

Remote sensing by visible light has also been used to estimate shallow water bathymetry, where acoustic methods become limited. Bathymetry has been quantified using passive methods, which measure only the natural light reflecting from the seafloor, and active methods, which use lasers to measure the distance to the seafloor.

Passive ocean color remote sensing

Of the sunlight that hits the ocean surface, only a small percentage is scattered back out again and can be remotely detected by aircraft or satellite. Passive ocean color sensors measure this small amount of solar radiation that has entered the water column and been scattered back out.[9] In waters that are shallow and clear enough for light to reach the bottom (called "optically shallow"), the color of the seafloor also contributes to the water leaving radiance in a way that depends on the bathymetry, substances in the water column, and bottom composition. In the clearest natural waters, bright sandy bottoms can be detected in 30 m or more. In most coastal areas, water is clarity is reduced by algae, sediment and other colored substances and generally depths less than 10 m can be measured. As the water becomes more turbid with sediment or dense phytoplankton, much of the light entering the water column is absorbed in the top layer of the water and the sea becomes "optically deep" at only a few meters in depth.

In sufficiently shallow and clear water, the magnitude and spectral quality of light reflected from the seafloor can be interpreted from remotely sensed ocean color. Methods to retrieve shallow water bathymetry from passive ocean color reflectance measurements have been either empirical[10] or radiometrically based using iterative modeling or look-up

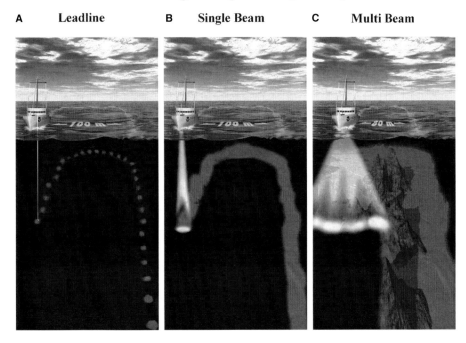

Bottom Coverage Comparison by Survey Method

A Leadline   B Single Beam   C Multi Beam

Fig. 2 Illustration of different methods for measuring bathymetry from (**A**) line and sinker technology; (**B**) echo sounding using single-beam acoustics; and (**C**) multibeam acoustics providing the highest spatial resolution of seafloor features.

tables.[11] Many of the empirical approaches use far red or near-infrared wavelengths where water is highly absorbing and the signal is less influenced by phytoplankton, sediments, and other absorbing and scattering material in the water column.[12] Radiometrically based approaches generally use most of the visible spectrum and require many wavelengths of data (i.e., hyperspectral) to retrieve bathymetry. In addition, the radiometric methods also solve for the color of the seafloor (i.e., benthic reflectance), as this is a key component of the water-leaving radiance signal. Using passive ocean color, bathymetry has been mapped with high spatial resolution from aircraft[12] to regional scales from satellites[13] (Fig. 3).

Because ocean color sensors are already up in space for oceanographic research with data freely available,[14] the use of satellites to derive shallow water bathymetry is quite effective and cost-efficient compared to acoustic methods. However, cloud cover masks the ocean color and only clear sky imagery can be analyzed. In addition, only a low percentage of the incident light that reaches the ocean is reflected back to a satellite. When viewed remotely, the ocean color is observed through a thick atmosphere containing gases and aerosols which also reflect sunlight back to the sensor. The atmosphere contributes more photons to a satellite sensor than the ocean surface itself and accurate "atmospheric correction" is one of the many challenges to obtaining an accurate water-leaving radiance required for estimating shallow water bathymetry. If too much signal is removed from the atmospheric correction, the ocean will appear too dark and bathymetry can be overestimated and vice versa.[15] Most methods also require some a priori knowledge of the benthic reflectance in the region and such measurements are difficult to make and integrate over the appropriate scales required for remote sensing. Bathymetry estimated with passive ocean color methods are generally not accurate enough for navigational purposes, but they provide much better results than gravimetric measurements (see further) over shallow basins (Fig. 3C) and the spectral information can also provide an estimate of the seafloor composition.

Active lidar

At present, the most effective means to map shallow water bathymetry is with active lidar systems. Active sensors produce and sense their own stream of light (e.g., light detection and ranging, or LIDAR) and are generally flown on aircraft, although several space-based lidars have been launched. Similar to acoustic measurements, LIDAR uses the round-trip travel time of a pulse of light to estimate the distance to the seafloor. Aircraft-mounted lasers pulse a

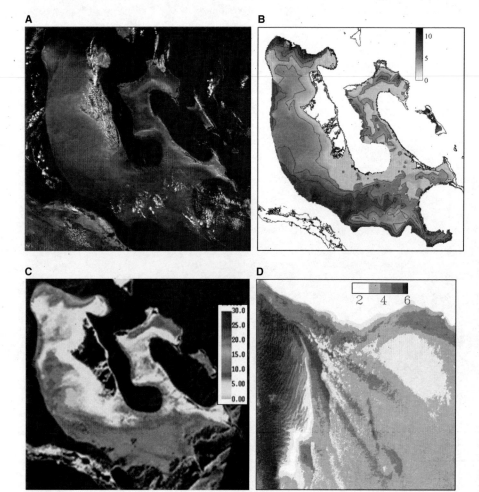

Fig. 3 The Great Bahama Bank is a large optically shallow bank that can be mapped with passive ocean color techniques. (**A**) A pseudo-true-color image from the MODIS Aqua sensor at 250 m resolution from 6 March 2004 shows the bright reflected color off of the shallow Bank; (**B**) bathymetry gridded using selected soundings from navigational chart;[26] (**C**) bathymetry modeled from the MERIS ocean color sensor;[13] (**D**) shallow water bathymetry mapped from a hyperspectral sensor on an aircraft for a small region south of Lee Stocking Island, Bahamas.[27] All bathymetry is in units of meters (m).

narrow, high-frequency laser beam toward the Earth and are capable of recording elevation measurements at rates of hundreds to thousands of pulses per second. Because they do not rely on sunlight, LIDAR systems can be operated at night.

Generally, two lasers are employed to estimate bathymetry: 1) an infrared laser (1064 nm), which does not penetrate water, is used to detect the sea surface and 2) a green laser (532 nm) is used to penetrate into the water column and provide a return signal from the seafloor.[16] In coastal waters, green light is the least absorbed and generally penetrates the deepest into the water column. Similar to passive ocean color measurements, the green laser is attenuated by the absorbing and scattering particles in the water column, and the maximum depth of measurement is determined by a combination of the optical properties of the water column and seafloor. In ideal conditions, depths up to 60 m have been measured with LIDAR systems,[17] but most applications are limited to depths <40 m.

Current LIDAR measurements with an elevation accuracy of 10–30 cm can provide point measurements from 0.1 to 8 pixels per m^2. They have been successfully used to map bathymetric features, beach erosion, coral reefs, and coastal vegetation. Unlike acoustic sensors on ships, the imagery is provided in fixed-width swaths independent of water depth and can provide high area coverage rates of up to 70 km^2 hr^{-1} (Fig. 4A). The method is particularly effective in shallow water where larger ships are excluded and acoustic surveying is least efficient. Another major advantage of LIDAR is that it can map land and water in the same mission and provide continuity of shoreline mapping. Such combined land and water measurements have been used by environmental managers to map coastlines and conduct city planning for storms and potential changes in sea level (Fig. 4B–D). Airborne systems are also useful in shallow waters with reefs or in regions where tidal flows make ship-based measurements difficult if not impossible. Often LIDAR systems are coupled with passive hyperspectral imagers to assess both the bathymetry and seafloor composition simultaneously.

Radar Systems

Marine radars can be used from satellite to detect large-scale changes in bathymetry that are related to sea-surface height. Imaging radar systems from ships and shore have been used to measure wave fields and deduce shallow water bathymetry from wave theory.

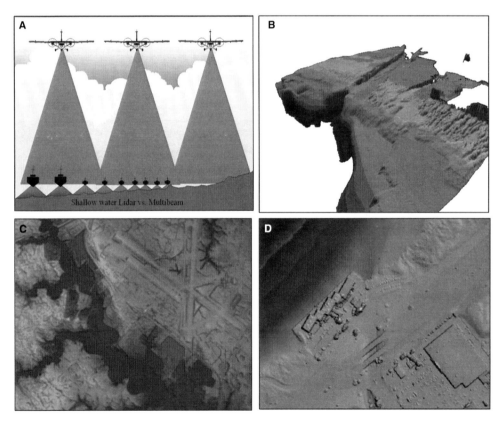

Fig. 4 (**A**) Airborne LIDAR provides a large sampling swatch for collecting bathymetry data compared to ship-based sonar systems and (**B**) can seamlessly blend water and land elevations in one image for precise coastline delineation. LIDAR data collected in (**C**) coastal Delaware can be used for mapping projected sea-level rise; and (**D**) Monterey Bay for evaluating erosion patterns along a seawall jutting into the coastal zone. Images in Panel A and Panel B provided courtesy of Optech Industries. Images from Panel C and Panel D are available from the NOAA Coastal Services Center Digital Coast project.

Satellite altimetry

The presence of bathymetric features, such as ridges and troughs, create changes in Earth's gravity field that produce small fluctuations in the height of the sea surface. The sea surface bulges slightly upward in response to a seamount, for example, as water is attracted to the greater mass (Fig. 5). Satellite-mounted radar altimeters orbiting the Earth can measure these slight variations in the sea surface-height by sending out radio wave pulses at high frequencies, usually in the range of 13 GHz, for determining sea-surface height variations. The radar pulse scatters off of the sea surface and the round-trip travel time of the signal is measured. If the satellite's position in orbit is well-characterized, then the round-trip travel time of the pulse can be related quite precisely to the height of the sea surface relative to the satellite sensor and can be estimated to within a few centimeters (Fig. 5). At wavelengths from 1–200 km, gravity anomaly variations are highly correlated with sea-floor topography and have been used to gravimetrically map bathymetry of the world oceans with a spatial resolution of ~10 km resolving features ~20 km in scale.[18]

Sea-surface heights are not impacted by bathymetry alone, but also by oceanographic factors such as waves, currents, and tides, as well as the underlying density of the seafloor. To effectively use altimetry for bathymetric purposes, the sea-surface height that would exist without any perturbing factors or "geoid" must be modeled (Fig. 5). Determining the geoid over the oceans is an active area of research. The footprint of the pulse must be large enough to average out the local irregularities in the surface due to

ocean waves. Tidal corrections must be applied. Corrections to the travel time of the pulse are also made for water vapor and electrons in the atmosphere. Profiles from many satellites, collected over many years, are combined to make high resolution images of the geoid showing major mass and density differences on the seafloor.

To produce reasonable estimates of bathymetry, the gravimetric anomaly measurements are combined with acoustically determined soundings to construct uniform grids of seafloor topography.[19,20] The methods work best in the deep-ocean basins, where sediments are thin and geological features can be mapped at coarser resolution. For example, gravimetric techniques are the primary method to locate seamounts >2000 m in height across the global ocean floor. However, errors arise in regions with sparse acoustic soundings and where thick sediments can bury the underlying "basement" topography. Gravimetric methods for estimating depth are of limited value over continental shelves, where sediments are thick and conventional bathymetric soundings are plentiful.

Comprehensive high-resolution bathymetry grids have been constructed from quality-controlled ship depth soundings with interpolation between sounding points guided by satellite-derived gravity data.[1,21] These data, freely available online, are used in a variety of applications, including tsunami path forecasting and ocean current models. However, such gridded datasets are not accurate enough to assess hazards to navigation, and in shallow water where other remote sensing techniques, such as LIDAR, would produce significantly better results.

Imaging radar systems

Wave theory has been used to deduce shallow bathymetry since the 1940s when sequences of aerial photographs were used by the Allied forces to map water depth along the Normandy coast of France.[22] As waves move into shallow water, their speed decreases and the wavelengths get shorter. Wave dispersion theory can be used to infer the bottom depth that would produce waves of a given speed (or celerity) and wavelength. Modern techniques rely on the same principles but use imaging radar systems to get accurate measurements of wavelength and speed of waves. The X-band radars, in particular, interact with the small sea-surface capillary waves and the backscattered signal can be used to resolve wave fields. A minimum degree of sea-surface roughness is needed to make such measurements with significant wave heights >1 m and wind speeds >3 ms⁻¹.[23] Water depth is calculated both from linear wave theory and from a nonlinear wave dispersion equation that approximates the effects of amplitude dispersion of the waves in the shallow water.[24]

Marine radar systems mounted on coastal stations have been used to infer nearshore bathymetry.[25] The technique was recently expanded for radar measurements collected on moving vessel.[23] Over 64 km² of coastal seas were

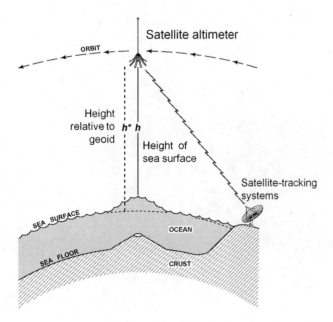

Fig. 5 Illustration of the use of radar altimetry to estimate large-scale bathymetry of the deep ocean floor.[18] The radar pulse provides the height of the sea surface to centimeter-scale, *h*, which is combined with the height relative to Earth's geoid, *h**, and acoustic soundings in order to estimate bathymetry.

mapped within a bay using radar data collected in 2 h using equipment already available on the ship.[23] The techniques were quite accurate down to 40–50 m water depth with a horizontal resolution of 50–100 m pixels. The practical limits of the technique vary by location and occur when the water depth is approximately one-quarter of the wavelength. Imaging radars measure a wide swath of sea in 360° around the vessel and, unlike acoustic systems, only a single pass is needed to estimate bathymetry over a wide area. Because ships already have radars for navigation, only a recorder is necessary to collect the data and the method is quite cost effective. The method does not resolve bathymetry at high enough resolution for most navigational purposes, but is especially useful for reconnaissance of coastal regions for military and scientific purposes and surveying migrating features such as interisland channels.

CONCLUSION

Knowledge of ocean bathymetry has progressed rapidly in the past century due to the advancement of techniques using acoustics, optics, and radar. The ocean has been mapped at a variety of spatial resolutions from kilometers to centimeters, but considerable work has yet to be done to accurately map the vast underwater landscape. More acoustic soundings are required to validate gravimetric bathymetry in remote regions of the world (e.g., the Southern Ocean), better characterization of the geoid is necessary for refining gravimetric techniques, more accurate tidal models are needed to assess nearshore bathymetry, and higher spatial resolution data are vital to assess shallow water bathymetry to delineate coastlines for storm surges and potential changes to sea level. The seafloor may be considered one of the last largely unexplored and most dynamic landscapes on Earth.

ACKNOWLEDGMENTS

Acknowledgments to Optic Industries, Dave Sandwell, and the U.S. National Oceanic and Atmospheric Association (NOAA) who provided figures for this entry. Thanks to Rick Stumpf, Jerry Mills, and Walter Smith for reviewing this entry. Funding was provided from the U.S. Office of Naval Research and U.S. National Aeronautics and Space Administration (Dierssen) and National Oceanographic and Atmospheric Administration (Theberge).

REFERENCES

1. Amante, C.; Eakins, B.W. ETOPO1 1 Arc-Minute Global Relief Model: Procedures, Data Sources and Analysis. *NOAA Technical Memorandum NESDIS NGDC-24,* 2009, 19.
2. Wang, Y.; Fang, G.; Wei, Z.; Wang, Y.; Wang, X. Accuracy assessment of global ocean tide models base on satellite altimetry. Adv. Earth Sci. **2010**, *25*, 353–359.
3. Stoker, J.M.; Tyler, D.J.; Turnipseed, D.P.; Van Wilson Jr, K.; Oimoen, M.J. Integrating disparate lidar datasets for a regional storm tide inundation analysis of Hurricane Katrina; J. Coast. Res. **2009**, *53*, 66–72.
4. Dierssen, H.M.; Theberge, A.E. Bathymetry: History of Seafloor Mapping. In *Encyclopedia of Ocean Sciences;* Taylor and Francis: New York, 2012.
5. Theberg, A. Appendix: The history of seafloor mapping. In *Ocean Globe*; Breman, J. Ed.; ESRI Press: 2010; pp 237–274.
6. Kvitek, R; Iampietro, P. In *Ocean Globe*; ESRI Press; Redlands, CA, J. Coast. Res. **2010**, *53*, 66–72.
7. Poppe, L; Polloni, C. Long Island Sound Environmental Studies. U.S. Geolog. Surv. Open File Rep. **1998**, *98*, 1.
8. Moline, M.A.; Woodruff, D.L.; Evans, N.R. Optical delineation of benthic habitat using an autonomous underwater vehicle. J. Field Robot. **2007**, *24*, 461–471.
9. Dierssen, H.M. Perspectives on empirical approaches for ocean color remote sensing of chlorophyll in a changing climate. Proc. Nat. Acad. Sci. **2010**, *107*, 17073.
10. Lyzenga, D.R. Passive remote sensing techniques for mapping water depth and bottom features. Appl. Opt. **1978**, *17*, 379–383.
11. Dekker, A.G. Intercomparison of shallow water bathymetry, hydro-optics, and benthos mapping techniques in Australian and Caribbean coastal environments. Limnol. Oceanogr. Methods **2011**, *9*, 396–425.
12. Dierssen, H.M.; Zimmerman, R.C.; Leathers, R.A.; Downes, T.V.; Davis, C.O. Ocean color remote sensing of seagrass and bathymetry in the Bahamas Banks by high-resolution airborne imagery. Limnol. Oceanogr. **2003**, *48*, 444–455.
13. Lee, Z.P.; Hu, C.; Casey, B.; Shang, S.; Dierssen, H.; Arnone, R. Global shallow-water high resolution bathymetry from ocean color satellites. EOS Trans. Amer. Geophy. Union **2010**, *91*, 429–430.
14. McClain, C.R. A decade of satellite ocean color observations. Annual Rev. Mar. Sci. **2009**, *1*, 19–42.
15. Dierssen, H.M.; Randolph, K. In *Encyclopedia of Sustainability Science and Technology*; Springer: Verlag Berlin Heidelberg 2013, http://www.springerreference.com/index/chapterdbid/310809.
16. Quadros, N.D.; Collier, P.A. *A New Approach to Delineating the Littoral Zone for an Australian Marine Cadastre*, 2010.
17. Wozencraft, J.; Millar, D. Airborne lidar and integrated technologies for coastal mapping and nautical charting. Mar. Techno. Soc. J. **2005**, *39*, 27–35.
18. Sandwell, D.T.; Smith, W.H.F. Marine gravity anomaly from Geosat and ERS 1 satellite altimetry. J. Geophys. Res. **1997**, *102*, 10039–10054.
19. Smith, W.H.F.; Sandwell, D.T. Bathymetric prediction from dense satellite altimetry and sparse shipboard bathymetry. J. Geophys. Res. All Ser. **1994**, *99*, 21.
20. Dixon, T.; Naraghi, M.; McNutt, M.; Smith, S. Bathymetric prediction from Seasat altimeter data. J. Geophys. Res. **1983**, *88*, 1563–1571.
21. GEBCO, *GEBCO_08 Gridded Bathymetry Data.* General Bathymetric Chart of the Oceans, 2008.
22. Hart, C.; Miskin, E. Developments in the method of determination of beach gradients by wave velocities. *Air Survey Research Paper No. 15, Directorate of Military Survey,* UK War Office, 1945.

23. Bell, P.S.; Osler, J.C. Mapping bathymetry using X-band marine radar data recorded from a moving vessel. Ocean Dyn. **2011**, *61*, 1–16.

24. Bell, P.; Williams, J.; Clark, S.; Morris, B.; Vila-Concej, A. Nested radar systems for remote coastal observations. J. Coast. Res. **2006**, *39*, 483–487.

25. Bell, P.S. Shallow water bathymetry derived from an analysis of X-band marine radar images of waves. Coast. Eng. **1999**, *37*, 513–527.

26. Dierssen, H.M. Benthic ecology from space: optics and net primary production in seagrass and benthic algae across the Great Bahama Bank. Mar. Ecol. Progress Ser. **2010**, *411*, 1–15.

27. Dierssen, H.M.; Zimmerman, R.C.; Leathers, R.A.; Downes, T.V.; Davis, C.O. Ocean color remote sensing of seagrass and bathymetry in the Bahamas Banks by high resolution airborne imagery. Limnol. Oceanogr. **2003**, *48*, 444–455.

Bathymetry: Features and Hypsography

Heidi M. Dierssen
*Department of Marine Sciences/Geography, University of Connecticut, Groton, Connecticut, U.S.A., and
Department of Biology, Norwegian University of Science and Technology, Trondheim, Norway*

Albert E. Theberge, Jr.
Captain, NOAA Corps, NOAA Central Library, Silver Spring, Maryland, U.S.A.

Abstract

The study of the distribution of elevations on the Earth or hypsography shows how more than half of the world ocean is dominated by deep basins (4–6.5 km) covered with abyssal plains and hills. Seafloor from 2 to 4 km depth is primarily comprised of oceanic ridge systems that spread over ~30% of the seafloor. The shallow seas and continental margins from sea level to 2 km depth cover the least amount of area and represent only ~15% of the seafloor. Superimposed on the primary features of the seafloor landscape are secondary features such as the median valley, seamounts, and submarine canyons. Fine-scale geomorphology of the underwater landscape and discontinuities in bathymetry are increasingly being used to assess distributions of marine organisms. New terminology and computational methods identify seafloor features that enhance marine biodiversity and allow for better management of vulnerable marine ecosystems.

INTRODUCTION

Up until the mid-20th century, the seafloor was thought to be featureless consisting of bits of rock and sediment eroded from the continents and flushed into the ocean reservoirs.[1] However, we now know that ocean bathymetry encompasses a varied seascape including vast mountain ranges, deep trenches, fracture zones extending for thousands of miles, the flattest plains on Earth, and a plethora of lesser meso- and microscale features ranging from individual seamounts to individual ripples and tidal channels. Comprehension of this landscape is a key component of the theory of plate tectonics, an intrinsic part of marine ecology, and a significant component of ocean circulation on global scales to individual estuaries. Indeed bathymetry and ocean science are intricately intertwined and one cannot be understood without the other.

Hypsography is the study of the distribution of bathymetric features across the Earth's surface and provides the percentage of the seafloor covered by large-scale features, such as continental shelves and abyssal plains. Fine-scale oceanic features can also be mapped at high resolution using new technology.[2] Assessing the geomorphology or shape of underwater landforms is now an important component in developing marine reserves to protect sensitive marine ecosystems. Many organisms aggregate at

discontinuities in the bathymetry that can also affect local eddies and current features that shape aspects of local water properties. Bathymetry of coastal estuaries and bays must also be evaluated in terms of tidal currents driven by the gravitational influence of the moon that can change bathymetry on hourly time scales. In a little over a century, the vast underwater landscapes from large- to small-scale features have been revealed, and this knowledge has contributed to the fundamental understanding of the Earth's processes.

BATHYMETRIC FEATURES AND HYPSOGRAPHY

The large-scale primary and secondary features of the deep seafloor and fine-scale features on the marine and coastal seafloor are discussed further.

Large-Scale Features

Covering 3.6×10^8 km², or 71% of the Earth's surface, the oceanic basins are dominated by three major physiographic features: 1) deep basins covered with abyssal plains and hills covering ~53% of the seafloor; 2) the world-girdling oceanic ridge system covering ~31% of the seafloor; and 3) continental margins and shallow seas comprised of

Encyclopedia of Natural Resources DOI: 10.1081/E-ENRW-120048589

continental shelf, slope and rise covering over 16% of the seafloor (Fig. 1). Undulating abyssal hills and gently sloping to flat abyssal plains lying between 4 and 6.5 km spread across the vast ocean floor. Very large features of significant lineal extent include the great *trenches* and *fracture zones* of the world ocean that can extend for thousands of miles either along the margins of ocean basins as in the case of trenches or across ocean basins as in the case of fracture zones. The ridges and rises associated with fractures zones predominantly occur at depths from 2 to 4 km across the world ocean (Fig. 1). Oceanic trenches from 6.5 to ~11 km in depth cover 0.2% of the world ocean. Continental margins are made up of continental shelves, continental slopes, and continental rises. *Continental shelves* border the continents and vary in size and shape across the world ocean. Comprised of continental crust, they are relatively flat containing thick layers of terrigenous sediments from riverine input. The widest continental shelves tend to occur along passive margins where considerable sediment can build up over time. In active continental margins with converging plates, there can be little to no

shelf at all. The depth limit describing the continental shelf is often considered to be <200 m, but deeper shelves extending to 500 m commonly occur in polar regions north and south of 60° latitude including Antarctica and Greenland. The *shelf break* indicates the point where the continental shelf abruptly ends and the slope steepens dramatically in what is referred to as the *continental slope*. At the base of the slope, the *continental rise* is characterized by a gentle slope of accumulated sediment that merges into the abyssal plain on passive continental margins.

Superimposed on the primary features of the seafloor landscape are secondary features such as the *median valley* associated with the mid-ocean ridge system, mountain ranges, and ridges not associated with the mid-ocean ridge system. The abyssal hills and plains are interrupted by a spectrum of larger hills, knolls, and cones with varying elevations culminating with the higher underwater mountains having relief >1000 m called *seamounts*. With over 100,000 estimated to occur on the seafloor, seamounts are typically formed from volcanic activity in association with diverging plates, hot spots, or converging plates.

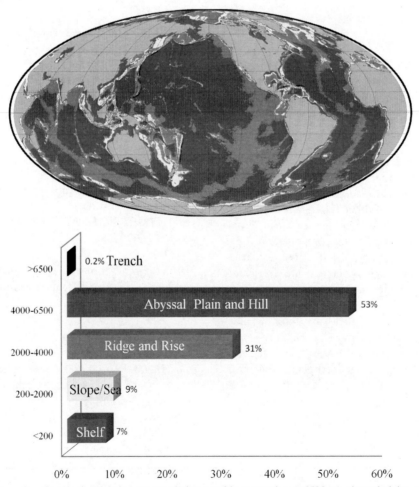

Fig. 1 Depth distribution of seafloor in the global ocean translating roughly to trenches (>6500 m), abyssal plains and hills (4000–6500 m), ridges and rises (2000–4000 m), continental slope and shallow seas (200–2000 m), and continental shelf (<200 m).
Source: Data from ETOPO1.[14]

Tablemounts or *guyots* are flat-topped seamounts formed by wave erosion which has been transported to deeper water and submerged on a moving plate. Volcanic activity forming mountains (either undersea or continental) commonly occurs on the back side of the trench feature where the downthrust subducting plate heats and melts. *Volcanic island arcs* are formed in this manner and run parallel to a trench at a distance of ~200 km from the trench axis. *Submarine canyons*, generally extensions of large rivers or formed by turbidity flows, periodically cut through the continental slope and transport considerable sediments from the shelf to the deep sea. Other secondary features include individual basins, troughs, deeps, holes, escarpments, benches, terraces, regional and local plains, great canyons, and lesser valleys found on the continental slopes of the world.

Virtually all of the primary and many of the secondary features of the world ocean were discovered by either line and sinker technology or by single-beam echo sounding.[1] However, because the introduction of multibeam sounding systems coupled with the global positioning system,[2] many additional deep-sea features have been discovered including hydrothermal vents and vent fields, mud volcanoes, mud wave and dune fields, individual lava flows, landslide scars, and diapiric structures including commercially valuable salt domes.

Comprehension of the significance and inter-relationships of the primary and secondary features of the seafloor has come about only since the formulation of the theory of plate tectonics. Plate tectonics describes the surface of the Earth in terms of numerous plates that either move away from each other at divergent plate boundaries, collide at convergent plate boundaries, or slide past each other along great faults known as transform faults. The median valleys of mid-oceanic ridges are the primary location of divergent boundaries, also known as seafloor spreading centers, where new seafloor is being produced from upwelling magma. Sites where plates collide with one plate being thrust under another consuming the seafloor, known as subduction zones, are marked by the great oceanic trenches, while colliding plates with no subduction form the great terrestrial mountain ranges of the world such as the Alpine-Himalayan belt of Europe and Asia. Sites where plates slide past one another are marked primarily by the numerous offsets on the mid-oceanic ridge system and are known as transform faults or fracture zones.

The presence of the abyssal plains and hills can also be explained from tectonic processes in combination with sedimentary processes. As new seafloor is formed and then pushed away from the median valley, parallel trains of abyssal hills are formed. With the passage of time, these hills are draped with sediment. When completely covered, the resulting surfaces are termed abyssal plains. Lineal chains of islands and seamounts can also be formed over long-lived stationary "hot spots" that provide conduits for magma to come to the surface. As the seafloor moves over these hot spots, lineal chains of islands and seamounts are formed trending in the direction of seafloor motion. The Hawaiian Islands and Emperor Seamount chains are examples of this phenomenon.

Fine-Scale Features

Marine geomorphology—the study of underwater landforms—is becoming an important component of management of marine ecosystems. Increasingly, the fine-scale ridges and troughs of the underwater landscape are being used to assess and define biological habitats for fish and other marine organisms. In Long Island Sound (Fig. 2A), the largest urban estuary in the United States, a concentrated mapping program has characterized the physical character of the seafloor using multibeam sonar for depth and "sidescan" sonar for a qualitative view of seafloor reflectivity characteristics.[3] Such high-resolution images showing sand ripples, uplifted ridges, and small depressions in bathymetry (Fig. 2B) are valuable for assessing benthic habitats, current modeling, and evaluating sedimentary processes.

Many species aggregate at locations with particular geomorphic features, such as reef promontories, uplifted ridges, and shelf edges, that are associated with abrupt discontinuities in surrounding structure.[4] When related to the seafloor, the terms "habitat heterogeneity" and "keystone structures" are applied to seafloor features that may enhance biodiversity by providing refugia from predation, enhanced or alternative food resources, spawning areas, stress gradients and substratum diversity.[5,6] Not only do undersea landforms provide three-dimensional structure, but they can also influence eddies and currents and create oceanographic conditions important for the growth and survival of marine communities.[7] Submarine canyons, for example, are sites of organic enrichment and benthic productivity due to sinking of particulate materials from the shores.[5]

Modern techniques allow for mapping of the seafloor with higher precision than ever before and assessment of benthic structures at meter scales or less. Because of linkages between bathymetric features, biodiversity, and biogeochemistry, new methods for classifying the seafloor have been developed. A benthic location can be classified according to a terrain ruggedness index[8] or more specifically the bathymetric position index (BPI), for example, which identifies the sloping nature of a region or "neighborhood." A positive BPI denotes ridge features; zero values indicate flat terrain or region with a constant slope; and negative values correspond to valleys. Bathymetric data tends to be spatially autocorrelated where locations closer together are more related than those further apart, and thereby the value of the BPI is related to the scale of data under investigation and the type of topography within the selected region.[9] Often BPIs are estimated at different scales to define different scales of features from small sand ripples to large-seascape features such as channels.

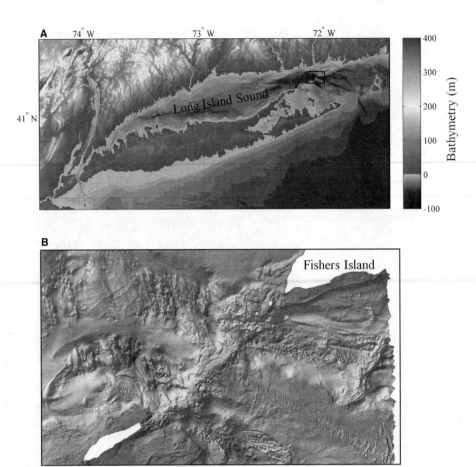

Fig. 2 (**A**) Bathymetry of Long Island Sound, the United States as mapped from NOAA digital high-resolution coastal gridded data (Amante[14]). (**B**) High-resolution bathymetry from multibeam sonar of the area around Fishers Island, outlined in black in Panel A, highlights a variety of detailed seafloor features (Poppe and Polloni[3]).
Source: Image courtesy of Long Island Sound Resource Center.

Fine-scale benthic features are also characterized by their "rugosity," which is a measure of the bottom complexity or "bumpiness." It is often parameterized as the surface area in relations to the planar area. A rugosity of 1 indicates a smooth, flat surface and values >1 indicate higher relief. Regions with higher rugosity have been linked to higher biodiversity.[10]

Ongoing efforts are underway to organize and define coastal and marine environments in meaningful units that can be compared across different temporal and spatial scales. The Coastal and Marine Ecological Classification Standard (CMECS) is one such effort that defines a habitat using five different components: water column, benthic biotic, surface geology, sub-benthic, and geoform component[11] (Fig. 3A). The geoform component classifies the major geomorphic or structural characteristics of the seafloor and is often assessed with bathymetric data. The classifiers tend from large to small scale and begin with coastal region, physiographic setting, geoform (natural), and subform, with an additional category called geoform (anthropogenic). The physiographic setting is defined from many of the primary oceanic features described earlier (e.g., mid-ocean ridge, abyssal plain, etc.)

and also includes the "coast" defined as the land–water interface. Geoform classifies features from centimeter to kilometer scale and has been separated into structures that are natural and anthropogenic in origin. Anthropogenic geoforms cover jetties, piers, artificial reefs, and other man-made structures and seafloor areas where human activities have changed the landscape (e.g., trawling and prop scars). Within the natural geoforms are definitions to describe marine features (e.g., seamounts, guyots, banks, boulder fields, and megaripples) and those structures in coastal regimes (e.g., fjord, beach, delta, coral reef, and lagoon). Finally, the subform category describes smaller features within the geoform and can include walls, escarpments, and also such factors as substrate orientation which can impact the availability of light and wave energy. Hence, many layers of complexity are required to describe the bottom topography itself, not to mention the additional components describing the type of substrate and biota associated with these features.

In a reef system, these techniques can be used to define zones corresponding to the geomorphology of the coral reef community. A recent analysis mapped the reef structures according to the following categories: shoreline intertidal,

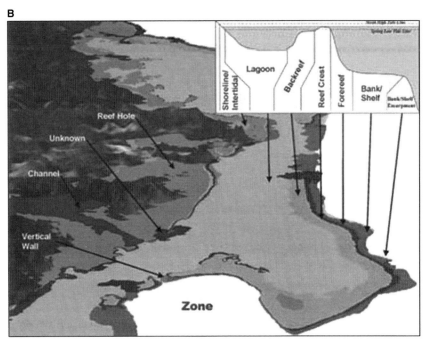

Fig. 3 (**A**) Geomorphologic classification scheme proposed by the Coastal and Marine Ecological Classification Standard (CMECS) categorize features from large (coastal region) to small (subform) spatial scales. (**B**) A classification of a coral reef demonstrating the different bathymetric features mapped from a bathymetric grid (Battista[12]).
Source: Image from NOAA, NOS, NCCOS Biogeography Branch.[12]

vertical wall, lagoon, back reef, reef flat, reef crest, fore reef, bank/shelf, bank/shelf escarpment, channel, and dredged area[12] (Fig. 3B). These geomorphologic zones can be comprised of different benthic substrate or cover types. The "lagoon," for example, may be covered by patch reefs, sand, and seagrass beds. Integrating and assessing the biota and geomorphology is important for managing these sensitive coastal resources.

HYPSOGRAPHY OF THE OCEAN

The bathymetry of each of the five ocean basins varies considerably (Fig. 4). Ridges and rises associated with seafloor spreading can be found in all five of the ocean basins. As mapped in the 1800s,[1] the Mid-Atlantic Ridge extends from the north of Iceland to the Southern Ocean boundary. Wide continental shelves (<200 m depth) are visible along much of the western Atlantic Ocean. The vast Pacific Ocean contains mostly broad expanses of deep ocean floor including abyssal hills, chains of seamounts, and lineal scars of great fracture zones. Prominent bathymetric features also occur along the margins of the Pacific Ocean. The Pacific Ocean is bounded by the "Ring of Fire" consisting primarily of subducting plates on the western, northern, and eastern margins with deep trenches and considerable earthquakes and volcanic activity. The southern boundary is defined by the Pacific–Antarctic Ridge and the southern portion of the East Pacific Rise. The Indian Ocean is crossed by ridges and volcanic island structures where the African, Indian, Australian, and Antarctic crustal plates all converge. With the exception of Western Australia, the Indian Ocean has

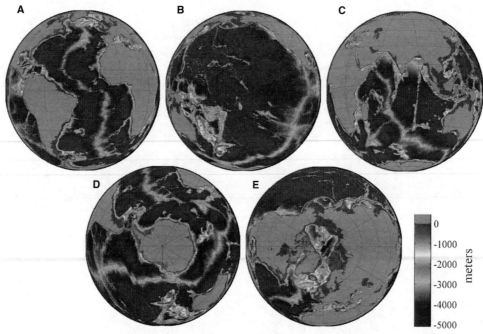

Fig. 4 Gridded bathymetric data are a compilation of satellite gravimetric data and sonar data from throughout the global ocean.[14] These plots show bathymetry (m) for each ocean basin in orthographic projection: (**A**) Atlantic, (**B**) Pacific, (**C**) Indian, (**D**) Southern, and (**E**) Arctic. Land surfaces are shown in gray.

limited continental shelf area. In contrast, the Arctic Ocean represents the shallowest of ocean basins and is characterized by broad expansive deep (500 m) shelves and shallow seas. The Southern Ocean also contains deep continental shelves (500 m), large expanses of abyssal plains, and prominent ridge features bordering the Pacific, Atlantic, and Indian Oceans.

Taken from the Greek word "hypsos," meaning height, hypsography is the study of the distribution of elevations on Earth and is used to relate bathymetry to the projected two-dimensional surface area of a region or feature. The "hypsographic curve" of Earth shows the cumulative percentage of the Earth surface area at different elevations from above and below sea level (Fig. 5). As shown, the ocean represents ~71% of the Earth's surface and most of this surface area is quite deeply submerged at close to 4 km below sea level. Based on our current and still limited knowledge of bathymetry, the average depth of the oceans is calculated to be 3.7 km. The prevalence of deep seafloor is related to the heavier elemental composition of oceanic crust that makes it much denser than terrestrial crust. Along the perimeters of the ocean basins, the ocean slopes rapidly and very little of the Earth's surface occurs between the continental shelves (<500 m) and the deep sea (4 km). Metaphorically, the ocean basins can be considered like deep bowls with thin rims and steep sides. However, the bowl bottoms are not smooth, but littered with abyssal hills and seamounts, as well as spreading ridges that appear as jagged seams laced across a baseball.

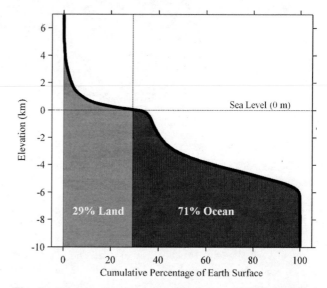

Fig. 5 Hypsographic curve modeled from ETOPO1 gridded bathymetric data[14] shows the cumulative percentage of Earth's surfaces above and below sea level. Land represents a smaller depth range and comprises 29% of Earth's surface; whereas, ocean has nearly an 11 km depth range and covers 71% of the Earth's surface. Over 50% of Earth falls between 4 and 6 km below sea level comprised predominantly of abyssal plains and hills.

CONCLUSION

Representing less than a third of the Earth's surface, the land is dwarfed by the enormity of ocean. The ocean has more relief with deep 11 km trenches compared to maximum peaks of 7 km on land. The large-scale features of the underwater landscape span the five ocean basins from the Southern Ocean northward to the Arctic Ocean. Describing and classifying coastal and marine geomorphology at fine scales is still an area of active research that requires more data and enhanced computational methods. Bathymetry is not static and the percentage of ocean is likely to increase in the next century as sea-level rises due to thermal expansion of ocean water and melting of land-based glaciers.[13] Ocean bathymetry will need to be revised particularly in the near-shore environment as the oceans get deeper and coastal regions become submerged. While islands and seamounts are continually being formed by volcanic activity, many islands will become submerged by tectonic activity and projected sea-level rise.

ACKNOWLEDGMENTS

Funding was provided from the U.S. Office of Naval Research, U.S. National Aeronautics and Space Administration (Dierssen), and U.S. National Oceanic and Atmospheric Administration (Theberge).

REFERENCES

1. Dierssen, H.M.; Theberge, A.E. Bathymetry: History of seafloor mapping. In *Encyclopedia of Ocean Sciences*; Taylor and Francis: New York, NY, 2012; 360 pp.
2. Dierssen, H.M.; Theberge, A.E. Bathymetry: Assessing methods. In *Encyclopedia of Ocean Sciences*; Taylor and Francis: New York, NY, 2012; 360 pp.
3. Poppe, L.; Polloni, C. Long Island Sound Environmental Studies. U.S. Geol. Surv. Open File Rep. **1998**, *98*, 1.
4. Heyman, W.D.; Wright, D.J. Marine Geomorphology in the Design of Marine Reserve Networks. Prof. Geogr. **2011**, *63*, 429–442.
5. Vetter, E.W.; Smith, C.R.; De Leo, F.C. Hawaiian hotspots: enhanced megafaunal abundance and diversity in submarine canyons on the oceanic islands of Hawaii. Mar. Ecol. **2010**, *31*, 183–199.
6. Drazen, J.C.; Goffredi, S.K.; Schlining, B.; Stakes, D.S. Aggregations of egg-brooding deep-sea fish and cephalopods on the Gorda Escarpment: a reproductive hot spot. Biol. Bul. **2003**, *205*, 1.
7. Ezer, T.; Heyman, W.D.; Houser, C.; Kjerfve, B. Modeling and observations of high-frequency flow variability and internal waves at a Caribbean reef spawning aggregation site. Ocean Dyn. **2010**, *61*(5), 1–18.
8. Riley, S.J.; DeGloria, S.D.; Elliot, R. A terrain ruggedness index that quantifies topographic heterogeneity. Intermt. J. sci. **1999**, *5*, 23–27.
9. Lundblad, E.; Miller, J.; Rooney, J.; Moews, M.; Chojnacki, J.; Weiss, J. Mapping Pacific Island Coral Reef Ecosystems with Multibeam and Optical Surveys. Coastal GeoTools. **2005**.
10. Lundblad, E.; Wright, D.J.; Naar, D.F.; Donahue, B.T.; Miller, J.; Larkin, E.M.; Rinehart, R.W. Classifying benthic terrains with multibeam bathymetry, bathymetric position and rugosity: Tutuila, American Samoa. Mar. Geod. **2006**, *29*, 89–111.
11. Federal Geographic Data Committee, Standards Working Group *Coastal and Marine Ecological Classification Standard Version 3.1 (Working Draft)* NOAA Coastal Services Center, 2010.
12. Battista, T.A.; Costa, B.M.; Anderson, S.M. in *NOAA Technical Memorandum NOS NCCOS 59, NOAA Center for Coastal Monitoring and Assessment, Biogeography Branch*; Silver Spring: MD, 2007.
13. IPCC, *Synthesis Report. Contribution of Working Groups I, II and III to the Fourth Assessment Report of the Intergovernmental Panel on Climate Change* Geneva, Switzerland, 2007.
14. Amante, C.; Eakins, B.W. ETOPO1 1 Arc-Minute Global Relief Model: Procedures, Data Sources and Analysis. *NOAA Technical Memorandum NESDIS NGDC-24,* 2009, 19 pp.

Aquaculture—
Bathymetry

Bathymetry: Seafloor Mapping History

Heidi M. Dierssen
Department of Marine Sciences/Geography, University of Connecticut, Groton, Connecticut, U.S.A., and Department of Biology, Norwegian University of Science and Technology, Trondheim, Norway

Albert E. Theberge, Jr.
Captain, NOAA Corps, NOAA Central Library, Silver Spring, Maryland, U.S.A.

Abstract

Terrestrial geography, including the outline of continents, islands, mountain chains, and great plains, was clearly known by the mid-19th century. However, the configuration of the seafloor was unknown and, even if there had been need, no technology was capable of accurately and quickly measuring the depths of the sea. This situation began changing in the 1840s, as both scientific and commercial interests began investigating the sea. Depth measuring technology progressed from point sounding line-and-sinker methods to primitive acoustic methods in the early 20th century. Line-and-sinker technology was capable of discovering most of the large features of the seafloor. Acoustic methods, by virtue of continual profiling with single-beam systems and with 100% bottom coverage capability with multibeam swath-mapping systems, have filled in the blanks of much of the detail of the world ocean. Our present world view and understanding of Earth's processes is a direct result of the mapping and understanding of the nature of the seafloor.

EARLY EFFORTS

The beginning of modern seafloor mapping coincided with the advent of systematic oceanographic observations (i.e., modern oceanography), deep-sea scientific dredging, and the commercial desire to lay deep-sea telegraph cables. Within a century, the concept of a featureless and static seafloor was shattered and the findings of a detailed ocean bathymetry were revealed (Fig. 1). Some of the first recorded measurements of bathymetry were made by the British explorer Sir James Clark Ross in 1840, by the U.S. Coast Survey beginning in 1845 with systematic studies of the Gulf Stream, and by the U.S. Navy, under the guidance of Matthew Fontaine Maury, beginning in 1849.[1] A weighted hemp or flax rope was dropped over the side of a vessel "lying to" (drifting) and the length of the line recorded once the sinker or lead weight reached the bottom. From a few such measurements, the first bathymetric map was produced and published by Maury in the 1853 Wind and Current Charts of the North Atlantic Ocean (Fig. 1A). Although this particular map was not very accurate, many important seafloor features were discovered from such measurements. For example, the first map showing the full extent of the Mid-Atlantic Ridge was produced in 1877 by Wyville Thomson from measurements made on HMS *Challenger* supplemented by additional soundings made by British vessels and those of the United States and other nations[1] (Fig. 1B). In 1875, HMS *Challenger* also discovered the first indication of

the Mariana Trench with a measured depth of 4475 fathoms although it was another 30 years before the true configuration of the trench was understood.[2] Just two years later in 1877, the German geographer, Augustus Petermann, produced the first bathymetric chart of the Pacific Ocean with many features including the "Challenger Tiefe" or "Challenger Deep," and the then deepest known spot in the ocean, the Tuscarora Deep, named after the *USS Tuscarora*, which had sounded in the Japan Trench in 1874.

PIANO-WIRE SOUNDING SYSTEM

The problems and inaccuracies inherent to making "line and sinker" measurements (e.g., angled line due to currents and vessel drifting, determining precisely when the sinker reached the bottom, etc.) led to the development of the piano-wire or Thomson sounding system in the 1870s by Sir William Thomson (later Lord Kelvin) (Fig. 2A). The piano-wire sounding system was a line-and-sinker technique but approximately three times faster than the old hemp rope system. Because of smaller cross-section of wire (vs. hemp rope) to be affected by surface and subsurface currents, less time to observe sounding, and exact indication of when the weight hit bottom, this method was approximately an order of magnitude more accurate than hemp rope sounding (10–20 m in 2000–3000 fathom vs. 100 m or more error with hemp rope sounding). It was the

Encyclopedia of Natural Resources DOI: 10.1081/E-ENRW-120047531

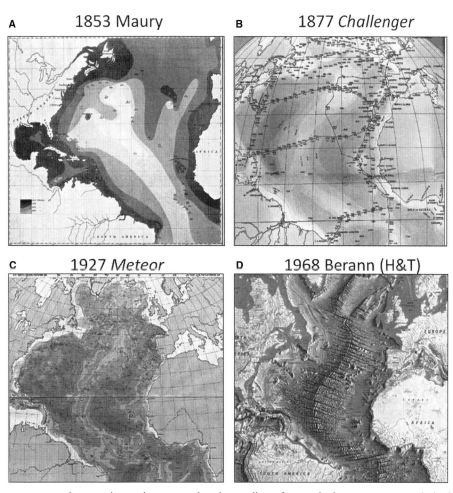

Fig. 1 In roughly one century, the mapping and conceptual understanding of ocean bathymetry were revolutionized as shown in the sequential maps of the North Atlantic. (**A**) First recorded bathymetric map created in 1853 by Maury in collaboration with the U.S. Navy only hints at the mid-ocean ridge. (**B**) Excerpt from the 1877 Thomson map based on the HMS *Challenger* measurements with line-and-sinker techniques shows the first continuous mapping of the mid-ocean ridge. (**C**) Echo sounding techniques allowed for increased frequency and a higher definition of the mid-ocean ridge, as shown in Theodor Stocks map from the *Meteor* cruises in 1927. (**D**) This 1968 Berann illustration, based on the Heezen and Tharp physiographic maps, outlines the ridge system in the North Atlantic (National Geographic Stock).

Thomson sounding machine and its variants, the Sigsbee Sounding Machine developed in the U.S. Coast and Geodetic Survey (USC&GS) (Fig. 2B) and the Lucas Sounding Machine developed by British surveyors, which outlined the great features of the world ocean prior to the introduction of acoustic sounding systems following the First World War.

The continental shelves and slopes, mid-ocean ridges, enclosed basins, and major trenches were discovered as a result of the nearly 20,000 soundings made in the deep ocean by the early 20th century. As a result of these discoveries, the 7th International Geographic Congress held in Berlin in 1899 appointed a committee on the nomenclature of undersea features and also formed a commission, under the chairmanship of Prince Albert I of Monaco, to publish a General Bathymetric Chart of the Oceans (GEBCO). This first GEBCO chart was published in 1905 and such charting has continued to modern day (Fig. 3).[3] From this early work, the significance of the seafloor features and their relationship to modern day plate tectonics began to unfold and in 1910 Frank Bursley Taylor wrote: "It is probably much nearer the truth to suppose that the Mid-Atlantic Ridge has remained unmoved, while the two continents on opposite sides of it have crept away in nearly parallel and opposite directions."[4] Shortly thereafter in 1912, Alfred Wegener first proposed the theory of continental drift. Another notable landmark from the piano-wire era was the construction in 1884 of the first 3D view of the seafloor from soundings made aboard the USC&GS steamer *Blake* in the Gulf of Mexico and western Atlantic Ocean (Fig. 3A). This ship also pioneered deep-sea anchoring and current measurement techniques while engaged in classic Gulf Stream studies.[1]

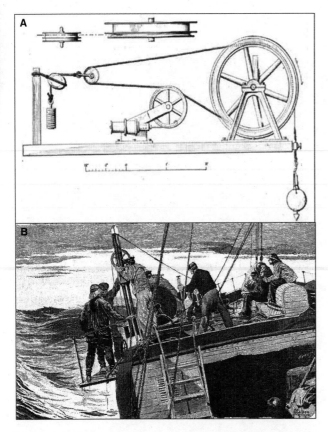

Fig. 2 (A) Diagram of the original version of the piano-wire sounding machine or Thompson Sounding Machine invented by Sir William Thomson (later Lord Kelvin) in 1872. (B) "The United States Fish Commission" by Richard Rathbun. Century Magazine, Vol. 43, issue 5. 1892. "Sounding the abyss with piano-wire." This image is among the most realistic representations of sounding with the Sigsbee Sounding Machine.

ACOUSTIC SOUNDING SYSTEMS

In the early 1900s, Submarine Signal Company, a forerunner of Raytheon Corporation, developed an underwater acoustic navigation system that was deployed from buoys and lightships for helping ships equipped with hydrophones to safely navigate to port during periods of reduced visibility. A similar system was also developed for ship-to-ship communication. Following the *Titanic* disaster, Reginald Fessenden of Submarine Signal Company developed an acoustic transducer that could both transmit and receive sound for the purpose of detecting objects in the water. During tests on the U.S. Revenue Cutter *Miami* in March 1914, reflections were obtained from an iceberg and, unexpectedly, from the bottom. Echo sounding was born. German, French, and American investigators modified and improved this technology for use in both outward looking antisubmarine warfare systems and downward-looking depth finding systems during the First World War. By 1922, the first truly functional acoustic depth measuring devices were in use making piano-wire sounding systems obsolete overnight.

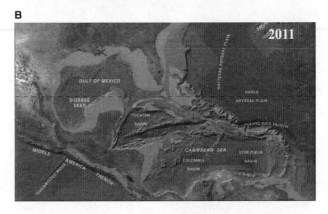

Fig. 3 Early mapping efforts along coastal margins were quite accurate, particularly as shown in (A) the first 3D rendition of ocean bathymetry of the Gulf of Mexico and Caribbean Sea produced in 1884 from the Blake soundings. (B) Same region highlighted from current bathymetric maps has many of the same features.
Source: Image reproduced from the GEBCO gridded bathymetric data.[3]

The first issue of the *International Hydrographic Review*, the publication of the newly formed International Hydrographic Organization, contained a profile of the Atlantic Ocean seafloor from Boston to Gibraltar obtained by a U.S. Navy-developed Hayes Sonic Depth Finder mounted on the U.S.S. *Stewart* in 1922 (Fig. 4A). This profile was derived from over 900 soundings taken during the transit proving both the efficacy of acoustic sounding (the word sounding does not refer to sound; it is derived from the Old French word *sonder* meaning, "to measure") and its ease of use and accuracy.

Overnight, echo sounding became the standard technique for observing bathymetry. Echo sounding determines bottom depth from measuring the time required for a sound pulse to be emitted from a transmitter, travel to the bottom and be reflected, and then travel back to a receiver unit. Dividing this time by 2 and multiplying times the velocity of sound in sea water gives the measure of the depth.[5] Contemporaneous with the *Stewart* work, the U.S. Navy also equipped U.S.S. *Corry* and

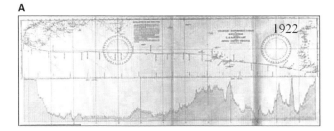

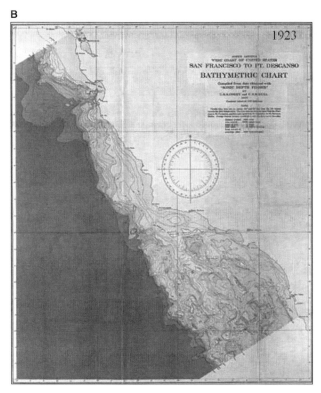

Fig. 4 (**A**) This profile illustrates the first acoustic line of soundings across the Atlantic Ocean obtained by the USS *Stewart* in 1922. (**B**) Published in 1923, Hydrographic Office Miscellaneous Chart 5194 showing the detailed coastal bathymetry of California from San Francisco Bay southward to San Diego, U.S.A. was the first bathymetric chart to be produced solely from acoustic soundings. NOAA/Department of Commerce.

U.S.S. *Hull* with echo sounders for conducting a survey of the California coast. This survey resulted in the first bathymetric map produced solely by acoustic technology (Fig. 4B). These successes were followed by the famous German Meteor Expedition (1925–1927), which resulted in over 67,000 soundings of the Atlantic Ocean. In addition to mapping the axis of the Mid-Atlantic Ridge (Fig. 1C), this expedition delineated for the first time the abyssal hills extending outward from the ridge axis. With the invention of the radio-acoustic ranging navigation system by the USC&GS in 1924, the position of a ship could be determined accurately and literally, millions of soundings were made on the continental shelf and slope of the U.S. prior to World War II.[1] This navigation system was

the first ever devised of sufficient accuracy for use in offshore bathymetric surveying operations. The combination of acoustic sounding and precise navigation led to the discovery of many mesoscale features that otherwise would have been impossible solely with the use of celestial navigation.

The 1930s saw a rapid advance of knowledge of the seafloor. The British Egyptian John Murray Mabahiss Expedition to the Indian Ocean discovered the first indications of the median valley of the mid-ocean ridge system; the German *Meteor* expeditions continued and, in 1937, the German oceanographer Gunter Dietrich discovered the median valley of the Mid-Atlantic Ridge; the USC&GS, in addition to its continental shelf and slope surveys, made a number of transects across the Gulf of Alaska discovering lineal chains of seamounts and also flat-topped seamounts, later termed guyots. The USC&GS also made detailed surveys of the bathymetry of the Aleutian Trench. As a result of its continental shelf and slope surveys, the C&GS discovered many large canyons on the east coast of the United States, the Mendocino Escarpment off the California coast which was the first indication of the great lineal features now known as fracture zones, and, of great commercial significance, salt domes on the Texas–Louisiana continental shelf and slope.

The exploration and mapping of the seafloor was interrupted by World War II except for isolated efforts such as the serendipitous discovery and mapping of seamounts and other features by Dr. Harry Hess (serving as a naval officer during the war) in the western Pacific Ocean. He discovered many flat-topped seamounts and called them "guyots." He coined the term "guyot" after Arnold Henry Guyot and in honor of the flat-topped building at Princeton University, which also gets its name from Arnold Henry Guyot.

COLD WAR YEARS AND BEYOND

Because of the rise of strategic submarine defense needs during the Cold War years, knowledge of bathymetry and other oceanographic parameters became critical. As a consequence, the world view we have of the seafloor today primarily stems from efforts of U.S. Navy-supported marine geologists and geophysicists, from government agencies, and major U.S. oceanographic institutions following World War II. Beginning in the late 1940s, ships from Columbia University's Lamont Geological Observatory, the Woods Hole Oceanographic Institution, and the Scripps Institution of Oceanography fanned out over the world ocean collecting bathymetry, geophysical data, and other oceanographic information. For the next quarter century, these institutions delineated the abyssal plains, the flattest surfaces on Earth; mapped the full extent of the world-girdling ridge system; defined the great fracture zones and their accompanying scars; surveyed hundreds of seamounts; and established that abyssal hills were the most abundant landform on Earth, particularly because of their widespread occurrence in the Pacific Ocean, where they covered approximately 85% of the seafloor.[6] Other nations and institutions took part in this endeavor. Notably the British *Challenger* II expedition, which established the Mariana Trench as the deepest spot in the ocean; the Swedish *Albatross* expedition; the Danish *Galathea* expedition; Japanese surveys of the northwest Pacific Ocean; and Seamap surveys by the USC&GS in the North Pacific Ocean all contributed to furthering knowledge of the seafloor and ocean in general. Both U.S. Navy and Russian surveys of most of the world ocean were also made, primarily in support of submarine warfare requirements, throughout most of the Cold War era but even today, most of those data remain classified.

Among the most widely known maps of the ocean basins are the iconic physiographic maps produced by Bruce Heezen and Marie Tharp, researchers at Columbia University's Lamont Geological Observatory, beginning in the early 1950s and subsequent collaborative illustrations of H.C. Berann in the 1960s and 1970s (Fig. 1D). These physiographic illustrations were based on data collected primarily from academic and military surveys conducted during the early Cold War era during which, due to U.S. national security restrictions, new bathymetric maps of many areas were "classified" and not available to the public.[7] The physiographic approach used by Heezen and Tharp portrayed physical features from an oblique perspective and an exaggerated vertical scale, and made detailed extrapolations between the soundings following from their burgeoning knowledge of geomorphology. Marie Tharp wrote: "I had a blank canvas to fill with extraordinary possibilities, a fascinating jigsaw puzzle to piece together: mapping the world's vast hidden seafloor. It was a once-in-a-lifetime—a once-in-the-history-of-the-world—opportunity for anyone, but especially for a woman in the 1940s."[8] During this same era, investigators from the Scripps Institution of Oceanography, such as H. W. Menard and T. E. Chase, were producing lesser-known atlases and physiographic diagrams of the Pacific Ocean basin.

CONCLUSION

The world view and regional views of bathymetry produced by these investigators were instrumental in helping form concepts of seafloor spreading, continental drift, and the theory of plate tectonics as we know them today. The advent of multibeam sonar in the civil mapping community, access to the remarkably accurate global positioning system, satellite altimetry remote sensing technology, and computer-aided analysis and interpolation has led to further advancements in bathymetric mapping in the last quarter of the 20th century. These advances have had implications in nearly every field of oceanography. The foundation for today's conceptual understanding of seafloor bathymetry, plate tectonics, and related bathymetric effects on many aspects of oceanography and marine ecology was laid down within one century by intensive effort from many talented individuals and institutions.

ACKNOWLEDGMENTS

Funding was received from the U.S. Office of Naval Research, U.S. National Aeronautics and Space Administration (Dierssen), and U.S. National Oceanic and Atmospheric Administration (Theberge).

REFERENCES

1. Theberge, A. Appendix: The History of Seafloor Mapping. In *Ocean Globe*; ESRI Press: Redlands, CA, Breman, J., Ed.; 2010; (ESRI Press): 237–274.
2. Theberge, A. Thirty Years of Discovering the Mariana Trench. Hydro, Int. **2008**, *12*, 38–41.
3. *GEBCO_08 Gridded Bathymetry Data,* General Bathymetric Chart of the Oceans, 2008.
4. Taylor, F.B. Bearing of the tertiary mountain belt on the origin of the Earth's plan. Bull. Geol. Soc. Am. **1910**, *21*, 179–226.
5. Dierssen, H.M.; Theberge, A.E. Bathymetry: Assessing Methods. *Encyclopedia of Ocean Sciences*; Taylor and Francis: New York, NY, 2012.
6. Dierssen, H.M.; Theberge, A.E. Bathymetry: Features and Hypsography. *Encyclopedia of Ocean Sciences*; Taylor and Francis: New York, NY, 2012.
7. Doel, R.E.; Levin, T.J.; Marker, M.K. Extending modern cartography to the ocean depths: military patronage, Cold War priorities, and the Heezen-Tharp mapping project, J. Hist. Geogr. **2006**, *32*, 605–626.
8. Tharp, M. Connect the dots: mapping the sea floor and discovering the Mid-Ocean Ridge. *Lamont-Doherty Earth Observatory of Columbia Twelve Perspectives on the First Fifty Years* 1999, *31–37.*

Coastal and Estuarine Waters: Light Behavior

Darryl J. Keith
Atlantic Ecology Division, U.S. Environmental Protection Agency (EPA), Narragansett, Rhode Island, U.S.A.

Abstract

Satellites and aircraft have proven to be valuable tools for providing synoptic and temporal views of coastal environments that cannot be obtained using traditional field-sampling techniques. Additionally, the form and magnitude from the visible portion of the electromagnetic spectra from these environments can be used to derive quantitative information on the types of substances present in the water and their concentrations. This entry will present an overview of the theory and techniques used to retrieve spectral information from nearshore coastal and estuarine waters using airborne and space-based platforms. The majority of the literature on remote sensing of the ocean has concentrated, understandably, on techniques and empirical models to retrieve information from the open ocean. However, remote sensing of the near-coastal ocean and estuaries is very important as 30–70% of the world's population lives in coastal and estuarine regions.[1,2]

INTRODUCTION

Water quality parameters (e.g., water clarity, chlorophyll concentration) are derived using remote-sensing systems from direct solar radiation entering the water column and either being absorbed or scattered within the water column. Some of the radiation is then reflected back to the atmosphere as water-leaving radiance. The ratio of the radiance reflected out to the direct solar radiation incident on the sea surface is termed reflectance. When the reflectance is passively recorded by a sensor, it is referred to as remote-sensing reflectance. The remote-sensing reflectance (R_{rs}) represents the ocean color signature of the water. R_{rs} can be used to deduce the concentrations of chlorophyll, colored dissolve organic matter (CDOM), or mineral particles within the near-surface water; the bottom depth and type in shallow waters; and other ecosystem information such as net primary production, phytoplankton functional groups, or phytoplankton physiological state. This entry discusses the inherent and apparent optical properties that affect the retrieval of color from coastal and estuarine waters. The most common example of remote sensing, and the one primarily discussed in this entry, is the use of sunlight that has been backscattered within the water and passively returned to the sensor.

Remote Sensing Basics

Coastal and inland waters contain a wide variety of optically active constituents that combine to create an optically complex system. Particulate and dissolved materials such as phytoplankton, detritus, suspended sediments, and CDOM vary over orders of magnitude in concentration to affect the underwater light regime and water quality.

Water quality parameters are derived using remote-sensing systems through direct solar radiation entering the water column and either being absorbed or scattered within the water column. Remote-sensing systems can be *active* or *passive*. *Active remote sensing* means that a signal of known characteristics (e.g., a laser beam) is sent from the sensor platform—an aircraft or satellite—to the sea surface, and the return signal is then detected after a time delay determined by the distance from the platform to the ocean and by the speed of light. The LIght Detection and Ranging (LIDAR) system that is usually flown on low-altitude aircraft and satellites is an example of an active system. In *passive remote sensing*, we simply observe the light that is naturally emitted or reflected by the water body. The night-time detection of bioluminescence from aircraft is an example of the use of emitted light at visible wavelengths. To derive water-quality parameters, coastal scientists primarily rely on passive remote-sensing systems. The most common example of passive remote sensing, and the one primarily discussed in this entry, is the use of sunlight that has been backscattered within the water and returned to the sensor. This light can be used to deduce the concentrations of chlorophyll, CDOM, or mineral particles within the near-surface water; the bottom depth and type in shallow waters; and other ecosystem information such as net primary production, phytoplankton functional groups, or phytoplankton physiological state.

OPTICAL PROPERTIES OF COASTAL AND ESTUARINE WATERS

The solar irradiance (E_d) that enters the Earth's atmosphere undergoes absorption and scattering by air molecules and aerosols. Radiance generated in the atmosphere by

Encyclopedia of Natural Resources DOI: 10.1081/E-ENRW-120049158

scattering (L_a) along the path between the sea surface and the sensor contains information about atmospheric aerosols and other atmospheric parameters. Aerosols are suspended liquid and particulate matter that include smoke, water, and hydrogen sulfate droplets, dust, ashes, pollen, spores, and other forms of atmospheric suspensions.[3] As a result, two radiative components reach the surface of the Earth: direct solar radiation that maintains the directionality of the sun's radiation as it exists outside the atmosphere and diffuse solar radiation (generally referred to as skylight or path radiance) that reaches the water surface from multiple angles due to molecular scattering (Rayleigh scattering) and aerosol scattering (Mie scattering) (Fig. 1). When direct sunlight (L_t) enters a water body, it is either absorbed or scattered (Fig. 1). The portion of L_t that is not absorbed by materials within the water column has its radiance altered, depending on the absorption and scattering properties of the water body and on the types and concentrations of the various constituents within the particular water body. The altered radiance (L_u) is then reflected (scattered) back out as water—leaving radiance (L_w). The total upwelling radiance (L_u) received by a sensor above the sea surface is the sum of the L_w that carries information about the water column, radiance reflected by the sea surface (L_r) contains information about the wave state of the sea surface, and radiance generated in the atmosphere by scattering (L_a). The ratio of the radiance reflected

out (in units of Watts/m²) to the solar irradiance (Watts/m²) is the *reflectance*. When the reflectance is passively recorded by a sensor, it is called remote sensing reflectance (R_{rs}, Fig. 2). The R_{rs} represents the "ocean color signature" of the water integrating the spectral absorption and backscattering properties of all the materials present in the water column.[4]

Where Do Estuarine and Coastal Waters Get Their Color Signature

The waters get their colors from the mixture of pigments in phytoplankton, CDOM, and minerals. If we are able to know how different substances (e.g., phytoplankton, suspended sediments) alter sunlight (either by wavelength-dependent absorption, scattering, or fluorescence), then we can attempt to deduce from the ocean color signature what substances must have been present in the water, and their concentrations. By working the analytical process "backward" in an inverse manner from the sensor to the estuary or coastal ocean, the remote-sensing community has been able to:

- Map chlorophyll concentrations
- Measure inherent optical properties (IOPs) such as absorption and backscatter
- Determine phytoplankton physiology, phenology, and functional groups
- Study ocean carbon fixation and cycling
- Monitor ecosystem changes resulting from climate change
- Map coral reefs, sea grass beds, and kelp forests
- Map shallow-water bathymetry and bottom type for military operations

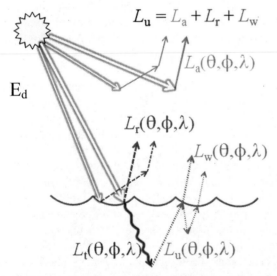

Fig. 1 Contributions to the total upwelling radiance above the sea surface, L_u, where θ = zenith angle, Φ = azimuth angle, and λ = wavelength. Thick arrows are the sun's unscattered beam; thin arrows are atmospheric path radiance L_a; dashed is surface-reflected radiance L_r; dotted is water-leaving radiance L_w. The curve arrow is total radiance (L_t) that enters the sea surface from direct sunlight. L_u, below the sea surface, is total upwelling radiance produced by scattering properties of the water column that contributes to the magnitude of L_w. E_d represents solar irradiance. Thick arrows represent single-scattering contributions; thin arrows illustrate multiple scattering contributions.
Source: Reproduced with permission from http://www.oceanopticsbook.info based on Mobley.[33]

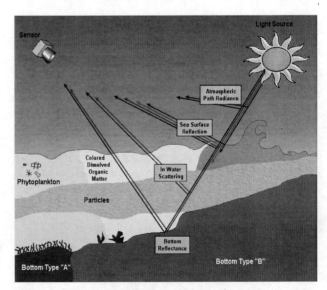

Fig. 2 Optical components and pathways of radiance and reflectance based on the absorption and scattering properties of coastal waters.
Source: Reproduced with permission from Arnone, R.A. from "Hyperspectral Imager for Coastal Ocean (HICO)" presented at the NASA Gulf Workshop in 2009.

- Monitor water quality for recreation and environmental protection
- Detect harmful algal blooms and pollution events

Ocean color remote sensing has completely revolutionized the ability of oceanographers, the environmental monitoring community, and coastal managers to understand estuaries and the coastal ocean at local to global spatial scales and daily to decadal temporal scales.

Inherent and apparent optical properties

The absorption and scattering properties of fresh and sea water are described by their IOPs. IOPs are properties of the medium and do not depend on the position of the sun or ambient light field in the water column. A volume of water has well-defined absorption and scattering properties whether or not there is any light there to be absorbed or scattered. Because of this basic principle, IOPs can be measured in a laboratory setting on a water sample, as well as in situ in an estuary or the ocean. The IOPs depend on the composition, morphology, and concentration of the particulate and dissolved substances in the medium. Composition refers to what materials make up the particle or dissolved substance, in particular the index of refraction of that material relative to that of the surrounding water. Different materials absorb light differently as a function of wavelength. Morphology refers to the sizes and shapes of particles. Particles with different shapes scatter light differently even if the particles have the same volume. Conversely, particles with different volumes scatter light differently even if they have the same shape. Concentration refers to the number of particles in a given volume of water, which is described by the particle size distribution. Given these complexities, it is not unusual that the magnitude of IOPs can vary by orders of magnitude.[5]

Apparent optical properties are those properties that depend both on the IOPs of the media as well as change in solar elevation (the angle between the horizon and the sun's disk, which alters the radiance distribution).[5,6] Apparent optical properties are R_{rs}, diffuse absorption, and diffuse backscattering.

Absorption Properties of Coastal and Estuarine Waters

There are only two things that can happen when light enters the water, it is either absorbed or scattered. Therefore, in order to understand the behavior of light as it passes into a water body, we need some measure of the extent to which water absorbs and scatters the incoming radiation. In estuarine and coastal waters, the total absorption of light is partitioned into absorption by the water itself, phytoplankton pigments, detritus particles, and CDOM (Figs. 3 and 4). Coastal and estuarine waters are considered to be optically complex because CDOM and mineral particles can make a significant (or even the dominant) contribution to water

color and brightness. Both CDOM and mineral particles absorb light most strongly in the blue part of the spectrum and exponentially decay with very low or no absorption in the red region (Fig. 3). When added to water (which absorbs mostly red light), they produce a green color that is easily confused with the effect of phytoplankton pigments.

Pure water

Pure water is a blue liquid whose color is derived from the fact that it absorbs very weakly in the blue and green regions of the spectrum.[7] However, absorption increases in the yellow and orange portions of the spectrum and is very significant in the red region (Fig. 3).[8]

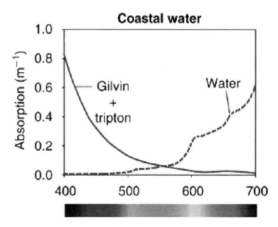

Fig. 3 Light absorption spectra of pure water (dashed line) and CDOM plus suspended matter (tripton).
Source: Reproduced with permission from http://www.oceanopticsbook.info based on Stomp et al.[34]

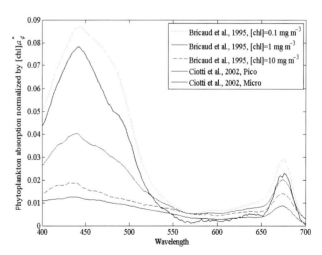

Fig. 4 Phytoplankton absorption spectra normalized by chlorophyll, which illustrates two dominant peaks: a primary peak in the blue region (at 440 nm) and a secondary absorption peak in the red region of the spectra (at 675 nm).
Source: Reproduced with permission from http://www.oceanopticsbook.info based on Bricaud et al.,[35] and Ciotti et al.[14]

Colored dissolved organic matter

Coastal waters get their color primarily due to the presence of CDOM (which is also known in scientific literature as yellow substance and gilvin), which imparts a brown to yellow-brown color to these waters. CDOM is the dissolved product of the decomposition of plant and animal matter from both terrestrial and marine sources that is composed of humic and fulvic acids. These acids form a water-soluble and chemically complex group of compounds termed "humic substances." As rainfall runoff and melting snow percolate through soils in coastal watersheds, humic substances are extracted and flow into rivers and streams and ultimately to estuaries and nearshore coastal waters.[4,7] Another likely origin for CDOM is from the decomposition of phytoplankton.

Tripton

Tripton (also known as suspended matter) includes both inorganic [e.g., mineral particles (colloids)] and organic (e.g., fecal pellets and cell fragments) materials that are suspended in estuarine and coastal waters. The composition of tripton generally reflects the geologic structure and composition of the adjacent watershed and coastal areas. Kirk[7] suggested that the yellow-brown color characteristic of suspended matter (tripton) from coastal waters is primarily due to particulate material either coated with humic compounds bound to particles or as free particles of humus. Characteristically, these substances do not absorb light strongly but scatter quite intensely. Spectrally, the absorption spectra for tripton is similar to that of CDOM with low absorption at the red region of the spectrum and rises increasingly as wavelength decreases toward the blue and ultraviolet end of the spectrum (Fig. 3). When light is transmitted through the water column, constituents (e.g., tripton, phytoplankton) in the water selectively absorb light at selective wavelengths. The resulting spectrum represents the unique absorption signature for that constituent. In the case of tripton, the spectral signature is dependent on particle shape, particle size distributions, and refractive index. Of all the optical constituents, detrital particles remain relatively understudied.[9,10]

Phytoplankton

Phytoplankton have a major effect on ocean color as they are a major absorber of light in estuarine, coastal, and open ocean environments. Phytoplankton possess chlorophyll, a pigment that allows them to harvest the sunlight and through the process of photosynthesis produce energy. Chlorophylls cause two dominant peaks in the absorption spectra: a primary peak in the blue region (at 440 nm) and a secondary absorption peak in the red region of the spectra

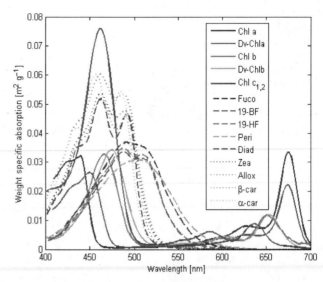

Fig. 5 Absorption spectra showing the primary and secondary absorption peaks for key phytoplankton pigment groups including Chlorophyll a (Chl a), Divinyl chlorophyll (Dv-Chl a), Chlorophyll b, Divinyl chlorophyll b (Dv-Chl b), Chlorophyll c1 and c2 (chl c1,2), 19′-Butanoyloxyfucoxanthin (19-BF), 19′-Hexanoxyfucoxanthin (19-HF), Fucoxanthin (Fuco), Peridinin (Peri), Diadino (Diad), Zeaxanthin (Zea), Alloxanthin (Allox), Beta-carotene (β-car), and Alpha-carotene (α-car).
Source: Reproduced with permission from http://www.oceanopticsbook.info based on Bidigare et al.[11]

(at 675 nm) (Fig. 4). Other pigments that are present (depending on species and taxa) will cause the broadening of the blue peak and the appearance of additional absorption maxima (Fig. 5). Spectra of phytoplankton absorption varies in magnitude and shape due to different cellular pigment compositions and pigment packaging.[11–16] Specific pigment-protein complexes present in the cell will cause changes in absorption spectra and magnitudes. Furthermore, the increase in cellular pigment concentration and cell size (i.e., packaging effect) will flatten the specific absorption spectra.[17–20] As the concentration of phytoplankton changes in a water body, changes in the different regions of the absorption spectra allow coastal scientists to create algorithms (step-by-step procedures) that can estimate chlorophyll abundance from R_{rs}.

In the open ocean, as phytoplankton concentrations increase, the water-leaving reflectance has been observed to "green" in a predictable fashion that allowed for the creation of empirical relationships from the ratio of blue-to-green wavelengths of light.[10] In contrast, in the productive turbid waters of estuarine and coastal environments, R_{rs} in the blue-green end of the spectrum are significantly affected by absorption due to CDOM, tripton, and phytoplankton pigments (Fig. 6). For this reason, satellite remote-sensing algorithms for chlorophyll developed from spectral data from the open ocean regularly fail in the coastal ocean and estuaries because they overestimate the chlorophyll

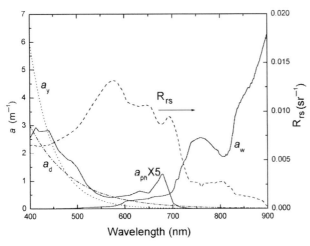

Fig. 6 Typical curve shapes for coastal remote-sensing reflectance (R_{rs}), phytoplankton pigment absorption (a_{ph}), C_{DOM} (a_y or a_{CDOM}), detrital absorption (a_d), and pure water absorption (a_w). **Source:** Reproduced with permission from the Journal of Coastal Research based on Sydor.[16]

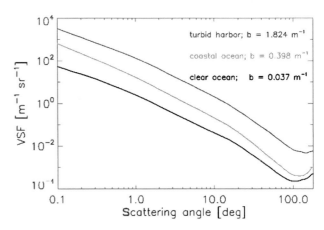

Fig. 7 Log-log plots of measured volume scattering function at 514 nm from the harbor at San Diego, San Pedro Channel off the California coast, and the Tongue of the Ocean, Bahama Islands. **Source:** Reproduced with permission from http://www.oceanopticsbook.info based on Petzold.[26]

content. In remote-sensing terms, these waters are classified as Case 2[21] as distinct from Case 1 open ocean waters.

To quantify chlorophyll concentrations in estuarine and coastal environments, algorithms have been developed which use information in spectral regions away from the influences of CDOM and tripton absorption and which are based on the properties of the phytoplankton reflectance peak in the red and near infrared portion of the spectrum near 700 nm.[22–24] The algorithms commonly ratio remotely sensed reflectances at 705 and 670 nm.[22,23,25]

Scattering Properties of Coastal and Estuarine Waters

Coastal and estuarine waters appear bright (highly reflective) when viewed from space due to the presence of sediments resuspended from the sea bottom by wave action, tidal currents, and storms; sediments brought in by river discharge; and phytoplankton. Scattering properties of phytoplankton are important since they are directly related to R_{rs} calculations. The manner in which incident light penetrates these waters depends not only on the scattering properties of the media but also on the angular distribution of the scattered flux resulting from the primary scattering process.[7] The angular distribution has a characteristic shape that is described in terms of the *volume scattering function* (VSF). The VSF describes the distribution of light scattered by a suspension of particles in a direction (forward or backward) at a specified wavelength (Fig. 7).[26] A measure of the overall magnitude of the scattered light is given by the scattering coefficient, which is the integral of the VSF over all angles. A quantitative definition for the scattering coefficient is found in Kirk.[7] Scattering and

backscattering coefficients of phytoplankton, as well as the VSF, are derived from either theoretical models (Mie theory) or direct measurements of the above-mentioned properties.[27–31] Coefficients are highly dependent on the size, shape, and refractive index of all components of the phytoplankton cell.[31,32]

CONCLUSION

Estuaries and coastal waters represent the dynamic interface between land and the open ocean. These waters are physically and biologically complex environments from which water quality information can be derived based on the absorption and scattering characteristics of light within this medium. All the apparent, inherent, and specific optical properties that were described in the preceding sections are wavelength dependent. It is the particular wavelength dependencies of these properties that define ocean color signatures in coastal and estuarine waters. The spectral qualities of the optical properties of pure water, CDOM, tripton, and phytoplankton determine the exact nature of the interdependence between individual constituents of the water and the R_{rs} signal that emerges from coastal and estuarine waters.

ACKNOWLEDGMENTS

I would like to thank Kenneth Rocha, Kristin Hychka, and Anne Kuhn for their review and comments of the manuscript. This is contribution number ORD-002860 of the Atlantic Ecology Division, National Health and Environmental Effects Research Laboratory, Office of Research and Development, U.S. Environmental Protection Agency. Any mention of trade names or commercial products does not constitute endorsement or recommendation for use.

REFERENCES

1. UNEP. *United Nations Environmental Programme Annual Report*; UNEP: Nairobi, Kenya, 2007.

2. Wilson, S.G.; Fischetti, T.R. *Coastline Population Trends in the United States: 1960 to 2008*; U.S. Census Bureau: Washington, D.C., 2010.

3. Sathyendranath, S., Ed.; IOCCG. *Remote Sensing of Ocean Colour in Coastal, and Other Optically-Comoplex Waters*, Reports of the International Ocean-Colour Coordinating Group, No. 3; IOCCG: Dartmouth, Canada, 2000.

4. Coble, P.; Hu, C.; Gould, R.W., Jr.; Change, G.; Wood, A.M. Colored dissolved organic matter in the coastal ocean – An optical tool for coastal zone environmental assessment and management. Oceanography **2004**, *17* (2), 50–59.

5. Mobley, C.D.; Stramski, D.; Bisset, W.P.; Boss, E. Optical modeling of ocean waters: Is the case-1 case-2 still useful? Oceanography **2004**, *17* (2), 60–67.

6. Prisendorfer, R.W. Application of radiative transfer theory to light measurements in the sea. Union Geod. Geophys. Inst. Monogr. **1961**, *10*, 11–30.

7. Kirk, J.T.O. *Light and Photosynthesis in Aquatic Ecosystems*; Cambridge University Press: Great Britain, 1994; p. 509.

8. Pope, R.M.; Fry, E.S. Absorption spectrum (380–700 nm) of pure water. II Integrating cavity measurements. Appl. Opt. **1997**, *36* (33), 8710–8723.

9. Bukata, R.P.; Jerome, J.H.; Kondratyev, K.Y.; Pozdnyakov, D.V. *Optical Properties and Remote Sensing of Inland and Coastal Waters*; CRC Press: Boca Raton, FL, 1995; pp. 362.

10. Schofield, O.; Arnone, R.; Bissett, W.P.; Dickey, T.D.; Davis, C.O.; Finkel, Z.; Oliver, M.; Moline, M.A. Watercolors in the coastal zone—What can we see? Oceanography **2004**, *17* (2), 24–31.

11. Bidigare, R.R.; Ondrusek, M.E.; Morrow, J.H.; Kiefer, D. In vivo absorption properties of algal pigments. In *Ocean Optics X*, Proceedings of SPIE, 1990; Vol. 1302, 290–302.

12. Bricaud, A.; Stramski, D. Spectral absorption coefficients of living phytoplankton and nonalgal biogenous matter: A comparison between the Peru upwelling area and the Sargasso Sea. Limnol. Oceanogr. **1990**, *35* (3), 562–582.

13. Hoepffner, N.; Sathyendranath, S. Effect of pigment composition on absorption properties of phytoplankton. Mar. Ecol. Prog. Ser. **1991**, *73* (1), 11–23.

14. Ciotti, A.M.; Lewis, M.R.; Cullen, J.J. Assessment of the relationships between dominant cell size in natural phytoplankton communities and the spectral shape of the absorption coefficient. Limnol. Oceanogr. **2002**, *47* (2), 404–417.

15. Bricaud, A.; Claustre, H.; Ras, J.; Oubelkheir, K. Natural variability of phytoplanktonic absorption in oceanic waters: Influence of the size structure of algal populations. J. Geophys. Res. Oceans **2004**, *109* (C11).

16. Sydor, M. Use of hyperspectral remote sensing reflectance in extracting the spectral volume absorption coefficient for phytoplankton in coastal water: Remote sensing relationships for the inherent optical properties of coastal water. J. Coast. Res. **2006**, *22* (3), 587–594.

17. Duysens, L.M.N. The flattening effect of the absorption spectra of suspensions as compared to that of solutions. Biochim. Biophys. Acta **1956**, *19* (1), 1–12.

18. Kirk, J.T.O. A theoretical analysis of the contribution of algal cells to the attenuation of light within natural waters, III. Cylindrical and spheroidal cells. New Phytol. **1976**, *77* (2), 341–358.

19. Morel, A.; Bricaud, A. Theoretical results concerning light-absorption in a discrete medium, and application to specific absorption of phytoplankton. Deep-Sea Res. A **1981**, *28* (11), 1375–1393.

20. Johnsen, G.; Nelson, N.B.; Jovine, R.V.M.; Prezelin, B.B. Chromoprotein-dependent and pigment-dependent modeling of spectral light-absorption in 2 dinoflagellates, prorocentrum-minimum and heterocapsa-pygmaea. Mar. Ecol. Prog. Ser. **1994**, *114* (3), 245–258.

21. Morel, A.; Prieur, L. Analysis of variations in ocean color. Limnol. Oceanogr. **1977**, *22* (4), 709–722.

22. Gitelson, A. The peak near 700 nm on radiance spectra of Algae and water - Relationships of its magnitude and position with chlorophyll concentration. Int. J. Remote Sens. **1992**, *13* (17), 3367–3373.

23. Gitelson, A.A.; Kondratyev, K.Y. On the mechanism of formation of maximum in the reflectance spectra near 700 nm and its application for remote monitoring of water quality. In *Transactions Doklady of the USSR Academy of Sciences: Earth Science Sections*; 1991; Vol. 306, 1–4.

24. Schalles, J.F. Optical remote sensing techniques to estimate phytoplankton chlorophyll a concentrations in coastal waters with varying suspended matter and CDOM concentrations. In *Remote Sensing of Aquatic Coastal Ecosystem Processes: Science and Management Applications*; Richardson, L., Ledrew, E., Eds.; Springer: the Netherlands, 2006; 27–79.

25. Dekker, A.G. *Detection of Optical Water Quality Parameters for Eutrophic Waters by High Resolution Remote Sensing, Ph.D. Thesis*; Vrije Universiteit: Amsterdam, 1993; p. 222.

26. Petzold, T.J. Volume scattering functions for selected ocean waters. In *Light in the Sea*; Tyler, J.E., Ed.; Hutchinson & Ross: Dowden, 1977; 150–174.

27. Bricaud, A.; Morel, A.; Prieur, L. Optical-efficiency factors of some phytoplankters. Limnol. Oceanogr. **1983**, *28* (5), 816–832.

28. Volten, H.; de Haan, J.F.; Hoovenier, J.W.; Schreurs, R.; Vassen, W. Laboratory measurements of angular distributions of light scattered by phytoplankton and silt. Limnol. Oceanogr. **1998**, *43* (6), 1180–1197.

29. Witkowski, K.; Krol, T.; Zielinski, A.; Kuten, E. A light-scattering matrix for unicellular marine phytoplankton. Limnol. Oceanogr. **1998**, *43*, 859–869.

30. Vaillancourt, R.D.; Brown, C.W.; Guillard, R.R.L.; Balch, W.M. Light backscattering properties of marine phytoplankton: Relationships to cell size, chemical composition and taxonomy. J. Plankton Res. **2004**, *26* (2), 191–212.

31. Sullivan, J.M.; Twardowski, M.S. Angular shape of the oceanic particulate volume scattering function in the backward direction. Appl. Opt. **2009**, *48* (35), 6811–6819.

32. Jonasz, M.; Fournier, G.R. *Light Scattering by Particles in Water: Theoretical and Experimental Foundations*; Academic Press, 2007.

33. Mobley, C. Overview of Optical Oceanography-Reflectances, 2010. http://www.oceanopticsbook.info.

Coastal—
Evaporation

34. Stomp, M.; Huisman, J.; Vörös, L.; Pick, F.R.; Laamanen, M.; Haverkamp, T.; Stal, L.J. Colorful coexistence of red and green picobacteria in lakes and seas. Ecol. Lett. **2007**, *10* (4), 290–298.

35. Bricaud, A.; Babin, M.; Morel, A.; Claustre, H. Variability in the chlorophyll-specific absorption coefficients of natural phytoplankton: analysis and paramentarization. J. Geophys. Res. **1995**, *100* (C7), 13211–13332.

Coastal and Estuarine Waters: Optical Sensors and Remote Sensing

Darryl J. Keith
Atlantic Ecology Division, U.S. Environmental Protection Agency, Narragansett, Rhode Island, U.S.A.

Abstract

Remote sensing science and technologies using space-based and airborne sensor systems have profoundly changed the practice of environmental monitoring and understanding of the dynamics of estuarine and coastal environments. This entry summarizes the types of remote sensors used to monitor coastal and estuarine waters. In this entry, comparisons are presented of the scanning systems used in ocean color remote sensing to collect radiance and reflectance data. The current suite of spaced-based as well as examples of aircraft-based scanning systems is presented. The entry concludes with a discussion of the challenges that the scientific and environmental communities face when incorporating remotely sensed data into monitoring strategies within the framework of the environmental laws and regulations.

INTRODUCTION

There are two basic types of optical multispectral or hyperspectral sensors used in imaging estuarine, coastal, and open ocean systems: whiskbroom (across-track scanning; Fig. 1) and pushbroom scanners (along track scanning; Fig. 2).

Scanners

Whiskbroom scanning (across-track scanning)

In a whiskbroom scanner system, a scan mirror rotates in front of a telescope to continuously sweep the Earth beneath the aircraft or satellite perpendicular to the direction of flight. Airborne scanners typically sweep large angles (between 90° and 120°), while satellites, because of their higher altitudes, only need to sweep fairly small angles (10–20°) to cover a broad region. At each ground cell or pixel (a two-dimensional array of individual picture elements arranged in rows or columns that has an intensity value and a location address), the incoming reflected or emitted radiation is detected independently and separated into ultraviolet, visible, near-infrared, and thermal spectral components based on their constituent wavelengths. Each pixel represents an area on the earth's surface. A bank of internal *detectors*, each sensitive to a specific range of wavelengths, detects and measures the energy for each spectral band and converts it into an electrical signal. The electrical signal is converted to digital data and recorded for subsequent computer processing. The ground resolution of each pixel is defined by the instantaneous field of view (IFOV) based on sensor characteristics and flight altitude. The sweep of the mirror (angular field of view) records a scan line of pixels. A scan line of pixels is equivalent to the imaged swath. As the platform moves forward over the Earth, successive scans build up a two-dimensional image of the earth's surface. A collection of scan lines in a single direction is generally defined as a mission flight line. For aircraft remote sensing, adjacent parallel flight lines are flown with adequate overlap to provide complete coverage of the desired area. Many well-known aircraft instruments are whiskbroom scanners such as the Airborne Ocean Color Imager (AOCI), the Calibrated Airborne Multispectral Scanner (CAMS), and the Airborne Visible/Infrared Imaging Spectrometer (AVIRIS).[1–3]

Pushbroom scanners (along-track scanning)

The pushbroom or along-track scanner is an imaging system that is "pushed" along in the flight track direction (i.e., along track) to map a scene. The system uses a linear array of solid semiconductive elements located at the focal plane of the image formed by the lens systems to collect spectral data. The semiconductive elements are individual detectors that measure the energy for a single pixel. A separate linear array is required to measure each spectral band or channel. For each scan line, the energy detected by each detector of each linear array is sampled electronically and digitally recorded. The size and IFOV of the detectors determine the spatial resolution of the system. Since all pixels for a given scan line are projected onto a detector array at the same time, the pushbroom design allows for a longer dwell-time over each pixel that increases the signal-to-noise performance. Common pushbroom systems for aircraft applications include the CASI (Compact Airborne Spectrographic Imager),[4–7] HYDICE (Hyperspectral Digital Imagery

Encyclopedia of Natural Resources DOI: 10.1081/E-ENRW-120051663

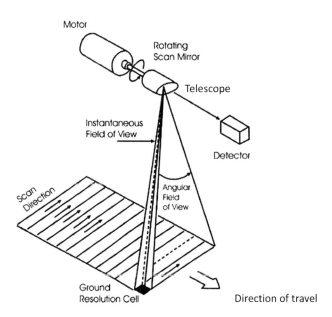

Fig. 1 Whiskbroom (across track) scanner system.
Source: Modified from Canadian Centre for Remote sensing.

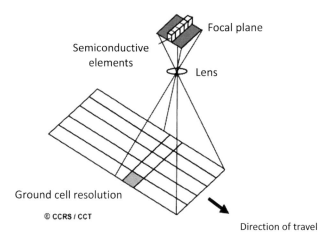

Fig. 2 Along-track scanner system.
Source: Modified from Canadian Centre for Remote Sensing.

Collection Experiment),[8] PHILLS (Hyperspectral Imager for Low Light Spectroscopy),[9] and HICO (Hyperspectral Imager for the Coastal Ocean)[10] sensors.

Sensors

Multispectral imaging sensors

Multispectral imaging is remote sensing that obtains optical representations in two or more ranges of frequencies or wavelengths. Multispectral imaging sensors capture image data from at least two or more wavelengths across the electromagnetic spectrum. On the sensor, each channel is sensitive to radiation within a narrow wavelength band resulting in a multilayer image (Fig. 3) that contains both the brightness and spectral (color) information of the pixels

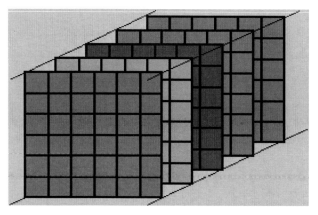

Fig. 3 An illustration of a theoretical multilayer image consisting of spectral data from five layers.
Source: Modified from the Centre for Remote Imaging, Sensing, and Processing (http:/www.crisp.nus.edu.sg).

sampled. In a multilayer image, data from each wavelength forms an image that carries some specific spectral information about the pixels in that image. By "stacking" images together from the same area, a multilayer image is formed composed of individual component images. Please note that multispectral images do not produce the spectrum of a pixel because wavelengths may be separated by filters or by the use of instruments that are sensitive to particular bands in the spectrum. Multispectral images are the main type of images acquired by most spaced-based or airborne radiometer systems. Radiometers are devices that measure the flux of electromagnetic radiation. Satellites may carry many radiometers in order to acquire data from selected portions of the electromagnetic spectrum. For example, one radiometer may acquire data from wavelengths in the red–green–blue (RGB) [700–400 nanometers (nm)] portion of the visible spectrum, a second radiometer may acquire data from wavelengths in the near infrared (700–3000 nm), and another might acquire data from mid-infrared to thermal region (greater than 3000 nm).

Hyperspectral imaging sensors

Hyperspectral systems offer the high spatial and spectral resolution needed to provide reflectance data from the numerous bays and estuaries that because of their spatial dimensions are smaller than the resolution capability of the multispectral ocean color satellite sensors. Hyperspectral sensors collect information from about a hundred or more contiguous spectral bands across the visible wavelengths from the near ultraviolet to near infrared regions of the electromagnetic spectrum. Information for these bands is combined to form a three-dimensional hyperspectral data cube (Fig. 4A) for processing and analysis. The precision of these sensors is typically measured in spectral resolution, which is the width of each band of the spectrum that is captured. Hyperspectral sensors usually have greater than 10 nm resolution. Hyperspectral sensors image narrow

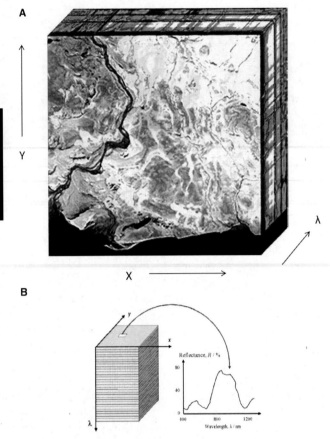

Fig. 4 (**A**) Two-dimensional projection of a hyperspectral data cube. The hyperspectral image data usually consists of over a hundred contiguous spectral bands, forming a three-dimensional (two spatial dimensions and one spectral dimension) image cube. (**B**) For these images, each pixel is associated with a complete spectrum of the imaged area. In this figure, X and Y represent location coordinates and λ represents all wavelengths associated with each pixel in the image.
Source: Modified from aviris.jpl.nasa.gov.

spectral bands over a continuous spectral range and produce the spectra of all pixels in the scene (Fig. 4B). Hyperspectral remote sensing is used in a wide array of applications. Although originally developed for mining, geology, and astronomy applications, hyperspectral imaging has been successfully used in oceanographic research for about 30 years.[11] Hyperspectral technology has expanded from hand-held radiometers to submerged sensors for measurements of inherent and apparent optical properties in estuarine/coastal and ocean waters.[11] Hyperspectral airborne and space-based detectors are now used to collect high spectral and spatial resolution measurements of radiance and reflectance from estuarine, coastal, and open ocean environments. The primary advantage to hyperspectral imaging is that each pixel has the entire spectrum acquired during overflights. The primary disadvantages are cost, complexity, and the need for fast computers, sensitive detectors, and large data-storage capacities for analyzing these data. Significant data storage capacity (exceeding

hundreds of megabytes) is necessary since hyperspectral cubes are large multidimensional datasets.

Platforms

Satellite and space-based remote sensing

Satellites have been an integral key to retrieving reflectance data from the coastal and open ocean. Remote sensing of ocean color began in 1978 with the launch of the Coastal Zone Color Scanner (CZCS). Through the years, governmental organizations around the world have developed increasingly advanced remote-sensing scanners that have provided higher spatial and temporal resolutions of the earth's surface. There are two types of orbits for Earth observation satellites, polar orbiting and geostationary. Polar-orbiting satellites typically operate at an altitude between 700 and 800 km, with a revisit time of 2–3 days, whereas geostationary satellites operate in time scales of hours, which could theoretically provide data on the diurnal variation in phytoplankton abundance and productivity. A summary of the operational characteristics of current satellite-based and other space-based systems that have provided data for coastal and estuarine monitoring is found in Table 1.

Aircraft remote sensing

Aircraft are now widely used to remotely sense estuaries and coastal waters. In general, an airborne system can provide considerably higher spatial resolution data (e.g., less than a meter to tens of meters) than space-based systems at relatively low costs and can be flown when atmospheric (i.e., cloud-free), environmental, and solar conditions are acceptable to study a specific phenomenon.[12] Virtually every class of aircraft from small single-engine propeller planes to large multi-engine commercial and specialized military platforms can be used to conduct remote sensing. Single-engine propeller planes are typically un-pressurized and operate at altitudes below 25,000 ft, with a range less than 1000 miles. Turbine or turbo-charged twin engine propeller aircraft operate at altitudes up to 35,000 ft. Commercial jets flying above 40,000 ft can map large areas (at lower resolutions) but are relatively expensive to operate. Special-purpose high-altitude platforms, such as the NASA ER-2 aircraft flying at 65,000 ft, collect spectral and image data at regional scales. The major benefits of airborne remote sensing, compared to satellite-based systems, are that the user can define the deployment schedule and operational characteristics of the remote-sensing system. The deployment can also be coordinated with a field program to acquire in situ measurements to monitor biological and physical processes that occur over temporal scales, and at spatial resolutions that cannot be sampled by most satellite instruments.[13,14]

Table 1 Operational characteristics of current ocean-color satellite sensors (IOCCG, 2012) http://www.ioccg.org/sensors/current.html

Sensor	Agency	Satellite	Launch date	Swath (km)	Spatial resolution (m)	Bands	Spectral coverage (nm)	Orbit
COCTS	CNSA (China)	HY-1B (China)	11 April 2007	2400	1100	10	402–12,500	Polar
CZI				500	250	4	433–695	
GOCI	KARI/KORDI (South Korea)	COMS	26 June 2010	2500	500	8	400–865	Geostationary
HICO	ONR and DOD Space Test Program	International Space Station	18 Sept. 2009	50 km Selected coastal scenes	100	124	380–1000	51.6°, 15.8 orbits p/d
MERIS	ESA	ENVISAT			300			
MERSI	CNSA (China)	FY-3A (China)	27 May 2008	2400	250/1000	20	402–2155	Polar
MERSI	CNSA (China)	FY-3B (China)	5 Nov. 2010	2400	250/1000	20	402–2155	Polar
MODIS-Aqua	NASA (USA)	Aqua (EOS-PM1)	4 May 2002	2330	250/500/1000	36	405–14,385	Polar
MODIS-Terra	NASA (USA)	Terra (EOS-AM1)	18 Dec. 1999	2330	250/500/1000	36	405–14,385	Polar
OCM-2	ISRO (India)	Oceansat-2 (India)	23 Sept. 2009	1420	360/4000	8	400–900	Polar
POLDER-3	CNES (France)	Parasol	18 Dec. 2004	2100	6000	9	443–1020	Polar
VIIRS	NOAA/NASA (USA)	NPP	28 Oct. 2011	3000	370/740	22	402–11,800	Polar

Post-flight data processing

Post-flight data processing is an extremely important process for removing atmospheric and seabed effects to convert the digital radiances into reflectance data. There are several commercially available software packages to process aircraft and satellite remote-sensing data such as ERDAS IMAGINE (Leica Geosystems GIS & Mapping) and ENVI (Research Systems International). Data are provided to a data user in a standard image processing format (e.g., ERDAS, ENVI, and GEOTIFF) or in a generic scientific data format such as HDF (Hierarchal Data Format). The HDF is an efficient structure for storing multiple sets of scientific, image, and ancillary data in a single data file. The data may be sent to a user as raw radiance files with no processing, image files with radiometric calibrations applied, or as radiometrically calibrated and atmospherically corrected digital image files georeferenced to a map projection. Complex atmospheric correction procedures and models [such as MODTRAN 4.0 (MODerate resolution TRANSsmittance)] are employed to compute the ocean color signal by determining the magnitude of and removing atmospheric scattering and absorption effects between the water surface and the sensor.[13,15,16] However, a commonly used and simple approach is the "clear water pixel" or "dark pixel" subtraction technique, which assumes that the sensor has a spectral band for which clear water is essentially a black body (i.e., no reflectance). Therefore, any radiance measured by the instrument in this band is due to atmospheric backscatter and can be subtracted from all pixels in the image.[17,18] In the shallow waters of estuarine and coastal systems, the seabed reflects part of the incident light in a way that is highly dependent on the bottom material and roughness. The reflected light is spectrally different than that of deep water, which allows scientists to obtain useful information about the nature of the seabed. The maximum depth at which a sensor receives any significant signal varies as a function of spectral wavelength and the clarity of the water. In some coastal waters, the bottom is detected to less than 10 m. In highly turbid waters, the bottom would not be visible as the depth of light penetration is less than a meter.[19] Once the data have been corrected for atmospheric and seabed effects, the standard method is to then georeference the imagery that links specific pixel locations in the image to their corresponding location on a mapped surface for which the mapped coordinates are well known.

Remote Sensing for Management of Coastal and Estuarine Environments

Increased population and development have contributed significantly to the environmental pressures in coastal regions. These pressures have resulted in substantial physical changes to the coastline, declines in water quality, and biological/chemical changes to waters with the addition of high volumes of nutrients (primarily nitrogen and phosphorous) from urban, nonpoint source runoff. Remote sensing provides near-synoptic, local, regional, and global views of the indicators of environmental condition and change that can be routinely monitored. Unfortunately, the application of ocean color remote sensing techniques and data to address coastal environmental issues has been limited to a small segment of the scientific community and there has not been a general transfer of the capability to the environmental management community for operational decision-making. Schaeffer et al.[20] indicated that four attributes may be responsible for the reluctance of the management community to incorporate remote sensing data into their decision-making: cost, data accuracy, satellite mission continuity, and obtaining upper level management approval to include remote sensing data in their work. Results from a survey of selected U.S. Environmental Protection Agency managers indicated there was an impression that all satellite imagery could only be obtained with a financial commitment.[20] Survey respondents did not know that most satellite data were available at no cost from NASA and the European Space Agency (ESA). Respondents wanted assurance that satellite products could be validated, with accuracy or error estimates identified, for their particular water body of interest.[20] Operational remote sensing of coastal and estuarine environments is in an evolutionary phase where the availability of real-time (or near real-time) data, coupled with advances in algorithm development and image processing, will soon revolutionize our ability to monitor coastal and estuarine systems. The ability to monitor and predict changes in biological and physical processes characteristic of the coastal environment, over large areas and long time periods, is only possible through satellite and aircraft observations. Even with uncertainties in accuracy estimates, the detection of change in a particular water body is possible if the data product was derived with a consistent methodology.[20–22] When incorporated into specific monitoring plans, remotely sensed information has been shown to provide an added value component to data analysis for better decision support.[23] For example, high-resolution optical systems on space-based platforms (e.g., Landsat and Satellite Pour l'Observation de la Terre (SPOT) satellites) provide a critical source of spatial information that gives coastal managers the ability to assess environmental conditions at a given point in time. When collected over multiple years, this information can be used to better understand the cumulative effects of human development, including impacts of changes in land cover on coastal water quality or ecosystem health.[24–27] The survey further indicated that mission continuity was identified as a critical element for acceptance as funding issues may jeopardize the life expectancy of satellite programs.[20] Finally, the need for experienced

personnel and to educate environmental managers was identified as an important requirement to create an organizational commitment to use remote sensing data.[20]

CONCLUSION

Optical sensors flown on space-based platforms and aircraft can remotely estimate the underwater light field of coastal and estuarine waters and link these optical properties to their in situ biological, chemical, and geological constituents. With this information, a wealth of new and exciting products that support coastal and estuarine monitoring are being created and could be effectively used to continually address new questions and challenges faced by decision-makers and managers.

ACKNOWLEDGMENTS

I would like to thank Kenneth Rocha, Kristin Hychka, and Anne Kuhn for their review and comments of the manuscript. This is contribution number ORD-002860 of the Atlantic Ecology Division, National Health and Environmental Effects Research Laboratory, Office of Research and Development, U.S. Environmental Protection Agency. Any mention of trade names or commercial products does not constitute endorsement or recommendation for use.

REFERENCES

1. Bagheri, S.; Peters, S.W.M. Retrieval of Marine Water Constituents Using Atmospherically Corrected AVIRIS Hyperspectral Data, 12th Aviris Workshop, JPL, Pasadena, CA, 2003.
2. Porter, W.M.; Enmark, H.T. A system overview of the Airborne Visible/Infrared Imaging Spectrometer (AVIRIS). *Proc SPIE, Imaging Spectrometer II*; San Diego, CA, 1987; Vol. 834, 22–31.
3. Richardson, L.L.; Buison, D.; Lui, C.J.; Ambrosia, V. The detection of algal photosynthetic accessory pigments using Airborne Visible-Infrared Imaging Spectrometer (AVIRIS) spectral data. Mar. Tech. Soc. J. **1994**, *28* (28), 10–21.
4. Anger, C.D.; Mah, S.; Babey, S.K. Technological enhancements to the Compact Airborne Spectrographic Imager (CASI). In *Proceedings of the First International Airborne Remote Sensing Conference and Exhibition*; Strasbourg, France, 1994; Vol. 2, 205–213.
5. Anger, C.D.; Achal, S.; Ivanco, T.; Mah, S.; Price, R.; Busler, J. Extended operational capabilities of CASI. In *Proceedings of the Second International Airborne Remote Sensing Conference and Exhibition-Technology, Measurement, and Analysis*, Environmental Research Institute of Michigan: Ann Arbor, MI, 1996; Vol. 2, 124–133.
6. Clark, C.D.; Ripley, H.T.; Green, E.P.; Edwards, A.J.; Mumby, P.J. Mapping and measurement of tropical coastal environments with hyperspectral and high spatial resolution data. Int. J. Remote Sens. **1997**, *18* (2), 237–242.
7. Hoogenboom, H.J.; Dekker, A.G.; De Haan, J.F. Retrieval of chlorophyll and suspended matter in inland waters from

CASI data by matrix inversion. Can. J. Remote Sens. **1998**, *24* (2):144–152.

8. Rickard, L.J.; Basedow, R.W.; Zalewski, E.F.; Silverglate, P.R.; Landers, M. HYDICE: An airborne system for hyperspectral imaging. In *Proc SPIE, Imaging Spec-55 trometry of the Terrestrial Environment*; Vane, G., Ed.; San Diego, CA, 1993; Vol. 1937, 173–179, doi: 10.1117/12.157055.

9. Davis, C.O.; Bowles, J.; Leathers, R.A.; Korwan, D.; Downes, V.; Snyder, W.A.; Rhea, W.J.; Chen, W.; Fisher, J.; Bissett, W.P.; Reisse, R.A. Ocean PHILLS hyperspectral imager: Design, characterization, and calibration. Opt. Expr. **2002**, *10* (4), 210–221.

10. Corson, M.; Davis, C.O. The Hyperspectral Imager for the Coastal Ocean (HICO) provides a new view of the Coastal Ocean from the International Space Station. AGU EOS, **2011**, *92* (19), 161–162.

11. Chang, G.; Mahoney, K.; Briggs-Whitmire, A.; Kohler, D.D.R.; Mobley, C.D.; Lewis, M.; Moline, M.A.; Boss, E.; Kim, M.; Philpot, W.; Dickey, T.D. The new age of hyperspectral oceanography. Oceanography **2004**, *17* (2), 16–23.

12. Myers, J.S.; Miller, R.L. Optical airborne remote sensing. In *Remote Sensing of Coastal Aquatic Environments Technologies, Techniques, and Applications (Remote Sensing and Digital Image Processing)*; Miller, R.L., Del Castillo, C.E., McKee, B., Eds.; Springer: Dordrecht, the Netherlands, 2005.

13. Miller, R.L.; Cruise, J.C.; Otero, E.; Lopez, J.M. Monitoring suspended particulate matter in puerto rico: Field measurements and remote sensing. Water Resour. Bull. **1994**, *30* (2), 271–282.

14. Miller, R.L.; Twardowski, M.S.; Moore, C.; Cassagrande, C. The Dolphin: Technology to support remote sensing algorithm development and applications. Backscatter **2003**, *14* (2), 8–12.

15. Richter, R.; Schläpfer, D. Geo-atmospheric processing of airborne imaging spectrometry data. Part 2: Atmospheric/Topographic correction. Int. J. Remote Sens. **2002**, *23* (13), 2631–2649.

16. Lavender, S.J.; Nagur, C.R.C. Mapping coastal waters with high resolution imagery: Atmospheric correction of multi-height airborne imagery. J. Opt. A Pure Appl. Opt. **2002**, *4* (4), S50–S55.

17. Gordon, H.R.; Morel, A. Remote assessment of ocean color for interpretation of satellite visible imagery: A review. In *Lecture Notes on Coastal and Estuarine Studies*; Springer-Verlag: Berlin, 1983; Vol. 4.

18. Siegel, D.A.; Wang, M.; Maritorena, S.; Robinson, W. Atmospheric correction of satellite ocean color imagery: the black pixel assumption. Appl. Opt. **2000**, *39* (21), 3582–3591.

19. Sathyendranath, S., Ed.; IOCCG. *Remote Sensing of Ocean Colour in Coastal, and Other Optically-Comoplex Waters*, Reports of the International Ocean-Colour Coordinating Group, No. 3; IOCCG: Dartmouth, Canada, 2000, 140 pp.

20. Schaeffer, B.A.; Schaeffer, K.G.; Keith, D.; Lunetta, R.S.; Conmy, R.; Gould, R. Barriers to adopting satellite remote sensing for water quality management. Int. J. Remote Sens. **2013**, *34* (21), 7534–7544.

21. Stumpf, R.P.; Culver, M.E.; Tester, P.A.; Tomlinson, M.; Kirkpatrick, G.J.; Pederson, B.A.; Truby, E.; Ransibrahmanakul, V.; Soracco, M. Monitoring *Karenia brevis* blooms in the Gulf of Mexico using satellite ocean color imagery and other data. Harmful Algae **2003**, *2* (2), 147–160.

22. Hu, C.; Muller-Karger, F.E.; Taylor, C.; Carder, K.L.; Kelbe, C.; Johns, E.; Heil, C.A. Red Tide detection and tracing using MODIS fluorescence data: A regional example in SW Florida coastal waters. Remote Sens. Environ. **2005**, *97* (3), 311–321.

23. Arnone, R.A.; Parsons, A.R. Real-time use of ocean color remote sensing for coastal monitoring. In *Remote Sensing of Coastal Aquatic Environments Technologies, Techniques, and Applications (Remote Sensing and Digital Image Processing)*; Miller, R.L., Del Castillo, C.E., McKee, B., Eds.; Springer: Dordrecht, the Netherlands, 2005; 317–337.

24. Wilson, E.H.; Sadler, S.A. Detection of forest type using multiple dates of Landsat TM imagery. Remote Sens. Environ. **2002**, *80* (3), 385–396.

25. Wang, Y.; Bonynge, G.; Nugranad, J.; Traber, M.; Ngusaru, A.; Tobey, J.; Hale, L.; Bowen, R.; Makota, V. Remote sensing of mangrove change along the Tanzania Coast. Mar. Geod. **2003**, *26* (1–2), 35–48.

26. Turner, W.; Spector, S.; Gardiner, N.; Fladeland, M.; Sterling, E.; Steininger, M. Remote sensing for biodiversity science and conservation. Trends Ecol. Evol. **2003**, *18* (3), 306–314.

27. Hurd, J.D.; Civco, D.L.; Wilson, E.M.; Arnold, C.L. Coastal area land-cover change analysis for connecticut. In *Remote Sensing of Coastal Environments*; Wang, Y., Ed.; CRC Press: Boca Raton, FL, 2010; 333–353.

Coastal Environments

Yuanzhi Zhang
Jinrong Hu
Yuen Yuen Research Centre for Satellite Remote Sensing, Institute of Space and Earth Information Science,
Chinese University of Hong Kong, Hong Kong, and Laboratory of Coastal Zone Studies,
Shenzhen Research Institute, Shenzhen, China

Abstract
Coasts are often highly scenic and contain abundant natural resources. Coastal environments contain a wide range of natural habitats such as sand dunes, barrier islands, tidal wetlands and marshes, mangrove forests, coral reefs, and submerged aquatic vegetation that provides food, shelter, and breeding grounds for terrestrial and marine species. However, coasts also face many environmental challenges including natural hazards and human-induced impacts. Coastal habitat loss caused by coastal land use, coastal pollution, sea-level rise, and other issues is becoming increasingly prominent.

INTRODUCTION

The coastal zone, broadly defined as near-coast waters and the adjacent land area, forms a dynamic interface of land and water of high ecological diversity and critical economic importance.[1] Coasts are often highly scenic and contain abundant natural resources. Coastal environments contain a wide range of natural habitats, such as sand dunes, barrier islands, tidal wetlands and marshes, mangrove forests, coral reefs, and submerged aquatic vegetation that provides food, shelter, and breeding grounds for terrestrial and marine species. Figure 1 shows a schematic view of coastal environments of coastal embayment. The majority of the world's population lives close to the sea. As many as 3 billion people (50% of the global total) live within 60 km of the shoreline.[2] The coastal zone is therefore perhaps the most critical zone for the world economy, culture, and future survival. However, the coasts also face many environmental challenges including natural hazards and human-induced impacts.

Coastal environments are characteristic of the coastal zones. This entry is designed as an introduction to the coastal environments to increase awareness and knowledge about coast. Beaches, barrier islands, salt marshes, mangrove swamps, estuaries, coral reefs are some examples of coastal environments and some typical coastal issues are described in this entry.

CHARACTERISTIC LANDFORMS AND ENVIRONMENTS OF COASTAL ZONE

The coastal areas are the most beautiful areas on Earth. For much of Earth's existence, coastal zones have been its most dynamic and changing areas. Coastal zones are continually changing because of the dynamic interaction between the oceans and the land. Coastal environments are also highly dynamic and complex. Currents, waves, and tides as the main coastal energy source are fundamental to defining coastal environments. Coastal regions have distinctive landforms that represent a balance among forces from the ocean, land, and atmosphere.[3] The shape of coastal landforms is a response of the materials that are available to the processes acting on them. In this section, some characteristic landforms and environments of coastal zone are introduced.

Beaches

A beach is a geological landform along the shoreline of an ocean, sea, lake, or a river. It usually consists of loose particles that are often composed of rock such as sand, gravel, shingle, pebbles, or cobblestones. The particles comprising the beach are occasionally biological in origin, such as mollusc shells or coralline algae. Beaches typically occur in areas along the coast where wave or current action deposits and reworks sediments.

Beaches are the result of wave action by which waves or currents move sand or other loose sediments of which the beach is made as these particles are held in suspension. Alternatively, sand may be moved by saltation (a bouncing movement of large particles). Beach materials come from erosion of rocks offshore, as well as from headland erosion and slumping producing deposits of scree.

Beaches also play an important role as a habitat for coastal plants and animals. Some small animals burrow into the sand and feed on material deposited by the waves.

Encyclopedia of Natural Resources DOI: 10.1081/E-ENRW-120047532

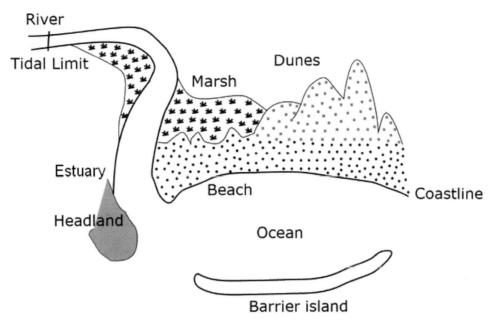

Fig. 1 A schematic coastal environment of a coastal embayment, which includes different types of coastal environments: Barrier island, beach, dunes, estuary, marsh, and headland.

Crabs, insects, and shorebirds feed on these beach dwellers. The endangered Piping Plover and some tern species rely on beaches for nesting. Sea turtles also lay their eggs on ocean beaches. Sea grasses and other beach plants grow on undisturbed areas of the beach and dunes.

Barrier Islands

Barrier islands, a coastal landform and a type of barrier system, are relatively narrow strips of sand that parallel the mainland coast. Barrier islands often form as chains of long, low, narrow offshore deposits of sand and sediment, running parallel to a coast but separated from it by bays, estuaries, or lagoons. Unlike stationary landforms, barrier islands build up, erode, migrate, and rebuild over time in response to waves, tides, currents, and other physical processes in the open ocean environment.

The world's barrier islands measure about 13,000 miles (21,000 km) in length.[4] They are found along all continents except Antarctica and in all oceans, and they make up roughly 10% of the Earth's continental shorelines. The northern hemisphere is home to 74% of these islands.

Barrier islands play an enormous role in mitigating ocean swells and other storm events for the water systems behind on the mainland side of the barrier island.[5] This effectively creates a unique environment of relatively low-energy brackish water. Multiple wetland systems, such as lagoons, estuaries, and/or marshes, can result from such conditions depending on the surroundings. Without barrier islands, these wetlands cannot exist and will be destroyed by daily ocean waves and tides as well as ocean storm events. One of the most prominent examples is that of the Louisiana barrier islands.

Salt Marshes

Salt marshes are coastal wetlands rich in marine life.[6] They are sometimes called tidal marshes, because they occur in the zone between low and high tides. Salt marshes are composed of a variety of plants: rushes, sedges, and grasses. Salt marsh plants cannot grow where waves are strong, but they thrive along low-energy coasts. They also occur in areas called estuaries, where freshwater from the land mixes with sea water. A distinctive feature of salt marshes is the color; the plants grow in various shades of gray, brown, and green.

Salt marshes serve many important functions.[7] They buffer stormy seas, slow shoreline erosion, and are able to absorb excess nutrients before they reach the oceans and estuaries. High concentrations of nutrients can cause oxygen levels low enough to harm wildlife, such as the "Dead Zone" in the Gulf of Mexico. Salt marshes also provide vital food and habitat for clams, crabs, and juvenile fish, as well as offering shelter and nesting sites for several species of migratory waterfowl.

Mangrove Forests

Mangroves are various kinds of trees growing up to a medium height and shrubs that grow in saline coastal sediment habitats in the tropics and subtropics, mainly between latitudes 25°N and 25°S.[8] Mangroves provide enormously important and economically valuable ecosystem services to coastal communities throughout the tropics. Their important ecosystems provide wood, food, fodder, medicine, and honey for humans and habitats for many animals such as crocodiles, snakes, tigers, deer, otters, dolphins, and birds. A wide range of fish and shellfish also depend on mangroves as the swamps help filter sediment and pollution from water upstream and stop it from disturbing the delicate balance of ecosystems like coral reefs.

Mangrove ecosystems should be better protected,[8] the UN's food agency has warned, as it published new figures showing that 20% of the world's mangrove area has been destroyed since 1980. The main causes of the destruction of mangrove swampland include population pressure, conversion for shrimp and fish farming, agriculture, infrastructure and tourism, as well as pollution and natural disasters.

Estuaries

An estuary is a partly enclosed coastal body of water with one or more rivers or streams flowing into it, and with a free connection to the open sea.[9] Estuaries are protected from the full force of the ocean by mudflats, sandspits, and barrier islands. Estuaries are sometimes called bays, lagoons, harbors, or sounds. All these examples are estuaries if fresh water mixes with salt water.

Estuaries play an important role in the economy. They attract tourists who like fishing, boating, and other water sports. They are an important part of the shipping industry because many industrial ports are located in estuaries. Estuaries are also a critical part of the commercial fishing industry. It is estimated that over 75% of all fish caught by commercial fishing operations lived in an estuary for at least part of their life cycle.

Coral Reefs

Reefs[10] built by coral and associated organisms occur extensively in tropical waters and are widespread between latitudes 30°N and 30°S in the western parts of the Pacific, Indian, and Atlantic oceans. They are well developed in the Caribbean Sea, the Red Sea, and Indonesia. The distribution [11] of coral reefs is influenced by environmental factors such as light (symbiotic zooxanthellae in corals require light to photosynthesize), sea-surface temperature, and carbonate saturation state (closely related to temperature). Corals are limited to waters, where sea-surface temperature rarely drops below 17–18°C or exceed 33–34°C for prolonged periods. Coral species diversity decreases with latitude in response to sea-surface temperature and currents of dispersal.[12] Today, richly diverse coral reefs are found in the tropics along coastlines, on the margins of volcanic islands, and on isolated coral atolls.

Coral reefs[13] are the most diverse and beautiful of all marine habitats; they are home to 25% of all marine species. However, many of the world's coral reefs have been severely damaged by natural processes and human activities in recent decades. Corals face serious risks from various diseases, including black-band, white-band, and yellow-band diseases that have been reported from many localities worldwide. Black-band disease is primarily caused by cyanobacteria, but the causes of white-band disease and yellow-band disease are unknown. When corals are stressed, they often expel the algal symbionts that are critical to their health in a process commonly known as coral bleaching. One known cause of coral bleaching is an increase in ocean temperatures. Regional increases in sea-surface temperatures occur during El Niño events, and ocean temperatures worldwide may be changing as a result of global warming.

The susceptibility of corals to disease may also be on the rise as a result of human activities. Many human activities are known to directly and indirectly harm coral reefs. Oil spills and pollutants can threaten entire reefs. Excessive nutrients from land sources, such as sewage outfall and agricultural fertilizers, promote the growth of algae that can smother corals. Such algae also thrive when fish that graze on them are overharvested. Other organisms harmful to corals, such as the crown-of-thorns starfish, multiply when the species that prey on them are removed. The collection of live corals and other reef organisms can directly degrade large areas of reef.

COASTAL HAZARDS

The highly variable environments [14] found along coastlines are shaped by a combination of different processes including waves, tides, storms, and in geologically recent times, humankind's changes to the shoreline environment. Most of the time, these processes operate at levels that are predictable, expected, and form part of the daily rhythm of life on the coasts, but other times these processes can impart dramatic changes very quickly, and sometimes pose great hazards to coastal residents. To appreciate and mitigate the hazards posed by living along coastlines, the processes that affect the interaction of the land, water, and atmosphere in this critical, dynamic, and ever-changing environment must be understood. There are several major factors that influence the development and potential of hazards along the coastline, including waves, tides, storms, and human-induced changes to the shoreline. These are each introduced in turn in this section.

Origin of Coastal Hazards

Waves

All coasts are affected to a certain degree by wave activity,[15] and waves provide the energy, which drives many of the coastal processes that create many of the world's most spectacular coasts. Waves are superficial undulations of the water surface produced by winds blowing over the sea.

Waves provide energy "powers" to many coastal processes. Coastlines are zones along which water is continually making changes. Waves can both erode rock and deposit sediment. Because of the continuous nature of ocean currents and waves, energy is constantly being expended along coastlines and they are thus dynamically changing systems, even over short (human) time scales.

Tides

Tides,[16] the rhythmic, twice-daily rise and fall of ocean waters, are caused by gravitational attraction between the Moon (and to a lesser degree, the Sun) and the Earth.

In some places, currents induced by tides are the most significant factor controlling development of the beach and shoreline environment. These places include tidal inlets, passages between islands and the mainland, and areas with exceptionally large tidal ranges. Tides are responsible for depositing deltas on the lagoonal and oceanward sides of tidal inlets, and for moving large amounts of sediment in regions with high tidal ranges. Tides also affect erosion and coastal sediment processes, since they control to what height waves can influence the land. Storm surges (high energy waves) that occur during spring tides can be particularly damaging to the shoreline and/or human property.

Storms

Storms[17] can cause some of the most dramatic and rapid changes to the coastal zones and represent one of the major, most unpredictable hazards to people living along coastline. Storms include hurricanes, which form in the late summer and fall, and extra tropical lows, which form in the late fall through spring. High winds blowing over the surface of the water during storms bring more energy to the coastline and can cause more rapid rates of erosion. One of the most famous hurricanes that had a significant impact on coastal communities in the United States was Hurricane Katrina in August 2005. According to the statistics, around 80% of New Orleans and some areas of neighboring communities were flooded, after which it took several weeks for the land surface to show up again. The hurricane left about 1800 people dead, which made it one of the deadliest U.S. tropical storms. The property damage was estimated over $100 billion, which was considered as the costliest hurricane in the USA.

Tsunami

Tsunami[18] is a series of traveling ocean waves of extremely long length generated by disturbances associated primarily with earthquakes occurring below or near the ocean floor. In the deep ocean, their length from wave crest to wave crest may be a hundred miles or more but with a wave height of only a few feet or less. They cannot be felt aboard ships nor can they be seen from the air in the open ocean. In deep water, the waves may reach speeds exceeding

500 miles per hour. Large tsunamis have been known to rise over 100 feet, while tsunamis 10–20 feet high can be very destructive and cause many deaths and injuries.

Tsunamis are a threat to life and property to anyone living near the ocean. For example, in 1992 and 1993 over 2000 people were killed by tsunamis occurring in Nicaragua, Indonesia, and Japan. Property damage was nearly $1 billion. The 1960 Chile earthquake generated a Pacific-wide tsunami that caused widespread death and destruction in Chile, Hawaii, Japan, and other areas in the Pacific. Two of the most famous tsunamis in recent years are the 2004 Indian Ocean tsunami and March 11, 2011 tsunami in northeastern Japan. The December 26, 2004 Indian Ocean tsunami was caused by an earthquake that was thought to have had the energy of 23,000 atomic bombs. By the end of the day, more than 150,000 people were dead or missing and millions more were homeless in 11 countries, making it perhaps the most destructive tsunami in history. On March 11, 2011, a devastating tsunami with 33 feet high waves, triggered by the biggest earthquake on record in Japan, destroyed huge areas of the country's northeastern coast, causing more than 1000 deaths.

HUMAN-INDUCTED HAZARDS

The human influence on coastlines looms large. Man is a major factor in coastal changes at various scales. Coastal zones are relatively fragile ecosystems, and disordered urbanization and development of infrastructure, alone or in combination with uncoordinated industrial, tourism-related, fishing and agricultural activities, can lead to rapid degradation of coastal habitats and resources. Coastal land use, pollution, and sea-level rise will be introduced in this section.

Coastal Land-Use

Coastal zones are an important area for human habitation, industry, location of centers of energy production, military activities, fisheries, bird life, and recreation. The fast development of coastal regions[19] will inevitably lead to the creation of vast built-up areas (such as construction of ports and tourist facilities) at the expense of natural habitats (e.g., dunes, saltmarshes) and as a result, will damage or destroy a substantial part of the natural coastline's habitats. In France, for example, 15% of natural areas on the coast have disappeared since 1976 and are continuing to do so at the rate of 1% a year. Italy, which had around 700,000 ha of coastal marshes at the end of the last century, had no more than 192,000 ha in 1972 and has less than 100,000 ha today. Some estimates suggest that about one-third of the coastal dunes in northwestern Europe and three quarters in the western Mediterranean have disappeared. Such large-scale habitat destruction will inevitably lead to a decline in species distribution and abundance.

Coastal Pollution: Red Tide

Pollution of the coastal zone[20] is primarily a result of contaminant load being discharged into receiving waters, resulting in deleterious effects such as harm to plants and animals, hazards to human health, hindrance to marine activities (including fishing), impairment of quality for use of sea water, and reduction of amenities. Despite the ability of coastal zones to reduce the harmful effects of some contaminants, coasts are also vulnerable to pollution since wastewater is often discharged directly or indirectly into sheltered and shallow coastal waters with poor mixing.

The most important contaminants[20] in the coastal zone are organic matter, synthetic organic compounds (e.g., PCBs and pesticides such as DDT and residues), and microbial organisms, nutrients (mainly nitrogen and phosphorus), and so on. This is the main reason for the steadily increase of harmful algal bloom (HAB) in coastal waters in recent years. Algal blooms[21] may cause harm through the production of toxins or by their accumulated biomass, which can affect co-occurring organisms and alter food-web dynamics. Impacts include human illness and mortality following consumption of or indirect exposure to HAB toxins, substantial economic losses to coastal communities and commercial fisheries, and HAB-associated fish, bird, and mammal mortalities. For example, in recent years, the frequent toxic red tide in the Gulf of Mexico has increasingly becoming a threat to the sea turtle population.

Coastal Pollution: Oil Spill

In recent years, oil spills caused by human error or carelessness have become one of the most serious disasters for the coastal and ocean ecosystems. Oil spills often result in both immediate and long-term environmental damage. Some of the environmental damage caused by an oil spill can last for decades after the spill occurs. Here, we only point out their effects on coastal environments, for example, the 2010 Gulf of Mexico oil spill and the 2011 Bohai Bay oil spill. The 2010 Gulf of Mexico oil spill led to at least 2500 km^2 of water covered with oil, and a large number of fish, birds, marine life, and plants were seriously affected. In November 2010, the U.S. government reported that 6104 birds, 609 turtles, and 100 dolphins were killed by this oil spill. The China State Oceanic Administration reported that the 2011 Bohai Bay oil spill resulted in 5500 km^2 of water pollution, roughly equivalent to 7% of the Bohai Sea area.

Sea-Level Rise

World sea-level is known to be rising, threatening an increase in storm and flood damage along many low-lying, populated coasts. The global rise may be as much as 1 m by the year 2050. The reasons for this rise in sea-level are complex, but one factor seems to be an increase in atmospheric CO_2 and trace gases, leading to increased heat absorption, the so-called "Greenhouse Effect."[22] Anthropogenic CO_2 emissions are currently responsible for more than half of the enhanced greenhouse effect.[23]

CONCLUSIONS

Coastal environments contain a wide range of natural habitats such as beaches, barrier islands, salt marshes, mangrove forests, coral reefs, and submerged aquatic vegetation that provides food, shelter, and breeding grounds for terrestrial and marine species. The coast and its adjacent areas on and off shore are an important part of a local ecosystem as the mixture of fresh water and salt water in estuaries provides many nutrients for marine life. However, coasts also face many environmental challenges including natural hazards and human-induced impacts. Currently, more and more issues face coastal managers: coastal storms and coastal habitat loss caused by land use, pollution, sea-level rise, and some other issues are becoming increasingly prominent. Management and protection of coastal environments to achieve the sustainable development of coastal zone are arduous and long-term tasks for humankind.

The coast, which is shaped by a variety of different forces, is a complex dynamic system and the coastal environments are also diverse. Due to space limitations, this entry cannot include detailed descriptions for each coastal environment but hopefully it can help the reader gain a general understanding about coastal environments.

REFERENCES

1. McLean, R.F.; Tsyban, A.; Burkett, V.; Codignotto, J.O.; Forbes, D.L.; Mimura, N.; Beamish, R.J.; Ittekkot, V. Coastal zones and marine ecosystems. In *Climate Change 2001: Impacts, Adaptation and Vulnerability*; Cambridge University Press: United Kingdom, 2001.
2. Woodroffe, C.D. *Introduction. Coasts: Form, Process and Evolution*, 1st Ed.; Cambridge University Press: Cambridge, United Kingdom, 2003.
3. Kusky, T. M. *Introduction. The Coast: Hazardous Interactions within the Coastal Environment*; Facts on File, Inc.: New York, 2008.
4. Stutz, M.L.; Pilkey, O.H. Open-Ocean Barrier Islands: Global Influence of Climatic, Oceanographic, and Depositional Settings. J. Coast. Res. **2011**, *27* (2), 207–222.
5. Stone, G.W.; McBride, R.A. Louisiana barrier islands and their importance in wetland protection: forecasting shoreline change and subsequent response of wave climate. J. Coast. Res. **1998**, *14*, 900–915.
6. http://www.dep.state.fl.us/coastal/habitats/saltmarshes.htm (accessed March 2012).
7. http://water.epa.gov/type/wetlands/marsh.cfm (accessed March 2012).
8. http://www.guardian.co.uk/environment/2008/feb/01/endangeredhabitats.conservation (accessed March 2012).

9. http://www.nhptv.org/natureworks/ (accessed March 2012)

10. Bird, Eric C.F. *Coral Reefs and Atolls. Coasts: An Introduction to Coastal Geomorphology*, 3rd Ed.; Blackwell: New York, 1984; 252.

11. Woodroffe, C.D. *The Coastal Zone. Coasts: Form, Process and Evolution*; 1st Ed.; Cambridge University Press: Cambridge, United Kingdom, 2003; 190–191.

12. Yonge, C.M. The biology of reef-building corals. Scientific Reports of the Great Barrier Reef Expedition 1928-1929, British Museum (Natural History), **1940**, *1*, 353–891.

13. http://pubs.usgs.gov/fs/2002/fs025-02/ (accessed March 2012).

14. Kusky, T.M. *Introduction. The Coast: Hazardous Interactions within the Coastal Environment*; Facts on File, Inc. Press: New York, 2008; 34.

15. Bascom, W.N. Ocean waves. Scient.Am. **1959**, *201*, 74–84.

16. Skinner, B.J.; Porter S.C. *The Oceans and Their Margins. The Dynamic Earth: An Introduction to Physical Geology*, 4th Edition; John Wiley & Sons, Inc.: New York, 2000; 378–379.

17. Kusky, T.M. Introduction. *The Coast: Hazardous Interactions within the Coastal Environment*; Facts on File, Inc. Press: New York, 2008; 45.

18. http://www.nws.noaa.gov/om/brochures/tsunami.htm (accessed March 2012)

19. Gehu, J.M. European dune and shoreline vegetation. *Nature and Environment* Series No. 32. Council of Europe, Strasbourg, 1985.

20. Europe's Environment - The Dobris Assessment; European Environment Agency: 1995; Contamination and coastal pollution.

21. http://www.whoi.edu/website/redtide/home (accessed March 2012).

22. Carter, R.W.G. *Coastal Issues. Coastal Environments: An Introduction to the Physical, Ecological and Cultural Systems of Coastlines*; Academic Press: 1989.

23. *The Supplementary Report to the IPCC Scientific Assessment of Climate Change*; Houghton, J.T., Callander, B.A., and Varney, S.K., Eds.; Cambridge University Press, Cambridge, 1992.

Coastal—Evaporation

Coastal Natural Disasters: Tsunamis on the Sanriku Coast

Takehiko Takano
Tohoku Gakuin University, Sendai, Japan

Abstract

The Sanriku Coast was the most severely damaged region in Japan due to the huge tsunami that followed the 2011 East Japan Earthquake. In the modern age, the coast suffered from big tsunamis in 1896, 1933, and 1960. Physical measures such as the construction of massive seawalls and breakwaters, and further measures such as hazard map making, evacuation drills, or establishing monuments, were done in every community on the coast. The experiences of the Sanriku Coast provide important lessons on the effectiveness of countermeasures against natural disasters, including tsunamis, in coastal regions. This entry outlines the 2011 East Japan tsunami disaster and the countermeasures for tsunamis on the Sanriku Coast.

INTRODUCTION

The Sanriku Coast is located on the Pacific side of the northeastern part of Japan's main island. Its coastal area was severely damaged when a huge tsunami with the height of more than 10 m arrived after the powerful earthquake on March 11, 2011, which shocked the world. Although this devastating tsunami was among the most severe natural disasters in world history, it was not the first one on the Sanriku Coast. The severely affected coastal area from Aomori to Miyagi is known as the "Sanriku Coast" or simply "Sanriku." The name "Sanriku" was originally used in the Meiji Era for the whole region of ancient Rikuzen (presently Miyagi), Rikuchu (Iwate), and Rikuou (Aomori), all of which contain the Japanese character "Riku." However, after the severe tsunami disaster that occurred in 1896, the name "Sanriku" began to mean only its coastal area. It was an impressive case that a severe natural disaster changed the meaning of a region's name.

HISTORY OF TSUNAMI DISASTERS ON THE SANRIKU COAST

Throughout history, the Sanriku Coast frequently suffered from tsunamis. According to "History of Earthquake and Tsunami in Miyagi Prefecture," there was a record of 16 tsunamis during the Edo Era from 1600 to 1868. Each of the historically known tsunamis, such as Jogan in 869, Keicho in 1611, Meiji in 1896, and Showa in 1933, were accompanied by severe damage with more than a thousand casualties (Table 1).

Although there was a lack of scientific observation data for the pre-modern Jogan and Keicho tsunamis, their outlines were described in ancient documents telling that the Jogan tsunami reached inland to a castle town about 4 km from the shoreline.[1] The description indicated that 1,000 people drowned, which reflects how severe the tsunami impact was in an ancient age when the population was far fewer than that of today. The Keicho tsunami is also considered to be the biggest one in Japan's history, of which the inundation wave was estimated to reach more distant places from the shoreline than the tsunamis in 1896 and 1933 (Fig. 1).

After the Meiji Restoration in 1868, seismological and geophysical survey methods were introduced to Japan and reliable data was collected. The Japan Society of Civil Engineering published a comparative chart of the past major tsunamis in the modern age including the East Japan tsunami in 2011 (Fig. 2). Figure 2 clearly shows that the 2011 tsunami was more intense than any of the other tsunamis in the modern age.

2011 EAST JAPAN EARTHQUAKE AND THE SANRIKU COAST

On JST 14:46, March 11, 2011, a mega earthquake with a Richter scale magnitude of 9.0 severely shook the Pacific coast of East Japan with a JMA seismic intensity of 6 or more. The U.S. Geological Survey reported that "this magnitude places the earthquake as the fourth largest in the world since 1900 and the largest in Japan since modern instrumental recordings began 130 years ago."[2] It was one of the most intense natural phenomena in Japan's history, and could be rated as a "millennium" class disaster. Among the damages, 15,872 people were killed, 2,777 people were missing, 129,577 houses were entirely destroyed, and another 266,101 houses were severely damaged.[3] Almost every type of consequence associated with an earthquake disaster followed, including lifeline cut, infrastructure damages, collapse of residential and agricultural lands, and liquefaction in lowland areas.

The most severely damaged area was the Pacific coast of the northeastern part of the main island, where most of

Encyclopedia of Natural Resources DOI: 10.1081/E-ENRW-120048592

Table 1 Major tsunami disasters affected the Sanriku Coast

Year	Name	Magnitude	Killed and lost lives	Tsunami reach (m)
869	Jogan	8.3*	1,000*	
1611	Keicho	8.9*	7,800*	
1896	Meiji	8.2	21,959	38.2
1933	Showa	8.1	3,064	28.7
2011	East Japan	9.0	18,649	40.5

*Estimated value in the cases for Jogan and Keicho.
Source: Data from Japan Meteorological Agency, Central Disaster Prevention Council.

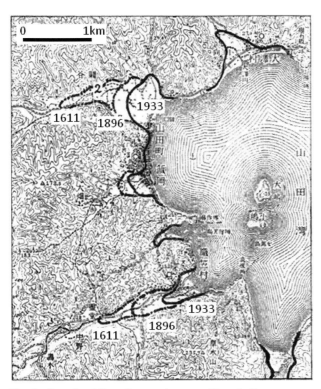

Fig. 1 Inundated line of tsunami wave at Yamada Bay.
Source: Adapted from Imamura.[10]

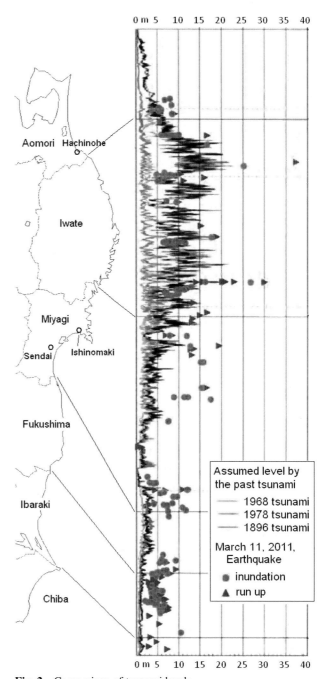

Fig. 2 Comparison of tsunami levels.
Source: Data from http://www.hkd.mlit.go.jp/zigyoka/z_kowan/jishintsunami/pdf/110801_2_1.pdf (partly modified by author)

the human casualties occurred (Fig. 3). The coastal areas of Iwate and Miyagi prefectures (Fig. 4) were heavily impacted, because the huge tsunami arrived about 50 min after the earthquake. Countless fishing boats and aquaculture facilities were washed away and destroyed.

MECHANISM OF TSUNAMI ON THE SANRIKU COAST

Tectonic Movement around the Japan Trench

Tsunamis along the Sanriku Coast are usually caused by tectonic movement of the plates around the Japan Trench, where the Pacific plate is creeping beneath the North American plate at a rate of 83 mm/yr.[4] The plate is actually considered to be divided into several micro plates and their activities can cause earthquakes and tsunamis. Smoothly descending micro plates are unlikely to cause major earthquakes and tsunamis. However, at the place where the micro plate strongly clings to the upper North American plates on which the Japanese archipelago lies, a sudden thrust of the plate can happen periodically. In the case where several micro plates move correlatively together, a greater magnitude of earthquake and an intensive scale of tsunami tend to occur.

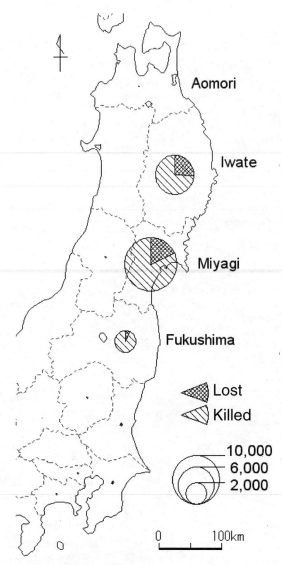

Fig. 3 Casualties by prefecture.
Source: Data from National Police Agency, October 11, 2011.

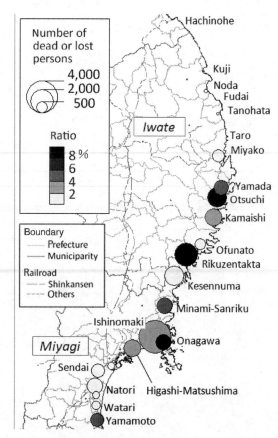

Fig. 4 Casualties by municipality.
Source: Data from National Police Agency, October 11, 2011.

The actual scale of a tsunami depends on which micro plates move. In the 2011 East Japan Earthquake, it was estimated that many micro plates offshore of East Japan (Fig. 5) moved correlatively in only about 3 minutes, which caused a huge tsunami to arrive not only to the Sanriku Coast but also to the entire Pacific shore of East Japan.

Tsunamis from Distant Places

Many of the tsunamis that affect the Sanriku Coast are caused not only in and near the Japan Trench, but also distant places where the plates meet around the Pacific Ocean. Trenches such as Kamchatka, Aleutian, Peru, and Chile are possible places. In the early morning of May 24, 1960, a tsunami with a height of 2–6 m struck the Sanriku Coast without any registered earthquake. It was caused by a M 9.5 mega-earthquake that occurred at the southern part of the Chile

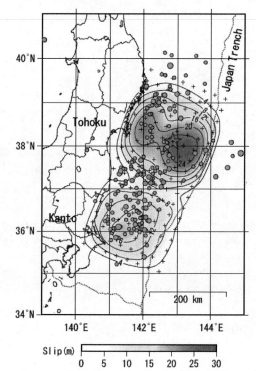

Fig. 5 Hypo-central area of 2011 East Japan Earthquake.
Source: Technical Report of The Japan Meteorological Agency, No.133 (2012) (http://www.jma.go.jp/jma/kishou/books/gizyutu/133/gizyutu_133.html)

Trench the previous day. The report indicated that 142 human lives were lost, 46,000 houses were washed away or destroyed, and fishing boats and aquaculture facilities were swept out.

Shape of Shoreline

The irregular shape of the Sanriku Coast is well known as a typical example of a "Ria type" coast, where narrow peninsulas and inlets appear in turn (Fig. 6). Ria is the regional name of Northwestern Spain of which the shape is saw-tooth-like. Such a coastal shape can easily increase the height and power of a tsunami especially at the bottom of inlets, where settlements are often located in the narrow coastal flat lands. That came true in the 2011 tsunami disaster that followed the East Japan Earthquake.

COUNTERMEASURES AGAINST TSUNAMI DISASTER

Some of the suggested countermeasures against tsunami disaster are summarized as follows.

Fig. 6 "Ria type" of coastline in the Sanriku Coast (Otsuchi and Kamaishi).
Source: Data from Geospatial Information Authority of Japan, http://www.gsi.go.jp/common/000059842.pdf

Seawalls, Breakwaters, and Water Gates

Seawalls, breakwaters, and water gates are the direct countermeasures to prevent or decrease the tsunami impact. Such structures have been constructed in almost every coastal settlement in Sanriku after suffering severe damages from two major tsunami disasters in 1896 and 1933. The fishery town of Taro is well known for its huge seawalls with a height of 10 m and total distance of about 2 km (Fig. 7). The 1,350 m long wall (shown by the bold line in Fig. 7) was constructed by 1958 and surrounds the old settlement near the western hill side. This was planned to divert tsunami flows eastward from the settlement. However, as the town expanded outside the original seawall, two other seawalls, shown by broken and dotted lines in Fig. 7, were constructed by 1978. In addition, breakwaters were constructed at the mouth of the port. The great tsunami of 2011 arrived higher than those seawalls, which swept away both the new settlement outside the original wall as well as the old settlements. Even so, the seawalls were effective in delaying the arrival of the tsunami into the settlement area.[5]

In Fudai village, about 1,500 people lived along the narrow flat land near the coast (Fig. 8). The 15.5 m high seawall was constructed by 1967 at a fishing settlement of Otanabu, and a water gate with the same height was constructed by 1984 at the river mouth of Fudai (Fig. 9). Both structures perfectly protected the inland settlement from the invasion of the 2011 tsunami. The height of the 15.5 m seawall was determined by a former chief of the village who believed villagers stating that the 1896 tsunami reached a height of about 15 m.

Fig. 7 Taro town. Bold line: constructed by 1958; broken line and dotted line: constructed by 1978.
Source: Data from Google Earth, July 20, 2009.

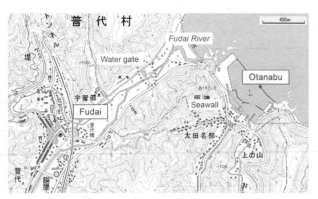

Fig. 8 Topographical map of Fudai village.
Source: Data from Geospatial Information Authority of Japan.

Fig. 9 The 15.5 m seawall at Otanabu port. The 2011 tsunami reached from the shore (left side), destroyed the factory (flat roof), but was interrupted by the seawall (upper right) before it could invade the settlement.
Source: Data from http://www.yomiuri.co.jp/national/news/20110403-OYT1T00599.htm

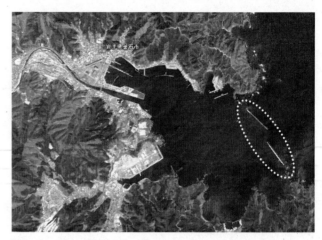

Fig. 10 Kamaishi bay.
Source: Data from Google Earth, July 20, 2009.

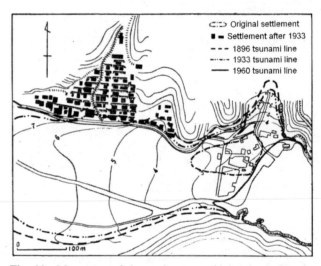

Fig. 11 Movement of the settlement to higher land after the 1933 tsunami in Toni village.
Source: Data from http://dil.bosai.go.jp/disaster/2011eq311/pdf/gsi_chileeq1961_all.pdf (partly modified by the author).

Fig. 12 Toni village just after the 2011 tsunami.
Source: Data from Google Earth, April 4, 2011.

The city of Kamaishi is known as the birthplace of Japan's modern iron manufacturing and is a major base port of the fishery on the Sanriku Coast. At the mouth of Kamaishi Bay with the maximum depth of 61 m, the world's biggest breakwater was constructed from 1978 to 2009 (Fig. 10). The height is 4 m above sea level, and its total length of the north and south parts is 1,660 m. By the 2011 tsunami, about half of the 66 caissons were subsided below sea level. However, it proved that every caisson kept its original shape and that the tsunami level was decreased less than 7 m in the central part of the city. Where no breakwater existed, the tsunami level was estimated to reach up to about 13 m.[6]

Relocation to Higher Ground

An option to avoid tsunami invasion would be the relocation of settlements to higher places away from the shore side. On the Sanriku Coast, relocation projects were done in many settlements after severe tsunami disasters in 1896 and 1933.[7] Figure 11 shows a small fishery village named "Toni," which was relocated from the shore side to a higher ground inland after the 1933 tsunami. The 2011 tsunami flooded and destroyed newly developed houses in lower land, but the relocated older settlement had little damage (Fig. 12).

Fig. 13 Rikuzentakata and its coastal forest before and after the 2011 tsunami.
Source: Data from Google Earth, July 20, 2010 – April 1, 2011.

Ground Level Up by Reclamation

Though relocation is surely an effective measure against tsunami invasion, it is difficult to realize in larger port towns and settlements with no available higher grounds to relocate. In such cases, "ground level up" by reclamation is a possible alternative measure. Needless to say, its cost may increase with the area and height of the ground to be uplifted, which may limit the feasible uplift level lower than the ideal safer level expected from past tsunamis. However, ground level up is an effective measure to reduce the height and pressure of tsunami flow to some extent, which could save the properties and lives of the inhabitants.

Coastal Forests

Coastal forests were created in many sandy shores on the Sanriku Coast. They were not originally created for the tsunami, but to protect rice fields from cold and salty winds coming from the north Pacific mainly after the Edo Era. The species of trees planted for the coastal forests were usually red pine or black pine. Figure 13 shows a coastal forest well known as "Takata Matsubara," where more than 70,000 pine trees were planted in the seventeenth and eighteenth centuries. "Matsubara" means "pine forest." In the 1933 tsunami disaster, it became known that the forest could decrease the tsunami's power and coastal trees could save lives of drowning people. As a result, coastal forests were established in many coastal areas in Sanriku. However, the 2011 tsunami was so powerful that it almost destroyed the entire coastal forest (Fig. 13, right), and only one pine tree was left, which gave hope to the people of the devastated town (Fig. 14).

It is concluded that the pine forest would not be effective against tsunamis higher than 3 m. The trees would fall down by a tsunami higher than 5 m.[8] After the 2011 disaster,

Fig. 14 Pine tree of hope (Rikuzentakata). This tree died due to salinity and was cut down, and returned to the original place as a monument after antiseptic treatment.
Source: Photographed by the author on July 2, 2013.

activities to create new coastal forests more defensive against tsunamis have been started by the Forestry Agency[9] and other interested ecologist groups.

Preparative Activities

The physical and biological measures would not be sufficient against the unexpected scale of tsunamis. Preparative activities of the people such as hazard map making, evacuation drills, and leaving lessons for next generations by the monuments and folktales are important. The Sanriku Coast experienced smaller scale tsunamis other than the big tsunamis in 1896, 1933, and 1960. Historically, a major earthquake has occurred offshore the Miyagi Prefecture about every 30 years. Such a situation increased people's awareness in Sanriku about the earthquake and tsunami possibly coming in the future. The installation of new caution signs with past tsunami levels (Fig. 15) and detailed and effective evacuation maps and practice drills have been performed by almost every community in the Sanriku Coast. The unexpected disaster in Sumatra and the Indian Ocean in December 2004 has further increased the awareness of tsunamis.

Having experienced the 2011 tsunami, planning activities to create villages or towns safer from coastal disasters have started in every Sanriku community, using almost all possible methods fitting local contexts. Those included the preservation of ruined structures by the tsunami as a memorial for the disaster (Fig. 16). However, negotiation and agreement among the local people are necessary before the decision to preserve is made, because such monuments can serve as a traumatic reminder of the disaster.

CONCLUSION

The Sanriku Coast is a rare region in the world that has experienced severe tsunami disasters cyclically. We can learn much about coastal natural disasters and the effectiveness of countermeasures from the experiences of the Sanriku Coast.

Fig. 15 Tsunami warning sign in Taro town.
Source: Data from http://www.bo-sai.co.jp/tunamihyoujiban.htm

Fig. 16 Ruined building in Minami-Sanriku town, Miyagi Prefecture. This building was the Disaster Prevention Office in town. Forty-two persons were lost in the 2011 tsunami, which reached up to 2 m higher than the roof top of the building. Many people requested to preserve the ruin as a memorial for the disaster.
Source: Photo by the author.

REFERENCES

1. Disaster Prevention Council: Tsunami Disasters in Sanriku Region (in Japanese), http://www.bousai.go.jp/kyoiku/kyokun/kyoukunnokeishou/rep/1896-meiji-sanrikuJISHINTSU-NAMI/pdf/1896-meiji-sanrikuJISHINTSUNAMI_05_chap1.pdf
2. USGS Updates Magnitude of Japan's 2011 Tohoku Earthquake to 9.0, http://www.usgs.gov/newsroom/article.asp?ID=2727&from=rss_home#.UC86EKCa2VA (accessed March 2011).
3. National Police Agency, http://www.npa.go.jp/archive/keibi/biki/higaijokyo.pdf (accessed October 2012).
4. USGS. Poster of the Great Tohoku Earthquake (northeast Honshu, Japan) of March 11, 2011 - Magnitude 9.0, http://earthquake.usgs.gov/earthquakes/eqarchives/poster/2011/20110311.php (accessed November 2012).
5. http://www.gyokou.or.jp/100sen/100img/02tohoku/021.pdf; http://www.ajg.or.jp/disaster/files/201105_taro.pdf.
6. http://www.pa.thr.mlit.go.jp/kamaishi/bousai/bouhatei/bouhatei-kamaishi.html; http://www.nikkei.com/article/DGXNASFK3100F_R30C11A3000000/.
7. Shudo, N. *History of Tsunami in Sanriku Region*; Part 5, Move to higher place (in Japanese)
8. Geographical Survey Institute, 1960: "Report on the Survey of the Abnormal Tidal Waves Tsunami Caused by the Chilean Earthquake on May 24, 1960" (http://tsunami.dbms.cs.gunma-u.ac.jp/xml_tsunami/xmltext.php). Shudo, N. *History of Tsunami in Sanriku Region*; Part 4, Takata pine forest (in Japanese).
9. Forestry Agency, 2011: On the reclamation of disaster preventive coastal forest (in Japanese), http://www.rinya.maff.go.jp/j/tisan/tisan/pdf/chuukanhoukoku.pdf
10. Imamura, A. *Past Tsunamis of the Sanriku District*; Vol. 1, Bulletin of the Earthquake Research Institute, University of Tokyo: 1933; pp. 1–16 (in Japanese).

Coral Reef: Biology and History

Graham E. Forrester
Department of Natural Resources Science, University of Rhode Island, Kingston, Rhode Island, U.S.A.

Abstract

Coral reefs are built from the secretion of calcium carbonate skeletons by corals and calcareous algae. Reef growth depends on the symbiosis between coral animals, and single-celled algae in the genus *Symbiodinium*. These limestone reefs are colonized by a tremendous variety of organisms to create an ecosystem that is the most biological diverse, and one of the most productive, in the marine environment. Coral reefs have a dynamic history spanning roughly 250 million years, which reflects the influences of biological evolution, continental drift, and sea level change.

INTRODUCTION

Although corals had been explored from above the surface for hundreds of years, scientific studies of coral reefs proliferated tremendously once scuba equipment became widely available in the 1960s. Like the millions of visitors who visit coral reefs each year, scientists have been captivated by the extraordinary beauty, abundance, and sheer variety of life to be found on coral reefs. This entry summarizes some of the discoveries made by those who have ventured underwater to learn more about these spectacular places.

CORALS AND SYMBIODINIUM

Corals belong to the class Anthozoa, within the phylum Cnidaria. The members of this group that build reefs (called hermatypic corals) are primarily members of the order Scleractinia. Their body form is called a polyp, and consists of a ring of tentacles surrounding a mouth. The mouth forms the only opening to the body cavity (coelenteron) (Fig. 1). Most corals on the reef are colonial; they grow as interconnected groups of genetically identical polyps that spread in a single layer over the reef surface. It is the external skeleton of calcium carbonate (limestone) secreted by hermatypic scleractinian corals that slowly builds up under the veneer of living tissue to form reefs (Fig. 1).

A key feature of most hermatypic corals is their mutualistic relationship with single-celled dinoflagellate algae in the genus *Symbiodinium* (sometimes referred to colloquially by the more general term zooxanthellae). The algal cells live in the tissues of the polyp and their photosynthesis produces energy that appears to increase the polyp's rate of growth, reproduction, and skeletal deposition. Some *Symbiodinium* can also live on the bottom separate from coral polyps, but those living inside polyps appear to benefit from an increased supply of nutrients and carbon dioxide, protection

from UV damage, and maintenance within a relatively constant set of ambient conditions.[1] It is this symbiosis between *Symbiodinium* and coral polyps that allows the high primary productivity and enhanced calcification making reef growth possible. While hermatypic corals acquire much of the energy they need from their symbiotic algae, most also capture prey by extending their tentacles (Fig. 1). Feeding mostly occurs at night, and typical prey includes zooplankton, bacteria, and small organic particles.

CORAL REPRODUCTION

Some corals can reproduce asexually, which occurs when a colony fragments to form genetically identical daughter colonies.[2] Corals display varied patterns of sexual reproduction and, while most can produce male and female gametes at the same time (simultaneous hermaphrodites), others are sequential hermaphrodites or have separate sexes (gonochores). Gametes may be fertilized internally so that the ciliated larvae develop internally, but the majority of species broadcast eggs and sperm into the water column to be fertilized.[3] Many broadcast spawning species only reproduce on a few nights each year, and the timing of these events is often predictable based on lunar cycles. On the Great Barrier Reef, spawning of up to 100 species is synchronized, and the clouds of gametes form massive slicks that are visible from the air.[4] How this precisely coordinated timing is achieved is not completely understood, but appears to involve a combination of responses to environmental signals, genetic precision, and possibly communication among neighboring corals.[5]

REEF GROWTH

Coral reefs are biogenic habitats, so-called because these rocky reefs that rise from the seabed are built by living

Encyclopedia of Natural Resources DOI: 10.1081/E-ENRW-120047534

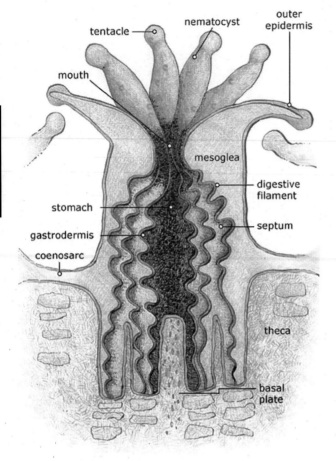

Fig. 1 Cross-section of a coral polyp.
Source: Wikipedia: http://en.wikipedia.org/wiki/Coral.

organisms. The calcium carbonate skeletons deposited by growing corals and calcareous algae gradually accrete reefs that can form enormous limestone landscapes underwater.[6,7] Reef-building corals and algae are therefore called "ecosystem engineers"[8] because they create a place to live for so many other species. Coral growth rates are highly variable, for example, the linear extension rate of two major Caribbean corals ranged from 0.6 to 0.9 cm/year for *Montastraea annularis* and 5–10 cm/year for *Acropora palmata*.[9] Reef growth is also fostered by the deposition of calcium and silica from sponges, which, along with calcareous algae also have an important role in cementing the various calcium carbonate skeletal deposits into a solid coherent matrix.[6]

The growth of reefs is continually opposed by processes that erode the reef to create coral rubble and sand. Most erosion is caused by organisms living in and around the reef and so is called bioerosion.[10] Some bioeroders are mobile species that roam above the reef, such as large parrotfish that will bite chunks out of both live corals and dead corals covered by algae. Other important external bioeroders include sea urchins and molluscs, which erode the reef as they scrape its surface to consume algae. Another group of bioeroding species lives with the reef

itself. These internal bioeroders include microscopic algae, fungi, and bacteria as well as larger organisms like sponges and molluscs that bore into the limestone. The rate of erosion is highly variable, but can be very damaging. Sponges in the family Clionaidae can, for example, remove up to 2.3 kg calcium carbonate from a square meter of reef per year.[11]

REEF HISTORY

Anthozoan corals have been forming extensive reefs in shallow coastal waters since the lower Cambrian period over 500 million years ago. The principal Paleozoic reef-forming corals were in the orders Tabulata and Rugosa, which deposited skeletons of calcite. These two groups are now extinct, and most reefs formed in the Mesozoic era are dominated by Scleractinian corals. Scleractinian corals first appeared in the fossil record roughly 240 million years ago and deposit aragonite skeletons. Over geologic time, coral reefs have undergone mass extinctions and regenerations with the movement of continents, the rise and fall of sea level, as well as with changes in ocean acidity that may have triggered shifts in dominance between scleractinian corals and other groups such as rudist bivalves. Although some of the coral reefs we can see today date back at least 50 million years, others are surprisingly recent. The earliest limestone that forms the Great Barrier Reef in Australia was laid down in the last 600,000 years and most of the extant reef was formed since the last glacial maximum 20,000 years ago, when sea level was 120 m lower than what it is today.[12]

REEF DISTRIBUTION AND LIMITS TO GROWTH

Large Scale

Currently, coral reefs are estimated to cover about 284,000 km² of the sea bed, of which most (261,000 km²) are in the Indo-Pacific region. Reefs cover much smaller areas of the Caribbean (21,600 km²) and Eastern Pacific (1600 km²). Coral reefs are found in tropical waters, mostly within 30° north or south of the equator (Fig. 2). They occur where temperatures are warm, growing best between 26 and 28°C. Hermatypic corals are generally unable to form reefs when temperatures regularly drop below 18°C or rise above 36°C, which largely explains their absence from the tropical coasts of West Africa and the Americas where upwelling brings cool water to the surface. Coral reefs form in shallow areas where sufficient light penetrates to support photosynthesis by zooxanthellae. Turbidity of the water thus also plays a role in limiting the distribution of hermatypic corals primarily by reducing the light available for photosynthesis. Corals are also restricted by their tolerance for salinity levels typical of seawater (25–42 PSU) and their requirement for enough dissolved calcium

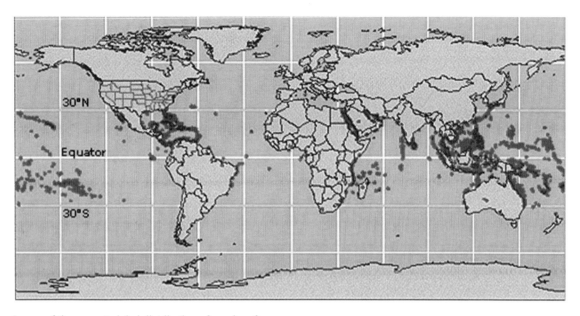

Fig. 2 A map of the current global distribution of coral reefs.
Source: National Ocean Service Education, National Oceanic and Atmospheric Administration (NOAA) http://oceanservice.noaa.gov/education/kits/corals/media/.

carbonate in the surrounding seawater to precipitate their skeletons (aragonite saturation greater than 3.2).[13]

Most of our knowledge of coral reefs comes from studying reefs less than 30 m deep, the comfortable limit for conventional scuba diving. In recent years, however, advances in diving technology and underwater robotics have fueled a growing appreciation of the importance of mesophotic coral reefs, which extend from 30 to 40 m to about 150 m, the lower limit of light penetration sufficient for photosynthesis by zooxanthellae.[14]

Local Distribution

Charles Darwin devised a simple and enduring classification of reefs, dividing them into fringing reefs, barrier reefs, and atolls.[15] Fringing reefs grow on the sloping perimeter of most tropical islands and along many coastlines. Here, corals colonize rocky substrata and form a layer of reef that grows upward toward the water's surface and may spread offshore if conditions permit. Barrier reefs, in contrast, are separated from the coastline or island landmass by a lagoon at least 10 m deep and typically 1–10 km wide. The best known examples are the world's two largest reef systems, the Great Barrier Reef (over 2900 reefs spanning 2600 km off the coast of Australia), and the Mesoamerican Barrier Reef (stretching roughly 1000 km from Mexico to Honduras). Another reef type, not classified by Darwin, is the patch reef, which as their name suggests grow as small isolated outcrops within lagoons. The third of Darwin's reef types is the atoll, which is a ring of reef that surrounds a lagoon. Underlying the lagoon is a volcanic basement. Atolls originally develop as fringing reefs

around volcanic islands. Over time, as the island gradually collapses or sea level rises, the ring of reef is eventually all that remains.

CONCLUSION

Based on the partnership between coral polyps and single-celled algae, scleractinian corals have been forming limestone reefs for over 200 million years. Their current distribution reflects this long history of evolution in response to changing water conditions and the configuration of ocean basins. Their narrow tolerance of temperature and salinity, and their need for light and dissolved calcium carbonate, restrict corals to warm shallow seas. Here, continued reef growth depends on the balance between the deposition of limestone by corals and erosive processes that break up the reef.

REFERENCES

1. Muller-Parker, G.; D'Elia, C.F. Interactions between corals and their symbiotic algae. In *Interactions between Corals and Their Symbiotic Algae. Life and Death of Coral Reefs*; Birkeland, C., Ed.; Chapman & Hall: New York, 1997; 96–113.
2. Highsmith, R.C. Reproduction by fragmentation in corals. Mar. Ecol. Prog. Ser. **1982**, *7* (2), 207–226.
3. Szmant, A.M. Reproductive ecology of caribbean reef corals. Coral Reefs **1986**, *5* (1), 43–53.
4. Harrison, P.l.; Babcock, R.C.; Bull, G.D.; Oliver, J.K.; Wallace, C.C.; Willis, B.L. Mass spawning in tropical reef corals. Science **1984**, *223* (4641), 1186–1189.

5. Levitan, D.R.; Fogarty, N.D.; Jara, J.; Lotterhos, K.E.; Knowlton, N. Genetic, spatial, and temporal components of precise spawning synchrony in reef building corals of the *Montastraea annularis* species complex. Evol. **2011**, *65* (5), 1254–1270.

6. Adey, W.H. Review-coral reefs: Algal structured and mediated ecosystems in shallow, turbulent, alkaline waters. J. Phycol. **1998**, *34* (3), 393–406.

7. Veron, J.E.N. *Corals of the World*; Australian Institute of Marine Science: Townsville, Australia, 2000; 490.

8. Jones, C.G.; Lawton, J.H.; Shachak, M. Organisms as ecosystem engineers. Oikos **1994**, *69* (3), 373–386.

9. Gladfelter, E.H.; Monahan, R.K.; Gladfelter, W.B. Growth rates of five reef-building corals in the northeastern caribbean. Bull. Mar. Sci. **1978**, *28* (4), 728–734.

10. Glynn, P.W. Bioerosion and coral reef growth: A dynamic balance. In *Bioerosion and Coral Reef Growth: A Dynamic Balance. Life and Death of Coral Reefs*; Birkeland, C. Ed.; Chapman & Hall: New York, 1997; 68–95.

11. Neumann, A.C. Observations on coastal erosion in bermuda and measurements of the boring rate of the sponge, cliona lampa. Limnol. Oceanogr. **1966**, *11* (1), 92–108.

12. Hopley, D.; Smithers, S.G.; Parnell, A.K.E. *The Geomorphology of the Great Barrier Reef: Development, Diversity, and Change*; Cambridge University Press: Cambridge, 2007; 532.

13. Kleypas, J.A.; McManus, J.W.; Menez, L.A.B. Environmental limits to coral reef development: Where do we draw the line? Am. Zool. **1999**, *39* (1), 146–159.

14. Lesser, M.P.; Slattery, M.; Leichter, J.J. Ecology of mesophotic coral reefs. J. Exp. Mar. Biol. Ecol. **2009**, *375* (1–2), 1–8.

15. Darwin, C.R. The structure and distribution of coral reefs. Being the first part of the geology of the voyage of the *Beagle*, under the command of Capt. R.N. Fitzroy, During the years 1832 to 1836. Smith Elder and Company: London.

Coral Reef: Ecology and Conservation

Graham E. Forrester
Department of Natural Resources Science, University of Rhode Island, Kingston, Rhode Island, U.S.A.

Abstract

Modern coral reefs are found in shallow, nutrient-poor waters within 30° north or south of the equator. Although they underlay only 0.1% of the ocean's surface area, coral reefs provide food for millions of people, income from tourism, and other critical ecosystem services. Over the past two hundred years, coral reefs have changed dramatically from local human influences such as run-off, pollution, and overfishing. More recently, they have come under severe threat from climate changes, which have immediate effects on corals via rising temperatures and the increasing acidity of seawater.

INTRODUCTION

Humans have a long history of interaction with coral reefs. Many tropical coastal communities have depended on coral reefs as a source of food, materials, and shoreline protection for millennia. More recently, millions of visitors have been drawn to snorkel and dive amidst their beautiful underwater scenery. Scientific studies of coral reefs proliferated tremendously once scuba equipment became widely available in the 1960s and have focused on the spectacular biodiversity supported by coral reefs. This entry summarizes our knowledge of reef ecology and the effects humans are having on this ecosystem.

BIODIVERSITY OF CORAL REEFS

Coral reefs are often called "rainforests of the sea" because both ecosystems have a complex biogenic habitat, are highly productive, and support spectacularly high biological diversity.[1–3] Globally, coral reefs harbor more species than any other marine ecosystem. Estimates of the total number of reef-associated species range from 600,000 to more than 9 million species.[4] This biodiversity reaches its highest level in the Indo-Australian archipelago (the "coral triangle"), where there are over 500 species of stony (scleractinian) corals and 2600 fishes. Diversity gradually declines as one moves away from this area across the western Pacific and Indian Oceans to the eastern Pacific, which supports less than 50 species of coral and 350 fishes.[5–7] The diversity of most invertebrates, seaweeds, and other reef-dwelling organisms also follows this pattern. For example, there are around 50 species of reef-associated mantis shrimp (members of the order Stomatopoda) in the coral triangle, but less than 5 in the eastern Pacific.[8] The gradual closure of the Central American Seaway that separated the eastern Pacific and Caribbean Sea

roughly 3.4 million years ago led to the evolution of a distinct ecosystem of fauna and flora in the Caribbean reef. Consequently, the two areas now have no hermatypic (reef-building) corals in common[9] and the Caribbean today has lower diversity than any of the major Indo-Pacific regions, with about 62 species of scleractinian coral.[10] These patterns of diversity appear to result from the interplay of global, regional, and local factors, which combine to generate the extraordinary biological richness we see on coral reefs today.[3,11,12]

THE FUNCTIONING OF REEF ECOSYSTEMS

Coral reefs are highly productive ecosystems, particularly in comparison to the clear oligotrophic waters that surround them.[13] Globally, primary producers on coral reefs are estimated to fix roughly 700×10^{12} g carbon/year.[14] The unicellular algae in the genus *Symbiodinium* that reside within coral polyps are key primary producers on coral reefs and provide energy to corals in the form of reduced carbon. Photosynthesis by *Symbiodinium* accounts for a substantial portion of the carbon fixation that occurs on reefs. Up to half of the carbon fixed by *Symbiodinium* is exuded by the coral as mucus, some of which dissolves to provide food for bacteria in the water column and most of the remainder eventually reaches the sediments.[15] Relatively little of the carbon fixed with coral polyps is, however, transferred directly up the food chain[14] because, with few notable exceptions, such as the crown-of-thorns starfish (*Acanthaster planci*) plus some butterfly fishes and parrot fishes, relatively few species consume coral polyps directly. Instead, turfs of filamentous and fleshy seaweeds are more important as the base of the reef food chain because they use sunlight to fix carbon, but are heavily grazed by a variety of herbivorous fishes and invertebrates. Reef food chains are also subsidized by offshore production,

Coastal—
Evaporation

from a constant supply of zooplankton that are delivered to reefs by currents and consumed by resident planktivores.[16] While the total amount of carbon dioxide fixed by all photosynthetic organisms in the reef is high (between 2–3–6.0 g carbon/m^2/day), the net production of organic matter by the community is much lower (0.01–0.29 g carbon/m^2/day), which indicates that most of this material is retained within reefs. Because of this strong competition for food, and because the surrounding waters are nutrient poor, it is also assumed that nutrients are efficiently recycled with coral reef ecosystems, but the mechanisms for this are still uncertain.

THE VALUE OF REEFS TO HUMANS

Coral reef ecosystems provide a number of services to humans including the provision of raw materials, protection of coastlines, food from fish and shellfish, processing of nutrients, and venues for recreation and tourism.[17] Although coral reefs are estimated to underlay only about 0.1% of the global surface area of the ocean, they are of tremendous importance to the people who live along coasts fringed by reefs.[6] More than 100 countries have coral reefs along part of their coastline and tens of millions of people in these countries depend partly on food from coral reefs.[18] Globally, species harvested from coral reefs account for 9–12% of fisheries, but in some areas they assume a much larger role in the provision of protein. For example, on many Pacific Islands and in coastal areas of Southeast Asia, the catch from coral reefs makes up at least 25% of the total fish catch.[19] Other important goods derived from coral reefs include live fish for the aquarium trade; coral limestone that is mined to produce sand, lime, mortar, and cement for building, as well as seaweed harvested to make agar; and shells that are collected for jewelry.[19] Coral reefs are also valuable to local economies by attracting tourists and supporting their recreational activity. The average economic value of reef-related tourism is estimated to be US$184 per visitor, but estimates vary widely and appear to depend on factors such as the total number of visitors and the number of reefs accessible to divers.[20] The high biological diversity of coral reefs has also made them a priority target for extraction of chemical compounds from tissues to be screened for medical applications.[21]

Other important ecosystem services provided by coral reefs have less direct, but no less important, value for coastal communities. For example, coral reefs are important to the cycling of organic and inorganic nutrients in tropical seas, including the transfer of surplus nitrogen produced by benthic microbes to the open water ecosystem.[19] Coral reefs also dissipate wave energy and thus buffer shorelines from the potentially devastating effects of storms and tsunamis. By dissipating waves and currents, coral reefs also create sheltered conditions inshore that promote the development of mangrove forests and sea-grass beds. These habitats in turn provide their own benefits to people living nearby, and in combination with coral reefs constitute a set of interlinked and mutually supporting habitats whose combined value is greater than that of each in isolation.[17,19]

EFFECTS OF HUMANS ON REEFS

Despite their a esthetic, cultural, and economic values, coral reefs have been substantially impacted by human activity over the past few hundred years (Fig. 1). Based on expert opinion and global monitoring efforts, 19% of the area of modern coral reefs has been effectively lost, and a further 15% is in serious threat of loss within 10–20 years.[22] One leading coral reef biologist has argued that within as little as 30–40 years reefs may bear so little resemblance to those historically present that they may become the first

Fig. 1 Illustration of the loss of coral on Caribbean reefs in the past 30 years.[31] (**A**) a Caribbean reef with roughly 40% of its surface covered with live coral (typical in the 1970s). (**B**) A similar Caribbean reef with less than 4% of its surface covered with live coral (typical since 2000).
Source: Photographs by G. Forrester.

major ecosystem to be effectively eliminated by human activity.[23]

These declines have a number of causes, some of which act in combination. Although people have fished on coral reefs for millennia, historical analyses suggest that fishing began to cause major changes to coral reef ecosystems in the 19th and 20th centuries.[24] Many large carnivores and herbivores, such as monk seals,[25] sea turtles,[26] dugongs, and manatees, had already become rare by the beginning of the 20th century. Likewise, large fishes, such as sharks,[27] groupers, snappers, and bumphead wrasses,[28,29] appear to have been depleted by overfishing before underwater research on coral reefs began in earnest in the 1970s. The use of dynamite, drive nets, and poisons like cyanide to catch fish is particularly destructive and has a major impact on some pacific reefs because these methods destroy the underlying reef and many animals in addition to their intended targets.[30]

More intensive scientific scrutiny over the past 30 years has revealed progressive declines in the cover of corals[31] (Fig. 1) and dramatic "phase shifts" from coral-dominated to seaweed-dominated states.[32] Some of these declines are attributable to localized effects of pollution and sedimentation, which coincide with the growth of human populations along tropical coastlines. In other cases, the loss of herbivores through overfishing and disease has allowed seaweeds to flourish and made it more like that they will outcompete corals for space on the reef. In Jamaica, this phase-shift from coral- to seaweed dominance was triggered by hurricane damage, which caused widespread coral mortality. Whereas historically corals could recover from storm-related mortality, the absence of herbivores allowed seaweeds to take over the reef and prevent corals from repopulating.[32]

More recently, the most important recent threats to coral reefs have come from the emergence of new diseases and from the effects of global climate change, though there are signs that invasive species will have increasing impacts in the future. By 1965, scientists had identified just one coral disease. Since then, the number of reported diseases has grown exponentially and 18 different diseases had been reported by 2004.[33] Some new diseases have emerged through the expansion of pathogens to new areas, followed by the infection of new host species.[34] Similarly, the expansion of other species to areas outside their historical range has strong effects when those species become "invasive" and disrupt the resident community. Some coral reefs in Hawaii have been overgrown by invasive seaweeds, but perhaps the most notably damaging invasive species yet is the Pacific lionfish, which has spread rapidly around the Caribbean over the past 20 years. Lionfish prey on smaller reef fishes, and are shockingly efficient predators in their new environment.[35] Climate change has exerted strong effects on coral reefs because even modest increases in seawater temperature can induce coral bleaching, a generalized stress response

in which the zooxanthellae are lost from the coral polyp.[36] As the ocean warms, episodes of higher than normal seawater temperature are becoming increasingly common and have triggered massive bleaching events, notably the global bleaching event of 1998 and the Caribbean-wide event of 2005.[37,38] Although bleaching-related stress can kill corals directly, some major bleaching events have coincided with epidemics of coral disease suggesting a synergistic effect where temperature stress and bleaching make corals more susceptible to disease.[39] Another threat to corals from climate change comes from the acidification of seawater. Elevated carbon dioxide emissions into the atmosphere have caused an increased flux of carbon dioxide into the ocean. Through a series of chemical reactions, the pH of surface waters has declined along with the availability of carbonate ions. As a result, the ability of corals and calcifying macroalgae to deposit calcium carbonate skeletons and form reefs will be progressively compromised and by 2050, the rate of calcium carbonate deposition may be reduced by 10–50% compared to preindustrial rates.[40]

CONCLUSION

Arresting the declines in diversity and abundance of reef-dwelling species, and restoring some of the ecosystem functions they provide, will require a concerted effort.[41] Local impacts, such as those from overfishing, can be addressed through marine protected areas and improved management of human activities in coastal regions. The growing impacts of climate change, though, require a worldwide commitment to reduce carbon emissions.

REFERENCES

1. Stehli, F.G.; Wells, J.W. Diversity and age patterns in hermatypic corals. Syst. Biol. **1971**, *20* (2), 115–126.
2. Huston, M.A. Patterns of species diversity on coral reefs. Annu. Rev. Ecol Syst. **1985**, *16*, 149–177.
3. Bellwood, D.R.; Hughes, T.P. Regional-scale assembly rules and biodiversity of coral reefs. Science **2001**, *292* (5521), 1532–1535.
4. Reaka-Kudla, M.L.; Wilson, D.E; Wilson, E.O. *Biodiversity 2*; Joseph Henry Press: Washington, DC, 1997; 83–108 pp.
5. Veron, J.E.N. Corals in space and time: *The Biogeography and Evolution of the Scleractinia*. Ithaca: Cornell University Press, 1995: 321 pp.
6. Spalding, M.; Ravilious, C.; Green, E.P. *World Atlas of Coral Reefs*. Berkeley: University of California Press, 2001: 421 pp.
7. Zapata, F.A.; Robertson, D.R. How many species of shore fishes are there in the tropical eastern Pacific? J. Biogeogr. **2007**, *34* (1), 38–51.
8. Reaka, M.L.; Rodgers, P.J.; Kudla, A.U. Patterns of biodiversity and endemism on Indo-West pacific coral reefs. Proc. Nat. Acad. Sci. **2008**, *105* (Supplement 1), 11474–11481.

9. Veron, J.E.N. *Corals of the World*; Australian Institute of Marine Science. John Edward Norwood: Townsville, Australia, 2000.

10. Miloslavich, P.; Diaz, J.M.; Klein, E.; Alvarado, J.J.; Diaz, C.; Gobin, J.; Escobar-Briones, E.; Cruz-Motta, J.J.; Weil, E.; Cortes, J.; Bastidas, A.C.; Robertson, R.; Zapata, F.; Ortiz, M. Marine biodiversity in the Caribbean: Regional estimates and distribution patterns. Plos. One **2010**, *5* (8), e11916.

11. Pandolfi, J.M. Coral community dynamics at multiple scales. Coral Reefs **2002**, *21* (1), 13–23.

12. Connolly, S.R.; Hughes, T.P.; Bellwood, D.R.; Karlson, R.H. Community structure of corals and reef fishes at multiple scales. Science **2005**, *309* (5739), 1363–1365.

13. Crossland, C.J.; Hatcher, B.G.; Smith, S.V. Role of coral reefs in global ocean production. Coral reefs **1991**, *10* (2), 55–64.

14. Hatcher, B.G. Coral reef primary productivity: A beggar's banquet. Trends Ecol. Evol. **1988**, *3* (5), 106–111.

15. Wild, C.; Huettel, M.; Klueter, A.; Kremb, S.G.; Rasheed, M.Y.M.; Jorgensen, B.B. Coral mucus functions as an energy carrier and particle trap in the reef ecosystem. Nature **2004**, *428* (6978), 66–70.

16. Hamner, W.M.; Jones, M.S.; Carleton, J.H.; Hauri, I.R.; Williams, D.M. Zooplankton, planktivorous fish, and water currents on a windward reef face: Great barrier reef, Australia. Bull. Mar. Sci. **1998**, *42* (3), 459-479.

17. Barbier, E.B.; Hacker, S.D.; Kennedy, C.; Koch, E.W.; Stier, A.C.; Silliman, B.R. The value of estuarine and coastal ecosystem services. Ecol. Monogr. **2011**, *81* (2), 169–193.

18. Salvat, B. Coral reefs—a challenging ecosystem for human societies. Global Environ. Change-Hum. Policy Dimens. **1992**, *2* (1), 12–18.

19. Moberg, F.; Folke, C. Ecological goods and services of coral reef ecosystems. Ecol. Eco. **1999**, *29* (2), 215–233.

20. Brander, L.M.; Van Beukering, P.; Cesar, H.S.J. The recreational value of coral reefs: A meta-analysis. Ecol. Eco. **2007**, *63* (1), 209–218.

21. Quinn, R.J.; De Almeida Leone, P.; Guymer, G.; Hooper, J.N.A. Australian biodiversity via its plants and marine organisms. A high-throughput screening approach to drug discovery. Pure Appl. Chem. **2002**, *74* (519–526), 2002.

22. Wilkinson, C. *Status of Coral Reefs of the World*; Global Coral Reef Monitoring Network and Reef and Rainforest Research Center: Townsville, Australia, 2008; 296 pp.

23. Sale, P.F. *Our Dying Planet: An Ecologist's View of the Crisis We Face*; University of California Press: Berkeley, c2011; 339 pp.

24. Pandolfi, J.M; Bradbury, R.H.; Sala, E.; Hughes, T.P.; Bjorndal, K.A.; Cooke, R.G.; McArdle, D.; McClenachan, L.; Newman, M.J.H.; Paredes, G.; Warner, R.R.; Jackson, J.B.C. Global trajectories of the long-term decline of coral reef ecosystems. Science **2003**, *301* (5635), 955–958.

25. McClenachan, L.; Cooper. A.B. Extinction rate, historical population structure and ecological role of the Caribbean monk seal. Proc Royal Soc. B-Biol. Sci. **2008**, *275* (1641), 1351–1358.

26. McClenachan, L.; Jackson, J.B.C.; Newman, M.J.H. Conservation implications of historic sea turtle nesting beach loss. Frontiers Ecol Environ. **2006**, *4* (6), 290–296.

27. Ward-Paige, C.A.; Mora, C.; Lotze, H.K.; Pattengill-Semmens, C.; McClenachan, L.; Arias-Castro, E.; Myers, R.A. Large-scale absence of sharks on reefs in the greater Caribbean: A footprint of human pressures. Plos One **2010**, *5* (8), e11968.

28. Sadovy, Y.; Kulbicki, M.; Labrosse, P.; Letourneur, Y.; Lokani, P.; Donaldson, T.J. The humphead wrasse, *Cheilinus undulatus*: Synopsis of a threatened and poorly known giant coral reef fish. Rev. Fish Biol. Fisheries **2003**, *13* (3), 327–364.

29. McClenachan, L. Documenting loss of large trophy fish from the Florida Keys with historical photographs. Conservation Biol **2009**, *23* (3), 636–643.

30. McManus. J.W. Tropical marine fisheries and the future of coral reefs: A brief review with emphasis on Southeast Asia. Coral Reefs **1997**, *16* (S121–S127).

31. Gardner, T.A.; Cote, I.M.; Gill, J.A.; Grant, A.; Watkinson, A.R. Long-term region-wide declines in Caribbean corals. Science **2003**, *301* (5635), 958–960.

32. Hughes, T.P. Catastrophes, phase shifts, and larg-scale degradation of a Caribbean coral reef. Science **1994**, *265* (1547–1551), 5178.

33. Sutherland, K.P.; Porter, J.W.; Torres, C, Disease and immunity in Caribbean and Indo-Pacific zooxanthellate corals. Mar. Ecol.-Prog. Ser. **2004**, *266* (273–302).

34. Harvell, C.D.; Kim, K; Burkholder, J.M.; Colwell, R.R.; Epstein, P.R.; Grimes, D.J.; Hofmann, E.E.; Lipp, E.K.; Osterhaus, A.D.M.E.; Overstreet, R.M.; Porter, W; Smith, G.W.; Vasta, G.R. Emerging marine diseases - climate links and anthropogenic factors. Science **1999**, *285* (5433), 1505–1510.

35. Albins, M.A.; Hixon, M.A. Invasive Indo-Pacific lionfish *Pterois volitans* reduce recruitment of Atlantic coral-reef fishes. Mar. Ecol. Pro. Ser. **2008**, *367* (233–238).

36. Douglas, A.E. Coral bleaching - how and why? Mar. Pollut. Bull. **2003**, *46* (4), 385–392.

37. Wilkinson, C.R. *Status of Coral Reefs of the World*; Institute of Marine Science: Cape Ferguson, Australia: Australian, 2000; 411–430.

38. Wilkinson, C; Souter, D. *Status of Caribbean Coral Reefs after Bleaching and Hurricanes*. Global Coral Reef Monitoring Network, and Reef and Rainforest Research Centre: Townsville, Australia, 2005; 152.

39. Bruno, J.F.; Selig, E.R.; Casey, K.S.; Page, C.A.; Willis, B.L.; Harvell, C.D.; Sweatman, H; Melendy, A.M. Thermal stress and coral cover as drivers of coral disease outbreaks. Plos Biology **2007**, *5* (6), 1220–1227.

40. Kleypas, J.A.; Yates, K.K. Coral reefs and ocean acidification. Oceanography **2009**, *22* (4), 108–117.

41. Cote, I.M., Reynolds, J.D. *Coral Reef Conservation*; Cambridge University Press: Cambridge.

Drainage and Water Quality

James L. Baker
Department of Agricultural and Biosystems Engineering, Iowa State University, Ames, Iowa, U.S.A.

Abstract

Artificial subsurface drainage is required on many agricultural lands to remove excess precipitation and/or irrigation water in order to provide a suitable soil environment for plant growth and a soil surface capable of physically supporting necessary traffic, e.g., for tillage, planting, and harvesting. Although this drainage makes otherwise wet soils very productive, subsurface drainage alters the time and route by which excess water reaches surface waters and can carry nutrients, pesticides, bacteria, and suspended solids to surface waters and cause nonpoint source pollution problems. The net water quality impact of subsurface drainage considered here is determined by comparison to the same cropping system not having subsurface drainage (not on the fact that the existence of adequate drainage will affect land use). The degree of pollutant transport with surface runoff and subsurface drainage is determined by the product of the volumes of water and the pollutant concentrations in the water. These are both influenced by environmental conditions, pollutant properties, and management factors, and their interactions with subsurface drainage are discussed.

ENVIRONMENTAL CONDITIONS

Soils/Hydrology

Whether water added to the soil surface as precipitation or irrigation infiltrates or becomes surface runoff is critical to water quality. Topography/slope, soil moisture content, texture, and soil structure, including the existence of preferential flow paths or "macropores," can affect both the rate and route of water infiltration (as shown in Fig. 1). While different drain spacings are used to provide a desired "drainage coefficient" (e.g., 0.5 in./day) based on the internal drainage characteristics of the subsoils, the conditions of the surface soil will determine what percentage of applied water (rain or irrigation) will infiltrate. In general, the existence of subsurface drainage increases the volume of infiltration and thus decreases the volume of surface runoff (and pollutant loss with that runoff), increases shallow percolation, and lowers water tables. This effect will be more pronounced for "lighter" soils that have higher infiltration rates and lower water holding capacities. These changes affect the final quality of cropland drainage because of differences in time and type of soil–water–chemical interactions. Surface runoff allows water to come in contact only with the surface soil and materials such as crop residue and surface-applied fertilizers, manures, and pesticides present on it for short periods of time. On the other hand, water that infiltrates and percolates through the soil comes in intimate contact with all the soils in the profile to at least the depth of the tile drain, and generally, for much longer periods of time. The soil residence time is shorter for that portion of infiltrated water intercepted by the tile drains than for water that must flow through the underground strata to appear as base flow.

Climate/Precipitation

The timing, amount, and intensity of water inputs, including precipitation and irrigation, relative to evapotranspiration (ET) determine the timing and amount of excess water that will leave agricultural land as surface runoff and subsurface drainage. Inputs at low intensities generally will totally infiltrate, but as intensities increase past the rate of infiltration, surface runoff begins. As just discussed, because decreased antecedent soil moisture contents, which result from the existence of artificial subsurface drainage, increase infiltration rates, the volume of surface runoff generally is decreased when subsurface drainage exists. This effect will be more pronounced for areas where input intensities often exceed the rates of infiltration.

For a given crop and climatic region, the amount of ET is relatively constant, so the volume of subsurface drainage is very much dependent on the amount of input water above that value. As an example, in a 4-year Iowa tile drainage experiment,[1] during a low rainfall year, only a trace of subsurface flow occurred; in a wet year (116 cm of precipitation vs. the average of 80 cm), there was 29 cm of flow; and the overall average flow for the four years was 15 cm.

Encyclopedia of Natural Resources DOI: 10.1081/E-ENRW-120010233

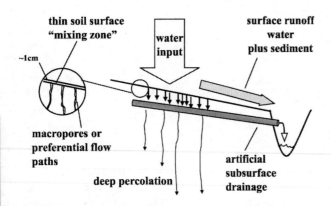

Fig. 1 Schematic of transport processes.

POLLUTANT PROPERTIES

Persistence

Pollutants that have a limited existence in agricultural lands because of plant uptake/chemical transformation (i.e., nutrients), degradation (i.e., pesticides), and die-off (i.e., micro-organisms) have different potentials for off-site transport with water based on their persistence. Because surface runoff is much more of an immediate process than subsurface drainage, concentrations of nonpersistent pollutants in subsurface drainage are usually lower than in surface runoff. This effect will be more pronounced, the lesser the persistence is. However, the "route" of infiltration, where in some cases the drainage water moves quickly through the soil profile because of macropores (see Fig. 1), plays a role and can somewhat negate the expected dissipation effect.

Adsorption/Filtration

Based on their chemical properties and/or their physical size, some pollutants transported down into and through the soil with subsurface drainage are removed from the flow stream by the soil. In particular, pollutants that are positively charged (e.g., ammonium-nitrogen (NH_4–N)) and larger, less soluble, organic compounds (e.g., some pesticides) can be attenuated by adsorption to soil clay and organic matter. Inorganic P ions can be removed by complexation or precipitation with soil cations. Microorganisms and sediment can be filtered out by small soil pores. Thus, in general, with the exception of soluble salts of nitrate–nitrogen (NO_3–N), sulfate (SO_4), and chloride (Cl) anions, pollutant concentrations in subsurface drainage are lower than in surface runoff. This effect will be more pronounced, the greater the interaction between the pollutant and soil is.[2]

For sediment itself, because soil erosion is dependent on the erosive ability and transport capacity of surface runoff, reducing the runoff flow rate and volume with subsurface drainage reduces sediment loss. For example, in a 6-year study in the lower Mississippi Valley, surface runoff volumes from plots on a clay loam soil with subsurface drainage were 34% less than for plots without subsurface drainage, and the corresponding soil loss of 3500 kg/ha/yr represented a decrease of 30%. For the plots with subsurface drainage, both sediment concentrations and losses in subsurface drainage were about one-tenth those in surface runoff.[3]

For nitrogen (N), loss from poorly drained soils is usually much less than that from soils with improved drainage systems.[4] While significant N can be transported with sediment (often sediment has at least 1000 ppm N), land needing subsurface drainage is usually not highly susceptible to erosion, and N loss is dominated by soluble inorganic-N loss. In a 3-year tile-drained watershed study in northeast Iowa,[5] NO_3–N losses in solution represented over 85% of the total N losses, including NH_4–N, organic-N in solution, and N associated with sediment. In a 5-year study in east-central Iowa,[6,7] where nutrient concentrations in both surface runoff and subsurface drainage from cropland were monitored, NO_3–N concentrations in subsurface drainage averaged about 12 mg/L and 2–3 times those in surface runoff. While NH_4–N concentrations were usually 2–10 times higher in surface runoff, on an absolute scale, the concentrations were overall much lower than those for NO_3–N and constituted only a fraction of the total N loss.

For phosphorus (P), transport is primarily in surface runoff with sediment (often sediment has at least 500 ppm P) and dissolved in surface runoff. For conventionally tilled cropland, about 75–90% of P transported in surface runoff is with sediment. In areas where soil erosion is minimal, soluble P in surface runoff water can dominate transport. The soluble P in surface runoff (and subsurface drainage) is regulated by adsorption/desorption characteristics of soil. Therefore, with P in surface soils generally much higher than in subsoils, subsurface drainage usually has much lower soluble P concentrations than in surface runoff.

For pesticides, because of their adsorption characteristics, like with P, concentrations in surface runoff water are usually much greater than in subsurface drainage. However, this effect is even more dramatic for pesticides because unlike P, pesticides have a limited persistence, which decreases their potential for movement with subsurface flows with long travel times. In a series of studies on herbicides in surface and subsurface drainage,[8–10] atrazine concentrations in May (the period of application) were 75 µg/L in surface runoff for plots with surface drainage only, and were 51 µg/L and 1 µg/L in surface runoff and subsurface drainage, respectively, for plots that also had subsurface drainage. Not only were there much lower atrazine concentrations in subsurface drainage water, but the presence of the subsurface drains delayed and reduced surface runoff such that concentrations in surface runoff from the plots with subsurface drainage were reduced about one-third.

For bacteria, a review by Crane et al.[11] showed that fecal coliform counts in surface runoff from manured lands often were greater than 10,000/100 mL. In a comparison of surface runoff and subsurface drainage, Culley and

Phillips[12] found similar counts (>10,000/100 mL) for fecal coliform in surface runoff from both manured and fertilized plots, but with much, much lower counts (<5/100 mL) in subsurface drainage.

MANAGEMENT PRACTICES

Tillage Systems

The hydrologic interactions between tillage systems and subsurface drainage that affects water quality involves how tillage affects the timing, route, and volume of infiltration (and hence the relative volumes of subsurface drainage and surface runoff). In a review of the hydrologic effects of conservation tillage, Baker[13] noted that changing the relative volumes of subsurface drainage and surface runoff also can affect chemical concentrations in those carriers. In general, increasing infiltration increases the time for beginning of surface runoff, which in turn reduces the concentrations of chemicals at the soil surface (shown as a thin mixing zone in Fig. 1) and therefore in surface runoff. However, the effect of conservation tillage on infiltration is time dependent. For the first storm after any tillage there is usually less runoff from the tilled soil, although on an annual basis, conservation (or less) tillage often results in lower total surface runoff volumes.

Cropping

As with tillage, the hydrologic interactions between cropping systems and subsurface drainage that affects water quality involves how cropping affects the timing, route, and volume of infiltration (and hence the relative volumes of subsurface drainage and surface runoff). A major difference between perennial crops such as forages, and row-crops such as corn and soybeans, is the volume and timing of ET demands. In general, with higher, more consistent ET, perennial crops would have lower total drainage volumes. A bigger effect of cropping would be the effect of needed chemical applications and their potential losses. For example, the large amounts of N needed (added, recycled, and/or fixed) in a continuous corn or corn–soybean rotation means there is usually high NO_3–N concentrations in the soil profile, and hence in subsurface drainage when it occurs. However, for grasses and alfalfa. NO_3–N concentrations in subsurface drainage are much lower.[14,15]

Controlled Drainage

In areas where subsurface drainage exists, controlling the timing of outflows has been suggested as one method to reduce chemical losses. This controlled drainage could reduce losses by reducing both subsurface volumes and chemical concentrations. The potential for reduced concentrations is probably the greatest for NO_3–N, where the process of denitrification would reduce NO_3 to N gases in the soil profile where high water tables and the presence of organic matter drive the system anaerobic. The results summarized from 125 site-years of data from North Carolina[16] showed that controlled drainage reduced subsurface drainage volumes an average of 30% compared to uncontrolled drainage systems. Reductions in N and P lost with subsurface drainage were 45% and 35%, respectively. While almost all the reduction of P loss was due to decrease in drainage volume; for N, reductions in NO_3–N concentrations also contributed to the reduction.

CONCLUSION

The total effect of surface drainage on surface water resources receiving drainage from agricultural lands involves the relative volumes of surface runoff and subsurface drainage, and the relative concentrations of sediment, nutrients, pesticides, and bacteria. In general, the existence of subsurface drainage increases infiltration rates which delays and reduces the volume of surface runoff. For pollutants lost mostly with surface runoff, which include sediment, NH_4–N, P, pesticides, and bacteria, not only is the volume of the carrier reduced, but also the concentrations. This is because delayed runoff and more water moving through the surface-mixing zone reduce the amounts of contaminants at the soil surface available to interact with added water and overland flow. Thus the only real water quality negative to subsurface drainage is the increased volume of water moving thorough the soil profile carrying the soluble unadsorbed NO_3–N anion. The use of improved in-field N management in the way of rate, method, and timing of N applications has some potential to reduce this problem[17] However, other practices such as controlled drainage or construction/reconstruction of wetlands may be needed to provide some NO_3–N reduction treatment. Alternatively, reducing the amount of row-crops grown on subsurface-drained lands could have a large impact although the current economics of doing that would be quite negative.

REFERENCES

1. Baker, J.L.; Campbell, K.L.; Johnson, H.P.; Hanway, J.J. Nitrate, phosphorus and sulfate in subsurface drainage water. J. Environ. Qual. **1975**, *4*, 406–412.
2. Gilliam, J.W.; Baker, J.L.; Reddy, K.R. Water quality effects of drainage in humid regions. *Drainage Monograph*; Am. Soc. Agron.: Madison, WI, 1999.
3. Bengston, R.L.; Carter, C.E.; Morris, H.F.; Bartkiewicz, S.A. The influence of subsurface drainage practices on nitrogen and phosphorus losses in a warm, humid climate. Trans. ASAE **1988**, *31*, 729–733.
4. Gambrell, R.P.; Gilliam, J.W.; Weed, S.B. Nitrogen losses from soils of the North Carolina coastal plain. J. Environ. Qual. **1975**, *4*, 317–323.

5. Baker, J.L.; Melvin, S.W.; Agua, M.M.; Rodecap, J. *Collection of Water Quality Data for Modeling the Upper Maquoketa River Watershed*; Paper No. 99-2222; ASAE: St. Joseph, MI, 1999.

6. Johnson, H.P.; Baker, J.L. *Field-to-Stream Transport of Agricultural Chemicals and Sediment in an Iowa Watershed: Part I. Data Base for Model Testing (1976–1978)*; Rep. No. EPA-600/S3-82-032; USEPA Environ. Res. Lab.: Athens, GA, 1982.

7. Johnson, H.P.; Baker, J.L. *Field-to-Stream Transport of Agricultural Chemicals and Sediment in an Iowa Watershed: Part II. Data Base for Model Testing (1979–1980)*; Rep. No. EPA-600/S3-84-055; USEPA Environ. Res. Lab.: Athens, GA, 1984.

8. Southwick, L.M.; Willis, G.H.; Bengston, R.L.; Lormand, T.J. Effect of subsurface drainage on runoff losses of atrazine and metolachlor in Southern Louisiana. Bull. Environ. Contam. Toxicol. **1990**, *45*, 113–119.

9. Southwick, L.M.; Willis, G.A.; Bengston, R.L.; Lormand, T.J. Atrazine and metolacholor in subsurface drain water in Louisiana. J. Irrig. Drain. Eng. **1990**, *116*, 16–23.

10. Bengston, R.L.; Southwick, J.M.; Willis, G.H.; Carter, C.E. The influence of subsurface drainage practices on herbicide losses. Trans. ASAE **1990**, *33*, 415–418.

11. Crane, S.R.; Moore, J.A.; Grismer, M.E.; Miner, J.R. Bacterial pollution from agricultural source: A review. Trans. ASAE **1983**, *26*, 858–866, 872.

12. Culley, J.L.B.; Phillips, P.A. Bacteriological quality of surface and subsurface runoff from manured sandy clay loam soil. J. Environ. Qual. **1982**, *11*, 155–158.

13. Baker, J.L. Hydrologic effects of conservation tillage and their importance relative to water quality. In *Effects of Conservation Tillage and Groundwater Quality*; Logan, T.J., Davidson, J.M., Baker, J.L., Overcash, M.R., Eds.; Lewis Publishers, Inc.: Chelsea, MI, 1987; Chap. 6.

14. Baker, J.L.; Melvin, S.W. Chemical management, status, and findings. *Agricultural Drainage Well Research and Demonstration Project—Annual Rep. and Project Summary*; Iowa Dep. of Agric. and Land Stewardship and Iowa State University: Ames, 1994; 27–60.

15. Weed, D.A.J.; Kanwar, R.S. Nitrate and water present in and flowing from root-zone soil. J. Environ. Qual. **1996**, *25*, 709–719.

16. Evans, R.O.; Skaggs, R.W.; Gilliam, J.W. Management practice effects on water quality. Proc. Irrig. Drain. 1990 Natl. Conf. ASCE Irrig. Div., Durango, CO, 1990; 182–191.

17. Baker, J.L. Limitations of improved nitrogen management to reduce nitrate leaching and increase use efficiency. Optimizing nitrogen management in food and energy production and environmental protection: proceedings of the 2nd international nitrogen conference on science and policy. Sci. World **2001**, *1*(S2), 10–16.

Endorheic Lake Dynamics: Remote Sensing

Yongwei Sheng
Department of Geography, University of California, Los Angeles, Los Angeles, California, U.S.A.

Abstract

Extensively distributed in arid and semiarid regions, endorheic lakes are sensitive and vulnerable to climate change and human activities. They are crucial water resources to humans and animals and carry important mineral resources. Their dynamics serve as a proxy indicator of climate change. There is a pressing need to assess endorheic lake dynamics under global warming and anthropogenic pressures. After decades of rapid development, remote sensing provides effective and efficient tools for lake dynamics monitoring at regional scales. This entry first introduces endorheic basins and endorheic lakes and their values to humans as well as their relationship to climate change, outlines the global distribution of major endorheic basins and lakes, and finally discusses remote sensing of lake area changes using optical sensors, lake level change monitoring using satellite radar and lidar altimetry, and paleolake change recovery using remote sensing and topographic data. The dramatically changing endorheic lakes in the Great Siling Co basin on the Tibetan Plateau are monitored using various remote sensing techniques as a case study.

INTRODUCTION

Atmospheric precipitation over much of the Earth's land eventually returns to the ocean through rivers, whereas precipitation falling into some interesting areas does not flow back into the ocean. These areas are endorheic basins. An endorheic basin, also known as a closed basin, is a drainage basin that retains water within it, prohibiting water outflow to external rivers or the ocean.[1] The retained water accumulates in the basin bottom, forming endorheic lakes. Endorheic lakes are water bodies whose water does not flow into the ocean. Endorheic lakes may include both through-flow lakes and terminal lakes in an endorheic basin. The water in through-flow lakes eventually flows to its final destination: terminal lakes. As endorheic lakes do not have outlets for the water to flow out of their basins, the only pathways for water to leave the drainage system are evaporation and seepage, either evaporating into the air or seeping into the ground. The water falling as precipitation in the basin dissolves and carries minerals during the course of transporting to its destinations—endorheic lakes, and the water in the lakes evaporates, leaving a high concentration of minerals and other inflow erosion products in the lakes. This process over time can render endorheic lakes rather saline and sensitive to environmental pollutions. Endorheic lakes include some of the world's large lakes, such as the largest, the Caspian Sea and the Aral Sea in Asia, the Lake Chad in Africa, the Great Salt Lake in North America, and Lake Eyre in Australia. They largely are saline or salt lakes located in dry climates.

Endorheic lakes are important natural resources. First of all, they are crucial water resources in the arid environment that they belong to, feeding people and animals and irrigating agriculture. They are also valuable industrial mineral resources. The minerals dissolved in the water are concentrated during the natural evaporation process and deposit in lakes. Zabuye Caka, for example, is a hypersaline lake on the western Tibetan Plateau with a salinity concentration of ~300 g/L as measured in 2011.[2] Lithium carbonate was discovered in this lake in the 1980s,[3] and it now is China's major source of lithium, which is a main component of lithium batteries extensively used in electronic devices.

Closed to major rivers and oceans, these lakes are particularly sensitive and vulnerable to climate change and human activities. Endorheic lakes are among the most sensitive recorders of past precipitation–evaporation balance due to changes in lake size, water level, and chemistry.[4] The size and water level of an endorheic lake are largely determined by the balance between precipitation in its basin and evaporation from the lake. Most endorheic lakes are rather sensitive to precipitation. Under global warming, precipitation, however, will likely become more unstable and variable in the future,[5] which will reinforce the vulnerability of endorheic lakes in arid regions. Water level, area, and volume of endorheic lakes fluctuate in response to changes in the evaporation and precipitation rates within their basins. Therefore, monitoring of these lake changes is not only important for hydrological and economic purposes but can also provide a climate record, a proxy indicator of climate change.[6,7]

Endorheic regions have been home to human cultures for thousands of years, and the population continues to increase in the areas.[8] However, population growth demands surface

Encyclopedia of Natural Resources DOI: 10.1081/E-ENRW-120049159

and ground water for irrigation. As a consequence, many endorheic lakes are under great pressure from human water demands, leading to their shrinkage and even desiccation. Rapid shrinkage of endorheic lakes may result in severe environmental problems. The environmental disaster caused by the Aral Sea's shrinkage presents a classic example. The Aral Sea, formerly the world's fourth largest lake in 1960 with an area of 68,000 km^2 only behind the Caspian Sea, Lake Superior, and Lake Victoria, has been dramatically shrinking since the 1960s due to the diversion of the Sya Darya and Amu Darya rivers for agriculture irrigation. Its lake level dropped nearly 13 m, its area decreased by 40%, and its salinity rose from 10 to 27 g/L between 1960 and 1987, which is the most rapid and pronounced recession in history, turning the Aral Sea into a residual brine lake.[9] The current Aral Sea comprises two separate bodies of water: the small North Aral Sea and the relatively large South Aral Sea, with the total area under 20% of its original size. The Aral is estimated to contain ~10 gigaton (GT) salts. An estimated 43 million tons of salts are annually transported by winds from the Aral's dried lakebed into adjacent areas and deposited as aerosols by rain and dew over 150,000 km^2. The wind blows salts and dust to travel a thousand kilometers to the Black Sea coast and even to the Russian Arctic shore. The biological diversity and productivity has steeply declined. Twenty of 24 native fish species disappeared, and the commercial catch is out of business.[9] In addition, desert animals have suffered from the greatly increased mineral content in their drinking water, and native plant communities have degraded, causing desertification in this region. The land around the Aral Sea is heavily polluted, and the people living in the area are suffering from a lack of fresh water and from health problems, including high rates of cancer and respiratory illnesses.

Many other endorheic lakes like the Aral Sea are under stress due to climate change and human activities. In the era of global change, it is of crucial importance to inventory and monitor endorheic lakes in the context of global warming. However, the abundance, distribution, and dynamics of these lakes are not well understood. There is a pressing need to assess endorheic lake dynamics under global warming and anthropogenic pressures and their potential impacts on water resources.

GLOBAL DISTRIBUTION OF ENDORHEIC LAKES

Endorheic lakes and basins occur in arid to semihumid climates but are commonly found in semiarid regions. Figure 1 shows the distribution of major endorheic basins (darker regions). These basins were extracted from global watershed data sets as lake basins closed to major rivers and oceans. Approximately 20% of the Earth's land drains to endorheic lakes. Continents vary in their concentration of endorheic regions due to conditions of geography and climate. In general, endorheic basins are extensively distributed in western and central Asia, northern Africa, central Australia, the Altiplano Plateau of South America, and the Great Basin of North America. Western and Central Asia are home to the world's largest group of endorheic lakes, including the largest, the Caspian Sea, the Dead Sea, and the Aral Sea. The Tibetan Plateau in the southern part of the region also contains a large number of endorheic lakes at an altitude of ~4500 m. Africa hosts three major groups of

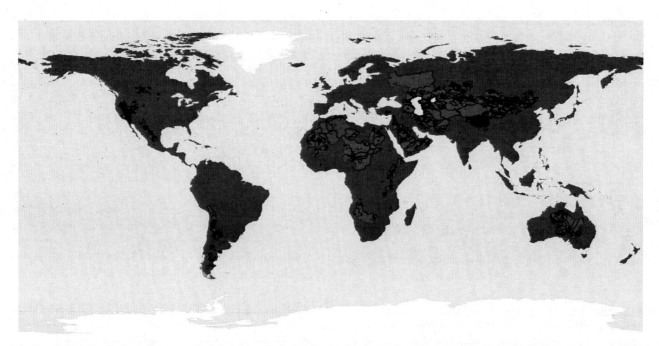

Fig. 1   Global distribution of major endorheic basins. Endorheic basins (darker regions) are extensively distributed in western and central Asia, northern Africa, central Australia, the Great Basin of North America, and the Altiplano Plateau of South America.

endorheic lakes in the Sahara Desert, the East African Rift, and the Kalahari Desert. Lake Chad in the southern Sahara Desert continues to shrink due to the decreased precipitation and increased irrigation pressure. The Great Basin in the United States contains a large region of endorheic basins, in which the Great Salt Lake is the largest endorheic lake in North America. The Valley of Mexico in Central America also contains a number of dried lakes. The Altiplano basin in South America includes a number of closed lakes such as Lake Poopó. Central Australia has many endorheic drainages, including the highly variant Lake Eyre.

REMOTE SENSING MONITORING OF ENDORHEIC LAKES

Lake area, water level, and salinity are measurable indicators of lake dynamics. However, field-based investigation of endorheic lakes is challenging due to the low accessibility and the harsh environment. These lakes are seldom observed, let alone monitored for dynamics. Spaceborne remote sensing provides a feasible tool for multitemporal lake mapping and monitoring at local, regional, and global spatial scales and monthly-to-decadal temporal scales. Harris[10] and Birkett[11] discussed the need for simultaneous measure of level and area of endorheic lakes and promoted remote sensing of areas and levels for large endorheic lakes from satellite images and radar altimeters.

Many remote sensing satellites that are currently in space can be used in mapping and monitoring lake dynamics. At high–medium resolutions, Landsat, *Satellite Pour l'Observation de la Terre* (SPOT), and Terra/Advanced Space-borne Thermal Emission and Reflection Radiometer (ASTER) images are widely used in detecting subtle lake area changes.[12–14] Coarse spatial resolution satellite sensors normally are able to monitor large lakes at much higher temporal resolutions to catch even seasonal variability. National Oceanographic and Atmospheric Administration/ Advanced Very High Resolution Radiometer (NOAA/ AVHRR; ~1 km resolution) and Terra/Aqua Moderate-Resolution Imaging Spectroradiometer (MODIS; 0.25 to 1 km resolutions) are used to monitor lake area dynamics as frequently as every two days, even to catch changes within closed basins as a result of possible changes in regional rainfall patterns.[15,16]

Satellite radar and laser altimetry technology provides a precise and efficient tool for measuring and monitoring water level of large inland lakes at 0.5 dm accuracy.[11,17] The widely used radar altimetry missions include Topex/ Poseidon (1992–2005), Jason-1 (2001 to present), Jason-2 (2008 to present), and Envisat (2002 to present), providing measurements since the early 1990s.[18] The only available spaceborne lidar mission Ice, Cloud, and land Elevation Satellite (ICESat, launched in January 2003) collected elevation data between 2003 and 2009 using the laser altimetry sensor Geoscience Laser Altimeter System (GLAS). Though the primary goal of the ICESat mission is to measure ice sheet mass balance and cloud heights, ICESat data have been found successful in monitoring water levels for oceans, lakes, rivers, and wetlands.[19] The combined available satellite altimetry missions have accumulated about a 20-year archive of lake level measurements.

Two water salinity satellites have been recently in operation in space, including Soil Moisture and Ocean Salinity (SMOS) launched in November 2009 by European Space Agency (ESA)[20] and Aquarius launched in June 2011 by the National Aeronautics and Space Administration (NASA).[21] As ocean salinity is crucial to the understanding of the role of the ocean in climate through the global water cycle, both missions were designed for sea surface salinity measurements; however, they are not able to resolve endorheic lakes at their current resolutions of 100–200 km.

The subsequent sections focus on the Great Siling Co (Co: lake in Tibet) basin as a specific case study to illustrate how remote sensing can be used to monitor dynamics for endorheic lakes. At a total area of 58,570 km², the Great Siling Co basin is the largest endorheic basin on the central Tibetan Plateau. The great basin consists of the primary Siling Co watershed (34,638 km²), the Gyaring Co–Urru Co watershed (15,838 km²), the Ringco Ogma Co watershed (3,364 km²), the Pangkog Co watershed (2,658 km²), and the Namka Co (2,072 km²) (Fig. 2). This region is cold all year round with an annual mean temperature of –2°C. The annual evaporation reaches 2,167 mm while the mean annual precipitation is only ~290 mm, 90% of which falls in the months of June to September.[22] Centered at (31.8°N, 89°E), Siling Co is the primary saline lake in the Great Siling Co basin with a salinity of ~18.5 g/L. The lake has been expanding dramatically over the past decades. The lake was only 1,640 km² in 1976 but expanded to its current size at 2,335 km², and its water level reached 4,544.5 m in 2009. In addition to atmospheric precipitation, it is fed by glacier meltwater from Mount Geladaindong in the far northeast via the Zagya Zangbo River entering the lake from the north and from Mount Gyagang in the south via the Zagen Zangbo River entering the lake from the west through Gyaring Co and Urru Co.

Lake Area Change

The Landsat program is particularly suitable and irreplaceable for detailed lake inventory and lake dynamics monitoring at regional and even global scales not only because of the recent release of the entire Landsat data archive for free public access but also due to the fact that the Landsat program provides the longest high-resolution image records of the Earth.[13] Images acquired by the early Multispectral Scanner (MSS; provided in 60 m resolution) sensors onboard Landsat 1–5, the later Thematic Mapper (TM, 30 m) onboard Landsat 4–5, and the recent Enhanced Thematic Mapper plus (ETM+, 30 m) onboard Landsat 7 have

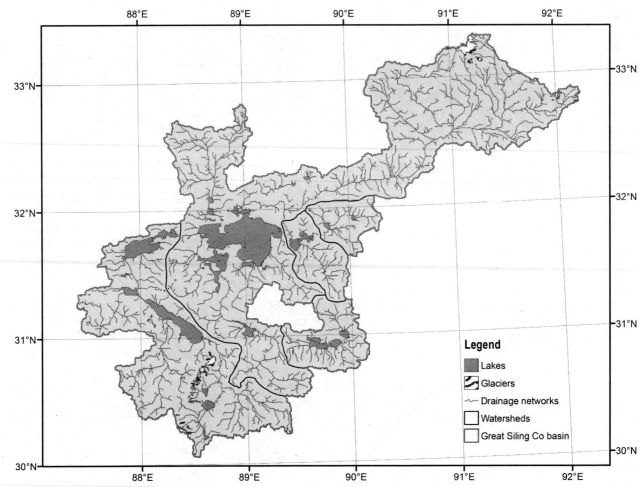

Fig. 2 Great Siling Co basin and lakes. The basin consists of the Siling Co watershed, the Gyaring Co–Urru Co watershed, the Ringco Ogma Co watershed, the Pangkog Co watershed, and the Namka Co.

continuously monitored the Earth in the past ~40 years repeatedly every ~16 days. With the recent Landsat 8 launched in February 2013 carrying the Operational Land Imager (OLI; also in 30 m resolution) sensor, the Landsat mission is expected to continue into the next decade.

Decadal lake area changes in the Great Siling Co basin were monitored using 1976 MSS, 1990 TM, 2000 ETM+, and 2009 TM images acquired between October and December when seasonal variability is minimal. Theoretically, lakes and other surface water bodies are readily identified in Landsat images owing to their very low reflectance in near infrared channels of Landsat sensors. However, the variant mineral content in endorheic lakes complicates the lake water spectral responses and imposes challenges to remote sensing lake mapping. The normalized difference water index (NDWI), derived using green and near infrared bands, can enhance water signature and depress disturbing factors and has been widely used in water body delineation from satellite images.[23] NDWI is less sensitive to mineral content in lake water and demonstrates better capability in delineating water bodies from land than

individual bands. However, lakes cannot be properly delineated in the image using a single threshold to segment an NDWI image due to the various conditions of the lakes.

Lakes in the Great Siling Co basin were mapped and monitored using an automated hierarchical lake mapping approach. The method simulates a human operator's procedures in delineating lakes and segments the image at both global and local levels.[13] In the global level of segmentation, lakes are first located by segmenting the entire NDWI image using a conservative threshold. Then, each identified lake is treated as an object and is further refined only in the area surrounding the lake (i.e., a buffer zone) for a fine adjustment at the local level. This hierarchical method can effectively map from Landsat imagery lakes under various complex water conditions in the Great Siling Co basin.

The monitoring results in Fig. 3 show that Siling Co has experienced significant change over the past ~35 years. It used to be the second largest lake on the central plateau (just after Num Co) in 1976 with an area of ~1,641 km², but it expanded steadily afterward (1990 and 2000) and became the largest

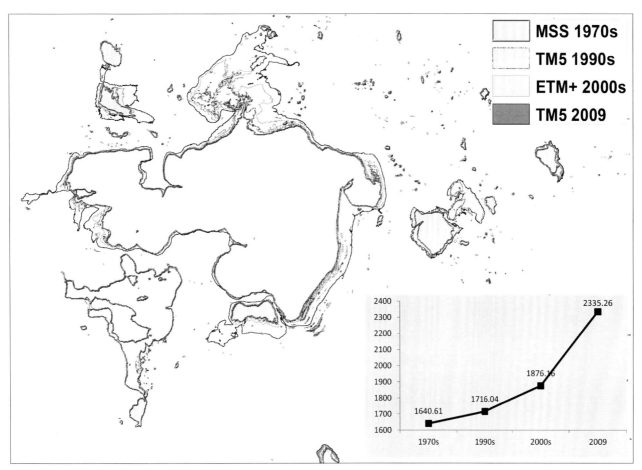

Fig. 3 Decadal area changes of Siling Co as monitored by remote sensing. Since 1976, Siling Co has expanded steadily (1990 and 2000) and became the largest lake in 2009 with an area of 2,335 km², a 42% expansion from the 1970s.

lake in 2009 with an area of 2,335 km², a 42% expansion from the 1970s. Moreover, the lake has been growing at an accelerating rate: 0.33%/year between 1976 and 1990, 0.93%/year in the 1990s, and 2.7%/year in the recent decade. The increased precipitation and the accelerated glacier melting in response to the rising temperature are responsible for the observed lake expansions.[13]

Lake Level Change

ICESat/GLAS collects elevation data with a high precision (~3 cm) every ~170 m along the spacecraft's ground track with small footprints (50–90 m), compared to Jason's ~5 km footprints.[19] As such, GLAS is able to measure more precisely water levels for smaller inland lakes in endorheic regions than the satellite radar altimeters.

The water level changes of Siling Co were monitored using ICESat/GLAS global land surface altimetry product, GLA14 delivered by National Snow and Ice Data Center (NSIDC), covering the operation period from February 2003 to October 2009. Because Siling Co was quite dynamic during this period, its varying extents were mapped using available Landsat images to screen GLAS

laser points in water level calculation. As such, a high-resolution lake level series was established for Siling Co. Figure 4 shows the lake level changes of Siling Co derived from GLAS data and the corresponding lake area changes (black dots) extracted from Landsat. The lake area measured by Landsat in 2002, 2005, 2006, 2007, and 2009 synchronizes rather well with the GLAS-detected lake level changes. Over the monitoring period, the lake level rose steadily with three rapid ~1 m jumps in the summers of 2003, 2004, and 2005. Siling Co expanded by ~15.2% from 2009 km² in December 2002 to 2314 km² in July 2009 with a ~4.5 m level rise. Combining the area and level changes, the total volume change of Siling Co between 2003 and 2009 is estimated at 9.7 billion m³.

A major issue with the current lake level remote sensing is that the tracks of satellite radar and laser altimeters are rather sparse. Envisat have a 35-day temporal resolution and 70 km intertrack spacing at the equator, while Topex/Poseidon, Jason-1, and Jason-2 have a 10-day orbital cycle and 350 km equatorial intertrack spacing.[18] ICESat has a 91-day repeat orbit with a 33-day subcycle. This 33-day scenario has a repeat track spacing of ~80 km at the equator and ~20 km at 75° latitude.[19] As a result, the current lake level data sets

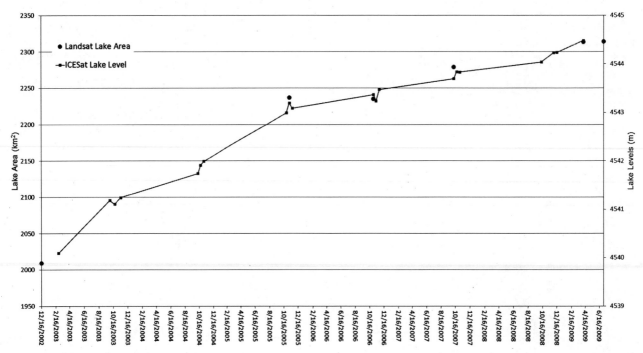

Fig. 4 Level change of Siling Co detected by Ice, Cloud, and land Elevation Satellite (ICESat)/Geoscience Laser Altimeter System (GLAS). The lake level rose steadily up to 4.5 m from 2003 to 2009, with three rapid ~1 m jumps in the summers of 2003, 2004, and 2005. The level changes synchronize rather well with the lake area change (black dots) detected by Landsat images.

are only able to measure hundreds of large lakes (such as Siling Co) occurring under the satellite tracks. The future Surface Water Ocean Topography mission (SWOT) planned to launch in 2020 by NASA will simultaneously monitor both lake area and level changes in nearly global coverage.[25] Flying two radar antennas, the SWOT satellite will measure surface elevations using radar interferometry along a 120 km wide swath. The radar sensors can penetrate both clouds and darkness and provide all-weather lake observations at high resolutions, overcoming the cloud-free limitation of its optical counterparts. The SWOT mission will periodically monitor lake areas at the spatial scale of decameters and the vertical water level change in centimeter accuracy. With the measurements of both lake surface area and water level changes, the SWOT mission will be able to monitor lake water storage changes, which is revolutionary in endorheic lake dynamics monitoring.

Lake Water Salinity Remote Sensing

Salinity is an important water parameter of endorheic lakes. Water salinity content in endorheic lakes changes along with their area and level dynamics. It is desirable to measure lake salinity using remote sensing. Figure 5 shows the Landsat/ETM+ image acquired in the western Tibetan Plateau on October 28, 2000, on which lakes appear in different colors, largely indicating water salinity content. Taro Co in the south is a freshwater lake with a low salinity of 0.77 g/L and appears completely dark on the image. The bluish lake north of Taro Co is Zabuye Caka (the major

lithium mineral site in China) in two parts: the North Zabuye Caka and the South Zabuye Caka. The dark blue north lake is a brine lake with a salinity of 393 g/L. The south lake appears blue, light blue, and white on the image from east to west. The blue part is brine waters at high salinity (~440 g/L) while the saturated salt deposits appear light blue, and the playa appears cyanish white on the image. The small playa lake in a valley north of the North Zabuye Caka is also surrounded by cyan-white lacustrine deposits. As this figure suggests, the optical image is able to separate these endorheic lakes into fresh water lakes, saline lakes, salt lakes, and playa lakes. However, quantitative retrieval of lake salinity using optical satellite remote sensing is far from robust and practical.

The only direct way to remotely assess water salinity from space seems to be through the use of passive microwaves. The dielectric constant of saltwater is a function of its salinity and temperature and directly impacts sea surface emissivity. Currently, both the SMOS and Aquarius satellite missions have only been able to derive sea surface salinity. Aquarius is an L-band radiometer and scatterometer instrument combination designed to map the salinity field at the surface of the ocean from space. The instrument is designed to provide global sea surface salinity maps on a monthly basis with a spatial resolution of 150 km.[21] SMOS carries an L-band interferometric radiometer, which receives the radiation emitted from the Earth's surface and converts it to salinity in the surface waters of the oceans at a resolution of 100–200 km.[20] At such coarse resolutions, inland lakes cannot be resolved. In addition, endorheic basins are not as

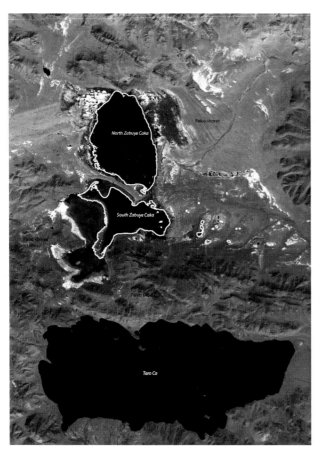

Fig. 5 Appearance of lakes with various salinity content on satellite imagery. The Landsat/Enhanced Thematic Mapper plus (ETM+) image acquired on October 28, 2000 illustrates the appearance of the lakes with varying salt content. Freshwater Taro Co appears rather dark, while Zabuye Caka shows the color tones of white outline black, dark gray, and white corresponding to brine water, salt deposits, and dry playas, respectively, along the west shore. In addition, Zabuye Caka is surrounded by abandoned paleo-shorelines, attesting to the existence of larger lakes in ancient time.

uniform as the ocean surface and the salinity range of endorheic lakes is much higher and more variant than that of the ocean. All these leave satellite remote sensing of inland lake salinity a currently unsolved problem.

Paleolake Change

Paleo-changes of endorheic lakes with paleo-shoreline relics in their basins can be recovered through integration of high-resolution imagery and digital elevation models (DEMs). Many endorheic lakes (e.g., Zabuye Caka in Fig. 5) are surrounded by paleo-shorelines, attesting to a substantially moister condition in the past. The paleolake extent is critical for us to understand the paleoclimatic and paleohydrologic environments in the basin. Aerial photographs, satellite images, and terrain data have been used to reconstruct paleolake extents for Lake Chad in Africa[24] and Lake Eyre in Australia.[26] Integrating high-resolution satellite imagery

and DEMs, Sheng (2009) developed a semiautomated tool for paleolake extent recovery at regional scales.[27]

Paleo-shorelines are found extensively on Landsat imagery in the Great Siling Co basin. With the help of the interactive paleolake recovery tool, paleolake extent was recovered in the basin using Landsat ETM+ images and Shuttle Radar Topography Mission (SRTM) DEMs (Fig. 6). Four paleolakes were recovered in the basin, occupying at least a total area of 10,221 km². These paleolakes, after shrinkage, broke into 95 contemporary lakes with a total area of 3,759 km². The total area shrinkage and water loss are estimated at 6,462 km² and 372 billion m³, respectively. Nearly two-thirds of the lake area has disappeared since the great lake period, which occurred ~11 ka BP in the basin according to optically stimulated luminescence (OSL) dating results from paleo-shore sand samples collected in the field.

Paleo Siling Co alone used to occupy an area of 7,665 km² in early Holocene and subsequently shrank as the climate became drier, evolving into 56 contemporary lakes including the current Siling Co and the sizable Pangkog Co, Urru Co, and Qiagui Co. The lake was recovered at the water level of 4,600 m on satellite images with field validation in 2009, which was ~56 m above the present-day water level. The recovery reveals that the lake shrinkage that occurred in Holocene led to a loss of ~310 billion m³ water and a disappearance of ~64% of the original lake extent. The water level of Paleo Gyaring Co is found to have dropped by ~14 m with a loss of 11.8 billion m³ of water and 33% of lake area. Paleo Yueqia Co and Paleo Ringco Ogma, the two paleolakes on highlands in the basin, suffered even more significant shrinkage. Paleo Yueqia Co experienced a 34 m water level drop, losing 18.6 billion m³ of water and 79% of lake area, while Paleo Ringco Ogma experienced a 26 m water level drop, losing 31.5 billion m³ of water and 77% of lake area. In summary, paleolakes in the Great Siling Co basin had shrunk in Holocene up to ~79% in area with an up-to-56 m level drop.

CONCLUSION

Endorheic lakes have been changing dramatically due to the climate change and human activities, for example, the rapid shrinkage of the Aral Sea since the 1960s and the accelerated expansion of Siling Co in the past four decades. Spaceborne remote sensing provides diversified means to monitor lake area changes, level changes, and thus the lake volume changes at regional and global scales. Future satellite missions are expected to simultaneously monitor lake area and level dynamics for a majority of endorheic lakes at high resolutions. Paleolake extents can be recovered through the integration of remote sensing and digital elevation data to reveal paleo-endorheic lake changes. Though water salinity satellite missions have recently been in operation, lake salinity remote sensing from space is still not practical in its current stage. Various remote sensing tools

Coastal—
Evaporation

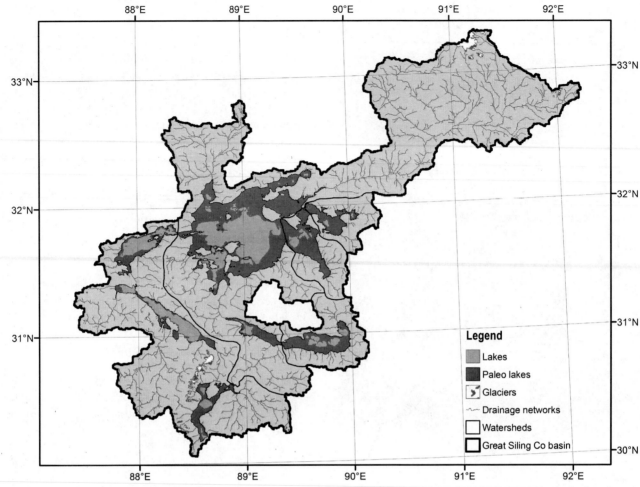

Fig. 6 Paleolake recovery in Great Siling Co basin. The extents of four paleolakes (in dark gray) were recovered in the basin using current Landsat images and digital elevation data. They had shrunk in Holocene up to ~79% in area with an up-to-56 m level drop in the basin.

have demonstrated their success in monitoring lake dynamics in the case study of the Great Siling Co basin. Siling Co and other endorheic lakes in the basin had dramatically shrunk in Holocene but have been significantly expanding in recent decades in response to the increased precipitation and the accelerated melting of glaciers in the basin.

ACKNOWLEDGMENTS

This work was supported in part by NASA New Investigator Program under Grant NNX06AE58G and in part by the U.S. Geological Survey (USGS) Landsat Science Team Program Grant G12PC00071.

REFERENCES

1. Hammer, U.T. *Saline Lake Ecosystems of the World*; Springer: Boston, 1986.

2. Yuan, F.; Sheng, Y.; Yao, T.; Fan, C.; Li, J.; Zhao, H.; Lei, Y. Evaporative enrichment of oxygen-18 and deuterium in lake waters on the Tibetan Plateau. J. Paleolimnol. **2011**, *46* (2), 291–307.

3. Zheng, M.; Liu, X. Hydrochemistry of salt lakes of the Qinghai-Tibet Plateau, China. Aquat. Geochem. **2009**, *15*, 293–320.

4. Smol, J.P. *Pollution of Lakes and Rivers: A Paleoenvironmental Perspective*; 2nd Ed.; Blackwell: Malden, MA, USA, 2008.

5. IPCC. *Climate Change 2007: The Physical Science Basis*; Cambridge University Press: Cambridge, 2007.

6. Street-Perrott, E.A.; Harrison, S.P. Lake levels and climate reconstruction. In *Paleoclimate Analysis and Modelling*; Hecht, A.D., Ed.; Wiley: New York, 1985; 291–340.

7. Mason, I.M.; Guzkowska, M.A.J.; Rapley C.G.; Street-Perrott, F.A. The response of lake levels and areas to climatic change. Clim. Change **1994**, *27* (2), 161–197.

8. Shen, Y.; Chen, Y. Global perspective on hydrology, water balance, and water resources management in arid basins. Hydrol. Process. **2010**, *24*, 129–135.

9. Micklin, P.P. Disiccation of the Aral Sea: A water management disaster in the Soviet Union. Science **1988**, *241* (4870), 1170–1176.

10. Harris, A.R. Time series remote sensing of a climatically sensitive lake. Remote Sensing Environ. **1994**, *50*, 83–94.

11. Birkett, C.M. The contribution of Topex/Poseidon to the global monitoring of climatically sensitive lakes. J. Geophys. Research-Oceans, **1995**, *100* (C2), 25179–25204.

12. Millington, A.C.; Drake, N.A.; Townshend, J.R.G.; Quarmby, N.A.; Settle, J.J.; Reading, A.J. Monitoring salt playa dynamics using Thematic Mapper data. IEEE Trans. Geosci. Remote Sensing **1989**, *27* (6), 754–761.

13. Sheng, Y.; Li, J. Satellite-observed endorheic lake dynamics across the Tibetan Plateau between circa 1976 and 2000. In *Remote Sensing of Protected Lands*; Wang, Y., Ed.; CRC Press: New York, 2011; 305–319.

14. Sivanpillai, R.; Miller, S.N. Improvements in mapping water bodies using ASTER data. Ecol. Inform. **2010**, *5* (1), 73–78.

15. Bryant, R.G. Application of AVHRR to monitoring a climatically sensitive playa, case study: Chott el Djerid, Southern Tunisia. Earth Surf. Process. Landforms **1999**, *24*, 283–302.

16. McCullough, I.M.; Loftin, C.S.; Sader, S.A. High-frequency remote monitoring of large lakes with MODIS 500 m imagery. Remote Sensing Environ. **2012**, *124*, 234–241.

17. Shum, C.; Yi, Y.; Cheng, K.; Kuo, C.; Braun, A.; Calmant, S.; Chambers, D. Calibration of Jason-1 Altimeter over Lake Erie. Mar. Geod. **2003**, *26* (3–4), 335–354.

18. Cretaux, J.F.; Jelinski, W.; Calmant, S.; Kouraev, A.; Vuglinski, V.; Bergé-Nguyen, M.; Gennero, M.C.; Nino, F.; Abarca Del Rio, R.; Cazenave, A.; Maisongrande, P. SOLS: A lake database to monitor in the near real time water level and storage variations from remote sensing data. Adv. Space Res. **2011**, *47*, 1497–1507.

19. Urban, T.J.; Schutz, B.E.; Neuenschwander, A.L. A survey of ICESat coastal altimetry applications: Continental coast, open ocean island, and inland river. Terr. Atmos. Ocean Sci. **2008**, *19* (1–2), 1–19.

20. Kerr, Y.H.; Waldteufel, P.; Wigneron, J.P.; Delwart, S.; Cabot, F.; Boutin, J.; Escorihuela, M.J.; Font, J.; Reul, N.; Gruhier, C.; Juglea, S.E.; Drinkwater, M.R.; Hahne, A.; Martin-Neira, M.; Mecklenburg, S. The SMOS mission: New tool for monitoring key elements of the global water cycle. Proc. IEEE **2010**, *98* (5), 666–687.

21. Le Vine, D.M.; Lagerloef, G.S.; Torrusio, S.E. Aquarius and remote sensing of sea surface salinity from space. Proc. IEEE **2010**, *98* (5), 688–703.

22. Wang, S.; Dou, H. *Catalog of China's Lakes*; Science Press: Beijing, China, 1998.

23. McFeeters, S.K. The use of the normalized difference water index (NDWI) in the delineation of open water features. Int. J. Remote Sensing **1996**, *17* (7), 1425–1432.

24. Leblanc, M.; Favreau, G.; Maley, J.; Nazoumou, Y.; Leduc, C.; Stagnitti, F.; van Oevelen, P.J.; Delclaux, F.; Lemoalle, J. Reconstruction of Megalake Chad using shuttle radar topographic mission data. Palaeogeogr. Palaeoclimatol. Palaeoecol. **2006**, *239* (1/2), 16–27.

25. Lee, H.; Durand, M.; Jungab, H.C.; Alsdorf, D.; Shum, C.K.; Sheng, Y. Characterization of surface water storage changes in Arctic lakes using simulated SWOT measurements. Int. J. Remote Sensing **2010**, *31* (14), 3931–3953.

26. DeVogel, S.B.; Magee, J.W.; Manley, W.F.; Miller, G.H. A GIS-based reconstruction of late quaternary paleohydrology: Lake Eyre, arid central Australia. Palaeogeogr. Palaeoclimatol. Palaeoecol. **2004**, *204* (1/2), 1–13.

27. Sheng, Y. PaleoLakeR: A semiautomated tool for regional-scale paleolake recovery using geospatial information technologies. IEEE Geosci. Remote Sensing Lett. **2009**, *6* (4), 797–801.

Coastal—
Evaporation

Eutrophication

Kyle D. Hoagland
School of Natural Resources, University of Nebraska, Lincoln, Nebraska, U.S.A.

Thomas G. Franti
Department of Biological Systems Engineering, University of Nebraska, Lincoln, Nebraska, U.S.A.

Abstract

Eutrophication is a natural process involving the nutrient enrichment of surface water and the subsequent impacts on water quality and the aquatic ecosystem. However, as a natural process it takes generations, or even thousands of years, for eutrophication to cause significant changes. The acceleration of the process, known as cultural eutrophication, is of the greatest concern to water quality and the health of aquatic systems.

INTRODUCTION

Eutrophication is the nutrient enrichment of surface water and the subsequent impacts on water quality and the aquatic ecosystem. The over abundance of plant nutrients, usually nitrogen and phosphorus—but sometimes silicon, potassium, calcium, iron, or manganese—creates the conditions for excessive plant growth.[1] Algal blooms are an example of excessive growth caused by over supply of nutrients. A body of water is classified by its trophic state based on the amount of nutrients supplied to it. An oligotrophic state is low in nutrients, a mesotrophic state is intermediate, and a eutrophic or hypereutrophic state is high in nutrients.[2] Eutrophication is a natural process. However, as a natural process it takes generations, or even thousands of years, for eutrophication to cause significant changes. It is the acceleration of the process, known as cultural eutrophication, that is of the greatest concern to water quality and the health of aquatic systems.

CULTURAL EUTROPHICATION

Cultural eutrophication is the result of excess nutrients—primarily nitrogen and phosphorus—delivered to rivers, lakes, and estuaries by the activities of humans. Nutrient loading can come from both point sources and diffuse sources. Point sources of nutrients are generally urban sewage treatment or industrial water treatment. Diffuse sources (or non-point sources) of nutrients include agriculture, deforestation, and urban lawn runoff. The use of commercial fertilizer in crop production can lead to losses of nutrients to surface water. Concentrated animal feeding operations are also a potential source of excess nutrients.[2,3] Finally, deforestation and subsequent soil erosion can deliver excess nutrients, sediment, and organic matter to surface water, creating the conditions for eutrophication.

The natural aging process of lakes in some climates can lead from an oligotrophic state, with clean open water supporting usually cold water fish species, to a eutrophic state of warmer, shallow water supporting different species of plants and fish, and ultimately, to a filled lake closed over by a bog or fen. This process usually takes thousands of years. In contrast, cultural eutrophication can occur in a matter of one or two generations.

The last 50 years of the twentieth century saw the greatest impact of eutrophication on the world's waters. Cultural eutrophication has accelerated the lake aging process, reduced water quality, and impacted aquatic plant and animal populations, at economic costs. These impacts are being felt worldwide.[1]

Modern society's use of phosphorus additives in detergents has lead to excessive phosphorus loading to surface waters.[1] Modern agriculture's use of commercial fertilizer and fertilizer's relative abundance, low cost, and over use, has lead to excessive nitrogen loading to surface water. Agriculture also contributes excessive phosphorus from fertilizer and from intensive livestock operations and land application of manure.[3,4] Urban sources, such as sewage treatment plants, storm water runoff, industrial water treatment facilities, and food processing contribute excess nutrients. Sewage treatment loading to waters, including treated gray water, is generally considered the greater source of phosphorus. The greater contributor of nitrogen is agricultural crop production. Other significant agricultural sources are animal confinement operations, forestry, and atmospheric inputs.

RATE OF EUTROPHICATION

The rate of eutrophication is controlled by the rate at which nitrogen and phosphorus are delivered to a body of water. It is generally accepted that phosphorous is the limiting

Encyclopedia of Natural Resources DOI: 10.1081/E-ENRW-120010332

nutrient for algal growth in lakes.[1,5] Another limiting factor for algal growth is light. As excess phosphorus enters a lake it can trigger high levels of algal growth. Excessive growth, or blooms, can reduce water clarity. Aquatic macrophytes may also thrive under these conditions. This process can increase the productivity of a lake to an extent. When it becomes excessive, the process of cultural eutrophication has begun.

The nutrients supplied to an aquatic system is the most important factor that determines the species and amount of plant material, which in turn controls available oxygen and the animal species that thrive.[1] As plants and macrophytes grow and die, organic matter accumulates on the lake bottom. Decomposition of excessive amounts of organic matter can consume available oxygen and create anoxic conditions. An anoxic condition drives away plant and animal species dependent on oxygen. Slowly a lake's original community changes.

This anoxic state creates another problem if a lake becomes stratified. Temperature differences in the water at the lake's surface and at depth can create thermal stratification. In this condition the cold, low oxygen water in the hypolimnion, or lower layer, does not mix with the epilimnion, the warmer surface water. In this condition the anoxic conditions in the hypolimnion creates a condition where the sediment releases phosphorus, which is transported upward to the epilimnion and contributes to increased algal growth. This internal cycling of phosphorus continues the eutrophication process even if no additional phosphorus enters the lake.[5,6]

EFFECTS OF EUTROPHICATION

There are various biological effects of eutrophication. Besides the problems of algal blooms and anoxic conditions previously described, there are changes in water temperature, reduction in water clarity, increased macrophyte production, population shifts in both plants and animals, and accelerated aging of lakes. Fish kills and the coastal "red tides" and "brown tides" are also potential environmental impacts from eutrophication.[5,7]

There are many impacts on human use of water from eutrophication. Potable water—water used for human consumption—can be significantly degraded by eutrophication. Besides the excess nutrients themselves, there is excess algae and other plant growth which can contribute to unwanted odors and tastes. Thus, additional water treatment is needed to make the water drinkable.[1,4]

The decline of commercial and recreational fisheries is also an indirect result of eutrophication. Fish populations may move away, or over time shift to less desirable fish species as water temperature, clarity, and quality change.[1] There are also negative impacts on recreational use of surface waters such as boating and swimming as a result of floating plant growth, smell, and the overall loss of aesthetic quality.[4] Finally, the general decline of an aquatic ecosystem and the loss of biodiversity are impacts of eutrophication.[1,3,7]

Some of these impacts can cause significant economic losses as well. Water treatment costs rise as water quality decreases. Algal blooms can contribute to shutting down water treatment plants. Advanced water treatment of municipal water to remove nutrients (such as alum addition)—hence removing a contributing factor to eutrophication—adds another expense to water treatment. Loss of commercial fisheries is an economic impact to a region. Finally, loss of recreation income from tourists, boaters, and recreational fisherman can have significant impacts on a local economy.[4,7]

REDUCTION AND MANAGEMENT

Two approaches can be taken to the reduction and management of eutrophication. 1) reduction of nutrient loads, and 2) managing the existing high-nutrient state.[1] Reduction of loading is clearly the more robust approach. Reduced loads are necessary if long-term improvement is expected. However, because of the ecosystem changes brought on by eutrophication and the potential problem of phosphorus cycling previously described, water quality and ecosystem improvements may not respond quickly to reduced loading.[3] A combination of reduced loads and in-lake controls may be needed.

Reducing nutrient loads requires reduction of both point and diffuse sources. To reduce point sources, sewage and industrial water treatment plants need advanced water treatment to remove nutrients.[1,7] Advanced water treatment is expensive, but is the only way to reduce these point source nutrients. Most diffuse sources contributing nutrients to surface water come from agricultural operations. Loading is dependent upon the type of crops grown, soil type, climate, cultural practices, fertilizer use, and whether animal waste management practices are sufficient to reduce loading. Nutrient reduction from these sources will require improved management of fertilizer and animal waste, and continued reduction of soil erosion throughout the watershed.[1]

There are several methods used to manage nutrient levels in lakes. These fall into two categories: those that 1) remove nutrients; or 2) manage nutrient levels without nutrient removal. Methods to remove nutrients include: 1) lake flushing; 2) hypolimnetic water withdrawal; 3) sediment removal (e.g., dredging); and 4) nutrient inactivation by precipitation (e.g., alum treatment). Methods to manage nutrients without removing them include: 1) artificial mixing and/or aeration; 2) dilution by the addition of water lower in nutrients; 3) bottom sealing to prevent internal nutrient cycling; 4) manipulation of lake biological communities (e.g., selective fish harvesting or the introduction of fish predators such as largemouth bass, walleye, or brown trout); 5) introduction of biological controls for unwanted macrophyte growth (e.g., weevils or grass carp); and 6) herbicide or other chemical treatments (e.g., copper sulfate) of excessive algal or macrophyte growth.[1,6]

Clearly the best way to reduce or reverse eutrophication is to reduce nutrient loading, that is, targeting the source of the problem.[1] This long-term solution involves participation and management by people throughout the entire watershed. In-lake nutrient management can be done, but may require annual inputs and regular management.

REFERENCES

1. Harper, D. *Eutrophication of Freshwaters: Principles, Problems and Restoration*; Chapman & Hall: London, 1992.
2. Smith, V.H.; Tilman, G.D.; Nekola, J.C. Eutrophication: impacts of excess nutrient inputs on freshwater, marine, and terrestrial ecosystems. Environ. Pollut. **1999**, *100*, 179–196.
3. Carpenter, S.R.; Caraco, N.F.; Correll, D.L.; Howarth, R.W.; Sharpley, A.N.; Smith, V.H. Nonpoint pollution of surface waters with phosphorus and nitrogen. Ecol. Appl. **1998**, *8* (3), 559–568.
4. Daniel, T.C.; Sharpley, A.N.; Lemunyon, J.L. Agricultural phosphorus and eutrophication: a symposium overview. J. Environ. Qual. **1998**, *27* (4), 251–257.
5. Correll, D.L. The role of phosphorus in the eutrophication of receiving waters: a review. J. Environ. Qual. **1998**, *27* (2), 261–266.
6. Cooke, G.D.; Welch, E.B.; Peterson, S.A.; Newroth, P.R. *Lake and Reservoir Restoration*; Butterworths Publishers: Stoneham, MA, 1986.
7. Klapper, H. *Control of Eutrophication in Inland Waters*; Ellis Horwood Limited: Chichester, West Sussex, England, 1991.

Evaporation and Energy Balance

Matthias Langensiepen
Department of Modeling Plant Systems, Humboldt University of Berlin, Berlin, Germany

Abstract

Solar radiation is the sole energy input of the Earth–atmosphere system. Its partitioning into surface fluxes of latent and sensible heat is determined by the physical properties and availability of surface water, the "evaporative demand" of the atmosphere, and the nature of the surface. Evaporation is the connecting link between the system's energy and water balances. The quantification of a system's energy balance requires a definition of its boundary conditions. They consist of the spatial and temporal dimensions of the system and its exchange surfaces, their physical transport properties, the energy states across the system boundaries, and possible modes of energy transfer.

INTRODUCTION

Various forms of energy drive water transport through the hydrological cycle. Radiant energy, originating from the sun, provides the input energy for the cycle. Once matter absorbs this energy, it is converted into sensible heat that elevates the temperature of the air and the ground, and latent heat that causes evaporation, driving thereby the cycle against the pull of gravity. Further transport is generated by kinetic energy and pressure energy of the moving air masses. Translocation of vapor is accompanied by continuous interchanges among radiant, thermal, kinetic, and pressure energy. Large amounts of latent heat are released when water condenses in the clouds and falls as precipitation on the Earth's surface. It carries kinetic energy while flowing through watersheds. Vertical movement and percolation through the Earth's crust finally causes changes in potential and pressure energies.

The first law of thermodynamics states that energy is neither created nor destroyed, only converted from one form into another. This effectively means that the input and output energies of a completely defined system must balance. Storage effects may temporarily disturb this equilibrium condition. The energy balance must thus be expressed in its most general form as

Energy Input = Energy Output + Energy Storage

The water balance of the Earth–atmosphere system can be treated analogically as the mass of water is conserved at all times. Evaporation is the connecting link between the system's water and energy balances. It is a surface process, which takes place at the lower boundary of the atmosphere and is an important component of the surface energy balance (see Fig. 1):

$$R_n = LE + H + G - A + S - L_p F_p$$

where R_n is the flux of net allwave-radiation, L the latent heat, E the evaporation rate, H the flux of sensible heat, G the heat flux at the lower boundary of the surface, and A the energy advected to the surface when the ground properties have horizontal discontinuities. The energy balance is sometimes parameterized for a volume of surface material (for example water body, soil, or canopy volume). As the solar energy input undergoes diurnal and annual fluctuations, heat storage S may become an important component of the balance when it is applied at time intervals shorter than the fluctuation period. When the layer includes vegetation, biochemical energy storage due to photosynthesis can also be considered. L_p is then the thermal conversion factor of carbon dioxide, and F_p is the flux of CO_2.

Shortwave radiation from the sun is the sole energy input of the Earth–atmosphere system. Its net amount available for heat conversion is related to geographical location, time, atmospheric transparency, atmospheric path length, geometrical distribution of the surface elements, and their optical properties. Complementary longwave radiation exchange is governed by surface to air temperature differences and cloudiness. Net shortwave and longwave radiation form the net allwave radiation R_n.

The partitioning of R_n into the remaining terms of the surface energy balance determines the rate of surface evaporation and depends on the availability of surface water.

ENERGY BALANCE AND WATER AVAILABILITY

Unlimited Water Availability

When water availability is unlimited on a large scale, such as in oceans, vertical temperature gradients within the atmosphere tend to be very close to the adiabatic value, and most of the available energy (R_n) is diverted to latent heat $(L_v E)$ from moisture flux at the surface. Wind

Encyclopedia of Natural Resources DOI: 10.1081/E-ENRW-120010307

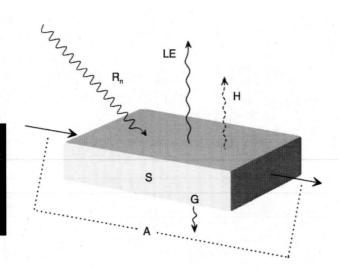

Fig. 1 Schematic illustration of the surface energy balance. (R_n = net radiation, LE = latent heat flux, H = sensible heat flux, G = ground heat flux, S = heat storage, A = advection. All symbols are expressed in $W\,m^{-2}$.)

gradients near the surface are typically very steep under such conditions and quickly approach values that remain nearly constant throughout the convective boundary layer. Vertical motion is damped out by strong subsidence inversion at the upper boundary of this well-mixed layer. Heat Storage (S) has a dominant effect on the diurnal course of the ocean's energy balance leaving only little energy for transport ($L_v E + H$) into the air until late in the afternoon. This situation is reversed during the night, where heat released from the ocean surface becomes the major source of energy. Large-scale advection (A) partly disturbs the thermal inertia of oceans, which has considerable effects on the global weather systems (Gulfstream, Humboldt-Stream, El Nino). The smaller the water volume becomes the more it is likely that its thermal inertia is disturbed by local advection due to horizontal discontinuities of thermal surface properties. Since radiation can penetrate into water bodies it is also possible that the underlying floor becomes a source of heat.

Snow Covered and Frozen Surface Layers

When water is bound in snow covered or frozen surface layers, energy partitioning is affected by the penetration of shortwave radiation, phase changes, and internal distribution of water. Net radiation gain (R_n) is commonly restricted by a high surface albedo. The optical depth of snow and ice also affect radiation absorption and penetration. The available energy is mainly partitioned between storage energy (S) and energy required to allow water to change between frozen and liquid states (L_f). Only little energy is consumed by atmospheric transport ($LE + H$). Phase changes of water within the layer (freezing, melting, condensation,

evaporation, sublimation) are accompanied by the continuous consumption or liberation of energy. The internal partitioning of available energy is thus influenced by the physical states of water: Water has a high specific heat of $4216\,J\,kg^{-1}\,K^{-1}$ at $0°C$ due to a strong intermolecular bonding force. Fifteen percent of the hydrogen bonds break when water changes from a solid to liquid state. The energy required to effect this change is $0.334\,MJ\,kg^{-1}$ and is called the latent heat of fusion L_f. Nearly 7.5 times as much energy is required at this temperature level to cause water to further change from a liquid to a gaseous state. The corresponding energy is called latent heat of vaporization L_v and is temperature dependant ($2.5\,MJ\,kg^{-1}$ at $0°C$, $2.45\,MJ\,kg^{-1}$ at $20°C$, and $2.41\,MJ\,kg^{-1}$ at $40°C$). In the event that water changes directly from a solid to a gaseous state (sublimation) the required latent heat of sublimation L_s is the algebraic sum of L_f and L_v. Freezing or condensation liberates energy, the amount depending on the corresponding phase shift. When the surface layer is below the freezing point and the sky is clear, net radiation can become negative under conditions of decreased radiation availability (high latitudes). It becomes positive, however, when the sky window is obstructed by clouds or surface emission is exceeded by incoming radiation. When the surface melts, both radiation and convection act as energy sources, sometimes accompanied by additional heat input from rainfall. Surface temperatures change only little during this process, because most energy is stored as latent heat of fusion.

Water Scarcity

When water is scarce, as is the case in deserts, most of the available energy (R_n) is consumed by surface heating, which can be sensed as a rise in surface temperature. Sensible heat (H) dissipation from dry surfaces lowers the density of air increasing its instability and tendency to rise. The instable air parcels form plumes (thermals) that progressively cool down as they mix with the surrounding air and are finally capped off by the inversion layer. Additional air is entrained from the top of the capping inversion layer and dragged to the ground by sinking motion of the cooling air masses. The height of the inversion layer is dependant on the amount of energy available for surface heating. At night and during early morning, winds in deserts are light, turbulence is low, air is stable or neutral, and the inversion layer is close to the ground. Net radiation (R_n) is partitioned into surface heat-flux (G) and heat storage (S) under such conditions. However, low thermal admittance of the barren dry soil diverts a major portion of the available energy to sensible heat (H), increasing air instability and turbulence. They promote the build up of miniature whirlwinds known as dust devils. The situation is reversed during afternoons, where sinking radiation energy input stabilizes air masses. High differences between day and night temperatures are a consequence of lacking water, the magnitude depending on the diurnal evolution of net radiation (R_n). Sloping terrain and thermal surface heterogeneities induce horizontal heat transport, known as advection (A). It causes the buildup of wind gusts and turbulence, which act as kinetic energy sources in soil erosion. Advection also plays a significant role in the energy balance of wet surface islands in dry areas (Oasis effect). The evaporative demand of the atmosphere is generally high under conditions of elevated air temperature and limited water availability.

Vegetation Control of Water Availability

When vegetation cover controls water availability, energy partitioning is affected by the physiological state of the plants. The sites of regulation are *stomata* (from Greek "mouth"), tiny pores serving as pathways between the plant interior and the atmosphere. Each pore is surrounded by a pair of specialized cells (guard cells), which control its aperture and respond to plant internal and external signals. Light, vapor pressure deficit, and water potential are the principal controlling signals. Carbon dioxide, hormones (abscisic acid and cytokinins), and photosynthetic assimilation capacity have so far been detected as additional regulating factors. Signals and plant responses are acting in an integrated manner and form the canopy resistance against water loss. Development and growth determine the evolution of plant stand architecture and hence the spatial distribution of exchange surfaces. The more the surfaces are vertically exposed against airflow, the higher is their capacity to absorb momentum. In neutral transport conditions, the logarithmic portion of the wind-profile above a canopy extrapolates downward to a height where wind speed becomes zero. This level is called *zero plane displacement* and is defined as the average height of mass and heat exchange within a canopy volume. This height changes in accordance with foliage density distribution, form drag, and wind speed. The type of surface vegetation cover thus influences the magnitude of heat and mass exchange. When determining the energy balance of a plant stand, two sources of water have to be considered, canopy and soil. If the vegetation cover is sparse or is at an early development stage, significant portions of the available energy (R_n) can reach the soil level. In this case, the availability of water depends on biological factors and the soil hydraulic properties (water retention, hydraulic conductivity, and soil water diffusivity). The partitioning of available energy (R_n) into the heat terms of the energy balance is largely determined by the water status of the soil–plant system. Latent heat (L_vE) from transpiration is the major energy sink when soil water is abundantly available. In case radiation reaches the canopy floor, latent heat (L_vE) from soil evaporation as well as soil heat flux (G) are additional sinks of energy. Advection (A) may become an

additional source of energy in hot climates. Energy storage due to photosynthesis (L_pF_p) is very small in comparison with the other components of the energy balance and is therefore often neglected. Heat storage (S) becomes important in massive canopies like forests. When water becomes limited, surface regulation restricts latent heat loss (L_vE) and sensible heat (H) becomes the principal sink of energy causing rises in surface temperature. Plants have flexible capabilities to optimize production in response to such conditions.

DETERMINATION OF THE SURFACE ENERGY BALANCE

Model determinations of the surface energy balance are commonly carried out with the combination equation, which emphasizes the mutual relation between latent and sensible heat fluxes. Practical methods assume equality, either between scalars and momentum (aerodynamic method) or between the eddy diffusivities for heat and vapor. The ratio of sensible to latent heat is then proportional to the ratio of air temperature over vapor concentration (Bowen ratio $\beta = H/L_vE$). Instrumentation can be categorized in accordance to their application. Surface parameters are commonly measured with net-radiometers (R_n), heat flow sensors (G), and lysimeters (L_vE). Gradient measurements above exchange surfaces involve determinations of wind speed (anemometers), air temperature (thermometers), air humidity (hygrometers, psychrometers), and CO_2 (infrared gas analyzers). Sonic anemometers, quartz thermometers, Lyman-alpha, and Krypton hygrometers are applied with the eddy correlation method. Remote sensors can be used to deduce turbulence parameters, heat and momentum fluxes from backscattered or forward-propagated signals (sodars, radars, and lidars).

CONCLUSION

The first law of thermodynamics states that the input and output energies of any given system must balance. Solar radiation is the sole energy input of the Earth–atmosphere system. Its partitioning into surface fluxes of latent and sensible heat is determined by the physical properties and availability of surface water, the "evaporative demand" of the atmosphere, and the nature of the surface. Evaporation is the connecting link between the system's energy and water balances. The quantification of a system's energy balance requires a definition of its boundary conditions. They consist of the spatial and temporal dimensions of the system and its exchange surfaces, their physical transport properties, the energy states across the system boundaries, and possible modes of energy transfer.

Coastal— Evaporation

BIBLIOGRAPHY

1. Brutsaert, W. *Evaporation into the Atmosphere*, 1st Ed.; Kluwer: Dordrecht, 1982; 299 pp.
2. Campbell, G.S.; Norman, J. *Introduction to Environmental Biophysics*, 2nd Ed.; Springer: Berlin, 1998; 286 pp.
3. Hillel, D. *Environmental Soil Physics*, 1st Ed.; Academic Press: San Diego, 1998; 771 pp.
4. Jones, H.G. *Plants and Microclimate*, 2nd Ed.; Cambridge University Press: Cambridge, 1992; 428 pp.
5. Kaimal, J.C.; Finnigan, J.J. *Atmospheric Boundary Layer Flows: Their Structure and Measurement*, 1st Ed.; Oxford University Press: Oxford, 1994; 289 pp.
6. Oke, T.R. *Boundary Layer Climates*, 2nd Ed.; Routledge: London, 1987; 435 pp.
7. Ross, J. *The Radiation Regime and Architecture of Plant Stands*, 1st Ed.; Junk: Den Haag, 1982; 391 pp.
8. Zeiger, E.; Farquhar, G.D.; Cowan, I.R. *Stomatal Function*, 1st Ed.; Stanford University Press: Stanford, 1987.

Evaporation: Lakes and Large Bodies of Water

John Borrelli
Department of Civil Engineering, Texas Tech University, Lubbock, Texas, U.S.A.

Abstract
Evaporation is a major component of the hydrologic cycle, second only to precipitation. As such, precise documentation of evaporation from lakes and other water bodies is required for wise management of our water resources. There is a necessity to accurately measure evaporation rates and provide numerical models for estimating evaporation from numerous lakes and reservoirs where direct measurement is too costly to undertake.

INTRODUCTION

The conversion of water from liquid state to vapor state is called evaporation. Evaporation requires energy—approximately 540 cal/cm^3 of water ($\approx$2.45 MJ/kg). Research has shown that the rate of evaporation is primarily a function of temperature, solar energy, wind velocity, vapor pressure deficit, and advected energy. The energy for evaporation comes predominately from solar radiation and wind. Evaporation is a major component of the hydrologic cycle, second only to precipitation. As such, precise documentation of evaporation from lakes and other water bodies is required for wise management of our water resources.

Annual lake evaporation across the United States has been estimated to range from 60 cm/yr to over 200 cm/yr.[1] The annual evaporation rates for several typical lakes vary from 51 cm/yr for Hungary Horse Reservoir (cool northern climate) to 223 cm/yr for Lake Mead (desert southwest) (Table 1). Consequently, there is a necessity to accurately measure evaporation rates and provide numerical models for estimating evaporation from numerous lakes and reservoirs where direct measurement is too costly to undertake.

TECHNIQUES FOR MEASURING LAKE EVAPORATION

There are three widely accepted methods for measuring the evaporation rates of lakes: a water budget, an energy budget, and the eddy correlation method. The water budget and the energy budget require a considerable amount of investment in personnel, instruments, and time. As a result, these methods are applied sparingly to calibrate numerical models.[2] With today's dependable computer technology, the eddy correlation method has become widely used in recent years. Most studies will employ two or all of the methods.

Water Budget

If all components of the water budget could be measured accurately, it is the only method that directly measures evaporation. The water budget for a lake is as follows:

$$Evap = \left[\left(SW_{in} - SW_{out} + GW_{in} - GW_{out} + S_b - S_e\right)\big/\text{Area} + \text{PPT}\right]\big/\text{Time}$$

where *Evap* [LT^{-1}] is evaporation, SW_{in} [L^3] is surface water inflow, SW_{out} [L^3] is surface water outflow, GW_{in} [L^3] is groundwater inflow, GW_{out} [L^3] is groundwater outflow, S_b [L^3] is lake storage at the beginning of the time period, S_e [L^3] is the lake storage at the end of the time period, Area [L^2] is the surface area of the lake, PPT [L] is precipitation, and Time [T] is the time period over which the measurements are made.[3] Evaporation is the residual of several measured terms and contains the errors included in the measurement of all those terms. Precipitation, for example, can have a bias error of up to 20% due to wind currents around the orifice of a rain gauge.[4] To use the water budget to measure evaporation, the inflow and outflow from the lake must be relatively small compared to the storage; otherwise, the errors in measurement will dominate the determination of evaporation. Overall, the error of measurement is $\pm$5–10%.

Energy Budget

The energy budget uses the conservation of energy principle to determine net transfer of energy into and out of a lake. Like the water budget, the evaporation rate is computed as the residual of all other terms; thus, it will contain residual measurement errors. Sturrock, Winter, and Rosenberry[5]

Encyclopedia of Natural Resources DOI: 10.1081/E-ENRW-120010365

Table 1 Annual evaporation from lakes

Lake	Annual evaporation (cm)	Longitude	Latitude	Area (ha)	Average depth (m)
Pyramid Lake[2]	128	119°40′	40°00′	46,640	61
Salton Sea[2]	179	116°10′	33°05′	88,100	8
Lake Ontario[2]	73	77°00′	44°00′	1,940,000	86
Hyco Lake[2]	94	79°05′	36°15′	1,760	6
Hungary Horse Reservoir[2]	51	113°55′	46°00′	9,700	15
Lake Kerr[2]	118	81°50′	29°20′	1,040	5
Lake Mead[2]	223	114°30′	36°05′	51,400	54
Lake Okeechobee[9]	147	80°55′	27°00′	182,130	3
Amistad Reservoir[2]	203	101°20′	29°20′	27,900	16
Great Salt Lake[2]	101	112°30′	41°00′	388,900	10

used the following energy budget equation in the study of Williams Lake:

$$Q_x = Q_s - Q_r + Q_a - Q_{ar} - Q_{bs} + Q_v - Q_e - Q_h - Q_w + Q_b$$

where Q_x is the change in energy content of the body of water, Q_s is incoming short-wave radiation, Q_r is reflected short-wave radiation, Q_a is incoming long-wave radiation, Q_{ar} is reflected long-wave radiation, Q_{bs} is long-wave radiation emitted from the body of water, Q_v is net energy advected to the body of water, Q_e is energy used for evaporation, Q_h is energy conducted from the water as sensible heat, Q_w is energy advected from the body of water by the evaporated water, and Q_b is heat transfer to the water from the bottom sediments. All terms are expressed in W/m².

Eddy Correlation

At the surface of the water, the water vapor in the air is nearly saturated. As air moves across the surface, small eddies transport the water vapor vertically at a net air movement of zero in the vertical direction. With current instrumentation, it is now possible to measure the vertical flux of water vapor or evaporation above the surface of a lake. The eddy correlation method directly measures the evaporative flux as presented by Shuttleworth[6] in the following formula:

$$E = 86.4 \rho_a \overline{w'q'}$$

where E is the evaporation rate (mm/day), ρ_a is the air density (g/m³), w' is the vertical wind velocity (m/sec¹), and q' is the specific humidity (g of water/g of air). The overbar denotes a mean value over a specific interval and the prime denotes an instantaneous deviation from the mean. Kizer and Elliot[7] provide a complete procedure to measure and calculate all terms needed to use the eddy correlation method. The accuracy for the eddy correlation measurements is 5–10%.[6] This compares favorably with the

energy and water budget methods, which have the same range of accuracy. Measurements are taken at a point but are used to represent a large area of a lake. This causes some error because there are different microclimates over a large lake.

ESTIMATION OF EVAPORATION

Evaporation cannot be measured at all lakes and reservoirs by using the methods described above. Thus, researchers have developed several equations that use climatological data for estimating evaporation. The most widely used equation is the modified Penman equation that was originally developed for evaporation as well as to estimate evapotranspiration from vegetation.[8] The modified Penman equation requires data on wind, net solar radiation, humidity, and temperature. There are many equations called modified Penman. The following is a good example of a modified Penman equation:[6]

$$E_p = \frac{\Delta}{\Delta + \gamma}(R_n + A_h) + \frac{\gamma}{\Delta + \gamma} \frac{6.43(1 + 0.536\,U_2)D}{\gamma}$$

where E_p is estimated potential evaporation (mm/d), R_n is net radiation exchange for the free water surface (mm/d), A_h is significant energy advected to the water body (mm/d), U_2 is wind speed at 2 m (m/sec), D is vapor pressure deficit (kPa), λ is latent heat of vaporization (MJ/kg), Δ the gradient of the saturation vapor–temperature curve (kPa/°C), and γ is the psychrometric constant (kPa/°C). Please refer to Shuttleworth[6] for details on the calculation of different variables.

Investigators have found that the modified Penman equation (not necessarily the same modifications as above) has estimated evaporation within the accuracy of measured evaporation rates.[9–11] The modified Penman equation does not take into account the heat stored in a lake, which can be significant. The Penman equation will overpredict evaporation during warmer months and underpredict evaporation during the colder months.[11] On an annual basis, the modified Penman has proven reliable over a wide range of locations and climatic conditions.

Pan evaporation rates have been widely used to estimate lake evaporation. Kohler, Nordenson, and Fox[12] reported on an extensive study at Lake Hefner in Oklahoma comparing lake evaporation and pan evaporation. They reported that the annual ratio for a U.S. Weather Bureau Class A pan evaporation to lake evaporation was 0.7. This proportional constant is called the pan coefficient. The USGS[13] reported that monthly pan coefficients varied from 0.13 in February to 1.32 in November. Annual pan coefficients have been reported as low as 0.51 at Lake Mead[14] to 0.75 at Lake Okeechobee.[9] Evaporation pans provide reliable results if several stations are used. However, pan evaporation records are often erratic and often trend downward with time because of environmental changes of surroundings and poor maintenance of the pan.

There are many other equations that have been developed to estimate evaporation. They include mass-transfer equations,[5] radiation equation,[9] temperature equations,[10] etc. The applicability of these equations is generally limited to their use in environments similar to those in which the equations were calibrated.

REFERENCES

1. Farnsworth, R.K.; Thompson, E.S.; Peck, E.L. *Evaporation Atlas for the Contiguous 48 United States*, NOAA Technical Report NWS-33; National Weather Service Reports: Washington, DC, 1982; 1–28.
2. Anderson, M.E.; Jobson, H.E. Comparison of techniques for estimating annual lake evaporation using climatological data. Water Resour. Res. **1982**, *18*(3), 630–636.
3. Rose, W.J.; Robertson, D.M. *Hydrology, Water Quality, and Phosphorus Loading of Kirby Lake, Barron County, Wisconsin*; Fact Sheet FS-066-98; U.S. Geological Survey, U.S. Department of the Interior: Washington, DC, 1998; 1–4.
4. Wanielista, M.; Kersten, R.; Eaglin, R. *Hydrology— Water Quantity and Quality Control*; John Wiley & Sons, Inc.: New York, 1997; 68–117.
5. Sturrock, A.M.; Winter, T.C.; Rosenberry, D.O. Energy budget evaporation from Williams Lake: a closed lake in North Central Minnesota. Water Resour. Res. **1992**, *28*(6), 1605–1617.
6. Shuttleworth, W.J. Evaporation. In *Handbook of Hydrology*; David, R.M., Ed.; McGraw-Hill, Inc.: New York, 1993; 4.1–4.53.
7. Kizer, M.A.; Elliott, R.L. *Eddy Correlation Measurement in Crop Water Use*. The 1987 International Winter Meeting of the American Society of Agricultural Engineers, Chicago, IL, Dec 15–18, 1987; American Society of Agricultural Engineers: St. Joseph, MI; Paper No. 87-2504, 1–14.
8. Penman, H.L. Natural evaporation from open water, bare soil, and grass. Proc. R. Soc., Ser. **A 1948**, *193*, 120–145.
9. Abtew, W. Evaporation estimation for Lake Okeechobee in South Florida. J. Irrig. Drain. Eng., Am. Soc. Civil Eng. **2001**, *127*(3), 140–147.
10. Hill, R.W. Consumptive Use of Irrigated Crops in Utah. Research Report 145; Utah Agricultural Experiment Station: Utah State University: Logan, Utah, 1994; 1–361.
11. Vardavas, I.M.; Fountoulakis, A. Estimation of lake evaporation from standard meteorological measurements: application to four Australian lakes in different climatic regions. Ecol. Model. **1996**, *84*, 139–150.
12. Kohler, M.A.; Nordenson, T.J.; Fox, W.W. *Evaporation from Pans and Lakes*, Research Paper No. 38; Weather Bureau, U.S. Department of Commerce: Washington, DC, 1955–20; 1–20.
13. U.S. Geological Survery (USGS). *Water-Loss Investigations: Lake Hefner Studies*, USGS Professional Paper 269; U.S. Geological Survey, Department of the Interior: Washington, DC, 1954; 1–158.
14. Hughes, G.H. *Analysis of Techniques Used to Measure Evaporation from Salton Sea, California*. Geological Survey Professional Paper 272-H. U.S. Geological Survey: Washington, DC, 1967; 151–176.

Coastal— Evaporation

Field Water Supply and Balance

Jean L. Steiner
Agricultural Research Service, U.S. Department of Agriculture (USDA-ARS), El Reno, Oklahoma, U.S.A.

Abstract
Field water supply has been a major focus of agricultural research and management. The soil water balance is a widely used method of tracking soil water supply in a field. This approach provided some of the earliest information available about the amount of water required to produce a crop, the relationship between water use and plant production, and water stress impacts on plant water use, and remains an important approach to research and management today.

INTRODUCTION

The soil water balance (Fig. 1) can be given as:

$$SW_t = SW_i + P + I - R - E - T - D \qquad (1)$$

where SW is soil water content within a defined root zone, t and i subscripts represent the end and beginning of a time period, respectively, P is precipitation, I is irrigation, R is runoff, D is drainage below the root zone, E is evaporation from the soil surface, and T is transpiration, with all terms in the same units and over the time period defined by the t and i subscripts. The R term might be modified to include any horizontal movement of surface water or shallow water table flow, which can be either imported to or exported from the defined soil volume. In most circumstances, the D term is downward flux below the root zone, but can be defined to include vertical flux across the bottom of the root zone that could include upward movement from a shallow water table to deep rooted plants. The E term can be considered to include water evaporated from any wetted surface (e.g., ponded water, wetted plants, evaporation or sublimation of accumulated snow), as well as evaporation of water from the soil profile. Frequently, the soil water balance is used to determine terms of Eq. (1) (e.g., $E+T$, $SW_t - SW_i$,) by measuring or estimating the remaining terms. Infiltration is often estimated by measuring P and R, if the other terms can be considered negligible during the precipitation event. Gardner[1] provides additional detail about the soil water balance.

WATER BALANCE COMPONENTS

Soil Water Content

Many different methods have been used to measure soil water content. A direct method that has been used since the early days of soil and agricultural research and remains common today is the gravimetric method. For the gravimetric method, soil samples, often cores, are collected and the water content is determined by weighing the sample before and after oven drying to determine the quantity of water lost by evaporation. Gravimetric sampling offers the benefit of providing a direct measurement of soil water content using simple equipment. However, it is time consuming and cannot be used to provide repeated measurement at the same location because it is destructive sampling. Given the high degree of spatial variability in most field soils, this limits the ability to determine temporal changes in soil moisture precisely, limiting the application of the soil water balance to relatively longer time periods. Methods that provide repeated measures of soil water content at the same location, such as using neutron probe or other technologies, reduce the problem of spatial variability and allow the soil water balance to be applied over shorter time periods.

Precipitation and Irrigation

Water is added to the system through precipitation or irrigation. Since precipitation is often highly variable, the rain gauge should be as near the site of the investigation as possible and should use standard weather gauges, properly sited away from tall buildings or vegetation that can distort rainfall catch. At field scales, irrigation applications are not totally uniform, so the gross irrigation amount may have to be adjusted by efficiency and uniformity factors to determine the net input to the soil water balance. However, well-managed, modern irrigation techniques such as low pressure applicators on center pivots and drip or subsurface irrigation methods, can provide very uniform distributions of water in a field with high efficiency.

Horizontal Movement of Water

The horizontal surface movement of water is largely runoff. In some situations, there may be a net gain through

Encyclopedia of Natural Resources DOI: 10.1081/E-ENRW-120010155

Field—Low

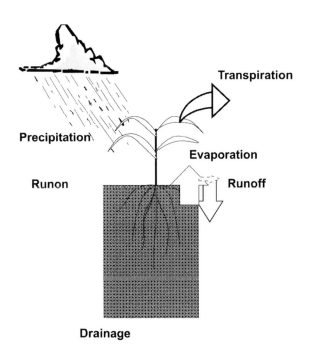

Fig. 1 Soil water balance for an agricultural plant.

run-on of water from another portion of the landscape. However, there is normally a net loss to the water balance of a particular field when horizontal movement occurs. The proportion of precipitation that runs off depends on soil properties, vegetation, and topography, as well as the intensity and duration of rainfall. In some soil water balance calculations, runoff is assumed to be negligible because of level topography, or in some cases berms are constructed around small experimental areas to restrict horizontal movement. When a large amount of precipitation falls as snow, there can be considerable movement caused by the wind and drifting. In some field situations, there is a substantial horizontal movement of water below the surface, making it difficult to use the water balance method.

Evaporation and Transpiration

Because it is difficult in field studies to separate loss of water by evaporation from soil and transpiration from plants, these terms are often linked into a single term, usually called evapotranspiration, E_t. From a management perspective, options for influencing soil evaporation and transpiration are different, so the terms are presented separately in Eq. (1).

Energy Balance

The energy balance approach, often used to estimate E_t, was developed for a plant–soil system under conditions where transpiration dominated the total loss. This approach has been successful because a large portion of energy that enters the Earth's atmosphere is required to transform water from liquid to vapor (the evaporation process). The combination

evaporation equation derived by Penman[2] and Monteith[3] can be expressed as

$$E_t = \left\{ \Delta\left(R_n + G\right) + \left[\rho C_p \left(e_s - e_a\right)/r_{av} \right] \right\} / L\left[\Delta + \gamma\left(1 + r_s/r_{av}\right) \right] \tag{2}$$

where Δ is the slope of the saturation vapor pressure curve, R_n is net radiation, G is soil heat flux, ρ is air density, C_p is the specific heat of dry air, e_s and e_a are the vapor pressures at the evaporating surface (assumed to be the saturation vapor pressure at the surface temperature) and of the atmosphere, respectively; r_{av} is the aerodynamic resistance, L is the latent heat of vaporization, γ is the psychrometric constant, and r_s is surface resistance. Because it is difficult to measure some of these terms, particularly the surface resistance term, methods have been developed to evaluate the equation for potential evaporation conditions using standard weather data, applying assumptions about a well-watered, vegetated surface; and then relating the potential evaporation to actual evaporation for a particular vegetative surface (such as early or late in the season with low vegetative cover or under water-stressed conditions) using crop coefficients and other adjustments. These methods are discussed thoroughly in Allen et al.[4] and Allen.[5]

Radiation Balance and Soil Heat Flux

Net radiation is the balance resulting from the incoming and outgoing fluxes of short and long-wave radiation and is affected by several aspects of the soil and plant cover at the surface.[6] Short-wave radiation is strongly affected by surface roughness, color, soil water content, and solar angle. For soils and plants, emitted long-wave radiation is largely determined by surface temperature, so wetter soils (which are generally cooler) have less outgoing long-wave radiation than drier, warmer soils. Incident long-wave radiation is influenced by sky conditions (cloudiness, cloud type, etc.), and it will generally decline with increased sky cover. As vegetative cover increases, the effect of soil conditions on net radiation decreases to negligible levels. For calculation of short-term evaporation, soil heat flux is sometimes taken as a fraction of net radiation (~10%). For bare soils, such as early or late in a growing season, the fraction can be much higher. For longer time periods, changes in air temperature can be used to estimate the flux of heat into or out of the soil over the period.[4] As vegetative cover or crop residue cover increases, the flow of heat in the soil is reduced because temperature gradients in the soil decrease as the soil is shaded.

Drainage of Water across the Lower Boundary

In some cropping systems, the flux across the lower boundary of the root zone can be considered to be negligible. This is true primarily in semiarid or arid climatic regimes and in soils with a high water holding capacity and/or a low hydraulic conductivity, such as clay loams, silty clay loams, or in some cases silt loams and loams. It may also be true where there is a restrictive layer in the soil profile that prevents or slows the flow of water below the root zone. However, when considering year round water balances, soil water is often lost below the root zone during at least part of the annual cycle in periods of high precipitation or low evapotranspiration. Water losses below the root zone can be measured using lysimeters, or can be calculated using measurements of soil water tension or content over time along with knowledge of soil hydraulic properties. In some situations, water can move upward from a shallow water table into the root zone of deep rooted plants.

APPLICATIONS

Irrigation Scheduling

Irrigated agriculture is one of the most intensive forms of agriculture. Having control of the water supply to ensure adequate water for crop growth allows a producer to invest more in other inputs that ensure a high yield and quality of the crop. However, irrigation applications can be expensive and excessive water application can result in loss of nutrients and other production inputs as well as causing environmental problems. Therefore, it is important to apply enough, but not too much, irrigation water. One way to do this is to monitor soil water content during the growing season and apply knowledge of the soil water balance to guide timing and amount of irrigation applications. Some irrigation scheduling models use weather data to simulate the evapotranspiration and maintain a soil water budget to predict changes in the soil water content. Depending on the rate of water use by the crop and the amount of soil water storage capacity, the producer can project the upcoming needs for irrigation applications.

Rainfed Cropping

Rainfed cropping is subject to great risks because of the high variability of rainfall in most agricultural regions. In many regions, water stored in the soil at planting time is an important component of the seasonal water supply. If there is a large amount of water stored at planting time, that stored water provides a buffer against dry periods during the growing season, and the producer might plan for an average or good yield level and invest in inputs to support those yield levels. If the water storage at planting is low, then the risk of crop losses due to growing season drought is high and the producer may decide to reduce or delay investment in some inputs until later in the season when more is known about growing season precipitation and forecasts. Analysis of long-term climatological records using a soil water balance approach can be used to evaluate alternative crops or

rotations for a region. In some cases, high soil water levels at the end of the growing season will result in a low faction of off-season precipitation being stored for the next crop with greater losses to percolation (with some possible nutrient leaching) or runoff (with possible greater erosion).

Plant Growth and Natural Resource Modeling

Soil water balance calculations are an integral part of plant growth and hydrologic models. Many of these models have been developed to operate at a daily time step and have been applied to a wide range of analyses. Some examples of such models that are available for downloading from the internet include crop growth, erosion, and hydrology models developed at the Grassland Soil and Water Research Laboratory at Temple, Texas, http://arsserv0.tamu.edu/intro.htm; the soil organic matter model, CENTURY, developed at the Natural Resource Ecology Laboratory at Colorado State University, http://www.nrel.colostate.edu/projects/century5/; water balance, irrigation management, and soils models developed by Dr. J. T. Ritchie and colleagues, http://nowlin.css.msu.edu/; the Decision Support System for Agrotechnology Transfer, DSSAT, which is a series of crop models and associated weather, crop, and soil data bases, http://icasanet.org/dssat/ and a suite of models, ranging from nutrient management and water quality models to an operational tool for whole farm/ranch strategic planning developed by the Great Plains System Research Unit at Fort Collins, Colorado, http://gpsr.ars.usda.gov/products/. These models, and many more, include a soil water balance as an integral part of the system.

CONCLUSION

The soil water balance approach has played an important role in improving our understanding and management of plant, water, and soil resources. The field soil water balance involves accounting for inputs of water to the system, such as precipitation and irrigation, as well as water leaving the system via evapotranspiration, runoff, and drainage below the root zone. The way a field is managed can have a large impact on the magnitude of the components of the water balance, such as runoff and drainage, as well as patterns of evapotranspiration and partitioning of the water loss into soil evaporation and transpiration. In agriculture, increasing the amount of transpiration increases productivity of the system. Soil water balance approaches can be applied to irrigated and rainfed agriculture and are an integral part of all plant growth and natural resource model, so understanding the basic concepts and principles is important for sound water management.

REFERENCES

1. Gardner, W.R. Soil properties and efficient water use: an overview. In *Limitations to Efficient Water Use in Crop*

Production; Taylor, H.M., Jordon, W.R., Sinclair, T.R., Eds.; Am. Soc. Agron., Crop Sci. Soc. Am., Soil Sci. Soc. Am.: Madison, WI, 1983; 45–71.

2. Penman, H.L. Natural evaporation from open water, bare soil, and grass. Proc. R. Soc. Lond. **1948**, 120–145.

3. Monteith, J.L. *Evaporation and Environment*, Symp. Soc. Exp. Biol.; Academic Press: New York, 1965; Vol. 19, 205–234.

4. Allen, R.G.; Pereira, L.S.; Raes, R.; Smith, M. *Crop Evapotranspiration—Guidelines for Computing Crop Water Requirements—FAO Irrigation and Drainage Paper 56*; Food and Agriculture Organization of the United Nations: Rome, 1998; 300.

5. Allen, R.G. Using the FAO-56 dual crop coefficient method over an irrigated region as part of an evapotranspiration intercomparison study. J. Hydrol. **2000**, *229*, 27–41.

6. Steiner, J.L. Crop residue effects on water conservation. In *Managing Agricultural Residues*; Unger, P.W., Ed.; Lewis Publishers: Boca Raton, Florida, 1994; 41–76.

Field—Low

Fisheries: Conservation and Management

Steven X. Cadrin
Department of Fisheries Oceanography, University of Massachusetts, Fairhaven, Massachusetts, U.S.A.

Abstract

Fisheries can be effectively managed to conserve ecosystem resources and achieve societal objectives such as seafood production, recreational opportunities, and sustenance of fishing communities. Fisheries science involves monitoring fisheries and fishery resources, assessing status of the resource and fishery, and predicting the response of the resource to alternative management decisions. Trends in fisheries management include ending overfishing, application of precautionary approaches, management strategy evaluation, and consideration of ecosystems. Maximum sustainable yield has played a central role in the history of fishery management and continues to be an organizing concept in fisheries science.

INTRODUCTION

Fisheries involve the utilization of living aquatic resources by individuals, communities, businesses, and societies. Fisheries provide opportunities for recreation, subsistence fishing, commercial harvesting, seafood processing and distribution, and sale for consumption. Advanced fishing technologies have the potential to deplete natural resources and alter aquatic ecosystems, requiring management of fishing activities to achieve societal objectives. Fisheries science is a multidisciplinary field involving limnology, oceanography, biology, engineering, economics, anthropology, and international policy. Effective fisheries management can achieve sustainable fisheries and productive ecosystems.

The goal of this entry is to briefly summarize the field of fisheries science, with a focus on stock assessment and fishery management. Similar to many disciplines, a description of the historical foundations of fisheries science helps to understand the basis of fishery conservation and management. Several recent trends in fisheries science and management are expected to continue in the future.

SUSTAINABLE FISHERIES

In 1882, Thomas Henry Huxley provoked the scientific community with his confident affirmation that "… *probably all the great sea fisheries are inexhaustible…*".[1] The subsequent debate on the exhaustibility of fishery resources was central to the formation of the International Council for the Exploration of the Seas and its initial task of considering the possibility of overfishing.[2–4] In the early 1900s, overfishing was defined as a magnitude of fishery removals that exceeds the fished population's capacity to replace itself, and sustainable yield was defined as a harvest that is less than the population's biological production (Fig. 1).[5] Each component of biological production (somatic growth, reproduction, and survival) is density dependent (i.e., their contribution to production depends on how many fish are in the population); so, sustainable yield is also a function of population size, and there is an intermediate population size that can produce maximum sustainable yield (MSY).[6] Given that population size is expected to decrease as harvest rate increases, there is also a harvest rate, or fishing mortality, associated with MSY (F_{msy}; Fig. 2).

Single-Species Stock Assessment and Fishery Management

Stock assessment involves fitting population dynamics models to fisheries information to estimate stock size and fishing mortality. The primary information used in stock assessment is from monitoring of the fishery removals and surveys of the resource. Surveys are statistically designed to represent trends in stock abundance or biomass as well as demographic information (e.g., age, size, gender, or spatial distributions) and life history traits (e.g., size and maturity at age, food habits). Fishery monitoring involves all components of catch, including landings and discarded catch, as well as fishing effort, catch rates, and demographic distributions of catch.

Fishery production can be modeled as aggregate biomass dynamics[7–9] or demographic processes to explicitly account for mortality, growth, and reproduction.[10–12] Most stock assessment models assume that the fishery exploits a unit stock that is more influenced by internal dynamics than interactions with adjacent stocks, but many management units also reflect practical considerations of jurisdiction and fishing grounds.[13]

Encyclopedia of Natural Resources DOI: 10.1081/E-ENRW-120047584

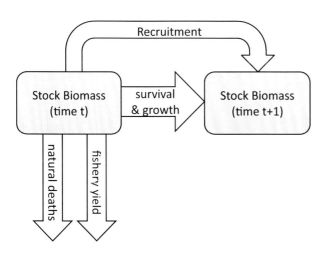

Fig. 1 Schematic representation of the components of biological production (survival, somatic growth, and recruitment) and fishery yield.

Models with size and age structure are considered the most informative approach to stock assessment.[14,15] Each component of production can be used to model population dynamics, by fitting a numerical model to observed data from fisheries and fishery resources to estimate parameters for fishery management (e.g., historical and current stock size and fishing mortality). Demographic approaches include length-based models,[16] stage-based models,[17] or more complex size and age structure.[15] Models that sufficiently represent population and fishery dynamics can be used to predict how the fishery resource will respond to alternative management actions.[14] Long-term projection can be used to derive the fishing mortality that produces maximum average yield and associated MSY reference points.[18]

Age-based survival is modeled using exponential mortality rates (Eq. 1).

$$N_{t+1} = N_t e^{-(F+M)} \qquad (1)$$

N_t: abundance at time t

F: fishing mortality rate

M: natural mortality rate

Somatic growth is either derived empirically, using samples of size at age, or theoretically modeled (Eq. 2).[19]

$$L_a = L_\infty \left[1 - e^{-k(a-a_0)} \right] \qquad (2)$$

L_t: length at age a

L_∞: asymptotic size

k: growth coefficient

a_0: age at length = 0

Reproduction is typically considered to be a theoretical expectation in which recruitment (the abundance of young fish) is a function of reproductive potential or the spawning stock biomass (Eq. 3).

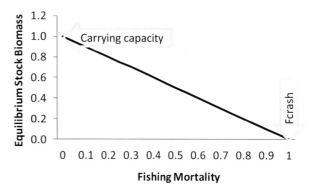

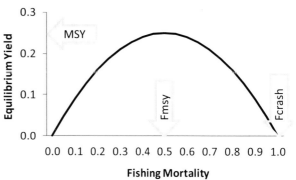

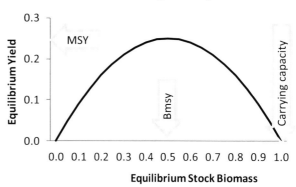

Fig. 2 Equilibrium expectations of fishing, with reduction of stock biomass from the unfished carrying capacity (upper panel), maximum sustainable yield (MSY) at intermediate fishing mortality rate (Fmsy) and equilibrium stock size (Bmsy) (middle and lower panels, respectively).

Ricker stock—recruit relationship:

$$N_{a,t+a} = aS_t e^{-\beta S_t}$$

Beverton—Holt stock—recruit relationship: $\qquad (3)$

$$N_{a,t+a} = \frac{aS_t}{b+S_t}$$

$N_{a,t}$: abundance at age of recruitment, a

S_t: biomass of spawners in time t

Recruitment can also be derived through more empirical relationships.[20–22]

Age-based methods are used to model population replacement as a function of fishing mortality, and similar properties of sustainable yield demonstrated by biomass dynamics (Fig. 2) are also found with demographic models so that MSY and associated fishing mortality and population size can be estimated.[20]

Fishery Management Tactics and Strategies

Governance of fisheries systems varies widely from "top-down" regulation by governments for fishery resources that are under their jurisdiction to "bottom-up" systems that involve co-operative behavior of fishermen to achieve sustainable fisheries and responsible fishing practices.[23] Fisheries can be managed through a large variety of regulatory approaches. "Input controls" include limited entry (e.g., fishing licenses), limiting fishing effort (e.g., limited days' fishing, seasonal or annual area closures), and restrictions on fishing gear specifications (e.g., mesh sizes, hook sizes). "Output controls" involve limiting the total fishery removals (e.g., total allowable catch, annual catch limits, daily catch limits, minimum legal size of fish). In limited-entry fisheries with output controls, individuals are allocated a portion of the total allowable catch each year, and some individual allocations are transferable to others (i.e., individual transferrable quotas). Management tactics are typically tailored to the type of fishery (e.g., small-scale artisanal vs. large-scale industrial), the availability of fishery-monitoring information, and management objectives. Effective fishery management requires enforcement of regulations to avoid illegal and unreported fishing.

Fishery production can also be enhanced through protection of essential fish habitat, mitigation of incidental bycatch, as well as culture and stocking of young fish. Habitat conservation extends to regulation of human uses of aquatic ecosystems beyond fisheries (e.g., coastal development, fish passage in rivers, pollution), and marine protected areas are designed to protect habitats from multiple human activities. Bycatch in mixed-species fisheries can be mitigated by using more selective fishing gears or bycatch avoidance through spatiotemporal fishing patterns.[24] Aquaculture programs can also rear fish to marketable sizes and contribute to seafood production.[25] Stocking programs and aquaculture can be productive, but may also have the unintended consequences of introducing nonnative species, which have caused major perturbations of many aquatic ecosystems.[26]

Many fishery management systems attempt to achieve optimum yield by limiting fisheries to the fishing mortality that produces MSY (Fmsy).[18] Simple economics can be coupled with population dynamics models to incorporate costs and derive revenue and profit as well as maximum economic yield (Fig. 3).[27] These biological and economic reference points can form fishery management objectives. Fishery management can also be aimed at other utilities, such as employment and sustaining fishing communities.

As a result of environmental variability and limited sampling of fishery resources, fisheries science is inherently

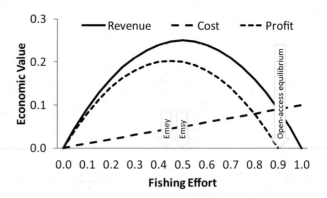

Fig. 3 Revenue, cost, and profit as a function of fishing effort to illustrate the open-access equilibrium expectation as well as the fishing effort associated with maximum sustainable yield (Emsy) and maximum economic yield (Emey).

uncertain. Recognition of it as an uncertain science led to the widespread adoption of a precautionary approach to fishery management in the 1990s that involves quantification of uncertainty in estimates (e.g., overfishing limits), derivation of risk-averse targets,[28] and risk-based management decisions.[29,30] Precautionary approaches involve an evaluation of trade-offs between the costs and benefits of conservation and utilization.

Performance of alternative management tactics and strategies for achieving stated objectives can be evaluated through adaptive learning.[31] Best practices or management procedures can also be determined using simulation techniques that allow management strategy evaluation before decisions are implemented.[32] In some situations, simple harvest control rules based on survey or fishery-monitoring information perform better for achieving objectives than more complicated stock assessment and management procedures.[33] A recent trend in fisheries is seafood certification to inform consumers about sustainability of fisheries to promote best practices in fishery management.

Ecosystem Approaches to Fishery Management

The single-species convention for stock assessment and fishery management developed during the industrialization of fishing technology and widespread depletion of fishery resources in the 20th century. Although overfishing persists in many fisheries, the implementation of management strategies to end overfishing have effectively reduced fishing mortality to within-sustainable limits in many regions and rebuilt many depleted stocks. The successes of single-species management strategies have reduced the effect of fishing on many populations and increased the influence of abiotic factors (e.g., climate change)[34] and biological interactions (e.g., predation and competition) with other species, prompting a transition to multispecies and ecosystem models. Ecological approaches to fisheries science and management involve the incremental inclusion of abiotic or biotic factors in single-species population dynamics models[35] or the holistic modeling of energy through aquatic ecosystems.[36] The integration of ecosystem components as well as the consideration of all human uses of aquatic systems promoted the development of integrated ecosystem assessments.[37] Ecosystem approaches to management require confrontation of many trade-offs for multiple ecosystem utilities.[38]

CONCLUSION

Overfishing persists in many fisheries, but fishery conservation and management have effectively ended overfishing in many regional management systems and achieved rebuilding toward the production of MSY. The concept of MSY encompasses all components of biological production, density dependence in the regulation of populations, and fishery dynamics. Bioeconomic extensions of MSY include broader definitions of optimum yield. The concept of MSY has endured many failures and critiques to be reconceived in the context of non-equilibrium dynamics, uncertainty, risk, and the application of precautionary and ecosystem approaches.[18] MSY is expected to continue to be an organizing concept in the future of fisheries conservation and management.

ACKNOWLEDGMENTS

I thank my mentors, peers, and students for the many lessons reflected in this brief summary of the diverse and challenging field of fisheries science.

REFERENCES

1. Huxley, T.H.H. Inaugural Address. Fisheries Exhibition, London. 1883; Blinderman, C., Joyce, D. 1998. The Huxley File. http://aleph0.clarku.edu/huxley/SM5/fish.html.
2. Sinclair, M. *Marine Populations: An Essay on Population Regulation and Speciation*; Washington Sea Grant Program: Seattle, 1988.
3. Smith, T.D. Stock assessment methods: the first fifty years. In *Fish Population Dynamics*, 2nd edn.; Gulland, J.A. Ed.; Wiley: New York, 1988; 1–34.
4. Smith, T.D. *Scaling Fisheries, The Science of Measuring the Effects of Fishing, 1855–1955*; Cambridge University Press: Cambridge, England, 1994.
5. Russell, E.S. Some theoretical considerations on the "overfishing" problem. Conseil Permanent International pour l'Exploration de la Mer **1931**, *6*, 3–20.
6. Graham, M. Modern theory of exploiting a fishery and application to North Sea trawling. J. Cons. Int. Explor. Mer. **1935**, *10*, 264–274.
7. Schaefer, M.B. Some considerations of population dynamics and economics in relation to the management of the commercial marine fisheries. J. Fish. Res. Bd. Can. **1957**, *14*, 669–681.
8. Polachek, T.; Hilborn, R.; Punt, A.E. Fitting surplus production models: comparing methods and measuring uncertainty. Can. J. Fish. Aquat. Sci. **1993**, *50*, 2597–2607.
9. Prager, M.H. A suite of extensions to a nonequilibrium surplus-production model. Fish. Bull. **1994**, *92*, 374–389.
10. Thompson, W.F.; Bell, F.H. Effect of changes in intensity upon total yield and yield per unit of gear. Rep. Int. Fish. Comm. **1934**, *8*, 7–49.
11. Ricker, W.E. Stock and recruitment. J. Fish. Res. Bd. Can., **1954**, *11*, 559–623.
12. Beverton, R.J.H.; Holt, S.H. On the dynamics of exploited fish populations. Fish. Invest., Ser. II, Mar. Fish., G. B. Minist. Agric. Fish. Food, **1957**, *19*, 1–533.
13. Cadrin, S.X.; Friedland, K.D.; Waldman, J. *Stock Identification Methods: Applications in Fishery Science*; Elsevier Academic Press: San Diego, 2005.
14. Hilborn, R.; Walters, C.J. *Quantitative Fisheries Stock Assessment: Choice, Dynamics & Uncertainty*; Chapman and Hall: New York, 1992.
15. Quinn, T.J. II; Deriso, R.B. *Quantitative Fish Dynamics*; Oxford University Press: New York, 1999.
16. Fournier, D.A.; Sibert, J.R.; Majkowski, J.; Hampton, J. MULTIFAN: a likelihood-based method for estimating growth parameters and age composition from multiple length frequency data sets illustrated using data for southern bluefin tuna (*Thunnus maccoyii*). Can. J. Fish. Aquat. Sci. **1990**, *47*, 301–317.
17. Collie, J.S.; Sissenwine, M.P. Estimating population size from relative abundance data measured with error. Can. J. Fish. Aquat. Sci., **1983**, *40*, 1871–1879.
18. Mace, P.M. A new role for MSY in single-species and ecosystem approaches to fisheries stock assessment and management. Fish and Fisheries, **2001**, *2*, 2–32.
19. Bertalanffy, L. von; A quantitative theory of organic growth (Inquiries on growth laws. II). Human Biol. **1938**, *10*, 181–213.
20. Shepherd, J.G. A versatile new stock–recruitment relationship for fisheries, and the construction of sustainable yield curves. J. Cons. Int. Explor. Mer. **1982**, *40*, 67–75.
21. Barrowman, N.J.; Myers, R.A. Still more spawner–recruitment curves: the hockey stick and its generalizations. Can. J. Fish. Aquat. Sci., **2000**, *57*, 665–676.
22. Getz, W.M.; Swartzman, G.L. A probability transition matrix model for yield estimation in fisheries with highly variable recruitment. Can. J. Fish. Aquat. Sci., **1981**, *38*, 847–855.
23. Ostrom, E. *Governing the Commons: The Evolution of Institutions for Collective Action*. Cambridge University Press: New York, 1990.
24. Alverson, D.L.; Freeburg, M.H.; Murawski, S.A.; Pope, J.G. A global assessment of fisheries bycatch and discards. FAO Fish. Tech. Paper **1994**, *339*.
25. Nash, C. *The History of Aquaculture*; John Wiley and Sons: New York, 2011.
26. Naylor, R.L.; Williams, S.L.; Strong, D.R. Aquaculture – A Gateway For Exotic Species. Science, **2001**, *294*, 1655–1656.

27. Gordon, H.S. An economic approach to the optimum utilization of fishery resources. J. Fish. Res. Bd. Can. **1953**, *10*, 442–457.

28. Garcia, S.M. The precautionary approach to fisheries and implications for fishery research, technology and management: an updated review. FAO Fisheries Tech. Paper, **1996**, *350* (2), 1–76.

29. Smith, S.J.; J; Hunt, J.J.; Rivard, D. Risk evaluation and biological reference points for fisheries management. Can. Spec. Publ. Fish. Aquat. Sci. **1993**, *120*.

30. Prager, M.H.; Shertzer, K.W. Deriving Acceptable Biological Catch from the Overfishing Limit: Implications for Assessment Models. N. Am. J. Fish. Manag., **2010**, *30*, 289–294.

31. Walters, C. *Adaptive Management of Renewable Resources*; Macmillan: New York, 1986.

32. Cooke, J.G. Improvement of fishery-management advice through simulation testing of harvest algorithms. ICES J. Mar. Sci. **1999**, *56*, 797–810.

33. Butterworth, D.S.; Punt, A.E. Experiences in the evaluation and implementation of management procedures. ICES J. Mar. Sci. **1999**, *56*, 985–998.

34. Cushing, D.H. *Fisheries Biology: A Study in Population Dynamics*; 2nd ed.; University of Wisconsin Press: Madison, 1982.

35. Rothschild, B.J. *Dynamics of Marine Fish Populations*; Harvard University Press: New York. 1986.

36. Steele, J.H.; Ruzicka, J.J. Constructing end-to-end models using ECOPATH data. J. Mar. Syst. **2011**, *87*, 227–238.

37. Levin P.S.; Fogarty, M.J.; Murawski, S.A.; Fluharty, D. Integrated ecosystem assessments: developing the scientific basis for ecosystem-based management of the ocean. Public Libr. Sci. (PLOS) Biol. **2009**, *7* (1), 23–28.

38. Link, J.S. *Ecosystem-Based Fisheries Management: Confronting Tradeoffs*; Cambridge University Press: New York, 2010.

Field—Low

Floodplain Management

French Wetmore

French & Associates, Ltd., Park Forest, Illinois, U.S.A.

Abstract

Throughout time, floods have altered the landscape. Flooding is a natural process and floodplains are created and altered by that process. Floodplains have also been altered by human development, with consequences to those who live in them. During the early settlement of the United States, locations near water provided required access to transportation, a water supply, and water power. These areas had fertile soils, making them prime agricultural lands. In recent decades, development along waterways and shorelines has been spurred by the recreational value of these sites. The result has been an increasing level of damage and destruction wrought by the natural forces of flooding on human development. Flooding has become the nation's number one natural hazard. It affects more property each year and has accounted for over 70% of the Presidential disaster declarations since 1970.

HISTORICAL APPROACHES

During the 1920s, the insurance industry concluded that flood insurance could not be a profitable venture because the only people who would want flood coverage would be those who lived in floodplains. As they were sure to be flooded, the rates would be too high to attract customers. Unlike other hazards, such as wind and hail, where the risk can be spread, private industry opted out of playing a role in flood protection.

With the great Mississippi River flood of 1927, the federal government became a major player in flooding. As defined by several Flood Control Acts, the role of government agencies was to build massive flood control structures to control the great rivers, protect coastal areas, and prevent flash flooding.

Until the 1960s, such structural flood control projects were seen as the primary way to reduce flood losses. In some areas, they still are. However, starting in the 1960s, people questioned the effectiveness of this single solution. Disaster relief expenses were going up, making all taxpayers pay more to provide relief to those with property in floodplains. Studies during the 1960s concluded that flood losses were increasing, in spite of the number of flood control structures that had been built.

One of the main reasons structural flood control projects failed to reduce flood losses was that people continued to build in floodplains. In response, federal, state, and local agencies began to develop policies and programs with a "non-structural" emphasis, ones that did not prescribe projects to control or redirect the path of floods.

A milestone in this effort was the creation of the National Flood Insurance Program (NFIP) in 1968. The NFIP is based on a mutual agreement between the Federal government [represented by the Federal Emergency Management Agency (FEMA)] and local governments. Federally guaranteed flood insurance is made available in those communities that agree to regulate development in their mapped floodplains.

If the communities do their part in making sure future floodplain development meets certain criteria, FEMA will provide flood insurance for properties in the community. The Federal government is willing to support insurance because, over time, local practices will reduce the exposure to flood damage.

Also during the 1960s and 1970s, interest increased in protecting and restoring the environment, including the natural resources and functions of floodplains. Coordinating flood loss reduction programs with environmental protection and watershed management programs has since become a major goal of federal, state, and local programs. This evolution is shown graphically in Fig. 1. Now, we no longer depend solely on structural projects to control floodwater. Instead of "flood control," we now speak of "floodplain management."

FLOODPLAIN MANAGEMENT

Floodplain management is officially defined by the Federal Government's *Unified National Program for Floodplain Management* as "a decision-making process that aims to achieve the wise use of the nation's floodplains." (see FEMA,[1] p. 8) "Wise use" means both reduced flood losses and protection of the natural resources and functions of floodplains. This is accomplished through different tools, including, but not limited to:

- Floodplain mapping.
- Land use regulations.

Encyclopedia of Natural Resources DOI: 10.1081/E-ENRW-120010108

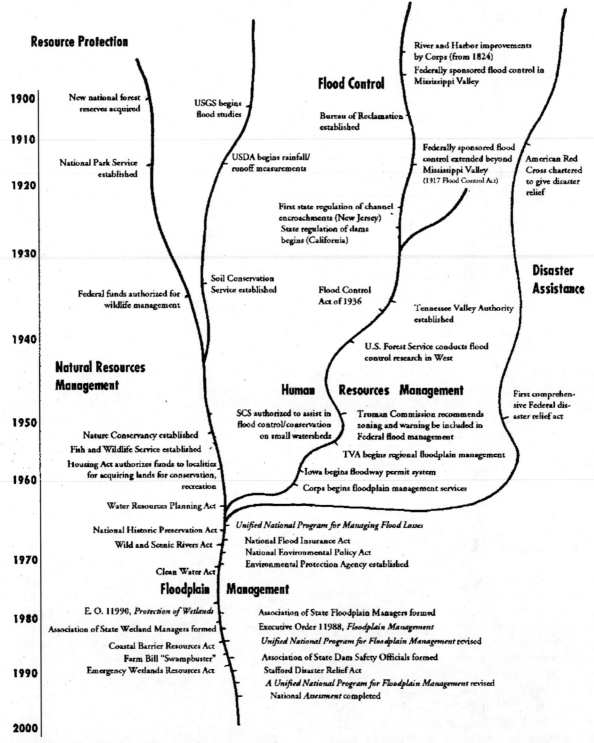

Resource Protection

1900 — New national forest reserves acquired

USGS begins flood studies

Flood Control

River and Harbor improvements by Corps (from 1824)
Federally sponsored flood control in Mississippi Valley

Bureau of Reclamation established

1910 —

National Park Service established

USDA begins rainfall/runoff measurements

Federally sponsored flood control extended beyond Mississippi Valley (1917 Flood Control Act)

American Red Cross chartered to give disaster relief

1920 —

First state regulation of channel encroachments (New Jersey)
State regulation of dams begins (California)

1930 —

Disaster Assistance

Federal funds authorized for wildlife management

Soil Conservation Service established

Flood Control Act of 1936

Tennessee Valley Authority established

1940 —

U.S. Forest Service conducts flood control research in West

Natural Resources Management

Human   Resources   Management

SCS authorized to assist in flood control/conservation on small watersheds

Truman Commission recommends zoning and warning be included in Federal flood management

First comprehensive Federal disaster relief act

1950 —

Nature Conservancy established
Fish and Wildlife Service established
Housing Act authorizes funds to localities for acquiring lands for conservation, recreation

TVA begins regional floodplain management
Iowa begins floodway permit system
Corps begins floodplain management services

1960 —

Water Resources Planning Act

Unified National Program for Managing Flood Losses

National Historic Preservation Act
Wild and Scenic Rivers Act

National Flood Insurance Act
National Environmental Policy Act
Environmental Protection Agency established

1970 —

Clean Water Act

Floodplain   Management

E. O. 11990, *Protection of Wetlands*

Association of State Floodplain Managers formed

1980 —

Association of State Wetland Managers formed

Executive Order 11988, *Floodplain Management*

Coastal Barrier Resources Act

Unified National Program for Floodplain Management revised

Farm Bill "Swampbuster"

Association of State Dam Safety Officials formed

Emergency Wetlands Resources Act

Stafford Disaster Relief Act

1990 —

A Unified National Program for Floodplain Management revised
National *Assessment* completed

2000 —

Fig. 1   Evolution of floodplain management in the United States.

- Preservation of floodprone open space.
- Flood control (levees, reservoirs, channel modifications, etc.)
- Acquiring and clearing damaged or damage-prone areas.
- Floodproofing buildings to reduce their susceptibility to damage by floodwaters.
- Flood insurance.

- Water quality best management practices.
- Flood warning and response.
- Wetland protection programs.
- Public information.

There are a variety of Federal, state, and local programs that administer these tools. Private organizations and property owners also have roles.

THE NATIONAL FLOOD INSURANCE PROGRAM

The nation's focal floodplain management program is the NFIP. It has prepared floodplain maps for 22,000 communities. FEMA sets the minimum land use development standards that participating communities must administer within the floodplains designated on their Flood Insurance Rate Maps. These standards are summarized in Fig. 2.

While participation is voluntary, communities that decide not to join or not to enforce those regulations do not receive Federal financial assistance for insurable buildings in their floodplains. Rather than face the loss of Federal aid (including VA home loans, HUD housing help, and disaster assistance), just about every community with a significant flood problem has joined. By 2002, 19,700 cities and counties were participating.

Within participating communities, Federal law requires the purchase of a flood insurance policy as a condition of receiving Federal aid, including mortgages and home improvement loans from Federally regulated or insured lenders. This requirement, coupled with personal experiences with flooding, has convinced over four million property owners to buy flood insurance. Unfortunately, it is estimated that only half of the properties in the FEMA mapped floodplains are insured.

The National Flood Insurance Program (NFIP) is administered by the Federal Emergency Management Agency (FEMA). As a condition of making flood insurance available for their residents, communities that participate in the NFIP agree to regulate new construction in the area subject to inundation by the 100-year (base) flood.

There are four major floodplain regulatory requirements. Additional floodplain regulatory requirements may be set by state and local law.

1. All development in the 100-year floodplain must have a permit from the community. The NFIP regulations define "development" as any manmade change to improved or unimproved real estate, including but not limited to buildings or other structures, mining, dredging, filling, grading, paving, excavation or drilling operations or storage of equipment or materials.

2. Development should not be allowed in the floodway. The NFIP regulations define the floodway as the channel of a river or other watercourse and the adjacent land areas that must be reserved in order to discharge the base flood without cumulatively increasing the water surface elevation more than one foot. The floodway is usually the most hazardous area of a riverine floodplain and the most sensitive to development. At a minimum, no development in the floodway may cause an obstruction to flood flows. Generally an engineering study must be performed to determine whether an obstruction will be created.

3. New buildings may be built in the floodplain, but they must be protected from damage by the base flood. In riverine floodplains, the lowest floor of residential buildings must be elevated to or above the base flood elevation (BFE). Nonresidential buildings must be either elevated or floodproofed.

4. Under the NFIP, a "substantially improved" building is treated as a new building. The NFIP regulations define "substantial improvement" as any reconstruction, rehabilitation, addition, or other improvement of a structure, the cost of which equals or exceeds 50 percent of the market value of the structure before the start of construction of the improvement. This requirement also applies to buildings that are substantially damaged.

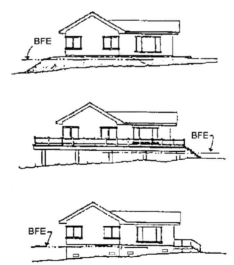

Communities are encouraged to adopt local ordinances that are more comprehensive or provide more protection than the Federal criteria. This is especially important in areas with older Flood Insurance Rate Maps that may not reflect the current hazard. Such ordinances could include prohibiting certain types of highly damage-prone uses from the floodway or requiring that structures be elevated 1 or more feet above the BFE. The NFIP's Community Rating System provides insurance premium credits to recognize the additional flood protection benefit of higher regulatory standards.

Fig. 2 Minimum National Flood Insurance Program regulatory requirements.

Field—Low

OTHER FEDERAL PROGRAMS

FEMA administers other floodplain management programs, including:

- Disaster assistance programs that help flooded communities and property owners recover after a flood.
- Mitigation assistance programs that fund local projects to acquire and clear floodprone properties.
- Research and technical assistance activities in the fields of mapping, planning, mitigation, and floodproofing.
- The National Dam Safety Program which assists state programs that regulate dams (dam failures were a factor in three of the four largest killer floods since 1970).

The U.S. Army Corps of Engineers is the second largest participant in Federal floodplain management programs. While it is best known as the builder of structural flood control projects, it has its own authority to regulate new development in navigable waterways and wetlands. It is also the leader in the technical aspects of floodproofing and river basin planning.

The U.S. Department of Agriculture's Natural Resources Conservation Service has a role in planning and building flood control projects, similar to the Corps,' but limited to smaller watersheds. Through local soil and water conservation districts, NRCS staff can be valuable advisors to local officials reviewing floodplain or watershed development proposals.

Just as rivers traverse many lands, floodplain management pervades many government programs. Other agencies with floodplain management responsibilities include:

- Tennessee Valley Authority (where floodplain management got its start)
- Bureau of Reclamation (water control projects in the west)
- U.S. Geological Survey (river data and mapping)
- Environmental Protection Agency (water quality programs)
- Small Business Administration (disaster assistance for private property owners)
- National Oceanic and Atmospheric Administration (coastal zone policies)
- National Weather Service (the lead in flood warning programs)

OTHER PROGRAMS

State and local agencies are also into a variety of floodplain management activities. Their regulatory programs often exceed the NFIP requirements. Many states set additional minimum standards for mapping, floodplain and wetland regulations and water quality. Some state agencies require their own permits, in addition to local permits, for new construction on waterways, lakes, shorelines, and floodplains.

In addition to being the lead regulators, most flood control projects are built and operated by local governments: cities, towns, counties, and special districts. The trend at the local level is toward special purpose authorities at the county or multicounty level to tackle problems holistically at the watershed level.

Private organizations have become more directly involved, too. Groups like the Nature Conservancy and land trusts work to preserve floodprone areas that have natural benefits. Others, like the National Wildlife Federation and American Rivers, are active on the political scene, reminding government agencies of their responsibilities and working to strengthen or expand their programs.

Over time, the distinction between what is done by what level of government has blurred. There are more and more cooperative and coordinated approaches, especially with increased non-federal cost sharing requirements and regional and river basin organizations. A recent example of this is FEMA's Cooperating Technical Partners program where a state or local government can contribute to the cost of floodplain mapping and have a say on the techniques and standards used to prepare their Flood Insurance Rate Maps.

Another reason for the blurring of the distinction is the increased professionalization of the field. Most people active in floodplain management are members of the Association of State Floodplain Managers. Private practitioners and staff from all levels of government work together on solving common problems, rather than debating authorities or funding. There is also a new program that certifies floodplain managers. In less than 3 years, over 1000 professionals have earned the right to put "CFM" after their names.

PROGRESS

The impact of these efforts can be measured in three ways: threat to life, property damage, and the environment. Statistics have shown that the loss of life due to floods decreased during the last century, primarily due to better warning and public information programs.

Progress in the other two fields has not been as encouraging. Property damage is still increasing, although at a slower rate than if there were no NFIP and other floodplain management efforts (Fig. 3). It is harder to see improvements in water quality and habitat protection, but it is generally concluded that while things are better than if there were no programs, we have a long way to go.

AGRICULTURAL CONCERNS

Farmers, ranchers, and other agricultural interests are likely to be involved in floodplain management in several different ways. First, as landowners, their freedom to

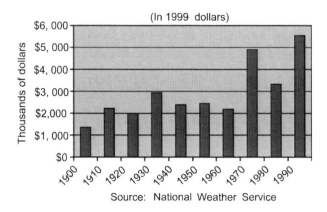

Fig. 3 Dollar damage caused by flooding.

develop the floodplain portions of their properties may be limited by floodplain management or wetland regulations.

Federal, state, and/or local regulations require permits for the following:

- Regrading in the floodway.
- Construction of a levee.
- Modifications to a channel.
- Filling in a wetland.
- Construction of a new building in the floodplain.

This is the controversial part of floodplain management: activities on one's own property are subject to government restrictions in order to prevent diverting flood flows to other properties or adversely affecting wetlands or habitat or to reduce government disaster response and assistance expenses. While many state laws exempt some agricultural activities from local zoning or building codes, FEMA has ensured that in every state, agricultural buildings will be regulated as a condition for a city or county to participate in the NFIP.

A loan or Federal financial assistance to purchase, improve or repair a building in the floodplain will likely be accompanied by a requirement to purchase a flood insurance policy on that building. However, by taking certain protection measures, such as elevating the building above flood levels, insurance premiums can be reduced.

Federal and state programs are not all about restrictive regulations. Federal disaster assistance, flood insurance and crop insurance can come to one's aid after a flood.

After the Great Flood of 1993 in the Mississippi River basin, many farmers accepted Federal funds to set aside wetlands and marginal farmland as a start to allowing Mother Nature to reclaim the natural floodplains.

Hopefully, farmers, ranchers, and other agricultural interests will become involved in floodplain management activities voluntarily and in a broader extent. They can reduce their own exposure to flood losses, help their communities and neighbors protect themselves, and improve their environment. Good places to learn more are the following websites:

- FEMA— http://www.FEMA.gov
- Association of State Floodplain Managers— http://www.floods.org

Both have links to other agencies and organizations. The latter has links to state floodplain management associations.

REFERENCE

1. *A Unified National Program for Floodplain Management*; FEMA 248, Federal Interagency Floodplain Management Task Force: Washington, DC, 1994.

BIBLIOGRAPHY

1. *Addressing Your Community's Flood Problems: A Guide for Elected Officials*; Association of State Floodplain Managers, Inc.: Madison, WI, 1996.
2. *Answers to Questions about the National Flood Insurance Program*; FEMA-387, http://www.fema.gov/nfip/qanda.htm (accessed February 2002) Federal Emergency Management Agency: Washington, DC 2001.
3. *Floodplain Management in the United States: An Assessment Report*; FIA-18, Federal Interagency Floodplain Management Task Force: Washington, DC, 1992.
4. *National Flood Programs in Review*, http://www.floods.org/PDF%20files/2000-fpm.pdf (accessed February 2002) Association of State Floodplain Managers: Madison, WI, 2000.
5. *Using Multi-Objective Management to Reduce Flood Losses in Your Watershed*; Association of State Floodplain Managers, Inc.: Madison, WI, 1996.

Groundwater Contamination

Charles W. Fetter, Jr.
C. W. Fetter, Jr. Associates, Oshkosh, Wisconsin, U.S.A.

Abstract

Contamination can be defined as the presence of a biological or chemical agent in groundwater in such a concentration that it renders water unfit for a particular use. Contaminants can be from both anthropogenic and natural sources.

INTRODUCTION

Contamination can be defined as the presence of a biological or chemical agent in groundwater in such a concentration that it renders water unfit for a particular use.[1] Agricultural uses of water include domestic drinking water, stock watering, and irrigation. Water that is contaminated for purposes of drinking might be perfectly suitable for use in irrigation.

Contaminants can be from both anthropogenic and natural sources, for example, arsenic. Arsenic found in groundwater in northeastern Wisconsin comes from a naturally occurring mineral, arsenopyrite, present in aquifer. Arsenic has also become a contaminant in groundwater due to use its use in agriculture as a pesticide as well as industrial sites where arsenic was used as a wood preservative.[1] The drinking water standard for arsenic in the United States for many years was $50\,\mu g/L$ (micrograms per liter). However, in May 2000 the U.S. Environmental Protection Agency reviewed the standard and it was lowered to $10\,\mu g/L$.

TYPES OF CONTAMINANTS

Groundwater contaminants fall into two broad categories, biological and chemical. Biological contaminants include bacteria, viruses, and protozoa. Chemical contaminants can be classified as organic or inorganic. Organic chemicals are based on a framework of carbon and hydrogen atoms. Inorganic compounds include all other chemicals, although some will have carbon present in an inorganic form, such as carbonate (CO_3^{2-}) and bicarbonate (HCO_3^-).

Organic chemicals include fuels and most pesticides. Fuels such as gasoline and diesel are composed of hundreds of different organic chemicals in varying proportions depending upon the source, and their composition will vary depending upon the season. Fuels do not mix with groundwater, rather if present in the ground they will float on the water table. They are sometimes referred to as Light Non-Aqueous Phase Liquid (LNAPL) as they are less dense than water. However, some of the chemicals that comprise gasoline and diesel will separate from fuel into a dissolved form in the groundwater. The most soluble of these chemicals are benzene, toluene, ethylbenzene, and xylenes. They are referred to by acronym BTEX.[2] Some organic pesticides may be soluble in water as they may be mixed with water prior to application to a field.

Inorganic chemicals found in groundwater are salts that dissociate into cations and anions when in contact with water. The cations include heavy metals such as iron, lead, manganese, cadmium, chromium, zinc, and mercury. The anions include nitrate (NO_3^-), nitrite (NO_2^-), sulfate (SO_4^{2-}), fluoride (F^-), chloride (Cl^-), arsenate (AsO_4^{3-}) and arsenite (AsO_3^{3-}).

SOURCES OF CONTAMINATION

Sources of contamination can be divided into point sources and non-point sources. As the name implies, point sources can be traced to a very specific location. An example of a point source might be a septic tank, a landfill, or a pesticide mixing area. Non-point sources are dispersed across the landscape. Fertilizer and pesticides applied to fields are examples of non-point sources.

Human and animal wastes are sources of potential groundwater contamination due to the presence of bacteria and viruses as well as nitrogen compounds. One chemical compound frequently found in groundwater in rural areas is nitrate. This can come from cesspools and septic tanks, barnyards, manure spread as fertilizer, and chemical fertilizers. Nitrate and nitrite in drinking water in excess of $10\,mg/L$ (milligrams per liter) as nitrogen have been implicated in infant methamoglobanemia or "blue baby syndrome." Another salt found in animal waste is chloride.

Encyclopedia of Natural Resources DOI: 10.1081/E-ENRW-120010305

This will impart a salty taste in drinking water if present in amounts in excess of 250 mg/L.

Pesticides are also a potential source of groundwater contamination. They can be found concentrated in areas where pesticides are mixed or equipment is washed. Likewise pesticides can be a non-point source of contamination when they are spread on a field. For example, atrazine has been found in groundwater in Wisconsin as a result of use on corn crops. Not only can pesticides occur in the environment, but breakdown products called metabolites can also occur in groundwater.

When water is used for irrigation, some will evaporate. This will concentrate the soluble salts in the remaining water, which will drain down to the water table. As a result, toxic salts may build up in the soil and groundwater. This situation has developed in some areas of California with selenium.

Fuels used on the farm can leak from underground storage tanks resulting in the formation of a pool of LNAPL on the water table below the tank and dissolved BTEX chemicals in the groundwater. The federal drinking water standard for benzene in the United States is 5 µg/L. In some states the groundwater standard is even lower, 1 µg/L.

Chemicals used for degreasing equipment can also contaminate groundwater if improperly disposed.[3] Many degreasers contain chlorinated organic compounds such as trichloroethylene (TCE) and 1,1,1-trichlorethane (TCA). These liquids are denser than water and mix poorly with water. They are referred to by acronym DNAPL. If disposed into the environment, for example by spilling on soil, they can migrate vertically to the water table and then sink below the water table into the underlying aquifers. These compounds are sparingly soluble in water, but even small amounts are dangerous. The federal drinking water standard for trichloroethylene in the United States is 5 µg/L. In some states the groundwater standard is even lower, 1 µg/L.

Chemicals used in wood preservatives are also potential groundwater contaminants. These include creosote and CCA (copper, chromium, arsenic). Treated wood itself would most likely not contaminate groundwater, but spilled or improperly disposed wood-treating chemicals could contaminate the groundwater.

EFFECT OF CLIMATE

In humid climates, the water table may be close to the surface and frequent rains can leach contaminants from the soil and transport them down to the water table. If the climate is more arid, contaminants in the soil zone are less likely to be transported to the water table, which itself is likely to be deeper than in a corresponding area that is more humid. However, evaporation of irrigation water in arid climates may result in a build up of soluble salts in the soil and the excess irrigation water that may eventually reach the water table.

TRANSPORT OF CONTAMINANTS

Dissolved contaminants are carried by flowing groundwater through a process called *advection*. If the contaminant is *conservative*, it will move at the same rate as the groundwater in which it is dissolved. An example of a dissolved salt that is conservative is chloride. Water flowing through an aquifer will not all be moving at the same rate. Groundwater moves through pores and cracks in the ground. Some of these openings in the ground are larger than others, and water in the larger openings will be moving faster than water in the smaller openings. As a result, the faster moving water will spread out in front of the rest of the mass of the water. If a contaminant is present in a low concentration, the closer the contaminant gets to the moving front the faster moving water mixes with uncontaminated water. This process is called *longitudinal dispersion*. Through dispersion and diffusion a *plume* of groundwater contamination is formed. This plume is nothing more than a contiguous zone where the contaminant is present in the groundwater. If there is an ongoing source of contamination at the start of the plume, the greatest concentration of the contaminant will be found there and the concentration will decrease in the direction of the groundwater flow. The contaminant plume will extend along the direction of groundwater flow, but also spread sideways through a process called *lateral dispersion*. This is due to the flowing groundwater taking branching pathways.[4]

Non-aqueous phase liquids also have the potential to move through the soil and underlying aquifers. Their movement is dependent upon the ability of the non-aqueous phase liquid to overcome capillary forces and displace air in the pores above the water table and water in the pores of the earth below the water table.

FATE OF CONTAMINANTS

Biological agents are particles of protoplasm. As such they can travel through large pores and cracks in the earth, but not small ones. Fine-grained soils can remove bacteria and viruses by *filtration*, usually within a few hundred meters or less of the source. Some aquifers such as coarse gravel, fractured rock, and carbonate rock have larger openings. Bacteria and viruses can travel for significant distances in such aquifers.

Ionic substances can be removed from groundwater by *ion exchange*. In this phenomenon, ions such as sodium and calcium, which are loosely bound to clay particles can be exchanged for other cations, such as lead, mercury, cadmium, and manganese. The heavy metal contaminants will thus be removed from the groundwater. The ability of a soil to remove contaminants by ion exchange is measured as the ion-exchange capacity of the soil.

Dissolved organic compounds can be removed from groundwater by *adsorption* onto organic matter contained in the soil or rock. The rate of adsorption is inversely proportional to the water solubility of the organic compound. Those that have a low solubility are more tightly bound to the soil organic matter than those that are more soluble. This propensity to be absorbed is measured by a property known as the octanol–water partition coefficient. The other important factor is the percentage of organic matter in the soil. Obviously, the greater the percentage of organic matter, the more of dissolved organic compounds it can absorb.

Finally, many of the dissolved organic compounds can potentially be broken down into simpler compounds by the action of microbes in the soil and aquifer. This process is known as *biodegradation*. The components of petroleum based fuels can be degraded by soil bacteria. The end result is either carbon dioxide or methane, depending upon the presence or absence of dissolved oxygen in the aquifer. BTEX compounds are most readily degraded under aerobic conditions, i.e., with dissolved oxygen present. However, under certain geochemical conditions in the aquifer they can also be degraded in the absence of oxygen, but at a slower rate. Many other organic chemicals dissolved in groundwater, such as the chlorinated solvents, can be degraded either biologically or abiotically under the right geochemical conditions.[5]

REFERENCES

1. Fetter, C.W., Jr. *Contaminant Hydrogeology*, 2nd Ed.; Prentice-Hall, Inc.: Upper Saddle River, NJ, 1999; 500 pp.
2. Bedient, P.B.; Rifai, H.S.; Newell, C.J. *Ground Water Contamination, Transport and Remediation*, 2nd Ed.; Prentice-Hall PTR: Upper Saddle River, NJ, 1999; 604 pp.
3. Pankow, J.F.; Cherry, J.A. *Dense Chlorinated Solvents*; Waterloo Press: Portland, Oregon, 1996; 522 pp.
4. Charbeneau, R.J. *Groundwater Hydraulics and Pollutant Transport*; Prentice-Hall, Inc.: Upper Saddle River, NJ, 2000; 593 pp.
5. Chapelle, F.H. *Ground-Water Microbiology and Geochemistry*, 2nd Ed.; Wiley and Sons, Inc.: New York, 2000; 475 pp.

Groundwater: World Resources

Lucila Candela

Department of Geotechnical Engineering and Geo-Sciences, Technical University of Catalonia (UPC), Barcelona, Spain

Abstract

Groundwater constitutes an important source of water supply for domestic, industrial, and irrigation uses in different countries due to its availability, which is not subject to multiannual and seasonal fluctuations. Groundwater runoff data are widely used to characterize regional groundwater resources and are an important component of the hydrologic cycle and environment.

INTRODUCTION

Groundwater constitutes an important source of water supply for domestic, industrial, and irrigation uses in different countries due to its availability, which is not subject to multiannual and seasonal fluctuations. At present it is the main source of domestic water supply in most European countries, the United States, Australia, and some countries of Asia and Africa both in large and small towns and in rural areas.[1,2] Groundwater is used for irrigation in about one-third of all irrigated lands.[3]

Natural groundwater resources are understood to be the total amount of recharge (replenishment) of groundwater under natural conditions as a result of infiltration of precipitation, seepage from rivers and lakes, leakage from overlying and underlying aquifers, and inflow from adjacent areas. In some cases, the average annual recharge of aquifers, evaluated from average annual precipitation, equals groundwater runoff. Natural groundwater resources may be equated to groundwater discharge (runoff) when the evaporation from the water table may be ignored or estimated separately.[4] Under this assumption, groundwater runoff data are widely used to characterize regional groundwater resources and are an important component of the hydrologic cycle and environment.[5–7]

The role of groundwater in the water balance and water resources of regions is quantitatively characterized by the groundwater runoff/precipitation ratio or groundwater recharge. The runoff/precipitation ratio is extremely variable depending on meteorological factors, composition of rocks, etc. Distribution of groundwater recharge to river/total river runoff ratios shows the effect of geographical and altitudinal zonality. The quantity of recharge ranges widely. Analysis of conditions of generation of groundwater resources within continents shows that this global process depends on a complex combination of various natural factors. Principal amongst these are precipitation, vegetation, soil type and geology, and the hydrogeological features of the area.

In regions where aquifers are mainly composed of sands, specific groundwater discharge values are twice as large as in regions where the percentage of sands in aquifers is small. In this regard distribution of specific values of groundwater discharge on a global scale is subject to latitudinal zonality. Values generally increase from subartic regions to medium-latitude zones, in humid tropics and tropics, and decrease in semiarid and arid regions. Large groundwater discharge values may be found in karst limestones (up to $20\,L\,sec^{-1}\,km^{-2}$), sand quaternary deposits (up to $18\,L\,sec^{-1}\,km^{-2}$) or highly fractured rocks (up to $10\,L\,sec^{-1}\,km^{-2}$), although values are normally dependent on topographic elevations and annual precipitation. Marine sandy and clayey sediments show minimal discharge values ($0.1\,L\,sec^{-1}\,km^{-2}$ and smaller).[4]

The main task of areal hydrogeological subdivision when compiling groundwater runoff and resources maps is to distinguish territories which are sufficiently uniform in terms of groundwater distribution and particularities of groundwater generation.[8]

AQUIFER TYPES

The principal aquifers are found in six types of permeable geologic materials: unconsolidated deposits of sand and gravel; semiconsolidated sand; sandstone; carbonate rocks interbedded sandstone and carbonate rocks; and basalt and other types of volcanic rocks.[8] Large areas of the world are underlain by crystalline rocks permeable only where they are fractured or weathered, and generally yield only small amounts of water to wells. In many places, they are the only source of water supply. However, because these rocks extend over large areas, important volumes of groundwater are withdrawn from them.

Encyclopedia of Natural Resources DOI: 10.1081/E-ENRW-120010218

Unconsolidated Sand and Gravel Aquifers

Unconsolidated sand and gravel aquifers are characterized by intergranular porosity and all contain water primarily under unconfined or water-table conditions, but locally confined conditions may exist where aquifers contain beds of low permeability. Different categories can be distinguished, which occupy different geologic settings. The sediments are mostly alluvial deposits, but locally may include windblown sand, coarse-grained glacial outwash, and fluvial sediments deposited by streams recharge. Large areas of the world are covered with sediments deposited during several advances and retreats of continental glaciers. The glacial sand and gravel deposits form numerous local but productive aquifers.

Aquifers commonly receive direct recharge from precipitation and streamflow infiltration. Regional movement is down the valley in the direction of stream flow, lake or playa (located in the center of the basin). Basins in arid regions might contain deposits of salt, anhydrite, gypsum or borate produced by evaporation or mineralized water in their central parts. Also, much of the infiltrating water is lost by transpiration by riparian vegetation.

Consolidated/Fractured Sedimentary Aquifers

Aquifers in sandstones are more widespread than those in all other kinds of consolidated rocks. Sandstone retain some primary porosity unless cementation has filled all the pores, but most of the porosity in these consolidated rocks consists of secondary openings such as joints, fractures, and bedding planes. The water is not highly mineralized in areas where the aquifer outcrops or are buried to shallow depths, but mineralization generally increases as the water moves downgrading toward the structural basin.

Carbonate Rock Aquifers

The water-yielding properties of carbonate rocks are highly variable; some yield almost no water and are considered to be confining units, whereas others are among the most productive aquifers known. The original texture and porosity of carbonate deposits can range from 1% to more than 50%. Recharge water enters the aquifer through sinkholes, swallow holes, and sinking streams, some of which terminate at large depressions called blind valleys.

Basaltic and Other Volcanic-Rock Aquifers

Volcanic rocks have a wide range of chemical, mineralogic structural, and hydraulic properties due largely to rock type. Unaltered pyroclastic deposits have porosity and permeability characteristics like those of poorly sorted sediments; rhyolites have low permeability except where they are fractured.

AVAILABILITY AND USE

Except for widely scattered places, existing data are not uniformly distributed in space and time because hydrologic investigations have been mostly conducted in areas where water supply or water quality problems existed, or where large quantities of groundwater were withdrawn. Long-term hydrologic records are rare and usually collected only during the course of a study or perhaps for a few years after the study has ended. No systematic investigation on groundwater resources and exploitation in many regions of the world have been conducted.[3,9]

The annual groundwater use for the world as a whole can be placed at $750–800 \times 10^9 \, m^3$, a modest value when compared to overall water availability (Tables 1 and 2). But an overwhelming majority of the world's cities and towns depend on groundwater for municipal water supplies. Over 35 countries of the world use more than $1 \times 10^9 \, m^3$ of groundwater annually.[10] Because of spatial imbalances in the occurrence of groundwater and the pattern of demand, massive problems of groundwater overexploitation are found in areas where high population exist or under intensive agriculture development.

GROUNDWATER RESOURCES DISTRIBUTION

Europe

All types of aquifers are currently exploited: large well fields in artesian basins of platform type, such as Paris and London; river valleys (France, Volga region); cones and intermontane depressions (Italy, Switzerland).[11,12] In many cases their exploitation is accompanied by the generation of large and deep cones of depression.

Table 1 Annual groundwater recharge and withdrawals in the world

World region	Average annual recharge (km³ yr⁻¹)[a]	Annual groundwater withdrawals (km³ yr⁻¹)[b]
Asia	2505	352
Europe	1368	78
Middle East and N. Africa	137	75
Sub-Saharan Africa	1548	9
North America	1884	110
C. America and Caribbean	344	29
South America	3693	14
Oceania	270	2

[a]Amount of water that is estimated to annually infiltrate into aquifers. It would represent the amount of water that could be annually withdrawn.
[b]Abstractions from aquifers. These data are scarce and not currently available for all countries in each region.
Source: WRI,[24] Dzhamalov & Zektser.[8]

Groundwater runoff in Europe is quite irregular, depending on the geostructural, climatic, and orographic conditions and the flow media generation: karst, porous fractured, and porous. Specific discharge values distribution is governed by the geological structure.

According to the available data (Table 1), groundwater use estimation in Europe is 78 km³ yr⁻¹, which constitutes 21% of the total water consumption. Urban and rural population constitute the most important water consumers, accounting for 56% of the total water consumption. Groundwater is the main source for public water supply (more than 70% of total resources), especially on islands and some European countries like Denmark. More than 90% of big cities and towns are exclusively supplied by groundwater (among them Berlin, Rome). Although groundwater is mainly used for irrigation in southern countries like Spain with values ranging between 0.7 km³ yr⁻¹ and 5 km³ yr⁻¹, other European countries, like the Netherlands, may also use it during dry years.

Africa

Africa is one of the regions of the world facing serious water shortages because of greater disparities in water availability and use, and because water resources are unevenly distributed. Groundwater, first considered as a main resource for water in urban, rural areas, and mining, especially in coastal areas and arid regions, is now tending to be extended to the most isolated desert and tropical regions. In Libya, groundwater accounts for 95% of country's freshwater withdrawals, while in some areas of North Africa it is a significant source for irrigated agriculture. In many parts of the continent, groundwater resources have not yet been fully explored and tapped. According to the geographic and climatic homogeneity, which has a direct influence on water resources, Africa can be divided into several regions: Northern, Sudano-Sahelian, Gulf of Guinea, Central, Eastern, Indian Ocean Islands and Southern. This vast territory can be subdivided into a number of large aquifer systems subject to very varied climatic conditions.[13,14]

Basement rocks cover most of the central territory and aquifers are not very productive except in few cases. Sedimentary formations of sandstones overlaying the basement areas may constitute good aquifers, such the Karoo basin. The coastal sedimentary basins are the most productive aquifers, being intensely exploited along the shoreline.

Table 2 Groundwater use in selected areas of the world

Country	Annual recharge (km³ yr⁻¹)	Groundwater use (%)
Russian Federation	900	<1
China	800	10
India	450	30

Source: From http://www.fao.org[3]

Alluvials are among the most important and also serve large populations, especially in Northern Africa. Karstified limestones of North-West Africa and Madagascar can yield flow rates up to 100 m³ hr⁻¹. Also large fossil aquifers are present in the Saharan and Nubian deserts made of sedimentary basins and being largely exploited.

South America, Central America, and the Caribbean

This area extending from the Central America Isthmus to South America has the most abundant river flow. Groundwater is unevenly distributed in quantity, but quality is usually good for domestic and industrial supply, presently the highest priority. Total water withdrawal from the aquifers is difficult to estimate because most comes from uncontrolled private and public wells. Based on UN estimates, 50–60% of total population domestic and industrial supply is from groundwater. Water withdrawn can be estimated at between 12 km³ yr⁻¹ and 14 km³ yr⁻¹, very low in comparison with the estimated renewable resources. Groundwater reserves estimation is 238,000 km³, discharge to rivers being 3898 km³ yr⁻¹. Discharge values are high in the humid equatorial zone and minimum in the Atlantic Andean Cordillera and northeast Brazil.

According to geologic and tectonic features, four major water-bearing domains can be distinguished:[15,16] superficial deposits; deep aquifers in sedimentary basins; folded mountain chains; and precambrian basement bedrocks. Vast areas of South America are composed of Precambrian crystalline rocks which are not highly productive unless weathered or intensively fractured. The hydrogeologic map of South America[16] shows 16 hydrogeological provinces with similar characteristics including the previously-mentioned water-bearing domains. Some of the formations' resources are considered as the most important water-bearing formations, such as the Amazon Sedimentary Basin (32,500 km³), Parnaíba-Maranhao (17,000 km³), and the Paraná Sedimentary basin, where the Guarani aquifer extending over 1,500,000 km² has 50,000 km³ of storage.

North America

Groundwater is an important source of water in the United States and Mexico, but it represents less than 5% of Canada's total water use. About 22% of the total water use in the United States (290 × 10⁶ m³ day⁻¹) is supplied by groundwater; about 50% of the U.S. population depends on groundwater for domestic uses and also major cities and metropolitan areas and irrigation has made the High Plains one of the most important agricultural areas. Half of the U.S. population draws its domestic water supply from groundwater.[1] In Mexico, where desert and semiarid conditions prevail over two-thirds of the country, groundwater is widely used. Urban areas of Mexico use groundwater as their sole or principal source.

Unconsolidated sand and gravel are the most widespread aquifers, with intergranular porosity, and water primarily under water-table conditions.[17,18] Some unconsolidated aquifers have supplied large amounts of water for irrigation, like the High Plains aquifer (56×10^6 m^3 day^{-1} withdrawn from the aquifer for irrigation in 1990); in the United States, about 20% of the groundwater withdrawn is derived from the High Plains aquifer.

Carbonate rock aquifers are most extensive in eastern United States and in the Bahamas, western Canada and Yucatan (Mexico), and some of them are considered among the most productive aquifers known. Most of them consist of limestone but dolomite and marble locally yield water. More than 13×10^6 m^3 day^{-1} (1990 data) were withdrawn from the Floridan aquifer system, the sole source of water supply for the city of Miami.

Oceania

While groundwater resources in New Zealand, Pacific Islands, and New Guinea are difficult to quantify due to the limited information, the aridity of much of the Australian continent is a significant factor in the occurrence and assessment of groundwater resources.[19] A large part of western and central Australia is arid, with a mean annual rainfall below 250 mm. Total amount of groundwater used in Australia is estimated at 2460×10^6 m^3 in 1983 from more than 500,000 tube-wells, 14% of the total amount of water used.[20] The greatest concentrations are near Perth, Adelaide, South Australia, western Victoria, and on the central Queensland coast. Surficial aquifers are the most important sources for irrigation, urban, and industrial supply. Fractured rock aquifers of igneous and metamorphic rocks are of relative importance, although they may locally provide high groundwater yields.

Most highly productive aquifers are the surficial sedimentary aquifers associated with inland or eastern coastal rivers, up to 100 m thick. Also sand dunes, coastal, and deltaic alluvium sediments form important aquifers along the east coast and in central Queensland. Australia's main arid-zone irrigation scheme is based on groundwater extracted from sands and gravel of Central Australia. Several large deep sedimentary basin aquifers (Amadeus, Canning, Great Artesian, Murray, Otway, Perth, Eucla, Officer), extending over more than 24,000 km^2, constitute a reliable source of old, good quality groundwater. The Great Artesian basin covers 1.7×10^6 km^2, is up to 300 m thick and is one of the largest basins in the world.[21] More than 20,000 non-flowing and more than 4000 flowing artesian wells have been drilled. Individual well flows exceeding 100 L sec^{-1} have been recorded. Diffuse natural discharge from the Great Artesian Basin has been calculated to be about 1.4×10^6 m^3 day^{-1}.

Asia and Middle East

Continental Asia is an immense and complex geographical area of great extremes. Some parts of China and India are among the most populated in the world while the deserts of central Asia and the interior high plateau are extremely thinly populated. Most of the land of the Arabian Peninsula and in central and eastern Iran is a desert, reflecting different patterns of groundwater use.

In Asia and Middle East groundwater has been developed since ancient times, especially in arid regions where no other source of water supply is permanently available.[22,23] Large-scale developments are found in northern and coastal areas of China, where artesian aquifers and the loess and karst areas of the central south are tapped for urban and industrial supply. Groundwater irrigation distribution by subregions is, according to AQUASTAT,[3] in the Arabian peninsula 96.6%; Middle East 18.2%; and Central Asia 34% (although for Bangladesh it represents 69%).

Groundwater exists in the area as semiconfined, unconfined shallow, and deep aquifers. Recharge is faster in the Middle East countries although the aquifers of the Arabian Peninsula contain much larger reserves. The interior arid regions of the Middle East countries may include geological formations which can be considered as aquicludes or aquitards. Among the groundwater-bearing zones, the most important are alluvial of large rivers, vast, complex sedimentary formations holding artesian water and sedimentary basins in coastal areas (Israel). Some carbonate basins of importance are also present in the Mediterranean area (Lebanon, Syria) and Pakistan. Weathered crystalline rocks and lava flows constitute important aquifers in peninsular India and Northern Syria.

Although groundwater quality is suitable for irrigation and domestic uses, salinization of groundwater has occurred in several areas of the Indus Plain and Pakistan.

REFERENCES

1. http://www.iwmi.cgiar.org/pubs/WWVisn/GrWater.htm
2. http://www.grida.no/geo2000/english
3. http://www.fao.org/ag/agl/aglw/Aquastatweb/Main/html/aquastat.htm
4. Zekster, I.S.; Dzhamalov, R.G. *Role of Ground Water in the Hydrological Cycle and in Continental Water Balance*; IHP-UNESCO: Paris, 1988; 133 pp.
5. Van der Leeden, F.; Troise, F.L.; Todd, K.D. *Water Encyclopedia*; Lewis Publishers: Chelsea, Michigan, 1990; 808 pp.
6. Gleick, P. *Water in Crisis: A Guide to the World's Fresh Water Resources*; Oxford University Press: New York, 1993; 1–493.
7. UN. *Comprehensive Assessment of the Freshwater Resources of the World: Report of the Secretary General*; United Nations, Commission on Sustainable Development, April 7–15, 1997; E/CN.17/1997/9, 1997; 1–35.

8.   Dzhamalov, R.G.; Zektser, I. *Digital World Map of Hydrogeological Conditions and Groundwater Flow*, CDrom Version 2.0; HydroScience Press: 1999.

9.   http://www.wri.org/water/index.html.

10. Llamas, R.; Back, W.; Margat, J. Groundwater use: equilibrium between social benefits and potential environment costs. Appl. Hydrol. **1992**, *1*, 3–14.

11. UN. *Groundwater in the Western Hemisphere*; Natural Resources/Water Series No. 4; United Nations, Department for Technical Co-operation for Development: New York, 1976; 337 pp.

12. UN. *Groundwater in Western and Central Europe*; Natural Resources/Water Series No. 27; United Nations, Department for Technical Co-operation for Development: New York, 1991; 363 pp.

13. UN. *Les Eaux souterraines de l'Afrique Orientale, Centrale et Australe*, Ressources naturelles/Serie eau No. 19; United Nations, Department for Technical Co-operation for Development: New York, 1988; 341 pp.

14. UN. *Groundwater in North and West Africa*; Natural Resources/Water Series No. 18; United Nations, Department for Technical Co-operation for Development: New York, 1988; 405 pp.

15. Rebouças, A.C. Desarrollo y tendencias de la hidrogeología en America Latina. In *Hidrogeología, Estado Actual y Perspectivas*; Anguita, Aparicio, Candela, Zurbano, Eds.; CIMNE: Barcelona, 1991; 429–453.

16. UNESCO-IHP/DNPM/CPRM. *Hydrogeological Map of South America 1: 5,000,000. Exploratory tex*; Rio de Janeiro, 1996; 212 pp.

17. http://www.water.usgs.gov/ogw.

18. http://capp.water.usgs.gov/gwa/ch_h/H-text1.html.

19. UN. *Groundwater in the Pacific Region*; Natural Resources/Water Series No. 12; United Nations, Department for Technical Co-operation for Development: New York, 1983; 289 pp.

20. DPIE. *Review of Australia's Water Resources and Water Use. Vol. 1. Water Resources Data Set*; Dep. of Primary Industries and Energy, Australian Government Publishing Service: Canberra, 1987; 158 pp.

21. Habermehl, M.A. The great artesian basin Australia. BRM J. Aust. Geol. Geophys. **1980**, *5*, 9–38.

22. UN. *Groundwater in the Mediterranean and Western Asia*; Natural Resources/Water Series No. 9; United Nations, Department for Technical Co-operation for Development: New York, 1982; 230 pp.

23. UN. *Groundwater in Continental Asia (Central, Eastern, Southern, South-Eastern Asia)*; Natural Resources/Water Series No. 15; United Nations, Department for Technical Co-operation for Development: New York, 1986; 391 pp.

24. WRI. *Environmental Data Tables*; World Resources 2000–2001: 2000.

Field—Low

Hydrologic Cycle

John Van Brahana
Division of Geology, Department of Geosciences, University of Arkansas, Fayetteville, Arkansas, U.S.A.

Abstract
The hydrologic cycle describes the dynamic, water-circulation system of the Earth. Water we see today is the same water that was originally derived from degassing of volcanoes as the Earth cooled from the molten mass that was our primordial planet several billion years ago. This water has been continuously recycled by natural processes, changing from liquid to solid or vapor and then back again, moving and flowing endlessly in response to the physical and chemical conditions of the environment of our planet.

PROCESSES AND PATHWAYS

The dominant processes of the hydrologic cycle, and the pathways along which we can trace water movement, include the following (Fig. 1): evaporation from oceans and open bodies of water on the Earth's surface into the atmosphere; evapotranspiration by plants of soil water into atmospheric vapor; condensation of water vapor into liquid or solid particles (clouds); precipitation as the condensed water or ice falls from the atmosphere back to the surface; infiltration of the water into subsurface (soil and groundwater) reservoirs; baseflow contribution to streams from groundwater; streamflow recharging of groundwater; streamflow runoff if the precipitation rate exceeds the infiltration rate; surface and subsurface flow back to oceans or intermediate reservoirs; and storage.[1–10]

In one sense, the hydrologic cycle is one of the most basic concepts of water science, yet in detail the concept is complex because it involves all forms of water of the hydrosphere, and it is affected by many influencing factors that are not always obvious. Because this circulation of water is intimately tied to energy transfer, it is helpful to start with the basic physics of the forces that drive this seemingly endless flow of water on our planet.

ENERGY SOURCES

The underlying source of energy that drives the movement of water throughout the hydrosphere of the Earth is solar radiation—thermal energy from our sun. Solar energy heats water, causes it to evaporate and change state from liquid to a gas, and in so doing, facilitates its movement through the atmosphere in response to wind and pressure changes. Every gram of liquid water at its point of vaporization requires an input of 540 cal of thermal energy to convert it to a gas. Worldwide, water vapor represents a huge source of energy storage and transport in a hydrologic

link between atmosphere, oceans, and continents, which we call "climate." The residence time of water vapor in the atmosphere is short, usually no more than several days or weeks, until it condenses and falls as precipitation. In undergoing condensation to a liquid, the energy stored in the vapor is released.

As a liquid, water is controlled by gravity. If it can move, it will, always move downhill to a lower potential energy state, and always along the path of least resistance, down the steepest gradient. As water moves, it expends energy. Fast-flowing runoff, particularly in streams, is the single most dominant agent of erosion of the surface of our planet. Glaciers likewise are effective at sculpting the land surface, but owing to their limited occurrence on only 10% of the continents, their impact is not nearly as widespread as that of flowing streams. Thus, erosion and the Earth's landforms are intimately tied to the hydrologic cycle. Ultimately, water reaches the lowest accessible level possible, which for most places on the Earth is sea level. Internally drained basins that are isolated from the oceans by mountain ranges and other high divides may exist below sea level (i.e., Death Valley in California; Caspian Sea in Kazakhstan; Dead Sea in Jordan), but these represent local base-level conditions rather than regional or global conditions. These areas of internal drainage are typically formed by tensional tectonics, where blocks of rock are down-faulted (grabens) due to forces that tend to pull the continents apart. Water drains into these depressions under the force of gravity from the surrounding highlands, and escapes only by evaporation.

HYDROLOGIC RESERVOIRS AND IMPLICATIONS FOR HUMANS

Oceans form the largest of our hydrologic reservoirs (Table 1), covering about 70% of the Earth's surface, and including about 96% of all of its water.[8,11,12] Ocean water

Encyclopedia of Natural Resources DOI: 10.1081/E-ENRW-120010048

Field—Low

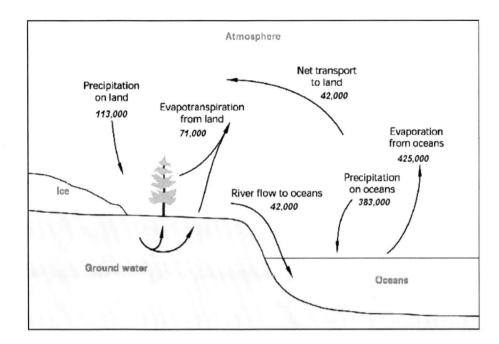

Fig. 1 Quantitative representation of processes in the hydrologic cycle showing transfer rates between reservoirs in units of cubic kilometers per year. **Source:** Adapted from Winter, et al.[5]

Table 1 Comparisons of quantities and percentages of Earth's water in the major storage reservoirs of the hydrologic cycle, based on estimates from UNESCO and NRC

Storage reservoirs of the hydrologic cycle	UNESCO		NRC	
	Volume, in $10^6 km^3$	%	Volume, in $10^6 km^3$	%
Oceans	1,338	96.5	1,400	95.96
Icecaps and glaciers	24.3	1.73	43.4	2.97
Groundwater	23.4	1.69	15.3	1.05
Lakes	0.176	0.013	0.125	0.009
Atmospheric water	0.0129	0.001	0.0155	0.001
Soil moisture	0.0165	0.0012	0.065	0.004
Rivers	0.0021	0.0002	0.0017	0.00012
Biologic water	0.001	0.0001	0.002	0.00013
Total	1,386		1,459	

Source: From U.S.S.R.[8,12]

unfortunately is saline and non-potable (undrinkable), containing about 35,000 mg L^{-1} of dissolved solids,[13] much too salty for human consumption. Of the remaining 3–4% of the Earth's water found in reservoirs on the continents, approximately 1% of this (principally saline lakes or deep, saline groundwater) also is non-potable.[6,8,11–13] Thus, about 97% of all the water on this Earth is too mineralized for humans to drink without expensive desalinization. Of the approximately 3% of the total water that is fresh, the largest percentage, estimated as 1.7–2.97% by different experts, is stored in icecaps and glaciers, far removed from most of the Earth's population and its water needs. Thus, the freshwater needs of the world are served by a fraction of 1% of the total hydrologic budget, primarily water in storage and transit as shallow groundwater, freshwater

lakes, soil moisture, water in manmade reservoirs, and rivers. These data are synthesized from several comprehensive studies, and although the values do not match exactly, they generally do not vary by more than 1% or 2%.[8,12]

As an integrated Earth system, the hydrologic cycle has no discrete beginning or ending point, but from a consumptive human point of view, the oceans are the major source of water, the atmosphere is the deliverer, and the land is the user. In this system, no water is lost or gained, but the amount of water available to the user may fluctuate mostly because of variations in the delivering agent. In the geologic past, large alterations in the cyclic roles of the atmosphere and the oceans produced deserts and glaciation across entire continents. During the last ice age, the colder climate resulted in a greater percentage of water being stored as snow and ice, with a decreased percentage of water being stored in the oceans. Scientists see evidence that major sea-level declines corresponded with maximum glacial development, and in fact, point to "drowned" valleys (e.g., Chesapeake Bay in the United States, and the fiords in the Scandanavian countries) that were eroded and formed when sea level was much lower, and have since been inundated and flooded by rising sea level as the glacial ice melted.

Historically, freshwater in the hydrologic cycle has been enough to serve human needs, but exponential population growth and water usage in regions with little freshwater are posing ever-increasing political and planning problems. For these areas, freshwater has to be imported from great distances, at great expense, and to the detriment of other regions that need the water for their own use. Many of the most pressing problems of the 21st century will be related to obtaining freshwater for the world's expanding population.

REFERENCES

1. http://observe.arc.nasa.gov/nasa/earth/hydrocycle/ hydro2.html (accessed July 2002).
2. http://www.uwsp.edu/geo/faculty/ritter/geog101/modules/ hydrosphere/hydrosphere_title_page.html (accessed July 2002).
3. http://ww2010.atmos.uiuc.edu/(Gh)/guides/mtr/hyd/home. rxml (accessed July 2002).
4. http://ga.water.usgs.gov/edu/mearth.html (accessed July 2002).
5. Winter, T.C.; Harvey, J.W.; Franke, O.L.; Alley, W.M. *Ground Water and Surface Water—A Single Resource*; U.S. Geological Survey Circular 1139: Denver, Colorado, 1999; 1–79.
6. Maidment, D.R.; Ed. *Handbook of Hydrology*; McGraw-Hill, Inc.; 1993; 1-1–1-15.
7. Driscoll, F.G. *Groundwater and Wells*, 2nd Ed.; Johnson Division: St. Paul, MN, 1986, Chapter 1.
8. U.S.S.R. Committee for the International Hydrological Decade, World Water Balance and Water Resources of the Earth, English translation. *Studies and Reports in Hydrology*; UNESCO: Paris, 1978; Vol. 25.
9. World Meteorological Organization–UNESCO. *International Glossary of Hydrology*, 2nd Ed.; WMO Report No. 385; Geneva, Switzerland, 1992.
10. Wilson, W.E.; Moore, J.E.; Eds. *Glossary of Hydrology*; American Geological Institute, 1998.
11. Bras, R.L. *Hydrology*; Addison-Wesley: Reading, MA, 1990.
12. National Research Council (NRC). *Global Change in the Geosphere–Biosphere*; National Academy Press: Washington, DC, 1986.
13. Hem, J.D. *Study and Interpretation of the Chemical Characteristics of Natural Water*; U.S. Geological Survey Water Supply Paper 2254; Washington, DC, 1992.

Field—Low

Hydrologic Modeling of Extreme Events

Jazuri Abdullah
Jaehoon Kim
Pierre Y. Julien
Department of Civil and Environmental Engineering, Engineering Research Center, Colorado State University, Fort Collins, Colorado, U.S.A.

Abstract

The TREX watershed model has become a very powerful tool to simulate floods from extreme rainstorms. Malaysia experiences some of the world's most devastating floods and the Semenyih watershed was selected for this study. Approximately half of this 236 km² watershed is covered with forest under rapid development for agriculture and urbanization. The TREX model has been calibrated using precipitation data from April 13, 2003 and has been validated with several other storms. The model is numerically very stable and matches both the peak discharge and the time to peak very well. The potential effects of the world's greatest precipitation at 68 mm/hr for 16 hours have been simulated to represent the case of long monsoon precipitation on the Semenyih watershed. The model results provide a dynamic illustration of the magnitude and duration of the vast flooding area with flow depths in excess of 6.0 m which would occur in the relatively flat and wide agricultural and urbanized areas. These flood mapping results should facilitate the analysis of flood hazards and disaster prevention.

INTRODUCTION

The simulation of extreme floods is one of the main concerns for public safety and hydrologic engineering.[1] In the analysis of extreme floods, the concepts of Probable Maximum Flood and paleohydrology are very useful.[2] In the United States, the models HEC-1 and HEC-HMS are used by the U.S. Army Corps of Engineers, whereas the Flood Hydrograph and Runoff (FHAR) model is promoted by the U.S. Bureau of Reclamation. The major advances in Geographical Information System (GIS)-based computer models in the past two decades offer a new perspective on the simulation of extreme flood events. We explore the use of a physically based and distributed model to simulate floods on a watershed under monsoon precipitation in South-East Asia.

The objectives of this research are to test the potential of GIS-based rainfall-runoff modeling to simulate extreme floods; and demonstrate modeling applicability through calibration, validation, and simulation of extreme floods. The Two-dimensional Runoff, Erosion and EXport (TREX) model was selected for the analysis and is tested at Sungai Semenyih in Malaysia where the 236 km² watershed can be subjected to some of the highest precipitation levels ever recorded world-wide.

TREX MODEL

The physically based TREX model is fully distributed and simulates rainfall-runoff, sediment transport, and chemical fate and transport at the watershed scale.[3] TREX uses a GIS-based raster format combining the CASC2D surface runoff model,[4–7] and the water quality WASP/IPX model.[8] TREX is classified as an event model as it simulates the Hortonian overland flow and the watershed responses from a single storm with no soil infiltration capacity recovery between events. The selection of the computational time step was done by satisfying the Courant Condition.

Upland Processes

The hydrologic processes in the model are as follows: 1) rainfall, interception, and surface storage; 2) infiltration and transmission loss; and 3) overland and channel flow. The model state variables are water depth in the overland plane and stream channels. Rainfall can be uniform or distributed in both time and space and can also be specified using several grid-based formats for point rain gauges or radar rainfall data.[9] Rainfall can be representing as the gross volume of water that reached the near surface as:

$$\frac{\partial V_g}{\partial t} = i_g \, A_s \tag{1}$$

where V_g = gross rainfall water volume [L³]; i_g = gross rainfall rate [L/T]; A_s = surface area over which rainfall occurs [L³]; and t = time [T]. The model calculates the volume of interception as:

$$V_i = \left(S_i + Et_r \right) A_s \tag{2}$$

Encyclopedia of Natural Resources DOI: 10.1081/E-ENRW-120047597

where V_i = interception volume [L^3]; S_i = interception capacity of projected canopy per unit area [L^3/L^2]; E = evaporation rate [L/T]; and t_r = precipitation event duration [T].

Infiltration and transmission loss rates are simulated using the Green and Ampt relationship[10]:

$$f = K_h \left(1 + \frac{(H_w + H_c)(1 - S_e)\theta_e}{F} \right) \tag{3}$$

where: f = infiltration or transmission loss rate [L/T]; K_h = effective hydraulic conductivity [L/T]; H_w = hydrostatic pressure head (pond water depth) [L]; H_c = capillary pressure (suction) head at the wetting front [L]; θ_e = $(\varphi - \theta_r)$ effective soil porosity [-]; φ = total soil porosity [-]; θ_r = residual moisture content [-]; S_e = effective saturation [-]; and F = cumulative infiltrated depth (depth to wetting front) [L].

Overland Flow

Overland flow is two dimensional and simulated using the diffusive wave approximation. Flow occurs when the water depth exceeds the storage depth:[4,10]

$$\frac{\partial h}{\partial t} + \frac{\partial q_x}{\partial x} + \frac{\partial q_y}{\partial y} = i_n - f + W = i_e \tag{4}$$

$$q_x = a_x \, h^\beta \tag{4a}$$

$$q_y = a_y \, h^\beta \tag{4b}$$

where: h = surface water depth [L]; q_x, q_y = unit flow discharge in the x- or y-direction = Q_x / B_x, Q_y / B_y [L^2/T]; Q_x, Q_y = flow in the x- or y-direction [L^3/T]; B_x, B_y = flow width in the x- or y-direction [L]; i_n = net precipitation rate (gross rainfall minus interception) [L/T]; W = flow point source (unit discharge) [L/T]; i_e = excess precipitation rate [L/T]; a_x, a_y = resistance coefficient for flow in the x- or y-direction = $|S_{fx}|^{1/2}/n_M$, $|S_{fy}|^{1/2}/n_M$ [L$^{1/3}$/T]; n_M = Manning roughness coefficient [T/L$^{1/3}$]; S_{fx}, S_{fy} = $S_{0x} - dh/dx$, $S_{0y} - dh/dy$ friction slope (energy grade line) in the x- or y-direction [-]; S_{0x}, S_{0y} = ground surface slope in the x- or y-direction [-] and β = exponent of the stage-discharge relationship. To solve the resistance coefficient equations for a_x and a_y, the absolute value of S_f is used and the sign of S_f indicates the flow direction.

Channel Flow

Channel flow is one dimensional and simulated using the diffusive wave approximation. Flow occurs when the water depth exceeds the storage depth and the friction slope is not zero:[4,10]

$$\frac{\partial A_c}{\partial t} + \frac{\partial Q_x}{\partial x} = q_l + W \tag{5}$$

$$Q_x = \frac{1}{n_M} A_c \, R_h^{2/3} \, |S_{fx}|^{1/2} \tag{6}$$

where A_c = cross-sectional area of flow [L^2]; Q = total discharge [L^3/T]; q_l = lateral unit flow (into or out of the channel) [L^2/T]; W = unit discharge from/to a point source/sink (including direct rainfall to the channel) [L^2/T]; R_h = hydraulic radius = A_c/P_c [L]; P_c = wetted perimeter of channel [L]. To solve Eq. (5), the absolute value of S_f is used and the sign indicates the flow direction.

In floodplain areas, water and any transported constituents are transferred between the overland plane and channel network based on the difference in water surface elevations. Floodplain transfers are bi-directional. Water and transported constituents move into stream channels by overland flow and can return to the overland plane when water levels in the stream exceed bank height. Similarly, materials can be moved from the sediment bed and can be delivered to the land surface by floodwaters.[11,12]

SITE DESCRIPTION AND MODEL CALIBRATION

Site Description

Located in the state of Selangor in Malaysia, the Semenyih watershed covers 236 km^2 (Fig. 1). The lowest elevation at the outlet is 40 m above sea level, whereas the highest point reaches 1100 m at the upstream end of the watershed. The average terrain slope is about 45% and ranges between 4% and 85% with very steep mountains overhanging flat and wide valleys (Fig. 2a). The length and width of the watershed are 17.7 km and 23.9 km, respectively. The average normal depth of the main channel in Sungai Semenyih ranges between 0.8 m and 2.49 m. The watershed is partly used for agriculture and the urbanization through residential and industrial development has rapidly transformed the area in the past 20 yr.

Located near the equator, the watershed climate is categorized as equatorial, being hot and humid throughout the year. The average rainfall precipitation reaches 2500 mm/yr and the average temperature is 27°C. Influenced by the southwest and northwest monsoons, the study area falls into the west coast rainfall region, where June and July are the driest months and November is the wettest.

Model Organization and Parameterization

The Semenyih watershed was simulated using the TREX watershed model. To resolve surface topography, the watershed was discretized at a 90 m × 90 m grid scale. This should yield satisfactory results according to the grid size analysis of

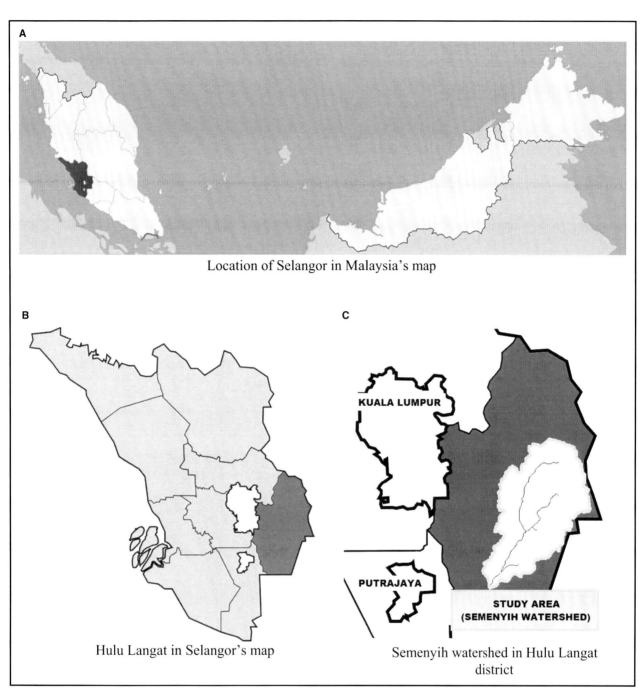

A

Location of Selangor in Malaysia's map

B

Hulu Langat in Selangor's map

C

KUALA LUMPUR

PUTRAJAYA

STUDY AREA
(SEMENYIH WATERSHED)

Semenyih watershed in Hulu Langat
district

Fig. 1 Semenyih study watershed location map.

Molnar and Julien.[13] The Digital Elevation Model (DEM) data for the site (Fig. 2A) were obtained from the Department of Surveying and Mapping Malaysia (DSMM). The resulting rectangular raster grid has 265 rows and 197 columns. Within this raster grid, the watershed area is defined by 29,139 cells. The DEM also allowed a delineation of the channel network with the watershed. The channel network includes seven links (reaches) and 399 nodes for a total stream length of approximately 36 km. Four soil types (Fig. 2B) and six land uses (Fig. 2C) were incorporated into the raster-based GIS representation of the watershed for the determination of hydraulic conductivity and surface roughness.

Calibration and Validation

The April 13, 2003, storm was used for calibration using the precipitation and flow records collected by Department of Irrigation and Drainage (DID). The calibration procedure focused on properly simulating peak flow, discharge volume, and time to peak at the outlet. Model parameters subject to calibration were simply the effective hydraulic conductivity (K_h), and roughness (Manning n). The values of the calibrated parameters are summarized in Table 1. The values of hydraulic conductivity K_h were adjusted during calibration to achieve good agreement between

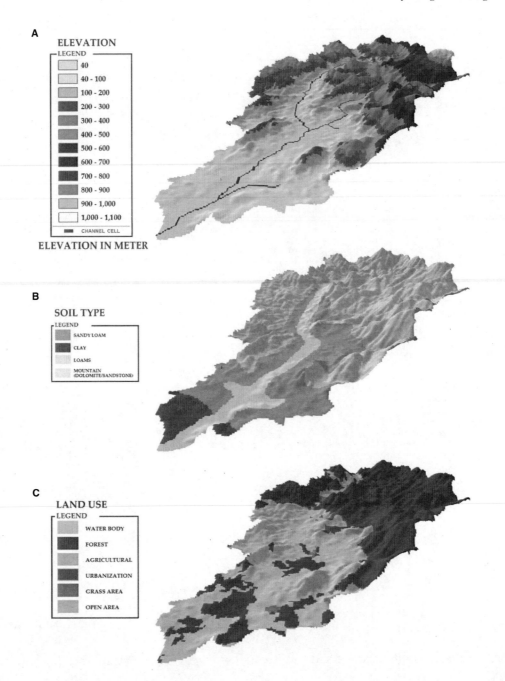

Fig. 2   Sungai Semenyih 90 m TREX GIS data layers: (**A**) elevation grid in meter and channels cells; (**B**) soil-type classes; (**C**) land use classes.

measured and simulated runoff. The antecedent moisture condition for the watershed was assumed to be fully dry.

The Nash–Sutcliffe Efficiency Coefficient (NSEC) and Percent BIAS (PBIAS) were used to evaluate model performance. The value of discharges and volumes from the simulated result were evaluated using these methods, respectively. The value of NSEC can be between -∞ and 1.0, with NSEC = 1.0 being the optimal value. Generally, NSEC values between 0.0 and 1.0 are acceptable levels of performance, whereas values less than 0.0 indicate unacceptable performance. PBIAS measures the tendency of the simulated data

to be larger or smaller than the observed data.[14] Positive and negative values indicate a model underestimated and overestimated, respectively. The optimal value of PBIAS is 0.0. In this study, the classifications for NSEC and PBIAS defined by Moriasi et al.[15] were used to determine whether the simulation results were good or very good.

The TREX model parameterization for the calibration on April 13, 2003, shows that peak flow, flow volume, and time to peak were all accurately simulated at the outlet (Fig. 3). A calibrated channel Manning n (0.04) was within the range proposed by Zakaria et al.[16] The Relative Percent

Table 1 Summary of model parameter values for Semenyih watershed

Parameter	Range	Application
Hydraulic	1.5×10^{-8}–6.1×10^{-7}	Sandy Loam
Conductivity	3.7×10^{-8}–6.9×10^{-8}	Loam
K_h (m/s)	1.7×10^{-9}–1.3×10^{-8}	Clay
	3.2×10^{-8}–3.2×10^{-10}	Mountain–Limestone
	0.05–0.18	Agricultural
	0.05–0.15	Urbanization
Manning's n (s/m$^{1/3}$)	0.20–0.40	Forest
	0.05–0.20	Grass area
	0.05–0.15	Open area
	0.02–0.04	Channel bed

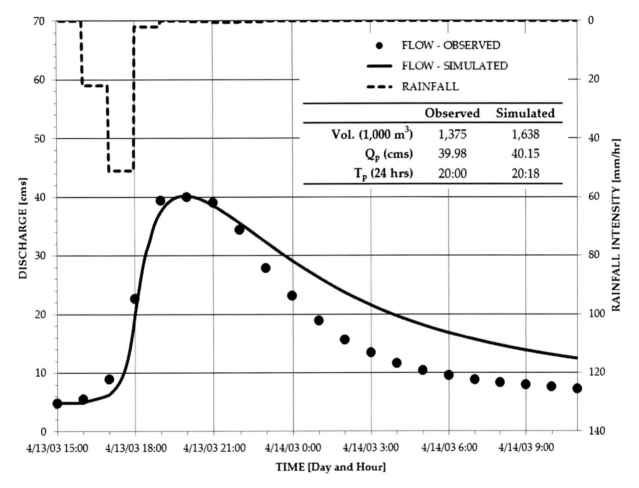

Fig. 3 April 13, 2003, hydrograph and TREX calibration

Difference (RPD) for peak flow, total volume, and time to peak was 0.4, 19.1, and 1.5, respectively. The NSEC and PBIAS value peak discharge and total volume were 0.8% and −19.3%, respectively, and indicate a very high simulation performance. For the validation events, the peak discharge found to be near-perfectly simulated with an average RPD value as low as 0.28%. The simulated hydrographs generally had the same shape as the observed hydrographs with an average total flow volume RPD of 32%. The peak discharge from measured and simulated for calibrated and validated model is presented in Fig. 4. The model is

numerically very stable and matches the peak discharge and time to peak. Performance evaluations for these simulations are presented in Table 2.

EXTREME EVENT SIMULATION

The importance of Probable Maximum Precipitation (PMP) is always highlighted and emphasized in terms of public safety and hazard downstream of any major river regulating structures, especially if it is located upstream of populated areas. In the case presented in this analysis, the

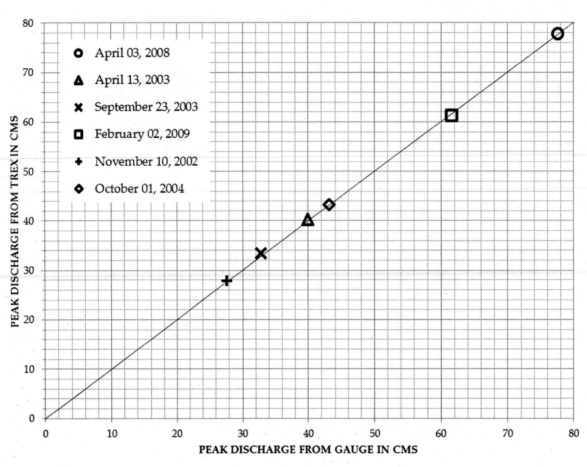

Fig. 4 Simulated and measured water discharge for the model calibration and validation events

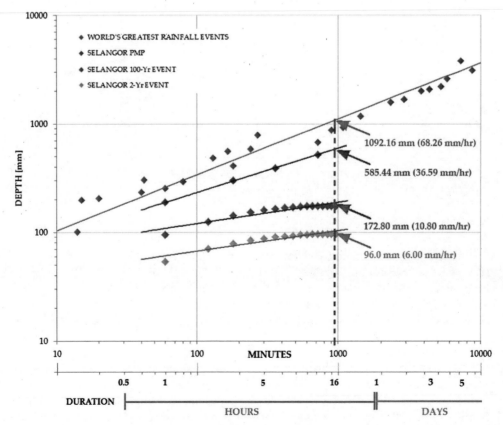

Fig. 5 Comparison between world's greatest rainfall events and Selangor extreme events.

Table 2 Hydrologic model performance evaluation summary

	Calibration										
	Peak flow (m³/sec)			**Total volume (× 1000 m³)**			**Time to peak (hr)**			**Performance evaluation**	
Event	**Observation**	**Simulated**	**RPD (%)**	**Observation**	**Simulated**	**RPD (%)**	**Observation**	**Simulated**	**RPD (%)**	**NSEC**	**PBIAS (%)**
April 13, 2003	39.98	40.15	0.4	1375	1638	19.1	20:00	20:18	1.5	0.8	−19.3

	Validation										
	Peak flow (m³/sec)			**Total volume (× 1000 m³)**			**Time to peak (hr)**			**Performance evaluation**	
Event	**Observation**	**Simulated**	**RPD (%)**	**Observation**	**Simulated**	**RPD (%)**	**Observation**	**Simulated**	**RPD (%)**	**NSEC**	**PBIAS (%)**
April 3, 2008	77.58	77.77	0.2	2939	3052	3.9	23:00	23:54	3.9	1.0	−7.6
September 23, 2003	32.83	33.37	1.6	590	950	61.2	7:00	7:42	10.0	0.1	−57.7
February 23, 2003	61.59	61.23	−0.6	1924	2530	31.5	22:00	22:45	3.4	0.4	−31.7
November 10, 2002	27.71	27.74	0.1	947	1277	34.9	0:00	0:42	2.9	0.8	−25.9
October 1, 2004	43.12	43.18	0.1	1236	1590	28.7	19:00	19:21	1.8	0.8	−28.9

Abbreviations: RPD, relative percent different; NSEC, Nash Sutcliffe Efficiency Coefficient; PBIAS, percent BIAS.

Table 3 Summary of intensity and peak discharge at Semenyih using 2 yr, 100 yr, PMP, and world's greatest precipitation

		Input		**Results**		
Events		**Depth (mm)**	**Intensity (mm/hr)**	**Peak discharge (cm)**	**Time to peak (24-hr)**	**Runoff coefficient**
Semenyih	2-Yr	96	6	136	18:12	0.3
	100-Yr	173	11	223	23:36	0.3
	PMP	585	37	1484	19:06	0.6
World's Precipitation		1092	68	3686	17:30	0.8

world's greatest precipitation (WGP) events are used as input to the calibrated TREX model to determine the magnitude of extreme floods. A 16 hr precipitation duration obtained from Jennings[17] with an intensity of 68.24 mm/hr has been applied (Fig. 5). The precipitation duration was selected at 16 hr because precipitation records for shorter duration can be determined with a sufficient level of certainty in this monsoon climate according to MSMA[18] (Urban Stormwater Management Manual). Table 3 shows the summary of the input for TREX. The 2 yr, the 100 yr[18] and also the PMP[19] conditions for Malaysia have also been applied in the model for a comparison of the peak discharge between the four simulated storms. The watershed runoff coefficient, $C = Q/iA$ (where Q = peak discharge $[L^3T^{-1}]$; i = rainfall intensity $[LT^{-1}]$; and A = watershed area $[L^2]$), increases in magnitude as intensity and peak discharge increased. The C values for the 2 yr and 100 yr

storm events were lower because the infiltration and flow characteristics in the flat areas were dominant compared with steep areas. The C values for the PMP and WGP storm events were closer to unity because the flat areas became fully saturated and almost impervious.

Figure 6 shows the results from the WGP event. Figure 6A represents the watershed starting to receive precipitation at time 01:39 hours. The WGP simulation result indicates that the lowlands would be flooded with a water depth as high a 6 m (Fig. 6B). The DID website defines three water levels in relation to floods in Malaysia: 1) the "normal" water level corresponds to water depths less than 4.49 m; 2) the "alert" level corresponds to water depths between 4.49 m and 6.09 m; and 3) the "danger" level is reached when the flow depth exceeds 6.09 m. This "danger" condition is reached during our simulation as indicated in red color on Figure 6C. Flooding is expected to be widespread in the low valley

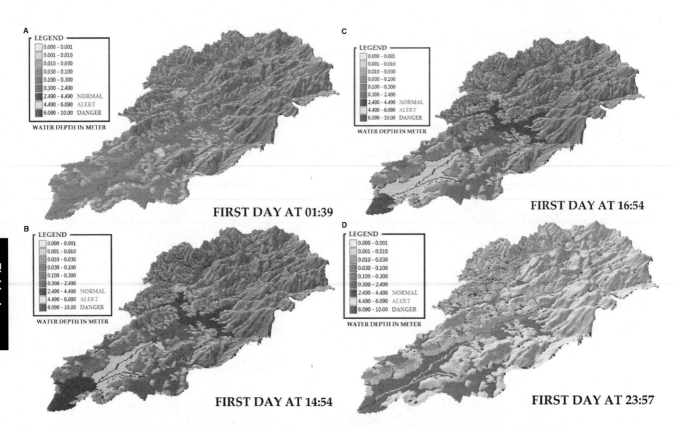

Fig. 6   Spatial distribution of water depth (m) with world's greatest precipitation event (68.24 mm/hr at 16 hr duration).

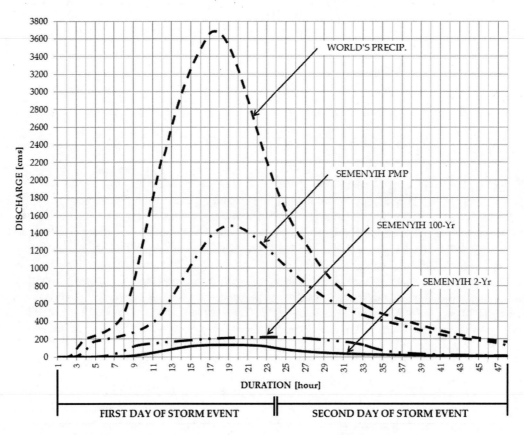

Fig. 7   TREX hydrograph with world's greatest precipitation event (68.24 mm/hr at 16 hr duration).

area. In this simulation, the "alert" condition (yellow in color) started around 11:00 hours and submerged about 50% of lowest area and lasted approximately 4 hr until 14:54 hours during the first day of the storm event. The flood took about 6 hr to recede and drain most of the surface water as shown in Fig. 6D.

The hydrograph (Fig. 7) for Semenyih watershed demonstrates that the TREX model can be effectively used to simulate and estimate the magnitude of flooding during extreme storms. The TREX result based on WGP input data are compared with floods of lesser magnitude, including the 2 yr, the 100 yr flood, as well as the PMP condition for Malaysia. A summary of the conditions and simulation results for these floods is presented in Table 3.

CONCLUSION

The model TREX has been successfully applied for the simulation of extreme floods on the 236 km^2 of Semenyih watershed in Malaysia. The April 13, 2003, storm was used to calibrate the model, whereas several other storm events served for the model validation. The calibrated TREX model can then simulate the hydrograph using the world's greatest precipitation (68.24 mm/hr for 16 hr duration). TREX successfully simulates extreme hydrographs for comparison with 2 yr and 100 yr flood events. TREX also provides visual flood hazard maps for this site.

ACKNOWLEDGMENTS

Financial support to the first author was granted through the Ministry of Higher Education (MOHE), Malaysia. Data for the Semenyih watershed application were provided by Mohd Rozi Talib (Ex-DSMM); Nor Haslinda Mohamed Yusop and Arshad Mohd Isa from DSMM; and Lizawati Turi, Abu Salim Abd. Aziz, Khairul Fadzilah Mohd Omar, Mohd Shawal Abd. Wahid, and Azmi Md. Jafri from DID. We also appreciate the assistance of Prof. Junaidah Ariffin, Joe Nyuin, Azmi Ibrahim, and Norizan Ismail from Universiti Teknologi MARA, Nur Shazwani Muhammad and Othman Jaafar from National University of Malaysia to provide us some useful information regarding Semenyih watershed. Additional support from Mark Velleux (HydroQual, New Jersey) and John England (U.S. Bureau of Reclamation) in using TREX software is also gratefully acknowledged.

REFERENCES

1. England, J.F. Jr.; Velleux, M.L; Julien, P.Y. Two-dimensional simulations of extreme floods on a large watershed. J. Hydrol. **2007**, *347* (12), 229241.

2. England, J.F. Jr.; Godaire, J.E.; Klinger, R.E; Bauer, T.R.; Julien, P.Y. Paleohydrologic bounds and extreme flood frequency of the upper Arkansas river, J. Geomorphol. **2010**, *124*, 1–16.

3. Velleux, M.L.; England, J.F. Jr.; Julien, P.Y. *TREX Watershed Modeling Framework User's Manual: Model Theory and Description*; Department of Civil Engineering; Colorado State University, Fort Collins, Colorado, 2006a; 83.

4. Julien, P.Y.; Saghafian, B.; Ogden, F.L. Raster-based hydrologic modeling of spatially-varied surface runoff. Water Resour. Bull. AWRA **1995**, *31* (3), 523–536.

5. Johnson, B.E.; Julien, P.Y.; Molnar, D.K.; Watson, C.C. The two-dimensional upland erosion model CASC2D-SED. J. Am. Water Resour. Assoc. AWRA **2000**, *36* (1), 31–42.

6. Ogden, F.L.; Julien, P.Y. CASC2D: A two-dimensional, physically-based, Hortonian hydrologic model. In *Mathematical Models of Small Watershed Hydrology and Applications*; Singh, V.P.; Frevert, D. Eds.; Water Resources Publications: Littleton, CO, 2002; 69–112.

7. Julien, P.Y.; Rojas, R. Upland erosion modeling with CASC2D-SED. Int. J. Sediment Res. **2002**, *17* (4), 265–274.

8. Velleux, M.L.; Julien, P.Y.; Rojas-Sanchez, R.; Clements, W.H.; England J.F. Simulation of metals transport and toxicity at a mine-impacted watershed: California Gulch, Colorado. Environ. Sci. Technol. **2006b**, *40* (22), 6996–7004.

9. Jorgeson, J.; Julien, P.Y. Peak flow forecasting with CASC2D and radar data, Water International. IWRA, **2005**, *30* (1), 40–49.

10. Julien, P.Y. *River Mechanics*; Cambridge University Press: Cambridge, 2002; 434.

11. Julien, P.Y. *Erosion and Sedimentation*; Cambridge University Press: Cambridge, 2nd Ed., 2010; 371.

12. Velleux, M.L.; England, J.F. Jr.; Julien, P.Y. TREX: Spatially distributed model to assess watershed contaminant transport and fate. J. Sci. Total Envir. **2008**, *404*, 113–128.

13. Molnar, D.K.; Julien, P.Y. Grid size effects on surface runoff modeling. J. Hydrol. Eng., ASCE, **2000**, *5* (1), 8–16.

14. Gupta, H.V.; Sorooshian, S.; Yapo, P.O. Status of automatic calibration for hydrologic models: comparison with multi-level expert calibration. J. Hydrol. Eng., ASCE, **1999**, *4* (2), 135–143.

15. Moriasi, D.N.; Arnold, J.G.; Van-Liew, M.W.; Bingner, R.L.; Harmel, R.D.; Veith, T.L. Model evaluation guidelines for systematic quantification of accuracy in watershed simulations. Am. Soc. Agric. Biol. Eng. **2007**, *50* (3), 885–900

16. Zakaria, N.A.; Azamathulla, H.M.; Chang, C.K.; Ab. Ghani, A. Gene expression programming for total bed material load estimation–a case study, Sci. Total Env. **2010**, *408*, 5078–5085.

17. Jennings, A. H. *World's Greatest Observed Point Rainfalls*. Hydrometeorological Section, U. S. Weather Bureau: Washington D.C., 1950; 4–5.

18. DID. Urban Stormwater Management Manual (MSMA Manual), 2000, 13 (3).

19. Hii, C.P.; Heng, H.H. Probable maximum precipitation derivation in Malaysia: Review and Comparison. Int. J. Hydro-Clim. Eng. **2010**, 37–72.

Field—Low

Hydrology: Environmental

Thomas Boving
Department of Geosciences/Department of Civil and Environmental Engineering, University of Rhode Island, Kingston, Rhode Island, U.S.A.

Abstract

The science of environmental hydrology focuses on the spatial and temporal variability of water quality and its evolution in the hydrosphere, lithosphere, and atmosphere. It encompasses the occurrence, distribution, and variability of water quality in ground water, surface water, the unsaturated zone, and the atmosphere. The principal scientific challenge of environmental hydrology is to describe in quantitative terms the complex chemical, physical, and microbiological processes that influence the solute transport in surface and subsurface water. Environmental hydrology, therefore, is a multidisciplinary science. In addition, a practicing environmental hydrologist must be cognizant about current environmental laws and regulations pertaining to their work. While environmental hydrology primarily focuses on water quality, its study cannot be isolated from studying the quantity of water interacting with the environment. Water scarcity is among the main challenges humanity faces in the 21st century and environmental hydrologists have to play an important role addressing this problem.

INTRODUCTION

Life on Earth depends on water that is both sufficient in quantity and quality. Anthropogenic activity, however, has degraded or depleted many water resources and humanity is now faced with remediating and restoring this essential natural resource. This endeavor can only be successful if scientists and engineers understand the complex connections between the various compartments of the environment that influence the availability and quality of water. This entry is about environmental hydrology. It describes how this branch of science stands at the nexus of competing interest for Earth's limited water resources and what role environmental hydrology can play in alleviating current and future ecohydrological stresses.

HISTORY

The science of hydrology, in general, focuses on the water on Earth, where it occurs, its circulation and distribution in biotic and abiotic systems, including its chemical and physical properties and interactions with the environment. Because water is vital for sustaining life on Earth, its study must be conducted in an environmental continuum[1] in which water interacts with the solid earth and the atmosphere in a system of complex and continuous reactions. This environmental continuum approach is emphasized in what is referred to as environmental hydrology.

The term environmental hydrology has been widely used since the early 1970s when hydrogeologists began debating and systematically investigating the effects of widespread pollution of the aqueous environment.[2] In those days, the public was confronted with massively polluted sites, such as the Cuyahoga River in Ohio, which became known as "the river that caught fire" or the notorious Love Canal site in Niagara Falls, NY, where an inadequate cover on a chemical company's dumpsite allowed a deadly brew of toxins to leak into a suburban neighborhood. The Love Canal became the site of one of the worst environmental disasters in American history, forcing 900 families to vacate their homes and causing damage worth over hundred million dollars.

As it became obvious that Love Canal was not an isolated case and that various types of toxic substances were dumped all over the environment, President Carter declared that uncovering these dumpsites was "one of the grimmest discoveries of the modern era."[3] The legacy of these sites eventually gave rise to the field of Environmental Hydrogeology, which represents a branch of hydrology that is primarily concerned about water quality. Environmental hydrology may be defined as the science dealing with the spatial and temporal variability of water quality and its evolution in the hydrosphere, lithosphere, and atmosphere. Atmospheric water must be considered a part of environmental hydrology because water cycling through the atmosphere impacts the water quality as demonstrated by the deposition of air-borne methyl mercury into surface water [4] or acid precipitation.[5] Hence, environmental hydrology

Encyclopedia of Natural Resources DOI: 10.1081/E-ENRW-120047594

encompasses the occurrence, distribution, and variability of water quality in ground water, surface water, the unsaturated (vadose) zone, and the atmosphere.[1] Further, there is mounting evidence that climate change is having a dramatic impact on the health of the ecohydrological environment,[6] which is why the hydrological consequences of climate change should also be covered under the umbrella of environmental hydrology.

Environmental hydrology is a multidisciplinary science, drawing knowledge from several branches of engineering and science, including chemistry, physics, biology, meteorology, geography, ecology, computer science, and mathematics, as well as agriculture and forestry. In addition, practicing environmental hydrologists are expected to have a working knowledge of the applicable environmental laws as well as an awareness of the potential sociopolitical implications of their work.

Many countries around the world have adopted environmental standards and regulations originally formulated by the U.S. Environmental Protection Agency (EPA). The EPA was proposed by President Nixon and charged by U.S. Congress on December 3, 1970 with protecting human health and the environment. The EPA enforces environmental regulations passed into law by congress. One of the earliest environmental laws is the Clean Water Act (CWA) of 1972. The CWA is the cornerstone of surface water quality protection in the United States and employs a variety of regulatory and nonregulatory tools to sharply reduce direct pollutant discharges into waterways, finance municipal wastewater treatment facilities, and manage polluted runoff.[7] It is noteworthy that CWA does not deal directly with ground water or with water quantity issues. In 1974, the Safe Drinking Water Act was passed to protect the drinking water supply by establishing enforceable national drinking water standards (e.g., maximum contaminant level). Probably the most important sets of environmental rules and regulations pertaining to the practicing Environmental Hydrologist are the Resources Conservation and Recovery Act (RCRA) and the Comprehensive Environmental Response Compensation and Liability Act (CERCLA), which were passed into law in 1976 and 1980, respectively. Under RCRA, the EPA has the authority to control hazardous waste from the "cradle-to-grave," including supervision of the generation, transportation, treatment, storage, and disposal of hazardous waste. CERCLA, also known as "Superfund law," provides broad federal authority to respond directly to releases or potential releases of hazardous substances that may endanger public health or the environment. Under CERLA, the EPA determines which hazardous waste sites warrant further investigation and are eligible for cleanup funding from the Superfund program. These sites are then placed on the National Priority List (NPL). As of February 2012, there were 1293 NPL sites.[8]

ENVIRONMENTAL HYDROLOGY AND WATER QUALITY

Environmental hydrology utilizes tools such as geographic information systems, hydraulic and transport/fate models, and global circulation models to solve complex environmental and ecological problems. Environmental hydrology research relies on studies and data generated in the field, including geophysical surveys, water and soil samples analysis, or tracer tests; in the laboratory, such as bench scale and column tests or isotherm experiments; or three dimensional pilot-scale test setups at dedicated research centers such as the Borden site in Canada [9] or the VEGAS installation in Germany.[10]

The information and data gathered in the field and laboratory have been used very successfully to predict the transport and fate of contaminants and to aid in the remediation of hazardous waste sites. The theory and concepts of contaminant transport and fate, including their modeling, are summarized in a number of noteworthy textbooks.[11–15] Chemical and physical properties of environmental pollutants are readily available in published format, such as[16–19] and on the internet.[20] Further, there are several textbooks specifically aimed at environmental hydrology, for example.[1,21]

A principal challenge of environmental hydrology science is to describe in quantitative terms the complex chemical, physical, and microbiological processes that influence the solute transport in surface and subsurface. The aqueous phase is the principal carrier of dissolved and suspended contaminants. In the absence of any interactions between the solute and the porous matrix, the advection–dispersion equation[11,22] can be utilized to describe solute transport in saturated media in terms of mixing and dilution. However, many contaminants interact with the porous matrix (sorption, reaction) or other compartments of the environment, including the atmosphere (volatilization) or biosphere (mineralization) or simply decay abiologically. Also, if nonaqueous phase liquids, such as petroleum products or chlorinated solvents are present, multiphase flow phenomena have to be considered as well.[11,23] To account for the various processes that influence the fate and transport of solutes in the environment, the advection–dispersion equation has to be expanded for the appropriate processes and solved numerically using a variety of powerful computer models for either saturated or unsaturated flow conditions. In a review of existing modeling software, over 250 codes were identified and rated for their ability to model the fate and transport of organic contaminants in the environment.[24] Environmental hydrologists commonly use these models as aides in designing remediation schemes or for predicting the response of hydrologic systems to changing environmental conditions. However, great care has to be taken to correctly set up these computer models and interpret their results.

Field—Low

ENVIRONMENTAL HYDROLOGY AND WATER QUANTITY

While environmental hydrology primarily focuses on water quality, its study cannot be isolated from understanding the quantity of water interacting with the environment. Many places on Earth already face severe water shortages; as the world's population continues to increase, the demand for water increases accordingly. Water scarcity is among the main problems many societies face in the 21st century.[25] For instance, as of 2004 it was estimated that 1.1 billion people lacked access to improved water supplies and 2.6 billion people lack adequate sanitation.[26] In many locations, trying to satisfy the demand for water resulted in falling water tables, depleted aquifers, or sea-water intrusion into overpumped coastal aquifers.[27] The ineffective use of water for irrigation has contributed to water depletion and, locally, has led to salinization of previously fertile farmland[28] and desertification.[29] In addition, the densely populated urban areas in the developing world will make up an estimated 81% of all urban inhabitants by 2030.[30] Supplying the growing number of urban and rural population with clean water and ensuring the effective treatment and disposal of wastewater will remain an enormous challenge for this and following generations of environmental hydrologists.

The consequences of climate change further stress the world's water resources. In 2007, the Intergovernmental Panel on Climate Change came to the conclusion that "all regions of the world show an overall net negative impact of climate change on water resources and freshwater ecosystems. Areas in which runoff is projected to decline are likely to face a reduction in the value of the services provided by water resources. The beneficial impacts of increased annual runoff in other areas are likely to be tempered in some areas by negative effects of increased precipitation variability and seasonal runoff shifts on water supply, water quality and flood risks."[6] Changes in the timing, intensity, and duration of precipitation together with higher water temperatures can affect water quality. Higher temperatures reduce dissolved oxygen levels, which can have an effect on aquatic life. Where stream flow and lake levels fall, there will be less dilution of pollutants; however, increased frequency and intensity of rainfall will produce more pollution and sedimentation due to runoff.[6] Today, environmental hydrologists are faced with trying to understand how a changing climate is affecting the water quality and availability and how the negative consequences of climate change can be mitigated.

Environmental problems have forced people to either vacate water-depleted, drought-stricken areas or pushed them into consuming polluted water. Besides pollution caused by the discharge of chemicals, water borne diseases like diarrhea pose a major threat to human health.

Waterborne diseases are caused by pathogenic microorganisms (viruses, bacteria, protozoa), which can be transmitted when contaminated water is consumed. The World Health Organization estimates that diarrheal disease alone is responsible for the deaths of 1.8 million people every year and is especially affecting children in developing countries.[26] The socioeconomic impacts of limited access to clean water and basic sanitation are substantial, estimated at annual losses of 6.4%, 5.2%, and 7.2% of GDP in India, Ghana, and Cambodia, respectively.[31] In 2000, the international community committed to the Millennium Development Goals (MDGs) with the aim to halve the proportion of people without access to clean water and basic sanitation by 2015. Currently, the world is on track to meet the water supply MDG but there are many regions and countries still lagging behind, particularly in sub-Saharan Africa.

Only about 3% of all the water on Earth is fresh water. Two-thirds of it is frozen, one-third is groundwater and a mere 0.3% remains as accessible surface water. Of the available renewable freshwater resources, about 54% is already appropriated by humanity for various uses: irrigated agriculture (70%), industry (22%), and domestic use (8%). By 2025, water withdrawals are projected to increase by 50% and 18% in the developing and developed world, respectively.[31] Water scarcity, commonly defined as water supplies of less than 1000 m^3 per person per year, is already a reality in much of North Africa and Arabia. By 2025, the United Nations expects absolute water scarcity to extend into Pakistan and India, affecting more than 1.8 billion people.[32] These statistics and predictions underline the need for integrated, cooperative solutions and adaptive water governance and management. This is particularly urgent in the 263 river basins, which are shared by two or more states, and in which nearly half the territory and population of the world are located.[33] In the absence of strong institutions and agreements, changes within a basin, such as dam construction or water diversion, can lead to transboundary tensions. When major water projects proceed without regional collaboration, they can become a point of conflict, heightening regional instability. Some of the world's largest watersheds and rivers, such as the Nile, the Jordan, the Colorado, and the Danube, may serve as examples. Given their expertise in dealing with complex water quantity and quality issues, environmental hydrologists are well placed to serve society as scientific advisors to political decision makers and water resources managers.

CONCLUSION

Environmental hydrology became a branch of hydrology in the early 1970s in response to ubiquitous environmental pollution problems. Since then it has evolved into a multidisciplinary science concerned not only with water quality but also the quantity of water on Earth. The underlying theoretical framework that describes the fate and transport of pollutants in surface and subsurface water is well developed, although there is plenty of room for future breakthroughs. Practicing environmental hydrologists perform their services within a legal framework, which grew out of the need for remediating, at great cost, a large number of hazardous waste sites. The prospect for employment in this area is likely to remain high for years to come.

The future of environmental hydrology is directly connected to how our society decides to deal with water scarcity and climate change. For billions of humans, safe and accessible water already is or shortly will be the natural resource that limits their standard of living and dictates where they can or cannot live. Now and in the future, environmental hydrologists have to play an important role in alleviating this global water crisis by making available to society their unique expertise regarding the sustainable management of the world's stressed water resources. This not only puts a great responsibility on the shoulders of currently practicing environmental hydrologists but also poses a tremendous economic opportunity for the next generation of young water scientists and professionals.

ACKNOWLEDGMENTS

Over the past few decades, many people have contributed to the field of environmental hydrology and a great number of excellent scientific articles, textbooks, and other materials have been published. Unfortunately, only a limited few could be cited in this entry. The selection of these sources, including emphasizing specific areas of research, is admittedly subject to this author's background and research interest. Many thanks to Ms. Laura Schifman for her critical review.

REFERENCES

1. Singh, V.P. What is environmental hydrology? In: *Environmental Hydrology* (Chap. 1). Singh, V.P., Ed.; Springer: 1995, Reprinted 2010.
2. Pettyjohn, W.A. *Water Quality in a Stressed Environment: Readings in Environmental Hydrology*; Burgess Pub. Co.: Minneapolis, 1972.
3. Beck, E.C. 1979. The Love Canal tragedy. EPA Journal - January 1979. http://www.epa.gov/aboutepa/history/topics/lovecanal/01.html (accessed September 2011).
4. Ullrich, S.; Tanton, T.; Abdrashitova, S. Mercury in the aquatic environment: a review of factors affecting methylation. Crit. Rev. Environ. Sci. Technol. **2001**, *31*, 241–293.
5. Likens, G. E.; Bormann F. H. Acid rain: a serious regional environmental problem. Science **1974**, *184* (4142),1176–1179.
6. Intergovernmental Panel on Climate Change (IPCC). Climate Change. Impacts, Adaptation, and Vulnerability. Contribution of Working Group II to the Fourth Assessment Report of the Intergovernmental Panel on Climate Change; Parry, M.L., Canziani, O.F., Palutikof, J.P., van der Linden, P.J., and Hanson, C.E. Eds.; Cambridge University Press: Cambridge, United Kingdom, 2007.
7. United States Environmental Protection Agency, 2008. Introduction to the Clean Water Act. http://www.epa.gov/owow/watershed/wacademy/acad2000/cwa/ (accessed September 2011).
8. United States Environmental Protection Agency. National Priorities List (NPL), http://www.epa.gov/superfund/sites/npl/ (accessed September 2011).
9. Sudicky, E.A.; Illman, W.A. Lessons learned from a suite of CFB Borden experiments. Rev. paper Ground Water **2011**, *49* (5), 630–648.
10. Koschitzky, H.P.; Braun, J. *VEGAS: Versuchseinrichtung zur Grundwasser- und Altlastensanierung.* Wasser und Abfall, Boden, Altlasten, Umweltschutz, **2008**, *10*, 54–55.
11. Bear, J., *Dynamics of Fluids in Porous Media*; Dover Publications: 1988.
12. Domenico, P.A.; Schwartz, F.W. *Physical and Chemical Hydrogeology*; 2nd Ed., Wiley: 1997.
13. Schwarzenbach, R.P.; Gschwend, P.M.; Imboden, D.M. *Environmental Organic Chemistry*; 2nd Ed., Wiley-Intersciences: 2002.
14. Fetter, C.W. *Contaminant Hydrogeology*; 2nd Ed., Waveland Press: 2008.
15. Bear, J.; Cheng, A.H.-D. *Modeling Groundwater Flow and Contaminant Transport: Theory and Applications of Transport in Porous Media*, Springer: 2010.
16. Yaws, C.L. *Chemical Properties Handbook: Physical, Thermodynamics, Environmental Transport, Safety & Health Related Properties for Organic & Inorganic Chemicals*; McGraw-Hill Professional: 1998.
17. Mackay, D.; Shiu, W.Y.; Ma, K.C. *Illustrated Handbook of Physical-Chemical Properties and Environmental Fate of Organic Chemicals*; 2nd Ed.; Vol. 1–4, CRC Press; 2006.
18. Montgomery, J.H. *Groundwater Chemicals Desk Reference*; 4th Ed., CRC Press: 2007.
19. Verschueren, K. *Handbook of Environmental Data on Organic Chemicals*; 5th Ed., Vol. 1–4, Wiley: 2008.
20. United States Environmental Protection Agency. Environmental Databases. http://www.epa.gov/opp00001/science/efed_databasesdescription.htm (accessed September 2011)
21. Ward, A.D.; Trimble, S.W. *Environmental Hydrology*; 2nd Ed., Lewis Publishers: 2003.
22. Freeze, A.R.; Cherry, J.A. *Groundwater*; Prentice Hall: 1979.
23. Brennen, C.E. *Fundamentals of Multiphase Flow*; Cambridge University Press: 2005.
24. MDH Engineering solutions corporation. Evaluation of computer models for predicting the fate and transport of hydrocarbons in soil and groundwater. The Water Research Users Group, Alberta Environment, Canada, 2005.
25. United Nation Development Program (UNDP). Human Development Report: Water Scarcity. Chapter 4. http://hdr.undp.org/en/media/HDR_2006_Chapter_4.pdf (accessed 09/12/2011).
26. United Nations Children's Fund (UNICEF) & World Health Organization (WHO). Meeting the MDG Drinking Water

Field—Low

and Sanitation Target: A Mid-Term Assessment of Progress. UNICEF/WHO, Geneva, Switzerland, 2004.

27. Barlow, P.M.; Wild, E.C. *Bibliography on the Occurrence and Intrusion of Saltwater in Aquifers along the Atlantic Coast of the United States*; U.S. Geological Survey: 2002.

28. United States Department of Agriculture (USDA). Soil quality resources concerns: Salinization. United States Natural Resources Conservation Service - Soil quality information sheet. http://soils.usda.gov/sqi/publications/files/salinzation.pdf (accessed September 2011).

29. United Nations International Strategy for Disaster Reduction (UNISDR). Drought, desertification and water scarcity. http://www.unisdr.org/we/inform/publications/3815 (accessed October 2013).

30. World Water Council. World water day 2011. http://www.world-watercouncil.org/index.php?id=32 (accessed September 2011).

31. United Nation Development Program (UNDP). Water and Ocean Governance. http://www.undp.org/water (accessed September 2011).

32. United Nation. Coping with water scarcity. ftp://ftp.fao.org/agl/aglw/docs/waterscarcity.pdf (accessed September 2011).

33. United Nations Educational, Scientific and Cultural Organization (UNESCO). From potential conflict to co-operation potential: Water for peace. Prevention and resolution of water related conflicts. http://www.unesco.org/water/wwap/pccp/pubs/brochures.shtml (accessed September 2011).

Field—Low

Hydrology: Urban

C. Rhett Jackson
Daniel B. Warnell School of Forest Resources, University of Georgia, Athens, Georgia, U.S.A.

Abstract

Urbanization reduces the cumulative volume of water storage provided by soils and vegetation, increases the fraction of rainfall that becomes surface runoff during storms, creates structures to hasten the movement of runoff to streams, reduces evaporation, and increases watershed yield. Protecting the hydrology and water quality of streams in newly urbanizing basins requires comprehensive implementation of structural and non-structural mitigation including but not limited to retention of natural vegetation, minimization of impervious surfaces, low impact site design, rain gardens, detention and infiltration facilities, and water quality treatment. The maximum possible extent of urbanization that can be accomodated while protecting local streams is uncertain.

INTRODUCTION

Since the passage of the 1972 Clean Water Act, the water quality of most United States rivers and streams has improved markedly, but urban streams have been an exception. Given the importance of flow to channel geomorphology and aquatic biota, the permanent and severe hydrologic changes wrought by urbanization make it difficult to restore good water quality conditions to previously urbanized streams.

ALTERATION OF HYDROLOGIC PROCESSES

Urbanization—broadly defined as all aspects of converting natural or agricultural land to residential, commercial, and institutional land—radically alters the hydrology of local streams, principally by reducing infiltration rates over most of the landscape. The infiltration rate is the depth of water per unit of time that can enter the soil from the surface. When infiltration rates are lowered, rainfall that formerly infiltrated into the ground and reached streams by relatively slow subsurface pathways instead runs over the ground surface, picks up pollutants spilled or applied to the ground or paved surfaces, typically enters engineered conveyance systems, and rapidly reaches streams during storms.

The influence of urbanization on local hydrologic systems can only be understood in comparison to the hydrology of natural watersheds. In a forest or grassland, soil structure and hydrologic behavior are strongly influenced by biological activity, the presence of leaf litter, and carbon accumulation from plant matter. Root growth, root decay,

cracking due to freeze/thaw and wetting/drying processes, animal burrowing, the windthrow of weak trees, subsurface erosion, and other natural processes all increase soil porosity (the ratio of void space to total soil volume), the number and size of macropores, and the conductivity of the soil to water. Leaf litter on the soil surface dissipates raindrop energy and allows rainfall to drip into the soil. Relatively high organic contents of natural soils increase the stability of soil aggregates, and stable aggregates prevent soil crusting during rainfall, reduce detachment of small soil particles, and maintain high surface infiltration rates. For all of these reasons, almost all rainfall infiltrates into the ground surface except during extremely intense rainfall events. When rainfall rates exceed infiltration rates, excess precipitation runs downhill over the ground surface. This type of surface runoff is called Horton overland flow and is rare on natural upland soils.

During the process of urbanization, lands are cleared of native vegetation; slopes and soils are graded to improve the topography for building; impervious surfaces (rooftops, parking lots, roads) are constructed; and typically curbs, gutters, and storm drains are created to hasten the flow of surface water off the landscape (Fig. 1). The process of clearing and grading compacts the soil, reducing porosity and macropore density and thus decreasing infiltration rates. Urban soils typically lack litter layers and have low carbon contents, so they are susceptible to surface sealing during rainfall. Horton overland flow occurs at lower rainfall rates on these urban soils. Impervious surfaces have infiltration rates approaching zero and thus convert nearly 100% of incident rainfall into overland flow.

Forests and grasslands have high leaf area indices—the ratio of vegetative surface area to the underlying ground

Encyclopedia of Natural Resources DOI: 10.1081/E-ENRW-120042089

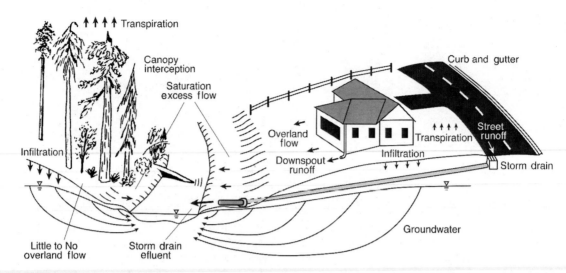

Fig. 1 Schematic illustrating multiple hydrologic changes wrought by urbanization including loss of forest cover, grading of soils, construction of impervious surfaces and stormwater conveyance facilities, occurrence of frequent overland flow, and reduced interception, infiltration, and evaporation.

area—and therefore a significant portion of rainfall is captured by the canopy and evaporated before ever reaching the ground. Urbanization typically reduces leaf area indices, increasing the amount of rainfall reaching the ground. Additionally, vegetative canopies provide temporary storage of rainfall, reducing precipitation intensities at the ground surface. Between storms, some water held by soils in the natural landscape is used by plants for transpiration and the rest moves by subsurface pathways to the water table and to streams, lakes, or wetlands at the base of the slope. Depending on hillslope position and the structure of the soils, travel times through these subsurface flow paths range from days to years, as opposed to minutes and hours for Horton overland flow. Because of interception and transpiration, a significant portion of rainfall on natural watersheds evaporates before reaching surface waters. The fraction of rainfall that becomes runoff is called the yield of a watershed, and yields increase in urbanized areas because of increased Horton overland flow and decreased transpiration. Furthermore, because of reduced evaporative cooling, reductions in soil water storage and canopy cover also increase urban heat island effects.

Where water tables occur near the ground surface, typically at the base of slopes and around the margins of streams, wetlands, and lakes, water tables may rise to the surface during rainfall. When the soil saturates from below, additional rainfall runs over the ground surface as saturation excess flow.[1] The areas that produce saturation excess flow are called variable source areas because they expand during wet periods and contract during dry periods.[2] These areas, along with direct precipitation on channels, are responsible for stormflow response in natural basins. Urbanization has relatively little effect on variable source areas but adds impervious surfaces and compacted soils to these rapidly responding areas. Thus, in relatively small

basins, stormflow volumes and peak flow rates increase almost linearly with increasing impervious surface coverage and also increase strongly with increasing coverage of compacted soil.

HYDROGRAPH EFFECTS

Within relatively small basins, the hydrologic alteration caused by urbanization can increase peak flow rates more than five times over their pre-development levels—with greater increases for frequent storm events and smaller increases for low frequency events[3]—and therefore increase flooding costs and problems. Stormflow volumes and cumulative durations of high flows also increase.[4] The additional erosive power leads to cycles of channel erosion and degradation[5,6] and to channel widening and simplification.[4,7] Channel erosion associated with flow increases in urban basins can contribute two-thirds of the sediment yield of an urban watershed.[8] These hydrologic and geomorphic alterations, along with changes in water chemistry associated with urbanization, are manifested in reduced diversity of aquatic organisms.[9,10] In addition to its direct hydrologic effects, increased percentages of impervious surfaces have been associated with increased nutrient loads and bacteria concentrations, higher summer stream temperatures, and reduced plant diversity in wetlands receiving urban runoff.[10]

Urbanization alters streamflow response to larger storms much less than the response to smaller storms because, in undeveloped watersheds, variable source areas expand and produce surface runoff over large areas during very large storms. The additive effect of impervious surfaces becomes less important during such storms (Fig. 2). Furthermore, the effects of peak flow alteration diminish going downstream as hydrograph phasing and channel routing progressively

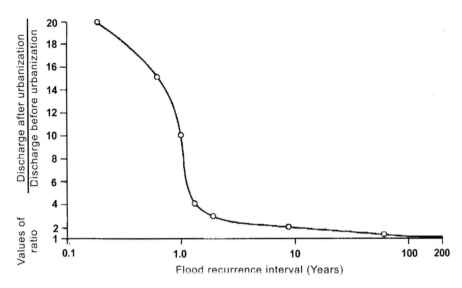

Fig. 2 Relationship between the relative increase in flood discharge due to urbanization (for watersheds with 20% pavement) and the flood recurrence interval. Urbanization has a very large effect on frequent floods and progressively less effect on large, low-frequency floods.
Source: Graph taken from Hollis.[3]

dampen flood flows as basin area increases. A city located on a large river has little effect on the large river flows; rather, it affects the tributaries draining the city into the river and also affects the river's chemical and biological quality.

Stormflows in urban streams also increase in frequency. In natural basins, when soils are dry, rainfall events are absorbed by the soils and streams respond very little. In most of the humid United States, natural streams experience very few large flows in late summer and early fall when soils are dry. Impervious surfaces provide no such buffering and produce stormflows during all rainfall events regardless of soil moisture conditions. Cumulative hydrologic alterations of an urban basin can also dramatically affect wetland hydropatterns (the time series of water levels), causing shifts in the composition of vegetative and amphibian communities. Urbanization reduces recharge rates over much of the landscape, thus reducing local sources of stream baseflow; but urban areas often import water from other basins, and groundwater and baseflows in urban streams may be supplemented by leakage from the water and sewage distribution system, lawn watering, and car washing. Thus, different studies have found both increases and decreases in baseflows as a result of urbanization.

The hydrograph of Peachtree Creek, Atlanta, when compared to the hydrograph of a nearby forested stream, clearly illustrates the multiple hydrograph effects of urbanization (Fig. 3). Peachtree Creek begins in the highly developed eastern suburbs of Atlanta and flows through Atlanta's urban core, draining 86.8 mi². Compared to a nearby forested stream on a unit area basis, Peachtree Creek has larger peak flows, greater stormflow volumes, more frequent high flow events, and greater total streamflow.

MITIGATION

Mitigating the potential hydrologic effects of urbanization involves maintaining infiltration rates and soil storage to

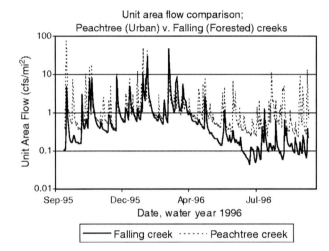

Fig. 3 Comparison of unit area runoff (cubic feet per second per square mile) between a highly urbanized stream (Peachtree Creek in Atlanta, Georgia) and a nearby forested stream (Falling Creek near Juliette, Georgia). Data are United States Geological Survey daily flow measurements shown for water year 1996 (October 1, 1995 through September 30, 1996).

the degree possible and creating artificial storage for runoff from impervious surfaces and compacted soils. The complete suite of stormwater mitigation strategies includes clustering development to reduce impervious surfaces and maintain forest cover, using pervious pavements where feasible,[11] employing low-impact development principles at the site scale (e.g., downspout infiltration, rain gardens, infiltration/biofiltration, and swales), incorporating green roof technology into commercial developments, constructing infiltration or detention basins, and maintaining riparian buffers around streams.[12,13] Detention facilities collect water from developed areas and release it at slower rates than it enters, whereas infiltration facilities are sited over soils that allow infiltration of a substantial portion of the collected waters. Riparian buffers provide little hydrologic benefit, but they increase the physical and biological resiliency of channels to hydrologic alteration. Structural solutions alone have been inadequate to mitigate urbanizations hydrologic alterations,[12] and protecting urban streams requires full integration of structural, non-structural, and site design strategies.[12,14]

In conventional structural stormwater mitigation, simple hydrologic models of runoff from single rain events are used to estimate stormflow peaks and volumes for pre- and post-development conditions. Then, a hydraulic routing model is used to design detention or infiltration ponds to reduce post-development peak flow rates to pre-development levels for various storms of interest, typically the 2-year, 10-year, and 25-year 24-hour rainstorms. There are two main problems with this design protocol: (1) matching only the peak flow rates still allows the duration of high flows to increase, thus the so-called peak matching design standards do not protect receiving streams from increased erosion[4] and (2) single rainfall event models are inherently incapable of matching peak flow rates because they do not account for problems associated with successive storms.[15] To protect receiving channels, detention and infiltration ponds should be sized to match durations of erosive flows. Development and acceptance of continuous rainfall/runoff models for stormwater design will result in better stormwater mitigation design and analysis.[15]

Because it takes years of post-development monitoring to evaluate hydrologic and water quality responses of streams to urbanization and because developments incorporating all modern concepts of stormwater mitigation have only recently been built, the cumulative effectiveness of modern stormwater management techniques for protecting the hydrology and water quality of urbanizing watersheds is unknown.

CONCLUSIONS

Urbanization reduces the volume of water storage provided by soils and vegetation, increases the fraction of rainfall that becomes surface runoff during storms, hastens the movement of runoff to streams, reduces evaporation, and increases watershed yield. Protecting the hydrology and water quality of streams in newly urbanizing basins requires comprehensive implementation of structural and non-structural mitigation, including but not limited to the retention of natural vegetation, the minimization of impervious surfaces, low-impact site designs, rain gardens, detention and infiltration facilities, and water quality treatments. However, the maximum possible extent of

urbanization that can be accommodated while protecting the hydrology, geomorphology, and biota of receiving waters is uncertain.

REFERENCES

1. Dunne, T.; Moore, T.R.; Taylor, C.H. Recognition and prediction of runoff producing zones in humid regions. Hydrol. Sci. Bull. **1975**, *20*(3), 305–327.
2. Hewlett, J.D.; Hibbert, A.R. Factors affecting the response of small watersheds to precipitation in humid areas. In *Forest Hydrology*; Sopper, W.E., Lull, H.W., Eds.; Pergamon Press: New York, 1967; 275–290.
3. Hollis, G.E. The effects of urbanization on floods of different recurrence intervals. Water Resour. Res. **1975**, *11*, 431–435.
4. Booth, D.B.; Jackson, C.R. Urbanization of aquatic systems—degradation thresholds, stormwater detention, and the limits of mitigation. J. Am. Water Resour. Assoc. **1997**, *33*(5), 1077–1090.
5. Wolman, M.G. A cycle of sedimentation and erosion in urban river channels. Geogr. Ann. **1967**, *49A*, 385–395.
6. Wolman, M.G.; Schick, A. Effects of construction on fluvial sediment, urban and suburban areas of Maryland. Water Resour. Res. **1967**, *3*, 451–464.
7. Hammer, T.R. Stream and channel enlargement due to urbanization. Water Resour. Res. **1972**, *8*, 1530–1540.
8. Trimble, S.W. Contribution of stream channel erosion to sediment yield from an urbanizing watershed. Science **1997**, *278*, 1442–1444.
9. Klein, R. Urbanization and stream quality impairment. Water Resour. Bull. **1979**, *15*, 948–963.
10. Schueler, T. The importance of imperviousness. Watershed Prot. Tech. **1995**, *1*(3), 100–111.
11. Ferguson, Bruce, K. *Porous Pavements*; CRC Press: Boca Raton, FL, 2005.
12. Booth, D.B.; Hartley, D.; Jackson, C.R. Forest cover, impervious surface area, and mitigation of stormwater impacts in King County, WA. J. Am. Water Resour. Assoc. **2002**, *38*(3), 835–846.
13. McCuen, R.H. Smart growth: hydrologic perspective. ASCE J. Prof. Issues Eng. Edu. Pract. **2003**, *129*, 151–154.
14. Walsh, C.J.; Roy, A.H.; Feminella, J.W.; Cottingham, P.D.; Groffman, P.M.; Morgan, R.P., II. The urban stream syndrome: current knowledge and the search for a cure. J. N. Am. Benthol. Soc. **2005**, *24*, 706–723.
15. Jackson, C.R.; Burges, S.J.; Liang, X.; et al. Development and application of simplified continuous hydrologic modeling for drainage design and analysis. In *Land Use and Watersheds: Human Influence on Hydrology and Geomorphology in Urban and Forest Areas*; Wigmosta, M.S., Burges, S.J. et al., Eds.; Water Science and Application; American Geophysical Union: Washington, DC, 2001; Vol. 2, 39–58.

Impervious Surface Area: Effects

Carl L. Zimmerman
Department of Urban and Environmental Policy and Planning, Tufts University, Medford, Massachusetts, U.S.A.

Daniel L. Civco
Department of Natural Resources and the Environment, University of Connecticut, Storrs, Connecticut, U.S.A.

Abstract

Humans are changing the world on a global scale by altering land use across the planet. Areas that consist of rooftops, driveways, pavements, and other anthropogenic materials are called impervious cover (IC). Measures of IC are often utilized to represent land cover changes and the environmental impacts of urbanization. The presence of IC in a watershed can have profound impacts on the local receiving waters through a variety of changes to the hydrologic cycle. Because IC can be mapped and analyzed with remote sensing and Geographic Information System (GIS) technologies, areas of IC can be compared across time. New low-impact development (LID) methods are becoming available that ameliorate some of the environmental impacts of IC.

INTRODUCTION

In this section, the concept of impervious cover (IC) as it relates to human activities will be introduced. The reader will learn how IC impacts and alters the environment, and an introduction on the measurement of IC at different scales will be provided. Finally, the entry will conclude with a discussion on a few methods that may ameliorate the impacts of IC on the environment.

LAND COVER CHANGE AND IMPERVIOUS COVER

The influence and impact of humans on the world has become an important area of research for scientists and planners. Some scientists are suggesting that we have entered a new geological epoch called the Anthropocene where human activities have become the dominant controller of the planet's geologic, climate, and biological systems.[1] An important component of our influence on the world is how we use and manipulate the landscape. Humans have managed and manipulated the environments in which we live for thousands of years to improve access to food and improve living conditions; however, the scale of anthropogenic land-use change has accelerated and altered in the past decade both in North America and in many developing regions undergoing rapid economic and population growth, such as India and China.

One of the most obvious changes in our environment is the rapid growth of urban areas. Accompanying the land development process is the engineered infrastructure such as roads, highways, and parking, and buildings with rooftops. Typically, these anthropogenic materials are made of

materials such as concrete, tar, or asphalt and interfere with the natural infiltration process because of their lack of permeability. Collectively, these areas are called impervious cover (IC).

Surprisingly, the relationship between population growth and IC coverage is not proportional. The expansion of IC in metropolitan areas of the United States such as Washington D.C., Houston, Texas, or Las Vegas, Nevada, has rapidly outstripped population growth.[2] In rapidly developing countries such as China and India, IC has expanded explosively in cities because of rapid industrialization and immigration. Even outside urban centers, IC has become more widespread and dispersed in areas that previously were dominated by other land cover types such as agriculture, forest, or grassland.[3]

As early as the 1930s,[4] aerial photography and mapping of land use and environmental variables was occurring at county or state scales. Later, multiple methods for understanding and mapping changes to the Earth's surface from anthropogenic activities such as urban growth, sprawl, and economic development emerged after the launch of Earth resource satellites such as Landsat in the 1970s and early 1980s. These images changed the way both citizens and scientists perceived the world. The medium-resolution, multi-spectral sensors on remote sensing satellites allowed for continental-scale classification of land cover such as the biological (e.g., forest) and physical (e.g., water) materials on the surface of the Earth, which can be captured within the field of view of a sensor.[5] As early as 1961, a definition of land cover included human influences on the planet's surface. Land cover was defined as "the vegetational and *artificial construction* covering the land surface."[6] These areas of "artificial construction" are made of man-made

Encyclopedia of Natural Resources DOI: 10.1081/E-ENRW-120047599

materials and are collectively known as IC. IC impacts the infiltration and runoff components of the water cycle.

Environmental Impacts

The process of land development places buildings and roads on areas that were typically utilized for agricultural land uses or covered by pervious land cover types such as grassland or forest. The alteration of a predevelopment area without IC to an area that has been developed adds more civil engineering infrastructure, such as stormwater drainage systems, and greater amounts of IC in the form or buildings and roads, one parcel or project at a time.[7] While a plethora of environmental impacts occurs in these areas, the sources of the environmental impacts are generally related to the alteration of the local hydrologic cycle. IC disrupts the natural infiltration of precipitation process in most landscape types and greatly increases the rate and amount of runoff in watersheds, which causes a number of cascading impacts (Fig. 1). The conversion of pervious natural land cover to IC in urbanizing areas generates the greatest runoff depth of any land cover type and introduces affiliated hydraulically efficient stormwater drainage systems found in roads and other transportation systems that rapidly remove water and interrupt infiltration. The flow pathway of water is simplified and shortened, and water is routed quickly away from homes and expensive land development projects.[8]

The resulting impacts of IC and development on the environment begin to occur at very low percentages of IC, typically in the 5% to 15% range of IC within a watershed.[9] "Vertical" hydrologic processes like infiltration and groundwater storage are reduced, which decrease year-round base flow in local streams and rivers. Water that previously infiltrated, stays at the surface that can, in some cases, exponentially increase runoff depth and volume with very high flow rates into receiving waters.[10] The hydrologic changes are related to structural changes in the local stream channels.

Fig. 1 Impervious cover without the infiltration of water.

They become rapidly incised with disproportionate deepening and increases in the cross-sectional area,[11] and the local riparian corridors and stream channels of the local receiving water become unstable because of the alteration of flow rate and volume. The channels erode and release a great deal of sediment and nutrients into the waters. Changes in the water quality impact stream ecology.

Fish populations of streams are negatively impacted and altered by the hydrologic, geomorphological, and water quality changes associated with IC especially when they are directly connected to receiving waters. Storm sewers and IC areas adjacent to streams play a critical role in delivering a wide variety of unattenuated pollutants into streams and rivers that degrade the quality of local habitat.[12] In the state of Connecticut, one study[13] found a wide number of nutrients and processes related to runoff from IC that degraded the quality of macroinvertebrate habitat when the IC percentage within watersheds reached 12%. Other possible impacts include those on coastal waters from fast-growing urban areas such as Tampa Bay, Florida, that could impact coastal water quality[14] and extreme heat events that have been related to the IC associated with sprawling urbanizing areas in the United States.[15] Finally, IC plays a role in fragmenting forested and other natural areas. One study found that approximately 62% of forest patches in the lower 48 states are located within 164 yards (150 m) of the nearest edge, and this edge is often related to some sort of IC-containing land-use area such as roads or developments.[16]

Measuring and Mapping IC

Professionals in land planning and development are using the percentage of IC within watersheds as a metric for evaluating the environmental impact of urbanization upon local receiving waters.[17] This metric is typically called percentage of total impervious area (%IC). Rather than representing one specific environmental impact or process, the percentage of IC within a watershed represents a whole collection of covarying pollutants associated with IC. Methods of determining %IC are derived from planning and geospatial data and are typically determined at either coarse planning scales or much finer site scales. Methods of determining %IC are implemented in a Geographic Information System (GIS) software package. These methods include:[18]

1. Deriving percent impervious surface coefficients from land cover (Fig. 2)
2. Modeling IC at watershed or census tract levels
3. Use of "subpixel" modeling to create spatially explicit per-pixel predictions of IC
4. Spatially explicit digitizing of IC into vector layers (Fig. 3)

The two primary data types used for mapping of IC are vector and raster data, and these data types are typically processed in a GIS software package. The vector data type is more commonly utilized in urban planning and is usually derived from engineering or surveying sources. It uses vertices located in a geographic coordinate system to define the perimeter of polygons of uniform IC surfaces. Typically, this data type has been utilized to define IC spatially at a site or local scale. This information can be acquired from construction documents, imported from computer-assisted design (CAD) maps, or created using different forms of digitizing. Accuracy is determined by the density of digitized points and by the quality of the source maps.

The other data type that is typically used to map and measure IC is the raster data type which is typically derived

Fig. 2 Impervious surfaces derived through on-screen digitizing using high-resolution aerial imagery.

Fig. 3 Percent impervious cover derived from processing of 30 m resolution Landsat satellite remote sensing data.
Source: Adapted from the United States Geological Survey National Land Cover Dataset (USGS NLCD).

from remote sensing imagery. A raster data type is organized similar to the pixels within a digital photograph and each pixel, also called a cell, is the same size. In the case of a Landsat image, the cell is 30 m × 30 m and incorporates reflectance information of the land cover acquired from the satellite for a specific location. The pixel does not explicitly define the IC. The analyst utilizes remote sensing software and interprets IC land use for specific locations based on the unique spectral characteristics of pavement, rooftops, roads, and other anthropogenic land cover. There are three advantages of the raster data sets over other types of environmental and geospatial data. The first advantage is that every location has data; so it is easy to get an overview of a larger area such as a watershed, and no gaps exist in the data. The second advantage is that satellite imagery data is in the raster format; so multidate comparisons of land cover change are simplified. The third advantage is that raster data is particularly well suited for integration with other environmental data and for environmental modeling of dynamic systems.

Extending the Utility of IC

Significant research efforts by geographers, ecologists, and hydrologists have gone toward extending the utility of land cover in modeling biological, hydrological, and ecological aspects of the environment. Typically run in a GIS, the goal is often to integrate different types and sources of collateral information with land cover data. As the resources for these professions are limited, the ability to scale up environmental models using land cover gives them potential to look at larger areas and problems that could not be investigated by field work alone. Many types of scientific modeling, planning, and environmental research have utilized IC in hydrologic modeling,[9] research on climate change, investigations into urban sprawl,[3] fragmentation of the habitat,[16] and land-use change and assessment research.[2] Landscape ecologists[19] and urban planning professionals have utilized IC to create spatial metrics that evaluate relationships between different types of land cover.[3]

One example of the extension of IC land cover data type is the application to some research related to water quality and stream ecology. By itself, a raster grid cell identified as being IC lacks the knowledge of its context in the urban or residential landscape, does not identify critical stormwater infrastructure, and does not identify flow paths to local rivers, streams, and ponds. By adding information about the connection or linkages and the location of stormwater infrastructure or flow patterns, a new type of IC assessment model can be created called directly connected impervious cover.[17] These hydrologically linked IC areas in this new type of model identify those areas that run off directly into local water resources, and hence are the most likely to carry nutrients directly into a water body without attenuation. Those impervious land cover areas that drain toward pervious areas and allow runoff to infiltrate are called unconnected impervious areas. Directly connected impervious areas have been found to be a better predictor of water quality, impacts on stream ecology, and fish assemblages than IC alone. Other researchers are extending the utility of land cover data by creating integrated land-planning models that evaluate environmental and hydrologic impacts within urban watersheds under different planning scenarios.[20]

CONCLUSION

Urbanization and growth of cities will continue to alter the environment both within and beyond urban centers as rapid economic growth occurs in many previously undeveloped nations throughout the world. Measuring and mapping of IC are improving and are becoming scalable, thanks to the availability of satellite data, the use of GIS software, and rapid improvements in computing hardware. Areas of IC growth can be measured, and potential environmental changes can be analyzed. New integrated landscape models using land cover such as IC are being explored to extend the utility of this data type for understanding the environment.

Fortunately, many new "greener" development methods are becoming practical and low-impact development (LID) methods are beginning to be widely used. These methods utilize pervious materials such as pavers, reroute stormwater runoff from IC to locally distributed infiltration areas such as rain gardens, and employ innovative approaches of routing stormwater runoff (Fig. 4). All these methods are intended to lessen or ameliorate the impacts of IC and its impact on the local hydrologic cycle.

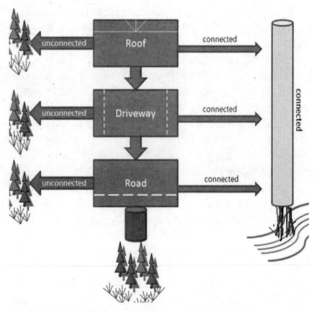

Fig. 4 Model of IC connection in residential area.

ACKNOWLEDGMENTS

The authors would like to thank the University of Connecticut NEMO program (Nonpoint Education for Municipal Officials) for the utilization of their figures.

REFERENCES

1. Kolbert, E. Enter the anthropocene: The age of man. Nat. Geogr. Mag. **2011**, 60–85.
2. Jantz, P.; Goetz, S.; Jantz, C. Urbanization and the loss of resource lands within the Chesapeake Bay watershed. Environ. Manag. **2005**, *36* (6), 808–825.
3. Angel, S.; Parent, J.; Civco, D.; Biel, A. The dimensions of global urban expansion: Estimates and projections for all countries, 2000–2050. Prog. Plan. **2011**, *75* (2), 53–108.
4. Rice, S. MAGIC Connecticut Historical Aerial Photography. Found at the University of Connecticut Libraries, Magic. http://magic.lib.uconn.edu/historical_aerial_indexes.html (accessed December 2013).
5. Arnoff, S. *Remote Sensing for GIS Managers*; ESRI Press: Redlands, CA, 2005; 1–524.
6. Anderson, J.; Hardy, E.; Roach, J.; Witmer, R. *A land use and land cover classification system for use with remote sensor data*, Paper 964; United States Geological Survey: Washington DC, 2006; 1–41.
7. Endreny, T. Land use and land cover effects on runoff processes: urban and suburban development. In *The Encyclopedia of Hydrologic Science*; Anderson, M., McDonnel, J., Eds.; John Wiley and Sons, 2005; Vol. 126, 1–29.
8. Endreny, T.; Thomas, K. Improving estimates of simulated runoff quality and quantity using road enhance road-enhanced land cover data. J. Hydraulic Eng. **2009**, *14* (4), 346–351.
9. Schueler, T.; Fraley-McNeal, L.; Cappiella, K. Is impervious cover still important? Review of recent research. J. Hydrol. Eng. **2009**, *14* (4), 309–315.
10. Dietz, M.; Clausen, J. Stormwater runoff and export changes with development in a traditional and low impact subdivision. J. Environ. Manag. **2007**, *87* (4), 560–566.
11. Booth, D. Stream channel incision following drainage-basin urbanization. Water Resour. Bull. **1990**, *26* (3), 407–417.
12. Walsh, C.; Papas, P; Crowther, P.; Sim, T.; Yoo, J. Stormwater drainage pipes as a threat to a stream dwelling amphipod of conservation significance, *Austrogammarus australis*, in southeastern Australia. Biodivers. Conserv. **2004**, *13*, 781–793.
13. Bellucci, C. Stormwater and aquatic life: Making the connection between impervious cover and aquatic life impairments. In *Water Environment Federation*, TMDL 2007 Conference Proceedings, 2007; 1003–1018.
14. Xian, G.; Crane, M. Assessment of urban growth in the Tampa Bay watershed using remotely sensing data. Remote Sensing Environ. **2005**, *97*, 203–215.
15. Stone, B.; Hess, J.; Frumkin, H. Urban form and extreme heat events: Are sprawling cities more vulnerable to climate change than compact cities? Environ. Health Perspect. **2010**, *118* (10), 1425–1428.
16. Riitters, K.; Wickham, J.; O'Neill, R.; Jones, K.; Smith, E.; Coulston, J.; Wade, T.; Smith, J. Fragmentation of continental United States forests. Ecosystems **2002**, *5*, 815–822.

17. Brabec, E.; Schulte, S.; Richards, P. Impervious surfaces and water quality: A review of current literature and its implications for watershed planning. J. Plan. Lit. **2002**, *16* (4), 1–17.

18. Chabaeva, A.; Civco, D.; Hurd, J. An assessment of impervious surface estimation techniques. J. Hydrol. Eng. **2009**, *14* (4), 377–387.

19. Moilanen, A.; Hanski, I. On the use of connectivity measures in spatial ecology. Oikos **2001**, *95* (1), 147–151.

20. Hong, B.; Limburg, K.; Hall, M.; Mountrakis, G.; Groffman, P.; Hyde, K.; Luo, L.; Kelly, V.; Myers, S. An integrated monitoring/modeling framework for assessing human-nature interactions in urbanizing watersheds: Wappinger and Onondaga Creek watersheds, New York, USA. Environ. Model. Software **2011**, *32*, 1–15.

BIBLIOGRAPHY

1. Skidmore, E. Environmental Modeling with GIS and Remote Sensing; Taylor and Francis: New York, 2002: 268 p.

2. Weng, Q. (Ed.). Remote sensing of impervious surfaces; CRC Press/Taylor and Francis Group: New York, 2008: 454 p.

Field—Low

Infiltration Systems and Nitrate Removal

Jose A. Amador

Laboratory of Soil Ecology and Microbiology, University of Rhode Island, Kingston, Rhode Island, U.S.A.

Abstract

The physicochemical properties of nitrate (NO_3^-) are presented in the context of its environmental fate, along with effects of excess levels on aquatic ecosystems and human and animal health. Two common infiltration-based approaches to removing NO_3^- from contaminated water—onsite wastewater treatment systems and permeable reactive barriers—are presented. The general mechanisms involved in NO_3^- removal in these systems as well as the specific microbial processes resulting in net loss of NO_3^- are discussed.

INTRODUCTION

Nitrate (NO_3^-) is an inorganic form of nitrogen commonly found in soil, ground, and surface water and precipitation. It is highly water soluble and anionic, properties that make it very mobile in the environment. Along with ammonia (NH_3), NO_3^- is the main form of N available to plants, and it is the preferred form of N for many terrestrial and aquatic plant species. Nitrate is also a source of N for microbial biomass and can serve as a terminal electron acceptor for denitrifying microorganisms that reduce NO_3^- to N_2O (nitrous oxide) and N_2 (nitrogen gas), resulting in a loss of N to the atmosphere.

Its function as an N source for organisms and as a terminal electron acceptor, and its mobility in the landscape, make NO_3^- an important component of the nitrogen cycle of terrestrial, freshwater, and marine ecosystems. These properties also translate into a high potential for pollution when NO_3^- is found in excess quantities. For example, nitrogen is often the *limiting nutrient* for primary production in coastal marine ecosystems. Because of its high mobility, NO_3^- can be the main form of inorganic N reaching these areas. Elevated NO_3^- levels can cause *eutrophication*, resulting in excessive algal growth, depletion of dissolved oxygen associated with decomposition of algal biomass, and fish kills. Excess NO_3^- levels in drinking water (10 mg NO_3^--N/L) from ground and surface sources are considered problematic because of negative effects on human and animal health.

The large-scale industrial synthesis of NH_3 from N_2 gas to manufacture fertilizers has resulted in a human-induced alteration of the global N cycle and the enrichment of natural and man-made ecosystems with nitrogen. Elevated NO_3^- levels in terrestrial and aquatic ecosystems can result from direct inputs, with major sources of NO_3^- including inorganic fertilizers and runoff. Some sources of N, such as wastewater, animal manure, and organic and inorganic fertilizers, may not contain much NO_3^-, but the nitrogen found

in them can be readily converted to it. Thus, organic N present in amino acids, amino sugars, nucleic acids, and proteins can be released as ammonia (NH_3) by microbial processes known as *ammonification*. Whether ammonia is present in the original input or results from ammonification, it can be oxidized to NO_3 via a process known as *nitrification*, by specific groups of autotrophic aerobic bacteria and archaea.

The ubiquity, mobility, and potential negative effects on the health of ecosystems, animals, and humans often require the removal of NO_3^- from water. One approach to removing nitrate present in water involves the infiltration of contaminated water into soil or other porous media with adequate metabolic capacity to transform this contaminant to gaseous forms that are readily lost to the atmosphere. We present two common examples of this approach to NO_3^- removal using infiltrations systems: 1) onsite wastewater treatment systems (OWTS); and 2) permeable reactive barriers (PRB). Because these systems rely on microbial processes for the removal of NO_3^-, a brief introduction to these processes is given first.

MICROBIAL REMOVAL PROCESSES

Denitrification

Most approaches for the removal of nitrate in infiltration systems take advantage of the fact that NO_3^- can be used as an electron acceptor by microorganisms through a process known as *denitrification*. It results in net loss of N from the system through the production of a gaseous form of nitrogen, namely, N_2 and N_2O. The general equation for the process is:

$$NO_3^- \rightarrow NO \rightarrow N_2O \uparrow \rightarrow N_2 \uparrow$$

The ability to reduce NO_3^- to N_2O and N_2 is mainly associated with bacteria, although fungi have also been

Encyclopedia of Natural Resources DOI: 10.1081/E-ENRW-120047600

shown to carry out this process. Denitrifying bacteria are facultative anaerobes that use NO_3^- as an alternate electron acceptor when O_2 is not available. The majority are heterotrophs that use organic carbon compounds as electron donors.

Nitrification

As mentioned previously, oxidation of ammonia to nitrate, or nitrification, is carried out by both bacteria and archaea. The general equation for the process is:

$$NH_3 + O_2 \rightarrow NO_2^- + O_2 \rightarrow NO_3^-$$

Nitrifying organisms can produce nitric oxide (NO) and N_2O when the levels of O_2 are low, resulting in net loss of N from the system. The relative importance of nitrification to gaseous N losses in environmental settings is still under debate. The amount of N lost as N gases via nitrification at low O_2 levels can be substantial relative to losses via denitrification,[1] and thus nitrification may also be important in N removal in infiltration systems.

Uptake

Microorganisms can reduce NO_3^- through a series of biochemical reactions to NH_3, which is then used in the synthesis of amino acids and nucleotides (*assimilatory nitrate reduction*). The resulting microbial biomass results in the storage of N in organic forms (e.g., proteins, nucleic acids) in microbial cells. If the microbial biomass is not removed from the system, the N associated with dead biomass can be released as NH_3 by ammonification, with the resulting NH_3 subject to oxidation to NO_3^-. Thus, although microorganisms take up NO_3^-, this process does not result in net NO_3^- removal.

ONSITE WASTEWATER TREATMENT SYSTEMS

Conventional Systems

Nearly a quarter of the households in the U.S.A. rely on OWTS for treatment and disposal of domestic wastewater.[2] In conventional OWTS, wastewater from the household flows by gravity to a septic tank, where the settling of solids and some removal of carbon and nutrients takes place (Fig. 1). The high biochemical oxygen demand in the septic tank results in anaerobic conditions. The anaerobic septic tank effluent (STE) flows by gravity to a distribution box and subsequently to a series of underground trenches (known as the drainfield, leachfield, or soil absorption area) where STE infiltrates into the soil. The infiltrative surface of the drainfield is generally at least 150 cm below the ground surface, and the separation distance between the infiltrative surface and the seasonal high water table generally ranges from 30 to 120 cm.[3] Nitrogen removal in a conventional

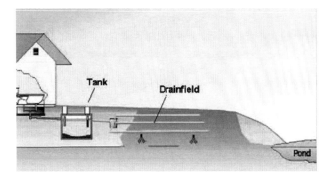

Fig. 1 Schematic diagram of a conventional onsite wastewater treatment system, or OWTS. Wastewater drains by gravity from home to the septic tank, where the removal of solids and some nutrients takes place, and on to the drainfield, where the removal of pathogenic organisms is the main function. Very little nitrogen is removed in the drainfield of a conventional OWTS, and NO_3^- leaching to groundwater can be a significant problem.
Source: Adapted from Joubert et al.[5]

OWTS is relatively low (0–30%), because these systems are not designed for N removal.[3,4] Because the anaerobic conditions prevalent in the septic tank preclude the oxidation of ammonia to nitrate, inorganic N in STE is present only as NH_3. However, as STE flows and infiltrates into the soil, O_2 present in the atmosphere of the trench can support some level of nitrification. The resulting NO_3^- may be removed via denitrification in the leachfield; however, a considerable amount of N still leaches from these systems, mostly as NH_3. This ammonia can be readily oxidized to NO_3^- once it reaches the vadose zone, posing a threat to groundwater quality.

Advanced Systems

The conventional OWTS design can be modified to address the poor N-removal efficiency and minimize NO_3^- leaching. These modified systems, often referred to as advanced treatment systems (ATS), are designed to promote the nitrification of ammonia in the STE, followed by denitrification before water is disposed of in the drainfield (Fig. 2). They are often installed in areas sensitive to nitrogen pollution, on lots that present site constraints, or in places where consistent system performance is desirable.[5] Although ATS designs vary, they involve the same approach: nitrification of ammonia, followed by denitrification. In order to promote nitrification, O_2 is introduced into the STE. This can be accomplished by passing STE through a filter filled with sand or other inert porous media, where O_2 can be entrained in the STE as it moves through the filter. Media filters provide habitat for microorganisms, in the form of particle surfaces, as well as aerobic conditions, since they are not saturated. Alternatively, aeration can be accomplished by bubbling air into a separate chamber through which STE flows. To promote denitrification of NO_3^- resulting from the aeration step, the nitrified STE is returned to the septic

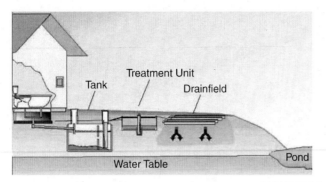

Fig. 2 Schematic diagram of an advanced onsite wastewater treatment system, or ATS. In systems designed for N removal, the additional treatment unit is used to convert NH_3 in STE to NO_3^-, which can be subsequently denitrified to N_2O and N_2 in the septic tank. This results in net loss of N from the ATS to the atmosphere, preventing NO_3^- contamination of groundwater.
Source: Adapted from Joubert et al.[5]

tank, where the anaerobic conditions and high levels of organic carbon support denitrification. A somewhat different ATS design relies on these two microbial processes taking place directly in the drainfield.[6] The first step involves accumulating STE in the septic tank or pump chamber, which is followed by aeration of the drainfield soil for a short period of time to promote nitrification. STE is then applied to the drainfield soil to displace oxygen and provide a carbon source, promoting denitrification.

Nitrogen removal efficiencies greater than 50% can be accomplished using ATS,[5] making them a highly effective technology for the removal of N from infiltrating wastewater, and thus the prevention of groundwater contamination with NO_3^-.

PERMEABLE REACTIVE BARRIERS

The construction of permeable reactive barriers (PRBs) involves the placement of a layer of porous media perpendicular to the flow path of contaminated water (Fig. 3). The porous medium can react directly with the contaminant to transform it, or support microbial processes that result in contaminant transformation. PRBs may be configured to treat water as it infiltrates the soil, with the barrier placed parallel to and below the soil surface, or as water moves horizontally in the vadose and saturated zones, in which case a trench is dug and filled with appropriate reactive media. PRBs have been employed successfully to remove a wide variety of organic and inorganic compounds, primarily from groundwater.[7]

PRBs have been used to effectively remove nitrate present in surface, drainage, and groundwater.[8–10] PRBs used for NO_3^- removal generally contain a medium that supports denitrification, the main mechanism by which NO_3^- is removed from contaminated water. The reactive medium in the trench is usually an inexpensive, easily available,

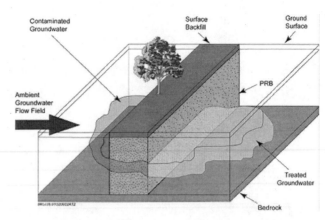

Fig. 3 Schematic diagram of a permeable reactive barrier, or PRB. For the removal of NO_3^- from groundwater, the PRB usually consists of a porous organic material that serves as habitat and a source of carbon for denitrifying bacteria, which convert NO_3^- to N_2O and N_2.
Source: From Los Alamos National Laboratory (© 2011 Los Alamos National Security, LLC All rights reserved) as included in Environment at LANL: Monitoring, Compliance and Risk Reduction. http://www.lanl.gov/environment/all/overview.shtml (accessed May 2012).

porous organic material that is relatively rigid (to prevent collapse) and has a sufficiently high hydraulic conductivity to allow water to flow relatively unimpeded. Commonly used barrier materials include wood chips, peat, and straw. The porous, organic material serves as a source of organic carbon for denitrification, promotes the establishment of anaerobic conditions for denitrification, and serves as a physical support for the bacteria involved in denitrification. The trench may be placed near the soil surface, to allow the infiltration of NO_3^- containing runoff, or at an appropriate depth to allow for the interception of drainage water or groundwater flow.

Removal rates between 50% and nearly 100% of initial NO_3^- concentration[10] have been reported for these systems. PRBs represent a relatively inexpensive, passive approach to the remediation of NO_3^--contaminated water that can provide long-term treatment with minimal need for maintenance or replenishment of reactive media.

CONCLUSION

Soil infiltration systems rely on the microbial reduction of nitrate to gaseous forms of N—denitrification—for the removal of this contaminant from ground, surface, and wastewater. Advanced OWTS involve a two-step approach, promoting the nitrification of ammonia to nitrate in wastewater prior to removal via denitrification. PRB rely on denitrification to remediate ground and surface water contaminated with nitrate. The barrier material supports nitrate removal by providing habitat and organic carbon compounds for denitrifying microorganisms.

REFERENCES

1. Kool, D.M.; Dolfing, J.; Wrage, N.; Van Groenigen, J.W. Nitrifier denitrification as a distinct and significant source of nitrous oxide from soil. Soil Biol. Biochem. **2010**, *43* (1), 174–178.

2. United States Census Bureau. American housing survey for the United States, Washington, DC, http:// www.census.gov/hhes/www/housing/ahs/ahs99/tab1a4.html (accessed May 2012).

3. U.S. Environmental Protection Agency. *Onsite Wastewater Treatment Systems Manual*; U.S. government: Washington, DC, 2002.

4. Kaplan, O.B. *Septic Systems Handbook*; Lewis Publishers: Chelsea, MI, 1987.

5. Joubert, L.; Loomis, G; Dow, D.; Gold, A.; Brennan, D.; Jobin, J. *Choosing a Wastewater Treatment System*; University of Rhode Island: Cooperative Extension, Kingston, RI, 2005.

6. Amador, J.A.; Loomis, G.W.; Kalen, D.; Patenaude, E.L.; Gorres, J.H.; Potts, D.A. Evaluation of Leachfield Aeration Technology for Improvement of Water Quality and Hydraulic Functions in Onside Wastewater Treatment Systems. Final Report to the Cooperative Institute for Coastal and Estuarine Environmental Technology (CICEET); NOAA/University of New Hampshire: Durham, NH, 2007.

7. Blowes, D.W.; Ptacek, C.J.; Brenner, S.G.; McRaea, C.W.T.; Bennetta, T.A.; Puls, R.W. Treatment of inorganic contaminants using permeable reactive barriers. J. Contam. Hydrol. **2000**, *45* (1–2), 123–137.

8. Ergas, S.J.; Sengupta, S.; Siegel, R.; Pandit, A.; Yao, Y.; Yuan, X. Performance of nitrogen removing bioretention systems for control of agricultural runoff. J. Environ. Eng. **2010**, *136* (10), 1105–1112.

9. van Driel, P.W.; Robertson, W.D.; Merkley, L.C. Denitrification of agricultural drainage using wood-based reactors. Trans. ASABE **2006**, *49* (2), 565–573.

10. Robertson, W.D.; Blowes, D.W.; Ptacek, C.J.; Cherry, J.A. Long-term performance of in situ reactive barriers for nitrate remediation. Ground Water **2000**, *38* (5), 689–695.

Field—Low

Irrigation: River Flow Impact

Robert W. Hill
Biological and Irrigation Engineering Department, Utah State University, Logan, Utah, U.S.A.

Ivan A. Walter
Ivan's Engineering, Inc., Denver, Colorado, U.S.A.

Abstract

The practice of irrigation necessitates developing a water source, conveying the water to the field, application of the water to the soil and collection and reuse or disposal of tailwater and subsurface drainage. These processes alter river basin hydrology and water quality in space and time. To sum up the effect of irrigation on a watershed in a word, it would be: *depletion*.

INTRODUCTION

In hydrologic studies it is common engineering practice to quantify the impact upon the stream(s) from which the irrigation water is diverted. The impact upon the stream is actually of two kinds: 1) diversions that decrease the streamflow and 2) return flows that increase the streamflow. The engineering term used to describe the overall impact is "streamflow depletion" which means the net reduction in streamflow resulting from diversion to irrigation uses. Actual stream depletions are a function of many factors including the amount and timing of diversions, the type of diversion structure (well vs. ditch), crops grown, soil type, depth to groundwater, irrigation method, irrigation efficiency, properties of the alluvial aquifer, area irrigated, and evapotranspiration of precipitation, groundwater, and irrigation water.

Depletion

Depletion, in this context, is the consumptive abstraction of water from the hydrologic system as a result of irrigation. It is in addition to consumptive water use that would have occurred in the unmodified natural situation. As an example, waters of the Bear River Basin of southern Idaho, northern Utah, and western Wyoming, because they are an interstate system, are administered by a federally established commission under the authority of the Bear River Compact.[1] Depletion is the basis, in the compact, for allocating Bear River water use among the three states. It is defined by a "Commission Approved Procedure" which includes consideration of land use and incorporates an equation for estimating depletion based on evapotranspiration. In a study for the commission, Hill[5] defined crop depletion as:

$$Dpl = Et - SMco - Pef \qquad (1)$$

where Dpl is estimated depletion for a given site or sub-basin; Et is calculated crop water use; SMco is moisture which is "carried over" from the previous non-growing season (October 1–April 30) as stored soil water in the root zone available for crop water use subsequent to May 1; and Pef is an estimate of that portion of precipitation measured at an NWS station during May–September, which could be used by crops.

The carry-over soil moisture (SMco) was estimated by assuming that 67% of adjusted precipitation from October through April could be stored in the root zone. If this exceeded 75% of the available soil water-holding capacity of the average root zone in the sub-basin, the excess was considered as lost to drainage or runoff and not available for crop use. Growing season precipitation was considered to be 80% effective in contributing to crop water use. The effectiveness factor of 80% allowed for precipitation depths throughout a sub-basin that might differ from NWS rain–gage amounts. It also included a reduction for mismatches in timing between rainfall events and irrigation scheduling.

HYDROGRAPH MODIFICATION

Diversion of significant amounts of water from rivers and streams for irrigation and subsequent return flows alters the shape and timing of downstream hydrographs. In watersheds where mountain snowmelt provides the irrigation supply, such as in the western United States, diversion during the spring runoff attenuates the peak flow rate while later return flows extend the flow duration into late summer and early fall.

Encyclopedia of Natural Resources DOI: 10.1081/E-ENRW-120010370

Reservoir Storage

Storage of water in reservoirs can significantly modify the natural stream hydrograph depending on the timing and quantity of the storage right. Irrigators with junior rights may only be able to store during time periods with low irrigation demand, such as during the winter, or during peak flow periods. Reductions of stream flow during the winter time may have considerable impact on downstream in-stream flows. Whereas, storage during periods of peak runoff may not affect minimum in-stream flow needs, but could deposit considerable amounts of sediment in the reservoir.

Irrigation Return Flows

Irrigation return flows are comprised of surface runoff and/ or subsurface drainage that becomes available for subsequent rediversion from either a surface stream or a groundwater aquifer downstream (hydrologically) of the initial use. Reusable return flow can be estimated as irrigation diversion minus crop-related depletions minus additional abstractions. Additional abstractions include incidental consumptive use from water surfaces as in open drains, along with non-crop vegetation. The timing of return flow varies from nearly instantaneous (recaptured tailwater) to delays of weeks and months or perhaps longer with deep percolation subsurface drainage. In a hydrologic model study of the Bear River Basin[4] delay times between diversion and subsequent appearance of the return flow at the next downstream river gage varied from 1.5 months to as long as 6 months. The delay appeared to be related to sub-basin shape and size.

Irrigation Methods

Four general irrigation methods are used: surface, subsurface, sprinkler, and trickle (also known as low flow or drip). Surface methods include wild or controlled flooding, furrow, border-strip, and ponded water (basin, paddy, or low-head bubbler). Hand move, wheel move, and center pivot are examples of sprinkler irrigation. Trickle irrigation includes point source emitters, microspray, bubbler, and linesource drip tape (above or below ground). Whereas the efficiency of surface irrigation is dependent upon the skills and experience of the irrigator, the performance of trickle and sprinkler systems is more dependent on the design. Generally, the more control that the system design (hardware) has on the irrigation system performance, the higher the application efficiency (E_a) can be. Thus, typical wheel move sprinklers have higher E_a values than surface irrigation, but lower values than for center pivots or trickle, assuming better than average management practices for each method.

The impact on river flows can be quite different among the various irrigation methods. The nature of furrow and border surface irrigation generally produces tail water run-off, which can be immediately recaptured and reused, as well as deep percolation, which may not be available for reuse until after a period of time. Tailwater is essentially eliminated and deep percolation reduced with sprinklers (Fig. 1) compared to conventional surface irrigation. Whereas, with drip methods, deep percolation can be further reduced. The reduction of deep percolation implies increased salt concentration in the root zone leachate, but, perhaps significant reduction in salt pick-up potential from geologic conditions.

Irrigation Efficiencies

Although a full discussion of the several variations of irrigation efficiency is beyond the scope herein, two terms will be defined and discussed. More complete discussions relating to irrigation efficiencies and water requirements are given elsewhere.[6-9,13] Keller and Bliesner[9] give a particularly thorough presentation of distribution uniformity and efficiencies.

Application efficiency (E_a):

$$E_a = 100 \times \frac{\text{Volume of water stored in the root zone} \left(V_s\right)}{\text{Volume of water delivered to farm or field}(V_f)}$$

Distribution uniformity:

The distribution uniformity is a measure of how evenly the on-farm irrigation system distributes the water across the field. The definition of DU is:

$$DU = 100 \times \frac{\text{Average of the lowest 25\% of infiltrated water depth}}{\text{Average of all infiltrated water depths across the field}}$$

On-farm or field application efficiencies can be affected by the distribution uniformity and vary widely for both surface and sprinkle irrigation methods. This is largely due to difference in management practices, appropriateness of design in matching the site conditions (slope, soils, and wind), and the degree of maintenance. In addition, for a given system uniformity, the higher the proportion of the field that is adequately irrigated (i.e., infiltrated water refills the soil water deficit) the lower will be the application efficiency. This is due to greater deep percolation losses in the over-irrigated portions of the distribution pattern. See Table 1 for an example of this phenomenon.

The E_a for a particular field may vary greatly during the season. Cultivation practices, microconsolidation of the soil surface and vegetation will alter surface irrigation efficiency

Surface Irrigation Water Budget

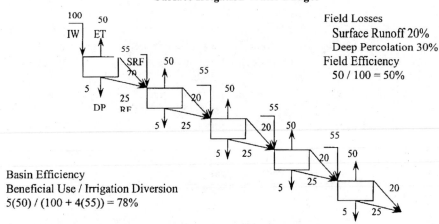

Field Losses
  Surface Runoff 20%
  Deep Percolation 30%
Field Efficiency
  50 / 100 = 50%

Basin Efficiency
Beneficial Use / Irrigation Diversion
5(50) / (100 + 4(55)) = 78%

IW – Irrigation Water Supply
ET – Evapotranspiration (beneficial use) from irrigation water
DP – Deep Percolation below root zone to subsurface water
SRF – Surface Return Flow
RF – Return Flow of subsurface water
EV – Evaporation from droplets in air and wind drift losses with sprinklers

Sprinkler Irrigation Water Budget

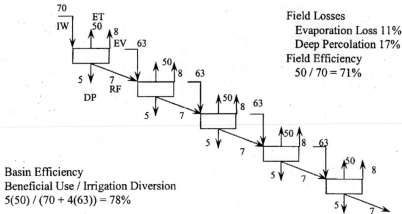

Field Losses
  Evaporation Loss 11%
  Deep Percolation 17%
Field Efficiency
  50 / 70 = 71%

Basin Efficiency
Beneficial Use / Irrigation Diversion
5(50) / (70 + 4(63)) = 78%

Fig. 1 Comparison of basin efficiencies between surface and sprinkler irrigation methods with four return flow reuse cycles.

Table 1   Values determined in Utah field evaluations

	Observed		
Method	High (%)	Low (%)	Typical (%)
Surface irrigation			
E_a	72	24	50
Tailwater	55	5	20
Deep percolation	65	20	30
Sprinkler irrigation			
E_a	84	52	70
Evaporation	45	8	12
Deep percolation	37	8	18

both up and down from the seasonal average. Seasonal and diurnal variations in wind, humidity, and temperature will also affect sprinkle application efficiencies.

BASIN IRRIGATION EFFICIENCY

The actual irrigation efficiency realized for several successive downstream fields where capture and reuse of return flows is experienced is higher than the E_a of an individual field. This notion of "Basin Irrigation Efficiency"[12,13] is illustrated in Fig. 1. This simple example comparison of surface and sprinkle methods assumes four reuse cycles. In each of the five "fields" Et is assumed to be 50 units. The surface runoff is captured for reuse on the next field. All of the irrigation-related evaporation is assumed "lost" as well as 5 units of deep percolation. After the fifth field, all surface and subsurface flows are lost. The basin efficiency for surface is 78%, which is the same as for sprinkle. The surface irrigation basin efficiency increase is dependent upon the surface return flow reuse, which is 20 units in this example. However, the depletion is greater for sprinkler due to the extra evaporation. In a Colorado field study, Walter and

Altenhofen[11] found a progressive increase in irrigation efficiencies from field (average E_a of 45%), to farm, to efficiency of ditch or sectors (average of 83%). This was due to the reuse of tailwater (10–20% of delivery) and deep percolation (46% of delivery).

ENVIRONMENTAL CONCERNS

The process of evapotranspiration, or crop water use, extracts pure water from the soil water reservoir, which leaves behind the dissolved solids (salts) contained in the applied irrigation water. The "evapoconcentration" of salts is an inevitable result of irrigation for crop production. As stated by Bishop and Peterson:[2]

"…Other uses add something to the water, but irrigation basically takes some of the water away, concentrating the residual salts. Irrigation may also add substances by leaching natural salts or other materials from the soil or washing them from the surface. Irrigation return flow is a process by which the concentrated salts and other substances are conveyed from agricultural lands to the common stream or the underground water supply…".

Water Quality Implications for Agriculture

Irrigated agriculture is dependent upon adequate, reasonably good quality water supplies. As the level of salt increases in an irrigation source, the quality of water for plant growth decreases. Since all irrigation waters contain a mixture of natural salt, irrigated soils will contain a similar mix to that in the applied water, but generally at a higher concentration. This necessitates applying extra irrigation water, or taking advantage of non-growing season precipitation, to leach the salts below the root zone.

Salt Loading Pick-Up

Water percolating below the root zone or leaking from canals and ditches may "pick-up" additional salts from mineral weathering or from salt–bearing geologic formations (such as the Mancos shale of western Colorado and eastern Utah). This salt pick-up will increase the salt load of return flows and consequently increase the salinity of receiving waters.

In the Colorado River Basin in the United States and Mexico salinity is a concern because of its adverse effects on agricultural, municipal, and industrial users.[10] The Salinity Control Act of 1974 (Public Law 93-320) created the Colorado River Basin Salinity Control Program to develop projects to reduce salt loading to the Colorado River. Salinity control projects include lining open canals and laterals (or replacing with pipe) and installing sprinklers in place of surface irrigation for the purpose of decreasing salt loading caused by canal leakage and irrigated crop deep percolation. Selenium in irrigation return flow has become a concern[3] and may also be reduced by salinity reduction projects.

In-Stream Flow Requirements

Diversions in some reaches in some western United States streams are "dried up" immediately downstream of diversion structures during times of peak irrigation demand. This condition eliminates any use of the reach for fisheries and other uses which depend on in-stream flow. In some instances, negotiated agreements with senior water rights users have allowed for bypass of minimal amounts of water to sustain the fishery or habitat, and for control of tailwater runoff to reduce agricultural related chemicals in the receiving water.

REFERENCES

1. Bear River Commission. *Amended Bear River Compact*, S.B. 255, Utah Legislature Session, 1979.
2. Bishop, A.A.; Peterson, H.B. (team leaders). Characteristics and Pollution Problems of Irrigation Return Flow. Final Report Project 14-12-408, Fed. Water Pollution Control Adm. U.S. Dept. of Interior (Ada, Oklahoma). Utah State University Foundation, Logan, Utah. May 1969.
3. Butler, D.L. *Effects of Piping Irrigation Laterals on Selenium and Salt Loads, Montrose Arroyo Basin, Western Colorado*; U.S. Geological Survey Water-Resources Investigations Report 01-4204 (in cooperation with the U.S. Bureau of Reclamation): Denver, Colorado, 2001.
4. Hill, R.W.; Israelsen, E.K.; Huber, A.L.; Riley, J.P. *A Hydrologic Model of the Bear River Basin*; PRWG72-1, Utah Water Resource Laboratory, Utah State University: Logan, Utah, 1970.
5. Hill, R.W.; Brockway, C.E.; Burman, R.D.; Allen, L.N.; Robison, C.W. *Duty of Water under the Bear River Compact: Field Verification of Empirical Methods for Estimating Depletion. Final Report*; Utah Agriculture Experiment Station Research Report No. 125, Utah State University: Logan, Utah, January 1989.
6. Hoffman, G.J.; Howell, T.; Solomon, K. *Management of Farm Irrigation Systems*; The American Society of Agricultural Engineers: St. Joseph, MI, 1990.
7. Jensen, M.E. *Design and Operation of Farm Irrigation Systems*; ASAE Monograph, American Society of Agricultural Engineers: St. Joseph, MI, 1983.
8. Jensen, M.E.; Burman, R.D.; Allen, R.G.; Eds. *Evapotranspiration and Irrigation Water Requirements*; ASCE Manual No. 70, American Society of Civil Engineers: New York, NY, 1990.
9. Keller, J.; Bliesner, R.D. *Sprinkle and Trickle Irrigation*; Van Nostrand Reinhold: New York, NY, 1990.
10. U.S. Department of the Interior. *Quality of Water— Colorado River Basin: Bureau of Reclamation, Upper Colorado Region*; Progress Report no 19; Salt Lake City, Utah, 1999.
11. Walter, I.A.; Altenhofen, J. Irrigation efficiency studies— northern colorado. Proceedings USCID Water Management

Field—Low

Seminar. Sacramento, CA Oct 5–7; USCID: Denver, CO, 1995.

12. Willardson, L.S.; Wagenet, R.J. *Basin-Wide Impacts of Irrigation Efficiency*; Proceedings of an ASCE Specialty Conference on Advances in Irrigation and Drainage, Jackson, WY, 1983.

13. Willardson, L.S.; Allen, R.G.; Frederiksen, H.D. *Elimination of Irrigation Efficiencies*; Proceedings 13th Technical Conference, USCID: Denver, CO, 1994.

Low-Impact Development

Curtis H. Hinman
School of Agricultural, Human and Natural Resource Sciences, Washington State University, Tacoma, Washington, U.S.A.

Derek B. Booth
Center for Water and Watershed Studies, University of Washington, Seattle, Washington, U.S.A.

Abstract

Low-Impact Development uses on-site natural features and small-scale hydrologic controls to manage runoff, closely mimicking predevelopment watershed hydrologic functions. LID can fully control flows associated with low-intensity storms, improve water-quality treatment by increasing storm-water contact with soils and vegetation, and significantly reduce flows from larger-volume storms on sites with relatively permeable soils. LID is a significant conceptual shift from conventional storm-water management; its broader application should encourage designers to incorporate native hydrologic processes as important organizing principles in the development of the urban and suburban landscape.

INTRODUCTION

Low-Impact Development (LID) is a strategy for storm-water management that uses on-site natural features integrated with engineered, small-scale hydrologic controls to manage runoff by maintaining or closely mimicking predevelopment watershed hydrologic functions.[1] Planning for LID is most effective at the scale of an entire subdivision or watershed; engineering and site design elements, however, are implemented at the scale of individual parcels, lots, or structures. In combination, these actions seek to store, infiltrate, evaporate, or otherwise slowly release storm-water runoff in a close approximation of the rates and processes of the predevelopment hydrologic regime.

TYPICAL CHACTERISTICS OF LOW-IMPACT DEVELOPMENT

Most applications of LID have several common components (Fig. 1):

- Preserving elements of the natural hydrologic system that are already achieving effective storm-water management, recognized by assessment of a site's watercourses and soils; channels and wetlands, particularly with areas of overbank inundation; highly infiltrative soils with undisturbed vegetative cover; and intact mature forest canopy.
- Minimizing the generation of overland flow by limiting areas of vegetation clearing and soil compaction; incorporating elements of urban design such as narrowed streets, structures with small footprints (and greater height, as needed), use of permeable pavements as a substitution for asphalt/concrete surfaces for vehicles or pedestrians;[3] and using soil amendments in disturbed areas to increase infiltration capacity.
- Storing runoff with slow or delayed release, such as in cisterns or distributed bioretention cells, across intentionally roughened landscaped areas or on vegetated roofing systems ("green roofs").[4,5] Runoff storage in LID differs from traditional storm-water management, notably the latter's use of detention ponds, primarily by its scale—small and distributed in LID, large and centralized in traditional approaches.

Native soils play a critical role in storage and conveyance of runoff, particularly in humid regions. In such regions, one to several meters of soil, generally high in organic material and relatively permeable, overlay less permeable substrates of largely unweathered geologic materials. While water is held in this soil layer, solar radiation and air movement provide energy to evaporate surface soil moisture and contribute to the overall evapotranspiration component of the water balance. Water not evaporated, transpired, or held interstitially moves slowly downslope or down gradient as shallow subsurface flow (also called interflow) over many hours, days, or weeks before discharging to streams or other surface-water bodies. In arid regions with relatively lower organic content soils and vegetation cover, precipitation events can produce rapid overland flow response naturally; however, the principles of LID remain—retain native soils, vegetation, topography, and hydrologic regime to preserve aquatic ecosystem structure and function.

The transition from a native landscape to a built environment increases the coverage of impervious surface from roads, parking areas, sidewalk, rooftops, and landscaping.

Encyclopedia of Natural Resources DOI: 10.1081/E-ENRW-120028721

Field—Low

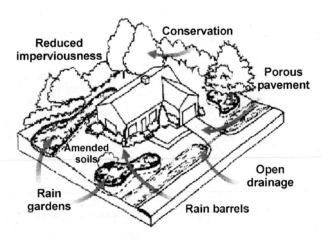

Fig. 1 Schematic of typical LID applications at the scale of a residential building lot.
Source: Modified from Puget Sound Action Team.[2]

The upper soil layers that evaporate, store, or infiltrate storm water are compacted or covered altogether. As a result, the watershed area contributing direct overland flow to streams, lakes, and wetlands increases; hence, precipitation will reach receiving waters much more rapidly and in greater volumes.[6]

Typical storm-water management focuses on flood control and thus emphasizes the efficient collection and rapid conveyance of precipitation away from residential and commercial development, commonly to central control ponds. Several factors have led to this approach: Storm water has been perceived as a liability and applications have evolved from wastewater technology, hard conveyance structures and central control ponds are considered reliable and relatively simple to maintain, and the conveyance and collection approach is relatively simple to model for regulatory requirements. Although newer conveyance and pond strategies, if properly designed and maintained, can manage predevelopment peak flows under some conditions, a number of problems (e.g., increased runoff volume and extended flow durations, conversion of dispersed flow to point discharges) will continue to challenge this traditional strategy of storm-water management.[7]

LIMITATIONS

Although the goal of LID to mimic the predevelopment hydrologic regime is laudable, it cannot feasibly be achieved everywhere or at all times. The hydrologic system evolved from, and is dependent on, the characteristics of undisturbed watersheds—mature vegetative cover, uncompacted soils, ungullied hillslopes—whose function cannot be expected to remain unchanged where half or more of the landscape has been appropriated for human uses. Thus the

objectives of any given LID must be strategically chosen, recognizing both the opportunities and the limitations of any given site. The limitations are not simply those of subdivision design and development density; they are also limitations of site topography, soil permeability and depth, and groundwater movement. They are likely to be most prominent during periods of extended rainfall, where the distributed on-site infiltration reservoirs common to most LID designs will experience their highest water levels and soil layers approach or reach full saturation. Under such conditions, the downstream impacts of uncontrolled run-off—be it flooding, channel erosion, or aquatic habitat disruption—could be as severe as with conventional storm-water control. Regulatory requirements, typical zoning and housing type, and costs of sophisticated control technology required on sites with higher development density and soils with low infiltration rates also create significant challenges for reducing or eliminating hydrologic impacts from development sites.

The potential failure of LID to control all flows from all storms, no matter how severe, is not an indictment of the approach as a whole. Indeed, LID probably represents the best opportunity to maintain a "natural," or at least minimally disrupted, flow regime in an urban watershed. Where downstream flooding or channel erosion are concerns, however, a comprehensive drainage design will need to consider more traditional storm-water management approaches (such as detention ponds or bypass pipelines) in addition to LID applications. Although such ponds would be smaller, they may still be required to achieve human or ecosystem protection at storm recurrence intervals that may be relatively long but still fall within regulatory thresholds.

SITE DESIGN AND MANAGEMENT OBJECTIVES FOR LOW-IMPACT DEVELOPMENT

The goals for LID are achieved through the following site objectives:

- Maximize retention of mature vegetation cover and restore disturbed vegetation to intercept, evaporate, and transpire precipitation.
- Preserve permeable, native soil and enhance disturbed soils to store and infiltrate storm flows.
- Retain and incorporate topographic site features that slow, store, and infiltrate storm water.
- Retain and incorporate natural drainage features and patterns.
- Locate buildings and roads away from critical areas and soils that provide effective infiltration.
- Minimize total impervious area and eliminate effective impervious surfaces ("effective" impervious surfaces

are the subset of the total imperviousness that has a direct hydraulic connection to the stream or wetland).

- Manage storm water as close to its origin as possible by utilizing small-scale, distributed hydrologic controls.
- Create a hydrologically rough landscape that slows storm flows and increases the time of concentration.
- Increase reliability of the storm-water management system by providing multiple or redundant points of control.
- Integrate storm-water controls into the development design and utilize the controls as amenities, creating a multifunctional landscape.
- Utilize an interdisciplinary approach that incorporates planners, engineers, landscape architects, and architects working together from the initial phases of the project.
- Reduce the reliance on traditional conveyance and pond technologies.
- Provide community education and promote community participation in the protection of LID systems and receiving waters.

These objectives can be grouped into five basic elements that constitute a "complete" LID design:[8]

1. Conservation measures.
2. Minimization techniques.
3. Flow attenuation.
4. Distributed integrated management practices.
5. Pollution prevention measures.

Although these five elements can be applied to any development, the manner in which they are used must be determined by the local climate and soils.

Conservation Measures

Conservation measures maintain as much of the natural landscape as possible. This includes retaining forests and other native vegetation, not filling wetlands, and providing buffers around wetlands and streams. These natural areas can then provide passive storm-water management opportunities, and they also double as open space.

Minimization Techniques

Minimization techniques reduce impacts of development on the hydrologic regime by reducing the amount of disturbance when preparing a site for development. Instead of grading an entire development site, only the lots and roads are graded while the rest of the ground is left undisturbed. Impervious surfaces are limited to areas where they are absolutely required. Cluster design is used to decrease the amount of the site developed and increase the amount of

open space. Graded soils with high infiltration-capacity soils are stockpiled and reused.

Flow Attenuation

By slowing runoff velocity, the opportunity for storm-water infiltration increases and the magnitude of peak discharges decline. Whereas traditional storm-water management directs water from a site as quickly as possible, LID holds runoff on-site as long as possible without causing flooding or other potential problems.

Distributed Integrated Management Practices

LID incorporates a range of integrated best management practices throughout a site, commonly in sequence. An example of this is connecting a bioretention area to a natural area by conveyance through a grass swale. During high flow events, excess storm water is given the opportunity to continue to infiltrate while flowing from the site.

Pollution Prevention Measures

Pollution prevention measures are accomplished through a variety of source control, rather than treatment, approaches. For example, community outreach activities, such as the publication of educational materials, can control pollution by not allowing contaminants to enter the watershed or to be released in the first place. This strategy complements other elements of LID design that also reduce the downstream delivery of pollutants by minimizing storm-water runoff volumes and by promoting filtration through soil.

CONCLUSION

Low-impact development uses on-site natural features and small-scale hydrologic controls to manage runoff, closely mimicking predevelopment watershed hydrologic functions. Challenges for more widespread use of this approach include uncertainty in its application on relatively non-infiltrative soils, its role in mitigating high-intensity and (or) large-volume storms, and its construction in new or previously developed areas where high urban densities are desired or already present (Fig. 2). Although additional research will be needed to characterize fully the performance of LID practices across different physical settings and development scenarios, available information indicates that LID can fully control flows associated with low-intensity storms, improve water-quality treatment by increasing storm-water contact with soils and vegetation, and significantly reduce flows from larger-volume storms

Field—Low

Field—Low

Fig. 2 Retrofitted residential block in north Seattle, displaying a variety of LID techniques: limited impervious area, amended soils, rain gardens, and bioswales. (Photo courtesy of Seattle Public Utilities.)
Source: Adapted from Seattle Public Utilities Natural Drainage Systems Program.[9]

on sites with relatively permeable soils. LID is a significant conceptual shift from conventional storm-water management; its broader application should encourage designers to incorporate native hydrologic processes as important organizing principles in the development of the urban and suburban landscape.

REFERENCES

1. Environmental Protection Agency. *Low Impact Development—A Literature Review*: EPA-841-B-00-005; 2000; 35 pp. (available at http://www.epa.gov/owow/ nps/lid/lid.pdf).
2. Puget Sound Action Team. *Low Impact Development (Brochure)*; Olympia: Washington, 2001; 4 pp. (available at http://www.psat.wa.gov/Programs/LID/lid_cd/brochure.pdf).
3. Brattebo, B.O.; Booth, D.B. Long-term stormwater quantity and quality performance of permeable pavement systems. Water Res. **2003**, *37*, 4369–4376.
4. Konrad, C.P.; Burges, S.J. Hydrologic mitigation using on-site residential storm-water detention. J. Water Resour. Plan. Manage. **2001**, *127*, 99–107.
5. *Penn State Center for Green Roof Research*; College of Agriculture, Pennsylvania State University: State College, PA, 2002. (available at http://hortweb.cas.psu. edu/research/greenroofcenter/).
6. Booth, D.B. Urbanization and the natural drainage system—impacts, solutions, and prognoses. Northwest Environ. J. 1991, *7*, 93–118.
7. Booth, D.B.; Jackson, C.R. Urbanization of aquatic systems—degradation thresholds, stormwater detention, and the limits of mitigation. Water Resour. Bull. 1997, *33*, 1077–1090.
8. Coffman, L.S. Using low-impact development in stormwater management. Water Environ. Res. Foundation, Progress Newslett. Winter 2001, *12*(1). (available at http://www.werf.org/press/winter01/01w_low.cfm).
9. *Seattle Public Utilities Natural Drainage Systems Program*. (available at http://www.ci.seattle.wa.us/util/ NaturalSystems/default.htm).

Marine Benthic Productivity

Bayden D. Russell
Sean D. Connell
Southern Seas Ecology Laboratories, School of Earth and Environmental Science, University of Adelaide, Adelaide, South Australia, Australia

Abstract

Primary productivity is the production of energy and biomass through the process of photosynthesis. In marine ecosystems, the major primary producers are algae, from phytoplankton in the open oceans to algal forests in shallow coastal waters. Primary production underpins all marine ecosystems because the growth of primary producers supplies the energy that supports entire food webs. The amount of this productivity that enters these ecosystems is largely determined by environmental conditions such as light and temperature, and resource availability such as carbon dioxide and nutrients. The relative dominance of different species often varies as a function of abiotic conditions, with some species being more common through greater net productivity. When abiotic conditions change from their long-term average conditions, primary productivity and growth among the alternate species of primary producers can also change, potentially leading to a shift in the dominant species. Such "phase-shifts" often lead to substantial changes in the productivity and species diversity of a system.

INTRODUCTION

Primary productivity is a term that refers to the production of energy through the process of photosynthesis. In marine ecosystems, the major primary producers are algae and plants, from phytoplankton in the open oceans to algal forests and seagrass meadows in shallow coastal waters. As a biological process, primary production underpins all marine ecosystems, because the growth of primary producers supplies the energy that supports entire food webs. Yet, the amount of biomass that is produced, and therefore the quantity that enters the ecosystem, is largely determined by abiotic conditions such as light, resource availability [mostly carbon dioxide (CO_2) and nutrients], and temperature. Therefore, a proper understanding of the role of primary productivity in the functioning of marine systems requires an understanding of how changing abiotic conditions, whether natural or anthropogenic in origin, will alter primary productivity.

This entry focuses on the primary productivity of benthic marine systems, in particular macroalgae that inhabit the rocky coastlines of the world. There are other excellent sources that discuss the processes and mechanisms of photosynthesis,[1] and primary productivity in phytoplankton[2] and seagrasses.[3]

BENTHIC PRIMARY PRODUCTIVITY

Macroalgae

Macroalgae, also known as seaweeds, are the predominant primary producers in benthic marine ecosystems. As light is one of the key requirements for photosynthesis, macroalgae are present from the intertidal to any depth that light can penetrate. All three of the major groups of algae, green (Chlorophyta), red (Rhodophyta), and brown (Phaeophyta), are represented in benthic habitats but as a group the brown algae are the most visually dominant, forming "forests" on many temperate rocky coastlines of the world. Globally, macroalgae produce more than 1 Pg of carbon per year that, relative to habitat available for them to inhabit, is equivalent to phytoplankton, which are generally considered to be more productive. This productivity is transmitted throughout coastal food webs and is responsible for many of the ecosystem services that human populations have come to rely on. However, primary productivity in these systems is strongly controlled by abiotic conditions such as light, temperature, and nutrient availability (both N and C), all of which are changing because of human activities.

Conditions Controlling Benthic Productivity

As with all photosynthetic organisms, light availability is the largest determinant of primary productivity; without light there can be no photosynthesis. This is not a simple linear relationship, however, as many algae have adapted to increase photosynthetic efficiency under low light conditions (e.g., red algae in deep waters). Much of this relationship is determined by the type and quantity of photosynthetic pigments, which varies greatly among species and within species under different environmental conditions.[4]

One of the factors that limits photosynthesis in terrestrial plants is the concentration or the availability of CO_2.

Encyclopedia of Natural Resources DOI: 10.1081/E-ENRW-120047577

In general, the productivity of terrestrial plants increases with increasing CO_2, especially in plants that utilize C3 photosynthesis.[5] This relationship does not necessarily hold for algae, however, because of the number of different mechanisms by which different algae acquire and assimilate inorganic carbon (C_i) as a substrate for photosynthesis. Most notably, CO_2 that is dissolved from the atmosphere into marine waters undergoes a series of chemical reactions, which means that it is not the most abundant source of C_i in marine waters. As such, many marine algae do not use CO_2 directly, but rather have carbon concentrating mechanisms (CCMs) that allow them to use bicarbonate as the dominant source of C_i for photosynthesis.[4,6] As bicarbonate is abundant in seawater, it is generally thought that photosynthesis of algae that use CCMs is carbon saturated at current CO_2 concentrations.[6,7] However, recent experiments have so far been inconclusive, with some species showing carbon saturation at current CO_2,[8,9] others demonstrating increased photosynthetic production with increasing CO_2,[10] and yet others switching between sources of carbon with greater CO_2 availability.[11,12] Regardless, the availability of resources is one of the largest factors controlling productivity in macroalgae (besides light availability) and as human activities change the availability of these resources (both C and N) the productivity of algae is likely to change.

ABIOTIC CONDITIONS AND PRIMARY PRODUCTIVITY

Resource Availability

The availability of resources has a fundamental role in regulating the productivity of individuals, species, and, ultimately, community structure and function.[13] In the context of primary productivity, the main resources of concern are light, the driver of photosynthesis, and nutrients such as carbon and nitrogen. The availability of these resources varies both spatially and temporally in most ecosystems, meaning that it is rare for primary producers to exist under their ideal conditions; that is, they are likely to be resource-limited.[14] Such resource limitation can be recognized as a change in the rate of processes in response to one resource, such as an increase in photosynthesis with the provision of CO_2. Yet, resource limitation is also partly determined by the ability of organisms to access available resources, meaning that the level of resource limitation will vary among organisms with different physiologies. For example, slower growing species of algae, such as kelp, tend to be able to store nutrients and so are less limited in periods of low nutrient availability. In contrast, fast-growing, ephemeral species of algae, such as filamentous turf-forming algae, cannot store nutrients but are well adapted to quickly acquire them from the environment. Biological communities are, therefore, generally comprised of functional groups experience diverse limitations, a condition that can determine the relative abundance of different primary producers in the community.[15]

The species-specific responses of marine algae to the enrichment of particular resources will manifest not only as changes in primary productivity, but also in the stoichiometry, or the ratio of different nutrients in their tissues.[16] The ratio of carbon to nitrogen in the tissues of primary producers, or C:N ratios, are often used to provide an index of the relative amounts of C and N available to algae.[17,18] High C:N ratios, or relatively more carbon than nitrogen, often indicates strong nitrogen limitation, while lower ratios often seen under nutrient enriched, or eutrophic, conditions generally indicate lower nitrogen limitation because there is a greater availability of N in the tissues. Therefore, the analysis of the tissue of different algae can be used as an indication of differential resource limitations experienced by different groups of algae under a variety of environmental conditions.

In addition to C:N ratio responses, the absolute nutrient content of tissues, primarily the % C and % N, provides insight into the availability of resources in the surrounding environment and the physiological processes by which resources are acquired by the primary producers. For example, the change in C:N ratio of a particular species of algae may indicate that it experiences nutrient limitation under oligotrophic (low nutrient) conditions but not under eutrophic (high nutrient) conditions. A response where the % N and not % C increases under enriched nutrient conditions would indicate that nutrient enrichment enables this alga to access and store more nitrogen but not carbon. Then, if under-elevated CO_2 conditions neither % C or % N change, this would be indicative of a species that is not carbon limited under natural CO_2 concentrations. By understanding these measures of resource limitation, and the mechanisms that underlay them, it is then possible to make informed predictions about changes that may occur when environmental conditions change and resource limitations are removed.

Primary Productivity under Changing Abiotic Conditions

It is becoming increasingly important to develop an understanding of the resource limitations experienced by primary producers as human activities are altering the availability of resources. Of particular concern is the potential for synergistic responses to increased resource availability, where the effects of increased carbon availability, through CO_2 emissions, may be amplified in places where human activities also increase nutrient loads (such as on urban and agricultural coastlines). As discussed above, the responses to increases in the availability of these resources will reflect the extent to which primary producers are carbon-limited as a consequence of the physiological mechanisms by which carbon is acquired for use in photosynthesis.[19–21] There are two key strategies of carbon uptake in marine algae, passive diffusion of CO_2 and active uptake via a carbon concentrating mechanism (CCM). While the majority of marine algae have CCMs that facilitate the active influx of CO_2 and/or HCO_3^- and elevate concentrations at the site of carbon fixation (i.e., Rubisco), some use dissolved CO_2 entering by diffusion and others are able to switch between these mechanisms.[22] Algae with CCMs are predicted to gain little benefit from elevated CO_2 in the world's oceans,[20] and as such will probably continue to grow at rates that are comparable to those seen currently. In contrast, algae that rely on diffusion of CO_2 are possibly carbon limited under current conditions and are likely to show increased primary productivity under elevated CO_2 conditions.[19] Species that can switch between these mechanisms are an interesting case because they are likely to increase productivity and growth under elevated CO_2 conditions even though they are not carbon limited at current CO_2 conditions; their increase in growth will be a result of the reduced energetic costs of using direct diffusion of CO_2 compared to CCMs, so their productivity can increase without an increase in photosynthesis per se. Unfortunately, recent work suggests that the taxa of algae that are more likely to benefit from increasing CO_2 availability, either because they do not possess CCMs or because they can switch to passive uptake of CO_2, are those taxa that also show rapid increases in growth in response to elevated nutrients,[23] possibly paving the way for a shift in relative abundance of algae along many coasts of the world.

LOSS OF BENTHIC HABITATS

In marine systems, the coastal zone is an area in which high productivity and species diversity coincide with human activities that alter abiotic conditions. Historically, this zone has been influenced by large-scale eutrophication from the run-off of nutrients from the land.[15,24,25] Increasingly, the oceans, and this area in particular, are set to be further influenced by the effects of a changing climatic condition,[26] with these waters absorbing approximately 30% of CO_2 emissions produced by humans globally. Therefore, primary productivity in marine systems is likely to increase into the future because of the increased availability of two of the essential resources, nitrogen and carbon.

This increased productivity is unlikely to be realized by all species, however. As discussed above, the faster-growing "weedy" species, such as filamentous algal turfs, are likely to benefit from these altered conditions to a greater extent than the slower-growing habitat formers such as kelps. Indeed, evidence to date suggests moderate increases of both nutrients[27] and CO_2 facilitate greater covers and biomass of turfs, potentially turning them from ephemeral to persistent habitats.[28–30] "Turfs" are a recurring component in the reporting of phase-shifts on coral reefs, seagrassmeadows, and in kelp forests and are likely to attract attention in climate studies. Turfs seem to become the dominant habitat by taking advantage of a pulse event (e.g., storms) into an ecological phase-shift under altered environmental conditions (e.g., greater nutrient availability). This "shifted" state will then often persist (e.g., see Fig. 1) because there is a positive feedback where turfs inhibit the recruitment of habitat formers (e.g., kelps), maintaining their dominance. However, turfs, like many fast-growing and ephemeral algae, lack the ability to store nutrients and have a growth strategy that allows them to exploit increases in resource availability, but collapse when resources are depleted.[17] Therefore, altered environmental conditions allow them to persist but the restoration of "natural" conditions by removing excess resources in the environment, such as by redirecting treated wastewater inland instead of into the sea, may aid in the recovery of these ecosystems.[31]

Fig. 1 Photos of a healthy kelp-dominated marine ecosystem (**A**) and a phase-shifted system dominated by algal turfs (**B**). In this case, the driver of the phase-shift was increased availability of carbon from subtidal CO_2 seeps that have similar CO_2 concentrations to those that are predicted globally by the year 2100.
Source: Photos by Bayden Russell.

CONCLUSION

In conclusion, primary productivity in marine ecosystems is largely controlled by abiotic conditions such as resource availability. Human activities increase the availability of resources, changing the conditions for productivity and growth. The resulting increase in productivity does not affect different taxa uniformly, however, with fast-growing "weedy" species often benefiting more than the slower-growing species. The ensuing change in relative dominance, from slower-growing, habitat-forming species to simpler and fast-growing species, causes a loss of habitat and, subsequently, species diversity. In some cases, this ecosystem shift may be able to be reversed, such as by limiting the input of nutrient-rich wastewater into the sea. It remains to be seen, however, whether this will be possible as global CO_2 emissions continue to increase, providing an almost unlimited source of carbon for enhanced primary productivity in the species of algae that maintain phase-shifts.

REFERENCES

1. Taiz, L.; Zeiger, E. *Plant Physiology*; Sinauer Associates: Sunderland, Mass, 2002.
2. Reynolds, C.S. Photosynthesis and carbon acquisition in phytoplankton; In *The Ecology of Phytoplankton*; Reynolds, C.S., Ed.; Cambridge University Press: Cambridge, 2006, 93–144.
3. Zimmerman, R.C. Light and photosynthesis in seagrass meadows; In *Seagrasses: Biology, Ecology and Conservation*; Larkum, A.W.D., Orth, R.J., Duarte, C., Eds.; Springer: Netherlands, 2006, 303–321.
4. Raven, J.A.; Hurd, C.L. Ecophysiology of photosynthesis in macroalgae. Photosynth. Res. **2012**, *113* (1–3), 105–125.
5. Ainsworth, E.A.; Long, S.P. What have we learned from 15 years of free-air CO_2 enrichment (FACE)? A meta-analytic review of the responses of photosynthesis, canopy properties and plant production to rising CO_2. New Phytol. **2005**, *165* (2), 351–371.
6. Beardall, J.; Beer, S.; Raven, J. A. Biodiversity of marine plants in an era of climate change: Some predictions based on physiological performance. Bot. Mar. **1998**, *41* (1–6), 113–123.
7. Gao, K.; McKinley, K.R. Use of macroalgae for marine biomass production and CO_2 remediation: A review. J. Appl. Phycol. **1994**, *6* (1), 45–60.
8. Beer, S.; Koch, E. Photosynthesis of marine macroalgae and seagrasses in globally changing CO_2 environments. Mar. Ecol.-Prog. Ser. **1996**, *141* (1996), 199–204.
9. Israel, A.; Hophy, M. Growth, photosynthetic properties and Rubisco activities and amounts of marine macroalgae grown under current and elevated seawater CO_2 concentrations. Glob. Change Biol. **2002**, *8* (9), 831–840.
10. Holbrook, G.P.; Beer, S.; Spencer, W.E.; Reiskind, J.B.; Davis, J.S.; Bowes, G. Photosynthesis in marine macroalgae: Evidence for carbon limitation. Can. J. Bot. **1988**, *66* (3), 577–582.
11. Johnston, A.M.; Raven, J.A. Effects of culture in high CO_2 on the photosynthetic physiology of *Fucus serratus*. Br. Phycol. J. **1990**, *25* (1), 75–82.
12. Schmid, R.; Forster, R.; Dring, M.J. Circadian rhythm and fast responses to blue-light of photosynthesis in *Ectocarpus* (Phaeophyta, Ectocarpales) II. Light and CO_2 dependence of photosynthesis. Planta. **1992**, *187* (1), 60–66.
13. Harpole, W.S.; Ngai, J.T.; Cleland, E.E.; Seabloom, E.W.; Borer, E.T.; Bracken, M.E.S.; Elser, J.J.; Gruner, D.S.; Hillebrand, H.; Shurin, J.B.; Smith, J.E. Nutrient co-limitation of primary producer communities. Ecol. Lett. **2011**, *14* (9), 852–862.
14. Andersen, T.; Pedersen, O. Interactions between light and CO_2 enhance the growth of *Riccia fluitans*. Hydrobiologia **2002**, *477* (1–3), 163–170.
15. Connell, S.D.; Russell, B.D.; Turner, D.J.; Shepherd, S.A.; Kildea, T.; Miller, D.; Airoldi, L.; Cheshire, A. Recovering a lost baseline: Missing kelp forests from a metropolitan coast. Mar. Ecol. Prog. Ser. **2008**, *360*, 63–72.
16. Elser, J.J.; Bracken, M.E.S.; Cleland, E.E.; Gruner, D.S.; Harpole, W.S.; Hillebrand, H.; Ngai, J.T.; Seabloom, E.W.; Shurin, J.B.; Smith, J.E. Global analysis of nitrogen and phosphorus limitation of primary producers in freshwater, marine and terrestrial ecosystems. Ecol. Lett. **2007**, *10* (12), 1135–1142.
17. Pedersen, M.F.; Borum, J. Nutrient control of algal growth in estuarine waters. Nutrient limitation and the importance of nitrogen requirements and nitrogen storage among phytoplankton and species of macroalgae. Mar. Eco. Prog. Ser. **1996**, *142* (1), 261–272.
18. Pedersen, M.F.; Borum, J. Nutrient control of estuarine macroalgae: Growth strategy and the balance between nitrogen requirements and uptake. Mar. Ecol.-Prog. Ser. **1997**, *161*, 155–163.
19. Kubler, J.E.; Johnston, A.M.; Raven, J.A. The effects of reduced and elevated CO_2 and O_2 on the seaweed *Lomentaria articulata*. Plant Cell Environ. **1999**, *22* (10), 1303–1310.
20. Hurd, C. L.; Hepburn, C.D.; Currie, K.I.; Raven, J.A.; Hunter, K.A. Testing the effects of ocean acidification on algal metabolism: Considerations for experimental designs. J. Phycol. **2009**, *45* (6), 1236–1251.
21. Hepburn, C.D.; Pritchard, D.W.; Cornwall, C.E.; McLeod, R.J.; Beardall, J.; Raven, J.A.; Hurd, C.L. Diversity of carbon use strategies in a kelp forest community: Implications for a high CO_2 ocean. Glob. Change Biol. **2011**, *17* (7), 2488–2497.
22. Raven, J.A.; Johnston, A.M.; Kubler, J.E.; Korb, R.; McInroy, S.G.; Handley, L.L.; Scrimgeour, C.M.; Walker, D.I.; Beardall, J.; Vanderklift, M.; Fredriksen, S.; Dunton, K.H. Mechanistic interpretation of carbon isotope discrimination by marine macroalgae and seagrasses. Funct. Plant Biol. **2002**, *29* (3), 355–378.
23. Falkenberg, L.J.; Russell, B.D.; Connell, S.D. Contrasting resource limitations between competing marine primary producers: Implications for associated communities under enriched CO_2 and nutrient regimes. Oecologia **2013b**, *172* (2), 575–583.
24. Eriksson, B.K.; Johansson, G.; Snoeijs, P. Long-term changes in the macroalgal vegetation of the inner Gullmar Fjord, Swedish Skagerrak coast. J. Phycol. **2002**, *38* (2), 284–296.

25. Gorman, D.; Connell, S.D. Recovering subtidal forests in human-dominated landscapes. J. Appl. Ecol. **2009**, *46* (6), 1258–1265.

26. Harley, C.D.G. Direct and indirect effects of ecosystem engineering and herbivory on intertidal community structure. Mar. Ecol. Prog. Ser. **2006**, *317*, 29–39.

27. Russell, B.D.; Connell, S.D. A novel interaction between nutrients and grazers alters relative dominance of marine habitats. Mar. Ecol. Prog. Ser. **2005**, *289*, 5–11.

28. Russell, B.D.; Thompson, J.I.; Falkenberg, L.J.; Connell, S.D. Synergistic effects of climate change and local stressors: CO_2 and nutrient driven change in subtidal rocky habitats. Glob. Change Biol. **2009**, *15* (9), 2153–2162.

29. Connell, S.D.; Russell, B.D. The direct effects of increasing CO_2 and temperature on non-calcifying organisms: Increasing the potential for phase shifts in kelp forests. Proc. R. Soc. B **2010**, *277* (1686), 1409–1415.

30. Russell, B.D.; Passarelli, C.A.; Connell, S.D. Forecasted CO_2 modifies the influence of light in shaping subtidal habitat. J. Phycol. **2011**, *47* (4), 744–752.

31. Falkenberg, L.J.; Connell, S. D.; Russell, B.D. Disrupting the effects of synergies among stressors: Improved water quality dampens the effects of future CO_2 on a marine habitat. J. Appl. Ecol. **2013a**, *50* (1), 51–58.

BIBLIOGRAPHY

1. Larkum, A.W.D.; Orth, R.J.; Duarte, C., Eds. *Seagrasses: Biology, Ecology and Conservation*; Springer: Netherlands, 2006.

2. Reynolds, C.S. Ed. *The Ecology of Phytoplankton*; Cambridge University Press: Cambridge, 2006.

Marine Mammals

Kathleen J. Vigness-Raposa
Marine Acoustics, Inc., Middletown, Rhode Island, U.S.A.

Abstract

Marine mammals are species that utilize the ocean environment for all or part of their life cycle. This includes whales, dolphins, porpoises, seals, sea lions, sea otters, walruses, polar bears, manatees, and dugongs. Land mammals began to live in the ocean over 50 million years ago. Remarkably, new species are still being discovered, both through direct observation and genetic analyses. Advances in research techniques continue to expand our understanding of the unique life histories of marine mammals.

INTRODUCTION

Marine mammals have captured the attention and imagination of humans for eons. They are fascinating creatures that have encouraged scientists to explore many aspects of their existence, from the evolutionary developments that led mammals into the marine environment to their complex social structures and remarkable acoustic abilities that allow them to survive underwater.

A marine mammal is defined as any mammal that is tied to the oceanic environment for all or part of its life cycle. This includes whales, dolphins, porpoises, seals, sea lions, sea otters, walruses, polar bears, manatees, and dugongs. These species occur in the mammalian order Cetacea (whales, dolphins, porpoises) and the order Sirenia (dugongs and manatees) and include many members of the order Carnivora (polar bear, sea otter, pinnipeds [seals and sea lions], and walruses). The total number of species is continually changing, as species relations are elucidated with modern genetic techniques, and hard-to-detect, elusive species are identified with new research methodologies or serendipitous observations.

This entry provides a brief overview of marine mammals, starting with the evolutionary history of marine mammals and recent phylogenetic discoveries, and concluding with recent advances in understanding these species with new research techniques. The more curious reader is referred to the References and Bibliography for more detailed works that expand on these topics, which are merely jumping-off points into the world of marine mammals.

EVOLUTIONARY BIOLOGY

Land mammals began to live in the ocean over 50 million years ago, and the results of millions of years of adaptation can be seen in four marine mammal groups: cetaceans, sirenians, pinnipeds, and sea otters. Each group is believed to descend from a different land mammal ancestor and experienced unique evolutionary development of marine adaptations. In addition, in the case of cetaceans and pinnipeds, there is continued controversy about the relationship among existing species and the identification of new species.

Cetaceans: Whales and Dolphins

The earliest marine mammals were predecessors of cetaceans, the artiodactyls (even-toed ungulates), most closely related to the hippopotamus,[1–3] which entered the marine environment approximately 53 million years ago. As the longest living marine mammals, cetaceans show the most pronounced adaptation to an exclusively oceanic existence. In order to breathe air most efficiently, their skulls have elongated, called "telescoping," resulting in the nostrils migrating to the top of the head. The long mobile neck, functional hind limbs, and most of the pelvic girdle of their terrestrial ancestors were lost as the body became more streamlined and torpedo-shaped, and horizontal tail fins, called "flukes," developed.

Another significant adaptation is the profound reliance on sound to acoustically sense their surroundings, communicate, locate food, and protect themselves underwater. Sound travels far greater distances than light underwater and it allows marine animals to gather information and communicate at great distances and from all directions. The reader is referred to the Discovery of Sound in the Sea website (http://www.dosits.org) for an extensive account of the ways in which marine mammals utilize sound.[4]

One unique adaptation of the suborder Mysticeti, comprised of the mysticetes or baleen whales, is the shift from heterodont dentition to rows of baleen plates.[5] Cetacean ancestors had teeth that were differentiated into incisors, canines, and grinding teeth (this is called heterodont dentition), but mysticetes evolved baleen plates, which are keratin structures made of the same tough protein as

Encyclopedia of Natural Resources DOI: 10.1081/E-ENRW-120047589

fingernails, hooves, and horns. The baleen plates are used to sieve great gulps of seawater in order to filter and consume plankton that could not be eaten by toothed whales.

Sirenians: Manatees and Dugongs

Sirenians are believed to have the same 50-million-year-old ancestor as elephants and hydraxes, forming an evolutionary lineage called Paenungulata.[6] Sirenians are the only extant herbivorous marine mammals and have experienced range declines and species extinctions due to oceanographic changes and human disturbance. All four extant species are considered at a high risk of extinction in the wild by the International Union for Conservation of Nature (http://www.iucnredlist.org). Similar to evolutionary adaptations of cetaceans, sirenians have lost their hind limbs and their tail has modified into a paddle used for propulsion; however, they have retained the use of their foreflippers for maneuvering underwater. They also utilize underwater sound for communication, and scientists are trying to use the vocalizations of manatees to warn boaters that animals are in the area. Fatal collisions with boats have contributed to the dwindling numbers of this endangered species.

Pinnipeds: Walruses, True Seals, Fur Seals, and Sea Lions

The earliest pinnipeds, from "pinna" meaning fin and "ped" meaning foot, were aquatic carnivores from 25 to 27 million years ago,[7] but there is continued debate about the exact origins. The traditional, biphyletic view of their evolution proposes that odobenids (walruses) and otariids (fur seals and sea lions) descended from an ursid (bear) ancestor and phocids (earless or true seals) descended from mustelids (weasels, skunks, otters).[8] However, recent morphological evidence and genetic data support the hypothesis that pinnipeds are monophyletic, having a single origin with ursids being the closest relatives.[7]

Pinnipeds are unique in that they are equally able to live in the ocean and on land. They have developed thick blubber and dense fur for marine survival, and the ability to see, hear, and move well both underwater and in the air, though in varying degrees based on the time they spend in each environment. Phocids, accounting for about 90% of pinnipeds, spend little time on land, with brief lactation periods lasting 4 days to several weeks. They have no external ears and their hindflippers cannot be turned forward under the body for mobility on land, resulting in an undulating movement to propel themselves across land. In water, they swim by moving their hindflippers and lower body in a side-to-side motion.[8] Otariids have visible external ears, pronounced sexual dimorphism, and hindflippers that can be turned under the body for moving on land. In water, they use their long foreflippers to swim. They have relatively

long lactation periods, from several months to 2 years, and highly polygynous breeding systems in which males defend territories. Odobenids are very similar to otariids, displaying pronounced sexual dimorphism, mobile hindflippers, and long lactation periods lasting 2 or more years; however, they have long tusks and no external ears. They "walk" like otariids when on land, but use their foreflippers to swim like phocids underwater.

Sea Otters

At least 6 of the 13 extant sea otter species are tied to the marine environment for all or part of their existence.[9] The sea otter (*Enhydra lutris*), the only fully marine otter, diverged recently from other mustelids, whereas the marine otter or chungungo (*Lontra felina*) separated by the late Miocene, about 5 to 11 million years ago. All marine species of otters are at high latitudes, with varying degrees of activity in the ocean. The sea otter conducts all activities at sea, including resting, breeding, birthing, and feeding. The chungungo only briefly enters the sea to forage, then returns to land to eat and rest. The remaining marine species exhibit similarly limited excursions into the oceanic environment.

PHYLOGENETIC DISCOVERIES

The ocean is a vastly unknown place and it is remarkable that animals as large as marine mammals have gone undiscovered, but new species continue to be discovered. In several cases, the new species exhibit cryptic behavior in unpopulated portions of the world where direct, at-sea observations are limited. For example, Shepherd's beaked whale (*Tasmacetus shepherdi*), assumed to have a circumpolar distribution in deep, cold temperate waters of the Southern Ocean, was only first reliably identified at sea in 2006.[10] It had been known earlier from stranded individuals, but a definitive description of its distinctive color pattern and external morphology were not completed until recently. Similarly, the external appearance of the Longman's beaked whale (*Indopacetus pacificus*), a deep, warm water species, was only described in 1999.[11,12] The Australian snubfin dolphin (*Orcaella heinsohni*), the first new dolphin species to be described in 56 years, is found off the northern coasts of Australia.[13] It was quickly followed by the discovery of the Burrunan dolphin (*Tursiops australis*), consisting of about 150 individuals in two locations in Victoria, Australia.[14]

The most common method for identifying new species is through DNA sequence data, a method called DNA taxonomy.[15,16] By comparing DNA samples to annotated and curated sets of reference nucleotide sequences from a comprehensive and representative range of cetaceans, insights into genetic diversity can be made. For example, the discovery of greater genetic diversity between minke whales

Marine—
Pollution

from the North Atlantic and the Antarctic than among all species within the genus *Balaenoptera* led to full species status for the Antarctic minke whale (*Balaenoptera bonaerensis*).[17] The genus *Tursiops* (bottlenose dolphin) has been particularly troublesome, with much variation in color patterns, body dimensions, and cranial structure that as many as 20 different species have been recognized over the years.[18] However, most recently, only the common bottlenose dolphin (*Tursiops truncatus*) was recognized until genetic analyses supported the separate classification of the Indo-Pacific bottlenose dolphin (*T. aduncus*), found mainly in the Indian Ocean and around Australia, China, and South Africa. Similarly, the genus *Delphinus* (common dolphin) had many defined species in the past, although they were subsequently considered variations of the single species *Delphinus delphis*.[19] At present, however, two different species have been defined with genetic analyses: a long-beaked form (*D. capensis*) and a short-beaked form (*D. delphis*). As genetic samples continue to be collected and archived, further elucidation of species differentiation will be made.

ADVANCES IN RESEARCH TECHNIQUES

Because of the difficulties in studying marine mammals, new research techniques are continually being developed and refined to extend our understanding of marine mammals. Since marine mammals spend a majority of their time underwater and out of visible view, technologies are being developed to capitalize on the underwater environment. One of the most significant advances has been the design of tags that attach to individuals. Traditional satellite tags include GPS sensors that transmit the location of an animal on a predefined schedule (e.g., one time per day). However, to document a location, the tag must be at the ocean surface to communicate with satellites in space, limiting their reliability with marine mammals. The next step in complexity is time-depth recorder tags that archive the animal's depth in the water column at fine-scale time steps. These tags must be retrieved to obtain the recorded data; however, they provided the first insight into animal behavior below the sea surface. Most recently, tags have been developed that include accelerometers that can measure heading, pitch, and roll, as well as animal depth and location.[20,21] In addition, they include hydrophones that can measure the sounds produced by the animals as well as those from the environment to which the animal is exposed. This type of tag has greatly extended the understanding of marine mammal behavior, providing insight into the rate at which marine mammals vocalize, their behaviors as they vocalize, and the multiple stressors they may encounter during a given time period.

Some of the advances in archival tags have allowed for advances in passive acoustics, in which underwater listening and recording devices are used to monitor the distribution and abundance of marine mammals.[22] Marine mammals have traditionally been surveyed with visual methods; however, visual surveys cannot be conducted at night or during poor weather conditions, and animals are not available for visual detection when they are underwater, which can occur up to 80% of the time. In addition, passive acoustic sensors can be towed behind a ship or mounted on an autonomous glider or other mobile platform to cover large areas, or placed at fixed locations to record in a specific area for potentially long periods of time. In fact, the use of gliders and other autonomous vehicles is increasing rapidly as cost-effective platforms for long-term oceanography studies. With information from archival tags on vocalization rates, data from fixed passive acoustic sensors can be used to estimate marine mammal densities.[23]

CONCLUSION

Since marine mammals utilize the ocean environment for all or part of their life cycle, they are an important component of ecosystem-based management systems as top-level trophic predators. Further understanding of species relations will inform managers as they attempt to understand the roles that marine mammals play within specific ecosystems. Recent advances in research methodologies have also helped scientists observe greater portions of the marine mammal life cycle and more accurately estimate the role they play in the greater ecosystem and their role as indicator species for the overall health of an ocean basin.

REFERENCES

1. Shimamura, M.; Yasue, H.; Ohshima, K.; Abe, H.; Kato, H.; Kishiro, T.; Goto, M.; Munechika, I.; Okada, N. Molecular evidence from retroposons that whales form a clade within even-toed ungulates. Nature **1997**, *388* (6643), 666–670.
2. Thewissen, J.G.M.; Williams, E.M.; Roe, L.J.; Hussain, S.T. Skeletons of terrestrial cetaceans and the relationship of whales to artiodactyls. Nature **2001**, *413* (6853), 277–281.
3. Price, S.A.; Bininda-Emonds, O.R.; Gittleman, J.L. A complete phylogeny of the whales, dolphins and even-toed hoofed mammals (Cetartiodactyla). Biol. Rev. Camb. Philos. Soc. **2005**, *80* (3), 445–473.
4. Scowcroft, G.; Vigness-Raposa, K.J.; Knowlton, C.; Morin, H. Discovery of Sound in the Sea, http://www.dosits.org (accessed July 2013).
5. Evans, P.G.H. *The Natural History of Whales and Dolphins*; Facts on File, Inc.: New York, 1987.
6. Domning, D.P. Sirenian evolution. In *Encyclopedia of Marine Mammals*, 2nd Ed.; Perrin, W.F., Wursig, B., Thewissen, J.G., Eds.; Academic Press: San Diego, 2009; 1016–1019.
7. Berta, A. Pinniped evolution. In *Encyclopedia of Marine Mammals*, 2nd Ed.; Perrin, W.F., Wursig, B., Thewissen, J.G., Eds.; Academic Press: San Diego, 2009; 861–868.
8. Riedman, M. *The Pinnipeds: Seals, Sea Lions, and Walruses*; University of California Press: Los Angeles, 1990.

9. Estes, J.A.; Bodkin, J.L.; Ben-David, M. Marine otters. In *Encyclopedia of Marine Mammals*, 2nd Ed.; Perrin, W.F., Wursig, B., Thewissen, J.G., Eds.; Academic Press: San Diego, 2009; 807–816.

10. Pitman, R.L.; van Helden, A.L.; Best, P.B.; Pym, A. Shepherd's Beaked Whale (*Tasmacetus shepherdi*): Information on appearance and biology based on strandings and at-sea observations. Mar. Mamm. Sci. **2006**, *22* (3), 744–755.

11. Pitman, R.L.; Palacios, D.M.; Brennan, P.L.R.; Brennan, B.J.; Balcomb, K.C.; Miyashita, T. Sightings and possible identity of a bottlenose whale in the tropical Indo-Pacific: *Indopacetus pacificus*? Mar. Mamm. Sci. **1999**, *15* (2), 531–549.

12. Dalebout, M.L.; Ross, G.J.B.; Baker, C.S.; Anderson, R.C.; Best, P.B.; Cockcroft, V.G.; Hinsz, H.L.; Peddemors, V.; Pitman, R.L. Appearance, distribution, and genetic distinctiveness of Longman's beaked whale, *Indopacetus pacificus*. Mar. Mamm. Sci. **2003**, *19* (3), 421–461.

13. Beasley, I.; Robertson, K.M.; Arnold, P. Description of a new dolphin, the Australian snubfin dolphin *Orcaella heinsohni* sp. N. (Cetacea, Delphinidae). Mar. Mamm. Sci. **2005**, *21* (3), 365–400.

14. Charlton-Robb, K.; Gershwin, L.-a.; Thompson, R.; Austin, J.; Owen, K.; McKechnie, S. A new dolphin species, the Burrunan dolphin *Tursiops australis* sp. nov., endemic to southern Australian coastal waters. PLoS ONE **2011**, *6* (9), e24047, doi:10.1371/journal.pone.0024047.

15. Ross, H.A.; Lento, G.M.; Dalebout, M.L.; Goode, M.; Ewing, G.; McLaren, P.; Rodrigo, A.G.; Lavery, S.; Baker, C.S. DNA surveillance: Web-based molecular identification of whales, dolphins and porpoises. J. Hered. **2003**, *94* (2), 111–114.

16. Dalebout, M.L.; Baker, C.S.; Steel, D.; Robertson, K.M.; Chivers, S.J.; Perrin, W.F.; Mead, J.G.; Grace, R.V.; Schofield, T.D. A divergent mtDNA lineage among *Mesoplodon* beaked whales: Molecular evidence for a new species in the tropical Pacific? Mar. Mamm. Sci. **2007**, *23* (4), 954–966.

17. Pastene, L.A.; Goto, M.; Kanda, N.; Zerbini, A.N.; Kerem, D.; Watanabe, K.; Bessho, Y.; Hasegawa, M.; Nielsen, R.; Larsen, F.; Palsboell, P.J. Radiation and speciation of pelagic organisms during periods of global warming: The case of the common minke whale, *Balaenoptera acutorostrata*. Mol. Ecol. **2007**, *16* (7), 1481–1495.

18. Natoli, A.; Peddemors, V.M.; Rus-Hoelzel, A.; Natoli, A. Population structure and speciation in the genus *Tursiops* based on microsatellite and mitochondrial DNA analyses. J. Evol. Biol. **2004**, *17* (2), 363–375.

19. Natoli, A.; Canadas, A.; Peddemors, V.M.; Aguilar, A.; Vaquero, C.; Fernandez-Piqueras, P.; Hoelzel, A.R. Phylogeography and alpha taxonomy of the common dolphin (*Delphinus* sp.). J. Evol. Biol. **2006**, *19* (3), 943–954.

20. Burgess, W.C.; Tyack, P.L.; Le Boeuf, B.J.; Costa, D.P. A programmable acoustic recording tag and first results from free-ranging northern elephant seals. Deep-Sea Res. II **1998**, *45* (7), 1327–1351.

21. Johnson, M.; Tyack, P.L. A digital acoustic recording tag for measuring the response of wild marine mammals to sound. IEEE J. Oceanic Eng. **2003**, *28* (1), 3–12.

22. Mellinger, D.K.; Stafford, K.M.; Moore, S.E.; Dziak, R.P.; Matsumoto, H. An overview of fixed passive acoustic observation methods for cetaceans. Oceanography **2007**, *20* (4), 36–45.

23. Ward, J.A.; Thomas, L.; Jarvis, S.; DiMarzio, N.; Moretti, D.; Marques, T.A.; Dunn, C.; Claridge, D.; Hartvig, E.; Tyack, P. Passive acoustic density estimation of sperm whales in the Tongue of the Ocean, Bahamas. Mar. Mamm. Sci. **2012**, *28* (4), E444–E455.

Marine—Pollution

Marine Protected Areas

Jennifer Caselle
Marine Science Institute, University of California, Santa Barbara, California, U.S.A.

Abstract

Globally, oceans are facing a large number of threats including overfishing, pollution, sedimentation, and climate change. As a result, marine protected areas (MPAs) are increasingly being implemented as part of an ecosystem-based management approach for conserving biodiversity and managing marine resources. MPAs are areas that limit or forbid various kinds of human uses. Considerable scientific research shows that MPAs increase the biomass, abundance, diversity, and size of marine species living within their borders. Increases in growth, reproduction, and biodiversity in an MPA can also replenish fished areas when larvae and adults move outside of the protected area, a process called spillover.

DEFINITION AND DESCRIPTION

Marine protected areas (MPAs) are defined as any intertidal or subtidal areas of the ocean, including all associated fauna, flora, and historical and cultural resources, that have been set aside for protection and management through legal or other means.[1] The general term "marine protected area" encompasses a range of human activities that might be allowed inside each MPA. Areas of the ocean that are completely protected from extractive activities and major human uses are referred to as marine reserves or no-take marine reserves. MPAs throughout the world may be termed nature reserves, marine parks and heritage sites and can be established to protect coral reefs, temperate kelp forests, seagrass beds, mangroves, bays and estuaries, as well as areas of the deep sea and open ocean. Data show that MPAs can provide a range of benefits for fisheries, local economies and the marine environment, leading to their increasingly popular use in marine management.

WHY DO WE NEED MPAS?

Globally, oceans are facing a large number of threats including overfishing, pollution, sedimentation, and climate change. Currently, there are no areas left in the oceans that are unimpacted by humans.[2] These changes are impairing the ocean's capacity to provide food, protect livelihoods, maintain water quality, and recover from environmental stress. These and other benefits, collectively called "ecosystem services," depend on healthy ocean ecosystems. The scale of most impacts to ocean ecosystems exceeds single habitats or species and as such, requires a more holistic, ecosystem-based approach to management. As a result, MPAs are increasingly being implemented as part of an ecosystem-based management approach for conserving biodiversity and managing marine resources. By protecting populations, habitats, and ecosystems within their borders, MPAs can provide a spatial refuge for the entire ecological system they contain. MPAs can provide a powerful buffer against a naturally fluctuating environment, catastrophes such as hurricanes, and the uncertainty inherent in establishing marine management schemes.

While the number of protected areas in the ocean is still far fewer than on land, the total ocean area protected has risen by >150% since 2003.[3] The total number of MPAs globally, as of 2010, stands at ~5880, covering >4.2 million km^2 of ocean. This figure equates to just over 1% of the marine area of the world. Although it is not possible to develop an exact account, fully protected, no-take areas cover only a small portion of MPAs worldwide.[3] Recently, very large protected areas have been implemented in locations such as the northwest Hawaiian Islands, the Phoenix Islands, the Marianas Trench, and the Chagos Archipelago. These large MPAs, together with Australia's Great Barrier Reef, contribute disproportionately to the total area of the ocean that is protected.

MPAs are one tool for managing ocean ecosystems, but they cannot protect the oceans from all human influences. MPAs alone may not address such pervasive problems as pollution and climate change, and they will most directly benefit fishes and invertebrates that do not move long distances. However, they can provide a range of benefits not seen with other management and conservation strategies.

BENEFITS OF PROTECTED AREAS

Considerable scientific research shows that MPAs and marine reserves increase the biomass, abundance, diversity, and size of marine species living within their borders

Encyclopedia of Natural Resources DOI: 10.1081/E-ENRW-120047590

(Fig. 1),[4] with stronger effects in fully protected marine reserves.[5] Generally, species that are subject to fishing pressure outside MPAs show the greatest responses to protection,[6] while other species may show no response or even decline. Such declines generally reflect interactions among species, where larger numbers of predators inside MPAs (i.e., generally those species most prone to overfishing) have cascading effects on lower trophic levels.[7] For example, in the California Channel Islands, the buildup of two urchin predators (CA sheephead and CA spiny lobster) inside a long-standing, fully protected marine reserve resulted in a decline in sea urchin abundance and a subsequent increase in kelp. Urchins are known to graze down giant kelp forests, often resulting in the formation of urchin barrens, which have lower productivity and biodiversity than healthy kelp forests. Indirect effects of the removal of fishing on sea urchin predators resulted in higher abundance of giant kelp and the absence of urchin barrens in the reserve.[8]

MPAs not only affect populations living within their borders but can also influence populations in adjacent areas. Increases in growth, reproduction, and biodiversity in an MPA can replenish fished areas when larvae and adults move outside of the protected area. For example, larger individuals produce exponentially more, and healthier offspring, and higher population densities improve the likelihood of reproductive success. By increasing reproductive output, MPAs can serve as a source for larvae that can restock the protected area itself as well as export larvae to adjacent areas open to fishing. Scientists are using genetic data, life-cycle information, computer models, and advanced tagging techniques to document the patterns of larval export from MPAs. As populations increase in MPAs, adult and juvenile organisms may also move from MPAs to open areas in a process called "spillover" (Fig. 2). Net emigration of adult and juvenile individuals from MPAs can result from density-dependent (e.g., competition for food or shelter, increased predation), or density-independent (e.g., shifts in home ranges, ontogenetic migrations) processes.[9] The rate of adult and juvenile spillover is hypothesized to increase with time after MPA establishment as populations become increasingly dense in the protected area.

FISHERIES AND SOCIAL BENEFITS

MPAs provide a number benefits to fisheries beyond the expected larval and adult spillover. These include (1) protecting genetic diversity, size distributions, sex ratios, or other stock characteristics; (2) reducing bycatch impacts on vulnerable species in multispecies assemblages; (3) rebuilding overfished stocks; (4) protecting critical habitats necessary for productive populations; (5) maintaining ecosystem processes necessary for productive populations; (6) providing a reference point to guide future management decisions; and (7) ensuring future catches even if management mistakes occur. MPAs also have many nonfisheries benefits, such as protecting biodiversity and ecosystem structure, serving as biological reference areas, providing nonconsumptive recreational activities such as ecotourism, and maintaining other ecosystem services such as shoreline protection, nutrient cycling, and climate control.

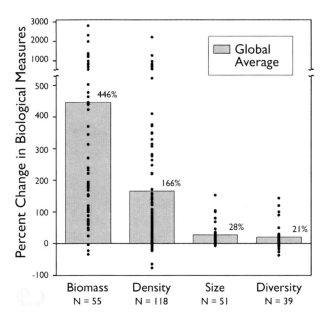

Fig. 1 Average changes in fishes, invertebrates, and algae within marine reserves around the world. Although changes varied among reserves (black dots), most reserves had positive changes. **Source:** Adapted from Lester et al.[4] PISCO Science of Marine Reserves.

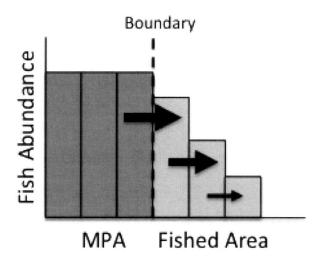

Fig. 2 Spillover from MPAs to fished areas may result from net emigration of adult and juvenile individuals from MPAs due to increased competition for resources or other density-dependent mechanisms (black arrows). Shading represents reduction in abundance outside of MPA due to fishing. **Source:** Adapted from Abesamis et al.[10]

Marine—
Pollution

MPA NETWORKS

By themselves, small MPAs do not tend to support populations that are large enough to sustain themselves. To ensure that young are available to replenish and sustain populations within MPAs, the protected area must be fairly large. However, in some regions, economic constraints may make it impractical to create a single large reserve or other type of MPA that can support viable populations. Establishing networks of several smaller MPAs can help reduce economic impacts without compromising conservation and fisheries benefits.

A network generally includes a set of multiple MPAs of different sizes, located in critical habitats, and connected by the dispersal of larvae and/or movement of juveniles and adults. The general objective of such networks is to maximize conservation and/or fisheries benefits from MPAs. A network can contain critical components of a particular habitat type, or portions of different kinds of habitats, that are critical to different life stages of organisms. In an effective network, organisms must be able to travel beyond the boundaries of a single protected area into other protected areas. To facilitate this movement, a network should be designed and implemented explicitly considering patterns of connectivity. By using different sizes and spacing of protected areas, a network can protect species with different characteristics. For example, a network of MPAs may include feeding habitats in open waters and breeding and nursery grounds in shallow bays. If MPAs protect these critical habitats, organisms that rely on them throughout their lifespans are likely to grow larger and have greater reproductive success. Networks that allow for redundancy in their protection of particular species and habitats also provide better fishery outcomes by protecting areas that are both sources and sinks of larvae. Networks can offer a better compromise between human use and conservation than single, large protected areas.

ACKNOWLEDGMENTS

Thanks to Kirsten Grorud-Colvert and the PISCO science of Marine Reserves team for thoughtful discussion and use of Fig. 1. This entry also benefitted from discussions with Alan Friedlander and Sarah Lester. This is contribution number 414 from PISCO, the Partnership for Interdisciplinary Studies of Coastal Oceans, funded primarily by the David and Lucile Packard Foundation and the Gordon and Betty Moore Foundation.

REFERENCES

1. Kelleher, G.E. *Guidelines for Marine Protected Areas*; IUCN: Gland, Switzerland and Cambridge, UK, 1999.
2. Halpern, B.S.; Walbridge, S.; Selkoe, K.A.; Kappel, C.V.; Micheli, F.; D'Agrosa, C.; Brumo, J.F.; Casey, K.S.; Ebert, C.; Fox, H.E.; Fujita, R.; Heinemann, D.; Lenihan, H.S.; madin, E.M.P.; Perry, M.T.; Selig, E.R.; Spalding, M.; Steneck, R.; Watson, R. A Global Map of Human Impact on Marine Ecosystems. Science **2008**, *319*, 948–952.
3. Toropova, C.; Meliane, I.; Laffoley, D.; Matthews, E.; Spalding, M. Eds.; *Global Ocean Protection: Present Status and Future Possibilities*; Brest, France: Agence des aires marines protégées, Gland, Switzerland, Washington, DC and New York, USA: IUCN WCPA, Cambridge, UK: UNEP-WCMC, Arlington, USA: TNC, Tokyo, Japan: UNU, New York, USA: WCS. 2010; 96 pp.
4. Lester, S.E.; Halpern, B.S.; Grorud-Colvert, K.; Lubchenco, J. Ruttenburg, B.I.; Gaines, S.D.; Airamé, S.; Warner, R.R. Biological effects within no-take marine reserves: a global synthesis. Mar. Ecol. Prog. Ser. **2009**, *384*, 33–46
5. Lester, S.E.; Halpern, B.S. Biological responses in marine no-take reserves versus partially protected areas. Mar. Ecol. Prog. Ser. **2008**, *367*, 49–56.
6. Hamilton, S.L.; Caselle, J.E.; Malone, D.P.; Carr, M.H. Incorporating biogeography into evaluations of the Channel Islands marine reserve network. Proc. Nat. Acad. Sci. **2010**, *107*, 18272–18277, http://www.pnas.org/cgi/doi/10.1073/pnas.0908091107.
7. Ling, S.D.; Johnson, C.R.; Frusher, S.D.; Ridgway, K.R. Overfishing reduces resilience of kelp beds to climate-driven catastrophic phase shift. Proc. Nat. Acad. Sci. **2009**, *106*, 22341–22345, http://www.pnas.org/cgi/doi/10.1073/pnas.0907529106.
8. Behrens, M.D.; Lafferty, K.D. Effects of marine reserves and urchin disease on southern Californian rocky reef communities. Mar. Ecol. Prog. Ser. **2004**, *279*, 129–139.
9. Goni, R.; Hilborn, R.; Diaz, D.; Mallol, S.; Adlerstein, S. Net contribution of spillover from a marine reserve to fishery catches. Mar. Ecol. Prog. Ser. **2010**, *400*, 233–243.
10. Abesamis, R.A.; Russ, G.R.; Alcala, A.C. Gradients of abundance of fish across no-take marine reserve boundaries: evidence from Philippine coral reefs. Aquat. Conserv. Mar. Freshw. Ecosyst. **2006**, *16*, 349–371.

Marine Resource Management

Richard Burroughs
Department of Marine Affairs, University of Rhode Island, Kingston, Rhode Island, U.S.A.

Abstract

Marine management addresses direct uses such as oil, fisheries, and mining as well as indirect and nonuse properties that include nutrient cycling, marine-protected areas, and cultural importance among other factors. The 1982 Law of the Sea Treaty followed by exclusive economic zones and regional arrangements created numerous jurisdictions for resource management. For each jurisdiction, decision makers select among policy instruments such as treaties, regulations, subsidies, sanctions, and property rights to shape human behavior consistent with evolving goals for the marine environment. At the resulting *resource management frontier,* the creation of new policy instruments incorporating more resources and values over larger geographic areas will shape future advances in this field.

INTRODUCTION

Marine resource management encompasses the uses and properties of the sea that reflect human needs and values. Primary elements include identifying the types of resources, specifying jurisdictions and goals, and selecting policy instruments to influence human behavior. The sections that follow explain each process.

MARINE RESOURCES AND VALUES

Marine resources consist of the non-material and material properties of marine environments. The interrelationship of natural systems and human systems starts with the values that people apply to the properties of marine environments (Table 1). Consumptive and nonconsumptive uses that involve physical interaction are referred to as *direct* use values.[1] Consumptive direct uses consist of oil drilling, mineral mining, fishing, and other extractive activities. Nonconsumptive direct uses include transportation, recreation, military, and research activities that depend on the ocean environment, but most often do not directly consume it. Measurable benefits grow from many properties of the marine environment. Nursery habitat supports fisheries. Salt marshes attenuate storm surges from hurricanes. Nutrient cycling can assimilate organic wastes in limited amounts. Each of these properties or processes has *indirect* benefits that help meet people's needs. *Nonuse* benefits include knowing about the existence of marine resources, enjoying the view, and reflecting on the culture that surrounds certain marine activities.

How people value benefits from the marine environment shapes decisions. Direct exploitation (Table 1) predominated in the past.[2,3] But now, the extents to which the physical beauty of nature inspires (attraction) or the power of the ocean instills fear (aversion) determine emotions that shape human involvement. In addition, individuals may seek to physically control the ocean (dominion), or they may have an emotional attachment to the sea that shapes sharing and cooperation (affection). And subtler values such as study of the ocean (reason), ethical concern and altruism (spirituality), and the use of nature in thought (symbolism) influence how people reach decisions concerning marine resources. All of these values can be reflected in decisions even though they rarely trade in markets.

Direct consumptive uses such as fisheries, oil and gas extraction, waste disposal, and mining dominate the entries in Table 1, and unsurprisingly have been the focus for most management initiatives. Historically, each triggered a sector-based management arrangement, if they were managed at all. Direct nonconsumptive uses such as transportation, recreation, and military activities require additional management activities often with separate authorizations.

While the history of marine resource management comprises the direct uses, the indirect uses and nonuse properties will shape the future. Without exception, they require management regimes that respect new and diverse values, and that operate in a holistic manner. Management of nutrient cycles in coastal waters, for example, requires the consideration of multiple sources (ocean dumping, sewage effluent, river contributions), some of which extend to land (agriculture) and air (fossil fuel combustion). Lack of careful attention to nutrient cycles renders the success of fisheries, recreation, aesthetics and other activities in doubt. Adapting current management institutions, which are

Encyclopedia of Natural Resources DOI: 10.1081/E-ENRW-120047578

Marine—
Pollution

Table 1 Direct, indirect, and nonuse of marine resources with examples of related values

Direct, Consumptive	
Use/property	Values
Fisheries	Exploitation, reason, affection
Aquaculture	Exploitation, reason
Oil and gas	Exploitation, dominion
Metallic minerals	Exploitation, dominion
Nonmetallic minerals	Exploitation, dominion
Water	Exploitation, symbolism
Ocean thermal energy conversion	Exploitation, reason
Hydrokinetic, wind	Exploitation, reason
Waste assimilation, accumulation	Exploitation, dominion

Direct, Nonconsumptive	
Use/property	Values
Transportation	Exploitation
Military	Exploitation, dominion
Recreation	Attraction, affection
Research	Reason, symbolism
Education	Reason, spirituality
Tourism	Attraction, affection, dominion

Indirect	
Use/property	Values
Nutrient cycling	Aesthetic, reason
Habitat, nursery	Reason, attraction
Sediment stability	Reason, symbolism
Flood control	Exploitation, reason

Non-Use	
Use/property	Values
Cultural	Affection, aesthetic
Existence	Symbolism, spirituality
Viewshed	Aesthetic
Biodiversity	Affection, attraction, reason

Source: Modified from Skinner,[27] Kellert,[2] Kellert,[3] Barbier.[1]

largely based on single extractive uses, to effectively address indirect and nonuse issues remains a central challenge. More holistic structures that consider all uses in a geographic area will produce more successful results.

Growing support for revised institutional structures is fueled by conflicts among some individual uses and the values that underlie them. Simultaneously using a single area, such as the Gulf of Mexico, for oil extraction, salt marsh restoration, and fishing leads to major conflicts when an oil spill such as Deepwater Horizon cripples, for a time, the biological systems. Promotion of one use at a specific geographic location may diminish or exclude other uses with single use primacy resulting. Decades-long use of urban waters for waste disposal, a trend that is just now being reversed, precluded their use for recreation. To the extent that adverse effects and/or incompatibility arise, institutional change is mandated. The need for change becomes acute as the possibility for resolving conflict through technological innovation and spatial separation is reduced.

JURISDICTIONS

Resources mentioned in Table 1 extend over various geographic areas—international, national, regional, and local. However, management occurs in defined jurisdictions. Therefore, specifying the geographic extent of a regime and identifying the resources covered determine a basis for management. The gradual division of the ocean into jurisdictions has occurred over many decades.

The ocean enclosure movement[4] expands national claims for area usually out to 200 miles and increases the number of resources under national control. In 1945 President Truman asserted that the United States had exclusive rights to exploit resources on and under its continental shelf;[5] and by 1983, the United States had claimed rights to an exclusive economic zone (EEZ) that extended to 200 miles from the shore.[6] This process, repeated by many other nations, transferred 39% of ocean space from high seas where open access predominated to jurisdictions where individual nations control most activities.[7]

The United Nations defined the area seaward of the EEZs as the common heritage of mankind[8] and directs the management of manganese nodules and polymetalic sulfides in that region. On a broad scale, these events signaled increasing jurisdictional clarity for marine resources, and over 150 nations participate in the Law of the Sea Convention.[9]

Where a nation's EEZ delimitation remains uncertain, much more remains to be done. For example, maritime boundaries among exclusive economic zones in the South China Sea remain in dispute, as do many aspects of delimiting control of and managing Arctic seas. Urgency for resolution arises because large quantities of oil and gas may be found in the disputed regions.

As freedom of the seas has given way to various forms of enclosure, the areal extent and the number of resources or uses encompassed in new management regimes has grown (Fig. 1). If the trend continues, coverage of more resources over greater areas seems likely. At the global scale, individual resources such as whales, biodiversity, straddling stocks, and ocean dumping are treated through separate international agreements and plot in the upper left hand area of Fig. 1. Regions such as the Mediterranean, the Baltic,

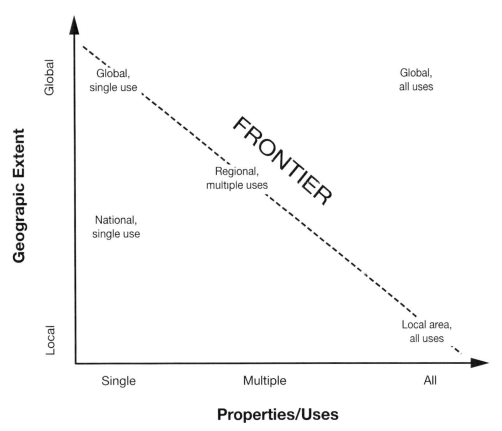

Fig. 1 Scales of management.
Source: Modified from Juda.[26]

Antarctica, and others have arrangements that cover multiple resources for part of the ocean and plot in the middle of the figure. To the lower right are local areas where some claim comprehensive management of all resources at least in theory. The *resource management frontier* runs from the upper left to the lower right. Most current debate exists along with this frontier and relates to the creation of new institutions to address natural and social issues that lie there.

DECISION PROCESSES

Decision-makers respond to a context and select policy instruments to shape human behavior utilizing a structured means for arriving at a solution. Context specifies the breadth of considerations that go into the decisions. Means for arriving at decisions, whether comprehensive and rational or circumscribed in numerous ways, portray the way that individuals process information. Stakeholders play an increasingly important role in making decisions.[10,11] Well-chosen instruments incorporated in resource management decisions create practical outcomes. Ultimately, environmental decisions can demonstrate success through monitored changes in natural systems that may reasonably be attributed to the program at hand.

Context

Initial efforts at resource management were directed toward individual resources or properties. For example, as offshore oil and gas extraction technologies improved in the 1950s,[12] U.S. law responded to establish decision processes. Later in the 1970s, recognition that oil and gas development affected other resources resulted in legally mandated projections of impacts through environmental statements. Over time, managing one sector, oil and gas, explicitly included other sectors such as fisheries and habitats. Soon thereafter, multiple use management in various forms developed protocols to deal with the interaction of several resources in one geographic area.

By the late 20th century, the decision-context embraced many of the elements of ecosystem-based management (EBM) as evidenced by consideration of interrelationships among resources as well as environmental impacts, and social values related to each. EBM has broadened the discussion about resources by adding properties beyond single consumptive uses and introduced the notion that natural systems have limits to the goods and services they can provide. For example, the U.S. Commission on Ocean Policy (USCOP) declared the need to manage across all ecosystem components and include humans as part of coastal ecosystems.[13] So, direct uses that produce revenue constitute an important but not sole consideration for management that ultimately must include indirect and nonuse values. Ecosystem services, a subsequent extension of EBM, define naturally functioning systems that provision (fish for example), regulate (climate), culturally enrich (recreation), and support (nutrient cycling) activities of great benefit to humans.[14]

Means

The means for reaching a resource management decision take many forms. When operating at the frontier portrayed in Fig. 1, decision-makers must create new resource management systems. Since there are few pat procedures, the selection of instruments for the first time requires extra creativity and diligence. This process has been identified as constitutive decision-making, which entails determining how policy should be made.[15,16] By indicating what considerations, which instruments, and who should be involved, decision-makers are setting the parameters for new law or guidance about how and when natural resources will be linked to society. Once completed a constitutive decision at the resource management frontier becomes the template for ordinary decisions which proceed within a known and more predictable system of authority and control. Ordinary decision processes can be found below and to the left of the frontier in Fig. 1 where specific procedures reduce but by no means eliminate controversy.

Constitutive decision-making takes three forms. First, a *rational* and comprehensive process assumes that the problem has been clearly defined and a goal established.[17] In this context, one need only select evaluation criteria, apply them to the policy alternatives, and recommend the solution or solutions that best meet the objectives.[18,19] By predicting the effects of alternative solutions and comparing across specific criteria, one identifies which policy shows the greatest promise and selects it.

Often however, selection does not present itself so cleanly, and this leads to a second approach to describe how decision makers act. Actual decision processes seldom match the ideal of rationality as described earlier.[20,21] Goals may not be clear, or they may be conflicting. This is often the case for marine resources. Information may not be readily available, or it may be so diverse that establishing an effective way to compare and utilize it is lacking. Complexities in goals and information cause decision makers to utilize an *incremental* or *bounded* approach. Organizations have to simplify complex decisions to make them tractable.[21] March and Simon observe that organizations establish boundaries of rationality by seeking only satisfactory not optimal results for a given criterion, and that they do this through actions that fit within a restricted

range of familiar situations. In this setting, actions are goal-oriented and adaptive. In addition, solutions are adopted through a series of small changes where each step is scrutinized for its relative effectiveness before the next step is taken.[20] The concept is that if one does not have all the information necessary, but action is mandated, then small actions rather than large ones are preferred. "Muddling through" in this manner copes with information inadequacies and goal deficiencies by limiting the decision-maker's liability should an initial solution prove inadequate.

Others identify a third path. They find that the problem and the preferences have not been clearly specified, that technological solutions remain unclear, and that participants' involvement and effort varies.[22] They envision participants dumping problems and solutions into a *garbage can,* and, after matching up a problem with a solution, removing both at the same time. At the moment when the organization must act, the decision-maker selects among problems, solutions, participants, and choice opportunities that are in the garbage can. One could envision harmful algal blooms or oil spills forcing decision-makers to hurriedly match existing technical solutions with the problem while the pressure for a resource management decision is high.

Instruments

Policy instruments (Table 2) enable society to shape human behavior consistent with common interests. Resource managers select instruments when they specify how to make operational decisions. Instruments define the types of actions that may be directed toward resolving marine resource problems. They vary in the extent to which government acts as opposed to other forces in society such as community values or markets. The instruments at the top of Table 2 depend almost exclusively on government action. In the middle of the table subsidies, economic frameworks, and information work with a government action supplying the impetus for nongovernmental entities to act. Voluntary agreements extend arrangements beyond the direct reach of government.

The last entry in Table 2, restructuring property rights, could significantly change the role of government. Over half a century ago Gordon[23] asserted that the nature of fisheries and their exploitation meant that successful fishermen must be lucky or in a fishery where social controls convert the open resource into property rights. If economic reasoning predominates, then success is equated to economic efficiency and that objective alone. But as shown here, many other policy instruments besides revising property rights may be used, and a great diversity of values beyond economic efficiency applies to the fisheries. Selective assessments in resource economics and law led proponents of the efficiency approach to define the

Table 2 Policy instruments and sample actions for marine resources

Instrument	Sample actions	Uses/Properties examples
International treaties	Establish jurisdictions, codify norms	Deep sea mining, biodiversity, ocean dumping
Regulation, sanctions	Change standards for permits, create performance-based approaches, fine	Mining, oil spills, water quality
Taxes, fees, charges	Establish, increase, decrease	Energy production, fishing licenses, beach access
Zoning	Exclusion, single use, multiple use	Marine protected areas, military areas, fish harvest exclusion zones, oil extraction areas
Subsidies, grants	Eligibility, adjust level	Aquaculture, agriculture
Economic activity framework	Competition/concentration	Oil, fisheries
Information, education, capacity building	Warn of hazards, exhort, train	Ecolabeling, management skills
Voluntary agreements	Coercion, norms, negotiation	Waste discharge, fisheries
Property rights, markets	Reallocate and/or redefine rights, establish tradable quotas, co-manage	Fisheries

Source: Modified from[9] Clark,[16] Sterner.[28]

problem in a manner that demanded changes in property rights as the solution.[24] However, that approach has substantial weaknesses of its own. Who holds the rights and how they handle them will determine success from the perspective of the society as a whole, yet those factors are omitted.[25] Optimizing fisheries decisions to favor one instrument and one value reflects neither the multiplicity of values within society nor the opportunity to reach them through various instruments.

CONCLUSIONS

The management of marine resources evolves with associated human needs and values. Technological developments enhance possibilities, but ultimately the limits inherent in the natural world control the extent to which society can intensify uses of marine resources. Initial controls began with a focus on single extractive uses and ultimately expanded to cover more uses over larger areas. Over the past half century, the understanding of interrelationships among properties of the marine environment and anthropogenic impacts have expanded at a time when indirect and nonuse activities gained in importance. Ironically, the society's reliance on marine resources grows at a time when decision processes lag in incorporating shifts in values and limits of the resources themselves.

In the future, the number of uses, properties, and values that influence decisions will likely increase, making resource management even more complicated. As uses and values proliferate, potential for conflict will follow. Calls for comprehensive management to reduce disputes will further focus innovation at the marine resource management frontier.

REFERENCES

1. Barbier, E. Progress and challenges in valuing coastal and marine ecosystem services. Rev. Environ. Econ. Policy **2012**, *6* (1), 1–19.
2. Kellert, S. *The Value of Life: Biological Diversity and Human Society*; Island Press: Washington, DC, 1996.
3. Kellert, S. *Birthright: People and Nature in the Modern World*; Yale University Press: New Haven, Connecticut, 2012.
4. Eckert, R. *The Enclosure of Ocean Resources: Economics and the Law of the Sea*; Hoover Institution Press: Stanford, California, 1979.
5. Truman, H. Policy of the United States with respect to the Natural Resources of the Subsoil and Sea Bed of the Continental Shelf: Proclamation 2667. Fed. Regist. **1945**, *10*, 12305.
6. Reagan, R. Exclusive Economic Zone of the United States: Proclamation 5030. Fed. Regist. **1983**, *48* (50), 10605–10606.
7. http://www.NationMaster.com/graph/geo_mar_cla_exc_eco_zon-maritime-claims-exclusive-economic-zone last visited May 26, 2012. This total, compiled from NationMaster, includes double counting where two or more nations have claimed the same space, and national boundaries of their respective EEZs have not been established.
8. http://www.un.org/Depts/los/convention_agreements/texts/unclos.unclos_e.pdf (accessed May 2012).
9. http://www.un.org/Depts/los/reference_files/status2010.pdf (accessed June 2012).
10. Burroughs, R. When stakeholders choose: Process, knowledge, and motivation in water quality decisions. Soc. Nat. Resour. **1999**, *12*, 797–809.
11. Dalton, T.; Forrester, G.; Pollnac, R. Participation, process quality, and performance of marine protected areas in the wider Caribbean. Environ. Manag. **2012**, *49*, 1224–1237.
12. Burroughs, R. *Coastal Governance*; Island Press: Washington, DC, 2011.

13. U. S. Commission on Ocean Policy. *An Ocean Blueprint for the 21st Century*: Final Report; U. S. Commission on Ocean Policy: Washington, DC, 2004.

14. United Nations Environment Program (UNEP). *Marine and Coastal Ecosystems and Human Well-Being: A Synthesis Report Based on Findings of the Millennium Ecosystem Assessment*; UNEP: Nairobi, Kenya, 2006.

15. Lasswell, H.*A Pre-View of Policy Sciences*; American Elsevier Publishing Co. Inc.: New York, 1971.

16. Clark, S. *The Policy Process: A Practical Guide for Natural Resource Professionals*; Yale University Press: New Haven, Connecticut, 2011.

17. Birkland, T. *An Introduction to the Policy Process: Theories, Concepts, and Models of Public Policy Making*; M.E. Sharpe: Armonk, New York, 2011.

18. Weimer, D.; Vining, *A. Policy Analysis: Concepts and practice*; Prentice Hall: Englewood Cliffs, New Jersey, 1992.

19. Bardach, E.*A Practical Guide for Policy Analysis: The Eightfold Path to More Effective Problem Solving*; CQ Press: Washington, DC, 2009.

20. Lindblom, C. The science of "muddling through." Public Adm. Rev. **1959**, *19* (2), 79–88.

21. March, J.; Simon, H. *Organizations*; John Wiley and Sons: New York, 1958.

22. Cohen, M.; March, J.; Olsen, J. A garbage can model of organizational choice. Adm. Sci. Q. **1972**, *17*, 1–25.

23. Gordon, H.S. The Economic theory of a common-property resource: The fishery. J. Political Econ. **1954**, *62*, 124–142.

24. Macinko, S.; Bromley, D. Property and fisheries for the twenty-first century: Seeking coherence from legal and economic doctrine. Vermont Law Rev. **2004**, *28*, 623–661.

25. Charles, A. Human rights and fishery rights in small-scale fisheries management. *In Small Scale Fisheries Management*; Pomeroy, R., Andrew, N., Eds.; CAB International: Wallingford, Oxfordshire, UK, 2011; 59–74.

26. Juda, L.; Burroughs, R. The prospects for comprehensive ocean management. Mar. Policy **1990**, *14* (1), 23–35.

27. Skinner, B.; Turekian, K. *Man and the Ocean*; Prentice-Hall Inc.: Englewood Cliffs, New Jersey, 1973.

28. Sterner, T. *Policy Instruments for Environmental and Natural Resource Management*; Resources for the Future: Washington, DC, 2003.

Marine—
Pollution

Maritime Transportation and Ports

Austin Becker
Emmett Interdisciplinary Program in Environment and Resources (E-IPER), Stanford University, Stanford, California, U.S.A.

Abstract

Ports and maritime transportation deliver low-cost transport and facilitate international trade. The vast majority of raw materials and manufactured goods traded are carried from port to port along the world's oceans and rivers; however, highly efficient ports and shipping also present environmental consequences for coastal, ocean, and fresh-water environments. This entry provides an overview of the impacts ports and shipping have on the environment, as well as a discussion of the unique problems that climate change poses to this critical sector of the global economy.

INTRODUCTION

Ports and maritime transportation form the backbone of the global economy by facilitating trade that provides civilization with the energy, goods, and materials upon which it depends. However, ports and shipping also bear responsibility for many negative environmental impacts, including oil spills, the introduction of invasive species, and air pollution. National and international policies have been created to address some of these issues, but problems do remain. In addition, ports and shipping contribute to global warming through CO_2 emissions. Climate change presents new challenges to this sector, due to both new emissions requirements and to the new environmental conditions that global warming will present. Ports and shipping will be on the frontline of impacts such as sea-level rise and increases in storminess. The industry may also benefit from certain changes such as the opening of the Northwest and Northeast passages due to melting sea ice.

THE ROLE OF PORTS AND SHIPPING

A global fleet of 50,000 commercial vessels transports 90% of the world's freight by volume (See Fig. 1).[1] Ports and maritime transportation play a major role on every scale of the economy. Today, over 3,000 ports around the world serve as transfer points for energy products (coal, oil, and gas), manufactured goods, and raw materials. Ports and shipping fulfill a wide variety of functions for the local, regional, and global economy. They provide jobs, facilitate trade, and serve as critical links between the hinterlands (region from which goods come from) and the forelands (the region to which goods are destined). Ports range in specialization from massive container ports (e.g., Los Angeles/Long Beach), to small niche ports that serve one type of freight (e.g., petroleum, coal, grain, or fishing).[2]

THE PORT'S LOCATION

Historically, many of the world's cities sprung from ports. Geographically, areas that connect oceans with rivers allow for economic development that results from the transshipment of products between ocean-going and riverine craft. Before the advent of rail and highway, goods would be transferred between larger ocean-going vessels and smaller river craft so that areas far inland could participate in global markets. During the Industrial Revolution, rivers also provided the power necessary for manufacturing and moving finished products and raw materials along those rivers to a port. This created an efficient means of connecting markets. Although most ports are located along the coast, many inland ports are strategically situated on lakes or along rivers. As cities grew around the trade center of the port, the port infrastructure itself often migrated away from the newly formed city center.[3]

TYPES OF PORTS

Ports can be categorized in numerous ways, but they are ultimately difficult to compare. Size may be measured by throughput, cargo value, land footprint, number of vessel calls, or other measures. Similarly, operation and ownership vary widely from port to port, with some being fully privatized and others being entirely public entities. Ports generally fall into one of four categories in terms of operations and management. "Service ports" are predominantly public. A public "port authority" owns the land and all assets and manages all cargo-handling operations. The "tool port" divides responsibility between the public port authority, which owns and maintains the infrastructure, and private firms, which handle the cargo. In a "landlord port," the public port authority owns the land and infrastructure, but leases it to private operating companies. Finally,

Encyclopedia of Natural Resources DOI: 10.1081/E-ENRW-120047591

Marine—
Pollution

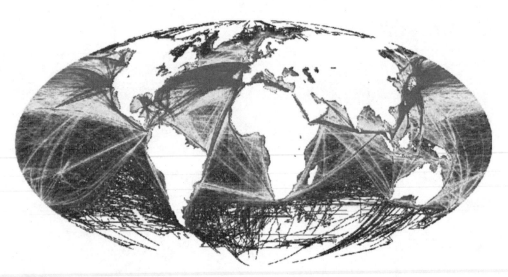

Fig. 1 Global maritime shipping routes.
Source: http://www.nceas.ucsb.edu/GlobalMarine/

the "private service port" is entirely owned and operated within the private sector.[4]

TYPES OF TRANSPORT

Maritime transportation generally falls into five main categories: bulk, break-bulk, ro–ro, containers, and passenger. Bulk shipping refers to freight carried directly in the hold of a ship without packaging. Examples include grain, coal, liquid petroleum products, and chemicals. Purpose-built ships carry these products between ports equipped to handle loading and offloading of raw materials. Break-bulk consists of cargo that has been "unitized" onto pallets or in barrels. Examples of break-bulk cargo include fruit and lumber. Ro–ro stands for "roll on/roll off" and includes vehicles and other equipment that usually operates under its own propulsion and is transferred on and off of ships via large ramps. Container shipping, technically a form of break bulk, became a global standard in the 1960s. Today, the international standard shipping container measures 20, 40, 45, 48, or 53 feet long. Regardless of the container size, throughput and capacity is usually expressed in twenty-foot equivalent units (TEUs). Containers revolutionized shipping because goods could be loaded from a source into a container, moved intermodally (on rail, truck, and ship), and unloaded at their destination. Finally, passenger ships are used to move people. This category of ships includes ferries, cruise ships, fishing vessels, and other commercial craft that carry people for pleasure or simply as a means of transport.

PORTS, SHIPPING, AND THE ENVIRONMENT

The strategically advantageous location of the coastal port from an economic standpoint often puts it in an estuarine area of critical importance from an ecological perspective. An "estuary" is defined as the part of a river's mouth where the river current meets the tide. Due to the abundance of nutrients and the mixing of salt and fresh water, these highly productive areas serve as breeding grounds for much of the world's marine life. The nature of a port's business puts it at odds with these ecological functions in a number of ways. Increasingly larger ships require ever-deeper channels so that they may access the port. The deepening of the channel, called dredging, displaces vast amounts of sediment, disturbing habitats, and stirring up contaminants that have settled to the bottom. Ships themselves can introduce new invasive species, which travel from one place to another on the ship's hull or in the ship's ballast water. Runoff from the port lands where industrial activity takes place and accidental spills also might degrade the water quality in an estuary. In the case of an extreme event such as a cyclone, or an earthquake, or a tsunami, the materials stored at the port also could be released into the environment. Port connections to other transportation modes often create a concentration of air and acoustic pollution, which combined with the buildings and other port operations and energy demands creates important implications for global climate change. Ports also contribute to other secondary environmental impacts, in that traffic congestion from trucks naturally increases around port operations. Trucks increase wear and tear on highways and also contribute to lower air quality and noise pollution.

Dredging

The dredging of navigation channels facilitates the movement of deep-draft vessels. Natural siltation and sedimentation slowly fills into these deep thoroughfares. The largest of container vessels today requires a channel that is around 52′ deep, though many ports have channels that are

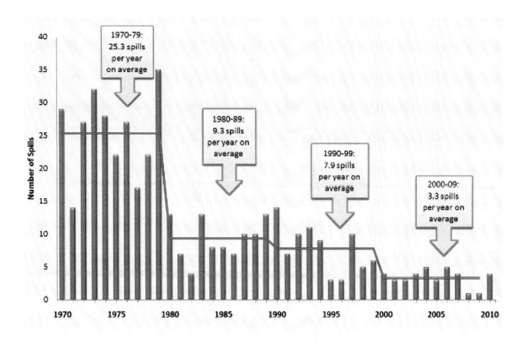

Fig. 2 Number of large (>700 tons) ship spills from 1970 to 2011.
Source: International Tanker Owners Pollution Federation Limited: Information Services - Data & Statistics – Stastics. http://www.itopf. com/information%2Dservices/data%2Dand%2Dstatistics/statistics/

substantially shallower. As vessels grow in size and ports expand to keep up with ever-increasing demand, port planners work to have their channels maintained at their authorized depths or deepened to meet the needs of larger ships. Dredging, however, creates environmental concerns. Dredging disturbs fragile habitats and, in many industrialized areas, stirs up contaminants that lay dormant under a layer of sediment. Disposing of contaminated sediment presents additional challenges. Often, sediment must be hauled by barge far offshore or disposed in special underwater cells and then capped with clean sediment. Dredging also facilitates the passage of larger ships, which can bring their own dangers to local ecosystems if an accident occurs.

Invasive Species and Ballast Water

Throughout history, ships have impacted the world's biodiversity through moving living organisms around the globe. Earthworms, for example, came to North America in ship's ballast in the 17th century.[5] Today, ballast water used to stabilize and adjust the ship's trim may be taken aboard in one ecosystem (fresh, brackish, or salt), transported thousands of miles, and discharged in another ecosystem. Many species survive the voyage and have no natural predators in the new environment. The introduction of zebra mussels (*Dreissena polymorpha*) from Europe to the Great Lakes in the mid-1980s continues to cause millions of dollars of damages annually. The resilient mussel colonizes and clogs power and water piping, as well as displacing other native species. The issues with mussels led to the adoption of

ballast water management standards in the United States. The International Maritime Organization (IMO) Ballast Water Convention was adopted in 2004 requiring both ships and ports to properly handle ballast water and minimize environmental impacts.[6]

Oil Spills

Spills of oil and other hazardous materials represent another negative impact of shipping. Although spill frequency has been vastly reduced over the past few decades (see Fig. 2), spills nevertheless represent a significant threat to coastal and ocean habitats. Oil spills may be caused by a breach in a ship's hull or by the discharging of oily bilge water. Some infamous spills include the Amoco Cadiz in 1978 in which 687 million gallons of oil polluted 200 miles of coastline. In 1991, 9 billion barrels of oil were spilled at the start of the Gulf War in Kuwait. In 1989, the Exxon Valdez spilled 10.8 million gallons of oil into Prince William Sound in Alaska. In 2002, the oil tanker Prestige broke apart spilling oil that washed up on thousands of beaches in France, Spain, and Portugal. Spills such as these kill huge numbers of fish and seabirds and effects can last for decades, despite advances in spill remediation technology and increased disaster preparedness planning. The International Convention for the Prevention of Pollution from Ships (MARPOL 73/78) was adopted as a measure to help prevent pollution of the marine environment by ships from operational or accidental causes.[7] Annex I of the convention addresses oil spills in particular and led to new requirements for double-hulled tankers. As seen in Fig. 2, the number of spills has sharply declined since the adoption of the MARPOL convention.

Other Sources of Pollution

Sewage, garbage, noxious liquids, and caustics have also presented environmental problems in the past. Through either accidental or operational discharges, these materials contributed to the degradation of marine habitats in both coastal and deep-water ocean areas. Deliberate ocean dumping of nonbiodegradable materials has led to the accumulation of plastics and other materials accumulating in great oceanic gyres. Seabirds, fish, marine mammals, and other marine life ingest or become ensnared in this waste. Toxic materials then bioaccumulate, working their way up the food chain. Annexes III – V of MARPOL 73/38 address this problem, though noncompliance remains a significant issue.[8]

Emissions (Port and Ship)

Historically, ships have burned the dirtiest type of fuel, bunker fuel. Pollutants from this source include sulfur oxides, nitrogen oxidescarbonaceous aerosols, and ozone.[9] Particulate-matter emissions for this fuel source have been associated with asthma, heart attacks, lung cancer, and other illnesses. Terminal operations, too, emit pollutants and new regulations are requiring ports to upgrade their equipment. "Cold ironing," for example, allows ships to utilize shore power rather than relying on their own shipboard power plants. This results in lower emissions at the port and provides the opportunity to utilize cleaner energy from the power utility.

PORTS, SHIPPING, AND CLIMATE CHANGE

The ports and shipping industries face new challenges from climate change. On the one hand, the emissions from ships and ports contribute significantly to global warming. Global marine shipping bears responsibility for 1.5–3% of total global CO_2 emissions. This is expected to double by 2050 if emissions-cutting technologies are not adopted.[10] On the other hand, maritime shipping produces less GHG emissions per unit shipped than any other transport mode (Fig. 3).[11] Both the vessel operators and the port facilities have initiated programs aimed to curb their contributions of greenhouse gasses to the atmosphere. The World Ports Climate Initiative formed in 2008 as a mechanism to assist ports combat climate change by initiating projects to reduce greenhouse gas emissions and improve air quality.[12] For example, the WPCI created a new Environmental Ship Index (ESI) scheme. The ESI creates an incentive for shipping companies to reduce the impacts of their vessels and

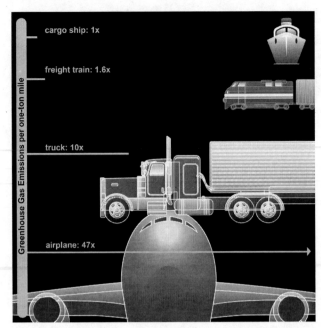

Fig. 3 Comparison of CO_2 emissions between different transport types.
Source: Christine Daniloff Creative Design Director, MIT News. http://web.mit.edu/newsoffice/2010/corporate-greenhouse-gas-1108.html

earn the right to claim a high standard for environmental responsibility and to fly a "clean ship" flag.

In addition, the maritime sectors will certainly be impacted by the many changes the planet will experience in the coming century.[13] Rising sea levels could inundate low-lying ports and will result in much higher storm surge levels than have ever been experience before. Sea level rise also will reduce effective under-bridge clearances (known as "air draft") and could impact other infrastructure that a port relies on (e.g., rail, highway, pipelines). Changes in cyclone behavior could mean more intense storms or more frequent storms, as well as shifting tracks that could result in future catastrophic events in areas that historically have not experienced cyclones. Shipping, too, could be impacted by these changes in storm patterns and have to adjust their routes accordingly to avoid storms.

Impacts on seaports are expected to be significant in many areas. Most ports, however, have not yet taken direct action to reduce their vulnerability.[14] Protecting infrastructure against a 1–2 meter rise in sea levels and more intense storms will require significant investment. Climate change will disproportionately affect ports and port-based economies, depending on their geographic location and the adaptive capacities of the ports themselves and the communities in which they are located. For example, ports in low-lying areas in a hurricane belt will face different physical challenges than those on emergent arctic coastlines where melting ice lowers the level of protection against storms. Ports in developing nations will have a different suite of options available to them than those in developed nations.[15,16] Ports located in estuaries that provide nursery environments for marine life have an even greater responsibility to protect coastal waters. The complexity and potential risks require the scientific community, policy makers, and the port authorities themselves to take an active role to better understand when and how to implement proactive adaptation strategies. Proper stakeholder engagement is also essential to ensure that solutions are sustainable not only from a financial point of view but also in consideration of environmental and social concerns.

To become more resilient to the impacts of climate change and to play a role in mitigating the acceleration of climate change, port decision makers will need to implement new strategies that range from policies (i.e., changing building codes), to design (i.e., creating new protective structures), to practices (i.e., emergency drills). The Port of Gulfport (MS) is currently in the midst of implementing a major resilience-building strategy that serves to illustrate the kind of actions that will inevitably be needed in other ports. After Hurricane Katrina, decision makers chose to elevate the entire port's footprint to 25′ above sea level in order to raise it out of the floodplain. Elevating, diking, or moving entire ports are some of the more drastic measures that will need to be considered.

PORT INDUSTRY AND ENVIRONMENTAL GOALS

The nature of the port industry and global supply chains makes environmental management more complicated, even with improved economic efficiencies. Ports have regionalized and grown in order to accommodate ever-increasing demands. While many of the negative externalities are concentrated in the immediate vicinity of the port itself, the region around the port is not immune to the environmental ramifications that result from the movement of such massive quantities of freight. The growth of ports has in turn spurred the need for larger highways, the development of warehouses, higher bridges to accommodate stacked containers on railways, and a whole host of support services for the port and ships. Termed, "logistics sprawl," the movement of these ancillary facilities away from the port itself and into the suburbs also spreads many of the environmental impacts such as truck emissions and runoff from industrial land use.[17] Finally, due to the global nature of the industry and competition within each link of the supply chain, improving environmental performance through collective actions on the local or regional level can be especially difficult.[18] Some firms, however, are beginning to improve environmental performance. Maersk's shipping, for example, recently began implementing a "fuel switch" program in which its ships will switch to low-sulfur diesel before entering near-shore waters.

THE FUTURE OF SHIPPING

As global population increases, concentrating especially in cities and coastal areas, and nations strive to improve their citizens' quality of life, international shipping will grow to meet new demands. Current forecasts project a doubling of cargo movement by 2040. Given its energy efficiency, maritime transport is likely to remain the predominant means for the shipment of freight. A number of new developments in shipping could produce some changes in routes. Short-sea shipping, already widely practiced in Europe, uses smaller ships to transport freight between areas normally served by road or rail. Though transit times may be longer, this reduces the number of trucks and trains and cuts back on overall emissions, traffic congestion, and wear-and-tear on infrastructure. As global warming continues to melt sea ice on the Arctic, new shipping lanes through the Northwest and Northeast Passages may open up, allowing ships to cut transit times down and avoid traveling through the heavily trafficked Panama and Suez Canals, and the Strait of Malacca.[19] However, it is important to note that increased traffic in the Arctic would also bring significant new environmental risks. Fragile Arctic ecosystems could be significantly damaged by contamination resulting from ship emissions and spills. The effectiveness of clean-up technologies for these environments remains uncertain and the infrastructure to support shipping does not yet exist.

CONCLUSION

Ports and maritime transportation fulfill a critical role in our global economy. Indeed, without ships and seaports, the world would be a very different place. From an environmental perspective, shipping remains one of the most efficient and clean methods of moving products on a unit weight per-distance basis. Shipping does, however, present numerous environmental concerns. Reducing particulate pollution, controlling CO_2 emissions, decreasing reliance on fossil fuels, improving design and operational standards, mitigating climate risk and managing the impacts of dredging will continue to challenge the industry for decades to come.

ACKNOWLEDGMENTS

Many thanks to Prof. Robert Desrosiers (Texas A&M), Nathan Chase (Arup), and the anonymous reviewers for insightful comments and suggestions.

REFERENCES

1. Shipping and World Trade: Key Facts, http://www.marisec.org/shippingfacts/worldtrade/
2. Hoyle, B.; Knowles, R. *Modern Transport Geography*; Belhaven Press: 1992.

Marine—
Pollution

3. Charmaz, K. Qualitative interviewing and grounded theory analysis. *Inside interviewing: New lenses, new concerns;* SAGE: 2003; 311–330.

4. Alderton, PM: *Port Management and Operations*; LLP: 2005; 255.

5. Mann, C.C. *Uncovering the New World Columbus Created*; Alfred a Knopf Inc: 2011.

6. GloBallast Partnerships - The New Convention, http://globallast.imo.org/index.asp?page=mepc.htm.

7. IMO International Convention for the Prevention of Pollution from Ships (MARPOL), http://www.imo.org/about/conventions/listofconventions/pages/international-convention-for-the-prevention-of-pollution-from-ships-(marpol).aspx.

8. Crist, P. Cost Savings Stemming from Non-compliance with International Environenmental Regulations in the Maritime Sector. In *Organization for Economic Co-operation and Development (OECD)*; Edited by Committee MT: 2003.

9. Patton, M.Q. *Qualitative Research and Evaluation Methods*; Sage Publications, Inc: 2002; 392 p.

10. Theys, C.; Notteboom, T.E.; Pallis, A.A.; De Langen, P.W. The economics behind the awarding of terminals in seaports: towards a research agenda. *International Handbook of Maritime Business;* Cheltenham: 2010; 232.

11. Simchi-Levi, D. *Operations Rules: Delivering Customer Value through Flexible Operations*; The MIT Press: 2010; 208 p.

12. World Ports Climate Initiative, http://www.wpci.nl/projects/projects_in_progress.php.

13. Hall, P.V. Seaports, urban sustainability, and paradigm shift. J. Urban Tech. **2007**, *14* (2), 87.

14. Becker, A.; Inoue, S.; Fischer, M.; Schwegler, B. Climate change impacts on international seaports: Knowledge, perceptions, and planning efforts among port adminstrators. J. Clim. Change **2011**.

15. Dasgupta, S.; Laplante, B.; Meisner, C.; Wheeler, D.; Yan, J. The impact of sea level rise on developing countries: a comparative analysis. Clim. Change **2008**, *93* (3–4), 379–388.

16. Nicholls, R.; Hanson, S.; Herweijer, C.; Patmore, N.; Hallegatte, S.; Corfee-Morlot, J.; Chateau, J.; Muir-Wood, R. Ranking Port Cities with High Exposure and Vulnerability to Climate Extremes: Exposure Estimates. In *OECD Environment Working Paper 1; ENV/WKP(2007)1.* Paris, France, OECD: 2007.

17. Dablanc, L.; Rakotonarivo, D. The impacts of logistics sprawl: How does the location of parcel transport terminals affect the energy efficiency of goods' movements in Paris and what can we do about it? Procedia-Social and Behavioral Sciences **2010**, *2* (3), 6087–6096.

18. Hall, P.V.; Jacobs, W. Shifting proximities: The maritime ports sector in an era of global supply chains. Reg. Stud. **2010**, *44* (9), 1103–1115.

19. Arup, *Climate Change and Ports*; Adaptation and Resileince of Assets and Operations. Internal White Paper. In.: 2011.

Marine—
Pollution

Markets, Trade, and Seafood

Frank Asche
Department of Industrial Economics, University of Stavanger, Stavanger, Norway

Cathy A. Roheim
Department of Agricultural Economics and Rural Sociology, University of Idaho, Moscow, Idaho, U.S.A.

Martin D. Smith
Nicholas School of the Environment and Department of Economics, Duke University, Durham, North Carolina, U.S.A.

Abstract

This entry describes the growth in seafood production and trade and the main factors causing these developments. We then review the leading economic research on the international seafood trade and markets with a focus on interactions of markets and the management of fisheries and aquaculture. Specific examples include the relationship between fisheries management institutions and international trade; the relationship between the value of seafood attributes and production practices; and the development of the Fish Price Index (FPI) by the UN Food and Agriculture Organization (FAO) to address food security concerns.

INTRODUCTION

Oceans, lakes, rivers, and other waterways provide about 16% of animal protein consumed globally. The fishing and aquaculture operations that produce this protein also support livelihoods for 8% of the world's population. Fully 39% of total seafood production (in live weight) enters into international trade, valued at US$102 billion per year in 2008.[1] Seafood from developing countries, in particular, constitutes the most valuable category of agricultural exports, totaling US$20 billion per year. By comparison, coffee exports are US$6 billion per year, and rubber, cocoa, bananas, meat, and tea are each less than US$5 billion per year.[2]

The amount of seafood traded internationally has grown significantly during the past decades. The technologies and logistics associated with globalization of food markets in general have influenced the seafood trade by creating access to new markets, lessening the importance of distance, and ultimately allowing producers to exploit new comparative advantages. These trends have greatly expanded aquaculture production, which accounts for much of the growth in the international seafood trade. Large wholesalers and retailers can now source fish from all over the world, providing new opportunities for some producers and new challenges for others. For instance, substitutes for Atlantic cod are no longer limited to regional substitutes from capture fisheries such as haddock and saithe, but now include distant capture fisheries such as Alaskan pollock and New Zealand hoki as well as farmed species such as pangasius and tilapia.

Seafood is distinguished from most other food industries because much of seafood production involves a close connection to an otherwise uncontrolled environment, and often institutions are insufficient to allow markets to organize production efficiently.[3] As the world´s last significant hunting-based industry, fisheries are limited by the supporting ecosystem and subject to the tragedy of the commons when poorly managed. There is little doubt that many of the world's fish stocks are degraded.[1] However, overfishing is just one of many issues that influence the size of fish stocks and the amount of seafood that enters international seafood trade. Habitat destruction from some types of fishing gear, bycatch (incidental catches of non-target species), and fishing that affects the size distributions of fish populations all influence the balance of ecosystems and indirectly the future availability of fishery resources. Similarly, most aquaculture production occurs in the open ocean or in ponds, and may lead to surrounding ecosystem effects. But aquaculture overall is more similar to livestock production in agriculture than fishing; it relies on the ecosystem for inputs but not for the product itself.[4] Aquaculture also is less exposed to commons problems. Hence, aquaculture faces weaker environmental and institutional barriers to further growth. Still, the continued growth of aquaculture production has implications for environmental health in producing countries if institutional safeguards are not embedded into the production systems. Aquaculture may have fewer commons problems than fisheries, but it must still resolve environmental externalities such that products are priced to reflect the true costs of production.[3]

Encyclopedia of Natural Resources DOI: 10.1081/E-ENRW-120047462

Interactions among domestic fisheries management, the international retail market, and a wide array of environmental concerns about aquaculture and fisheries could all affect the international seafood trade. In this entry, we first provide background on the growth in seafood production and trade and the main factors causing these developments. We then briefly review the economic literature on the international seafood trade and markets. Because space limitations preclude a complete review, we focus on state-of-the-art analysis of seafood markets with implications for management of fisheries and aquaculture that aim to reduce market inefficiencies.

PRODUCTION AND TRADE

The total supply of seafood increased from 71.7 million tonnes in 1976 to 145.1 million tonnes in 2009.[1] Seafood comes from two main modes of production—harvest from capture fisheries and aquaculture. Until the 1970s, aquaculture was unimportant. However, since then a virtual revolution has taken place. Figure 1 shows the changing production from wild fisheries and aquaculture. Aquaculture production has grown from ~3.5 million tonnes in 1970, or 5.1% of total seafood supply, to 60.5 million tonnes, or 40% of total

seafood supply in 2011.[5] The combined effect of increased productivity, increased market access and demand growth has made aquaculture the world's fastest growing animal-based food sector.[1] Capture fisheries production, on the other hand, has fluctuated between 90 and 100 million tonnes in annual landings with no particular trend. Fisheries supply is not expected to increase in the future, as the majority of fish stocks are either fully or overexploited.[1]

Seafood has long been traded internationally, but trade in seafood has increased dramatically in recent decades. The increase in seafood trade is largely attributable to growth in aquaculture productivity and increased exports from developing countries. International trade has increased much faster than total seafood production. From 1976 to 2008, the export volume of seafood increased from 7.9 million tonnes to 33.5 million tonnes, or almost four-fold. Adjusted for inflation, the export value during this period increased threefold. It is important to note that export quantities are not directly comparable to the production quantities because exports are measured in product weight. This can lead to dramatic differences. The fillet weight of tilapia, for instance, is only between 30% and 40% of the harvest weight.

A number of factors have caused the increased trade in seafood. Transportation and logistics have improved

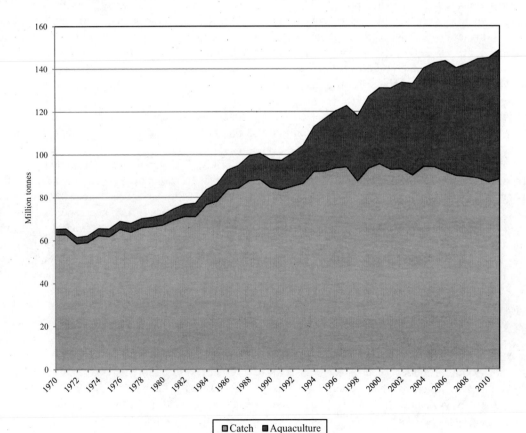

Fig. 1 Global Production from Capture Fisheries and Aquaculture: 1970–2008.
Source: FAO FishStat.[5]

significantly. Lower transportation costs have also given new producers access to the global market. Coastal nations imposed the 200-mile exclusive economic zones (EEZs) in the 1970s and 1980s that increased incentives for international trade. Peru, Ecuador, and Chile implemented EEZs as early as 1952. By the time the U.S. declared its 200-mile EEZ in 1976, 37 nations had already extended their jurisdiction, and by the mid-1980s, nearly all coastal nations had imposed EEZs.[6] Countries with considerable distant-water fishing fleets, such as Spain and Japan, were negatively affected, as other coastal nations expanded their domestic fleets to exploit the fisheries within their 200-mile EEZs. As a result, countries that relied on harvesting within the 200-mile EEZ of foreign nations had to increase their imports to maintain domestic consumption at the same levels.

When seafood export quantity increases fourfold and export value only threefold, the unit value decreases. Lower unit values suggest seafood's competitiveness as a food source has increased and is a key factor explaining increased trade. Successful aquaculture species such as salmon and shrimp illustrate this phenomenon. Real prices for each are less than one-third of what they were 25 years ago, and internationally traded quantities have grown at an annual rate of 16.8% and 13.6%, respectively, during the period from 1985 to 2008.[5] Total growth in quantities traded in salmon and shrimp for the period were 3014% and 1491%, respectively. The profitable expansion in the production of these species, despite decreasing prices, is due to a combination of lower production costs, improved production technologies, and lower distribution and logistics costs.

The trade patterns are widely different between exports and imports. The source of seafood exports were split almost equally between developing and developed countries in 2008, as shown in Fig. 2. The value share of exports going from developing countries has increased from 37% in 1976, to 50% in 2008.[5] Alternatively, developed countries comprised 78% of all imports in 2008. Even though that share has declined from 86% in 1976, most of the increased trade pattern in seafood is to developed countries, and a considerable share is exported from developing countries. Japan and the United States are the two largest importers. However, if the EU countries are aggregated, the European Union is clearly the largest market.

It is certainly not arbitrary that developed countries account for most of the imports and that the European Union, Japan, and the United States are the largest seafood importers. These regions are the world's wealthiest customers with the best ability to pay. Economic growth in China and Southeast Asia has led to substantial growth in seafood imports, with prospects for even further growth in demand.[7]

ECONOMICS OF SEAFOOD MARKETS AND TRADE

This section reviews the state of the art in the economic literature about markets, trade, and seafood. It covers the

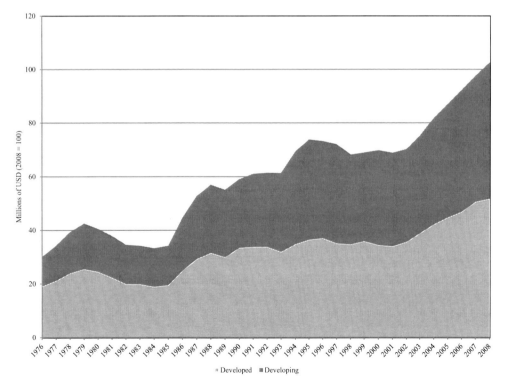

Fig. 2 Real Value of Seafood Exports from Developing and Developed Countries: 1976–2008.
Source: FAO FishStat.[5]

relationship between fisheries management and trade; the relationship between the value of seafood attributes and production practices value of seafood attributes; and the development of the Fish Price Index (FPI) by the UN Food and Agriculture Organization (FAO) to address food security concerns.

The Relationship between Fisheries Management and Trade

The impacts of trade liberalization will differ depending on several factors, including production method (capture or aquaculture) and domestic fisheries management policies. Modern theories of open access (and of fisheries management in general) are dynamic and account for the inseparable link between harvest now and the stock of fish in future periods. Economists began to grapple with the dynamic nature of fishery resources in the 1950s[8] and developed a fully dynamic theory of open access in the late 1960s.[9,10] These results have important implications for conventional models of international trade. Trade policies that decrease (or increase) prices or costs may have the opposite effect of what one might see in other resource or agricultural sectors depending on the level of exploitation of the fishery resource.[11] The basic intuition is that any short-run increase in harvest of the renewable resource sparked by trade liberalization will lead to long-run decreases in harvest when the resource is governed by open access.

A well-known theoretical analysis considers trade between two countries with equal natural resource endowments, but that differ based on their resource management institutions: one country with open-access management and another country with a conservationist approach.[12] By allowing for excess exploitation, open access can be a source of comparative advantage and lead to exporting the resource to the conservationist country. Society benefits for the conservationist country but declines for the open-access country in the long run. However, it is also possible that the well-managed resource in the conservationist country can lead to this country having a lower price. The result is that the conservationist country exports the resource, and both countries gain from trade. Brander and Taylor aptly summarize that "when a renewable resource is subject to open access, or something approaching it, then free trade may not be the tide that raises all boats. Improved management of renewable resources may be a necessary precondition for gains from trade."[13]

The effects of trade liberalization have also been considered under other fisheries management scenarios. Trade liberalization may negatively affect the resource under open access or even under limited access fisheries management regimes.[14] For the society at large, trade will always be beneficial if the resources are well managed, although there will be redistribution effects within each society.

International Markets: Seafood Safety

There are many food safety issues involved in the seafood markets and trade. Some stem from the food system and others arise from the natural environment from which the products are raised or captured. Health risks from seafood consumption are varied and include the potential for a wide range of episodic events—including bacterial and parasitic illnesses and histamine poisoning—as well as concerns about repeated exposure to high levels of mercury and PCBs. The economic issues which result are equally varied, complex, and important for public and private decision making (see Cato[15] for an overview). Some health issues stem from water quality problems, e.g., shellfish contamination from algal blooms or local waste water treatment difficulties. Other seafood-borne illnesses result from processing, handling, inadequate refrigeration, and spoilage. The public is concerned about the health benefits and risks involved in seafood consumption.[16] However, consumers have difficulty estimating health risks in the food they consumed even when given seafood advisories.[17] Consumers are flooded with information and misinformation related to the health risks of farmed seafood consumption, affecting both purchase decisions and consumer welfare.[16–19] Evidence is almost universal that information widely available to consumers is inadequate to promote informed consumption of farmed seafood.[19] Government agencies and nonprofit organizations often provide conflicting guidance, and research repeatedly demonstrates suboptimal consideration of long-term health risks and benefits related to farmed seafood.[19,20] Indeed, a recent National Academies study concludes that "research is needed to develop and evaluate more effective communication tools for use when conveying the health benefits and risks of seafood consumption."[19] Also there is some experimental evidence that public communication programs could reduce the uncertainty by providing the accurate health benefit and risk information.[21]

International Seafood Markets: Ecolabeling

Seafood ecolabeling programs that certify fisheries based on sustainability allow consumers to signal preferences for healthier global oceans each time they purchase labeled seafood. Ecolabels can thus create an economic incentive for environmental improvements.[3,22,23] Ecolabeling is an increasingly important tool used in the promotion of sustainable fishery products around the world. Whether the consumer is actually paying a price premium for ecolabeled products is of fundamental importance as it indicates a return on the investment of sustainable practices, providing an incentive for producers to undertake such practices. A portion of consumers in the United States and select European countries have a statistically significant preference for seafood with ecolabels relative to unlabeled

Marine—Pollution

seafood.[24–27] In a study of consumers' noneconomic motivations to purchase ecolabeled seafood in Europe, there is a relationship between consumers' preference for ecolabeled fish and stated beliefs about the level of fisheries regulation and fish stocks.[28] Moreover, French consumers' taste for ecolabeled seafood is a function of perceptions of the fishing industry.[29] And most notably, there is a statistically significant premium of 14.2% for ecolabeled frozen processed pollock in the UK market.[30] Altogether, these studies jointly imply the potential for successful market differentiation for ecolabeling programs and possible incentives for sustainable fisheries practices. Whether these modest price premiums translate into actual changes in fishing practices is still being debated and researched.

International Seafood Markets: Attributes

Consumers value a wide array of seafood attributes, ranging from species, taste, texture, color, freshness, nutritional content, and product form to less tangible attributes such as country of origin or the presence of an ecolabel.[25,27,31] Moreover, attributes of fish also have value for the producer.[32–40] Some individual attributes may be more highly valued than others, and particular combinations of attributes lead to higher or lower valued fish at the producer level. For example, quality (resulting from handling during catch) or size of fish may segment fish within a particular fishery into different fresh or processed product forms.[40] Results of several of these studies have shown that quality matters; in particular, that there is a higher return for the raw fish product that is of sufficiently high quality to be put into a higher valued market segment. Quality issues can thus be an additional source of economic inefficiency in fisheries if poor fish quality results from management systems that provide incentives for long trips and poor treatment of fish. Under regulated open access, a larger share of the raw fish product can flow toward the inferior low-value market rather than the high-value fresh market.[41] An alternative is for processors to add value by breading into fish sticks to cover poor fish quality and create some direct employment in fish processing.[40] However, this strategy may reduce the overall value of the fishery for coastal communities and the fishermen when the fish is relegated to lower-valued product forms.

Overall, globalization and the expansion of the international seafood trade have often led to market integration. Competition from aquaculture, in particular, has tended to decrease prices, an undeniably good outcome for consumers but a challenge to capture fisheries.[42] This competition creates an even greater sense of urgency for wild fisheries to adopt effective management institutions that organize production efficiently and produce sustainable profits. Market integration also reduces opportunities for price compensation from ecological and technological disasters. Hypoxia (low dissolved oxygen), for example, has decreased

profits for the shrimp fishery in North Carolina, which makes up a small portion of the U.S. total domestic shrimp production.[43] In the much larger Gulf of Mexico shrimp fishery, which also experiences hypoxia and was affected by the recent Deepwater Horizon oil spill, price data indicate that the market is integrated with imports; producer losses from recent supply disruptions likely have not been offset even partially by price increases.[44] Pushing back against market integration are attempts to segment the seafood market through country-of-origin labeling, ecolabeling, differentiating wild from farmed fish, and promotion of local seafood.

FAO Fish Price Index

Seafood is a critical contributor to livelihoods and food security.[3] As world food prices hit an all-time high in February 2011 and are still almost two and a half times those in 2000, there is an increased awareness that prices are an important indicator of food scarcity. The FAO, which has for a long time compiled prices for other major food categories, recently began tracking seafood prices, working together with a team of researchers to develop the FPI.[45] The FPI can facilitate understanding of seafood crises and may assist in averting them. The index uses a similar format as price indices for other foodstuffs, all of which are published in the FAO Food Outlook (http://www.fao.org/giews/english/fo/index.htm).

The FPI relies on trade statistics because seafood is a highly traded commodity, these data are easily updated, and trade data can provide a timely proxy for domestic seafood prices that are difficult to observe in many regions and costly to update with global coverage. Calculations of the extent of price competition in different countries support the plausibility of reliance on trade data. The FPI can also be separated into subindices according to production technology, fish species, or region, providing a valuable tool for understanding and tracking development in different segments of the seafood market. In Fig. 3, the FPI is shown, as well as separate indices for capture fisheries and aquaculture. Figure 3 suggests increased scarcity of capture fishery resources in recent years but growth in aquaculture that is keeping pace with demand.

CONCLUSIONS

The international seafood trade will likely continue to grow. As aquaculture production expands in developing countries, it will continue to be exported to developed countries and will likely experience growth in exports to consumers in developing countries. Production from capture fisheries is unlikely to grow overall, although improvements in fisheries management in some countries (and continued degradation in fisheries stocks in other countries) may change the

Marine—Pollution

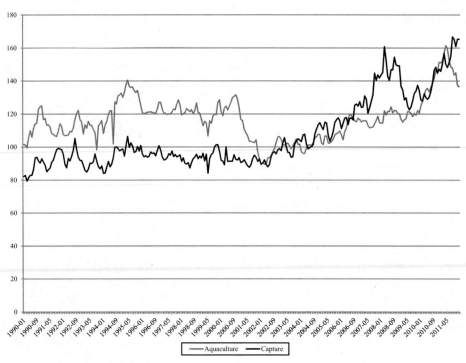

Fig. 3 FAO Fish Price Index, Monthly 1990–2011. Aquaculture v. Capture.
Source: Tverterås et al.[45]

relative mix of seafood from capture fisheries available on the international market. Increasing affluence of consumers in emerging markets and increasing populations will lead to continued demand growth. Thus, to promote a future in which the international markets maintain access to seafood and developing countries have seafood as a source of food security, a challenge will be to create economic incentives for sustainability. In other words, institutions are required that allow the market to operate efficiently such that natural resource prices reflect the costs of sustainability.

REFERENCES

1. Food and Agricultural Organization (FAO). *The State of the World Fisheries and Aquaculture*. Rome, 2010.

2. Food and Agricultural Organization (FAO). *The State of the World Fisheries and Aquaculture* 2008. Rome, 2009.

3. Smith, M.D.; Roheim, C.A.; Crowder, L.B.; Halpern, B.S.; Turnipseed, M.; Anderson, J.L.; Asche, F.; Bourillón, L.; Guttormsen, A.G.; Khan, A.; Liguori, L.A.; McNevin, A.; O'Connor, M.; Squires, D.; Tyedmers, P.; Brownstein, C.; Carden, K.; Klinger, D.H.; Sagarin, R.; Selkoe. K.A. Sustainability and Global Seafood. Science, **2010**, *327*, 784–786.

4. Asche, F. Farming the Sea. Mar. Resour. Econ., **2008**, *23* (4), 527–547.

5. Food and Agricultural Organization (FAO). 2012. Rome. FISHSTAT. http://www.fao.org/fishery/topic/16140/en

6. Anderson, J.L.; Asche, F.; Tveterås, S. World Fish Markets. In *Handbook of Marine Fisheries Conservation and Management*. Grafton, R.Q., Hilborn, R., Squires, D.,

7. Delgado, C.L.; Wada, N.; Rosegrant, M.; Meijer, S.; Ahmed, M. *Fish to 2020: Supply and Demand in Changing Global Markets;* International Food Policy Research Institute: Washington, DC and WorldFish Center, Penang, Malaysia, 2003.

8. Scott, A. The fishery: the objectives of sole ownership. J. Political Econ. **1955**, *63*, 116–124.

9. Smith, V.L. Economics of Production from Natural Resources. Am. Econ. Rev. **1968**, 409–431.

10. Smith, V.L. On Models of Commercial Fishing. J. Political Econ. **1969**, *77*, 181–198.

11. Anderson, J.L. *The International Seafood Trade*; Cambridge, Woodhead Publishing: 2003.

12. Brander, J.A.; Taylor, M.S. International Trade and Open-Access Renewable Resources: The Small Open Economy Case. Can. J. Econ. **1997a**, *30*, 526–552.

13. Brander, J.A.; Taylor, M.S. International Trade between Consumer and Conservationist Countries. Resour. Energy Econ. **1997b**, *19*, 267–298.

14. OECD. *Liberalising Fisheries Markets: Scope and Effects*; Paris: OECD, 2003.

15. Cato, J.C. *Seafood Safety - Economics of Hazard Analysis and Critical Control Point (HACCP) programmes*. FAO FISHERIES TECHNICAL PAPER – 381. Rome: Food and Agriculture Organization of the United Nations, 1998, http://www.fao.org/docrep/003/x0465e/x0465e00.htm

16. Roosen, J.; Marette, S.; Blancemanche, S. Does Health Information Matter for Modifying Consumption? A Field Experiment Measuring the Impact of Risk Information on Fish Consumption. Rev. Agric. Econ. **2009**, *31*, 2–20.

Tait, M. and Williams, M. Eds.; Oxford University Press: Oxford, 2010.

17. Shimshack J.P.; Ward, M.B.; Beatty, T.K.M. Mercury Advisories: Information, Educations, and Fish Consumption. J.Environ. Econ. Manag. **2007**, *53*, 158–179.

18. Roheim C.A. An Evaluation of Sustainable Seafood Guides: Implications for Environmental Groups. Mar. Resour. Econ. **2009**, *24*, 301–310.

19. Nesheim, M.C.; Yaktine, A.L., Eds.; *Seafood Choices: Balancing Benefits and Risks*; National Academy of Sciences, Committee on Nutrient Relationships in Seafood: Selections to Balance Benefits and Risks, Food and Nutrition Board, M.C. National Academies Press: Washington, DC, 2007.

20. Leiss, W.; Nicol, A.M. A Tale of Two Food Risks: BSE and Farmed Salmon in Canada, J. Risk Res. **2006**, *9*, 891–910.

21. Marette, S.; Roosen, J.; Blanchemanche, S. The Choice of Fish Species: An Experiment Measuring the Impact of Risk and Benefit Information. J. Agric. Resour. Econ. **2008**, *33*, 1–18.

22. Marine Stewardship Council (MSC). 2011. Marine Stewardship Council Annual Report 2010/2011. http://www.msc.org/documents/msc-brochures/annual-report-archive/annual-report-2010-11-english/view (accessed April 2012).

23. Roheim, C. The Economics of Ecolabelling, In *Seafood Ecolabelling: Principles and Practice*, Ward, T., Phillips, B. Eds.; Blackwell Publishing: Oxford, UK, 2008; p. 38–57.

24. Wessells, C.R., Johnston, R.; Donath, H. Assessing Consumer Preferences for Ecolabeled Seafood: The Influence of Species, Certifier, and Household Attributes. Am. J. Agric. Econ. **1999**, *81*, 1084–1089.

25. Johnston, R.J.; Wessells, C.R.; Donath, H.; Asche, F. Measuring Consumer Preferences for Ecolabeled Seafood: An International Comparison. J. Agric. Resour. Econ. **2001**, *26*, 20–39.

26. Johnston, R.J.; Roheim, C.A. A Battle of Taste and Environmental Convictions for Ecolabeled Seafood: A Contingent Ranking Experiment. J. Agric. Resour. Econ. **2006**, *31*, 283–300.

27. Jaffry, S.; Pickering, H.; Ghulam, Y.; Whitmarsh, D.; Wattage, P. Consumer Choices for Quality and Sustainability Labelled Seafood Products in the UK. Food Policy **2004**, *29*, 215–228.

28. Brécard, D.; Hlaimi, B.; Lucas, S.; Perraudeau, Y.; Salladarré, F. Determinants of demand for green products: An application to ecolabel demand for fish in Europe. Ecol. Econ. **2009**, *69*, 115–125.

29. Salladarré, F.; Guillotreau, P.; Perreudeau, Y.; Monfort, M.C. The Demand for Seafood Ecolabels in France. J. Agricu. Food Ind. Organ. **2010**, *8* (1), Article 10. doi: 10.2202/1542-0485.1308.

30. Roheim, C.; Asche, F.; Santos, J. The Elusive Price Premium for Ecolabeled Products: Evidence from Seafood in the U.K. Retail Sector, J. Agric. Econ. **2011**, *62*, 655–668.

31. Holland, D.; Wessells, C.R. Predicting Consumer Preferences for Fresh Salmon: The Influence of Safety Inspection and Production Method Attributes. Agricu. Resour. Econ. Rev. **1998**, *27* (1), 1–14.

32. Gates, J. Demand Price, Fish Size and the Price of Fish. Can. J. Agric. Econ. **1974**, *22* (1), 1–22.

33. Anderson, L. Optimal Intra- and Interseasonal Harvesting Strategies when Price Varies with Individual Size. *Mar. Resour. Econ.* **1989**, *6*, 145–162.

34. Larkin, S.; Sylvia, G. Intrinsic Fish Characteristics and Production Efficiency. Am. J. Agric. Econ. **1999**, *81* (1), 29–43.

35. McConnell, K.; Strand, I. Hedonic Prices for Fish: Tuna Prices in Hawaii. Am. J. Agric. Econ. **2000**, *82*, 133–144.

36. Carroll, M.; Anderson, J.; Martinez-Garmendia, J. Pricing U.S. North Atlantic Bluefin Tuna and Implications for Management. *Agribusiness* **2001**, *17*, 243–254.

37. Asche, F.; Hannesson, R. Allocation of Fish between Markets and Product Forms. Mar. Resour. Econ. **2002**, *17*, 225–238.

38. Fong, Q.S.W.; Anderson, J.L. International Shark Fin Markets and Shark Management: An Integrated Market Preference-Cohort Analysis of the Blacktip Shark (*Carcharhinus limbatus*). Ecol. Econ. **2002**, *40*, 117–130.

39. Kristofersson, D.; Rickertsen, K Efficient Estimation of Hedonic Inverse Input Demand Systems. Am. J. Agric. Econ. **2004**, *86* (4), 1127–1137.

40. Roheim, C.; Gardiner, L.; Asche, F. Value of Brands and Other Attributes: Hedonic Analysis of Retail Frozen Fish in the UK. Mar. Resour. Econ. **2007**, *22* (3), 239–253.

41. Homans, F.R.; Wilen, J.E. Markets and rent dissipation in regulated open access fisheries. J. Environ. Econ. Manag. **2005**, *49*, 381–404.

42. Anderson J.L. Aquaculture and the future: why fisheries economists should care. Mar. Resour. Econ. **2002**, *17*, 133–151.

43. Huang, L.; Nichols, L.A.B.; Craig, J.K.; Smith, M.D. Measuring welfare losses from hypoxia: the case of North Carolina brown shrimp. Mar. Resour. Econ. **2012**, *27*, 3–23.

44. Asche, F.; Bennear, L.S.; Oglend, A.; Smith, M.D. U.S. Shrimp Market Integration. *Marine Resource Economics*, **2012**, *27*, 181–192.

45. Tveterås, S.; Asche, F.; Bellemare, M.F.; Smith, M.D.; Guttormsen, A.G.; Lem, A.; Lien, K.; Vannuccini, S. Fish Is Food - The FAO's Fish Price Index. PLoS ONE. **2012**, *7* (5), e36731. doi:10.1371/journal.pone.0036731.

Marine—Pollution

Ocean–Atmosphere Interactions

Vasubandhu Misra
Department of Earth, Ocean and Atmospheric Science and Center for Ocean-Atmospheric Prediction Studies, Florida State University, Tallahassee, Florida, U.S.A.

Abstract
In this entry we define ocean–atmosphere interaction in terms of exchange of fluxes at the interface of ocean–atmosphere boundary and through compensatory dynamical circulations that maintain the observed climate of the planet. We describe the strong feedback between the atmosphere and ocean that is characterized in the fluxes, their global distribution of the fluxes, and diagnosis of the feedbacks.

INTRODUCTION

Ocean–atmosphere interaction commonly refers to the exchange of energy at the interface of the ocean and atmosphere. However, besides energy, there is also exchange of mass including that of fresh water (precipitation, runoff, melting of sea ice, evaporation) and of inert and sparingly soluble gases in seawater [e.g., oxygen (O_2), methane (CH_4), nitrous oxide (N_2O), carbon dioxide (CO_2)], which are quite significant even if they seem small on a unit area basis because of the extent of the ocean surface. We will, however, limit our discussion to the exchange of energy and freshwater. The energy exchange usually takes the form of exchange of heat (sensible heat), moisture (latent heat), and momentum (wind stress) at the overlapping boundary of the atmosphere and the ocean. The exchange of heat, moisture, and momentum are represented usually as fluxes, which are defined as the rate of exchange of energy per unit surface area of ocean–atmosphere interface. So the wind stress is the flux of horizontal momentum imparted by the atmospheric wind to the ocean. Similarly the latent heat flux refers to the rate at which energy associated with phase change occurs from the ocean to the atmosphere. Likewise the sensible heat flux refers to exchange of heat (other than that due to phase change of water) by conduction and/or convection. There is also heat flux from precipitation, which comes about as a result of the difference in temperature of the raindrops and the ocean surface. The heat flux from precipitation could become important in a relatively wet climate.

In the following sub-sections we discuss air–sea feedback, their global distribution, the common practices to diagnose this feedback, and the dynamical implications of ocean–atmosphere interactions, with concluding remarks in the final section.

AIR–SEA FEEDBACK

The surface energy budget of the ocean mixed layer, of which sensible heat flux, latent heat flux, precipitation heat flux are a part, dictates the temperature of the sea surface, which by virtue of its role in determining the stability of the lower atmosphere and the upper ocean dictates the fluxes across the interface. It may be noted that the surface mixed layer of the ocean refers to the depth up to which surface turbulence plays an important role in mixing so that the density is approximately the same as the surface. The entire mixed layer is active in transferring heat to the ocean–atmosphere interface. This forms a feedback loop between sea surface temperature (SST), stability of the lower atmosphere and the upper ocean, and the air–sea fluxes. Therefore determination of air–sea fluxes is a form of diagnosis of the coupled ocean–atmosphere processes. Similarly, the fresh water flux is part of the ocean surface salinity budget that dictates the stability of the upper ocean, which feeds back to the air–sea fluxes. For example, the fresh water flux from river discharge into the coastal ocean sometimes results in forming a relatively thin fresh water "lens" or an ocean barrier layer[1] that stabilizes the vertical column and inhibits vertical mixing in the upper ocean. This ocean barrier layer is then relatively easily modulated by the atmospheric variations, thus modulating the SST and, therefore, the air–sea fluxes. The influence of cloud radiation interaction on SST is another form of air–sea feedback. Several studies,[2–4] for example, note the existence of positive cloud feedback on SST in the summer time over the North Pacific (~35°N). This feedback refers to the reduction of SST under enhanced maritime stratiform clouds in the boreal summer season that reduces the downwelling shortwave flux and cools the mixed layer. In turn, the cooler SST favors further enhancement of the stratiform clouds affecting the stability of the atmospheric

Encyclopedia of Natural Resources DOI: 10.1081/E-ENRW-120048428

boundary layer. Wind Induced Surface Heat Exchange (WISHE) phenomenon is another form of air–sea feedback, which refers to the positive feedback between the surface (sensible and latent) heat fluxes with wind speed that results in increased fluxes leading to cooling of the SST as the wind speed increases. These negative correlations are apparent at large spatial scales across the subtropical oceans[5] as well as in mesoscale features (e.g., hurricanes).[6] The asymmetry of the inter-tropical convergence zone residing largely in the northern hemisphere in the eastern Pacific and in the eastern Atlantic Ocean was explained in terms of WISHE.[7]

DISTRIBUTION OF AIR–SEA FLUXES

Direct measurements of air–sea flux are few, limited both in time and in space. These measurements of air–sea fluxes are important, however, for developing, calibrating, and verifying the estimated air–sea flux from parameterization schemes.[8,9] The parameterization schemes for air–sea fluxes use state variables of the atmosphere and the ocean (e.g., wind speed, temperatures, humidity) to estimate the fluxes. A commonly used parameterization scheme for air–sea fluxes is the "bulk-aerodynamic" formula. This is based on the premise that wind stress is proportional to the mean wind shear computed between surface and 10 m above surface, and sensible heat flux and latent heat flux are proportional to the vertical temperature and moisture gradients computed between surface and 2 m above surface. As a result, air–sea fluxes have been computed globally and regionally from a variety of analyzed and regionally observed or analyzed atmospheric and oceanic states leading to a number of air–sea flux intercomparison studies.[10–12] These intercomparisons provide insight into the uncertainty of estimating air–sea fluxes as well as revealing salient differences in the state variables used in the parameterization scheme. For example, Smith et al.[10] found that in many regions of the planet the differences in surface air temperature and humidity amongst nine different products of air–sea fluxes had a more significant impact than the differences in the surface air wind speed (at 10 m) on the differences in the air–sea fluxes. Climatologically, large values of sensible heat flux are observed in the winter along the western boundary currents of the middle-latitude oceans, when cold continental air passes over the warm ocean currents (e.g., Gulf Stream). In the tropics and in the sub-tropical eastern oceans, the sensible heat flux is usually small. In the former region, climatologically, the wind speeds and the vertical temperature gradients between the surface and 2 m above the surface are weak. In the sub-tropical eastern oceans, with prevalence of upwelling, the SSTs are relatively cold leading to generally smaller sensible heat flux. The latent heat flux is observed climatologically to be large everywhere in the global oceans

relative to the sensible heat flux, with exceptions over polar oceans in the winter season. The ratio of sensible to latent heat flux, called the Bowen ratio, has a latitudinal gradient with a higher (smaller) ratio displayed in the polar (tropical) latitudes. This gradient of the Bowen ratio largely stems from the decreasing SST with latitude that affects the moisture-holding capacity of the overlying atmosphere and thereby affecting the latent heat flux.

The higher Bowen ratio may lead one to believe that air–sea coupling is weak in regions of cold SST. However, several studies[13,14] reveal that this is not the case. Using satellite-based (scatterometer) winds and SST that affords high space–time resolution, a robust relationship (positive correlation) between surface wind stress and SST is revealed along major ocean fronts that display strong SST gradients. Such a positive correlation between SST and surface wind speed is suggestive of a vertical momentum-mixing mechanism, which allows atmospheric wind to adjust to SST changes across major ocean fronts.[15,16] In other words, static stability of the lower atmosphere is reduced as the SST warms, which results in intensifying the turbulent mixing of the fast moving air aloft that causes the surface wind to accelerate. Such positive correlations have been observed in many regions of the global oceans at relatively small spatial scales.[14] On the other hand, negative correlations of wind speed with SST is suggestive of the so-called WISHE phenomenon discussed earlier.

DIAGNOSIS OF AIR–SEA FEEDBACK

Cayan[5] showed that vast regions of the middle-latitude ocean surface temperature variability is forced by the atmospheric variations. He showed this by comparing local, simultaneous correlations of the monthly mean values of latent heat and sensible heat fluxes with time tendency of SST that capture this interaction at spatial scales, which span a major portion of the North Atlantic and Pacific Oceans. The simultaneous correlation of SST with heat flux is shown to operate at a more local scale. Similarly, Wu et al.[17] showed that the atmospheric variability in the eastern equatorial Pacific is forced largely by the underlying SST variations, where simultaneous correlations of latent heat flux with tendency of SST is positive. This is in contrast to the tropical western Pacific Ocean where the correlations of latent heat flux with time tendency of SST is negative, suggesting that atmospheric variability is forcing the SST variability.

DYNAMICAL INTERACTIONS

Ocean–atmosphere interaction also happens through dynamics as illustrated in the case of Asian monsoon.[18] Here it is shown that in order to maintain the net heat balance of the coupled ocean–land–atmosphere Asian monsoon system, the ocean transports heat to the winter hemisphere of the Indian Ocean as comparable heat is released in the atmosphere of the summer hemisphere from the monsoonal convection. Similarly, Czaja and Marshall[19] demonstrate that atmospheric heat transport from tropical to higher latitudes is comparably higher to that by the oceans to maintain the observed latitudinal gradients of temperature and hence the general circulation of the Earth. In other words, in both these examples, the ocean and the atmosphere is acting in tandem to maintain the energy balance of the climate system of the planet, exhibiting a manifestation of the coupled ocean–atmosphere processes. Another prime example of ocean–atmosphere interaction is the natural variability of the El Niño and the Southern Oscillation (ENSO), one of the largest and most well-known interannual variations of the Earth's climate system, manifesting most prominently as the SST variability in the eastern equatorial Pacific Ocean. ENSO is a result of ocean dynamics and air–sea interaction processes,[20,21] which has a quasi-periodic oscillation with a period in the range of 2–7 years that affects the global climate variations.

CONCLUSIONS

Ocean–atmosphere interaction comprises a critical part of the Earth's physical climate system. ENSO variations that affect the global climate variability was first explained in terms of air–sea feedback in the equatorial eastern Pacific Ocean in the late 1960s. The first forays of predicting seasonal climate in the 1970s came from the hypothesis that through air–sea interaction, the persistent tropical SST anomalies would impart climate memory to the overlying atmosphere. Ever since, steady progress has been made in modeling (or parameterizing) and observing air–sea fluxes. Our growing understanding of the importance of air–sea interaction on observed climate variability has also led us to wean away slowly from a reductionist approach of developing component (such as atmosphere, land, ocean) models in isolation to coupled climate numerical models. The air–sea interaction happens not only at the interface of the atmosphere–ocean boundary through fluxes (defined as the amount of exchange of energy or mass per unit area) but also through compensatory dynamical circulations to maintain the observed climate of the planet.

ACKNOWLEDGMENTS

This work is supported by grants from NOAA, USGS, and CDC. However, the contents of the paper are solely the responsibility of the author and do not represent the views of the granting agencies.

REFERENCES

1. Lukas, R.; Lindstrom, E. The mixed layer of the western equatorial Pacific Ocean. J. Geophys. Res. **1991**, *96*, 3343–3457.
2. Norris, J.E.; Zhang, Y.; Wallace, J.M. Role of low clouds in summertime atmosphere-ocean interactions over the North Pacific. J. Clim. **1998**, *11*, 2482–2490.
3. Park, S.; Deser, C.; Alexander, M.A. Estimation of the surface heat flux response to sea surface temperature anomalies over the global oceans. J. Clim. **2005**, *18*, 4582–4599, DOI: 10.1175/JCLI3521.1.
4. Wu, R.; Kinter III, J.L. Atmosphere-ocean relationship in the midlatitude North Pacific: Seasonal dependence and east-west contrast. J. Geophys. Res. **2010**, *115*, D06101, DOI: 10.1029/2009JD012579.
5. Cayan, D.R. Latent and sensible hat flux anomalies over the northern oceans: Driving the sea surface temperature. J. Phys. Oceanogr. **1992**, *22*, 859–881.
6. Emanuel, K.A. An air-sea interaction theory for tropical cyclones. Part I. J. Atmos. Sci. **1958**, *43*, 585–604.
7. Xie, S.-P.; Philander, S.G.H. A coupled ocean-atmosphere model of relevance to the ITCZ in the eastern Pacific. Tellus. **1994**, *46A*, 340–350.
8. Weller, R.A.; Anderson, S.P. Temporal variability and mean values of the surface meteorology and air–sea fluxes in the western equatorial Pacific warm pool during TOGA COARE. J. Clim. **1996**, *9*, 1959–1990.
9. Fairall, C.W.; Bradley, E.F.; Rogers, D.P.; Edson, J.B.; Young, G.S. The parameterization of air–sea fluxes for Tropical Ocean-Global Atmosphere Coupled-Ocean Atmosphere Response Experiment. J. Geophys. Res. **1996**, *101*, 3734–3764.
10. Smith, S.R.; Hughes, P.J.; Bourassa, M.A. A comparison of nine monthly air-sea flux products. Int. J. Climatol. **2010**, *31*, 1002–1027.
11. Bourras, D. Comparison of five satellite-derived latent heat flux products to moored buoy data. J. Clim. **2006**, *19*, 6291–6313.
12. Chou, S.-H. A comparison of airborne eddy correlation and bulk aerodynamic methods for ocean-air turbulent fluxes during cold-air outbreaks. Bound-Lay. Meteorol. **1993**, *64*, 75–100.
13. Xie, S.–P. Satellite observation of cool ocean-atmosphere interaction. Bull. Amer. Soc. **2004**, 195–208
14. Chelton, D.B., Schlax, M.G.; Freilich, M.H.; Milliff, R.F. Satellite radar measurements reveal short-scale features in the wind stress field over the world ocean. Science **2004**, *303*, 978–983.
15. Wallace, J.M.; Mitchell, T.P.; Deser, C. The influence of sea surface temperature on surface wind in the eastern equatorial Pacific: Seasonal and interannual variability. J. Climat. **1989**, *2*, 1492–1499.
16. Hayes, S.P.; McPhaden, M.J.; Wallace, J.M. The influence of sea-surface temperature on surface wind in the eastern equatorial Pacific: Weekly to monthly variability. J. Clim. **1989**, *2*, 1500–1506.
17. Wu, R.; Kirtman, B.P.; Pegion, K. Local air–sea relationship in observations and model simulations. J. Clim. **2006**, *19*, 4914–4932.

18. Loschnigg, J.; Webster, P.J. A coupled ocean-atmosphere system of SST regulation for the Indian Ocean. J. Clim. **2000**, *13*, 3342–3360.

19. Czaja, A.; Marshall, J. The partitioning of poleward heat transport between the atmosphere and ocean. J. Atmos. Sci. **2006**, *63*, 1498–1511.

20. Cane, M.A.; Zebiak, S.E. A theory for El Niño and the Southern Oscillation. Science **1985**, *228*, 1085–1087

21. Philander, S.G. *El Niño, La Niña, and the Southern Oscillation*; Academic Press: 1990; 289pp.

**Marine—
Pollution**

Oceans: Observation and Prediction

Oscar Schofield
Josh Kohut
Grace Saba
Xu Yi
John Wilkin
Scott Glenn
Coastal Ocean Observation Laboratory, Institute of Marine and Coastal Sciences, School of Environmental and Biological Sciences, Rutgers University, New Brunswick, New Jersey, U.S.A.

Abstract

The ocean is a complex and difficult sample, and our understanding of how it operates and its future trajectory is poor. This has implications for humanity as increasing numbers are living in coastal zones. To better understand the ocean, it is necessary to adopt new sampling strategies that consist of coupling new observational technologies with numerical ocean models. New observational technologies are becoming available to oceanographers. Fixed assets include moorings, sea floor cables, and shore-based radars. These fixed assets are complemented with mobile platforms that provide spatial maps of subsurface data. The mobile platforms include profiling floats, gliders, and autonomous underwater vehicles. The observational data is complemented with numerical models. Many simulation models are becoming available to the community and the appropriate model is a function of the specific need of the user. Increasingly, observational data are used to constrain the models via data assimilation. The coupled observational and modeling networks will provide a critical tool to better understand the ocean.

INTRODUCTION

The oceans cover the majority of the Earth's surface and despite centuries of exploration they remain relatively unexplored. This gap of knowledge reflects the difficulty of collecting physical, chemical, and biological data in the ocean, as it is a harsh and unforgiving environment in which to operate. Despite centuries of ship-based exploration, the immense size and hazards associated with wind, waves, and storms limit the ability of humans to sustain a coherent global sampling network. Satellite and aircraft remote sensing approaches provide powerful tools to map global synoptic properties (Fig. 1); however, satellite systems largely provide information on the surface ocean. Fixed and mobile sensors deployed in the ocean can provide subsurface data, however, their numbers, while expanding, are limited and the technology still struggles with issues related to the onboard power availability and the number of available robust sensors. These sampling shortcomings have significant implications for human society especially as there is increasing evidence that the physics, chemistry, and biology of the ocean have changed over the last few decades. These changes reflect both natural cycles and the anthropogenic forcing from human activity.

Quantitatively understanding the relative importance of the natural and anthropogenic forcing in the ocean remains an open question, which needs to be resolved as the environmental impacts associated with human activity will increase, reflecting the growth of human populations.[1] The current projections suggest that human population growth at coastlines will be the most rapid and largest on the planet.[2] This will increase the importance of marine systems in national economies around the world making managing coastal systems critical. The close proximity of large populations will expose them to potential natural and man-made disasters associated with the oceans. These disasters include tsunamis, hurricanes, offshore industrial accidents, and human health issues such as outbreaks of waterborne disease. Our current capabilities to predict, respond to, manage, and mitigate these events is astonishingly poor. Improving our ability to observe and predict changes in the ocean will require technical improvements combined with an increased fundamental understanding of physical, chemical, and biological processes.

WHAT IS THE PATH FORWARD?

Improving our understanding and management of the ocean system will require an improved ability to map ocean properties in the present and improving our ability to forecast future ocean conditions. The ability to map ocean

Encyclopedia of Natural Resources DOI: 10.1081/E-ENRW-120048087

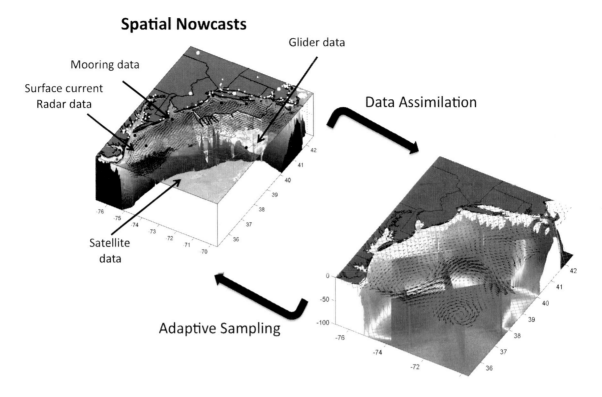

Spatial Nowcasts

Glider data

Mooring data

Surface current
Radar data

Data Assimilation

Satellite
data

Adaptive Sampling

4-D Numerical Forecasts

Fig. 1 An example of a coupled ocean observation and modelling network along the East coast of the United States. The spatial nowcast consists of observational data delivered in near real-time to shore. The nowcast consists of satellite data (sea surface temperature is shown), surface current radar (black arrows on the ocean), mooring data (black dots), subsurface data collected from gliders (water temperature shown here), and weather data collected by shore based stations (small white dots). The nowcast provides data to simulation models via data assimilation. The simulation model provides a 48-hour forecast, which is used to redistribute observational assets to provide an improved data set before the next forecast cycle.

properties will require a distributed portfolio of ocean infrastructure that will be linked together through an increasing number of ocean models.[3] The observation networks will collect quantitative data about the current status of the ocean. The forecasts are driven by numerical models that use current scientific understanding to project how the ocean will evolve. The combined observatory and numerical modelling capacity will improve our fundamental understanding and ability to respond to changes in the physics, chemistry, and biology of the marine systems.

The observations will assist the modelling efforts in several ways. Observations will provide data required to parameterize processes within the models. If the data are delivered in real-time they will be assimilated into the model to improve the predictive skill of the forecast. Finally, as new data are collected, they will be used to validate the predictive skill of the model. In turn the model forecasts will assist the observational efforts by providing forecasts that will allow scientists to adjust the spatial configuration and sampling rates of sensors to better sample future ocean conditions. These coupled systems are a rapidly maturing technology and builds off the more

mature science of weather forecasting, which has its roots in the early 19th century. The fundamental approaches are based on the seminal work of Lewis Fry Richardson who is considered the father of numerical weather prediction in the 1920s. These approaches are computationally intensive and it was not until the advent of electronic computers that the science moved forward to become an indispensable tool for humanity. Modern computer models use data as inputs collected from automated weather stations and weather buoys at sea. These instruments, observing practices and timing are standardized through the World Meteorological Organization.

Oceanographic efforts are evolving in a similar fashion where observations inform operational models. The weather and ocean models, most often run by federal agencies, provide forecasts that are used by scientists, the maritime industry, state and local communities. Most often they are used to issue warnings of unsafe conditions due to storms and high waves. The motivation for global standardized ocean forecast systems can be traced back to the sinking of the Titanic in 1912, which prompted the international community to call for development of systems

Marine—
Pollution

to improve safety at sea. Modern approaches and forecasting tools for the ocean did not mature until the 1980s and are rapidly evolving as the computing and ocean observation technologies are rapidly improving.

HOW ARE OBSERVATIONS MADE IN THE OCEANS?

Many platforms are available for making ocean measurements and, although the list below is not exhaustive, it provides a snapshot of the major platforms. The platforms carry sensors that can measure physical, chemical, and biological properties of the sea; however, most new novel sensors can only be carried on ships. A smaller number of sensors can be deployed on autonomous platforms and the discussion in this entry is focussed on those sensors that can be deployed on a variety of ocean platforms.

The most mature sensors are those that measure physical and geophysical variables, such as temperature, salinity, pressure, currents, waves, and seismic activity. Except for seismic variables, most of the physical sensors can be deployed on most of the platforms listed below. Many of the physical properties are the key variables in ocean numerical models. Currently, chemical sensors can measure dissolved gases (primarily oxygen) and dissolved organic material; however, recently the sensors capable of measuring nutrients (primarily nitrogen) are becoming commercially available. Biological sensors currently consist of optical and acoustic sensors. The optical sensors are used to provide information on the concentration, composition, and physiological state of the phytoplankton. Acoustic sensors can provide information on zooplankton to fish depending on the acoustic frequency band that is chosen.

Ships

The primary tool for oceanographers for centuries has been ships and will remain a central piece of infrastructure for the foreseeable future.[4] Ships are ideal as they are extremely flexible and allow teams to conduct experiments at sea. Ships are expensive to operate and must avoid hazardous conditions, such as storms, which limit the ability to make sustained measurements.

Satellites

Satellites are the most important oceanographic technology in modern times (beyond the ships).[5] Satellite observations have resulted in numerous advances in our fundamental understanding of the oceans[6] by resolving both global features associated with the mesoscale circulation of physical and biological properties. Satellite datum is fundamental to weather and ocean state prediction. Physical parameters available from space-based sensors include ocean surface temperature, wind speed and direction, sea surface height and topography, and sea ice distribution and thickness. Biological and chemical parameters can be derived from ocean colour radiometers.

High-Frequency Radar

High-frequency radar measures ocean surface current velocities over hundreds of square miles simultaneously. Each site measures the radial components of the ocean surface velocity directed towards or away from the site[7,8] and the estimated velocity components allow surface currents (upper meter of water column) to be estimated.[9] These systems are cost effective and have many applied uses.

Ocean Moorings

The modern ocean moorings grew out of the weather stations established in the 1940s. Since the 1960s, modern buoys have enabled a wide range of studies addressing the ocean's role in climate, weather, as well as providing insight into the biogeochemistry of the sea. Moorings provide the backbone to many of the global ocean networks studying ocean–atmosphere interactions and are the foundation for the global tsunami warning system network. They will continue to be a key element of ocean observing infrastructure for the foreseeable future.

Seafloor Cables

Scientists often require high bandwidth and power for sustained periods of time. Seafloor electro-optic cables provide a means for maintaining a sustained presence in the ocean. Cables have been deployed off the east and west coasts of the United States, Canada, Japan, and Europe. Many other countries are planning to deploy seafloor cables.

Drifters and Floats

Passive Lagrangian platforms are tools for creating surface and subsurface maps of ocean properties. These platforms are relatively inexpensive and thus allow thousands of these platforms to be deployed. Drifters have historically been a key tool for oceanography as evidenced by the important works by Benjamin Franklin[10] and Irving Langmuir.[11] The drifters can carry numerous sensors to create global maps of surface circulation. The first neutrally buoyant drifters were designed to observe subsurface currents.[12] The subsurface profiling drifters were enabled in the early 1990s with communication capabilities[13] and now anchor the international ARGO program, which has over 3000 floats deployed in the ocean.

Gliders are a type of autonomous underwater vehicle (Fig. 2) that use small changes in buoyancy in conjunction

Fig. 2 A Webb underwater glider being deployed in the Ross Sea Antarctica in February 2011.
Source: Photo credit Chris Linder.

with wings to convert vertical motion to horizontal motion, and thereby propel itself forward with very low-power consumption.[14] Gliders follow a saw-tooth path through the water, providing data on large temporal and spatial scales. They navigate with the help of periodic surfacing for Global Positioning System (GPS) fixes, pressure sensors, tilt sensors, and magnetic compasses. Using buoyancy-based propulsion, gliders have a significant range and duration, with missions lasting up to a year and covering over 3500 km of range.[15–17]

Propeller-driven autonomous underwater vehicles (AUVs) are powered by batteries or fuel cells and can operate in water as deep as 6000 m. Similar to gliders, AUVs relay data and mission information to shore via satellite. Between position fixes and for precise maneuvering, inertial navigation systems are often available onboard the AUV to measure the acceleration of the vehicle, and combined with Doppler velocity measurements, it is used to measure the rate of travel. A pressure sensor measures the vertical position. AUVs, unlike gliders, can move against most currents nominally at 3–5 knots, and can therefore systematically and synoptically survey a particular line, area, and/or volume.

WHAT NUMERICAL OCEAN MODELS ARE AVAILABLE?

Over the last 30 years, there have been significant developments in three-dimensional numerical models for the ocean.[18] Many models exist spanning from global ocean scales down to the scale of individual estuaries. Models vary in their coordinate system (linear, spherical, and others), resolution in space and time, complexity (i.e., number of state variables), and the parameterization of key processes within the model. There are several excellent texts that outline many of the details of numerical ocean modelling[19,20] and one key lesson is that the choice of particular modelling approach depends on its intended application and on the available computational resources. Although an exhaustive list of the ocean models is beyond the scope of this text, several classes are described in the following paragraphs.

Mechanistic models are simplified models used to study a specific process, and are used to provide insight into the underlying processes influencing the physics, chemistry, and biology of the ocean. These models are most often constructed as a learning tool in order to assess processes and feedbacks within marine systems.

Simulation models are complex and describe three-dimensional (3-D) ocean processes using the continuity and momentum equations. For this reason they are called the primitive equation models. These models can be used to simulate many processes, including ocean circulation, mixing, waves, and responses to external forces (such as storms). All these models are constructed using different assumptions. Additionally, the resolution of the models requires that the trade-offs of the computation burdens be measured against the processes that need to be simulated. For example, if the model must resolve mesoscale eddies, it will require the resolution of a few tenths of a degree of latitude and longitude. In contrast, most primitive equation climate models have much coarser horizontal resolution as they were designed to study large-scale hydrographic structure, climate dynamics, and water-mass formation over decadal time scales; however, for a specific question, there are climate models with sufficient resolution to resolve mesoscale eddies if one is willing to accept the computation cost. Simulations are also constructed for

Marine— Pollution

coastal systems and can resolve coastal currents, tides, and storm surges. Increasingly the biological and ocean chemistry models are being coupled into these 3-D simulation models. Although these biogeochemical models are rapidly improving, there is unfortunately no set of "primitive" equations yet capable for describing biological and chemical systems in the ocean; however, as these models evolve, they will be increasingly useful tools for managing living resources and water quality in the ocean.

Often models of varying resolutions are combined. Coarser-scale global or basin-scale models provide outer boundary inputs to higher resolution nested models, which allows a myriad of processes to be modelled with a lower computation burden but allow a range of processes to be simulated even if they require high resolution. This is often the case for coastal and continental shelf models. The advantage of this downscaling approach is that it allows basin scale models to resolve large-scale forcing that drives the regional to local-scale processes that are effectively modelled by a higher resolution model. The approach by which one links these models is a difficult problem and remains an area of active research.

Data assimilation is an approach by which model simulations are constrained by observations. For example, model calculations and observations of temperature and salinity can be compared, and then the model can be "adjusted" based on the mismatch. This is a difficult problem as 1) it represents an inverse problem (where a finite number of observations are used to estimate a continuous field), 2) many of the ocean processes of interest are non-linear, and 3) the observations and models both have unknown errors. Descriptions of data assimilation approaches for oceanography have been reviewed.[21,22] These approaches allow modellers to increase the forecast skill of their models by essentially keeping the models "on track" if the observations and data assimilation approaches can be provided in a timely fashion. Many in the ocean modelling community are focusing on using these approaches to increase model forecast skill as it determines how well these approaches will serve a wide range of science, commercial, and government needs.[23]

CONCLUSIONS

Ocean observation and modelling capabilities are rapidly diversifying and improving. These systems are increasingly linked by data assimilation approaches that when combined, provide a coupled observing and forecasting network. These approaches will increase the predictive skill of forecast models that in turn can serve a wide range of applications spanning from basic research to improving the efficiency of the maritime industry. The combined technologies will be critical to improving our understanding of the ocean today and the potential trajectory in the future.

ACKNOWLEDGMENTS

We acknowledge the support of the Office of Naval Research's Major University Research Program, the National Oceanic and Atmospheric Administration's Integrated Ocean Observing System, and the National Science Foundation's Ocean Observing Initiative.

REFERENCES

1. De Souza, R.; Williams, J.; Meyerson, F.A.B. Critical links: Population, health, and the environment. Popul. Bull. **2003**, *58* (3), 3–43.
2. Vitousek, P.M.; Mooney, H.A.; Lubchenco, J.; Melillo, J.M. Human domination of Earth's ecosystems. Science **1977**, *277*, 494–499, DOI: 10.1126/science.277.5325.494.
3. National Research Council. *Critical Infrastructure for Ocean Research and Societal Needs in 2030*; National Academy Press: Washington, D.C., 2011.
4. National Research Council. *Science at Sea: Meeting Future Oceanographic Goals with a Robust Academic Research Fleet;* National Academy Press: Washington, D.C., 2009.
5. Munk, W. Oceanography before, and after, the advent of satellites. In *Satellites, Oceanography and Society*, Halpern, D., Ed.; Elsevier Science: Amsterdam, 2000; 1–5.
6. Halpern, D., Ed. *Satellites, Oceanography and Society*; Elsevier Science: Amsterdam, 2000; 361 pp.
7. Crombie D.D. Doppler spectrum of sea echo at 13.56 Mc/s. Nature **1955**, *175*, 681–682.
8. Barrick, D.E.; Evens, M.W.; Weber, B.L. Ocean surface currents mapped by radar. Science **1977**, *198*, 138–144.
9. Stewart R.H.; Joy J.W. HF radio measurements of ocean surface currents. Deep Sea Res. **1974**, *21*, 1039–1049.
10. Franklin, B. Sundry marine observations. Trans. Am. Philos. Soc. **1785**, *1* (2), 294–329.
11. Langmuir, I. Surface motion of water induced by wind. Science **1938**, *87*, 119–123.
12. Swallow, J.C. A neutral-buoyancy float for measuring deep currents. Deep Sea Res. **1955**, *3*, 74–81.
13. Davis, R.E.; Webb, D.C.; Regier, L.A.; Dufour, J. The Autonomous Lagrangian Circulation Explorer (ALACE). J. Atmos. Ocean. Tech. **1992**, *9*, 264–285.
14. Rudnick, D.L.; Davis, R.E.; Eriksen, C.C.; Fratantoni, D.M.; Perry, M.J. Underwater gliders for ocean research. Mar. Technol. Soc. J. **2004**, *38*, 73–84.
15. Sherman, J.; Davis, R.E.; Owens, W.B.; Valdes, J. The autonomous underwater glider "Spray." IEEE J. Ocean. Eng. **2001**, *26*, 437–446.
16. Eriksen, C.C.; Osse, T.J.; Light, R.D.; Wen, T.; Lehman, T.W.; Sabin, P.L.; Ballard, J.W.; Chiodi, A.M. Seaglider: A long-range autonomous underwater vehicle for oceanographic research. IEEE J. Ocean. Eng. **2001**, *26*, 424–436.
17. Webb, D.C.; Simonetti, P.J.; Jones, C.P. SLOCUM: An underwater glider propelled by environmental energy. IEEE J. Ocean. Eng. **2001**, *26*, 447–452.
18. Bryan, K. A numerical method for the study of the circulation of the world ocean. J. Comput. Phys. **1969**, *4* (3), 347–376, DOI: 10.1016/0021-9991(69)90004-7.

Marine—
Pollution

19. Kantha, L.H.; Clayson, C.A. *Numerical Models of Oceans and Oceanic Processes*; International Geophysics Series, Elsevier: Amsterdam, 1995; 750 pp.

20. Haidvogel, D.; Beckmann, A. *Numerical Ocean Circulation Modeling*; Imperial College Press: London, 1999; 318 pp.

21. Wunsch, C. *The Ocean Circulation Inverse Problem*; Cambridge University Press: Cambridge, 1996; 442 pp.

22. Malanotte-Rizzoli P. *Modern Approaches to Data Assimilation in Ocean Modeling*; Elsevier Science: Amsterdam, Netherlands, 1996; 455 pp.

23. Pinardi, N.; Woods, J. *Ocean Forecasting: Conceptual Basis and Applications*; Springer Verlag: Berlin, Germany, 2002; 472 pp.

Marine— Pollution

Pesticide Contamination: Groundwater

Roy F. Spalding
Water Science Laboratory, University of Nebraska, Lincoln, Nebraska, U.S.A.

Abstract

Toxic pesticides and their transformation products are a concern from the standpoint of human health. They are also a risk to the environment in areas where contaminated groundwater enters surface water. There is a need to evaluate the environmental cost/benefit of safer product replacements used in conjunction with genetically altered crops. As new products are registered to replace more persistent and mobile pesticides, long-term fate studies on groundwater quality, including monitoring of the effect of transformation product, are necessary.

INTRODUCTION

Trace concentrations of most of the commonly used pesticides have been confirmed in groundwaters of the United States. Since groundwater is the source of 53% of the potable water, the more toxic pesticides and their transformation products are a concern from the standpoint of human health. Others are a risk to the environment in areas where contaminated groundwater enters surface water. Through toxicological testing, the USEPA has established Maximum Contaminant Levels (MCLs) or lifetime Health Advisory Levels (HALs) for several pesticides (Table 1).

The EPA also has a separate list of unregulated compounds, including newly registered pesticides and their transformation products, such as acetochlor and alachlor-ESA, that are presently being evaluated or being considered for toxicological evaluation. Based on the results of the EPAs National Pesticide Assessment,[2] 10.4% of 94,600 community systems contained detectable concentrations of at least one pesticide. Evaluation of these results led to an estimated 0.6% of rural domestic wells containing one or more pesticides above the MCL.

PESTICIDE USE

In the United States about 80% of pesticide usage is in agriculture. The remainder is used by industry, homeowners, and gardeners. About 500 million pounds of herbicide, 180 million pounds of insecticide, and 70 million pounds of fungicide were applied for agricultural purposes in 1993.[3] Several maps of the United States delineate usage patterns of several pesticides.[4] The majority of the triazine and amide herbicides are applied to fields in the north central corn belt states of Michigan, Wisconsin, Minnesota, Nebraska, Iowa, Illinois, Indiana, and Ohio. Commonly used organophosphorus insecticides are more heavily applied to fields in California and along the southeastern seaboard than in the northern corn belt. Carbamate and thiocarbamate pesticides are heavily used in potato growing areas of northern Maine, Idaho, the Delmarva Peninsula, and vegetable fields of California and the southeastern coastal states. Fungicide use is concentrated in high humidity and irrigated areas of the coastal states and to some extent along the Great Lakes and Mississippi River Valley. The fumigants carbon tetrachloride and ethylene dibromide (EDB) were used heavily in the past at grain storage elevators throughout the Midwest and elsewhere in the United States.

ASSOCIATED PESTICIDE BEHAVIOR IN SOILS AND WATER

Although pesticide use is a dominant factor in groundwater contamination, leaching variability among pesticides exhibiting similar behaviors is striking and explains why several heavily used pesticides seldom if ever are detected in groundwater. In general, pesticides within a class have similar chemical characteristics upon which soil leaching predictions can be made based on persistence, solubility, and mobility. Pesticide class relationships with soils and water transport described in the following text are detailed in Weber.[5] Individual frequencies of groundwater pesticide detection, in parenthesis next to commonly used products, are calculated from the Pesticide Groundwater Data Base (PGWDB)[4] and the National Water Quality Assessment (NAWQA) database.[6] High frequencies of detection identify those pesticides with a disposition to leach.

Encyclopedia of Natural Resources DOI: 10.1081/E-ENRW-120010368

Marine—Pollution

Table 1 U.S. maximum contaminant levels for drinking water

Organic chemical name	MCL (mg/L)	Organic chemical name	MCL (mg/L)	Organic chemical name	MCL (mg/L)
2,4,5-TP (Silvex)	0.05	Chlordane	0.002	Heptachlor	0.0004
2,4-D	0.07	Dalapon	0.2	Heptachlor Epoxide	0.0002
Alachlor	0.002	Dinoseb	0.007	Lindane	0.0002
Aldicarb	0.007	Diquat	0.02	Methoxychlor	0.04
Aldicarb sulfone	0.007	Endothall	0.1	Oxamyl (Vydate)	0.2
Aldicarb sulfoxide	0.004	Endrin	0.002	Picloram	0.5
Atrazine	0.003	Ethylene dibromide	0.00005	Simazine	0.004
Carbofuran	0.04	Glyphosate	0.7	Toxaphene	0.003
Carbon tetrachloride	0.005				

Source: Adapted from U.S. Environmental Protection Agency.[1]

Insecticides

Chlorinated hydrocarbons are one of the oldest chemical classes of insecticides. Some of the best-known compounds include aldrin, dieldrin, DDE, DDT, endrin, and toxaphene. Although banned since the 1960s, their extremely persistent nature precludes their detection in very trace quantities in groundwater of the upper Midwest. On the other hand, heavily used organophosphates like malathion, methylparathion, disulfoton, and others have been extensively surveyed during several groundwater monitoring studies and have not been detected. The organophosphate insecticides, parathion (not reported (NR), <1; % occurrence from PGWDB and % occurrence from NAWQA data), terbufos (<1, <1), fonofos (<1, <1), and chlorpyrifos (<1, <1), which are heavily used on corn and sorghum, were also seldom detected. Diazinon (1.1, 1.3), the common garden insecticide, is occasionally detected in groundwater. Generally, the organic phosphates are rapid degraders and are strongly retained on soils.

For the most part, carbamates and thiocarbamates are very sparingly soluble and exhibit low to moderate soil retention; however, a small number have high solubility and low soil retention. Most carbamates are characterized as having short longevities. Generally, pesticides in this group having half-lives of 30 days or more have the potential to leach. The thiocarbamates butylate (<1, <1) and EPTC (2.6, <1) are extensively used in agriculture and have relatively short half-lives. Aldicarb (<1, <1) and carbofuran (14.7, <1) are at the high end for solubility and longevity in their class. Their metabolites have been frequently detected beneath high use crops, such as potatoes in the potato growing regions of the United States.

The pyrethroid insecticides have low solubilities, short half-lives, and high soil retentions that make them unlikely to leach. Yet, permethrin (<1, <1) is occasionally detected in very trace quantities in groundwater.

Fungicides and Fumigants

Fungicides are non-volatile organometallic compounds with low aqueous solubility that inhibit growth of actinomycetes and many fungi. The best known fungicides zineb (not detected (ND), NR) and captan are zinc-based, and maneb (ND, NR) is manganese-based. Some, like bordeaux, are copper sulfate-based. Although their detection frequency is very low, fungicides have not been analyzed in many surveys.

Fumigants are very volatile halogenated compounds that generally are knifed below the soil surface. These compounds have high aqueous solubility and very low soil retention. The fumigants EDB and 1,3-dichloropropene have been frequently detected in the subsurface and in groundwater in high-use regions, such as California.[7] Ethylene dibromide and carbon tetrachloride were also used in grain storage facilities during the 1950s and 1960s. Spills, leaks, and improper handling resulted in 400 reported groundwater contamination sites in Kansas and Nebraska.

Herbicides

There are at least eight major chemical classes of herbicides. These include: quaternary N, basic, acidic, carboxylic acid, hydroxy and aminosulfonyl, amide and anilide, dinitroaniline, and phenylurea herbicides.[5] Several herbicide classes have similar behaviors with respect to soil and water.

Both quaternary N and dinitroaniline herbicides are very highly retained by soils and are not expected to be detected in groundwater. However, paraquat, pendimethalin, and trifluralin have been reported several times in groundwater. Their presence indicates that transport is dependent on factors not directly related to compound longevity, solubility, and mobility. Vertical transport by preferential flow through macropores is a commonly accepted mechanism used to explain these detections. In some instances, compounds have been described as preferentially transported attached to colloidal material.

Carboxylic, hydroxy, and aminosulfonyl acids, and thiocarbamate herbicides have very low to low soil retention and very short to moderate longevity. Thus, the more heavily used and persistent pesticides in these groups are the ones most generally detected in groundwater. They

Marine—Pollution

include the acids, dicamba (2.0, NR), picloram (2.5, <1), bromacil (1.8, 1.0), and dinoseb (1.4, <1).

Phenylurea herbicides have low to high soil retentions and short to moderate longevity. Linuron (16.7, <1) and diuron (<1, 1.9) are the most frequently detected in groundwater and both have moderately long half-lives ranging from 60 to 90 days.

Amide and anilide herbicides have low soil retention and short to moderate longevity. Several amide herbicides and their transformation products have been detected in groundwater. The commonly used amides in the Midwestern corn belt, namely alachlor (1.7, 2.7), metolachlor (<1, 12), propachlor (1.2, <1), and acetochlor (NR, <1), are the most frequent offenders because they are relatively persistent.

As suggested by the name, basic herbicides behave as bases. The group contains several subclasses including aniline, formamidine, imidazole, pyrimidine, thiadiazole, triazines, and triazole. Basic herbicides have low to high soil retention and very short to moderate longevity. Again, it is generally the most persistent and heavily used pesticides that are more frequently found in groundwater. The most frequently detected compounds in the group are the triazines, namely atrazine (5.6, 30), metribuzin (4.2, 1.9), cyanazine (2.0, 1.4), simazine (2.0, 14.8), and prometon (2.1, 11.6).

GROUNDWATER CONTAMINATION

It stands to reason that there are generally good associations between pesticide use and their detection in groundwater. Since groundwater flows very slowly at rates normally ranging from 0.1 ft/day to 3 ft/day, pesticide sources are generally very near the monitored well. Thus, high frequencies of triazine and acetamide detections are reported in the states of the northern corn belt. More fungicides and fumigants were detected in warm humid states of California and Florida where vegetable and fruit crops dominate the landscape. In an analysis of the 20 NAWQAs for pesticides, frequencies of pesticide detection in groundwater were significantly related to the estimated amount of agricultural use within a 1 km radius of the sampled site.[6] They also emphasized that pesticides were detected beneath both agricultural (60.4%) and urban areas (48.5%). Discontinued used pesticides have been detected numerous times in shallow aquifers.

In general, families of pesticides have similar chemical characteristics from which predictions have been made as to the product's potential for contamination of groundwater; however, differences in the leaching behavior of pesticides exhibiting similar chemistry can be appreciable and is the reason several heavily used pesticides are seldom, if ever, detected in groundwater.

MANAGEMENT OF POINT SOURCES OF GROUNDWATER CONTAMINATION

Important steps are being taken to reduce water quality pollution by pesticides occurring from spills and back siphoning events (point sources). Since it is easier to resolve point than non-point sources, laws have been enacted to eliminate contamination of surface water bodies, which may be in hydraulic contact with groundwater, from used pesticide containers and rinseate from chemical wash downs. Check valves are mandatory when pesticides are mixed and/or diluted and prevent backflow to groundwater. Soils at and adjacent to agrichemical supply facilities have been surveyed in several states and found to be highly contaminated with pesticide residues. The herbicides atrazine, alachlor, metolachlor, cyanazine, and metribuzin are the worst offenders from the standpoint of pesticide mass in the soils at sites in Wisconsin and Illinois.[4] Many of these sites and those in other states are now involved in soil cleanups, which are designed to protect underlying groundwater from further pollution.

MANAGEMENT OF NON-POINT SOURCES OF GROUNDWATER CONTAMINATION

Normal farm chemical applications of pesticides are generally considered potential non-point sources of groundwater contamination because they are dispersed over large areas ranging from fields to watersheds. Management strategies are in place to reduce leaching of field applied chemicals.[8] These strategies vary from regulatory restrictions to outright bans on application in areas deemed more vulnerable to leaching. Integrated pest management, fostered by the office of pesticide management at the USEPA, is designed to reduce chemical applications. The practice of banding applications has reduced amounts applied. Both target more efficient pesticides and genetically engineered plants sensitive only to specific herbicidal action have been and are being developed. These new pesticides and pesticide–plant combinations require less chemical than in the past, and the altered plants allow for pest control with more environmentally sensitive chemicals. The USEPA has announced a plan to reduce the mass of applied chemicals from commonly used triazines and amides that are frequently detected in groundwater.

Irrigation Management

Irrigation practices can influence pesticide leaching. Atrazine was vertically transported deeper and faster when using flood rather than sprinkler irrigation.[9] Sprinkler systems allow for much more uniform and efficient water management practices than furrow irrigation, and recent studies have shown that they reduce chemical leaching.[10] In Nebraska's Platte Valley[11] and in the Walnut Creek watershed in Iowa,[12] peak herbicide concentrations were strongly related to rapid flushing beneath drainage areas where surface water ponds during heavy rainfall events on the cropped fields. Application of excess irrigation water was also reported to increase herbicide leaching.[9,11]

FUTURE RESEARCH

More research is necessary to evaluate the health risks of transformation products from heavily used pesticides that are frequently detected in groundwater. Research needs to focus on precision application of pesticides to specific field problem areas as a potential mechanism to reduce chemical application.

There is a need to evaluate the environmental cost/benefit of safer product replacements used in conjunction with genetically altered crops. As new products are registered to replace more persistent and mobile pesticides, long-term fate studies, including the monitoring of the transformation product impact, on groundwater quality are necessary.

REFERENCES

1. U.S. Environmental Protection Agency. *Drinking Water Standards and Health Advisories*, EPA 822-B-00-001; USEPA Office of Water: Washington, DC, Summer, 2000.
2. U.S. Environmental Protection Agency. *National Survey of Pesticides in Drinking Water Wells: Phase I Report*, EPA 570/9-90-015; USEPA Office of Pesticide and Toxic Substance: Washington, DC, November, 1990.
3. Aspelin, A.L. *Pesticide Industry Sales and Usage, 1992 and 1993 Estimates*; Economic Analysis Branch Report 733-K-92-001; USEPA Office of Pesticide Programs, Biological and Economic Analysis Division, 1994; 1–37.
4. Barbash, J.E.; Resek, E.A. *Pesticides in Ground Water. Distribution, Trends, and Governing Factors*; Ann Arbor Press, Inc.: Chelsea, MI, 1996; 1–588.
5. Weber, J. Properties and behavior of pesticides in soils. In *Mechanisms of Pesticide Movement into Ground Water*; Honeycutt, R.C., Schabacker, D.J., Eds.; Lewis Publishers: London, 1994; 15–41.
6. Kolpin, D.; Barbash, J.E.; Gilliom, R.J. Pesticides in ground water of the United States, 1992–1996. Ground Water **2000**, *38*(6), 858–863.
7. Troiano, J.; Weaver, D.; Marade, J.; Spurlock, F.; Pepple, M.; Nordmark, C.; Bartkowiak, D. Summary of well water sampling in California to detect pesticide residues resulting from nonpoint-source applications. J. Environ. Qual. **2001**, *30*(2), 448–459.
8. Guyot, C. Strategies to minimize the pollution of water by *pesticides. In Pesticides in Ground and Surface Water*; Borner, H., Ed.; Springer: Berlin, 1994; 87–148.
9. Troiano, J.; Garretson, C.; Krauter, C.; Brownwell, J.; Huston, J. Influence of amount and method of irrigation water application on leaching of atrazine. J. Environ. Qual. **1993**, *22*, 290–298.
10. Spalding, R.F.; Watts, D.G.; Schepers, J.S.; Burbach, M.E.; Exner, M.E.; Poreda, R.J.; Martin, G.E. Controlling nitrate leaching in irrigated agriculture. J. Environ. Qual. **2001**, *30* (4), 1184–1194.
11. Spalding, R.F.; Watts, D.G.; Snow, D.D.; Cassada, D.A.; Exner, M.E.; Schepers, J.S. Herbicide loading to shallow ground water beneath Nebraska's management systems evaluation area. J. Environ. Qual. **2003**, *32*, 84–91.
12. Moorman, T.B.; Jaynes, D.B.; Cambardella, C.A.; Hatfield, J.L.; Pfeiffer, R.L.; Morrow, A.J. Water quality in walnut creek watershed: herbicides in soils, subsurface drainage and groundwater. J. Environ. Qual. **1999**, *28*(1), 35–45.

Marine—
Pollution

Pesticide Contamination: Surface Water

Sharon A. Clay
Plant Science Department, South Dakota State University, Brookings, South Dakota, U.S.A.

Abstract
Although other methods of pest control are important, pesticides continue to be an efficient, cost-effective method of management. The challenge is to select and apply pesticides in a manner that kills the target pest and does not harm non-target organisms or the environment. Sound and sensible management can reduce the probability of pesticides entering surface water.

INTRODUCTION

About 1 billion pounds of pesticide were applied in the United States in 1997.[1] This extensive use of pesticides has caused concern about pollution in the environment. Indeed, pesticides have been detected in groundwater, surface water, and rain. However, the amount found normally is less than 0.1% of the amount used and occurs in seasonal cycles. Management options that limit losses from target sites will be presented.

DETECTION FREQUENCY, CONCENTRATIONS, AND SEASONAL CYCLES OF PESTICIDES FOUND IN SURFACE WATER

Pesticide classification may be based on the type of pest controlled. Chemicals that control weeds (herbicides) accounted for 60% of the pesticide use in 1997.[1] Chemicals that control insects (insecticides) and plant diseases (fungicides) accounted for 13% and 7%, respectively, of the chemical use.[1] Rodenticides, fumigants, and nematicides represent other pesticide classes that account for the remaining 20% of use. In a typical year, agricultural lands receive about 80% of all pesticides applied while homeowners/gardeners (8%) and the commercial market (12%) apply the rest.

The United States Geological Survey (USGS) has sampled surface water for pesticide pollution since the late 1970s. Thousands of samples have been analyzed for about 100 pesticides or breakdown products (metabolites). Herbicides are the pesticide type most commonly found in rural streams.[2] Atrazine, a broadleaf herbicide ranked number one in total pounds applied in the United States from 1987 to 1997 (about 75 million lb/yr), has been detected most often.[1] Other herbicides found in surface water include simazine and cyanazine (used for broadleaf and grass control in corn) and alachlor and metolachlor (used for grass control in corn and soybean).[2] Most of these herbicides also ranked high in U.S. agricultural use between 1987 and 1997.[1] The amount of these herbicides found in streams has typically ranged from less than 1% of the amount applied (cyanazine, metolachlor, and alachlor) up to 3% (atrazine).[2] The five insecticides frequently detected were diazinon, carbaryl, chlorpyrifos, carbofuran, and malathion.[2]

Detection of cyanazine and alachlor in the U.S. environment will decline in the future. Cyanazine is no longer labeled for use in the United States as of 2002. Alachlor is being replaced by acetochlor. However, acetochlor was detected in surface water in its first year of general use (1994) with occurrence patterns similar to other herbicides of the same family.[3] The concentration detected was lower than alachlor due to lower application rates.

Pesticides have also been detected in urban streams.[4,5] In fact, the estimated contribution of insecticides to surface water contamination from urban and rural areas may be similar. Insecticides, such as malathion, carbaryl, and diazinon, are commonly used in urban settings for control of mosquitoes, turfgrass and garden insects, and termites.[4] Herbicides detected in urban streams were those commonly used for broadleaf weed control in lawns (i.e., 2,4-D, dicamba, and MCPA)[5] along with prometon and tebuthiuron, both used for total vegetation control in right-of-way areas.[4] Herbicides almost exclusively used in agricultural settings (atrazine, alachlor, metolachlor, and cyanazine) have also been detected in urban streams. Atmospheric depositions in sediment, rain, or snow or transport from agricultural watersheds upstream of urban settings are the most likely sources of these herbicides.

Pesticides in surface water have been detected throughout the year.[6] The greatest concentrations in rural streams are reported in the spring and early summer coinciding with agricultural applications.[7] The mean pesticide concentrations in most months are about 1 μg/L.[2] However, during peak usage, concentrations in some samples may

Encyclopedia of Natural Resources DOI: 10.1081/E-ENRW-120010192

Marine—Pollution

exceed 12 μg/L. In urban streams, the mean pesticide concentration is fairly stable (<0.5 μg/L) throughout the year[2] but can differ by area of the country. In the southern United States, pesticides may be applied both earlier and later in the year due to the longer growing season compared to the northern tier of states.

PESTICIDE TOXICITY IN SURFACE WATER

The U.S. Environmental Protection Agency[8] has established pesticide concentration criteria values for some pesticides for the protection of aquatic health. Pesticides can be toxic to aquatic invertebrates such as plankton[9] and vertebrates such as frogs and fish when above critical concentrations. The herbicides, atrazine [water quality criterion (WQC) – 1.8 μg/L] and trifluralin (WQC – 0.2 μg/L), and the insecticides, chlorpyrifos (WQC = 0.041 μg/L) and diazinon (WQC = 0.08 μg/L), were the pesticides that most often exceeded the aquatic health criteria in 37 rural streams.[2] In a survey of eight urban streams across the United States monitored from 1993 to 1995, 41 pesticides were detected. Simazine (WQC = 10 μg/L), prometon (no established criterion), atrazine, and diazinon were detected at all sites and 20 other pesticides were detected above their WQC in one or more of the streams.[4] The estimated number of days that the pesticide concentration exceeds the established standards varies by chemical.[2] Atrazine was estimated to exceed the standard in 15 rural streams from 1 day to 84 days (with an average of 36 days). In comparison, chlorpyrifos and azinphos-methyl (WQC = 0.01 μg/L) exceeded the standard in 8 streams, ranging from 1 day to 8 days (average of 3 days) and 1 day to 70 days (average of 13 days), respectively.

In addition, herbicides may be used to control plants in ponds and lakes. The reduction in vegetation may have an indirect effect on the number of weed-clinging invertebrates such as dragonflies or damselflies. However, removal of plants does not always decrease the numbers of aquatic organisms. For example, the number of aquatic organisms remain fairly constant or increase due to an increase in organisms that feed on plant debris after vegetation is controlled.[10]

FACTORS THAT AFFECT PESTICIDE MOVEMENT TO SURFACE WATER

Distance Between Application and Surface Water

The distance between pesticide application and surface water can greatly affect the likelihood of surface water contamination. In most cases, buffer zones of 50–100 ft are recommended on the pesticide label. Filter strips of grass or a dense cover crop at the field edge also reduces the pesticide concentration in the runoff in two ways.[11] First, filter strips slow water movement and allow soil particles to settle out of the runoff. Second, these areas are normally high in organic matter and can sorb the pesticide out of water. Herbicide concentrations have been reduced up to 64% after flowing through a filter strip compared to concentrations upstream of the filter strip area.[11]

Choice of Rate, Pesticide, and Application Timing

Some pesticides are much more vulnerable to movement into surface water. Pesticides that degrade quickly (days to a few weeks) are less likely to contaminate surface water than those that linger in the environment (months or longer). Applications to sandy soils with low organic matter are more likely to have runoff than loamy or clayey soils with more organic matter.

The higher the rate of application the more the pesticide is available for runoff. For example, banding herbicides to only row areas and using cultivation to control weeds in interrow areas is one technique to reduce application rates. The amount of herbicide applied depends on the bandwidth. If the bandwidth is 10″ on a 30″ row, the amount applied is reduced by two-thirds compared to a broadcast application rate where the entire area is covered. Banding has been reported to reduce the amount of herbicide in runoff water up to 70%.[12] Applying the pesticide as a split application also reduces the amount of chemical present at one time and can reduce the potential risk of surface water contamination.

Pesticides that are applied at ounces per acre rather than pounds per acre are available. Examples of herbicides include the sulfonylureas and imidiazilinones. The low application rates of these herbicides reduce the total amount of herbicide applied to an area and therefore reduce the risk of contamination.

Rainfall After Application

The amount, intensity, and time of rainfall after application are all factors that affect the total amount and pesticide concentration in runoff. Generally, a large amount of pesticide is removed with the first rainfall after application.[13] If the soil is bare and crusted or has little or no plant canopy, the amount of runoff will be very large. Storms with high intensity rains also have more runoff than if the rain is slow and steady. The amount of water already present in the soil is another factor that affects runoff. There is little or no infiltration on saturated or very wet soils whereas dry soils will allow more water to infiltrate before runoff occurs.

Marine—
Pollution

Drift, Volatilization, and Atmospheric Transport

Drift is the transport of spray applications in the wind. The smaller the droplet size, the farther the droplets can travel. Very fine particles may move for miles before deposition occurs. These drops could land in lakes, streams, or other off-site areas. Wind speed is also an important factor in drift. Most applications should not take place when wind speeds are over 10 mph. Pesticides should not be applied when weather conditions that cause an inversion (warm air over cold air) are present. Fine spray droplets move much farther when an atmospheric inversion exists.

There are some factors that can be used to limit drift. Applying pesticides under low pressure in high amounts of water will increase droplet size, thereby decreasing the number of droplets available for drift. Larger droplets will also lower the pesticide concentration of individual droplets. There are nozzles specifically designed to limit the amount of fine particles in a spray pattern. Antidrift chemicals are available to be mixed with pesticide application. These chemicals reduce the number of very fine droplets in the spray pattern. Lowering the spray boom limits the amount of spray subjected to the wind. Shields for the boom or individual nozzles can be attached to limit wind interception with spray patterns.

Pesticides can volatilize (change from a liquid to a gas) or sublime (change from solid to a gas) from the soil surface. The amount volatilized is a function of the vapor pressure of the chemical and the environmental conditions. For example, EPTC, a grass herbicide, is highly volatile with most of the herbicide lost to the atmosphere within a few hours if not incorporated into the soil.[14] In contrast, atrazine is not very volatile but under the right conditions, 2% of the applied chemical can be lost through this mechanism.[15] This amount is about one-half the loss due to surface runoff on a regional scale. A pesticide applied to warm moist soils on windy days will have a greater loss than if the same pesticide is applied to cool dry soils.[14]

Pesticides attached to very small soil particles are also moved into the atmosphere.[16] Soil particles and aggregates that are less than 1 mm in diameter are considered wind-erodible.[17] In the Great Plains area of the United States, the loss of soil due to wind erosion is greater than the amount lost due to water erosion.[18] The small size soil particles make up about 50% of the soil mass in the top 0.5″ of soil, if the area has been chisel plowed. The amount of herbicide found on these particles can range from 50% to 200% more than the amount found on larger aggregates one day after application.[19] If there are windy conditions within days after application, the amount of pesticide lost could be very high. Shallow incorporation of pesticides limits pesticide losses into the atmosphere.

Concentrations of pesticides in the atmosphere can decrease by several methods. Dilution, removal in rain, snow, or by dry deposition, and photochemical degradation are the three ways in which pesticide concentrations can be reduced.[20] However, pesticides can move long distances from their sites of application.[21] For example, surface water and rainfall has been monitored in the remote pristine area of Isle Royale National Park, MI, located in Lake Superior.[22] Atrazine was detected in rainfall from mid-May to early-June, corresponding to peak application timing of atrazine in the U.S. Midwestern Corn Belt. Surface water of the lakes in Isle Royale National Park also contained atrazine at trace (part per trillion) levels. These levels are not high enough to be toxic to organisms, but their presence in this remote area point to the need to use caution in the use and application of pesticides.

CONCLUSION

Although other methods of pest control are important, pesticides continue to be an efficient, cost-effective method of management. The challenge is to select and apply pesticides in a manner that kills the target pest and does not harm non-target organisms or the environment. Sound and sensible management can reduce the probability of pesticides entering surface water. Precision application to only those areas needing treatment is one method to reduce the total amount of pesticide applied. Other techniques include using split applications, banding, and planting filter strips along streams and water courses.

REFERENCES

1. Aspelin, A.L.; Grube, A.H. *Pesticides Industry Sales and Usage: 1996 and 1997 Market Estimates.* EPA/OPP. http://www.epa.gov/oppbead1/pestsales/97pestsales. Updated Jan 2000. (accessed September 2001).

2. Larson, S.J.; Gilliom, R.J.; Capel, P.D. *Pesticides in Streams of the United States—Initial Results from the National Water-Quality Assessment Program.* USGS/NAWQA, WRIR98-4222, 1999, http://www.water.wr.usgs.gov/pnsp/rep/wrir984222 (accessed September 2001).

3. Koplin, D.W.; Nations, B.K.; Goolsby, D.A.; Thurman, E.M. Acetochlor in the hydrologic system in the Midwestern United States, 1994. Environ. Sci. Technol. **1996**, *30*, 1459–1464.

4. Hoffman, R.S.; Capel, P.D.; Larson, S.J. Comparison of pesticides in eight U.S. urban streams. Environ. Toxicol. Chem. **2000**, *19*, 2249–2258.

5. Wotzka, P.J.; Lee, J.; Capel, P.D.; Lin, M. Pesticide concentrations and fluxes in an urban watershed. AWRA Technical Publication Series TPS-4; American Water Resources Association: Herndon, VA, 1994; 135–145.

6. Ferrari, M.J.; Ator, S.W.; Blomquist, J.D.; Dysart, J.E. *Pesticides in Surface Water of the Mid-Atlantic Region.*

USGS-WRIR97-4280, 1997, http://www.md.water.usgs. gov/publications/wrir-97-4280 (accessed February 2001).

7. Council for Agricultural Science and Technology. Issue paper 2, April, 1994. *Pesticides in Surface and Groundwater*. Revised May, 1997. CAST. http://www.castscience.org/ pwq_ip.htm (accessed February 2001). 818 Pesticide Contamination: Surface Water.

8. U.S. Environmental Protection Agency, *National Recommended Water Quality Criteria*, USEPA 822-Z-99-001; USEPA: Washington, DC, 1999.

9. Thompson, A.R.; Edwards, C.A. Effects of pesticides on nontarget invertebrates in freshwater and soil. In *Pesticides in Soil and Water*; Guenzi, W.D., Ahlrichs, J.L., Chesters, G., Bloodworth, M.E., Nash, R.G., Eds.; Soil Sci. Soc. Amer.: Madison, WI, 1974; 341–386.

10. Harp, G.L.; Campbell, R.S. Effects of the herbicide silves on benthos of a farm pond. J. Wildl. Manag. **1964**, *28*, 308–317.

11. Hall, J.K.; Hartwig, N.L.; Hoffman, L.K. Application mode and alternative cropping effects on atrazine losses from a hillside. J. Environ. Qual. **1983**, *18*, 439–445.

12. Gaynor, J.D.; van Wesenbeeck, I.J. Effects of band widths on atrazine, metribuzin, and metoachlor runoff. Weed Technol. **1995**, *9*, 107–112.

13. Ma, L.; Spalding, R.F. Herbicide mobility and variation in agricultural runoff in the beaver creek watershed in Nebraska. In *Herbicide Metabolites in Surface Water and Groundwater*; ACS Symposium Series 630; Meyer, M.T., Thurman, E.M., Eds.; American Chemical Society: Washington, DC, 1996; 226–236.

14. Spencer, W.F.; Cliath, M.M. Movement of pesticides from soil to the atmosphere. In *Long Range Transport of Pesticides*; Kurtz, D.A., Ed.; Lewis Publishers: Chelsea, MI, 1990; 1–16.

15. Nash, R.G.; Hill, B.D. Modeling pesticide volatilization and soil decline under controlled conditions. In *Long Range Transport of Pesticides*; Kurtz, D.A., Ed.; Lewis Publishers: Chelsea, MI, 1990; 17–28.

16. Ciba-Geigy Corporation. *Biological Assessment of Atrazine and Metolachlor in Rainfall*, Technical Paper 1; Bern, Switzerland, 1993; 4 pp.

17. Chepil, W.S. Dynamics of wind erosion: I. Nature of movement of soil by wind. Soil Sci. **1945**, *60*, 305–320.

18. United States Department Agriculture. *Summary Report: 1997 National Resources Inventory. Revised Dec. 2000; USDA Natural Resources Conservation Ser.: Washington, DC, 2000*, http://www.nhq.nrcs.usda.gov/NRI/1997/summary_report (accessed September 2001).

19. Clay, S.A.; DeSutter, T.M.; Clay, D.E. Herbicide concentration and dissipation from surface wind-erodible soil. Weed Sci. **2001**, *49*, 431–436.

20. Richards, R.P.; Kramer, J.W.; Baker, D.B.; Krieger, K.A. Pesticides in rainwater in the Northeastern United States. Nature **1987**, *327*, 129–131.

21. Kurtz, D.A. *Long Range Transport of Pesticides*; Lewis Publishers: Chelsea, MI, 1990; 462 pp.

22. Thurman, E.M.; Cromwell, A.E. Atmospheric transport, deposition, and fate of triazine herbicides and their metabolites in pristine areas of Isle Royale National Park. Environ. Sci. Technol. **2000**, *34*, 3079–3085.

Marine—
Pollution

Pollution: Nonpoint Source

Ravendra Naidu
Mallavarapu Megharaj
Peter Dillon
Rai Kookana
Ray Correll
Commonwealth Scientific and Industrial Research Organisation (CSIRO), Adelaide, South Australia, Australia

W. W. Wenzel
Institute of Soil Research, University of Natural Resources and Life Sciences, Vienna, Austria

Abstract

The biosphere is a life-supporting system to the living organisms. The detection of hazardous compounds in Antarctica, where these compounds were never used or no man has ever lived before, indicates how serious is the problem of long-range atmospheric transport and deposition of non-point source pollution. This ubiquitous pollution has had a global effect on our soils, which in turn has been affecting their biological health and productivity.

Marine—
Pollution

INTRODUCTION

Non-point source pollution (NPSP) has no obvious single point source discharge and is of diffuse nature (Table 1). An example of NPSP includes aerial transport and deposition of contaminants such as SO_2 from industrial emissions leading to acidification of soil and water bodies. Rain water in urban areas could also be a source of NPSP as it may concentrate organic and inorganic contaminants. Examples of such contaminants include polycyclic aromatic hydrocarbons, pesticides, polychlorinated biphenyls that could be present in urban air due to road traffic, domestic heating, industrial emissions, agricultural treatments, etc.[1–3] Other examples of NPSP include fertilizer (especially Cd, N, and P) and pesticide applications to improve crop yield. Use of industrial waste materials as soil amendments have been estimated to contaminate thousands of hectares of productive agricultural land in countries throughout the world.

CONTAMINANT INTERACTIONS

Non-point pollution is generally associated with low-level contamination spread at broad acre level. Under these circumstances, the major reaction controlling contaminant interactions are sorption–desorption processes, plant uptake, surface runoff, and leaching. However, certain contaminants, in particular, organic compounds are also subjected to voltalization, chemical, and biological degradation. Sorption–desorption and degradation (both biotic and abiotic) are the two most important processes controlling organic contaminant behavior in soils. These processes are influenced by both soil and solution properties of the environment. Such interactions also determine the bioavailability and/or transport of contaminants in soils. Where the contaminants are bioavailable, risk to surface and groundwater and soil, crop, and human health are enhanced.

IMPLICATIONS TO SOIL AND ENVIRONMENTAL QUALITY

Environmental contaminants can have a deleterious effect on non-target organisms and their beneficial activities. These effects could include a decline in primary production, decreased rate of organic matter break-down, and nutrient cycling as well as mineralization of harmful substances that in turn cause a loss of productivity of the ecosystems. Certain pollutants, even though present in very small concentrations in the soil and surrounding water, have potential to be taken up by various micro-organisms, plants, animals, and ultimately human beings. These pollutants may accumulate and concentrate in the food chain by

Encyclopedia of Natural Resources DOI: 10.1081/E-ENRW-120017970

Table 1 Industries, land uses, and associated chemicals contributing to non-point source pollution

Industry	Type of chemical	Associated chemicals
Agricultural activities	Metals/metalloid	Cadmium, mercury, arsenic, selenium
Non-metals	Nitrate, phosphate, borate	
Salinity/sodicity	Sodium, chloride, sulfate, magnesium, alkalinity	
Pesticides	Range of organic and inorganic pesticides including arsenic, copper, zinc, lead, sulfonylureas, organochlorine, organophosphates, etc., salt, geogenic contaminants (e.g., arsenic, selenium, etc.)	
Irrigation	Sodium, chloride, arsenic, selenium	
Automobile and industrial emissions	Dust	Lead, arsenic, copper, cadmium, zinc, etc.
Gas	Sulfur oxides, carbon oxides	
Metals	Lead and lead organic compounds	
Rainwater	Organics	Polyaromatic hydrocarbons, polychlorbiphenyls, etc.
Inorganic	Sulfur oxides, carbon oxides acidity, metals and metalloids	

Source: Adapted from Barzi, et al.[8]

several thousand times through a process referred to as biomagnification.

Urban sewage, because of its nutrient values and source of organic carbon in soils, is now increasingly being disposed to land. The contaminants present in sewage sludge (nutrients, heavy metals, organic compounds, and pathogens), if not managed properly, could potentially affect the environment adversely. Dumping of radioactive waste (e.g., radium, uranium, plutonium) onto soil is more complicated because these materials remain active for thousands of years in the soil and thus pose a continued threat to the future health of the ecosystem.

Industrial wastes, improper agricultural techniques, municipal wastes, and use of saline water for irrigation under high evaporative conditions result in the presence of excess soluble salts (predominantly Na and Cl ions) and metalloids such as Se and As in soils. Salinity and sodicity affect the vegetation by inhibiting seed germination, decreasing permeability of roots to water, and disrupting their functions such as photosynthesis, respiration, and synthesis of proteins and enzymes.

Some of the impacts of soil pollution migrate a long way from the source and can persist for some time. For example, suspended solids can increase water turbidity in streams, affecting benthic and pelagic aquatic ecosystems, filling reservoirs with unwanted silt, and requiring water treatment systems for potable water supplies. Phosphorus attached to soil particles, which are washed from a paddock into a stream, can dominate nutrient loads in streams and down-stream water bodies. Consequences include increases in algal biomass, reduced oxygen concentrations, impaired habitat for aquatic species, and even possible production of cyanobacterial toxins, with series impacts for

humans and livestock consuming the water. Where waters discharge into estuaries, N can be the limiting factor for eutrophication; estuaries of some catchments where fertilizer use is extensive have suffered from excessive sea grass and algal growth.

More insidious is the leaching of nutrients, agricultural chemicals, and hydrocarbons to groundwater. Incremental increases in concentrations in groundwater may be observed over long periods of time resulting in initially potable water becoming undrinkable and then some of the highest valued uses of the resource may be lost for decades. This problem is most severe on tropical islands with shallow relief and some deltaic arsenopyrite deposits, where wells cannot be deepened to avoid polluted groundwater because underlying groundwater is either saline or contains too much As.

SAMPLING FOR NON-POINT SOURCE POLLUTION

The sampling requirements of NPSP are quite different from those of the point source contamination. Typically, the sampling is required to give a good estimate of the mean level of pollution rather than to delineate areas of pollution. In such a situation, sampling is typically carried out on a regular square or a triangular grid. Furthermore, gains may be possible by using composite sampling.[4] However, if the pollution is patchy, other strategies may be used. One such strategy is to divide the area into remediation units, and to sample each of these. The possibility of movement of the pollutant from the soil to some receptor (or asset) is assessed, and the potential harm is quantified. This process requires an analysis of the bioavailability of the pollutant, pathway analysis, and the toxicological risk. The risk analysis is then assessed and decisions are then made as to how the risk should be managed.

MANAGEMENT AND/OR REMEDIATION OF NON-POINT SOURCE POLLUTION

The treatment strategies used for managing NPSP are generally those that modify the soil properties to decrease the bioavailable contaminant fraction. This is particularly so in the rural agricultural environment where soil–plant transfer of contaminants is of greatest concern. Soil amendments commonly used include those that change the ion-exchange characteristics of the colloid particles and those that enhance the ability of soils to sorb contaminants. An example of NPSP management includes the application of lime to immobilize metals because the solubility of most heavy metals decreases with increasing soil pH. However, this approach is not applicable to all metals, especially those that form oxyanions—the bioavailability of such species increases with increasing pH. Therefore, one of the prerequisites for remediating contaminated sites is a detailed assessment of the nature of contaminants present in the soil. The application of a modified aluminosilicate to a highly contaminated soil around a zinc smelter in Belgium was shown to reduce the bioavailability of metals thereby reducing the Zn phytotoxicity.[5] The simple addition of rock phosphates to form Pb phosphate has also been demonstrated to reduce the bioavailability of Pb in aqueous solutions and contaminated soils due to immobilization in the metal.[6] Nevertheless, there is concern over the long-term stability of the processes. The immobilization process appears attractive currently given that there are very few cheap and effective in situ remediation techniques for metal-contaminated soils. A novel, innovative approach is using higher plants to stabilize, extract, degrade, or volatilize inorganic and organic contaminants for in situ treatment (cleanup or containment) of polluted topsoils.[7]

PREVENTING WATER POLLUTION

The key to preventing water pollution from the soil zone is to manage the source of pollution. For example, nitrate pollution of groundwater will always occur if there is excess nitrate in the soil at a time when there is excess water leaching through the soil. This suggests that we should aim to reduce the nitrogen in the soil during wet seasons and the drainage through the soil. Local research may be needed to demonstrate the success of best management techniques in reducing nutrient, sediment, metal, and

chemical exports via surface runoff and infiltration to groundwater. Production figures from the same experiments may also convince local farmers of the benefits of maintaining nutrients and chemicals where needed by a crop rather than losing them off site, and facilitate uptake of best management practices.

GLOBAL CHALLENGES AND RESPONSIBILITY

The biosphere is a life-supporting system to the living organisms. Each species in this system has a role to play and thus every species is important and biological diversity is vital for ecosystem health and functioning. The detection of hazardous compounds in Antarctica, where these compounds were never used or no man has ever lived before, indicates how serious the problem of long-range atmospheric transport and deposition of these pollutants is. Clearly, pollution knows no boundaries. This ubiquitous pollution has had a global effect on our soils, which in turn has been affecting their biological health and productivity. Coupled with this, over 100,000 chemicals are being used in countries throughout the world. Recent focus has been on the endocrine disruptor chemicals that mimic natural hormones and do great harm to animal and human reproductive cycles.

These pollutants are only a few examples of contaminants that are found in the terrestrial environment.

REFERENCES

1. Chan, C.H.; Bruce, G.; Harrison, B. Wet deposition of organochlorine pesticides and polychlorinated biphenyls to the great lakes. J. Great Lakes Res. **1994**, *20*, 546–560.
2. Lodovici, M.; Dolara, P.; Taiti, S.; Del Carmine, P.; Bernardi, L.; Agati, L.; Ciappellano, S. Polycyclic aromatic hydrocarbons in the leaves of the evergreentree*Laurus Nobilis*. Sci. Total Environ. **1994**, *153*, 61–68.
3. Sweet, C.W.; Murphy, T.J.; Bannasch, J.H.; Kelsey, C.A.; Hong, J. Atmospheric deposition of PCBs into green bay. J. Great Lakes Res. **1993**, *18*, 109–128.
4. Patil, G.P.; Gore, S.D.; Johnson, G.D. *Manual on Statistical Design and Analysis with Composite Samples*; Technical Report No. 96-0501; EPA Observational Economy Series Center for Statistical Ecology and Environmental Statistics; Pennsylvania State University, 1996; Vol. 3.
5. Vangronsveld, J.; Van Assche, F.; Clijsters, H. Reclamation of a bare industrial area, contaminated by non-ferrous metals: in situ metal immobilisation and revegetation. Environ. Pollut. **1995**, *87*, 51–59.
6. Ma, Q.Y.; Logan, T.J.; Traina, S.J. Lead immobilisation from aqueous solutions and contaminated soils using phosphate rocks. Environ. Sci. Technol. **1995**, *29*, 1118–1126.

7. Wenzel, W.W.; Adriano, D.C.; Salt, D.; Smith, R. Phytoremediation: a plant-microbe based remediation system. In *Bioremediation of Contaminated Soils*; Soil Science Society of America Special Monograph No. 37, Adriano, D.C., Bollag, J.M., Frankenberger, W.T., Jr., Sims, W.R., Eds.; Soil Science Society of America: Madison, USA, 1999; 772 pp.

8. 8. Barzi, F.; Naidu, R.; McLaughlin, M.J. *Contaminants and the Australian Soil Environment. In Contaminants and the Soil Environment in the Australasia-Pacifi c Region*; Naidu, R., Kookana, R.S., Oliver, D., Rogers, S., McLaughlin, M.J., Eds.; Kluwer Academic Publishers: Dordrecht, the Netherlands, 1996; 451–484.

Marine— Pollution

Pollution: Point Source

Mallavarapu Megharaj
Peter Dillon
Ravendra Naidu
Rai Kookana
Ray Correll
Commonwealth Scientific and Industrial Research Organisation (CSIRO), Adelaide, South Australia, Australia

W. W. Wenzel
Institute of Soil Research, University of Natural Resources and Life Sciences, Vienna, Austria

Abstract

Environmental pollution is one of the foremost ecological challenges. Pollution is an offshoot of technological advancement and overexploitation of natural resources. From the standpoint of pollution, the term environment primarily includes air, land, and water components including landscapes, rivers, parks, and oceans. Pollution can be generally defined as an undesirable change in the natural quality of the environment that may adversely affect the well being of humans, other living organisms, or entire ecosystems either directly or indirectly. Although pollution is often the result of human activities (anthropogenic), it could also be due to natural sources such as volcanic eruptions emitting noxious gases, pedogenic processes, or natural change in the climate. Where pollution is localized it is described as point source (PS). Thus, PS pollution is a source of pollution with a clearly identifiable point of discharge that can be traced back to the specific source such as leakage of underground petroleum storage tanks or an industrial site.

INTRODUCTION

Some naturally occurring pollutants are termed geogenic contaminants and these include fluorine, selenium, arsenic, lead, chromium, fluoride, and radionuclides in the soil and water environment. Significant adverse impacts of geogenic contaminants (e.g., As) on environmental and human health have been recorded in Bangladesh, West Bengal, India, Vietnam, and China. More recently reported is the presence of geogenic Cd and the implications to crop quality in Norwegian soils.[1]

The terms contamination and pollution are often used interchangeably but erroneously. Contamination denotes the presence of a particular substance at a higher concentration than would occur naturally and this may or may not have harmful effects on human or the environment. Pollution refers not only to the presence of a substance at higher level than would normally occur but is also associated with some kind of adverse effect.

NATURE AND SOURCES OF CONTAMINANTS

The main activities contributing to PS pollution include industrial, mining, agricultural, and commercial activities as well as transport and services (Table 1). Uncontrolled mining, manufacturing, and disposal of wastes inevitably cause environmental pollution. Military land and land for recreational shooting are also important sites of PS contamination. The contaminants associated with such activities are listed in Table 1. Contamination at many of these sites appears to have resulted because of lax regulatory measures prior to the establishment of legislation protecting the environment.

CONTAMINANT INTERACTIONS IN SOIL AND WATER

Inorganic Chemicals

Inorganic contaminant interactions with colloid particulates include: adsorption–desorption at surface sites, precipitation, exchange with clay minerals, binding by organically coated particulate matter or organic colloidal material, or adsorption of contaminant ligand complexes. Depending on the nature of contaminants, these interactions are controlled by solution pH and ionic strength of soil solution, nature of the species, dominant cation, and inorganic and organic ligands present in the soil solution.[2]

Encyclopedia of Natural Resources DOI: 10.1081/E-ENRW-120010213

Marine—
Pollution

Table 1 Industries, land uses, and associated chemicals contributing to points, non-point source pollution

Industry	Type of chemical	Associated chemicals
Airports	Hydrocarbons	Aviation fuels
	Metals	Particularly aluminum, magnesium, and chromium
Asbestos production and disposal	Asbestos	
Battery manufacture and recycling	Metals	Lead, manganese, zinc, cadmium, nickel, cobalt, mercury, silver, and antimony
	Acids	Sulfuric acid
Breweries/distilleries	Alcohol	Ethanol, methanol, and esters
Chemicals manufacture and use	Acid/alkali	Mercury (chlor/alkali), sulfuric, hydrochloric and nitric acids, sodium and calcium hydroxides
	Adhesives/resins	Polyvinyl acetate, phenols, formaldehyde, acrylates, and phthalates
	Dyes	Chromium, titanium, cobalt, sulfur and nitrogen organic compounds, sulfates, and solvents
	Explosives	Acetone, nitric acid, ammonium nitrate, pentachlorophenol, ammonia, sulfuric acid, nitroglycerine, calcium cyanamide, lead, ethylene glycol, methanol, copper, aluminum, *bis*(2-ethylhexyl) adipate, dibutyl phthalate, sodium hydroxide, mercury, and silver
	Fertilizer	Calcium phosphate, calcium sulfate, nitrates, ammonium sulfate, carbonates, potassium, copper, magnesium, molybdenum, boron, and cadmium
	Flocculants	Aluminum
	Foam production	Urethane, formaldehyde, and styrene
	Fungicides	Carbamates, copper sulfate, copper chloride, sulfur, and chromium
	Herbicides	Ammonium thiocyanate, carbanates, organochlorines, organophosphates, arsenic, and mercury
	Paints	
	Heavy metals	Arsenic, barium, cadmium, chromium, cobalt, lead, manganese, mercury, selenium, and zinc
	General	Titanium dioxide
	Solvent	Toluene, oils natural (e.g., pine oil) or synthetic
	Pesticides	Arsenic, lead, organochlorines, and organophosphates
	Active ingredients	Sodium, tetraborate, carbamates, sulfur, and synthetic pyrethroids
	Solvents	Xylene, kerosene, methyl isobutyl ketone, amyl acetate, and chlorinated solvents
	Pharmacy	Dextrose and starch
	General/solvents	Acetone, cyclohexane, methylene chloride, ethyl acetate, butyl acetate, methanol, ethanol, isopropanol, butanol, pyridine methyl ethyl ketone, methyl isobutyl ketone, and tetrahydrofuran
	Photography	Hydroquinone, pheidom, sodium carbonate, sodium sulfite, potassium bromide, monomethyl paraaminophenol sulfates, ferricyanide, chromium, silver, thiocyanate, ammonium compounds, sulfur compounds, phosphate, phenylene diamine, ethyl alcohol, thiosulfates, and formaldehyde

Marine—Pollution

(Continued)

Table 1 (***Continued***)   Industries, land uses, and associated chemicals contributing to points, non-point source pollution

Industry	Type of chemical	Associated chemicals
	Plastics	Sulfates, carbonates, cadmium, solvents, acrylates, phthalates, and styrene
	Rubber	Carbon black
	Soap/detergent	
	General	Potassium compounds, phosphates, ammonia, alcohols, esters, sodium hydroxide, surfactants (sodium lauryl sulfate), and silicate compounds
	Acids	Sulfuric acid and stearic acid
	Oils	Palm, coconut, pine, and tea tree
	Solvents	
	General	Ammonia
	Hydrocarbons	e.g., BTEX (benzene, toluene, ethylbenzene, xylene)
	Chlorinated organics	e.g., trichloroethane, carbon tetrachloride, and methylene chloride
Defense works		See "Explosives" under "Chemicals Manufacture and Use, Foundries, Engine Works, and Service Stations"
Drum reconditioning		See "Chemicals Manufacture and Use"
Dry cleaning		Trichlorethylene and ethane
		Carbon tetrachloride
		Perchlorethylene
Electrical		PCBs (transformers and capacitors), solvents, tin, lead, and copper
Engine works	Hydrocarbons	
	Metals	
	Solvents	
	Acids/alkalis	
	Refrigerants	
	Antifreeze	Ethylene glycol, nitrates, phosphates, and silicates
Foundries	Metals	Particularly aluminum, manganese, iron, copper, nickel, chromium, zinc, cadmium and lead and oxides, chlorides, fluorides and sulfates of these metals
	Acids	Phenolics and amines
		Coke/graphite dust
Gas works	Inorganics	Ammonia, cyanide, nitrate, sulfide, and thiocyanate
	Metals	Aluminum, antimony, arsenic, barium, cadmium, chromium, copper, iron, lead, manganese, mercury, nickel, selenium, silver, vanadium, and zinc
	Semivolatiles	Benzene, ethylbenzene, toluene, total xylenes, coal tar, phenolics, and PAHs
Iron and steel works		Metals and oxides of iron, nickel, copper, chromium, magnesium and manganese, and graphite
Landfill sites		Methane, hydrogen sulfides, heavy metals, and complex acids
Marinas		Engine works, electroplating under metal treatment
	Antifouling paints	Copper, tributyltin (TBT)
Metal treatments	Electroplating metals	Nickel, chromium, zinc, aluminum, copper, lead, cadmium, and tin
	Acids	Sulfuric, hydrochloric, nitric, and phosphoric
	General	Sodium hydroxide, 1,1,1-trichloroethane, tetrachloroethylene, toluene, ethylene glycol, and cyanide compounds

(*Continued*)

Marine—
Pollution

Table 1 (*Continued*)   Industries, land uses, and associated chemicals contributing to points, non-point source pollution

Industry	Type of chemical	Associated chemicals
	Liquid carburizing baths	Sodium, cyanide, barium, chloride, potassium chloride, sodium chloride, sodium carbonate, and sodium cyanate
	Mining and extracting industries	Arsenic, mercury, and cyanides and also refer to "Explosives" under "Chemicals Manufacture and Use"
	Power stations	Asbestos, PCBs, fly ash, and metals
	Printing shops	Acids, alkalis, solvents, chromium (see "Photography" under "Chemicals Manufacture and Use")
Scrap yards		Hydrocarbons, metals, and solvents
	Service stations and fuel storage facilities	Aliphatic hydrocarbons
		BTEX (i.e., benzene, toluene, ethylbenzene, xylene)
		PAHs (e.g., benzo(a) pyrene)
		Phenols
		Lead
Sheep and cattle dips		Arsenic, organochlorines and organophosphates, carbamates, and synthetic pyrethroids
Smelting and refining		Metals and the fluorides, chlorides and oxides of copper, tin, silver, gold, selenium, lead, and aluminum
Tanning and associated trades	Metals	Chromium, manganese, and aluminum
	General	Ammonium sulfate, ammonia, ammonium nitrate, phenolics (creosote), formaldehyde, and tannic acid
Wood preservation	Metals	Chromium, copper, and arsenic
	General	Naphthalene, ammonia, pentachlorophenol, dibenzofuran, anthracene, biphenyl, ammonium sulfate, quinoline, boron, creosote, and organochlorine pesticides

Source: Adapted from Barzi et al.[11]

Organic Chemicals

The fate and behavior of organic compounds depend on a variety of processes including sorption–desorption, volatilization, chemical and biological degradation, plant uptake, surface runoff, and leaching. Sorption–desorption and degradation (both biotic and abiotic) are perhaps the two most important processes as the bulk of the chemicals is either sorbed by organic and inorganic soil constituents, and chemically or microbially transformed/degraded. The degradation is not always a detoxification process. This is because in some cases the transformation or degradation process leads to intermediate products that are more mobile, more persistent, or more toxic to non-target organisms. The relative importance of these processes is determined by the chemical nature of the compound.

IMPLICATIONS TO SOIL AND ENVIRONMENTAL QUALITY

A considerable amount of literature is available on the effects of contaminants on soil microorganisms and their functions in soil. The negative impacts of contaminants on microbial processes are important from the ecosystem point of view and any such effects could potentially result in a major ecological perturbation. Hence, it is most relevant to examine the effects of contaminants on microbial processes in combination with communities. The most commonly used indicators of metal effects on microflora in soil are: (1) soil respiration, (2) soil nitrification, (3) soil microbial biomass, and (4) soil enzymes.

Contaminants can reach the food chain by way of water, soil, plants, and animals. In addition to the food chain transfer, pollutants may also enter via direct consumption or dust inhalation of soil by children or animals. Accumulation of these pollutants can take place in certain target tissues of the organism depending on the solubility and nature of the compound. For example, DDT and PCBs accumulate in human adipose tissue. Consequently, several of these pollutants have the potential to cause serious abnormalities including cancer and reproductive impairments in animal and human systems.

SAMPLING FOR PS POLLUTION

The aims of the sampling system must be clearly defined before it can be optimized.[3] The type of decision may be to determine land use, how much of an area is to be remediated, or what type of remediation process is required.

Because sampling and the associated chemical and statistical analyses are expensive, careful planning of the sampling scheme is therefore a good investment. One of the best ways to achieve this is to use any ancillary data that are available. These data could be in the form of emission history from a stack, old photographs that give details of previous land uses, or agricultural records. Such data can at least give qualitative information.

As discussed before, PS pollution will typically be airborne from a stack, or waterborne from some effluent such as tannery waste, cattle dips, or mine waste. In many cases, the industry will have modified its emissions (e.g., cleaner production) or point of release (increased stack height), hence the current pattern of emission may not be closely related to the historic pattern of pollution. For example, liquid effluent may have been discharged previously into a bay, but that effluent may now be treated and perhaps discharged at some other point. Typically, the aim of a sampling scheme in these situations is to assess the maximum concentrations, the extent of the pollution, and the rate of decline in concentration from the PS. Often the sampling scheme will be used to produce maps of concentration isopleths of the pollutant.

The location of the sampling points would normally be concentrated towards the source of the pollution. A good scheme is to have sufficient samples to accurately assess the maximum pollution, and then space additional samples at increasing intervals. In most cases, the distribution of the pollutant will be asymmetric, with the maximum spread down the slope or down the prevailing wind. In such cases more samples should be placed in the direction of the expected gradient. This is a clear case of when ancillary data can be used effectively. A graph of concentration of the pollutant against the reciprocal of distance from the source is often informative.[4] Sampling depths will depend on both the nature of the pollution and the reason for the investigation. If the pollution is from dust and it is unlikely to be leached, only surface sampling will be required. An example of this is pollution from silver smelting in Wales.[5] In contrast, contamination from organic or mobile inorganic pollutants such as F compounds may migrate well down to the profile and deep sampling may be required.[6,7]

ASSESSMENT

In order to assess the impacts of pollution, reliable and effective monitoring techniques are important. Pollution can be assessed and monitored by chemical analyses, toxicity tests, and field surveys. Comparison of contaminant data with an uncontaminated reference site and available databases for baseline concentrations can be useful in establishing the extent of contamination. However, this may not always be possible in the field. Chemical analyses must be used in conjunction with biological assays to reveal site contamination and associated adverse effects. Toxicological assays can also reveal information about synergistic interactions of two or more contaminants present as mixtures in soil, which cannot be measured by chemical assays alone.

Microorganisms serve as rapid detectors of environmental pollution and are thus of importance as pollution indicators. The presence of pollutants can induce alteration of microbial communities and reduction of species diversity, inhibition of certain microbial processes (organic matter breakdown, mineralization of carbon and nitrogen, enzymatic activities, etc.). A measure of the functional diversity of the bacterial flora can be assessed using ecoplates (see http://www.biolog.com/section 4.html). It has been shown that algae are especially sensitive to various organic and inorganic pollutants and thus may serve as a good indicator of pollution.[8] A variety of toxicity tests involving microorganisms, invertebrates, vertebrates, and plants may be used with soil or water samples.[9]

MANAGEMENT AND/OR REMEDIATION OF PS POLLUTION

The major objective of any remediation process is to: (1) reduce the actual or potential environmental threat; and (2) reduce unacceptable risks to man, animals, and the environment to acceptable levels.[10] Therefore, strategies to either manage and/or remediate contaminated sites have been developed largely from application of stringent regulatory measures set up to safeguard ecosystem function as well as to minimize the potential adverse effects of toxic substances on animal and human health.

The available remediation technologies may be grouped into two categories: (1) ex situ techniques that require removal of the contaminated soil or groundwater for treatment either on-site or off-site; and (2) in situ techniques that attempt to remediate without excavation of contaminated soils. Generally, in situ techniques are favored over ex situ techniques because of: (1) reduced costs due to elimination or minimization of excavation, transportation to disposal sites, and sometimes treatment itself; (2) reduced health impacts on the public or the workers; and, (3) the potential for remediation of inaccessible sites, e.g., those located at greater depths or under buildings. Although in situ techniques have been successful with organic contaminated sites, the success of in situ strategies with metal contaminants has been limited. Given that organic and inorganic contaminants often occur as a mixture, a combination of more than one strategy is often required to either successfully remediate or manage metal contaminated soils.

GLOBAL CHALLENGES AND RESPONSIBILITY

The last 100 years have seen massive industrialization. Indeed such developments were coupled with the rapid increase in world population and the desire to enhance economy and food productivity. While industrialization has led to increased economic activity and much benefit to the human race, the lack of regulatory measures and appropriate waste management strategies until the early 1980s (including the use of agrochemicals) has resulted in contamination of our biosphere. Continued pollution of the environment through industrial emissions is of global concern. There is, therefore, a need for politicians, regulatory organizations, and scientists to work together to minimize environmental contamination and to remediate contaminated sites. The responsibility to check this pollution lies with every individual and country although the majority of this pollution is due to the industrialized nations. There is a clear need for better coordination of efforts in dealing with numerous forms of PS pollution problems that are being faced globally.

REFERENCES

1. Mehlum, H.K.; Arnesen, A.K.M.; Singh, B.R. Extractability and plant uptake of heavy metals in alum shale soils. Commun. Soil Sci. Plant Anal. **1998**, *29*, 183–198.

2. McBride, M.B. Reactions controlling heavy metal solubility in soils. Adv. Soil Sci. **1989**, *10*, 1–56.

3. Patil, G.P.; Gore, S.D.; Johnson, G.D. *EPA Observational Economy Series Volume 3: Manual on Statistical Design and Analysis with Composite Samples*; Technical Report No. 96-0501;Center for Statistical Ecology and Environmental Statistics: Pennsylvania State University, 1996.

4. Ward, T.J.; Correll, R.L. Estimating background concentrations of heavy metals in the marine environment, Proceedings of a Bioaccumulation Workshop: Assessment of the Distribution, Impacts and Bioaccumulation of Contaminants in Aquatic Environments, Sydney, 1990; Miskiewicz, A.G., Ed.; Water Board and Australian Marine Science Association: Sydney, 1992; 133–139.

5. Jones, K.C.; Davies, B.E.; Peterson, P.J. Silver in welsh soils: physical and chemical distribution studies. Geoderma **1986**, *37*, 157–174.

6. Barber, C.; Bates, L.; Barron, R.; Allison, H. Assessment of the relative vulnerability of groundwater to pollution: A review and background paper for the conference workshop on vulnerability assessment. J. Aust. Geol. Geophys. **1993**, *14*(2–3), 147–154. 880 Pollution: Point Source (PS)

7. Wenzel, W.W.; Blum, W.E.H. Effects of fluorine deposition on the chemistry of acid luvisols. Int. J. Environ. Anal. Chem. **1992**, *46*, 223–231.

8. Megharaj, M.; Singleton, I.; McClure, N.C. Effect of pentachlorophenol pollution towards microalgae and microbial activities in soil from a former timber processing facility. Bull. Environ. Contam. Toxicol. **1998**, *61*, 108–115.

9. Juhasz, A.L.; Megharaj, M.; Naidu, R. Bioavailability: the major challenge (constraint) to bioremediation of organically contaminated soils. In *Remediation Engineering of Contaminated Soils*; Wise, D., Trantolo, D.J., Cichon, E.J., Inyang, H.I., Stottmeister, U., Eds.; Marcel Dekker: New York, 2000; 217–241.

10. Wood, P.A. Remediation methods for contaminated sites. In *Contaminated Land and Its Reclamation*; Hester, R.E., Harrison, R.M., Eds.; Royal Society of Chemistry, Thomas Graham House: Cambridge, UK, 1997; 47–73.

11. Barzi, F.; Naidu, R.; McLaughlin, M.J. Contaminants and the Australian soil environment. In *Contaminants and the Soil Environment in the Australasia–Pacific Region*; Naidu, R., Kookana, R.S., Oliver, D., Rogers, S., McLaughlin, M.J., Eds.; Kluwer Academic Publishers: Dordrecht, the Netherlands, 1996; 451–484.

Marine— Pollution

Riparian Wetlands: Mapping

Kristen C. Hychka
Atlantic Ecology Division, Office of Research and Development, U.S. Environmental Protection Agency (EPA), Narragansett, Rhode Island, U.S.A.

Abstract

Riparian wetlands are critical systems that perform functions and provide services disproportionate to their extent in the landscape. Mapping wetlands allows for better planning, management, and modeling, but riparian wetlands present several challenges to effective mapping due to their morphology, size, forest cover, and relative dryness. There are three general approaches to mapping riparian wetlands. These include using on-site, remotely sensed, or ancillary data. These approaches all have trade-offs in terms of time, cost, accuracy (omission and commission), and repeatability. However, there are promising approaches to improve the accuracy of mapping riparian wetlands, particularly through radar, LiDAR intensity returns, multitemporal images, mechanistic modeling, and mixed method approaches.

INTRODUCTION

Riparian wetlands are critical systems that perform functions and provide services disproportionate to their extent in the landscape. Mapping wetlands allows for better planning, management, and modeling, but riparian wetlands present several challenges to effective mapping. However, there are promising approaches to improve the accuracy of mapping riparian wetlands.

CHARACTERISTICS AND SIGNIFICANCE OF RIPARIAN WETLANDS

Riparian ecosystems are biophysical communities that border rivers, streams, or lakes and are some of the most diverse and dynamic systems on Earth.[1,2] They encompass the stream channel and adjacent land that are influenced by flooding or elevated water tables and have soils that hold water[3] and are comprised of aquatic, wetland, and upland components. This entry focuses specifically on riparian wetlands associated with fluvial or lotic systems. A wetland is an area where soil formation and the composition of the biota of the system are primarily determined by consistent or periodic saturation or inundation of the soils.[4] Riparian wetlands are predominantly associated with headwater streams and often drier than other wetland types.[5] Though they have a range of cover types, riparian wetlands are predominantly forested,[5] as the majority of vegetated freshwater wetlands in the continental United States are forested[6] and greater than 80% of stream networks are comprised of small streams,[7,8] which are more likely to be forested than streams lower in the network.[9]

Riparian wetlands perform a suite of functions including sediment retention, provision of wildlife habitat, carbon sequestration, temperature regulation, biogeochemical processing, and are particularly critical for water quality and quantity regulation, because of their position in the landscape.[10–12] However, approximately 80% of riparian corridors in North America and Europe have been severely degraded in the past 200 years.[3] Though total wetland area has shown no significant change in the Continental United States recently, there have been significant losses of vegetated freshwater wetlands and particularly of forested wetlands.[6]

IMPORTANCE OF AND CHALLENGES IN MAPPING RIPARIAN WETLANDS

Mapping wetlands is important for spatial planning; water resources management;[13] and climatic, hydrologic, and coupled hydroclimatic models.[14] Also mapping extent, characteristics, and stressors associated with wetlands are important for monitoring, particularly in these rapidly changing systems.

The approaches to mapping wetlands outlined below apply to all wetland types, however riparian wetlands are especially "controversial wetlands," as they often function as wetlands but do not always meet regulatory definitions of wetlands[15] and are difficult wetlands to map for a number of reasons. First, their morphology is often thin and long, and therefore smaller than the detection limits of some mapping efforts.[16] They are also often small, because a large proportion of riparian wetlands are associated with headwater streams.[5] Additionally, they are generally forested,

Encyclopedia of Natural Resources DOI: 10.1081/E-ENRW-120048593

so they are particularly difficult to identify through optical remote sensing approaches often used to map wetlands.[17] Riparian wetlands are also often ecotones that do not have abrupt boundaries.[15] Some mapping approaches handle gradient boundaries or fuzzy classification in mapping, but traditionally wetlands are mapped with crisp boundaries between wetland types and between wetlands and uplands. Finally, riparian wetlands, particularly those associated with extreme headwaters, are often "drier end" wetlands,[5] so they are more difficult and less accurate to map when soil saturation and inundation are criteria in wetland mapping.[15]

There are extensive bodies of literature focusing on remote sensing of wetlands or of riparian systems, but much less attention has been paid to specifically riparian wetlands. This entry seeks to identify and summarize references that focus specifically on riparian wetlands and summarize some of the relevant information from more general papers on riparian systems or wetlands based on the characteristics of riparian wetlands.

MAPPING QUALITY AND STANDARDS

The output and accuracy of wetland maps vary greatly based not only on the characteristics of the wetland described above but also on the approach, classification, validation, and base data used. First, the criteria used to define a wetland vary. For example, wetlands mapped through the detection of standing water versus the presence of mapped hydric soils will result in very different wetland extents for the same area.[18] Also, classification approaches may be visual, supervised (commonly maximum likelihood), or unsupervised (commonly clustering)[17] and delineate wetlands based on crisp or fuzzy classes.[19] Validation can be spatial or aspatial (aggregated validation by region), and the accuracy of a map can also be evaluated in terms of errors of omission (producer's accuracy) and errors of commission (user's accuracy).[16] Finally, maps are generated from different types of base data and this is discussed in more detail in the coming sections.

MAPPING APPROACHES

For the reasons outlined earlier, some approaches to mapping riparian wetlands are more successful than others. There are three general approaches to mapping wetlands: on-site, remotely sensed, or ancillary data. Specific applications of these approaches have trade-offs in terms of time, cost, resolution, accuracy (omission and commission), and repeatability. It is important to know the strengths and weaknesses associated with an approach to mapping, particularly the type and amount of error associated with different mapping approaches.[18]

On-Site Mapping

On-site mapping typically follows ecological or jurisdictional protocol and involves identifying characteristics of vegetation, soils, and hydrology. Reviews of these approaches are covered in other texts.[20–22] On-site mapping provides the highest level of confidence in the data (low user's and producer's errors), but can be labor intensive and logistically difficult, so this approach is often used on small spatial extents for a single period.

Remote Sensing

Remote sensing approaches to mapping rely on the visual or automated classification of reflectance data from panchromatic, multi- or hyperspectral imagery, and other wavelengths including radar.

Pan-chromatic data are primarily in the visible light range and are either aerial photography or satellite imagery. Aerial photo interpretation was the first remote sensing approach to mapping wetlands and is still considered to be one of the most accurate approaches given cloud-free, high-resolution imagery collected at appropriate time of year based on phenology and hydroperiod.[16,23] Another advantage to using aerial photo interpretation is if older images are available, they can be used to create historical time series of wetland change. Optical images can have a wide range of horizontal spatial resolutions (sub-meter to hundreds of meters), which greatly influence the detection of small and, or narrow riparian wetlands.[18] Drawbacks to using this approach are that the interpretation is typically manual, so it is labor intensive and subjective, and forested wetlands are often underrepresented.[24] Automated methods for analyzing optical images are also important in detecting riparian wetlands; for example, leaf area index and other vegetation indices have also been used to map both the extent and the characteristics (such as biomass) of riparian wetlands.[25] Repeatability is an advantage of automated mapping over manual delineation. The National Wetlands Inventory of the United States primarily relies on the manual interpretation of color IR imagery to map wetlands.[6,23] This inventory, typical of optically derived wetlands maps, is fairly conservative and has high producer's accuracy, but does miss a lot of wetlands (low user's accuracy), particularly dry, narrow, or forested wetlands.[18,20]

Radar is a powerful tool in mapping riparian wetlands, because it is not constrained by weather conditions, does not require sunlight, and is useful in detecting both standing water and saturated soils.[26,27] Radar has been successfully used to detect soil saturation below herbaceous vegetation, but is less successful under woody vegetation, and consistently flooded wetlands are easier to detect than periodically flooded systems.[25,26] Radar data are available in a wide range of spatial resolutions, though most continental- or global-scale has horizontal resolution in hundreds of meters,[26] which is not accurate enough to detect small or narrow riparian wetlands.

LiDAR intensity returns have also been used to map inundation under canopy. In some cases this approach was better than topographic indices or reflectance in a panchromatic image[28] and particularly when used in combination with statistical approaches to account for spatial autocorrelation.[29] These studies were conducted primarily on isolated wetlands, but their approaches can be applied to mapping riparian wetlands. LiDAR data have particularly fine spatial resolution (meters to sub-meter),[30] though until recently the spatial extent of available LiDAR-derived data has been limited.

Additionally using a multitemporal approach, which relies on imagery from different times of the year, is also an effective way to accurately capture riparian wetland extent. Most commonly, this approach is used to determine vegetation from images during leaf off conditions to detect the ground below the canopy and to determine if vegetation is deciduous.[17,31] Also, it is useful in evaluating images from the wettest times of the year to determine soil saturation for drier systems and flooding in periodically flooded systems.[32]

Ancillary Data

Ancillary data are used to model processes or hydromorphic settings that support wetland formation. Some advantages to using an ancillary data approach to mapping riparian wetlands are that these can identify "wetland supportive settings" even in a disturbed landscape where wetlands have been drained and can be useful in scenario development. However, a key negative to using ancillary data is that though there may be a high user's accuracy there is often a low rate of producer's accuracy. Some of the approaches that use ancillary data include topographic indices, climato-topographic indices, and directly modeled hydrologic processes.

Many studies have used topographic indices to predict wetland occurrence primarily by using the terrain to model the occurrence of soil saturation.[28] There is a suite of topographic indices that are based on combinations of the contributing area and slope, and they differ mostly in their calculation of upslope contribution.[33] These approaches are helpful in identifying "cryptic" or forested riparian wetlands, as canopy cover has a smaller influence on the results compared to optical approaches.[34] Climato-topographic indices refine DEM-derived (digital elevation model) topographic indices by adding measured or modeled climate data, accounting for differences in the availability of water,[14,35] which is particularly important in large study areas that cross several climatic zones.[36] A drawback to both topographic and climato-topographic indices is that they do not work well in areas with complex geology where soil permeability is highly variable. An emerging approach to riparian wetland

mapping is to use mechanistic modeling or to directly model processes that drive wetland occurrence, such as groundwater level and flooding.[37,38] This is a particularly useful approach when making projections of wetland change under differing scenarios; however, this approach is data intensive and combines data from multiple sources, which can have unaccounted error propagation.[39,40] As with other mapping approaches, the spatial resolution of the base data (DEMs in this case) influences the detection limits of mapping, and the horizontal resolution of DEMs ranges widely (sub-meter to kilometers).[41]

Mixed Method Approaches

Because different sensors can detect different characteristics of wetlands, the use of a multisensor approach can be the most fruitful for accurately mapping riparian wetlands.[17,27,42] Using multiple sensors can draw on the strongest aspects of each approach, for example using radar to determine the extent and timing of flooding and optical data to determine vegetative characteristics.[43] Another promising approach to riparian wetland mapping is the coupled use of remotely sensed data with data from in situ sensors.[44,45] For example, in periodically flooded systems, in situ sensors can identify potentially useful imagery for mapping wetland extents. Using participatory sensing or local knowledge can also inform riparian wetland mapping. One study coupled remotely sensed data with local ecological knowledge to map isolated wetlands,[46] and the same approach could be used for riparian systems.

CONCLUSION

Though riparian wetlands are difficult to map because of their morphology, size, typical forest cover, and often drier hydrology, several mapping approaches and combinations of approaches have improved or shown promise in improving the accuracy of mapping riparian wetlands.

REFERENCES

1. Naiman, R.J.; DeCamps, H. The ecology of interfaces: riparian zones. Ann. Rev. Ecol. Syst. **1997**, *28* (1), 621–658.
2. Malanson, G.P. *Riparian Landscapes*; Great Britain: Cambridge University Press: Cambridge, 1993.
3. Naiman, R.J.; DeCamps, H.; Pollock, M. The role of riparian corridors in maintaining regional biodiversity. Ecol. Appl. **1993**, *3* (2), 209–212.
4. Cowardin, L.M.; Carter, V.; Golet, F.C. LaRoe, E.T. *Classification of Wetlands and Deepwater Habitats of the United States*; Washington, D.C., 1979.
5. Whigham, D.F. Ecological issues related to wetland preservation, restoration, creation and assessment. Sci. Total Environ. **1999**, *240* (1–3), 31–40.

6. Dahl, T.E. *Status and Trends of Wetlands in the Contermi-nous United States 2004 to 2009*; U.S. Department of the Interior, Fish and Wildlife Service: Washington, D.C. 2011; 108 pp.

7. Peterson, B.J.; Wollheim, W.M.; Mulholland, P. J.; Webster, J.R.; Meyer, J.L.; Tank, J.L.; Martí, E.; Bowden, W.B.; Valett, H.M. ; Hershey, A.E. ; McDowell, W.H.; Dodds, W.K.; Hamilton, S.K.; Gregory, S.; Morrall, D.D. Control of nitrogen export from watersheds by headwater streams. Science **2001**, *292* (5514), 86–90.

8. Naiman, R.J.; DeCamps, M.E. McClain, M.E. *Riparia: Ecology, Conservation, and Management of Streamside Communities*; Academic Press: Elsevier, 2005.

9. Vannote, R.L.; Minshall, G.W.; Cummins, K.W.; Sedell, J.R.; Cushing, C. E. The river continuum. Canadian J. Fisheries Aquat. Sci. **1980**, *37* (1), 130–137.

10. Mitsch, W.J. Gosselink, J.G. The value of wetlands: Importance of scale and landscape setting. Ecol. Econ. **2000**, *35* (1), 25–33.

11. Brinson, M.M. A *Hydrogeomorphic Classification for Wetlands*; WRP-DE-4, U.S. Army Corps of Engineers, Vicksburg, MS, USA 1993; 101 pp.

12. Mitsch, W.J. Landscape design and the role of created, restored, and natural riparian wetlands in controlling non-point source pollution. Ecol. Eng. **1992**, *1* (1–2), 27–47.

13. Baker, C.; Lawrence, R.; Montagne & D. Patten Mapping wetlands and riparian areas using Landsat ETM+ imagery and decision-tree-based models. Wetlands **2006**, *26* (2), 465–474.

14. Merot, P.; Squividant, H.; Aurousseau, P.; Hefting, M.; Burt, T.; Maitre, V.; Kruk, M.; Butturini, A.; Thenail, C.; Viaud, V. Testing a climato-topographic index for predicting wetlands distribution along an European climate gradient. Ecol. Model. **2003**, *163* (1–2), 51–71.

15. National Research Council. *Wetlands: Characteristics and Boundaries*; National Academy Press: Washington, D.C., 1995.

16. FGDC Wetlands Subcommittee. *Wetlands Mapping Standard*. Federal Geographic Data Committee: Reston, VA, 2009; 39 pp.

17. Ozesmi, S.L., Bauer, M. E. Satellite remote sensing of wetlands. Wetlands Ecol. Manag. **2002**, *10* (5), 381–402 http://dx.doi.org/10.1023/A:1020908432489

18. Shapiro, C. *Coordination and Integration of Wetland data for Status and Trends and Inventory Esitimates*; Federal Geographic Data Committee: Reston, VA, 1995.

19. Adam, E.; Mutanga, O.; Rugege, D. Multispectral and hyperspectral remote sensing for identification and mapping of wetland vegetation: a review. Wetlands Ecol. Manag. **2010**, *18* (3), 281–296.

20. Tiner, R.W. *Wetland Indicators: A Guide to Wetland Identification, Delineation, Classification, and Mapping*; CRC Press: Boca Raton, FL, 1999.

21. Richardson, J.L. Vepraskas, M.J. *Wetland soils: Genesis, Hydrology, Landscapes, and Classification*; Lewis Publishers: Boca Raton, FL, 2001.

22. Lyon, J.G.; Lyon, L. K. *Practical Handbook for Wetland Identification and Delineation,* 2nd Ed.; CRC Press, Taylor and Francis Group: Boca Raton, FL, 2011.

23. Dahl, T.E. *Status and Trends of Wetlands in the Contermi-nous United States 1986 to 1997*; U.S. Department of the Interior, Fish and Wildlife Service: Washington, D.C., 2000; 82 pp.

24. Yang, X. Integrated use of remote sensing and geographic information systems in riparian vegetation delineation and mapping. Int. J. Remote Sensing **2007**, *28* (2), 353–370.

25. Ramsey, E.W., III. Radar remote sensing of wetlands. In *Remote Sensing Change Detection: Environmental Monitoring Methods and Application;* Lunetta, R. S., Elvidge, C.D., Eds.; Sleeping Bear Press, Inc.: Chelsea, MI, 1998; 211–243 pp.

26. Henderson, F.M.; Lewis, A.J. Radar detection of wetland ecosystems: a review. Int. J. Remote Sensing **2008**, *29* (20), 5809–5835.

27. Bourgeau-Chavez, L.L.; Kasischke, E.S.; Brunzell, S.M.; Mudd, J.P.; Smith, K. B. ; Frick, A. L. Analysis of space-borne SAR data for wetland mapping in Virginia riparian ecosystems. Int. J. Remote Sensing **2001**, *22* (18), 3665–3687.

28. Lang, M.W.; McCarty, G. W. Lidar intensity for improved detection of inundation below the forest canopy. Wetlands **2009**, *29* (4), 1166–1178.

29. Julian, J.; Young, J.; Jones, J.; Snyder, C.; Wright, C. The use of local indicators of spatial association to improve LiDAR-derived predictions of potential amphibian breeding ponds. J. Geographical Syst. **2009**, *11* (1), 89–106.

30. Lefsky, M.A.; Cohen, W.B.; Parker, G.G; Harding, D. J. Lidar remote sensing for ecosystem studies. BioScience **2002**, *52* (1), 19–30.

31. Elvidge, C.D.; Miura, T.; Jansen, W.T.; Groeneveld, D.P.; Ray, J. Monitoring trends in wetland vegetaion using a landsat MSS time series. In *Remote Sensing Change Detection: Environmental Monitoring Methods and Application;* Lunetta, R.S., Elvidge, C.D., Eds.; Sleeping Bear Press, Inc.: Chelsea, MI, 1998; 191–210 pp.

32. Townsend, P.A.; Walsh, S. J. Remote Sensing of Forested Wetlands: Application of Multitemporal and Multispectral Satellite Imagery to Determine Plant Community Composition and Structure in Southeastern USA. Plant Ecol. **2001**, *157* (2), 129–149.

33. Sørensen, R.; Zinko, U.; Seibert, J. On the calculation of the topographic wetness index: evaluation of different methods based on field observations. Hydrol. Earth Syst. Sci. **2006**, *10* (1), 101–112.

34. Creed, I.F.; Sanford, S.E.; Beall, F.D.; Molot, L.A.; Dillon, P.J. Cryptic wetlands: integrating hidden wetlands in regression models of the export of dissolved organic carbon from forested landscapes. Hydrol. Processes **2003**, *17* (18), 3629–3648.

35. Rodhe, A. & J. Seibert. Wetland occurrence in relation to topography: a test of topographic indices as moisture indicators. Agric. Forest Meteorol. **1999**, *98–99*, 325–340.

36. Curie, F.; Gaillard, S.; Ducharne, A.; Bendjoudi, H. Geomorphological methods to characterise wetlands at the scale of the Seine watershed. Sci. Total Environ. **2007**, *375* (1–3), 59–68.

37. Murphy, P.; Ogilvie, J.; Connor, K.; & Arp, P. Mapping wetlands: A comparison of two different approaches for New Brunswick, Canada. Wetlands **2007**, *27* (4), 846–854.

38. Fan, Y.; Miguez-Macho, G. A simple hydrologic framework for simulating wetlands in climate and earth system models. Climate Dynamics **2011**, *37* (1–2), 253–278.

Riparian—
Wetlands

39. Leavesley, G.A.; Hay, L.E., Viger, R.J.; Markstrom, S. L. Use of a priori parameter-estimation methods to constrain calibration of distributed-parameter models. Water Sci. Appl. **2003**, *6*, 255–266.

40. Oreskes, N.; Shrader-Frechette, K. Belitz, K. Verification, validation, and confirmation of numerical models in the earth sciences. Science **1994**, *263* (5147), 641–646.

41. Wilson, J.P. ; Gallant, J. C. *Terrain Analysis: Principles and Applications*; John Wiley & Sons, Inc.: New York, NY, 2000.

42. Bourgeau-Chavez, L.L.; Riordan, K.; Powell, R.B.; Miller, N.; Barada, H. Improving Wetland Characterization with Multi-Sensor, Multi-Temporal SAR and Optical/Infrared Data Fusion. In *Advances in Geoscience and Remote Sensing*; Jedlovec, G., Ed.; InTech: 2009; 742 p.

43. Li, J. Chen, W. A rule-based method for mapping Canada's wetlands using optical, radar and DEM data. Int. J. Remote Sensing **2005**, *26* (22), 5051–5069.

44. Hart, J.K.; Martinez, K. Environmental Sensor Networks: A revolution in the earth system science? Earth Sci. Rev. **2006**, *78* (3–4), 177–191.

45. Quinn, N.W.T.; Ortega, R.; Rahilly, P. J. A.; Royer, C. W. Use of environmental sensors and sensor networks to develop water and salinity budgets for seasonal wetland real-time water quality management. Environ. Model. Soft. **2010**, *25* (9), 1045–1058.

46. Pitt, A.; Baldwin, R.; Lipscomb, D.; Brown, B.; Hawley, J.; Allard-Keese, C.; Leonard, P. The missing wetlands: using local ecological knowledge to find cryptic ecosystems. Biodivers. Conserv. **2012**, *21* (1), 51–63.

Riparian Zones: Groundwater Nitrate (NO$_3^-$) Cycling

D. Q. Kellogg
Department of Natural Resources Science, University of Rhode Island, Kingston, Rhode Island, U.S.A.

Abstract

Nitrate is the most mobile form of nitrogen in groundwater and can be transported long distances without transformation. Coastal estuaries are sensitive to excessive nitrogen loading from the watershed, with over-fertilization contributing to eutrophication and hypoxia, stressing marine life. Vegetated riparian zones are a common management practice used to reduce nitrate delivery to coastal waters. Because of their position between terrestrial and aquatic ecosystems, vegetated riparian zones provide opportunities for nitrate transformation before emerging to surface water. Riparian zones vary spatially and temporally and this variation affects the extent to which groundwater nitrate cycling occurs. Plant uptake, microbial immobilization and denitrification are major groundwater nitrate transformation pathways.

INTRODUCTION

Groundwater nitrate cycling in riparian zones is an important component in managing nitrogen delivery to surface waters. Humans have had an enormous effect on the global nitrogen (N) cycle, dramatically increasing nitrogen pools in atmospheric, terrestrial and aquatic ecosystems over the last century, largely through the production and use of mineral fertilizers, but also through the burning of fossil fuels and cultivation of nitrogen-fixing crops, such as legumes.[1] One effect has been the over-fertilization of estuarine ecosystems, which are generally nitrogen-limited. When nitrogen is added to these nitrogen-limited systems algae blooms are excessive and are not consumed, drifting to bottom waters where they are decomposed by microbes. Microbial activity depletes dissolved oxygen in the water column, and contributes to hypoxia (low oxygen) in coastal ecosystems, stressing marine life.[2] An example of this phenomenon is the so-called "dead zone" in the Gulf of Mexico, into which the Mississippi River flows, carrying nitrogen from distant reaches in the watershed to the outlet in the Gulf.[3] Other coastal estuaries around the world are also showing symptoms of over-fertilization. For example, in the U.S. the Chesapeake Bay is a large and economically important estuary on the east coast where nitrogen has been identified as a pollutant of concern, prompting aggressive nitrogen management throughout the watershed, i.e., the area of land that contributes water to a given water body.[4] Additionally, nitrate in groundwater contaminates drinking water supplies and is a human health hazard, regulated by the Clean Water Act. Groundwater nitrate is increasing in many areas of the United States and is being actively monitored by the United States Geological Survey.[5]

SOURCES OF NITRATE IN GROUNDWATER

Anthropogenic sources of nitrate to groundwater include mineral fertilizers applied to lawns and agricultural lands, on-site septic systems, the cultivation of nitrogen-fixing crops, and animal waste. Atmospheric deposition, as rain or particulates, carries N from the burning of fossil fuels and also contributes nitrogen to ecosystems.[6] About 78% of the earth's atmosphere is made up of nitrogen gas in the form of dinitrogen (N_2), a relatively inert gas due to its triple bond. Nitrogen is an essential nutrient to life, and most ecosystems are relatively N limited.[7] With the development of the Haber-Bosch process that takes N_2 out of the atmosphere to create mineral fertilizers, agricultural systems have expanded in size and production to meet the demand for food for the world's growing population.[8] As a result nitrogen delivery to terrestrial and aquatic ecosystems has increased, frequently over-fertilizing these systems. Nitrogen is found in both organic and inorganic forms. Inorganic, or mineral, nitrogen forms include ammonium (NH_4^+), nitrite (NO_2^-) and nitrate (NO_3^-). Ammonium and nitrate are "bioavailable" or accessible to plants and microbes, while nitrite is rarely found in high concentrations in nature and is toxic to plants. Nitrate is the most stable form of N in groundwater because of the presence of oxygen in most groundwaters. Due to its negative charge nitrate is also mobile and easily transported with groundwater. Fine soil particles (i.e., clays and humus) are almost always negatively charged and repel nitrate, except where soils are very acidic.[9,10]

RIPARIAN ZONES AND MANAGEMENT OF GROUNDWATER NITRATE

Riparian zones are the lands next to surface water (e.g., rivers, streams, lakes, reservoirs) and can play an important

Encyclopedia of Natural Resources DOI: 10.1081/E-ENRW-120047595

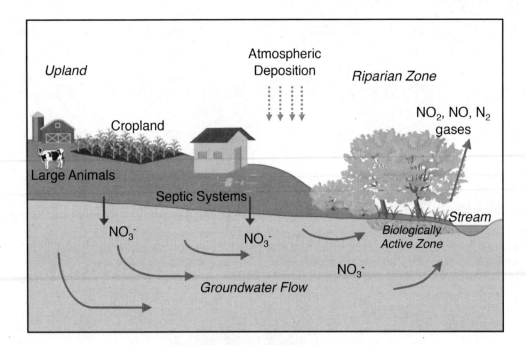

Fig. 1 Idealized riparian zone, with cropland, large animals, septic systems and atmospheric deposition as sources of nitrogen to groundwater. In humid climates groundwater flows toward surface water, carrying nitrate (NO₃⁻) to streams and ponds, and eventually to the coast.

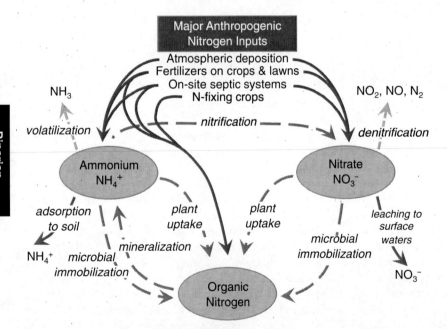

Fig. 2 Simplified nitrogen cycle in the subsurface. Anthropogenic (human) inputs include atmospheric deposition, fertilizers, on-site septic systems and N-fixing crops.

role in mediating nitrogen delivery to N-limited estuaries.[11] Situated between terrestrial and aquatic ecosystems (Fig. 1), riparian zones display gradients in water table, vegetation, soil wetness, soil carbon, temperature, oxygen, and many other biogeochemical characteristics, providing opportunities for biogeochemical transformations.[12]

GROUNDWATER NITRATE PATHWAYS FOR TRANSFORMATION

As groundwater flows toward surface water, interaction with shallow, biologically active soils provides opportunities for

nitrate transformations: plant uptake, microbial immobilization, and denitrification (Fig. 2). These shallow soils are rich in organic matter (mostly decomposing plant debris), providing essential elements for microbial activity and plant growth. Decomposition begins the cycle of converting organic nitrogen back to inorganic nitrogen, returning bioavailable nitrogen to the ecosystem as ammonium and nitrate.

Plant Uptake

Nitrogen is necessary for plants to thrive, providing the building blocks for chlorophyll and essential proteins that drive photosynthesis and growth. With the exception of nitrogen-fixing plants (e.g., legumes) that extract N$_2$ from the atmosphere, plants absorb mineral nitrogen as nitrate and ammonium from soil pore water.[10] During the growing season trees and grasses actively transpire, pulling groundwater up toward the soil surface and taking up nutrients, often limiting nitrate transport towards surface waters.[13] During the dormant season when plants are in senescence and temperatures are cooler, evaporative demand slows dramatically as does plant uptake of nutrients. Nitrogen may also be removed from riparian zones through active riparian zone management such as tree harvesting.

Microbial Immobilization

There are countless types of microbes found in soils and each plays a different role in ecosystem function, but all require nitrogen to grow and thrive. Microbial immobilization refers to the process by which microbes take up mineral nitrogen, transforming it to an organic form as it is used for biological function. When microbes die, they decompose and mineral N is released back to the ecosystem.

Decomposition

Decomposition of plant debris initiates the active cycling of nitrogen, with plant-based N undergoing transformations by soil invertebrates and microbial communities, resulting in organic matter and microbial biomass. These organic N pools can be stored for long periods in wetter soils. Wetter soils (also called hydric soils and frequently identified as wetlands) tend to be low in oxygen (anaerobic) because oxygen diffusion into water from the atmosphere is slow. Decomposition is slower in anaerobic soils, allowing organic matter to accumulate. When these soils are drained, nitrogen can be released as decomposition increases. Organic N is transformed to inorganic forms through mineralization of organic matter, releasing bioavailable nitrogen (NH$_4^+$ and NO$_3^-$) back to the ecosystem.[7] The process begins with the production of ammonium (NH$_4^+$). Ammonium may be taken up by plants, immobilized by microbes, volatilized as ammonia (NH$_3$), adsorbed to negatively charged soil particles, or oxidized to NO$_3^-$ (nitrification) in the presence of oxygen. Nitrate may be taken up by plants, immobilized by microbes, denitrified, or leached to surface waters. In some environments nitrate may also be adsorbed to positively charged soil particles.[14]

Denitrification

Denitrification is the subject of much study because it is the only process by which nitrogen is returned to the atmosphere and permanently removed from aquatic or terrestrial ecosystems. Denitrification is largely carried out by microbes that use the oxygen in the NO$_3^-$ for respiration, releasing dinitrogen (N$_2$), nitrous oxide (N$_2$O), and nitric oxide (NO) gases. Microbial denitrification requires low oxygen (anaerobic) conditions, a source of electron donors, most frequently in the form of organic carbon such as decomposing plant matter, and a supply of NO$_3^-$. In general, nitrate is denitrified in a series of steps:

$$2NO_3^- \rightarrow 2NO_3^- \rightarrow 2NO \rightarrow N_2O \rightarrow N_2.$$

The conditions and types of denitrifiers determine the specific mechanisms involved. The process does not always go to completion, which results in the production of NO and N$_2$O gases, both of which have negative environmental impacts. They contribute to acid rain and ground-level ozone, act as greenhouse gases in the upper atmosphere, and may contribute to the destruction of the protective ozone layer in the stratosphere.[10] Alternative electron donors that have been found to play a significant role in groundwater denitrification include inorganic sulfide (often in the form of pyrite, FeS$_2$), and reduced iron, Fe(II).[15] Other potentially important microbial transformations of groundwater nitrate are dissimilatory reduction of nitrate to ammonium (DNRA), and anaerobic ammonium oxidation (anammox).[16]

Riparian zones vary spatially in soils, slope, vegetation, and other physical characteristics affecting the extent to which they can remove groundwater nitrate through denitrification. Some characteristics vary seasonally, such as temperature and water table elevation. Some factors directly influence denitrification rates, while others control groundwater flowpaths and residence time that influence the extent to which nitrate-laden groundwater interacts with denitrifying microbes in the biologically active zone.

Factors that directly influence denitrification rates

Laboratory and field studies have explored the limiting factors of microbial denitrification. These include a supply of nitrate, low oxygen (anaerobic) conditions, and the availability of electron donors. Temperature and pH play a less critical role, but also influence denitrification rates.[17] The transformation of ammonium to nitrate requires oxygen, while denitrification requires anaerobic conditions. Riparian zones frequently provide the conditions necessary to support nitrification and denitrification because they are at the interface between terrestrial and aquatic systems and possess a gradient of biogeochemical conditions, such as oxygen status. Therefore these wetland riparian soils are most favorable for groundwater denitrification; however, not all riparian zones have wet soils. If a stream is deeply incised, then riparian soils tend to be dryer, keeping surface soils more aerated and limiting the accumulation of organic matter.[18] Groundwater denitrification tends to be patchy and occurs in hot spots in the subsurface[19] making field measurements of groundwater denitrification rates challenging.[20]

Factors that influence groundwater nitrate interaction with denitrifying microbes

Field studies have focused on the many physical factors that influence the interaction between nitrate-laden groundwater and denitrifying microbes, such as groundwater flow paths [21] and residence time. In humid climates, groundwater generally flows towards streams and ponds and supplies water to these systems when it is not actively raining (baseflow) (Fig. 1). Total groundwater nitrate removal by denitrification in riparian zones depends on slope, the depth of permeable soils in the upland, and the depth of permeable soils in the riparian area.[22] These factors determine the supply of nitrate to the riparian zone as well as the residence time within the riparian zone. Residence time can also be expressed in terms of riparian width or the distance necessary to remove a given percentage of nitrate. The longer the residence time (or greater the width) the greater the potential for nitrate removal. Highly permeable soils allow for a large volume of groundwater to move through the riparian zone and potentially a large supply of nitrate, but highly permeable soils also reduce the residence time. All of these factors explain why riparian nitrate removal is so variable across the landscape.[23] Regardless of the variability, vegetated riparian zones help to buffer aquatic ecosystems from nitrate loading and are an effective nitrogen management practice.[24] Urbanization and intensive agriculture have degraded riparian buffers in the past, prompting restoration efforts to improve water quality.[25,26]

CONCLUSIONS

Groundwater nitrate cycling in riparian zones affects nitrogen delivery to surface waters. In many environments nitrate is the most mobile form of nitrogen in groundwater and can be transported long distances without transformation. Coastal estuaries are sensitive to excessive nitrogen loading from the watershed, with over-fertilization contributing to eutrophication and hypoxia, stressing marine life. Vegetated riparian zones are a common management practice used to reduce nitrate delivery to coastal waters. Because of their position between terrestrial and aquatic ecosystems, vegetated riparian zones provide opportunities for nitrate transformation before emerging to surface water. Riparian zones vary spatially and temporally and this variation affects the extent to which groundwater nitrate cycling occurs. Plant uptake, microbial assimilation and denitrification are major groundwater nitrate transformation pathways. Denitrification is the only process that returns nitrogen to the atmosphere as N$_2$, N$_2$O and NO, with the latter two gases representing incomplete reduction to N$_2$ and having negative impacts on the environment. Riparian wetlands provide conditions that are particularly favorable to microbial denitrification. Nitrogen removal

may also occur through agroforestry management approaches such as selective tree harvesting. Mineralization of decomposing plants and microbes returns bioavailable inorganic nitrogen, as nitrate and ammonium, to the riparian ecosystem.

REFERENCES

1. Galloway, J.N.; Dentener, F.J.; Capone, D.G.; Boyer, E.W.; Howarth, R.W.; Seitzinger, S.P.; Asner, G.P.; Cleveland, C.C.; Green, P.A.; Holland, E.A.; Karl, D.M.; Michaels, A.F.; Porter, J.H.; Townsend, A.R.; Vorosmarty, C.J. Nitrogen cycles: past, present, and future. Biogeochemistry **2004**, *70*(2), 153–226.

2. Diaz, R.J.; Rosenberg, R. Spreading dead zones and consequences for marine ecosystems. Science **2008**, *321*, 926–929.

3. Rabalais, N.N.; Turner, R.E.; Wiseman, Jr., W.J. Hypoxia in the Gulf of Mexico. J. Environ. Qual. **2001**, *30* (2), 320–329.

4. Boesch, D.F.; Brinsfield, R.B.; Magnien, R.E. Chesapeake Bay eutrophication: Scientific understanding, ecosystem restoration, and challenges for agriculture. J. Environ. Qual. **2001**, *30*, 303–320.

5. Lindsey, B.D.; Rupert, M.G. Methods for evaluating temporal groundwater quality data and results of decadal-scale changes in chloride, dissolved solids, and nitrate concentrations in groundwater in the United States, 1988-2010. U.S. Geological Survey Scientific Investigations Report 2012-5049.

6. Vitousek, P.M.; Aber, J.; Howarth, R.W.; Likens, G.E.; Matson, P.A.; Schindler, D.W.; Schlesinger, W.H.; Tilman, G.D. *Human Alteration of the Global Nitrogen Cycle: Causes and Consequences*. Issues in Ecology 1997, Issue 1. Ecological Society of America, Washington, D.C.

7. Schlesinger, W.H. *Biogeochemistry: An Analysis of Global Change*; Academic Press: London, 1997.

8. Erisman, J.W.; Sutton, M.A.; Galloway, J.; Klimont, Z.; Winiwarter, W. How a century of ammonia synthesis changed the world. Nat. Geosci. **2008**, *1* (10), 636–639.

9. Freeze, R.A.; Cherry, J.A. *Groundwater*; Prentice-Hall, Inc.: Englewood Cliffs, NJ, 1979.

10. Brady, N.C.; Weil, R.R. *The Nature and Properties of Soils*; 12th edition. Prentice-Hall, Inc.: Upper Saddle River, N.J, 1999.

11. Hill, A.R. Nitrate removal in stream riparian zones. J. Environ. Qual. **1996**, *25* (4), 743–755.

12. McClain, M.E.; Boyer, E.W.; Dent, C.L.; Gergel, S.E.; Grimm, N.B.; Groffman, P.M.; Hart, S.C.; Harvey, J.W.; Johnston, C.A.; Mayorga, E.; McDowell, W.H.; Pinay, G. Biogeochemical hot spots and hot moments at the interface of terrestrial and aquatic ecosystems. Ecosystems **2003**, *6* (4), 301–312.

13. Kellogg, D.Q.; Gold, A.J.; Groffman, P.M.; Stolt, M.H.; Addy, K. Riparian ground-water flow patterns using flownet analysis: Evapotranspiration-induced upwelling and implications for N removal. J. Am. Water Res. Assoc. **2008**, *44* (4), 1024–1034.

14. Korom, S.F.; Seaman, J.C. When "conservative" anionic tracers aren't. Ground Water **2012**, doi: 10.1111/j.1745-6584.2012.00950.x.

Korom, S.F.; Schuh, W.M.; Tesfay, T.; Spencer, E.J. Aquifer denitrification and in situ mesocosms: Modeling electron donor contributions and measuring rates. J. Hydrol. **2012**, *432–433*, 112–126.

15. Bergin, A.J.; Hamilton, S.K. Have we overemphasized the role of denitrification in aquatic ecosystems? A review of nitrate removal pathways. Front. Ecol. Environ. **2007**, *5* (2), 89–96.

16. Korom, S.F. Natural denitrification in the saturated zone: A review. Water Resour. Res. **1992**, *28* (6), 1657–1668.

17. Gold, A.J.; Groffman, P.M.; Addy, K.; Kellogg, D.Q.; Stolt, M.; Rosenblatt, A.E. Landscape attributes as controls on ground water nitrate removal capacity of riparian zones. J. Am. Water Resour. Assoc. **2001**, *37* (6), 1457–1464.

18. Parkin, T.B. Soil microsites as a source of denitrification variability. Soil Sci. Soc. Am. J. **1986**, *51* (5), 1194–1199.

19. Addy, K; Kellogg, D.Q.; Gold, A.J.; Groffman, P.M.; Ferendo, G.; Sawyer, C. In situ push-pull method to determine ground water denitrification in riparian zones. J. Environ. Qual. **2003**, *31* (3), 1017–1024.

20. Speiran, G.K. Effects of groundwater-flow paths on nitrate concentrations across two riparian forest corridors. J. Am. Water Resour. Assoc. **2010**, *46* (2), 246–260.

21. Vidon, P.G.F.; Hill, A.R. Landscape controls on nitrate removal in stream riparian zones. Water Resour. Res. **2004**, *40*, W03201, DOI: 10.1029/2003WR002473.

22. Mayer, P.M.; Reynolds, S.K.; McCutchen, M.D.; Canfield, T.J. Meta-analysis of nitrogen removal in riparian buffers. J. Environ. Qual. **2007**, *36* (4), 1172–1180.

23. Lowrance, R.; Altier, L.S.; Newbold, D.; Schnabel, R.R.; Groffman, P.M.; Denver, J.M.; Correll, D.L.; Gilliam, J.W.; Robinson, J.L.; Brinsfield, R.B.; Staver, K.W.; Lucas, W.; Todd, A.H. Water quality functions of riparian forest buffers in Chesapeake Bay watersheds. Environ. Manage. **1997**, *21* (5), 687–712.

24. Mitsch, W.J.; Day, J.W.; Gilliam, J.W.; Groffman, P.M.; Hey, D.L.; Randall, G.W.; Wang, N. Reducing nitrogen loading to the Gulf of Mexico from the Mississippi River Basin: Strategies to counter a persistent ecological problem. BioScience **2001**, *51* (5), 373–388.

25. Weller, D.E.; Baker, M.E.; Jordan, T.E. Effects of riparian buffers on nitrate concentrations in watershed discharges: new models and management implications. Ecol. Appl. **2011**, *21* (5), 1679–1695.

Riparian— Wetlands

Stormwaters: Management

John C. Clausen
Department of Natural Resources and the Environment, University of Connecticut, Storrs, Connecticut, U.S.A.

Michael E. Dietz
Center for Land Use Education and Research (CLEAR), University of Connecticut, Storrs, Connecticut, U.S.A.

Abstract

Stormwater management is the handling and control of stormwater. Stormwater is caused by rainfall rates exceeding infiltration rates, and increases with urbanization as a result of impervious surfaces such as roads, parking lots, and buildings. Stormwater best management practices (BMPs) have been developed to reduce peak flow rates and control the volume of runoff. The effectiveness of BMPs in treating stormwater quality varies with the pollutant, the BMP, the land area contributing runoff to the BMP, and the time elapsed since the previous rainfall, rainfall intensity, and many other variables.

INTRODUCTION

Stormwater is excess water running off the surface of the land during and immediately after a period of rain.[1] It is also termed rainfall excess, representing water that is not infiltrated into the soil. Direct runoff and surface runoff are other synonyms. A related term is overland flow, which is water that travels over the surface of the land before it enters a water body, best management practice (BMP), or stormwater collection system.[1] Stormwater management is generally regarded as everything to do with the handling and control of stormwater. This entry emphasizes stormwater management from the perspectives of stormwater peak control and stormwater volume control. Water quality aspects of stormwater management are also explained. BMPs are briefly described and their pollutant removal effectiveness is summarized.

STORMWATER CAUSES

Stormwater is generally caused by rainfall rates exceeding the infiltration capacity of the soil's surface. The melting of snow can also result in stormwater. Rainfall excess is commonly shown graphically by comparing infiltration rates and rainfall rates (Fig. 1). Depression storage capacity is the volume of water required to fill surface depressions to their overflow levels.[1] If the rainfall intensity exceeds the infiltration capacity of the soil, this excess begins to fill depression storage. Infiltration is defined as the process of water entering the surface of the soil. As a rainstorm progresses, this rate typically declines over time, as shown in Fig. 1, due to spaces in the soil being filled with water. Saturated areas in watersheds, often near streams, are more likely to be a source of runoff as compared to other parts of the watershed. These have been termed saturated source areas.

Urbanization is known to increase stormwater runoff.[2] Stormwater is commonly thought to be caused only by impervious surfaces in urbanized areas, also called total impervious area.[3] Typical impervious areas include streets, roof tops, and parking lots. Some of these areas might drain to pervious areas, and therefore are not directly connected to downstream areas. However, soil alteration from construction and compaction can also cause stormwater runoff; thus, a surface like a lawn that is commonly thought to be pervious might generate significant stormwater runoff, especially if the rain storm has a high intensity. The term "effective impervious area" refers to impervious areas that are directly connected to downstream water bodies and can include the typical impervious areas and compacted or low-permeability soils.[3]

WATER QUALITY CHARACTERISTICS OF STORMWATER

The water quality characteristics of stormwater are a function of its source. For example, pollutants from land surface sources that are washed into surface water resources often have components that are particulate as well as dissolved. Subsurface sources of pollutants are primarily in the dissolved form. Nutrients, such as phosphorus, and metals are found mostly in particulate forms in stormwater runoff. Nitrogen in stormwater runoff can be in several forms, and the dominant form may change within a BMP due to nitrogen cycling. Treatment of the quality of stormwater runoff must consider the pollutants involved and their form,

Encyclopedia of Natural Resources DOI: 10.1081/E-ENRW-120047552

Riparian—
Wetlands

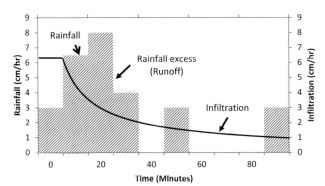

Fig. 1 Hypothetical rainfall, infiltration, and runoff during a rainstorm.

whether particulate or dissolved. Stormwater is known to contain increased concentrations of sediment, nutrients (nitrogen, phosphorus), metals (copper, lead, zinc), bacteria, and hydrocarbons. Increasingly, chloride and other deicing chemicals are found in stormwater runoff.

STORMWATER CONTROL

The control of stormwater is conducted both at a watershed scale and through the use of individual practices. The concept of low-impact development (LID) has emerged for primarily new residential developments. The individual practices are known as BMPs. Both approaches are discussed.

Low-Impact Development

The goal of LID is the preservation of the predevelopment hydrology of a site. More specifically, predevelopment runoff peak, volume, and timing are to be maintained. To accomplish these goals, careful planning needs to begin long before construction. Design concepts include preserving functional streams and wetlands, fitting a development to the terrain of the site, and preserving as much native vegetation and soils as possible. The amount of impervious surface that is directly connected to the stormwater conveyance system is typically minimized in a LID. Common practices for impervious area reduction include bioretention, on-site infiltration, pervious pavements, vegetated swales, rainwater harvesting, and vegetated or "green" roofs. Road and driveway sizes and lengths also can be minimized, to a point consistent with safety standards.

Best Management Practices

To control the adverse effects of urbanization on stormwater, BMPs have been developed. A stormwater BMP is a practice, either structural or nonstructural, to reduce nonpoint source pollution from stormwater. A report from the National Research Council[4] recommends use of the term stormwater control measure (SCM) rather than BMP. Initially, stormwater management practices were aimed to mitigate the increased peak flow rate that occurs from development. Practices were designed to treat the 100-year, 50-year, or 25-year 24-hour rainfall amount.[5] The 100-year, 24-hour storm is the rainfall amount, based on historical records, that is expected to occur once every 100 years for a 24-hour duration. The following practices are used to reduce the peak flow rate, thereby decreasing downstream flooding and bank scouring. Stormwater prevention is perhaps the most cost efficient mechanism for stormwater control. Approaches for stormwater prevention include education, proper land use planning and site design, reduction of impervious areas, and preservation of undisturbed soils.[4] Many BMPs require pretreatment and all require regular maintenance, further described later.

Practices to Reduce Peak Flow Rate

Common practices traditionally used to control peak flow rate associated with stormwater runoff are listed in Table 1. Other practices are possible, but may only provide peak runoff control for more frequent, smaller storms (e.g., 2-year, 10-year). The overall approach to peak control is to first provide available storage volume, and second to provide a control structure that lets the water out more slowly at a rate that mimics predevelopment conditions. Commonly used outlet control structures include weirs, pipes, and orifices. Dry detention basins were the first practices used to reduce peak flow rates. A dry detention basin is an excavated area with an outlet control structure, and it typically does not have a permanent pool of water. Stormwater inflow is stored in the basin and released over an extended period of time, usually 48 hours or less. However, these basins do not typically reduce the volume of stormwater leaving the basin. The bottom of the basin tends to become compacted and clogged with sediment, thus reducing infiltration into the soil. Extended detention basins are intended to store the water for longer time periods, which provides more time for water quality treatment (Fig. 2). These basins can be wet (ponded), dry, a combination, or include a wetland. Wet ponds are similar to dry ponds except that their outlet structure is elevated; therefore, the pond maintains a more permanent pool of water. Because wet ponds allow more time for particle settling, they provide additional water quality treatment.[6] Stormwater wetlands can be added to detention basin bottoms to enhance water quality treatment, at least temporarily.

Research on detention basins has shown that they have variable water quality benefits (Table 1). The existence of the permanent pool (Fig. 3) of water may help to settle out more suspended solids and pollutants attached to particulates as compared to detention basins, but their ability to retain nutrients such as N and P is variable, and generally poor (Table 1). These systems may also increase the

Fig. 2 Extended detention basin.

Fig. 3 Wet pond.

Table 1 Commonly used stormwater BMPs for peak control and their pollutant removal efficiencies

BMP	Pollutant removal (%)					Source
	Sediment	Nitrogen	Phosphorus	Metals	Pathogens	
Detention basins	30–65	15–45	15–45	15–45	<30	[20]
Extended detention basins	5–90	20–60	10–55	25–65	–	[21]
Wet ponds	−30–91	5–85	10–85	10–95	–	[21]

Table 2 Commonly used stormwater BMPs for volume control and their pollutant removal efficiencies

BMP	Pollutant removal (%)					Source
	Sediment	Nitrogen	Phosphorus	Metals	Pathogens	
Bioretention	54–99	30–99	−249–99	44–99	–	[10,11]
Green roofs	–	32	−191	−50–71	–	[17]
Grassed swales	30–88	15–45	15–45	2–67	<30	[18,20]
Infiltration basins	50–80	50–80	50–80	50–80	65–100	[20]
Infiltration trenches	50–80	50–80	15–45	50–80	65–99	[18,20]
Porous pavement	65–100	65–100	30–65	65–100	65–100	[20]
Sand filters	70	21	33	45	76	[19]

temperature of runoff waters. The addition of wetlands can improve treatment performance temporarily, but as the wetland vegetation dies and decomposes, additional N and P will be released and the wetland can become a source of N and P.

Practices to Reduce Runoff Volume

BMPs reduce runoff volume by infiltrating it into the soil, by enhancing evapotranspiration, or by water reuse (Table 2). Additional practices can provide some volume control on a more limited basis, such as grassed swales and buffer strips. BMPs with permanent pools generally provide little volume control because little or no infiltration occurs. Constructed wetlands and extended wet

detention basins typically have permanent pools of water and volume reduction is not significant. Various studies have shown that the volume of runoff leaving a site can be reduced to predevelopment levels, at least for smaller (<25 yr) storms.[7] Pollution reduction by volume control practices vary.

Bioretention

Bioretention is a vegetated depression in the landscape, which may have an engineered media that is designed to collect and infiltrate stormwater.[8] Unlike previously used "end-of-pipe" BMPs, one major advantage of bioretention is that it can be used in residential settings to treat runoff close to the source of generation. Bioretention areas, or rain gardens as they are commonly called in a residential setting, can be functional practices that are integrated into residential landscaping (Fig. 4).

Mechanisms for pollutant retention in bioretention include adsorption to mulch and soil media, exfiltration into native sub soils, plant uptake, and bacterial degradation. Laboratory research on bioretention has indicated the potential for excellent retention of several pollutants of concern such as N, P, copper, lead, and zinc.[9] More recent field research on bioretention has shown variable N and P removal, but high metals removal.[10,11] The variable N and P removal is due to the decomposition of dead plant material that releases nitrogen and phosphorus as well as differences in the absorptive capacity of the media. Sand, for example, does not retain P well. Chloride may not be well retained by this practice, but bacteria treatment has been successful.[11] Downstream thermal loading can be reduced simply by a reduction in volume transported. However, the effects of bioretention on stormwater temperatures have been mixed, showing little effect in one case [12] and reductions in another.[13] Bioretention use has expanded greatly throughout the Unites States since its introduction. Bioretention installations are also found in New Zealand and Australia.[14,15] By comparison, vegetated filter strips are shallow planted areas designed to spread stormwater out and slow it down. This allows more particles to settle and be filtered and more water to infiltrate into the soil.

Grassed swales

Vegetated swales can be used as an alternative to conventional curb and gutter conveyance systems (Fig. 5). Modern swales, also termed "bioswales,"[4] are designed differently from the older ditches that were just meant to convey runoff. Modern swales are engineered to provide storage and infiltration of stormwater, but they also will convey runoff from larger events to reduce flooding potential. However, pollutant removal is generally low in swales

(Table 2). Vegetated swales are similar to vegetated filter strips, but they are designed to convey stormwater as well. They may retain high percentages of suspended solids, N, and P, but the retention rates are variable (Table 2).

Green roofs

Vegetated or "green" roofs have been used for centuries in northern Europe. Benefits of green roofs include retention of rainfall, and in some cases, energy cost savings.[16] The traditional or "intensive" green roof consists of a thick soil layer capable of supporting grasses, plants, and sometimes even trees. As would be expected, the building structure below a green roof needs to be specially engineered to support its weight safely. An "extensive" green roof consists of a shallow media (10–15 cm) of expanded clays or other light materials. The advantage of the extensive system is that the weight is much less, so the system can be applied with less structural support. However, this type and thickness of media cannot support shrubs or trees. Typically, extensive green roofs are planted with *Sedum* species, which are highly tolerant of the harsh conditions that exist on a rooftop (Fig. 6). The amount of rainfall that is retained

Fig. 5 A grassed swale.

Fig. 4 Household rain garden.

Fig. 6 An extensive green roof.

by a green roof depends on the type of roof, the vegetation, as well as the regional climate. Retention can be a substantial portion of annual precipitation.[17] The green roof is an ideal BMP for highly developed urban areas where little greenspace exists for traditional BMPs, but open rooftops are plentiful.

Infiltration basins

Infiltration basins are large areas designed to infiltrate stormwater runoff. These types of basins are generally constructed in well-draining native soils, or soils that have been amended to encourage infiltration. They are designed to infiltrate most or all stormwater inflow, although they can be designed with storage and outlet control structures to provide more time for infiltration. An underdrain may also be installed to provide hydraulic relief if native soils have low infiltration capacity. Systems containing underdrains, however, typically provide little or no volume reduction. Due to the large volume of stormwater infiltrated with these systems, pollutant removal is generally high (Table 2).

Infiltration trenches

Stormwater runoff can also be piped to underground storage areas, called infiltration trenches. An infiltration trench is filled either with crushed stone or housed in other types of structures (concrete or plastic). The structure or the stone act as the storage reservoir so that part or all of the water can infiltrate into surrounding soils.[18] Solids should be removed through pretreatment to avoid clogging of these types of systems. Pollutant removal is dependent on design, but generally is high due to the large storage volume and potential recharge to groundwater (Table 2).

Pervious pavement

Several alternatives to traditional pavements have been developed: pervious interlocking concrete pavers (PICPs), plastic grid pavers, pervious asphalt, and pervious concrete. All are pervious and allow water to flow through the surface into the base materials. Although they differ in construction and appearance, all pervious pavements have one thing in common: the preparation of the base materials below the pavement is very different than with traditional pavement. For traditional pavement, a base of processed gravel, which contains a high proportion of fine material, gets compacted to provide a stable base for the pavement. Water flow through this compacted layer is slow. Base preparation for pervious pavements involves a crushed stone reservoir, a choker course of smaller aggregate, and then the pavement. The thickness of the reservoir depends on site needs, and in some cases this reservoir can be designed to act as a below-ground retention system to contain the 100-year storm.

Many types of PICPs now exist. In concept PICPs are similar to traditional concrete paver blocks; however, they are constructed in such a way that there are spaces between the blocks when they are installed. These voids are filled with crushed stone or gravel, and allow runoff to pass by the paver block into the subsurface layers (Fig. 7). Several types of plastic grid pavers also have been developed. They are constructed of various types of plastic, and are available in rolls or grids that lock together. The grids are highly pervious, and can be filled with topsoil and planted with turf, or filled with crushed aggregate.

Pervious asphalt has been in use for several decades (Fig. 8), although it was not initially used as an LID tool. It was originally termed open-graded friction course (OGFC), and was used on highways to improve traction and reduce road spray. The mix is similar to traditional asphalt, except the sand component is omitted. Numerous asphaltic binders have been developed to improve adhesion, and promote stability in hot conditions. One advantage to this type of pavement is that it can be applied with the same equipment that is used for traditional asphalt.

Fig. 7   Eco-stone® pavers.

Fig. 8   Pervious asphalt.

To avoid potential clogging of the pores in this material, winter sanding should be avoided. Periodic maintenance with vacuum suction or power washers may be required to remove fines that have been deposited on the surface.

Pervious concrete is a variation of traditional concrete, where the sand is omitted from the mix (Fig. 9). The ratio of Portland cement, water, and aggregate is very specific. The mix then gets rolled or tamped to a specific density. The goal is to make sure there is sufficient water in the mix to react with the Portland cement and bond the aggregates together, without sealing the surface. The application of this type of concrete is different from traditional concrete, and should only be performed by a contractor with previous installation experience. As with pervious asphalt, winter sand application should be avoided to reduce clogging potential, and just as with salt application on traditional concrete, care should be taken to use deicing agents that will not react with the concrete.

Rainwater collection/reuse

Rainwater collection has occurred for thousands of years to provide water for domestic uses. Modern rainwater harvesting is encouraged to reduce stormwater runoff volumes and provide some water for irrigation of landscaping plants. Rain barrels are the cheapest and easiest form of rainwater collection, although distribution of the water is limited to gravity draining (Fig. 10). Larger cisterns can be installed with filters and pressurized pump systems. Much larger volumes of water can be captured and utilized with a cistern, but installation is more costly. Typically, however, rainwater harvesting does not collect a large fraction of runoff.

Sand filters

Sand filters have been used for more than 100 years to treat water for drinking, but their use has expanded to stormwater. Sand filters are effective in removing several pollutants but provide little flow rate control.[19] Larger particles and the pollutants that are adsorbed to them can be physically strained and retained by the sand filter. Smaller particles and dissolved pollutants typically pass right through the filter without treatment. Most pollutants such as metals and P will occur in both particulate and dissolved form. Thus a sand filter can only remove a fraction of the total pollutant load. Others, such as N, and specifically nitrate nitrogen (NO_3–N), are almost all dissolved and essentially no removal will be achieved. Three designs of sand filters are in common practice: the Austin (Texas) sand filter, the Washington, DC, sand filter, and the Delaware sand filter.[19] The Austin filter is above ground or at grade while the Washington, DC, and Delaware filters are in below ground concrete tanks. The Austin sand filter is used for larger watersheds with mixed land uses up to 50 acres. The Washington, DC, and Delaware systems are used for smaller (1 acre) highly impervious watersheds. Experiments are being conducted on sand filters with peat and metal shavings for additional pollutant treatment.

Operation and maintenance

All stormwater BMPs require maintenance, the amount dependent on the actual practice, watershed inputs, and precipitation. They also have a design life, and require

Fig. 9 A pervious concrete sidewalk.

Fig. 10 Wooden rain barrel.

rejuvenation or replacement when that time is exceeded. Additionally, BMP pretreatment may extend the life of the practice. Because stormwater BMPs are found at the low end of the landscape and water flows into them, they are often collectors of trash and debris. Such debris can interfere with some treatment processes and should be regularly removed. Basins and ponds have outlets that can also clog and require regular maintenance. Forebays to such systems may prevent them from filling up with sediment, but these should be regularly cleaned. Bioretention systems may require media replacement, mulch renewal and plant maintenance. Excess mulching could fill up these BMPs. Infiltration BMPs and sand filters are prone to clogging. Operation and maintenance of sand filters is especially important. Clogging may occur within 3–5 years after construction, and the media may need replacement. Clogging may also occur with porous pavement and prevention of clogging is an important consideration.

Operation and maintenance are complicated by ownership of the BMP. Practices on private land may not receive the attention they are given in municipalities.

CONCLUSION

The management of stormwater is necessitated by the effects of urbanization on infiltration rates. Stormwater prevention is the first and perhaps best tool in reducing stormwater runoff. Several BMPs have been successfully applied in reducing the peak flow rate and the volume of runoff. Some of these practices work well for large rarer rain storms, whereas others only perform well for small frequent storms. Pollution retention by BMPs varies widely and depends on the pollutant and BMP type, the watershed land uses, the original design of the BMP, pretreatment, and the performance of regular maintenance.

REFERENCES

1. American Public Health Association. *Glossary: Water and Wastewater Control Engineering*; 1969; 387 pp.
2. Leopold, L.B. *Hydrology for Urban Land Planning – A Guidebook on the Hydrologic Effects of Urban Land Use.* U.S. Geological Survey Circular 554, 1968.
3. Booth, D.B.; Jackson, C.R. Urbanization of aquatic systems: degradation thresholds, stormwater detection, and the limits of mitigation. J. Am. Water Res. Assoc. **1997**, *33* (5), 1077–1090.
4. National Research Council. *Urban Stormwater Management in the United States*; The National Academies Press: Washington, DC. 2008.
5. USDA-Soil Conservation Service. *Urban Hydrology for Small Watersheds*; Technical Release 55: 1986.
6. Schueler, T.R. *Controlling Urban Runoff: A Practical Manual for Planning and Designing Urban BMPs.* Metropolitan Information Center. Metropolitan Washington Council of Governments. Washington, D.C. 1987.
7. Bedan, E.S.; Clausen, J.C. Stormwater runoff quality and quantity from traditional and low impact development watersheds. J. Amer. Water Res. Assoc. **2009**, *45*, 998–1008.
8. Clar, M.L.; Green, R. *Design Manual for Use of Bioretention in Stormwater Management*; Prince George's County, MD Department of Environmental Resources, Watershed Protection Branch, MD Department of Environmental Protection: Landover, MD. 1993.
9. Davis, A.P.; Shokouhian, M.; Sharma, H.; Minami, C. Laboratory study of biological retention for urban stormwater management. Water Environ. Res. **2001**, *73* (1), 5–14.
10. Dietz, M.E. Low impact development practices: a review of current research and recommendations for future directions. Water, Air, Soil Pollut. **2007**, *186*, 351–363.
11. Davis, A.P.; Hunt, W.F.; Traver, R.G.; Clar, M. Bioretention technology: overview of current practice and future needs. J. Environ. Eng. **2009**, *135* (3), 109–117.
12. Dietz, M.E.; Clausen, J.C. A field evaluation of rain garden flow and pollutant treatment. Water, Air, Soil Pollut. **2005**, *167*, 123–128.
13. Jones, M.P.; Hunt, W.F. Biotretention impact on runoff temperature in trout sensitive waters. J. Environ. Engr. **2009**, *135* (8), 577–585.
14. Read, J.; Wevill, T.; Fletcher, T.; Deletic, A. Variation among plant species in pollutant removal from stormwater in biofiltration systems. Water Res. **2008**, *42* (4–5), 893–902.
15. Trowsdale, S.A.; Simcock, R. Urban storm water treatment using bioretention. J. Hydro. **2011**, *397* (3–4), 167–174.
16. Hoffman, L. *Green Roofs: Ecological Design and Construction*; Schiffer Publishing Ltd.; PA. 2005.
17. Gregoire, B.G.; Clausen, J.C. Effect of a modular extensive green roof on stormwater runoff and water quality. Ecol. Engr. **2011**, *37*, 963–969.
18. ASCE. *Guide for Best Management Practice (BMP) Selection in Urban Developed Areas*; American Society of Civil Engineers: Reston, VA, 2001.
19. U.S. EPA. *Storm Water Technology Fact sheet: Sand Filters*; United States Environmental Protection Agency: Washington, D.C., EPA 832-F-99-007, 1999.
20. U.S. EPA. *Preliminary Data Summary of Urban Storm Water Best Management Practices*; United States Environmental Protection Agency: Washington, D.C., EPA-821-R-99-012, 1999.
21. U.S. EPA. *Guidance Specifying Management Measures for Sources of Nonpoint Pollution in Coastal Waters*; United States Environmental Protection Agency: EPA-840-B-92-002, 1993.

Stormwaters: Urban Modeling

Miguel A. Medina, Jr.
Department of Civil and Environmental Engineering, Duke University, Durham, North Carolina, U.S.A.

Cevza Melek Kazezyılmaz-Alhan
Department of Civil Engineering, Istanbul University, Istanbul, Turkey

Abstract

Modern urban water management has evolved from the need to address rapidly changing land-use that has local, regional and global hydrologic consequences. In an urban setting, water is concentrated into small natural channels, man-made open channels and storm sewers, natural and engineered swales, and detention structures. The complexity of the many interactions from runoff generated by precipitation over both pervious and impervious surfaces requires a comprehensive mathematical modeling approach, albeit limited by the availability of rainfall/runoff quantity and quality data within the study area. Quantitative methods incorporated into computer codes such as the comprehensive distributed routing U.S. EPA Storm Water Management Model (SWMM) are reviewed and applications are presented.

INTRODUCTION

Elements of the hydrologic processes that describe the transformation of precipitation into surface runoff are depicted in Figs. 1 and 2. Figure 1 illustrates the relationship among the rainfall process, infiltration to the subsurface, generation of overland flow, and flow into the man-made drainage system. Among the many variables that describe these processes mathematically are the width, area, percent imperviousness, ground slope, roughness parameters of the land cover for both impervious and pervious fractions, and several infiltration rate parameters that depend upon methods chosen. Eventually, the runoff makes its way into inlets of the man-made drainage system.

Figure 2 illustrates the transport of runoff through both natural and man-made conduits, as well as routing through a detention structure. Whether natural or man-made, the length, geometry, slope, and roughness characteristics must be specified for each conduit. Natural channels can be approximated with triangular, trapezoidal, and other cross-sectional shapes. For storage elements such as detention basins (there is also storage along a channel), stage-discharge curves and other geometric data are required. Although the focus of this entry is to demonstrate the physical basis for the hydrologic components represented, any mathematical model is an abstraction of the actual physical system it attempts to simulate. The limitations of numerical modeling and data requirements are discussed further, as well as the need for calibration. The authors recognize that there is more to urban water management than modeling. At the same time, management without application of these powerful tools is missing a planning tool that could be used very effectively to compare alternative quantity and quality control strategies.

OVERLAND FLOW

Overland flow is very thin sheet flow that develops after precipitation over a sloped land surface and infiltration to the subsurface. The depth and flow rate of this thin sheet of water depend, among others, on the rainfall intensity, ground surface characteristics (e.g., roughness, slope), and antecedent and immediate subsurface conditions (e.g., hydraulic conductivity, moisture). By neglecting the acceleration and pressure terms in the momentum equation for dynamic waves, the kinematic wave model[1] substitutes a steady uniform flow (stage–discharge) relationship for the momentum equation. However, unsteady flow is preserved through the continuity equation.

The dynamic wave equations for overland flow consist of continuity and momentum equations, as follows, respectively:[2]

$$\frac{\partial y}{\partial t} + V \frac{\partial y}{\partial x} + y \frac{\partial V}{\partial x} = i - f \tag{1}$$

$$\frac{\partial V}{\partial t} + V \frac{\partial V}{\partial x} + g \frac{\partial y}{\partial x} = g\left(S_o - S_f\right) + \left[\left(i - f\right)\frac{V}{y}\right] \tag{2}$$

where y is depth of water (L), S_f is friction slope (L/L), S_o is bed slope (L/L), V is water velocity (L/T), i is rainfall intensity (L/T), f is infiltration rate (L/T), g is acceleration

Encyclopedia of Natural Resources DOI: 10.1081/E-ENRW-120047560

Riparian—
Wetlands

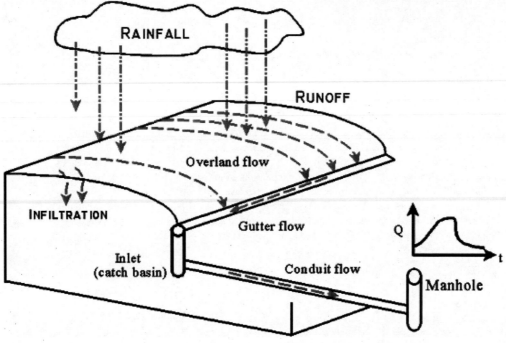

Subcatchment data: Width, Area, Percent Imperviousness, Ground Slope, Manning's *n* for impervious and pervious areas, Infiltration rate parameters

Fig. 1   Rainfall, infiltration, overland flow, and flow into drainage system.

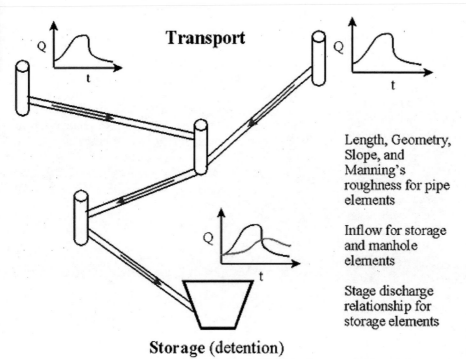

Length, Geometry, Slope, and Manning's roughness for pipe elements

Inflow for storage and manhole elements

Stage discharge relationship for storage elements

Fig. 2   Transport and storage components of the drainage system.

of gravity (L/T^2), t is time (T), and x is distance (L). These equations are for moderately wide overland flow and small bottom slope. An order-of-magnitude analysis of Eq. (2) shows that $(i - f)$ has a negligible effect on the flow dynamics.[2] The flow per unit width, q (L^2/T) is equal to Vy. Thus, Eq. (1) may be written as follows:

$$\frac{\partial q}{\partial x} + \frac{\partial y}{\partial t} = i - f \tag{3}$$

The simplest distributed routing method is the *kinematic wave* model, which neglects the local acceleration, convective acceleration, and pressure terms in the momentum equation (thus, friction and gravity forces essentially balance each other, $S_f = S_o$). The most common formulas relating water velocity, slope, and catchment roughness are the following:[3]

Manning formula: $V = \frac{k}{n} R^{2/3} \sqrt{S_o}$ (4)

Chezy formula: $V = C\sqrt{R S_o}$ (5)

where V is water velocity (L/T), n is the Manning friction coefficient (dimensionless), R is the hydraulic radius (L), C is the Chezy coefficient ($L^{1/2}/T$), $k = 1.486$ for English system units and 1.0 for the SI system, and the bed slope S_o replaces the friction slope S_f. For wide rectangular sections (e.g., overland flow), the hydraulic radius reduces to water depth y, and the momentum equation for kinematic waves reduces to

$$q = \alpha y^m \tag{6}$$

where, for *fully turbulent* flow:

using Manning: $\alpha = \frac{k}{n}\sqrt{S_o} \quad m = \frac{5}{3}$ (7)

using Chezy: $\alpha = C\sqrt{S_o} \quad m = \frac{3}{2}$ (8)

Values of exponent m for laminar and mixed laminar–turbulent conditions are tabulated by Ponce.[3] For overland flow problems using the kinematic wave, the continuity and momentum equations can be combined into one differential equation, as follows:

$$\left.\begin{array}{l}\frac{\partial y}{\partial t} + \frac{\partial q}{\partial x} = i - f \\ q = \alpha y^m\end{array}\right\} \Rightarrow \frac{\partial y}{\partial t} + \alpha\frac{\partial(y^m)}{\partial x} = i - f \tag{9}$$

CHANNEL FLOW

The distributed routing models allow us to compute flow rate and water-level variation through space and time: a major advantage over the lumped model in terms of the design criteria of any storage structure, such as a detention pond or reservoir. Similar to the overland flow problems,

by eliminating some terms in the momentum equation of the Saint-Venant equations, alternative distributed flow routing models are obtained. The dynamic wave equations for channel flow consist of continuity and momentum equations, as follows, respectively:[4]

$$\frac{\partial Q}{\partial x} + \frac{\partial A}{\partial t} = q(x,t) \tag{10}$$

$$\frac{1}{A}\frac{\partial Q}{\partial t} + \frac{1}{A}\frac{\partial}{\partial x}\left(\frac{Q^2}{A}\right) + g\frac{\partial y}{\partial x} - g(S_o - S_f) = 0 \tag{11}$$

where y is depth of water (L), S_f is friction slope (L/L), S_o is bed slope (L/L), Q is flow rate (L^3/T), q (x, t) is the net lateral inflow per unit length of channel (L^2/T), g is acceleration of gravity (L/T^2), A is the channel cross-sectional area (L^2), t is time (T), and x is distance (L). In Eq. (11), the first, second and third terms represent local acceleration, convective acceleration, and pressure force, respectively. The bed slope and friction slope in the last term represent the gravity force and friction force.

Again, the simplest distributed routing method is also the *kinematic wave* model, which neglects the local acceleration, convective acceleration and pressure terms in the momentum equation for dynamic waves. For open channel flows, the continuity and momentum equation and their combined form for kinematic waves are given as follows:

$$\left.\begin{array}{l}\frac{\partial A}{\partial t} + \frac{\partial Q}{\partial x} = q(x,t) \\ Q = \alpha A^m\end{array}\right\} \Rightarrow \frac{\partial A}{\partial t} + \alpha\frac{\partial(A^m)}{\partial x} = q(x,t) \tag{12}$$

where, for *fully turbulent* flow:

Manning: $\alpha = \frac{k}{n}\frac{\sqrt{S_o}}{P^{2/3}} \quad m = \frac{5}{3}$

Chezy: $\alpha = C\frac{\sqrt{S_o}}{\sqrt{P}} \quad m = \frac{3}{2}$ (13)

Here, A is the channel cross-sectional area (L^2), Q is the flow rate (L^3/T), n is the Manning friction coefficient (dimensionless), C is the Chezy coefficient ($L^{1/2}/T$), P is the wetted perimeter (L), S_o is the bed slope (L/L), t is time (T), q (x, t) is the net lateral inflow per unit length of channel, and x is the distance along the flow axis (L). As before, $k = 1.486$ for English system units and 1.0 for the SI system.

INFILTRATION

Most urban simulation models have used Horton's equation for prediction of infiltration capacity into the soil as a function of time,

$$f_{cap} = f_\infty + (f_0 - f_\infty)e^{-\alpha t} \tag{14}$$

where f_{cap} = infiltration capacity into soil, say in./hr;
f_∞ = minimum or ultimate value of f (at $t = \infty$), in./hr;
f_0 = maximum or initial value of f (at $t = 0$), in./hr;
t = time from beginning of storm, sec; and
α = decay coefficient, 1/time

The actual infiltration is given by the following equation:

$$f(t) = \min[f_{cap}(t), i(t)] \tag{15}$$

where f = actual infiltration into the soil, (say in./hr) and
i = rainfall intensity (say in./hr).

Equation 15 simply states that the actual infiltration will be the lesser of actual rainfall and available infiltration capacity.

Typical values for parameters f_0 and f_∞ are often greater than typical rainfall intensities. Thus, when Equation (14) is used such that f_{cap} is a function of time only, then f_{cap} will decrease even if rainfall intensities are very light. This results in a reduction in infiltration capacity regardless of the actual amount of entry of water into the soil. Thus, to correct this problem, the integrated form of Horton's equation may be used:

$$F(t_p) = \int_0^{t_p} f_{cap}\, dt = f_\infty t_p + \frac{f_0 - f_\infty}{\alpha}(1 - e^{-\alpha t_p}) \tag{16}$$

where F = cumulative infiltration at time t_p, (inches or mm). The true cumulative infiltration will be the following:

$$F(t) = \int_0^t f(\tau)\, d\tau \tag{17}$$

where f is given by Equation (15). Equations (16) and (17) may be used to define the equivalent time, t_p. That is, the actual cumulative infiltration given by Equation (17) is equated to the area under the Horton curve [given by Equation (16)] and the resulting equation is solved for t_p and serves as its definition.

$$F = f_\infty t_p + \frac{f_0 - f_\infty}{\alpha}(1 - e^{-\alpha t_p}) \tag{18}$$

Unfortunately, the equation cannot be solved explicitly for t_p: it must be done iteratively. Note that $t_p \leq t$ which states that the time t_p on the cumulative Horton curve will be less than or equal to the actual elapsed time. This also implies that available infiltration capacity $f_{cap}(t_p)$ will be greater than or equal to that given by Equation 14. Thus, f_{cap} will be a function of the actual water infiltrated and not a function of time only, with no other effects.

The error involved if the integrated form of Horton's model is not used is illustrated with a simple model, where the same Horton parameters and rainfall amounts are used, but the sequence of rainfall amounts is rearranged. The integrated form accounts for the changes while the original Horton model does not. Figure 3 shows a hyetograph with light rainfall early followed later by the higher rates. The original Horton model does poorly when light rainfall falls early, since it decays in time independently of accumulated infiltration. The integrated form of Horton's equation is applicable to urban environments, particularly for relatively short storm events. It should be noted that the Horton equation is an approximate solution to the *Richards*

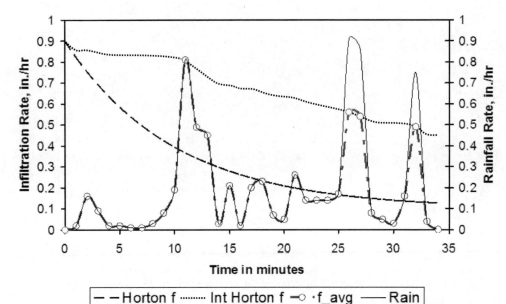

Fig. 3 Hyetograph with light rainfall during the earlier time steps.

equation: one-dimensional, unsteady, unsaturated flow—under several simplifying assumptions.

An alternative model is the Green–Ampt formulation, better suited for non-urban environments. Green and Ampt[5] derived the first physically based equation describing the infiltration of water into a soil. It has been refined and adapted for use in computer codes. A schematic representation of the Green–Ampt model is presented in Fig. 4. In Fig. 4, the wetting front has penetrated to a depth L in time t since infiltration began. Water is ponded to a small depth h_0 on the soil surface. The Green–Ampt equations for *cumulative infiltration F* and infiltration rate *f* are given by the following equations:

$$F(t) - \psi \, \Delta\theta \, \ln\left[1 + \frac{F(t)}{\psi \, \Delta\theta}\right] = Kt \tag{19}$$

$$f(t) = K\left[\frac{\psi \, \Delta\theta}{F(t)} + 1\right] \tag{20}$$

The parameters are defined as follows:
Ψ = wetting front soil suction head (cm)
θ = moisture content
K = hydraulic conductivity, (cm/h).

Equation 19 is commonly solved by Newton–Raphson iteration. Once F is found, then the infiltration rate is obtained. The parameters are well documented for a variety of soil classes (sand, sandy loam, loam, silt loam, sandy clay loam, clay loam, silty clay loam, sandy clay, and clay).

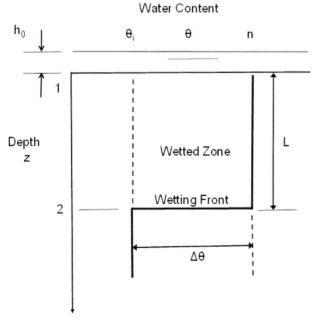

Fig. 4 Schematic for Green–Ampt equation.
Source: Adapted from Chow, Maidment et al.[4]

SURFACE WATER QUALITY

In surface water quality, pollutant accumulation and washoff processes are mathematically approximated. The assumptions of these processes are (i) the amount of pollutant, which can be removed during a storm event, depends on rainfall duration and initial quantity of pollutant available; (ii) chemical changes or biological degradation do not affect the pollutant decay during the runoff process; and (iii) the amount of pollutants percolating into the soil by infiltration is insignificant.[6] The exponential decay of solids build-up is generally represented as follows:

$$p_{n,j,t} = p_{\max_{n,j}}\left[1 - e^{-\lambda_{n,j}t}\right] \tag{21}$$

where $p_{n,j,t}$ is the loading of pollutant j on subcatchment n at time t [lb/acre (kg/ha)], $p_{\max n,j}$ is the maximum (asymptotic) loading of pollutant j on subcatchment n, [lb/acre (kg/ha)], $\lambda_{n,j}$ is power exponent coefficient for pollutant j on subcatchment n, [day^{-1}], and t is accumulation time [day]. The washoff is calculated by making it proportional to runoff rate at each time step of simulation as follows:

$$P_t = \frac{dP}{dt} = -ar^b P_o \tag{22}$$

where P_t is the pollutant load washed off at time t [lb/sec (kg/sec)], P_o is the load available for washoff at time t, [lb (kg)], r is the runoff rate, [in/hr (mm/hr)], a is the washoff coefficient, including conversion units, and b is a power constant. Pollutant transport in overland flow and through urban storage/treatment systems such as pipes, channels, detention basins, may be represented by the one-dimensional version of the classical advective–dispersive equation[7]

$$\frac{\partial C}{\partial t} = \frac{\partial}{\partial x}\left[E \frac{\partial C}{\partial x} - UC\right] \pm \sum_{i=1}^{n} S_i \tag{23}$$

where C is the concentration of pollutant [M/L^3], t is time [T], $-E\partial C/\partial x$ is the mass flux due to longitudinal dispersion along the flow axis (x-direction) [M/L^2], UC is mass flux due to advection by the fluid containing the mass of pollutant [M/L^{2}T], S_i is the sources or sinks of the substance C [M/L^{3}T], n is the number of sources or sinks, U is the flow velocity [L/T], and E is the longitudinal dispersion coefficient [L^2/T]. The mass balance equation for a well-mixed, variable volume unit is derived from Equation (23) and is given by Medina et al.:[7]

$$\frac{d(VC)}{dt} = I(t) \, C^I(t) - O(t) \, C(t) - K \, C(t) \, V(t) \tag{24}$$

Riparian— Wetlands

where V is the reservoir volume [L³], C^I is the influent pollutant concentration [M/L³], C is the effluent and reservoir pollutant concentration [M/L³], I is the inflow rate [L³/T], O is the outflow rate [L³/T], t is time [T], and K is the first-order decay coefficient [1/T]. This equation is solved numerically over time for concentration, using average values for quantities that might change over a time step, such as flow rate and volume.

ENVIRONMENTAL PROTECTION AGENCY STORM WATER MANAGEMENT MODEL (EPA SWMM)

EPA SWMM is a dynamic rainfall-runoff simulation model that can be used for single-storm event simulation or continuous simulation of multiple storms of a specific watershed. The model calculates the flow and water quality constituents of the watershed based on hydraulic parameters set by the user and provides graphs of simulation results such as flow hydrographs and water quality concentrations. The model is widely used to plan, analyze, and control storm water runoff; to design drainage system components; and to evaluate watershed management of both urban and nonurban areas.[6,8]

The runoff component of SWMM simulates surface runoff over subcatchment areas, as generated from rainfall input. Then, SWMM routes this runoff with the transport component through a conveyance system of pipes, channels, storage/treatment devices, pumps, and regulators. SWMM calculates the quantity and quality of runoff generated within each subcatchment, and the flow rate, flow depth, and quality of water in each pipe and channel during a simulation period comprised of multiple time steps. SWMM inputs include precipitation data, subcatchment delineation, pipe system characteristics, and soil properties. The watershed profile is formed by defining input data such as slope, area, and imperviousness of subcatchments; length, cross-section, height, and slope of conduits; porosity, hydraulic conductivity, and field capacity of aquifers, into the program. Outputs of the model include change of flow rate (hydrograph), change of concentration (pollutograph) through time and daily, monthly, annual, and total simulation summaries (for continuous simulation) and statistical frequency analysis. SWMM is also capable of viewing results in tabular and graphical form.

While calculating surface runoff, precipitation and flow from upstream subcatchments contribute to the inflow, infiltration and evaporation contribute to the outflow. For flow routing in conduits, steady flow, and kinematic, diffusion and dynamic wave routing options may be used. In addition, for infiltration calculation, three options exist, which are the Integrated Horton Method, the Green–Ampt Method, and the SCS Curve Number Method. EPA SWMM simulates surface runoff quality by using exponential, power, and saturation functions for buildup and exponential,

rating curve, and event mean concentration functions for washoff.

APPLICATION TO DUKE UNIVERSITY CAMPUS WATERSHEDS

The Duke University West Campus watershed is located in Durham, North Carolina, USA. Runoff from the West Campus of Duke University is simulated by using the EPA SWMM to predict surface runoff and quality and transport through the drainage system. The first formal application of EPA SWMM to the Duke Campus was completed by Mahi.[9]

Calibration and Discretization

In order to be able to use EPA SWMM for surface runoff and channel flow predictions, the model needed to be calibrated against measured rainfall and runoff data. A rain gage was placed on the roof of Hudson Hall (Pratt School of Engineering) and several storm events were measured in 15-min intervals. About 30 storms were recorded from 1994–1995: these were grouped into summer and winter storms for calibration purposes. The year 1997 was determined as the reference year to establish runoff *baseline conditions* (as approved by the City of Durham, NC). The last major calibration was conducted in 1996 (*and thus valid for 1997 baseline conditions*). In order to measure the flow rate, an aluminum compound weir was placed at the outlet of conduit 210, near node 345, a 1.83 m (6-ft.) diameter pipe (see Fig. 5). Typically, pressure transducers were located a few feet upstream of the weirs, transmitting date, time of day, and depth of flow data to electronic data loggers. Several model parameters (e.g., infiltration, depression storage, Manning's roughness, etc.) were adjusted to match predicted versus observed flows. An example of a winter storm calibration is shown in Fig. 6.

The discretization of the entire Duke University West Campus watershed using EPA SWMM is shown in Fig. 7: it illustrates the location of the weirs used for calibration (near node 345). The rapid urbanization (medical center) upstream of this measurement station over the past 5 years resulted in the decision to consider the design and construction of a large detention pond to control stormwater flowing off-campus (at the Erwin Road Outfall) into the City of Durham.

Modeling a Best Management Practice

Detention ponds are considered one of the most effective runoff control devices; in particular, to reduce peak flows, settle particulate matter, while also reducing some pollutant concentrations in the outflow. They are not as effective in reducing dissolved fractions of contaminants. SWMM5 can also simulate other BMPs; for example, low-impact development

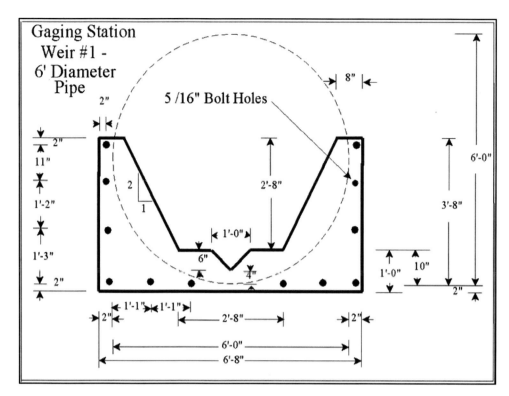

Fig. 5 Compound weir at node 345.

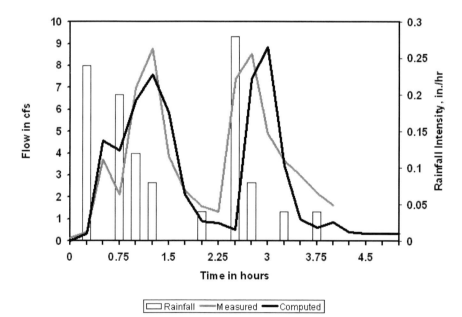

Fig. 6 Observed rainfall and runoff, computed flow, February 4, 1995.

runoff control options, such as bioretention cells, porous pavement, infiltration trenches, and vegetative swales. The effect of a large detention pond in capturing a relatively small storm (storm of February 15–17, 1995; for which runoff measurements were available) is presented in Fig. 8. Conduit

203 is upstream of the simulated detention pond. Conduit 202 is downstream from the pond and shows the pond storage effects on the hydrograph. The methodology described by Gironás et al.[10] includes a procedure for sizing the water quality control volume (WQCV). An actual pond is

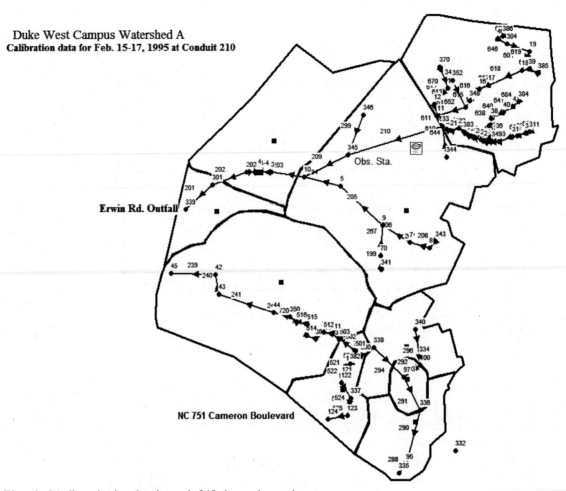

Fig. 7 Watershed A discretization showing node 345 observation station.

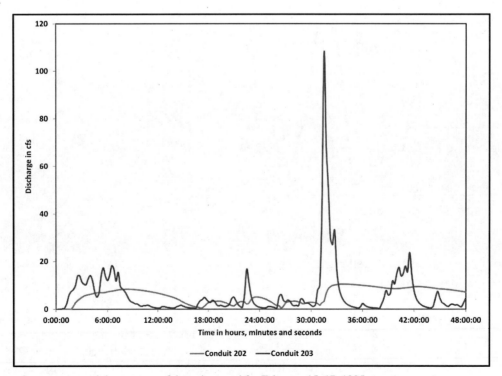

Fig. 8 Hydrographs upstream and downstream of detention pond for February 15–17, 1995 storm.

currently being designed by a consulting firm based on the location (between Conduit 202 and 203) and configurations simulated with SWMM5 and programs that incorporate the statistical properties of hourly rainfall recorded for the past 63 years, as recorded by National Climatic Data Center rain gauges near the campus.

APPLICATION TO SAZLIDERE WATERSHED IN ISTANBUL, TURKEY

Sazlıdere Watershed is located on the European Continental side of Istanbul in Turkey and Sazlıdere Lake is located downstream from the Sazlıdere Watershed. Surface runoff generated over the Sazlıdere Watershed (165 km² of drainage area) flows to Sazlıdere Dam Lake which was built up in 1996. Protecting and improving the Sazlıdere Watershed is of great importance as it supplies a major portion of the drinking water of Istanbul. There is a great potential of population growth, and therefore increases in the residential area. The hydrodynamic model of the Sazlıdere Watershed is depicted in Fig. 9, using the EPA SWMM. Part of the Sazlıdere watershed area, which feeds Türkköse Stream, is divided into 177 subcatchments by using the topographical map of the modeled area. A total number of 171 junctions and 173 conduits, which are ~44 km long in total, are defined for the hydrodynamic model. In order to calibrate and verify the hydrodynamic model, a rain gauge was set up on top of a building 4 stories high, located in the middle of the watershed in Haraççı, and a flow meter was set up close to the downstream segment of the Türkköse Stream (which is the main channel in the watershed). The locations of

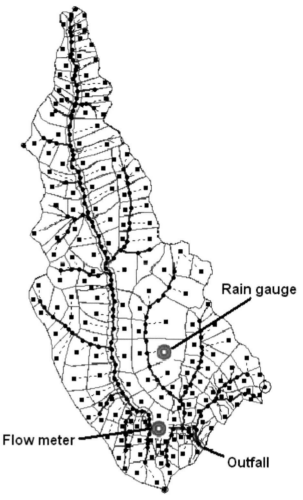

Fig. 9 Hydrodynamic model of Sazlıdere Watershed in EPA SWMM.

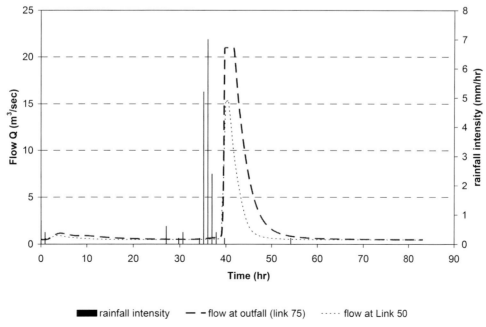

Fig. 10 Rainfall intensity and predicted flow rate versus time in Sazlıdere Watershed during December 26–29 rainfall event.

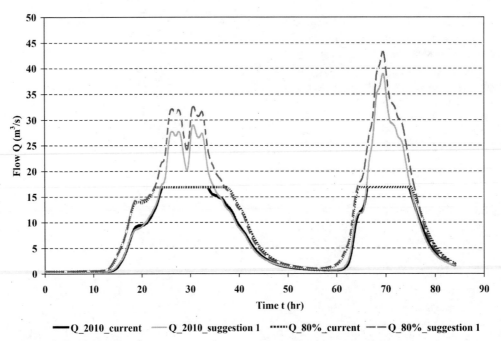

Fig. 11 Comparison of the hydrographs obtained for 10% and 80%, respectively, of urbanization for the cases with stream restoration and with no stream restoration.

the rain gauge and flow meter are indicated by a circle in the figure.

The flow rate was predicted during the December 26–29, 2009 rainfall event with the calibrated model at link 50, at the midpoint of the watershed where there is dense urbanization, and at link 75, which is located at a point downstream from the watershed. The rainfall data were measured at the field site. The predicted flow rates and the measured rainfall intensity are shown in Fig. 10. The peak flow reaches ~15 m³/sec near the urbanized area. On the other hand, the channel surcharges close to the outlet of the watershed for a duration of 2 hours and with a maximum flow of 21 m³/sec. In order to see the effect of extreme urbanization, the surface runoff and stream flow were also simulated during a rainfall event from December 10–13, 2009 for 80% of imperviousness on the catchment. Figure 11 shows the hydrographs for 10% and 80%, respectively, of urbanization for the cases with stream restoration and with no stream restoration. The calculated flow rates at the outfall for year 2010 (which corresponds to 10% urbanization) under current stream sections and under suggested stream restoration are labeled as "Q_2010_current" and "Q_2010_ suggestion 1," respectively. Under the current stream sections scenario, the channel cannot sustain the flow, and flooding occurs between the hours of 25–40 and 65–75. Therefore, a stream restoration is proposed by increasing the cross-sectional areas of the Türkköse Stream to 2.5 times its current value in order to prevent flood events on the catchment. When the restored stream sections are considered, we see that the stream will be able to carry the expected rising flow rate and no flood

will occur under the extreme urbanization with this suggested stream restoration.

REFERENCES

1. Lighthill, M.J.; Whitham, G.B. On kinematic waves. I. Flood movement in long rivers. Proc. R. Soc. Lond., London, Series A, **1955**, *229* (1178), 281–316.

2. Eagleson, P.S. *Dynamic Hydrology*; McGraw-Hill, Inc.: New York, 1970.

3. Ponce, V.M. *Engineering Hydrology: Principles and Practices*; Prentice Hall: Englewood Cliffs, New Jersey, 1989.

4. Chow, V.T.; Maidment, D.R.; Mays, L.W. *Applied Hydrology*; McGraw-Hill: New York, 0-07-010810-2, 1988.

5. Green, W.H.; Ampt, G.A. Studies on soil physics, 1, The flow of air and water through soils. J. Agr. Sci. **1911**, *4* (1), 1–24.

6. Huber, W.C.; Dickinson, R.E. *Storm Water Management Model*; Version 4, User's Manual, Athens, GA. Environmental Research Laboratory, Office of Research and Development, U.S. Environmental Protection Agency (EPA): 1988.

7. Medina, M.A.; Huber, W.C.; Heaney, J.P. Modeling stormwater storage/treatment transients: Theory, J. Environ. Eng. Div. ASCE, **1981**, *107* (EE4), 781–797.

8. Rossman, L.A. *Storm Water Management Model, User's Manual*; Version 5. Water Supply and Water Resources Division National Risk Management Research Laboratory, Cincinnati, Ohio, U.S. Environmental Protection Agency: 2010, EPA/600/R-05/040.

9. Mahi, H. *Stormwater Management and Modeling at Duke University*, M.S. Thesis, 1994.

10. Gironás, J.; Roesner, L.A.; Davis, J. *Storm Water Management Model Applications Manual*; Cincinnati, Ohio, 2009, EPA/600/R-09/077.

Streams: Perennial and Seasonal

Daniel von Schiller
Vicenç Acuña
Sergi Sabater
Catalan Institute for Water Research, Girona, Spain

Abstract
Perennial (permanently flowing) and seasonal (temporarily flowing) streams are diverse ecosystems that can be found on every continent. This entry provides an updated overview of perennial and seasonal streams, including their definition, classification, spatial zonation, their role as fluvial ecosystems, the ecosystem services they provide and the main issues related to their management and conservation.

INTRODUCTION

A stream is defined as a body of running water confined within a channel. Streams and rivers are analogous except that the latter are larger. Although they contain only a minor fraction of the total amount of water in the world (0.0001%), streams and rivers play a key role in the hydrologic cycle by connecting the atmosphere with the land and the ocean.[1]

Most streams originate when groundwater flow surfaces, usually at the lowest topographic areas of a valley.[2] Several connected streams build drainage or stream networks. Each stream has a drainage area, also referred to as watershed or catchment, which is the topographical region from which rainfall drains into a stream. The climate, geology, and vegetation of the drainage area exert a strong influence on the structure and function of streams.[3] For instance, stream flow or discharge (i.e., the volume of water passing through a channel cross-section per unit time) is mainly driven by the balance between precipitation and evapotranspiration within the drainage area, the second depending in turn on the vegetation, solar radiation, and temperature.

The aim of this entry is to provide an overview of perennial and seasonal streams considering their classification, spatial zonation, role as ecosystems, the ecosystem services they provide, as well as their management and conservation.

CLASSIFICATION

Perennial streams have permanent flow year round, whereas seasonal streams only flow at certain times (Fig. 1). The former are usually marked on topographic maps with a solid blue line, whereas a dashed blue line is used to indicate the latter. Depending on the length of the dry (flow cessation) period, seasonal streams are often referred to as temporary, intermittent, or ephemeral streams.[4] Seasonal streams are not restricted to headwaters, as they can also be found in the mid-reaches and lowlands of stream networks.[5]

In natural landscapes, the perennial or seasonal nature of a stream is mainly determined by the local climate and geology. Thus, seasonal streams tend to occur in areas where evapotranspiration exceeds precipitation for at least part of the year and where water infiltration rates through soils are high (e.g., areas with calcareous geology). In humanized landscapes, seasonal streams can also be artificially created through excessive water abstraction. In addition, because of climate change many naturally perennial streams are shifting to a seasonal hydrologic regime.[5]

In addition to classifying streams by the permanence of their flow, they are often classified by stream order. The most common method for ordering streams is the Strahler classification system (Fig. 2).[6] In this hierarchical method, the smallest streams are assigned first order. The order of the streams increases when two streams of the same order join. Streams of lower order joining a higher order stream do not change the order of the higher stream. Based on their order, streams are typically classified as small size (first to third order), medium size (fourth to sixth order), or large size (higher than seventh order). Within a drainage area, low-order streams are more frequent than high-order streams and comprise the longest part of the total stream network. Stream order is positively related to physical characteristics such as stream size, discharge, and drainage area. Therefore, it is a simple and informative classification system. Its major drawback is that it is often difficult to determine the perennial first-order streams on maps.

Streams are also classified by the landscape, predominant soil uses, or the vegetation of their drainage areas.[1]

Encyclopedia of Natural Resources DOI: 10.1081/E-ENRW-120047603

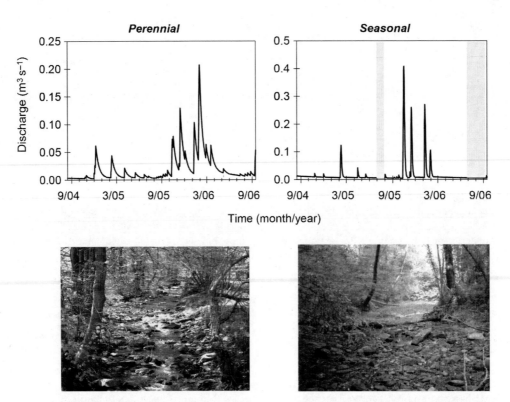

Fig. 1 Biannual hydrographs and pictures of a perennial and a seasonal stream from the Mediterranean region (NE Iberian Peninsula) during the dry summer period. The gray bars indicate the dry (flow cessation) periods in the seasonal stream.

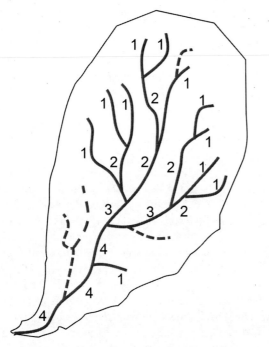

Fig. 2 Stream network illustrating the Strahler stream order within a fourth-order drainage area. Perennial streams are marked as solid lines. Seasonal streams are marked as dashed lines.

Accordingly, there are desert streams, forest streams, arctic streams, prairie streams, agricultural streams, urban streams, etc. This classification system is useful because the drainage area greatly affects the discharge, morphology, and chemistry of the draining streams. Other characteristics used to classify streams include their discharge patterns, geology, profile, gradient, water color, riparian vegetation, flora, fauna, etc.[7]

SPATIAL ZONATION

Stream networks can be divided into three zones on the basis of the dominant geomorphological processes dominating within each zone:[8] 1) erosion zone, characterized by steep v-shaped valleys, rapidly flowing water, and a high export of sediments; 2) transfer zone, characterized by gradual slope valleys, incipient meandering, and quick reception and delivery of sediments; and 3) deposition zone, with broad and flat valleys, developed stream meandering, and a high deposition of sediments. Nonetheless, rivers form part of a longitudinal continuum defined by changes in geomorphological, physicochemical, and hydrological conditions, where materials distribution and organisms play a significant role.

Streams are hierarchically organized systems that can be partitioned in progressively smaller and overlapping spatial units, i.e., drainage area, drainage network, segment, reach, habitat, and microhabitat.[9] Each spatial unit shows different patterns and processes that influence lower hierarchical units, but not vice versa. This hierarchy implies that many processes are unidirectional and that upstream sections always, at least partially, influence downstream sections. Humans usually view streams at the reach scale perspective, which has many characteristic components or parts (Fig. 3).

Riparian—
Wetlands

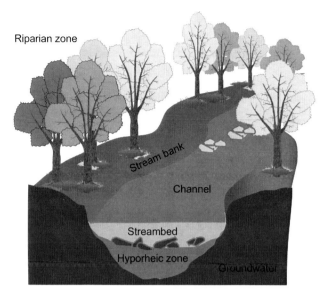

Fig. 3 Cross-section illustrating the main parts of a flowing stream reach.

THE FLUVIAL ECOSYSTEM

Streams have distinct physical, chemical, and biological characteristics when compared with other aquatic or terrestrial ecosystems. Generally, streams are defined by unidirectional flow, high biogeochemical reactivity, and unique biological communities.[10] Moreover, streams are especially open and dynamic ecosystems that have been described as having four dimensions.[11] The upstream–downstream connection constitutes the longitudinal dimension. The lateral dimension includes interactions between the channel and riparian/floodplain systems. The vertical dimension involves the relation between the channel and hyporheic/groundwater. The fourth dimension is time.

As any other ecosystem, streams integrate a variety of interactions between abiotic (non-living) factors and living organisms.[7] Spatial and temporal variation in flow shapes the morphology of the streambed and creates habitats for aquatic biota. Light is the main source of energy for primary production and its availability is primarily controlled by the riparian vegetation. Temperature is a key determinant of the functioning of many organisms and its variation is mainly driven by shading, climate, and water source. The substrata of streams are composed by organic and inorganic materials. Organic materials (e.g., leaves, wood) constitute a key source of energy for the heterotrophic organisms at the base of headwater stream food webs, and their amount is mostly driven by the presence of riparian vegetation. The size of the inorganic substrata is mainly controlled by the stream gradient and typically decreases downstream from headwaters to larger streams. Stream water chemistry is largely determined by the geology and land use of the drainage area and can vary dramatically along the stream network.

The biota inhabiting streams is diverse and adapted to the unique physical and chemical environment.[12] Microorganisms such as algae, bacteria, and fungi constitute the base of the stream food web, often forming biofilms on benthic substrata and/or plankton in the water column of larger streams. Larger algae, mosses, liverworts, and vascular plants contribute to the primary production of the stream and form microhabitats for fauna. Invertebrates (mostly insects) show a high diversity and act as key consumers and prey in streams. Vertebrates (mostly fishes) are at the top of the food web in most fluvial ecosystems, though they may be absent in seasonal streams. Both invertebrates and vertebrates show a large variety of feeding strategies (e.g., shredders, collector-gatherers, grazers, suspension feeders, predators) and occupy myriad habitats. In seasonal streams, the aquatic flora and fauna have developed adaptations (e.g., desiccation-resistant stages, propagules) to the frequent absence of flow.[4–5] Less is known, however, about the terrestrial biota that inhabits the dry streambeds of seasonal streams and the processes that occur in the absence of surface water.[13]

Energy sources in lotic ecosystems can be autochthonous (i.e., originated within the stream) or allochthonous (i.e., originated in the terrestrial environment and later transported to the stream). Autochthonous energy sources include the organic carbon compounds formed by the stream's primary producers through photosynthesis. Allochthonous energy sources include coarse (i.e., leafs, wood, and fruits), fine, or dissolved organic matter introduced from the terrestrial environment through surface or subsurface pathways. The changes in energy sources and the metabolic balance along the stream network are nicely illustrated by the river continuum concept:[14] In low-order streams, production is lower than ecosystem respiration (P/R ratio <1) and allochthonous coarse organic matter is the main source of energy. In mid-order streams, the P/R ratio approaches 1, indicating that much more energy is supplied by autochthonous primary production. In high-order streams, photosynthetic rates by phytoplankton become limited due to turbidity; thus, the P/R ratio decreases again. Headwaters and downstream sections are linked through the export of fine organic matter. The generality of the river continuum concept has been criticized mainly because it mostly applies for temperate climate regions, it does not explicitly account for seasonal streams, and it is based on pristine streams which are rarely found in our highly humanized landscapes. Nonetheless, it remains a useful conceptual model, which has been expanded by other conceptual models such as the serial discontinuity concept,[15] the flood pulse concept[16,17] or the riverine productivity model.[18]

On the whole, streams are not just transport conduits connecting land with the oceans. These ecosystems possess a high capacity to store, transform, and remove biologically reactive elements. Due to the presence of a strong unidirectional flow, in streams we speak of material spiraling instead of material cycling.[19] The study of the contribution of material spiraling in stream networks to the material budgets of whole drainage areas and global elemental budgets as well as the role of flow cessation periods in seasonal streams on these processes are considered important issues in current research.[13,20]

ECOSYSTEM SERVICES

Resources and processes supplied by stream ecosystems provide valuable benefits to humankind. The ecosystem services offered by streams can be divided into three categories:[21] 1) water supply; 2) supply of goods other than water; and 3) non-extractive or in-stream benefits (Table 1). In many regions, streams constitute a critical source of freshwater for irrigation, households, aquaculture, and industries. Goods other than water include important food resources such as fish, waterfowl, and shellfish. The natural functioning and geomorphology of streams helps mitigate the energy of floods, dilute and attenuate pollution, and fertilize floodplain soils, estuaries, and coastal areas near the river deltas. In addition, water from streams is used for hydroelectric power generation. Streams can also be used as transportation routes or as recreational sites for swimming, boating, fishing, etc. Moreover, streams provide a crucial habitat for the conservation of aquatic biodiversity and a series of important non-use values. Despite the absence of water during some periods, seasonal streams and their dry streambeds also have important human and ecological values mainly in terms of culture, recreation, and biodiversity conservation.[13] For instance, seasonal streams host a unique mixture of aquatic, terrestrial, and amphibious communities as a result of their dry and wet phases.

Table 1 Ecosystem services provided by streams

Water supply	Supply of goods other than water	Non-extractive or in-stream benefits
Irrigation	Fish	Flood control
Household	Waterfowl	Transportation
Industry	Shellfish	Hydroelectric power
Aquaculture		Pollution, dilution, and attenuation
		Soil fertilization
		Recreation
		Biodiversity
		Non-use values

Source: Based on Postel & Carpenter.[21]

MANAGEMENT AND CONSERVATION

Perennial and seasonal streams are highly susceptible to anthropogenic pressures.[7] Pollution (e.g., nutrients, heavy metals, and pharmaceuticals), originated from point (e.g., waste water treatment plant effluents) or diffuse (e.g., agricultural fertilizers) sources, represents a major impact that has deleterious effects on stream structure and function. Hydro-morphological modifications including damming, channelization, water abstraction, and removal of riparian vegetation strongly influence the flow, temperature, and sediment regime of streams among other factors. In addition, dams fragment stream networks, thus breaking the longitudinal connectivity characteristic of these ecosystems. Furthermore, humans have intentionally or unintentionally introduced a high number of non-native species into streams. These invasive organisms predate native species, compete with them for prey or habitat, alter stream habitats, and introduce harmful diseases.[12] Furthermore, changes in climate together with land use modifications are expected to cause an increase in the temporal and spatial extent of seasonal streams in many regions of the world.[5,13] Seasonal streams, mostly if water interruption is not natural, favor the disappearance of sensitive taxa and offer opportunities for invasive species.

In view of these major threads, the management and conservation of stream ecosystems and their associated values remain a major challenge. With this purpose, many countries have developed bodies of legislation such as the Clean Water Act in the U.S.A., the Water Framework Directive in the European Union, or the Water Act in Australia. These represent landmarks in sustainable management, but water security is often brought about by economic investment in water treatment rather than the prevention of impacts on freshwater ecosystems.[22] Unfortunately, many of these legislations fail to protect small low-order streams and are not yet well defined with seasonal streams.[10]

CONCLUSION

Streams are complex and diverse ecosystems. In a wide sense, streams can be divided into perennial (permanently flowing) or seasonal (temporarily flowing) streams, which show important commonalities but also differences in their structure, function, and dynamics. Moreover, streams are often characterized by their stream order as well as the type of landscape or geology of their drainage areas. Stream networks form a continuum from headwaters to the river mouth, which can be divided into different zones and hierarchical units on the basis of geomorphological and ecological attributes. Fluvial ecosystems are primarily characterized by unidirectional flow, high connectivity, unique biological assemblages, and high biogeochemical

Riparian—
Wetlands

reactivity. They have the ability to retain, transform, and remove important amounts of material during downstream transport and provide a series of highly valuable ecosystem services to humankind. Nonetheless, perennial and seasonal streams are highly susceptible to human impacts and their management and conservation remains a major challenge.

ACKNOWLEDGMENTS

We thank E. Martí for providing drawings of Fig. 3. This research was funded through the projects SCARCE (CON-SOLIDER-INGENIO CSD2009-00065) and CARBONET (CGL2011-30474-C02-01) from the Spanish Ministry of Science and Innovation. D. von Schiller was additionally supported by a Juan de la Cierva postdoctoral fellowship from the Spanish Ministry of Economy and Competitiveness (JCI-2010-06397).

REFERENCES

1. Dodds, W.K.; Whiles, M.R. *Freshwater Ecology: Concepts and Environmental Applications*; Academic Press: San Diego, CA, USA, 2010.
2. Charlton, R. *Fundamentals of Fluvial Geomorphology*; Routledge: London, UK, 2008.
3. Hynes, H.B.N. The stream and its valley. Verh. Internat. Verein. Limnol. **1975**, *19*, 1–15.
4. Williams, D.D. *The Biology of Temporary Waters*; Oxford University Press: New York, NY, USA, 2006.
5. Lake, P.S. *Drought and Aquatic Ecosystems: Effects and Responses*; Wiley-Blackwell: Oxford, UK, 2011.
6. Strahler, A.N. Hypsometric (area-altitude) analysis of erosional topography. Bull. Geol. Soc. Am. **1952**, *63* (11), 1117–1142.
7. Allan, J.D.; Castillo, M.M. *Stream Ecology: Structure and Function of Running Waters*; Springer: Dordrecht, the Netherlands, 2007.
8. Schumm, S.A. *The Fluvial System*; John Wiley & Sons: New York, USA, 1977.
9. Frissell, C.A.; Liss, W.L.; Warren, C.E.; Hurley, M.D. A hierarchical framework for stream habitat classification: Viewing streams in a watershed context. Environ. Manag. **1986**, *10* (2), 199–214.
10. Doyle, M.W.; Bernhardt, E.H. What is a stream? Environ. Sci. Technol. **2011**, *45* (2), 354–359.
11. Ward, J.V. The four-dimensional nature of lotic ecosystems. J. North Am. Benthol. Soc. **1989**, *8* (1), 2–8.
12. Giller, P.S.; Malmqvist, B. *The Biology of Streams and Rivers*; Oxford University Press: New York, 1998.
13. Steward, A.L.; von Schiller, D.; Tockner, K.; Marshall, J.C.; Bunn, S.E. When the river runs dry: Human and ecological values of dry riverbeds. Front. Ecol. Environ. **2012**, *10* (4), 202–209.
14. Vannote, R.L.; Minshall, G.W.; Cummins, K.W.; Sedell, J.R.; Cushing, C.E. The river continuum concept. Can. J. Fish. Aquat. Sci. **1980**, *37* (1), 130–137.
15. Ward, J.V.; Stanford, J.A. The serial discontinuity model: Extending the model to floodplain rivers. Reg. Rivers Res. Manag. **1995**, *10* (2–4), 159–168.
16. Junk, W.J.; Bayley, P.B.; Sparks, R.E. The flood pulse concept in river-floodplain systems. In *Proceedings of the International Large River Symposium*; Dodge, D.P., Ed.; Canadian Special Publications of Fisheries and Aquatic Sciences, 1989; 110–127.
17. Tockner, K.; Malard, F.; Ward, J.V. An extension of the flood pulse concept. Hydrol. Process. **2000**, *14* (16–17), 2861–2883.
18. Thorp, J.H.; Delong, M.D. The riverine productivity model: An heuristic view of carbon sources and organic processing in large river ecosystems. Oikos **1994**, *70* (2), 305–308.
19. Newbold, J.D. Cycles and spirals of nutrients. In *The Rivers Handbook*; Calow, P., Petts, G.E., Eds.; Blackwell Scientific: Oxford, UK, 1992; pp. 379–408.
20. Cole, J.J.; Prairie, Y.T.; Caraco, N.F.; McDowell, W.H.; Tranvik, L.J.; Striegl, R.G.; Duarte, C.M.; Kortelainen, P.; Downing, J.A.; Middelburg, J.J.; Melack, J. Plumbing the global carbon cycle: Integrating inland waters into the terrestrial carbon budget. Ecosystems **2007**, *10* (1), 171–184.
21. Postel, S.; Carpenter, S. Freshwater ecosystem services. In *Nature's Services*; Daily, G.C., Ed.; Island Press: Washington D.C., USA, 1997; pp. 195–214.
22. Vörösmarty, C.J.; McIntyre, P.B.; Gessner, M.O.; Dudgeon, D.; Prusevich, A. Green, P.; Glidden, S.; Bunn, S.E.; Sullivan, C.A.; Reidy Liermann, C.; Davies, P.M. Global threats to human water security and river biodiversity **2010**, *467* (7315), 555–561.

Riparian— Wetlands

Submerged Aquatic Vegetation: Seagrasses

Alyssa Novak
Frederick T. Short
University of New Hampshire, Durham, New Hampshire, U.S.A.

Abstract

Seagrasses are marine angiosperms found in shallow coastal waters along every continent except Antarctica. They form extensive meadows that provide important ecosystem services, including storing carbon, improving water quality, providing food and habitat, and acting as a biological indicator of coastal conditions. Seagrass meadows are rapidly declining primarily due to anthropogenic factors, including global climate change. Strong science-based management and regulatory strategies are needed to maintain and increase seagrass habitats, as well as build their resilience to stressors in a globally changing environment.

INTRODUCTION

Coastal waters are among the most productive yet highly threatened ecosystems in the world. They are vulnerable and at risk from development, overexploitation, physical alteration and destruction of habitat, as well as climate change related stress. Submerged aquatic vegetation (SAV) is an important component of coastal ecosystems. SAV is comprised of nonflowering and flowering macrophytes that grow completely underwater in freshwater, estuarine, and marine habitats. This entry focuses on seagrasses, an ecological group of flowering plants (angiosperms) that thrive in estuarine and shallow marine environments to 70 m. The entry provides an overview of seagrasses and their ecosystem services, evolutionary history, adaptations, classification, distribution, status and potential threats, as well as strategies to ensure their future existence in a globally changing environment.

AN ESSENTIAL NATURAL RESOURCE

Seagrasses are angiosperms that have adapted to exist fully submerged in estuarine and marine environments. They grow in shallow coastal waters and form extensive meadows that provide ecosystem services more valuable than saltmarshes, mangroves, and coral reefs.[1] Seagrass meadows exhibit high primary and secondary production and are considered a significant global carbon sink, a key to combatting global climate change.[2] While seagrasses comprise only 2% of the oceans' area, they are responsible for more than 15% of the carbon annually buried there, with some of the largest documented organic carbon stores found in sediment mattes produced by the Mediterranean seagrass *Posidonia oceanica* (Fig. 1).[2] Seagrasses also serve an important role in trophic transfer and export 24%

of their net production (0.6×10^{15} g C yr^{-1}) to adjacent ecosystems,[3] including the nutrient-poor deep sea.[4] In addition to their roles in primary production, carbon storage and export, seagrass meadows serve as filters and improve water quality and clarity through the direct trapping of suspended particles and the retention of organic matter.[5–7] Seagrass meadows are also hotspots of biodiversity, providing food and habitat to a variety of organisms including microbes, invertebrates, and vertebrates that are often endangered, such as dugongs, or commercially important, such as fish and shrimp.[8–11] Finally, seagrasses are considered biological indictors of coastal conditions because they are vulnerable and rapidly respond to anthropogenic stressors including eutrophication, sedimentation, oil spills, and commercial fishing.[12,13]

Origin, Adaptations, and Classification

Seagrasses evolved 40 million to 100 million years ago in the Late Cretaceous Period. They arose from terrestrial monocotyledons that reinvaded the sea and developed into three separate lineages: Cymodoceaceae complex, Hydrocharitaceae, and Zosteraceae.[14,15] The colonization of the sea required seagrasses to grow and reproduce while enduring the osmotic effects of salt water, changes in the availability of dissolved CO_2, changes in the intensity and quality of light, and the density and mechanical drag of water.[16–18] Seagrasses have a number of unique adaptations that allow for survival in these conditions, including a well-developed horizontal rhizome that anchors plants into the substrate and roots extending from the rhizome that assist in anchoring and nutrient uptake. Leaves are flexible and offer little resistance to wave action. They also function in nutrient uptake and as receptors of light, with the epidermis of blades serving as the main site for photosynthesis.

Encyclopedia of Natural Resources DOI: 10.1081/E-ENRW-120047540

Riparian—Wetlands

Fig. 1 An erosional escarpment in a *Posidonia oceanica* meadow with organic-rich soils in Calvi Bay, Corsica Island, France. The exposed face of the matte has a thickness of 2 m.
Source: Arnaud Abadie.

Tissues (aerenchyma) extending through the roots, rhizomes, and leaves facilitate internal gas and solute transport while regularly arranged air spaces (lacunae) give plants buoyancy. Finally, all species of seagrasses exhibit hydrophilous pollination while completely submerged except for *Enhalus acoroides* and *Ruppia* spp., which pollinate at the water surface.[16–20]

Seagrasses are classified as a functional rather than a taxonomic group of angiosperms. Historically, species designations were based on ecological, reproductive, and vegetative characteristics including leaf and flowering characteristics, vein numbers, fiber distributions, epidermal cells, and roots and rhizomes to create a complete taxonomic description.[19,20] Advances in technology now allow seagrass species designation to be based on genetic difference.[21] As of 2012, there are 72 seagrass species belonging to 6 families and 13 genera.[19,21,22] Five genera are placed in the family Cymodoceaceae (*Amphibolis, Cymodocea, Halodule, Syringodium,* and *Thalassodendron*), three in Hydrocharitaceae (*Enhalus, Halophila,* and *Thalassia*), one in Posidoniaceae (*Posidonia*), one in Ruppiaceae (*Ruppia*), one in Zannichelliaceae (*Lepilaena*), and two in Zosteraceae (*Phyllospadix* and *Zostera*; Table 1).[19,21,22] Although many species superficially resemble terrestrial grasses, seagrasses exhibit a diversity of size and morphologies and

may be as small as 1 cm (*Halophila minor*) or as long as 7 m (*Zostera caulescens*). Similarly, blades may be strap-like, cylindrical, ovate, or ovate-linear, with some species resembling ferns and others clover.[16,17,19]

Distribution

Seagrasses meadows are found in coastal waters along every continent except Antarctica, growing from the intertidal zone to depths between 60 and 70 m in the clearest ocean waters.[23] The dominant factor controlling distribution is light, with most seagrasses requiring a minimum of 10–25% of incident surface radiation for survival.[24] Other parameters that influence seagrass geographic and depth distribution include water clarity, temperature, salinity, current and wave patterns, nutrients, and substrate.[25]

The distribution of seagrass species across the globe is divided into six geographic bioregions (four temperate and two tropical), based on assemblages of taxonomic groups in temperate and tropical areas and the physical separation of oceans (Fig. 2).[26] Within each bioregion, seagrass species are further distributed according to the physical habitat (e.g., lagoon, estuary, surf-zone, back-reef, deep-water) and/or different successional roles. According to the bioregional model, there are approximately the same number of genera and species in temperate and tropical areas although the Tropical Indo-Pacific bioregion has the greatest seagrass species diversity. The most widely distributed species is *Ruppia maritima*, a species that can tolerate a wide range of salinities, and is found in both temperate and tropical bioregions in a variety of habitats.[26]

Status

Seagrass ecosystems are facing a global crisis.[13] There has been a 10-fold increase in reports of seagrass declines in recent decades and Waycott et al.[27] estimated that a minimum of 29% of the known global extent of seagrass area has been lost since 1879. Moreover, the International Union for Conservation of Nature (IUCN) Red List of Threatened Species recently classified 10 species of seagrass at elevated risk for extinction and 3 species as endangered (Fig. 2).[21]

Many stressors, including global climate change, have been responsible for the decline in the distribution of seagrasses. Threats may be localized or regional and some seagrass populations are being exposed to multiple stressors.[21,28] The greatest anthropogenic threats have been eutrophication and sedimentation from urban and agricultural runoff, as well as aquaculture practices.[6,10,27,28] Both eutrophication and sedimentation decrease the amount of light available to seagrasses for photosynthesis. Moreover, in systems with high nutrient loadings, epiphytes and fast-growing macroalgae outcompete seagrasses since they uptake nutrients more effectively and have relatively lower

Riparian—Wetlands

Table 1   List of the 72 seagrass species of the world[19,21]

Family	Genus: *Species*
Cymodocaceae	
	Amphibolis C. Agardh:
	Amphibolis antarctica (Labillardière) Sonder et Ascherson
	Amphibolis griffithii (J.M. Black) den Hartog
	Cymodoceaceae König in König et Sims:
	Cymodocea angustata Ostenfeld
	Cymodocea nodosa (Ucria) Ascherson
	Cymodocea rotundata Ehrenber et Hemprich ex Ascherson
	Cymodocea serrulata (R. Brown) Ascherson et Magnus
	Halodule Endlicher:
	Halodule beaudettei (den Hartog)
	Halodule bermudensis den Hartog
	Halodule emarginata den Hartog
	Halodule pinifolia (Miki) den Hartog
	Halodule uninervis (Forsskål) Ascherson
	Halodule wrightii Ascherson
	Syringodium Kützing in Hohenacker:
	Syringodium filiforme Kützing in Hohenacker
	Syringodium isoetifolium (Ascherson) Dandy
	Thaslassodendron den Hartog:
	Thalassodendron ciliatum (Forsskål) den Hartog
	Thalassodendron pachyrhizum den Hartog
Hydrocharitaceae	**Enhalus L.C. Richard:**
	Enhalus acoroides (Linnaeus *f.*) Royle
	Halophila Du Petit Thours:
	Halophila australis Doty et Stone
	Halophila baillonii Ascherson ex Dixie in J.D. Hooker
	Halophila beccarii Ascherson
	Halophila capricorni Larkum
	Halophila decipiens Ostenfeld
	Halophila engelmanni Ascherson
	Halophila euphlebia Makino
	Halophila hawaiiana Doty et Stone
	Halophila johnsonii Eiseman in Eiseman et McMillan
	Halophila minor (Zollinger) den Hartog
	Halophila nipponica Kuo
	Halophila ovalis (R. Brown) J.D. Hooker
	Halophila ovata Gaudichaud in Freycinet
	Halophila spinulosa (R. Brown) Ascherson
	Halophila stipulacea (Forsskål) den Hartog
	Halophila sulawesii Kuo
	Halophila tricostata Greenway

Table 1 (*Continued*)   List of the 72 seagrass species of the world[19,21]

Family	Genus: *Species*
Posidoniaceae	**Thalassia Banks ex König in König et Sims:**
	Thalassia hemprichii (Ehrenberg) Ascherson in Petermann
	Thalassia testudinum Banks ex König in König et Sims
	Posidonia König in König et Sims:
	Posidonia angustifolia Cambridge et Kuo
	Posidonia australis J.D. Hooker
	Posidonia coriacea Cambridge et Kuo
	Posidonia denhartogii Kuo et Cambridge
	Posidonia kirkmanii Kuo et Cambridge
	Posidonia oceanica (Linnaeus) Delile
	Posidonia ostenfeldii den Hartog
	Posidonia sinuosa Cambridge et Kuo
Ruppiaceae	**Ruppia Linnaeus:**
	Ruppia cirrhosa (Petagna) Grande
	Ruppia filifolia (Phil.) Skottsb.
	Ruppia maritima L.
	Ruppia megacarpa R. Mason
	Ruppia polycarpa R. Mason
	Ruppia tuberosa J.S. Davis & Toml.
Zannichelliaceae	**Lepilaena Frummond ex Harvey:**
	Lepilaena australis Harv.
	Lepilaena marina E.L Robertson
Zosteraceae	**Phyllospadix W.J. Hooker:**
	Phyllospadix iwatensis Makino
	Phyllospadix japanoicus Makino
	Phyllospadix scouleri W.J. Hooker
	Phyllospadix serrulatus Ruprecht ex Ascherson
	Phyllospadix torreyi S. Watson
	Zosetera Linnaeus:
	Zostera asiatica Miki
	Zostera caespitosa Miki
	Zostera capensis Setchell
	Zostera capricorni Ascherson
	Zostera caulescens Miki
	Zostera chilensis Kuo
	Zostera geojeensis Shin.
	Zostera japonica Ascherson et Graebner
	Zostera marina Linnaeus
	Zostera mulleri Irmisch ex Ascherson
	Zostera nigricaulis Kuo
	Zostera noltti Hornemann
	Zostera pacifica L.
	Zostera polychlamis Kuo
	Zostera tasmanica (Marten ex Ascherson) den Hartog
	Zostera nigricaulis
	Zostera noltti Hornemann
	Zostera pacifica S. Watson

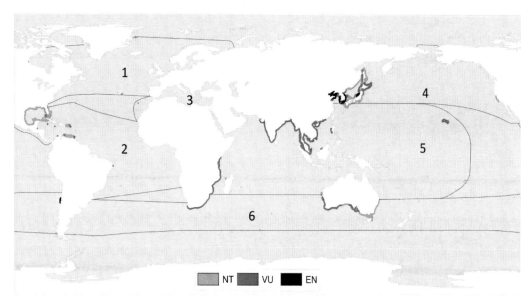

Fig. 2 Distribution of Near Threatened (NT), Vulnerable (VU), and Endangered (EN) seagrasses across the globe. Numbers 1–6 indicate bioregions.
Source: © Elsevier BV 2011 as included in Extinction risk assessment of the world's seagrass species, Biol Conserv Vol. 144, No. 7, Reprinted with permission.

light requirements to sustain growth.[29,30] The relationship between human watershed activity and seagrass declines is well documented: SeagrassNet, a global monitoring program, recently captured the dramatic decrease in percent cover and shoot density of seagrass meadows in Placencia, Belize as coastal housing and tourism development surged in the region.[31] Other anthropogenic activities that have had direct impacts on the distribution of seagrasses by reducing water clarity and/or uprooting plants include dredge and fill, land reclamation, and dock and jetty construction.[6,10,27]

The direct loss of seagrasses by organisms other than humans has also occurred through overgrazing (e.g., by dugongs, urchins, sea turtles), bioturbation, and disease.[10,13,31] For example, in the early 1930s, 90% of all *Zostera marina* (eelgrass) populations from Canada to North Carolina disappeared as a result of "wasting disease," caused by a marine slime mold-like protist, *Labyrinthula zosterae*.[32,33] Although eelgrass has since returned, the disease still affects eelgrass beds in North America and Europe and has been responsible for some recent losses.[34–36]

Threats to seagrasses from global climate change include increases in sea-surface temperature of the ocean, sea-level rise, CO_2 concentrations, storm events, and levels of ultraviolet-B (UV-B) radiation.[10,13,37] Recent field studies have shown that increased maximum annual seawater temperature in the Mediterranean has led to increased seagrass mortality.[38] Moreover, the Mediterranean Institute for Marine Studies expects *Posidonia oceanica*, the dominant seagrass in the Mediterranean, to decline by 90% and become a functionally extinct ecosystem by 2050 because of surface water warming.[39] The impacts of climate changes on other seagrass species are uncertain.[13,28] Scientist expect a shift in the geographic and depth distribution of species depending on their tolerance to different climate stressors, with some species becoming extinct.[10,37] Genetically diverse seagrass populations are also expected to have a higher chance of success than genetically conserved ones.[40]

Ensuring the Future of Seagrasses

The long-term survival of seagrasses in a globally changing environment depends upon reducing human impacts to the coastal oceans. Researchers and managers can support seagrass ecosystem resilience by continuing efforts to better understand the status and health of seagrasses and using the information gained to develop effective management strategies. Actions that support the maintenance of healthy seagrass ecosystems include establishing mapping and monitoring programs; improving habitat conditions; protecting populations in addition to restoring; and raising awareness about the value, status, and threats to seagrasses.[40–42]

Establishing mapping and monitoring programs

Mapping and monitoring programs provide crucial information on the health, status, and trends of populations, as well as environmental conditions. Aerial photography, remote sensing, and GPS can be used to delineate and map seagrass beds within a given area and to assess changes in seagrass abundance and distribution. Monitoring programs collect more detailed information on selected seagrass and environmental parameters. SeagrassNet, a global monitoring

Fig. 3 SeagrassNet sampling in *Thalassia testudinum* meadows at Rosario Island, Columbia.
Source: Fred Short.

network composed of scientists, managers, and stakeholders, uses a standardized protocol to detect changes in seagrass habitat, monitoring both seagrass parameters (e.g., percent cover, shoot density, canopy height) and environmental variables (e.g., changes in water clarity; Fig. 3).[43–44] The information is used to evaluate trends in the distribution, structure, and function of seagrasses habitats, identify potential threats, and adapt or implement management strategies.

Improving habitat conditions

Seagrass growing along coastlines that are stressed by human activities are the most vulnerable to generalized disturbances such as climate change. Improving habitat conditions in these areas can enhance seagrass health for better resistance or adaptation to future conditions. Poor water quality and clarity can be improved by encouraging land-use better practices such as developing coastal buffer zones to decrease nutrient and sediment run-off, reducing the use of fertilizers and pesticides, and increasing filtration of effluent. Moreover, the effects of disturbances such as dredging, fishing and boating activities on seagrass habitats can be minimized by establishing codes of conduct.[40–42] In systems where seagrass habitats are exposed to multiple

anthropogenic threats acting at broad scales, spatially explicit assessments can be conducted to identify "hotspots" for prioritizing seagrass management actions.[45]

Protecting resilient populations

Seagrass populations exhibit different responses to stressors, with some populations exhibiting higher tolerances to poor water quality,[46] increased temperatures[47] and/or elevated levels of UV-B.[48–50] Identifying and protecting seagrass communities that are potentially resilient to stressors are an important strategy because it supplies a source of seeds to repopulate following disturbances.[40] Priority should be given to resilient populations growing in different types of conditions and from a wide geographical range. Patterns of connectivity between seagrass beds and adjacent habitats such as salt marshes, mangroves, and coral reefs should also be identified to promote ecological linkages and shifts in species distribution.[40] Once resilient populations are identified, their status and health should be monitored and habitat conditions conducive for their growth maintained or improved if needed.

Restoring

Restoration of seagrass beds is considered to be a viable strategy for increasing seagrass habitats although transplanting techniques are often costly and not always successful.[51–53] Two factors critical to transplanting success are the selection of suitable transplant sites and robust donor populations.[52,54] Transplant sites that are conducive to establishment and growth have good water quality and clarity, appropriate substrate (type and size), low exposure (wind, waves, dessication), and are free of conflicting uses.[52] Donor populations that are from sites with similar conditions as transplant sites and have high genetic diversity increase transplanting success by contributing to seagrass productivity and recovery potential.[55–59] All transplanting efforts should be assessed through long-term mapping and monitoring programs.

Raising awareness

Seagrasses have received less attention from the media than other marine ecosystems despite their higher economic value.[13] Engaging local communities and stakeholders in seagrass programs is an essential conservation strategy that raises awareness about the importance of these valuable ecosystems and the threats to their survival. Building awareness of seagrasses ultimately improves coastal habitats and encourages habitat protection. Community monitoring programs (e.g., Seagrass-Watch) or community-based restoration efforts (e.g., TERFS) can reinforce the value of seagrass habitats while collecting information about the condition of seagrasses.[40,60–62]

Riparian—
Wetlands

CONCLUSION

Seagrasses are angiosperms adapted to exist fully submerged in estuarine and near-shore marine environments. They form extensive meadows that provide important ecological and economic services in coastal systems throughout the world by stabilizing and enriching sediments, trapping and cycling nutrients, maintaining water quality and clarity, and providing habitat for organisms in a vast food web. Moreover, seagrass meadows have the ability to mitigate climate change through carbon sequestration and storage. Despite their many values, seagrass meadows are rapidly declining and there is no evidence that seagrass recoveries are compensating for large-scale losses. The future of seagrass meadows is dependent upon immediate actions to implement strong science-based strategies that will maintain and increase seagrass habitats in a globally changing environment while considering the interaction between land and sea processes in policy, planning, and management decisions. Their survival also relies on increasing our understanding and awareness of these species, the relationships between species, their genetic variability, unique adaptive traits, and ability to naturally recolonize. Beyond research and monitoring, management is needed to secure the future of seagrass ecosystems so they continue to deliver important ecological and ecosystem services to the world's oceans and people.

ACKNOWLEDGMENTS

Thanks to Cathy Short for suggestions and editing and to two anonymous reviewers for their comments and suggestions.

REFERENCES

1. Costanza, R.; d'Arge, R.; de Groot, R; Farber, S.; Grasso, M; Hannon, B; Naeem, S; Limburg, K; Paruelo, J; O'Neill, R.V; Raskin, R; Sutton, P; van den Belt, M. The value of the world's ecosystem services and natural capital. Nature. **1997**, *387*, 253–260.
2. Fourqurean, J.W.; Duarte, C.M.; Kennedy, H.; Marba, N.; Holmer, M.; Mateo, M.A.; Apostolaki, E.T; Kendrick, G.A.; Krause-Jensen, D.; McGlathery, K.J.; Serrano, O. Seagrass ecosystems as a globally significant carbon stock. Nature Geosci. **2012**, *5*, 505–509.
3. Duarte, C.M.; Cebrian, J. The fate of autotrophic production in the sea. Limnol. Oceanogr. **1996**, *41*, 1758–1766.
4. Suchanek, T.H.; Williams, S.W.; Ogden, J.C.; Hubbard, D.K.; Gill, I.P. Utilization of shallow-water seagrass detritus by Caribbean deep-sea macrofauna: $\delta^{13}C$ evidence. Deep Sea Res. **1985**, *32*, 2201–2214.
5. Heck, K.L.; Able, K.W.; Roman, C.T.; Fahay, M.P. Composition abundance, biomass, and production of marcrofauna in a New England estuary: comparisons among eelgrass meadows and other nursery habitats. Estuaries **1995**, *18* (2), 379–389.
6. Short, F.T.; Wyllie-Echeverria, S. Natural and human-induced disturbance of seagrasses. Environ. Conserv. **1996**, *23* (1), 17–27.
7. Terrados, J.; Duarte, C.M.; Experimental evidence of reduced particle resuspension within a seagrass (*Posidonia oceanica* L.) meadow. Exper. Mar. Biol. Ecol. **2000**, *243*, 45–53.
8. Fry, B.; Parker, P.L. Animal diet in Texas seagrass meadows. C evidence for the importance of benthic plants. Estuar. Coast. Mar. Sci. **1979**, *8*, 499–509.
9. Hemminga, M.A.; Duarte, C.M. *Seagrass Ecology*; Cambridge University Press: Cambridge, UK: 1991.
10. Duarte, C.M. The future of seagrass meadows. Environ. Conserv. **2002**, *29* (2), 192–206.
11. Heck, K.L.; Hays, C.; Orth, R.J. A critical evaluation of the nursery role hypothesis for seagrass meadows. Mar. Ecol. Prog. Ser. **2003**, *253*, 123–136.
12. Bricker, S.B.; Ferreira, J.G.; Simas, T. An integrated methodology for assessment of estuarine trophic status. Ecol. Model **2003**, *169*, 39–60.
13. Orth, R.J.; Carruthers, T.J.B.; Dennison, W.C.; Duarte, C.M.; Fourqurean, J.W.; Heck, K.L. Jr.; Hughes, A.R.; Kendrick, G.A.; Kenworthy, W.J.; Olyarnik S.; Short, F.T.; Waycott, M.; Williams, S.L. A global crisis for seagrass ecosystems. BioScience **2006**, *56* (12), 987–996.
14. Les, D.H.; Cleland, M.A.; Waycott, M. Phylogenetic studies in the Alismatidae, II: Evolution of the marine angiosperms (seagrasses) and hydrophily. Syst. Bot. **1997**, *22*, 443–463.
15. Waycott, M.; Procaccini, G.; Les, D.H.; Reusch, T.B.H. Seagrass evolution, ecology and conservation: a genetic perspective. In *Seagrasses: Biology, Ecology and Conservation*; Larkum, A.W.D.; Orth, R.J.; Duarte, C.M.; Eds; Dordrecht, Springer: 2006; 25–50.
16. den Hartog, C. Seagrasses of the World. Verh. Kon. Ned. Akad. Wetens. Afd. Naturk. Ser. 2 59, 1-275+31 plates, 1970.
17. Phillips, R.C.; Meñez, E.G. Seagrasses. Smiths. Contr. Mar. Sci. Number 34, Smithsonian Institution Press: Washington, D.C., 1998.
18. Dawes, C.J. *Marine Botany*. 2nd edition.; John Wiley & Sons: New York, 1998.
19. Kuo, J.; den Hartog, C. Seagrass taxonomy and identification key. In *Global Seagrass Research Methods*; Short, F.T.; Coles, R.G., Eds; Amsterdam: Elsevier Science B.V, 2001; 31–58.
20. Kuo, J.; den Hartog, C. Morphology, anatomy, and ultra-structure. In *Seagrasses: Biology, Ecology and Conservation*; Larkum, A.W.D., Orth, R.J., Duarte, C.M., Eds; Berlin: Springer Verlag: 2006; 51–88.
21. Short, F.T.; Polidoro, B.; Livingstone, S.R.; Carpenter, K.E.; Bandeira, S.; Bujang, J.S.; Calumpong, H.P.; Carruthers, T.J.; Coles, R.G.; Dennison, W.C.; Erftemeijer, P.L.A; Fortes, M.D.; Freeman, A.S.; Jagtap, T.G.; Kamal, A.H.M.; Kendrick, G.A.; Kenworthy, W.J.; La Nafie, Y.A.; Nasution, I.M.; Orth, R.J.; Prathep, A.; Sanciangco, J.C.; van Tussenbroek, B.; Vergara, S.G.; Waycott, M.; Zieman, J.C. Extinction risk assessment of the world's seagrass species. Biol. Conserv. **2011**, *144*, 1961–1971.
22. Moore, K.A.; Short, F.T. *Zostera*: Biology, ecology and management. In *Seagrasses: Biology, Ecology and Conservation*; Larkum, A.W.D., Orth, R.J., Duarte, C.M., Eds; Berlin: Springer Verlag, 2006; 361–386.

Riparian—
Wetlands

23. Coles, R.G.; McKenzie, L.J.; De'ath, G.; Roelofs, A.J.; Lee Long, W. Spatial distribution of deepwater seagrass in the inter-reef lagoon of the Great Barrier Reef World Heritage Area. Mar. Ecol. Prog. Ser. **2009**, *392*, 57–68.

24. Dennison, W.C.; Orth, R.J.; Moore, K.A.; Stevenson, J.C.; Carter, V.; Kollar, S.; Bergstrom, P.W.; Batuik R.A. Assessing water quality with submersed aquatic vegetation. BioScience **1993**, *43*, 86–94.

25. Short, F.T.; Coles, R.G.; Pergent-Martini, C. Global seagrass distribution. In *Global Seagrass Research Methods*. Short, F.T., Coles, R.G., Eds; Amsterdam: Elsevier Science B.V, 2001; 5–30.

26. Short, F.T.; Carruthers, T.; Dennison, W.; Waycott, M. Global seagrass distribution and diversity: A bioregional model. J. Exper. Mar. Biol. Ecol. **2007**, *350*, 3–20.

27. Waycott, M.; Duarte, C.M.; Carruthers, T.J.B.; Orth, R.J.; Dennison, W.C.; Olyarnik, S.; Calladine, A.; Fourqurean, J.W.; Heck, K. Jr.; Hughes, A.R.; Kendrick, G.A.; Kenworthy, W.J.; Short, F.T; Williams, S.L. Accelerating loss of seagrasses across the globe threatens coastal ecosystems. Proc. Nat. Acad. Sci. **2009**, *106*, 12377–12381.

28. Grech, A.; Chartrand-Miller, K.; Erftemeijer, P.; Fonseca, M.; McKenzie, L.; Rasheed, R.; Taylor, H.; Coles, R.G. A comparison of threats, vulnerabilities and management opportunities in global seagrass bioregions. Environ. Res. Lett. 024006.

29. Harlin, M. M.; Thorne-Miller, B. Nutrient enrichment of seagrass beds in a Rhode Island coastal lagoon. Mar. Biol. **1981**, *65*, 221–229.

30. Twilley, R.R.; Kemp, W. M.; Staver, K.W.; Stevenson, J. C.; Boynton, W.R. Nutrient enrichment of estuarine submersed vascular plant communities. 1. Algal growth and effects on production of plants and associated communities. Mar. Ecol. Prog. Ser. **1985**, *23*, 179–191.

31. Short, F.T.; Koch, E.; Creed, J.C.; Magalhaes, K.M. SeagrassNet monitoring of habitat change across the Americas. Biol. Mar. Medit. **2006**, *13*, 272–276.

32. Rasmussen, E. The wasting disease of eelgrass (*Zostera marina*) and its effects on environmental factors and fauna. In *Seagrass Ecosystems*. McRoy, C.P., Helfferich, C., Eds; Marcel Dekker: New York, 1977; 1–51.

33. Ralph, P.J.; Short, F.T. Impact of the wasting disease pathogen, Labyrinthula zosterae, on the photobiology of eelgrass, *Zostera marina*. Mar. Ecol. Prog. Ser. **2002**, *228*, 265–271.

34. Short, F.T.; Mathieson, A.C.; Nelson, J.I. Recurrence of the eelgrass wasting disease at the border of New Hampshire and Maine, USA. Mar. Ecol. Prog. Ser. **1986**, *29*, 89–92.

35. Short, F.T.; Ibelings, B.W.; den Hartog, C. Comparison of a current eelgrass disease to the wasting disease of the 1930s. Aquat. Bot. **1988**, *30*, 295–304.

36. den Hartog, C. Wasting disease and other dynamic phenomena in *Zostera* beds. Aquat. Bot. **1987**, *27*, 3–14.

37. Short, F.T.; Neckles, H.A. The effects of global climate change on seagrasses. Aquat. Bot. **1999**, *63*, 169–196.

38. Marbà, N.; Duarte, C.M. Mediterranean warming triggers seagrass (*Posidonia oceanica*) shoot mortality. Global Change Biol. **2010**, *16*, 2366–2375.

39. Jordà, G.; Marbà, N. Duarte, C.M. Mediterranean seagrass vulnerable to regional climate warming. Nat. Clim. Change **2012**, doi: 10.1038/nclimate1533.

40. Björk, M.; Short F.T.; Mcleod E.; Beer, S. Managing Seagrasses for Resilience to Climate Change. IUCN: Gland, Switzerland, 2008.

41. Long, W.L.; Thom, R.M. Seagrass habitat conservation–Improving seagrass habitat quality. In *Global Seagrass Research Methods*. Short, F.T., Coles, R.G., Eds; Amsterdam: Elsevier Science B.V, 2001; 407–445.

42. Coles, R.C.; Fortes, M. Protecting seagrass—approaches and methods. In *Global Seagrass Research Methods*. Short, F.T., Coles, R.G., Eds; Amsterdam: Elsevier Science B.V, 2001; 445–463.

43. Short, F.T.; McKenzie, L.G.; Coles, R.G.; Vidler, K.P; Gaeckle, J.L. SeagrassNet Manual for Scientific Monitoring of Seagrass Habitat – Worldwide Edition. University of New Hampshire Publication: Durham, NH, USA, 2006.

44. http://www.SeagrassNet.org.

45. Grech, A.; Coles, R.G.; Marsh, H. A broad-scale assessment of the risk to coastal seagrasses from cumulative threats. Mar Policy, **2011**, *35* (5), 560–567.

46. Short, F.T.; Burdick, D.M.; Moore, G.E. The Eelgrass Resource of Southern New England and New York: Science in Support of Management and Restoration Success. The Nature Conservancy, 2012.

47. Franssen, S.U.; Gu, J; Bergmann, N.; Winters, G.; Klostermeier, U.C.; Rosenstiel, P.; Bornberg- Bauer, E.; Reusch, T.B.H. Transcriptomic resilience to global warming in the seagrass *Zostera marina*, a marine foundation species. Proc. Nat. Acad. Sci. **2011**, 19276–19281.

48. Trocine, R.P.; Rice, J.D.; Wells, G.N. Inhibition of seagrass photosynthesis by ultraviolet-B radiation. Plant Physiol. **1981**, *68*, 74–81.

49. Dawson, S.P.; Dennison, W.C. Effects of ultraviolet and photosynthetically active radiation on five seagrass species. Mar. Biol. **1996**, *124*, 629–638.

50. Novak, A.B; Short, F.T. UV-B induces leaf reddening and contributes to the maintenance of photosynthesis in the seagrass *Thalassia testudinum*. J. Exp. Mar. Biol. Ecol. 2011, 136–142.

51. Fonseca, M.S.; Kenworthy, W.J.; Thayer, G.W. Guidelines for conservation and restoration of seagrass in the United States and adjacent waters. NOAA/NMFS Coastal Ocean Program and Decision Analysis Series, No. 12. NOAA Coastal Ocean Office: Silver Spring, Maryland, 1998.

52. Short, F.T.; Davis, R.C.; Kopp, B.S.; Short, C.A.; Burdick, D.M. Site-selection model for optimal transplantation of eelgrass *Zostera marina* in northeastern U.S. Mar. Ecol. Prog. Ser. **2002**, *227*, 253–267.

53. Cunha, A. H.; Marbá, N. N.; van Katwijk, M. M.; Pickerell, C.; Henriques, M.; Bernard, G.; Ferreira, M. A.; Garcia, S.; Garmendia, J. M.; Manent, P. Changing paradigms in seagrass restoration. Rest. Ecol. **2012**, *20*, 427–430.

54. van Katwijk, M. M.; Bos, A.R.; de Jonge, V.N.; Hanssen, L.S.A.M.; Hermus, D.C.R.; de Jong, D.J. Guidelines for seagrass restoration: importance of habitat selection and donor population, spreading of risks, and ecosystem engineering effects. Mar. Pollut. Bull. **2009**, 179–188.

55. Procaccini, G.; Piazzi, L.; Genetic polymorphism and transplantation success in the Mediterranean seagrass *Posidonia oceanica*. Rest. Ecol. **2001**, *9*, 332–338.

56. Williams, S.L. Reduced genetic diversity in eelgrass transplantations affects both population growth and individual fitness. Ecol. Appl. **2001**, *11*, 1472–1488.

Riparian—
Wetlands

57.   Hughes, A.R.; Stachowicz, J.J. Genetic diversity enhances the resistance of a seagrass ecosystem to disturbance. Proc. Nat. Acad. Sci. **2004**, *101*, 8998–9002.

58.   Reusch, T.B.H.; Ehlers, A.; Hammerli, A.; Worm, B. Ecosystem recovery after climatic extremes enhanced by genotypic diversity. Proc. Nat. Acad. Sci. **2005**, *102*, 2826–2831.

59.   Reusch, T.B.H.; Hughes, A.R. The emerging role of genetic diversity for ecosystem functioning: Estuarine macrophytes as models. Estuar Coast. **2006**, *29*, 159–164.

60.   McKenzie, L.J.; Lee Long, W.J.; Coles, R.G.; Roder, C.A. Seagrass-Watch: Community based monitoring of seagrass resources. Biol. Mar. Medit. **2000**, *7* (2), 393–396.

61.   http://www.Seagrasswatch.org.

62.   Short, F.T.; Short, C.A.; Burdick-Whitney, C. *A Manual for Community-Based Eelgrass Restoration*. Sponsored by NOAA Restoration Center: University of New Hampshire, Durham, NH, 2002.

Riparian—
Wetlands

Surface Water: Nitrogen Enrichment

Tim P. Burt
Penny E. Widdison
Department of Geography, University of Durham, Durham, U.K.

Abstract

Nitrogen is essential for plant growth. In order for nitrogen to be used for plant growth, it must be "fixed." In the terrestrial nitrogen cycle, microbes break down organic matter to produce much of the available nitrogen in soils. Nitrate is completely soluble in water and it is not adsorbed to clay particles; therefore, it is vulnerable to being leached out of the soil by percolating rainfall or irrigation water.

INTRODUCTION

Nitrogen is essential for plant growth and comprises nearly 79% of the earth's atmosphere in the form of N_2 gas. In order for nitrogen to be used for plant growth, it must be "fixed" in the form of ammonium (NH_4) or nitrate (NO_3). In the terrestrial nitrogen cycle, microbes break down organic matter to produce much of the available nitrogen in soils. Nitrate is completely soluble in water and it is not adsorbed to clay particles; therefore, it is vulnerable to being leached out of the soil by percolating rainfall or irrigation water. Generally, the movement of nitrogen can be described in three ways: 1) upward, crop uptake and gaseous loss; 2) downward, as leaching to groundwater; and 3) lateral, via surface and subsurface flow to surface waters. The nitrogen cycle under arable soils is shown in Fig. 1.

NITRATE LEVELS IN SURFACE WATERS

The natural water quality of a river will be determined primarily by the catchment soil type and underlying geology to which water, falling on the catchment as rain, is exposed as it drains to the river. Climate provides an important context for nitrogen cycling by controlling the propensity for carbon and nitrogen to be stored within the catchment; thus, in the United Kingdom, upland soils tend to conserve organic matter as peat, whereas organic matter tends to decompose much more readily in lowland soils. Deviations from this baseline water quality are generally caused by the influence of human activities through point and diffuse pollution sources. Up to 40% of total nitrogen reaches the aquatic system through direct surface runoff or subsurface flow.[1] Nitrogen delivery to surface waters is further controlled by: 1) soil structure and type; 2) rainfall; 3) the amount of nitrate supplied by fertilizers; and 4) plant cover and root activity.[2]

In pristine river systems, the average level of nitrate is about 0.1 mg/L as nitrogen (mg/L N). However, in Western Europe, high atmospheric nitrogen deposition results in nitrogen levels of relatively unpolluted rivers to range from 0.1 to 0.5 mg/L.[3] In recent years, nitrate concentrations in European rivers have been rising (Fig. 2) and "No progress has been made in reducing the concentration of nitrates in Europe's rivers."[4] High rates of nitrogen input to rivers and coastal waters are not confined to Europe. In an average year, the Mississippi River discharges 1.57 million metric tons of nitrogen into the Gulf of Mexico.[5] About 7 million metric tons of nitrogen in commercial fertilizers are applied annually in the basin leading to nitrate concentrations in agricultural drains of 20–40 mg/L or more.[5] In the United States, in 1998, more than one-third of all river miles, lakes (excluding the Great Lakes), and estuaries did not support the uses for which they were designated under the Clean Water Act.[6] Table 1 illustrates N inputs to rivers and coasts in areas of America, Africa, and Asia.

It is now widely acknowledged that agriculture is the main source of N pollution in surface waters and groundwater in rural areas of Western Europe and the United States.[2,7,8] The U.K. House of Lords' report *Nitrate in Water*[9] commented on the conflicts that can arise when the use of land for farming comes into conflict with the use of land for water supply. Concern initially focused on alleged links between high nitrate concentrations in drinking water and two health problems in humans: the "blue-baby" syndrome methemoglobinemia and gastric cancer. Now, there are also concerns for environmental degradation. Nutrient enrichment in water bodies encourages the growth of aquatic plants (see Fig. 3). Reed beds and other marginal plants may be attractive on a small scale, but when these and, particularly, underwater plant growth are excessive, this can cause a narrowing of waterways and become a nuisance to recreational users of rivers and lakes. Furthermore, eutrophication (a group of effects caused by

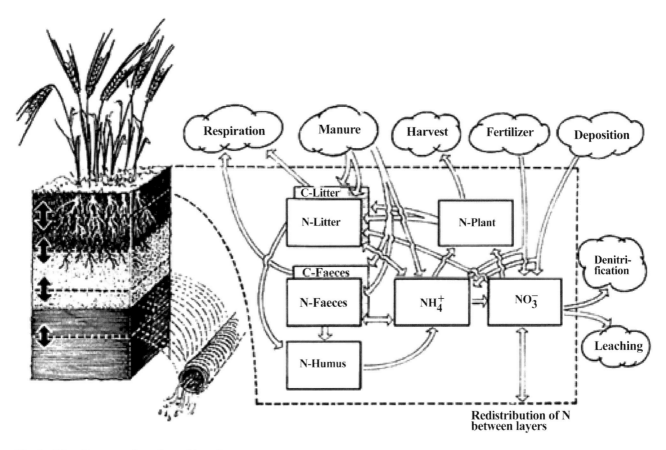

Fig. 1 The nitrogen cycle under arable soils.
Source: Adapted from Burt, et al.[2]

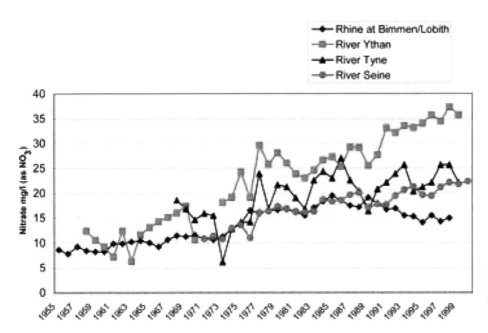

Fig. 2 Nitrate concentration in selected European rivers
Source: Adapted from European Environment Agency.[4]

nutrient enrichment of water bodies) can adversely affect the aquatic ecosystem. An algal bloom may cut out light to the subsurface, and when it dies back, decomposition uses the oxygen supply needed by other species. Some algae are toxic to fish, whilst others, for example, cyanobacterial species, are toxic to mammals including domestic pets.[10] Studies in Asia have demonstrated the link between increasing use of fertilizers and increasing incidence of algal blooms. Table 2 illustrates rates of fertilizer application for selected Asian countries. In some Chinese

Table 1 Nitrogen inputs to rivers and coastal waters

River	N inputs to rivers (kg^{-2} yr^{-1})	N exports to coastal waters (kg^{-2} yr^{-1})
Mississippi	7,489	597
Amazon	3,034	692
Nile	3,601	268
Zaire	3,427	632
Zambezi	3,175	330
Rhine	13,941	2,795
Po	9,060	1,840
Ganges	9,366	1,269
Chang Jiang	11,823	2,237
Juang He	5,159	214

Source: Adapted from Norse.[11]

Fig. 3 Choked watercourse, River Skerne, County Durham, August 2002.
Source: From P. Widdison.

Table 2 Average fertilizer use (kg/ha of cropland 2000)

Country	Average fertilizer use kg/ha (2000)
China	255.6
Japan	301.0
Korean Rep.	407.3
Vietnam	285.3

Source: From FAO.[17]

provinces, fertilizer application is greater than 400 kg/ha. This is usually applied as a single application, and with crop utilization efficiency as little as 30–40%, a high proportion is lost to rivers, lakes, and coastal waters.[11] The environmental impact at the regional level has led to a rise in the incidence of red tides (algal blooms). During the 1960s, less than 10 red tides per year were recorded, but in the late 1990s over 300 per year were recorded.[11]

The popular misconception that the nitrate problem is caused by farmers applying too much nitrate fertilizer is too simplistic. Nevertheless, there is now little doubt that the high concentrations of nitrate in fresh waters noted in recent years have mainly resulted from runoff from agricultural land and that the progressive intensification of agricultural practices, with increasing reliance on the use of nitrogenous fertilizer, has contributed significantly to this problem. Since 1945, agriculture in the industrialized world has become much more intensive. Fields are ploughed more frequently; more land is devoted to arable crops, most of which demand large amounts of fertilizer; grassland too receives large applications of fertilizer to ensure a high quality silage for winter feed; stocking densities in general are higher, leading to increased inputs of manure on grassland and problems of disposal of stored slurry; cattle often have direct access to water courses resulting in soil and bank erosion and direct contamination from animal waste; many low-lying fields are now underdrained, encouraging more productive use of the land and speeding the transport of leached nitrate to surface water courses. It is true that lowland rivers close to urban areas receive larger quantities of nitrogen from sewage effluent, but budgeting studies confirm that agriculture is the main source of nitrate in river water.[11,12]

Betton, Webb, and Walling[13] have mapped nitrate concentrations for mainland Britain. A marked northwest to southeast gradient is evident, reflecting relief, climatic conditions, and agricultural activity. Upland areas in the north and west are usually characterized by nitrate concentrations below 1 mg NO$_3$–N L^{-1}. This reflects the high rainfall and low temperatures of such areas: upland soils tend to conserve organic matter and mineralization rates are low. In contrast, a decreasing ratio of runoff to rainfall and an increasing intensity of agricultural land use towards the south and east of Britain result in higher mean concentrations of nitrate in river water. Many of the lowland rivers are characterized by concentrations above 5 mg NO$_3$–N L^{-1}; in East Anglia and parts of the Thames basin, mean nitrate concentrations in rivers are close to the E.C. limit of 11.3 mg NO$_3$–N L^{-1}, a level exceeded in some spring waters especially in the Jurassic limestones of the Cotswolds and Lincolnshire Wolds.[14]

The changing pattern of lowland agriculture since 1945 is reflected in long-term records of nitrate for surface and ground waters.[14] For both large and small rivers, there has been a relatively steady upward trend in nitrate concentrations, often of the order of 0.1–0.2 mg NO$_3$–N L^{-1}. Analyses for relatively short time series of just a few years[13] have shown that the upward trend may be interrupted, either because of climatic variability (drier years are associated with lower nitrate concentrations) or because of land use change. Nevertheless, statistical analysis of long time series shows that the main effect is a steady increase in nitrate levels over time that is independent of climate.[14] If trends continue, the mean nitrate concentration of many rivers in Europe will soon be above the E.C. limit; in many cases, this level is already exceeded during the winter when nitrate concentrations reach their maximum. In catchments where groundwater is the dominant discharge source, this long-term trend may be prolonged since it may take years for nitrate to percolate down to the saturated zone. In such basins, nitrate pollution may remain a problem for decades to come. In recent years, a number of options have been considered as a means of halting the upward trend.

LAND USE CONTROLS TO REDUCE NITROGEN ENRICHMENT TO SURFACE WATERS

Trends in water management in Europe include moves toward catchment-level management, improved intersectoral co-ordination and co-operation, and frameworks facilitating stakeholder participation. This approach is developed by the European Union in its Water Framework Directive, which sets targets for good ecological status for all types of surface water bodies and good quantitative status for groundwater.[3] More localized schemes, like the U.K. Nitrate Vulnerable Zones, involve greater restrictions on farming practice, such as restricting the amount and timing of organic and inorganic fertilizer application. The EU Common Agricultural Policy is to change the way payments are made to farmers. Single-farm payments will encourage farming in a more environment friendly way. Financial payments may be available to farmers for loss of income or for changing farming practice such as improving slurry storage and fencing off watercourses to restrict livestock access.[2] Much interest currently focuses on the use of riparian land as nitrate buffer zones.[15]

The terrestrial–aquatic ecotone (boundary zone) occupies the zone between the hillslope and the river channel, usually coinciding with the floodplain. Given their position, nearstream ecotones can potentially function as natural sinks for sediment and nutrients emanating from farmland. Observed denitrification rates in floodplain sediments may be sufficient to remove all nitrate from groundwater flowing under a riparian woodland, with a floodplain width of 30 m. Saturated, anoxic soils, rich in carbon, are exposed to nitrate-rich groundwater. Rates of denitrification are high within this zone since the nutrients required by denitrifying bacteria are abundant. Wetlands and wet meadows (defined as areas where the water table is at or above land surface for long enough each year to promote the formation of hydric soils and support the growth of

aquatic vegetation) also have potential as nitrogen sinks.[16] High production rates by wetland vegetation result in an abundance of carbon providing an organic substrate for bacterial processes. Wetland plants transport oxygen into anaerobic sediments, which can enhance denitrification leading to losses of nitrogen as N_2O or N_2 from wetland sediments.

The type of vegetation found on the floodplain controlling the efficiency of nitrate absorption is the subject of much debate (see, for example,[15]). Several studies have argued the presence of trees is crucial; yet others state the role of surface vegetation is secondary to the presence of saturated conditions together with a carbon-rich sediment. Denitrifying bacteria operate best at the junction of anaerobic/aerobic zones where both carbon and nitrate are abundant. It is clear that nitrate losses may be reduced by creating a nutrient-retention zone between the farmland and the river. Given that many floodplains around the world are part of an intensive agricultural system, creating permanently vegetated buffer strips between field and water courses is an idea that should be actively promoted. However, buffer strips will only be successful nutrient sinks if they are managed in an appropriate way. Underlying artificial drainage should be broken or blocked up to prevent a direct route to the watercourse for solutes, and grassland strips need maintenance to prevent them becoming choked with sediment and losing their sediment retention potential.

Solving the problem of nutrient enrichment of surface waters cannot be seen in the short term. Long-term land use change is needed. Taking farm land immediately adjacent to water courses out of production is one option that could go some way to allow modern agriculture and water supply to coexist in the same basin.

REFERENCES

1. Heathwaite, A.L. Nitrogen cycling in surface waters and lakes. In *Nitrate: Processes, Patterns and Management*; Burt, T.P., Heathwaite, A.L., Trudgill, S.T., Eds.; Wiley: 1993.
2. Burt, T.P.; Heathwaite, A.L.; Trudgill, S.T.; Eds. *Nitrate: Processes, Patterns and Management*. Wiley, 1993.
3. WHO. *Water and Health in Europe: A Joint Report from the European Environment Agency and the WHO Regional Office for Europe*; World Health Organisation: 2002.
4. European Environment Agency. *Nitrogen Concentrations in Rivers*; European Environment Agency: 2001.
5. United States Geological Service. Nitrogen in the Mississippi basin–estimating sources and predicting flux to the Gulf of Mexico. In *USGS Fact Sheet No. 135-00*; USGS: U.S.A., 2000.
6. Ribaudo, M. Non-point source pollution control policy in the USA. In *Environmental Policies for Agricultural Pollution Control*; Shortle, J.S., Abler, D.G., Eds.; CAB International: 2001.

Riparian—
Wetlands

7. Power, J.F.; Wiese, R.; Flowerday, D. Managing farming systems for nitrate control: a research review from management systems evaluation areas. J. Environ. Qual. **2001**, *30*, 1886–1880.

8. Royal Society. The *Nitrogen Cycle of the United Kingdom*; Royal Society: London, 1983.

9. House of Lords. *Nitrate in Water: Sixteenth Report from the European Communities*; HMSO: London, 1989.

10. Addiscott, T.M. Fertilizers and nitrate leaching. In *Agricultural Chemicals and the Environment*; Hester, R.E., Harrison, R.M., Eds.; Royal Society of Chemists, 1996.

11. Norse, D. Fertilisers and world food demand. Implications for environmental stress. IFA-FAO Agriculture Conference. Rome, 2003.

12. Burt, T.P.; Johnes, P.J. Managing water quality in agricultural catchments. Trans. Inst. Br. Geogr. **1997**, *22*(1).

13. Betton, C.; Webb, B.W.; Walling, D.E. *Recent Trends in NO3–N Concentration and Load in British rivers*; IAHS Publication, 1991; 203, 169–180.

14. Johnes, P.J.; Burt, T.P. Nitrate in surface waters. In *Nitrate: Processes, Patterns and Management*; Burt, T.P., Heathwaite, A.L., Trudgill, S.T., Eds.; Wiley: 1993; 269–317.

15. Haycock, N.E. et al., Eds. Buffer zones: their processes and potential in water protection. Quest Environmental: Harpenden.

16. Howard-Williams, C.; Downes, M.T. Nitrogen cycling in wetlands. In *Nitrate: Patterns, Processes and Management*; Burt, T.P., Heathwaite, A.L., Trudgill, S.T., Eds.; Wiley: 1993; 141–167.

17. Food and Agriculture Organisation. Earth Trends Data Tables: Agricultural Inputs 2003; http://earthtrends. wri.org.

Riparian—Wetlands

Surface Water: Nitrogen Fertilizer Pollution

Gregory McIsaac
Natural Resources and Environmental Sciences, University of Illinois, Urbana, Illinois, U.S.A.

Abstract

For centuries farmers practiced mixed cropping of leguminous and non-leguminous plants without knowing the basis of the empirically observed benefit from the practice. In 1838, it was first recognized that leguminous plants could utilize nitrogen (N_2) from the air. This entry explores this process.

INTRODUCTION

The use of industrially manufactured nitrogen (N) fertilizers increased rapidly in developed countries between 1960 and 1980. This facilitated a large increase in the production of feed and food grains (maize, wheat, and rice) per unit of cultivated land, but in some regions it also contributed to enrichment of surface and groundwater with various forms of nitrogen. Fertilizer, however, is not the only source of nitrogen that can cause contamination of surface waters. Biological nitrogen fixation, mineralization of soil organic nitrogen, and animal wastes can also contribute to nitrogen enrichment of water bodies. Additionally, under some conditions, nitrogen applied to the soil may be converted to gaseous or immobile forms of nitrogen that do not contribute to surface water contamination. Because of these various sources and transformations of nitrogen, the severity of surface water contamination by nitrogen fertilizer has been difficult to precisely quantify. Existing research indicates that the amount of contamination from fertilizer varies depending on the amount of fertilizer applied, and characteristics of the soils, crops, climate, and the receiving water bodies.

PROBLEMS CAUSED BY NITROGEN POLLUTION OF SURFACE WATERS

There are three water quality concerns associated with different forms of nitrogen. First, the combined concentrations of nitrate (NO_3^-) plus nitrite (NO_2^-) in excess of $10\,mg\,N\,L^-$ can contribute to methemoglobinanemia ("blue baby syndrome") in infants if ingested.[1] To guard against this, the U.S. Public Health Service limits nitrate plus nitrite concentration in public drinking water supplies to $10\,mg\,N\,L^-$. Secondly, unionized ammonia (NH_3) may be toxic to fish at concentrations as low as $0.02\,mg\,N\,L^-$. Finally, elevated total nitrogen concentrations (including nitrate, ammonia, and organic forms) in rivers can promote the process of cultural eutrophication in coastal waters, whereby increased production and decomposition of algae, leads to reduced oxygen concentrations. This, in turn, may reduce the abundance and diversity of marine life and may promote the outbreak of nuisance algae.[2]

SOURCES OF N POLLUTION

Nitrogen contamination may come from a variety of sources: municipal sewage, animal manure, atmospheric deposition, biological N fixation, soil organic N, and/or nitrogen fertilizers. The consequences of contamination in a specific water body will depend upon the amount of contamination from all sources and characteristics of the receiving waters. Shallow rivers, wetlands, lakes, and reservoirs, have some capacity to remove nitrogen by microbial denitrification. The susceptibility of estuaries and coastal waters to eutrophication depends on temperature, availability of phosphorus and silica for algae production, and the rate of water exchange with the open ocean.

FERTILIZERS

The contribution of inorganic fertilizer to surface water N contamination increased after 1960 as the widespread and intensive use of inorganic N fertilizers rapidly expanded.[3] The use of N fertilizer has allowed greater production of feed and food crops per unit area cultivated. In the United States, 75% of N fertilizer is applied to maize, while in other countries, N fertilizer is primarily used on wheat and rice. Prior to 1960, nitrogen for crop production was obtained primarily by using crop rotations that included legumes such as clover and alfalfa, which can establish a symbiotic relationship with soil bacteria that can convert atmospheric N_2 gas to biologically available forms of N.

Commercial nitrogen fertilizer is primarily manufactured as gaseous ammonia (NH_3), using the Haber–Bosch process in which gaseous nitrogen is reacted with gaseous hydrogen under pressure. The gaseous ammonia may be

Encyclopedia of Natural Resources DOI: 10.1081/E-ENRW-120010336

injected into the soil, which is a common fertilizer application practice in the United States. Additionally, a wide variety of granular and aqueous fertilizer products containing nitrogen are manufactured from manufactured ammonia.

WHAT HAPPENS WHEN FERTILIZER IS APPLIED TO SOIL?

Biochemical Processes

In the soil, ammonia reacts with water and is largely converted to ammonium (NH_4^+), which tends to be strongly adsorbed on soil particles. This adsorption inhibits the movement of ammonium through the soil. Ammonium is an energy rich substance and certain soil bacteria can utilize this energy by decomposing the ammonium to nitrate (NO_3^-). Unlike ammonium, nitrate is not adsorbed to soil particles and, therefore, moves readily with water in the soil. Nitrate that is not taken up by plant roots or soil microorganisms can be transported to groundwater and surface water by a variety of mechanisms.

HYDROLOGIC PROCESSES

Rainfall, snow melt, or irrigation water input to the soil periodically exceeds the water holding capacity of the soil in the root zone. Depending on the characteristics of the soil, this may lead to one or more of the following: 1) saturation of the root zone with water; 2) surface runoff; and 3) drainage of water through the soil profile to groundwater and/or surface water bodies. Each of these has different consequences for transport of nitrate to surface waters.

If the soil becomes saturated, oxygen may become scarce and in anoxic conditions, denitrifying bacteria may convert the nitrate to nitrogen gases (NO, N_2O, and N_2). Nitrogen converted to these gases becomes unavailable for plant uptake or for surface water contamination. Additionally, saturated soil during the growing season is harmful to many crops like maize that cannot tolerate low oxygen concentrations in the root zone for more than a few days.

Surface runoff has the capacity to transport soil, vegetation, and surface applied granular fertilizers from agricultural fields to surface water bodies. If a granular form of nitrogen fertilizer had been applied immediately prior to the event that caused the surface runoff to occur, nitrate and ammonia concentrations in runoff can be very high.[4] This does not appear to be a common phenomenon, however. Small rainfall events are much more common than large events that typically produce surface runoff. After granular fertilizer is applied, it is likely that a series of small rainfall events will dissolve the granules and move the nitrogen into the soil profile, where it is less likely to contaminate surface runoff. Surface runoff usually has a low nitrate concentration but it can be high in organic and particulate N derived from soil and vegetation.

Drainage of water through the soil profile to groundwater and surface water appears to be the hydrologic pathway that most frequently leads to problematic nitrate contamination of surface waters in agricultural watersheds. This can occur in two ways: by natural drainage where ground water contributes to stream flow and river flow, and by artificial subsurface drainage, where perforated pipes (sometimes called tile drains) have been buried in the soil for the purpose of removing water to reduce damage caused by saturated conditions and thereby enhance crop production (Fig. 1).

Artificial subsurface drainage improves aeration of the soil root zone and increases the length of time that machinery can be used on the soil.[6] It is a common practice in the north central United States, and in northwestern Europe, where flat and swampy land has been converted to cropland. The water removed from the soil by artificial drainage is usually directed to surface ditches, streams, and rivers. This water can have high concentrations of nitrogen, principally in the nitrate form, especially where nitrogen fertilizers are applied in excess to the amount necessary for crop production.[7,8] This nitrate can also be derived from microbial conversion of soil organic matter to inorganic N in the process of mineralization. Mineralized soil organic N may come from crop residues (unharvested leaves, stalks, and roots). Of course, some of the N in crop residues may have originated from fertilizer applied in previous years, but it can also derive from biological N fixation or animal manures applied on a field.

SPATIAL VARIABILITY

In most agricultural settings, commercial fertilizer provides only one source of N used for crop production. Animal manure, biological N fixation, mineralization from soil organic N, and deposition of N from the atmosphere can also contribute to soil fertility and surface water contamination. Because there are multiple sources and sinks of N in the soil, the relationship between N fertilizer application rate and nitrogen loss in drainage water is not always consistent across locations and across studies. If denitrification and plant and microbial uptake of N are large, nitrate concentrations in subsurface drainage may be low in spite of high fertilizer N inputs. If mineralization of soil organic matter is large, nitrogen in drainage water may be large without N fertilizer input. High rates of mineralization of soil organic N occur after the initial cultivation of virgin land, and after a leguminous forage crop such as alfalfa or clover is cultivated into the soil. Appropriate use of N fertilizer should take all of these N sources into account, as should studies examining the relationship between N fertilizer use and water quality.

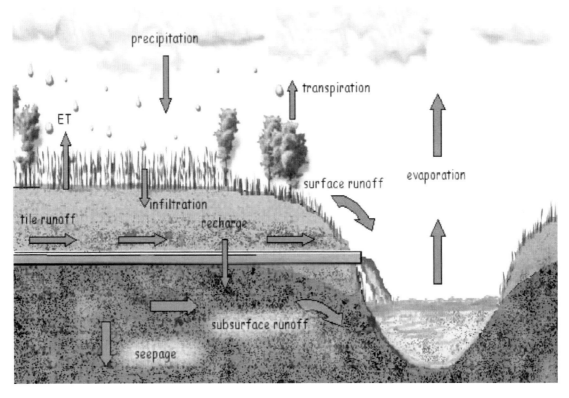

Fig. 1 Illustration of the hydrologic cycle with artificial subsurface drainage (tile runoff) contributing to surface channel flow.
Source: Adapted from Steinheimer, et al.[5]

WATERSHED SCALE ANALYSES

Regional Nitrogen Input–Output Analyses

Howarth et al.[10] developed an approach for estimating the net nitrogen inputs to a region N that is highly correlated with average nitrogen transport in the rivers draining temperate regions (Fig. 2). Net N input to a region was defined as sum of N in fertilizer used, biological N fixation of agricultural crops, oxidized N in atmospheric deposition in the region, and the N in food and feed imported to the region minus the N in food and feed exported from the region. This approach assumes that there is no net gain or loss of N from soil organic matter. This assumption appears to be reasonable in regions where most soils have been under continuous cultivation for 60 years or more, at which time, annual mineralization of soil organic N is roughly replaced by organic N returned to the soil in crop residues and microbial biomass.[11]

In temperate regions, riverine N transport is, on average 25% of the net N input to the region. The fate of the other 75% of the net N is unknown, but much of it is probably converted to gaseous forms of N by microbial denitrification. The high net N input in countries draining to the North Sea, most notably the Netherlands, is in part due to high density of domestic animals as well as use of N fertilizers. In tropical regions, riverine N flux is much greater than 25% of net N inputs, even in regions where little N fertilizer was used. The reasons for this are not precisely known but it is believed to be due, in part, to greater rates of biological N fixation in both cultivated and non-cultivated land in the tropics. This may also be due to the recent conversion of forest, wetlands, and grasslands to crop production, which leads to high rates of mineralization of soil organic N to nitrate which is highly mobile.

HYDROLOGIC PROCESS MODELS

The quantity of nitrate transported in rivers is also related to the quantity of water flowing in the rivers per unit of land area, which is also known as water yield. Caraco and Cole[12] demonstrated that riverine nitrate N transport in major rivers in the world was a function of water yield, fertilizer use, population density, and atmospheric deposition of oxides of N. Building on these results, McIsaac et al.[13] developed the following model of annual nitrate discharge in the Lower Mississippi River 1960–1998:

$$NF_m = 0.66 WY^{0.93} e^{\left(0.13 NNI^{2-5} + 0.06 NNI^{6-9}\right)} \tag{1}$$

where NF_m = annual nitrate N flux in Lower Mississippi River (kg N ha^{-1} yr^{-1}), NNI^{2-5} = average annual net N input

Riparian— Wetlands

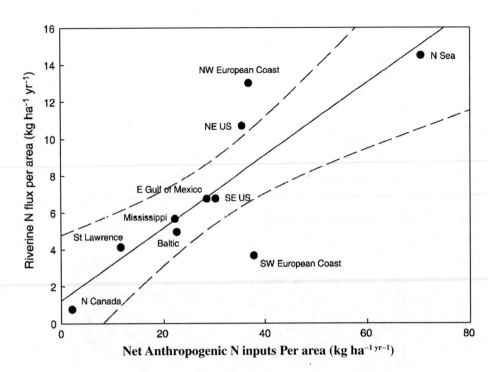

Fig. 2 Average annual riverine total N flux as a function of net anthropogenic N inputs to temperate regions draining to the North Atlantic Ocean.
Source: Adapted from Howarth, et al.[10]

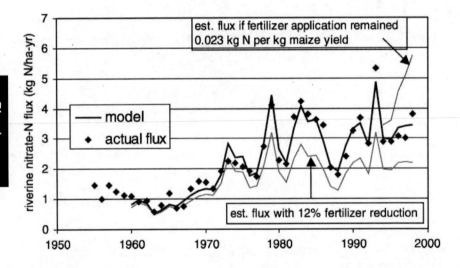

Fig. 3 Annual riverine nitrate flux in the Lower Mississippi River at St. Francisville, Louisiana, as determined from measurements (diamonds) as estimated from Eq. (1) (thick line), as estimated from Eq. (1) assuming a 12% reduction in N fertilizer input (thin lower line), and assuming fertilizer applications remained 0.023 kg N per bushel of harvested maize rather than declining to 0.018 kg N per bushel of harvested maize (thin upper line).
Source: Adapted from McIsaac.[13]

during the previous 2–5 years (kg N ha⁻¹ yr⁻¹), $NNI^{6-9} =$ average annual net N input during the previous 6–9 years (kg N ha⁻¹ yr⁻¹), and $WY =$ annual water yield (m yr⁻¹).

This equation accounted for 95% of the annual variation in nitrate flux in the Mississippi River from 1960 to 1998 and suggested that riverine nitrate in a given year was correlated with net N input averaged over the previous 2–9 years. Furthermore, calculations with the model suggest that if the N fertilizer use in the basin had been 12% lower than actual during this period, nitrate flux to the Gulf of Mexico would have been 33% less than observed (Fig. 3), assuming crop yields were not limited by N shortages.

THE ROLE OF FERTILIZER USE EFFICIENCY

The efficiency of fertilizer used for maize production in the major maize producing states (Illinois, Iowa, Indiana, Minnesota, and Nebraska) in the Mississippi River Basin increased between 1986 and 2000. Maize yields have increased about 20% from 1986 to 2000, while N fertilizer use has remained roughly constant. Between 1976 and 1986, an average of 0.023 kg N of fertilizer was applied for each kg of maize harvested. Between 1996 and 2000, an average of 0.018 kg N of fertilizer was applied per kg of maize harvested. If this improvement in fertilizer use efficiency had not occurred, nitrate flux in the Mississippi

River in 1996–1998 would have been about 50% greater than the measured flux, according to the model of McIsaac et al.[13] (Fig. 3).

Farmers face two major uncertainties when making fertilizer application decisions: they do not know what their yields will be nor how weather conditions might influence the availability of N fertilizer to the crop. The cost of nitrogen fertilizer has been relatively low in relation to the value of the increased yields, and consequently many farmers have believed that applying more N fertilizer than necessary provides "cheap insurance" against the uncertainties. In some instances, farmers did not consider N available from animal manure or from previously harvested legume crops like soybeans.

A number of factors are likely responsible for the increased fertilizer use efficiency. Research and outreach efforts have provided farmers with better information for making N fertilizer decisions. Water quality concerns have focused attention on the need for improved nutrient management. Weather during the 1990s was generally more favorable for corn production than the 1980s, when three major droughts occurred in the corn growing region of the Mississippi River Basin.

ADDITIONAL NEEDS AND APPROACHES FOR REDUCING NITROGEN TRANSPORT

A recent improvement in the efficiency of N fertilizer use has also been observed in wheat production in the United Kingdom and rice production in Japan.[3] However, improved fertilizer use efficiency alone may not be sufficient to address water quality problems in some settings. Jaynes et al.[8] reported that even with N fertilizer rates at recommended levels, nitrate concentration in tile drainage water sometimes exceeded the drinking water standard of 10 mg N L$^-$ in Iowa.

Zucker and Brown[9] recommended several additional practices that can reduce nitrogen contributions in tile drainage: water table management, treatment of drainage water in wetlands, and use of crop rotations that reduce N losses. Additional monitoring and documenting the changes in water quality associated with changing fertilizer management practices are needed to improve our understanding of the connections between N fertilizer use and water quality in different geographic settings.

CONCLUSIONS

In many settings nitrogen enrichment of surface water bodies has increased following the increased use of N fertilizers. The precise contribution of nitrogen fertilizers to surface water nitrogen has been difficult to quantify because there are multiple sources of nitrogen contributing

to most water bodies, and, depending on environmental conditions, a certain portion of soil nitrogen may be converted to gaseous or immobile forms. In general, however, agricultural regions with extensive artificial subsurface drainage systems or with sandy soils tend to have the most nitrogen enriched surface waters.

The efficiency of nitrogen fertilizer used for crop production increased in many areas in the 1990s and this has very likely limited or reduced the subsequent contamination of surface waters. Continued improvements in fertilizer use efficiency, and the use of wetlands for removing nitrogen from surface waters will help alleviate problems caused by nitrogen enrichment. Additional monitoring and research are needed to more precisely quantify how nitrogen management practices influence surface water nitrogen concentrations in different settings. A more precise understanding of the causal relationships between nitrogen inputs to the land and the contamination of surface waters could provide more effective guidance for management, policies, and programs intended to protect aquatic resources while maintaining optimal use of land resources.

REFERENCES

1. Skipton, S.; Hay, D. *Drinking Water: Nitrate and Methemoglobinemia ("Blue Baby" Syndrome)*; Publication G98-1369; University of Nebraska Cooperative Extension: Lincoln, NE, 1998, http://www.ianr.unl. edu/pubs/water/g1369.htm (accessed February 2002).
2. National Research Council. *Clean Coastal Waters: Understanding and Reducing the Effects of Nutrient Pollution*; National Academy Press: Washington, DC, USA, 2000.
3. Smil, V. *Enriching the Earth: Fritz Haber, Carl Bosch and the Transformation of World Food Production*; MIT Press: Cambridge, MA, USA, 2001.
4. Romkens, M.J.M.; Nelson, D.W.; Mannering, J.V. Nitrogen and phosphorus composition of surface runoff as affected by tillage method. J. Environ. Qual. **1973**, *2* (2), 292–295.
5. Steinheimer, T.R.; Scoggin, D.K.; Kramer, L.A. Agricultural chemical movement through a field-size water-shed in Iowa: subsurface hydrology and distribution of nitrate in groundwater. Environ. Sci. Technol. **1998**, *32*, 1039–1047.
6. Badiger, S. Integrated Numerical Modeling of Spatial and Time-Variant Hydrologic Response in Subsurface Drained Watersheds. Ph.D. thesis. Department of Agricultural Engineering, University of Illinois: Urbana, 2001.
7. Stevenson, F.J.; Ed. *Agricultural Drainage*; American Society of Agronomy (Monograph # 38): Madison, WI, USA, 1999.
8. Jaynes, D.B.; Colvin, T.S.; Karlen, D.L.; Cambardella, C.A.; Meek, D.W. Nitrate loss in subsurface drainage as affected by nitrogen fertilizer rate. J. Environ. Qual. **2001**, 30, 1305–1314.
9. Zucker, L.A.; Brown, L.C. *Agricultural Drainage: Water Quality Impacts and Subsurface Drainage Studies in the Midwest*; Ohio State University Extension Bulletin: 1998;

871, http://ohioline.osu.edu/b871/ (accessed February 2002).

10. Howarth, R.W.; Billen, G.; Swancy, D.; Townsend, A.; Jaworski, N.; Lajtha, K.; Downing, J.A.; Elmgren, R.; Caraco, N.; Jordan, T.; Berendse, F.; Freney, J.; Kudeyarov, V.; Murdoch, P.; Zhao-Liang, Z. Regional nitrogen budgets and riverine N&P fluxes for the drainages to the North Atlantic Ocean: natural and human influences. Biogeochemistry **1996**, 35, 75–139.

11. Stevenson, F.J.; Ed. Origin and distribution of nitrogen in the soil. *Nitrogen in Agricultural Soils*; Agronomy Monograph 22; American Society of Agronomy: Madison, WI, 1982; 1–42.

12. Caraco, N.F.; Cole, J.J. Human impact on nitrate export: an analysis using major world rivers. Ambio **1999**, 28, 167–170.

13. McIsaac, G.F.; David, M.B.; Gertner, G.Z.; Goolsby, D.A. Nitrate flux in the Mississippi River. Nature **2001**, *414*, 166–167.

Tourism Management: Marine and Coastal Recreation

C. Michael Hall
Department of Management, University of Canterbury, Christchurch, New Zealand,
Centre for Tourism, University of Eastern Finland, Savonlinna, and
Department of Geography, University of Oulu, Oulu, Finland

Abstract

Coastal and marine tourism is one of the most significant areas of global tourism activity. Ranging from beach tourism to whale watching, the coast is the focal point of many countries tourism industries. However, the significant population pressure placed on coastal tourism destinations as a result of increased visitor numbers can substantially affect infrastructure and resources as well as environmental services in general. This entry provides a definition and historical overview of coastal and marine tourism before reviewing the significance of coastal resort morphology on tourism's environmental impacts. It then reviews tourism activities in relation to one of the most import marine tourism environments—coral reefs—before briefly discussing the cruise industry. It concludes by noting tourism's role in coastal and marine area conservation and management.

INTRODUCTION

The coastal environment is a major focus for tourism and recreational activity in many parts of the world and is often associated with tourism images of "sun, sand and surf."[1] However, the development of beach resorts and the increasing popularity of coastal and marine tourism (e.g., fishing, scuba diving, whale watching, windsurfing, and yachting) have all placed increased pressure on coastal regions, an area in which use is often already highly concentrated in terms of agriculture, human settlements, fishing, and industry.

The concept of coastal tourism includes a range of tourism, leisure, and recreationally oriented activities that occur in the coastal zone and immediate offshore coastal waters. These include tourism-related development (accommodation, restaurants and food services, attractions, and second homes), and the infrastructure supporting coastal and marine tourism development (e.g., retail businesses, transport hubs, marinas, and activity suppliers). Also included are tourism activities such as recreational boating, coast- and marine-based ecotourism, cruises, swimming, recreational fishing, snorkeling, and diving. Marine tourism is closely related to coastal tourism, but it is more geographically expansive and also includes ocean-based tourism activities, such as cruise ships and yacht cruising.[2] The entry first discusses the historical development of coastal and marine tourism and the values associated with it before examining coastal resort morphology and the associated environmental impacts of coastal and marine tourism. The entry then discusses coastal and

marine tourism in the specific contexts of coral reefs, cruise tourism, and conservation.

HISTORICAL DEVELOPMENT

Although coastal areas are now synonymous with tourism and recreation, this has not always been the case. In western landscape traditions, natural coastal areas were often perceived negatively until the late 1700s when the Romantic Revolution transformed perceptions of landscape and seascape. Simultaneously, a belief developed in the healing and recuperative benefits of sea air and seawater. The period 1750–1840 witnessed a fundamental reassessment of coastal areas as leisure place. In that period, the beach developed as an activity space for recreation and tourism, with distinct cultural and social forms emerging in relation to resort fashions, tastes, and innovations. The development of piers, jetties, and promenades for formal recreational activities led to new ways of experiencing and appreciating the sea and development of positive coastal amenity values dates from this time.[3]

From the mid-19th century to the mid-20th century, the development of coastal resorts was closely linked to the expansion of rail and shipping transport networks. For example, for what was arguably the world's first modern seaside resort, Margate in southeast England, the visits by the wealthy gentry and upper middle classes, were replaced by lower middle class and working class tourists from the 1820s as a result of direct steamer access from London. This expansion of tourist numbers but change in their socioeconomic composition

Encyclopedia of Natural Resources DOI: 10.1081/E-ENRW-120047588

received further impetus when Margate was connected to London by railway after the 1840s. This particularly innovation, together with the growth of personal car access in the twentieth century, also meant that visitation could take the form of day excursions rather than overnight stays with consequent effects on the economic returns from tourism at the destination level as well as increased pressures on the transit routes to and from coastal resorts.[4]

Charting the history of seaside resorts in northern Europe and North America also highlights how continued changes in transport technology led to the decline of many traditional resorts as, for a similar cost, it became possible for tourists in major urban centers to travel to the Mediterranean or the Caribbean. This shift in coastal tourism activity has had significant implications for resource management in newer coastal resorts as well as for climate change given growth in emissions from aviation-related travel to coastal resorts. Such historical change also indicates that, like many industries, tourism also moves through development cycles and that resorts can enter into significant decline and even abandonment as tourist destinations.[5] This can therefore create new problems for such cities and towns as they seek to regenerate infrastructure and find new forms of economic development at a time of diminished returns from tourism. In some cases, this may mean focusing on new markets or attractions, for example, casinos in the case of Atlantic City, New Jersey, or cultural tourism in the case of Margate, that has few direct links to the water-based tourism activities. In older coastal resorts, the tourism base may also be replaced by a second-home and retirement home function that still includes a coastal leisure component.[2]

COASTAL RESORT MORPHOLOGY AND IMPACTS

Tourism urbanization is spatially and functionally different from other urban places.[6] In many coastal areas, tourism urbanization is highly linear with resultant impacts on the coastal environment. These are particularly pronounced in coastal resort areas in the Mediterranean where tourism and leisure-related urbanization is held to be primarily responsible for coastal urbanization. Of the 220 million people who visit the Mediterranean region each every year, over 100 million visit Mediterranean beaches.[7] In Italy, over 43% of the coastline is completely urbanized, 28% is partly urbanized, and less than 29% is still free of construction. There are only six stretches of coast over 20 km long that are free of construction and only 33 stretches between 10 and 20 km long without any construction.[8] Similarly, in Cyprus, 95% of the tourism industry is located within two kilometers of the coast,[9] placing the coastal environment as well as archeological heritage under extreme pressure. In Tunisia, it is expected that about 150 km of shoreline, over 13% of the total Tunisian coastline, will eventually be occupied by tourism and leisure facilities

and infrastructure.[10] In many coastal resorts, the inappropriate siting of tourist infrastructures on foredunes has accelerated beach erosion processes as a result of sediment processes being interrupted,[11] while tourism development has also altered the water dynamics of coastal regions.[10]

Island destinations are a special case of coastal and marine tourism particularly affected by tourism urbanization and the resource demands of tourism. The government of the Caribbean Island of Barbados estimates that 98% of renewable freshwater resources are already being used, and that industrial and commercial uses (including tourism) had increased from 20% in 1996 to 44% by 2007. Tourism operations (hotels, cruise ships, golf courses), which represented approximately one-sixth of total consumption in 1996, is projected to represent one-third of overall demand by 2016.[12]

Impacts of Seasonality

The seasonal nature of coastal tourism can place significant stress on infrastructure with populations frequently more than doubling in resort destinations during high season. Island destinations with already limited resources may be especially affected. For example, in the case of Anguilla in the Caribbean, annual tourist numbers are equivalent to a 30.5% increase in permanent population, while in the Cayman Islands it is equivalent to an 89% increase.[13] Such high levels of tourist visitation place extreme strain on water and sewage infrastructure.[14] In the Mediterranean, only 80% of the effluent of residents and tourists is collected in sewage systems, with the remainder being discharged directly or indirectly into the sea or to septic tanks. However, only 50% of the sewage networks are actually connected to wastewater treatment facilities with the rest being discharged into the sea.[15] The lack of appropriate sewage treatment leads to nutrient enrichment of littoral waters and the growth of algal blooms which can affect both coastal ecology as well as tourist perception of beaches and destinations.[16] The effects of seasonal tourism demands on infrastructural systems such as sewage and water supply, as well as resources means that in a number of coastal destinations local water supplies have to be supplemented by water imports. The Mediterranean is a coastal tourism region identified as being extremely vulnerable to water scarcity,[17] while small islands and regions such as Baja California Sur, Sharm El Sheikh, Egypt, Zanzibar, and Almería province, Spain, also have overused water capacity or are at high-risk of overusing their water capacity.[18]

Beach Environments

Beaches and wetland systems are the coastal environments most under threat from tourism urbanization and leisure activities. Visitor pressure increases dune degradation and vulnerability highlighting the need for close monitoring of impacts and changes in dune morphology as well as the development of appropriate management regimes.[11] Many significant wildlife habitats have been lost through resort development. This includes not only the construction of hotels and tourism infrastructure but also the development of activities, such as golf courses. The Parliamentary Assembly, Council of Europe [19] estimated there has been a reduction in the Mediterranean wetland area of approximately three million hectares by 93% since Roman times. Of which, a third has been lost since 1950. In the Asia-Pacific many coastal wetlands have been drained or substantially modified for resort development, although the loss of mangrove swamps has substantial long-term implications for increased coastal erosion especially as a result of increased storm events and sea-level rise while habitat loss affects local fish stocks given the nursery role of mangroves.[20]

CORAL REEFS

Coral reefs are one of the most important marine ecosystems for tourism. However, they are highly susceptible to urban run-off and sedimentation; over-fishing; souveniring of coral and shells; as well as over use by divers and snorkelers. Coral reefs and are also considered one of the most vulnerable to climate change and other anthropogenic environmental change,[21] including several climate change-related impacts: ocean acidification, coral bleaching, and for in-shore coral reefs, greater land runoff as a result of increased storm events.[20]

Coral bleaching can occur for multiple reasons, but temperature change is the primary cause, and acidification can be a strong contributor. Mass coral bleaching transforms large reef areas from a mosaic of color (if healthy) to a stark white. With very high confidence, the IPCC[22] concluded that a warming of 2°C above 1990 levels would result in mass mortality of coral reefs globally. However, the vulnerability of reefs to the impacts of climate change will vary spatially and temporally, with shallow reefs, reefs with species closest to their thermal maximum threshold, and those closest to sources of pollution or other human impacts the most vulnerable and where impacts will be visible to tourists the soonest. Nevertheless, even moderate further warming will consequently affect the attractiveness of coral reefs to tourists in some destinations. Further, coral reefs provide other important ecosystem services to the tourism sector, including as a fishery resource and coastal protection against storms. For destinations where reefs are the key attraction for tourists, the long-term damage arising from bleaching incidents will have important implications for the quality of dive tourist experiences, and therefore for the sustainability of tourism operations.[23] However, previous bleaching events suggest that the effects may be variable among different markets at different destinations.[20]

In the Caribbean, 76% of tourists at the diving destination of Bonaire (where 99% of respondents took at least one dive during their trip) indicated they would be unwilling to return for the same holiday price in the event that corals suffered "severe bleaching and mortality."[24] Andersson's[25] study of tourism perceptions at the African islands of Zanzibar and Mafia found that serious divers who visit Mafia were more aware of bleaching than the recreational divers who visit Zanzibar (62% versus 29%). When asked if they would be willing to dive on a bleached reef, 40% of respondents at Zanzibar and 33% at Mafia indicated that they would. In their study of Mauritius, Gössling et al.[26] found that there are significant differences between dive destinations. They found that the state of coral reefs was largely irrelevant to dive tourists and snorkellers, as long as a threshold level was not exceeded. This level was defined by visibility, abundance and variety of species, and the occurrence of algae or physically damaged corals. The results by Gössling et al.[26] are consistent with the findings of Main and Dearden[27] that 85% of recreational divers failed to perceive any damage to reefs in Phuket, Thailand after the 2004 Indian Ocean tsunami. Dearden and Manopawitr[28] even question whether future generations of divers will perceive environmental changes, such as bleaching and degraded reef conditions, in the same way as contemporary divers if they have no frame of reference of previous, more pristine conditions. Such results highlight how different scuba diving market segments will be differentially affected by the impacts of climate and other forms of environmental change depending on motivations, prior experiences, and budgets.

CRUISE TOURISM

The cruise sector is one of the fastest-growing areas of tourism with cruise demand in North America alone increasingly at an average annual rate of 7.6% between 1990 and 2010.[29] In 2010 14.89 million people travelled on CLIA member cruise lines. Cruising impacts the marine environment as a result of discharging untreated waters and other waste at sea,[30] the role of shipping as a vector of alien species,[31] and the high-energy intensity of cruise ships on a per-passenger basis.[32] Global ocean-going cruise emissions for 2005 were estimated at 34 $MtCO_2$, less than 5% of global shipping emissions.[33] However, this figure does not include the full range of tourist passenger vessels.[20] Marine transport is a major contributor to the spread of disease and biological invasion. Cruise ships and yachts, as well as site visitation by tourists, have been recognized as particularly significant avenues of species introduction via hull fouling[34] and ballast water.[35]

TOURISM AND COASTAL AND MARINE MANAGEMENT AND CONSERVATION

Although tourism is responsible for a range of impacts on coastal and marine environments, it also provides an economic justification for species and habitat conservation.

Riparian—Wetlands

For example, it is estimated that in 2008 more than 13 million people took whale watching tours in 119 countries worldwide, generating US$2.1 billion in total expenditures.[36] However, while individual species may receive legal protection the conservation of marine habitats lags far behind their terrestrial counterparts. As of 2010, the global total are for marine-protected areas (MPAs) was 4.7 million km^2, or 1.31% of the global ocean surface (3.21% of potential marine jurisdictions/EEZ areas, and 1.06% of off-shelf areas).[7] Only 12 out of 190 states and territories with marine jurisdictions have an MPA coverage of 10% or more in the areas under their jurisdiction. Even in island systems such as the Caribbean and the South Pacific, the proportion of protected marine area in both regions is much lower than the terrestrial area. In the Caribbean, Jamaica has the highest proportion of marine area set aside at 3.56% and in the Pacific, Palau has 8.74% of its marine territory as protected area.[13]

Given the large number of stakeholders in the coastal and marine areas, tourism is often but one component of multiple-objective spatial management and planning strategies. Marine spatial planning often aims to reduce the environmental impacts of tourism, among other marine area users, but simultaneously ensures that the resource base is sufficient to encourage further visitation, especially as tourism, fees, licenses, and income can assist in financing biodiversity conservation. Management and spatial planning strategies aim to reduce conflict between marine users via the concentration of compatible and separation of incompatible uses.[37] However, such integrated management approaches are harder to achieve in the coastal zone because of the larger numbers of stakeholders and population base that is experienced there, especially in urban areas. Nevertheless, in the Integrated Coastal Management Framework, tourism is usually recognized as one of the most important activities in coastal areas as part of their sustainable development.[38]

CONCLUSION

Tourism is a major economic activity in coastal and marine areas. Yet, in historical terms, the association of tourism with the coast and the sea is relatively recent. Nevertheless, in a short period of time, tourism has become an industry critical to many coastal regions and communities as well as generating income for coastal and marine conservation activities. Yet, at the same time, tourism has also been one of the biggest sources of environmental impact of coastal and near-shore areas particularly when seasonal tourist visitation overwhelms water and sewage infrastructure. Tourism is also a significant source of greenhouse gas emissions which may have long-term consequences for significant recreational resources such as coral reefs as well as the threat of sea level rise to coastal tourism resources and infrastructure. Integrated marine area spatial strategies and coastal zone management are the primary planning means to manage coastal and marine tourism although tourism is only one, sometimes conflicting, use of coastal areas. Yet, the income provided by tourism may provide some of the strongest economic justifications to conserve coastal and marine ecosystem services and resources.

ACKNOWLEDGMENTS

The support of Academy of Finland (project SA 255424) is gratefully acknowledged.

REFERENCES

1. Hall, C.M. Trends in coastal and marine tourism: the end of the last frontier? Ocean Coast Manage **2001**, *44*, 9–10, 601–618.
2. Hall, C.M.; Page, S. *The Geography of Tourism and Recreation*, 3rd Ed.; Routledge: London, 2006.
3. Towner, J. An *Historical Geography of Recreation and Tourism in the Western World 1540–1940*; Wiley: Chichester, 1996.
4. Walton, J. The *English Seaside Resort: A Social History 1750–1914*; Leicester University Press: Leicester, 1983.
5. Walton, J. *The British Seaside: Holidays and Resorts in the Twentieth Century*; Manchester University Press: Manchester, 2000.
6. Hall, C.M. Tourism urbanization and global environmental change. In *Tourism and Global Environmental Change: Ecological, Economic, Social and Political Interrelationships*; Gössling, S., Hall, C.M., Eds.; Routledge: London, 2006; 142–156 pp.
7. Toropova, C.; Meliane, I.; Laffoley, D.; Matthews, E.; Spalding, M. *Global Ocean Protection: Present Status and Future Possibilities*; IUCN: Gland, 2010.
8. World Wide Fund for Nature. *Tourism Threats in the Mediterranean*; WWF Mediterranean Programme: Rome, 2001.
9. Loizidou, X. Land use and coastal management in the Eastern Mediterranean: the Cyprus example. In *International Conference on the Sustainable Development of the Mediterranean and Black Sea Environment*; Thessaloniki: Greece; May 28–31, 2003. http://www.iasonnet.gr/abstracts/loizidou.html; 2003.
10. De Stefano, L. *Freshwater and Tourism in the Mediterranean*; WWF Mediterranean Programme: Rome, 2004.
11. Williams, A.; Micallef, A. *Beach Management: Principles and Practice*; Earthscan: London, 2009.
12. Emmanuel, K.; Spence, B. Climate change implications for water resource management in Barbados tourism. Worldwide Hospitality Tourism Themes **2009**, *1*, 252–268.
13. Hall, C.M. An island biogeographical approach to island tourism and biodiversity: An exploratory study of the Caribbean and Pacific Islands. Asia Pac. J. Tourism Res. **2010**, *15*, 383–399.
14. UN Division for Sustainable Development. *Trends in Sustainable Development: Small Island Developing States*; United Nations: New York, 2010.

15. Scoullos, M.J. Impact of anthropogenic activities in the coastal region of the Mediterranean Sea. In *International Conference on the Sustainable Development of the Mediterranean and Black Sea Environment*; Thessaloniki: Greece; May 28–31, 2003. http://www.iasonnet.gr/abstracts/Scoullos .pdf; 2003.

16. Bauer, M.; Hoagland, P.; Leschine, T.M.; Blount, B.G.; Pomeroy, C.M.; Lampl. L.L.; Scherer, C.W.; Ayres, D.l.; Tester, P.A.; Sengco, M.R.; Sellner, K.G.; Schumacker, J. The importance of human dimensions research in managing harmful algal blooms. Front Ecol. Environ. **2010**, *8*, 75–83.

17. Iglesias, A.; Garrote, L.; Flores, F.; Moneo, M. Challenges to manage the risk of water scarcity and climate change in the Mediterranean. Water Resource Manag. **2010**, *21*, 775–788.

18. Gössling, S.; Peeters, P.; Hall, C.M.; Ceron, J.P.; Dubois, G.; Lehmann, L.V.; Scott, D. Tourism and water use: Supply, demand, and security. An international review. Tourism Manage **2012**, *33*, 1–15.

19. Parliamentary Assembly, Council of Europe. *Erosion of the Mediterranean Coastline: Implications for Tourism, Doc. 9981 16 October 2003, Report Committee on Economic Affairs and Development*; Brussels: Council of Europe, 2003.

20. Scott, D.; Gössling, S.; Hall, C.M. *Tourism and Climate Change: Impacts, Adaptation and Mitigation*; Routledge: London, 2012.

21. Hughes, T.P.; Graham, N.A.J.; Jackson, J.B.C.; Mumby, P.J, Steneck, R.S. Rising to the challenge of sustaining coral reef resilience. Trends Ecol. Evol. **2010**, *25*, 633–642.

22. Schneider, S.H.; Semenov, S.; Patwardhan, A.; Burton, I.; Magadza, C.H.D.; Oppenheimer, M.; Pittock, A.B.; Smith, J.B.; Suarez, S.; Yamin, F. Assessing key vulnerabilities and the risk from climate change. In *Climate Change 2007: Impacts, Adaptation and Vulnerability. Contribution of Working Group II to the Fourth Assessment Report of the Intergovernmental Panel on Climate Change*; Parry, M.L.; Canziani, O.F.; Palutikof, J.P.; van der Linden, V.D.; Hanson, C.E., Eds.; Cambridge University Press: Cambridge, 2007; 779–810 pp.

23. Flugman, E., Mozumder, P.; Randhir, T. Facilitating adaptation to global climate change: perspectives from experts and decision makers serving the Florida Keys. Climatic Change **2011**, *112*, 1015–1035.

24. Uyarra, M.C.; Côté, I.; Gill, J.; Tinch, R.R.T.; Viner, D.; Watkinson, A.R. Island specific preferences of tourists for environmental features: implications of climate change for tourism-dependent states. Environ. Conserv. **2005**, *32*, 1, 11–19.

25. Andersson, J. The recreational costs of coral bleaching – a stated and revealed preference study of international tourists. Ecol. Econ. **2007**, *62*, 704–715.

26. Gössling, S.; Lindén, O.; Helmersson, J.; Liljenberg, J.; Quarm, S. Diving and global environmental change: a Mauritius case study. In *New Frontiers in Marine Tourism: Diving Experiences, Management and Sustainability*; Garrod, B., Gössling, S., Eds.; Amsterdam: Elsevier: Amsterdam, 2007; 67–92 pp.

27. Main, M.; Dearden, P. Tsunami impacts on Phuket's diving industry: geographical implications for marine conservation. Coastal Manag. **2007**, *35*, 4, 1–15.

28. Dearden, P.; Manopawitr, P. Climate change – coral reefs and dive tourism in South-east Asia. In *Disappearing Destinations*; Jones, A., Phillips, M., Eds.; CABI Publishing: Wallingford, 2011; 144–160 pp.

29. Cruise Lines International Association. 2011 CLIA Cruise Market Overview: Statistical Cruise Industry Data through 2010; Cruise Lines International Association: Fort Lauderdale, 2011.

30. Klein, R. The cruise sector and its environmental impact. In *Tourism and the Implications of Climate Change: Issues and Actions. Bridging Tourism Theory and Practice*; Schott, C., Ed.; Emerald Group Publishing: Bingley, **2011**; 113–130 pp.

31. Hall, C.M.; James, M.; Wilson, S. Biodiversity, biosecurity, and cruising in the Arctic and sub-Arctic. J. Herit. Tourism **2010**, *5*, 351–364.

32. Eijgelaar, E.; Thaper, C.; Peeters, P. Antarctic cruise tourism: the paradoxes of ambassadorship, "last chance tourism" and greenhouse gas emissions. J. Sustain Tour **2010**, *18*, 337–354.

33. World Economic Forum. *Towards a Low Carbon Travel & Tourism Sector*; World Economic Forum: Davos, 2009.

34. Drake, J.M.; Lodge, D.M. Hull fouling is a risk factor for intercontinental species exchange in aquatic ecosystems. Aquatic Invasions **2007**, *2*, 121–131.

35. Endresen, O.; Behrens, H.L.; Brynestad, S.; Bjørn Andersen, A.; Skjong, R. Challenges in global ballast water management. Mar. Pollut. Bull. **2004**, *48*, 615–623.

36. O'Connor, S.; Campbell, R.; Cortez, H.; Knowles, T. *Whale Watching Worldwide: Tourism Numbers, Expenditures and Expanding Economic Benefits, Prepared by Economists at Large*; International Fund for Animal Welfare: Yarmouth, MA, 2009.

37. Beck, M.W.; Ferdana, Z.; Kachmar, J.; Morrison, K.K.; Taylor, P. *Best Practices for Marine Spatial Planning*; The Nature Conservancy: Arlington, VA, 2009.

38. UN Environmental Programme. *Sustainable Coastal Tourism: An Integrated Planning and Management Approach*; UNEP: Milan, 2009.

BIBLIOGRAPHY

1. Dimmock. K.; Musa, G., Eds.; *Scuba Diving Tourism*; Routledge: London, 2013.

2. Dowling, R.K. *Cruise Ship Tourism: Issues, Impacts, Cases*; CABI Publishing: Wallingford, 2010.

3. Gössling, S.; Scott, D.; Hall, C.M., Ceron, J.P.; Dubois, G. Consumer behaviour and demand response of tourists to climate change. Ann. Tourism Res. **2012**, *39*, 36–58.

4. Hall, C.M.; Lew, A.A. *Understanding and Managing Tourism Impacts: An Integrated Approach*; Routledge: London, 2009.

Riparian—
Wetlands

Water Cycle: Ocean's Role

Don P. Chambers
College of Marine Science, University of South Florida, St. Petersburg, Florida, U.S.A.

Abstract

The oceans contain more than 95% of all the water on the planet and are the source of much of the water precipitated over land. Changes in the amount of water evaporated from the ocean and returned via precipitation and runoff from land can cause measurable changes in the sea surface salinity (SSS), sea level, and ocean mass. Measurements of these quantities, especially measurements from space-borne sensors, are yielding new insight into how the Earth's water cycle is changing.

INTRODUCTION

The water, or hydrological, cycle, describes the processes that move water from one area to another on the Earth's surface. A significant part of the water cycle involves the transformation of water from its liquid state on the Earth's surface to the gaseous state in the atmosphere (*evaporation*) and then back to its liquid state (*precipitation*). Although the amount of water that can be stored at any time in the atmosphere is only a fraction of the amount that is stored on land or in the ocean (Table 1),[1] it can transport water great distances, and therefore is a vital component of the water cycle. An additional component of the water cycle is the liquid water transported via rivers, or from melting of frozen water in glaciers and icesheets (*runoff*).

Because of the importance of water to human life, much of the study of the water cycle has focused on understanding measurements of precipitation, evaporation, and runoff over the land. However, when placed in perspective of the amounts of water available for evaporation and the relative amounts of water being exchanged with the atmosphere, it is clear that the ocean largely controls the water cycle.[2] The oceans contain ~1.3 billion cubic km (km³) of water, which is 51 times more than the amount of water stored as ice (mainly in ice sheets over Greenland and Antarctica), 85 times more than the amount of water stored in all the underground aquifers on land, 4,000 times more than the water stored in the soil and rivers/lakes, and more than 133,000 times more than the water stored in the atmosphere (Table 1).

The amount of water evaporated from the ocean every year is also much larger than the amount evaporated from land, by a factor of more than 5. Although about 90% of this returns relatively quickly as precipitation over the oceans, the other 10% is transported via the atmosphere to precipitate over land. Thus, precipitation over land exceeds evaporation, and this is only possible because of the water transported from the oceans. This excess land precipitation cannot be stored permanently on land, and most of it flows back into the ocean via rivers resulting in a near balance in the water mass storage and transport on global scales.[1,3]

However, the balance is not exact on time-scales from months to several years, and changes in the climate can move the balance father from equilibrium, especially regionally. Some climate models predict that in a warming climate, there will be an intensification of both precipitation and evaporation,[4] with a result that that wet areas become wetter and dry areas become dryer. Although several studies have attempted to quantify trends in specific components of the water cycle (e.g., precipitation, evaporation, or runoff) over large regions or even globally, assessing the change in the balance over time is difficult, primarily because estimates of the uncertainty of yearly means of precipitation, evaporation, and runoff are all of order 3000 km³, which means a balance cannot be determined to better than this.[1] Imbalances at this level in terms of the ocean volume (sea level) would imply more than 8 mm per year of sea-level rise, which is far larger than observed changes.[5] Because of this, most studies of the global water cycle budget assume a perfect closure, and fix one estimate (often runoff) to perfectly balance the other two.[1]

Because the ocean is a large component of the water cycle and is the primary source of cycling water, several important quantities of the ocean state can provide important clues to the size of water cycle variability and how it is changing. These are salinity, sea level, and ocean mass variations. The following sections will briefly describe how these are related to the water cycle and how measurements of changes in each quantity are providing important clues to how the water cycle is changing.

Encyclopedia of Natural Resources DOI: 10.1081/E-ENRW-120047581

Riparian—
Wetlands

Table 1 Estimated average volumes of water stored in various reservoirs on the Earth

Reservoir	Water volume storage (in millions of km³)
Oceans	1,335
Ice	26.4
Groundwater (aquifers)	15.3
Soil	0.12
Rivers/Lakes	0.18
Atmosphere	0.01

Source: Adapted from Trenberth, Smith, et al.[1]

OCEAN SALINITY

The ocean is saline due to dissolved salts that have accumulated in the oceans over time. Oceanographers can measure the ratio of mass of dissolved salts to the mass of water and express this in terms of salinity. The global average salinity of the ocean is 35 parts per thousand, but can locally be more or less than this. It is clear from comparing the pattern of mean ocean surface salinity with that from evaporation (E) minus precipitation (P) over the ocean (Fig. 1), that high salinity regions occur where there is excess evaporation (E–P positive) and low salinity regions occur where there is excess precipitation (E–P negative).

A Evaporation minus Precipitation

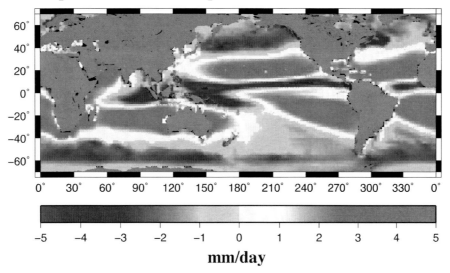

mm/day

B Sea Surface Salinity

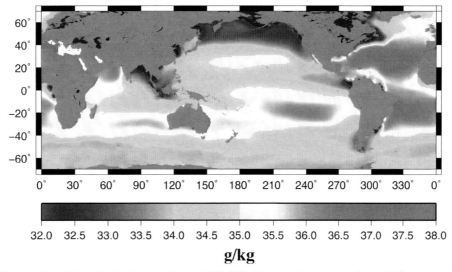

g/kg

Fig. 1 (**A**) Mean Evaporation - Precipitation in mm/day and (**B**) SSS. Evaporation data are from OAflux (http://oaflux.whoi.edu/),[6] averaged 1979–2009. Precipitation data from Global Precipitation Climatology Project (GPCP) (http://precip.gsfc.nasa.gov/),[7] averaged 1979–2009. Sea surface salinity data from World Ocean Atlas 2009 (http://www.nodc.noaa.gov/OC5/WOA09/pr_woa09.html).[8]

Riparian—Wetlands

Although the differences in E–P over the ocean basins are generally small for particular latitude bands, the difference in salinity between the difference basins are much larger. This is due mainly to how water is transported through the atmosphere and the humidity of air blowing across the oceans.[9] For example, the air over the narrow Atlantic tends to be dryer because it blows from the continents; therefore, the internal precipitation tends to be less than that of the Pacific.

While salts are being added to the ocean every year from river outflow, it is a fraction of the amount of total salts in the ocean, and would change the ocean salinity by <1 part in 17 million over the course of a year,[10] which is far too low to measure. Changes in precipitation and evaporation, along with river runoff and transport of different water masses within the ocean, however, can change the surface salinity locally by amounts that are several times larger than the measurement precision of current in situ instruments. Thus, changes in ocean surface salinity can provide insight into the changing water cycle.

Several studies have attempted to do this.[11,12] They find that since the 1950s, the SSS has increased in areas of high salinity and decreased in areas of low salinity. Although there is still uncertainty due to incomplete knowledge of oceanic transports of salt and mapping of sparse data, the results are consistent with the hypothesis of an increasing water cycle. Recently, two different ocean observing satellites have been launched to measure SSS from space for the first time: the soil moisture ocean salinity (SMOS) mission from the European Space Agency and the Aquarius/SAC-D mission from NASA and Argentina.[13] Although neither will match the precision of in situ salinity measurements, it is hoped that their near global coverage will provide better insight to into the changing global water cycle from an ocean perspective.

GLOBAL MEAN SEA LEVEL AND OCEAN MASS

The water transported between the ocean and continents in the water cycle will be reflected in both changes in global mean sea level (GMSL) and the mass of the ocean. If there is more water leaving the ocean than being added, GMSL and the global mass of the ocean will drop. If more water is entering the ocean either from increased precipitation or runoff from land, or less is leaving via decreased evaporation, then GMSL will increase. GMSL will also be affected by ocean warming/cooling,[14,15] and so will only provide limited information on changes in the global water cycle without other information. With measurements of in situ temperatures, it has been possible to use GMSL measurements to extract some information about changes in the water cycle, especially at seasonal periods.[14,16] Beginning in the late 1990s, it became evident that there are large changes in the global ocean mass on seasonal time-scales as water is cycled between the land and continents (Fig. 2). The ocean's mass drops during the early part of the year as the water is cycled from the ocean to the continents (around April). Six months later (around October), the ocean gains mass as water is cycled from the continents.

This has been confirmed with data from the Gravity Recovery and Climate Experiment (GRACE).[17] GRACE is a novel satellite mission that can measure changes in the Earth's gravity at monthly time-scales accurately enough to resolve mass changes over areas as small as river basins.[18] In addition to confirming the seasonal exchange of water mass between the oceans and continents as part of the water cycle, GRACE has also measured interannual variations in water cycling for the first time (Fig. 3). While ocean mass has been generally increasing since 2003 due to addition of previously frozen water from Greenland and

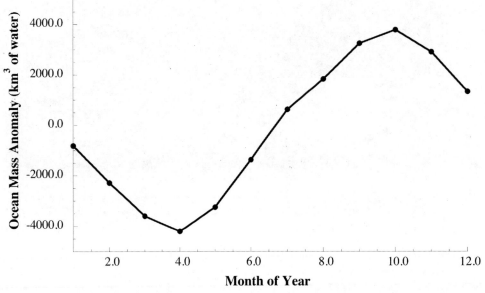

Fig. 2 Mean seasonal ocean mass variations in km³ of water, computed from 8 years of GRACE data.

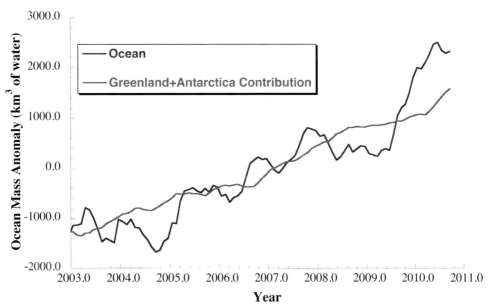

Fig. 3 Non-seasonal ocean mass variations from GRACE (dark line), along with contribution to ocean mass change from Greenland and Antarctica (gray line), both in km³ of water. Both time-series have been smoothed with a 6-month running mean filter.
Source: Chambers & Schröter,[19] Velicogna.[20]

Antarctica,[20] there are much larger year to year variations that are caused by cycling of water between the continents and ocean related to interannual imbalances in the water cycle. These variations can often account for much of the interannual GMSL variations.[21,22] For instance, there are significant deviations in GMSL associated with El Niño and La Niña events,[23,24] and evidence suggests that much of this is due to changes in the cycling of water between the ocean and continents.[25]

CONCLUSION

Most of the water that eventually returns to the Earth as precipitation is first evaporated from the oceans. Thus, the ocean has a vital role in the global water cycle. Understanding how the water cycle is changing is important for predicting future water resources for humanity.[2] However, while much has been learned about the relative sizes and variability of precipitation, evaporation, and water storage over large basins, much is still unknown. Because of its important role in the water cycle, ocean properties such as SSS, GMSL, and mean ocean mass are proving to be useful probes into the water cycle, and can aid us in understanding how the cycling of water is changing over time. Although results are still preliminary, new observations, especially from space-borne sensors, will likely improve our understanding in the future.

REFERENCES

1. Trenberth, K.E.; L. Smith; T. Qian; A. Dai; J. Fasullo. Estimates of the global water budget and its annual cycle using observational and model data. J. Hydrometeor. **2007**, *8*, 758–769.
2. Oki, T.; Entekhabi, D.; Harrold; T.I. The global water cycle. *The State of the Planet: Frontiers and Challenges in Geophysics,* Sparks, R.S.J., Hawkesworth, C.J. Eds.; Geophysical Monograph 150, IUGG: 2004; Vol. 19, 225–237, doi:10.1029/150GM1829.
3. Oki, T. Global water cycle. *Global Energy and Water Cycles*; Browning, K., Gurney, R., Eds.; Cambridge University Press: 10–27, 1999.
4. Held, I.M.; Soden, B.J. Robust responses of the hydrological cycle to global warming. J. Clim. **2006**, *19*, 5686–5699.
5. Cazenave, A.; Llovel, W. Contemporary sea level rise. Ann. Rev. Marine Sci. **2010**, *2*, 145–174 pp.
6. Yu, L.; Jin, X.; Weller, R.A. *Multidecade Global Flux Datasets from the Objectively Analyzed Air-Sea Fluxes (OAFlux) Project*; Latent and sensible heat fluxes, ocean evaporation, and related surface meteorological variables. Woods Hole Oceanographic Institution, OAFlux Project Technical Report. OA-2008-01, Woods Hole, Massachusetts, 2008; 64pp.
7. Adler, R.F.; Huffman, G.J.; Chang, A.; Ferraro, R.; Xie, P.; Janowiak, J.; Rudolf, B.; Schneider, U.; Curtis, S.; Bolvin, D.; Gruber, A.; Susskind, J.; Arkin. P.; Nelkin, E. The version 2 global precipitation climatology project (GPCP) monthly precipitation analysis (1979–Present). J. Hydrometeor. **2003**, *4*, 1147–1167.

Riparian—
Wetlands

8. Antonov, J.I.; Seidov, D.; Boyer, T.P.; Locarnini, R.A.; Mishonov, A.V.; Garcia, H.E.; Baranova, O.K.; Zweng, M.M.; Johnson, D.R. *World Ocean Atlas 2009, Volume 2: Salinity.* Levitus, S. Ed. NOAA Atlas NESDIS 69, U.S. Government Printing Office: Washington, D.C., 2010; 184 pp.

9. Schmitt, R.W. Salinity and the global water cycle. Oceanography, **2008**, *21* (1), 12–19.

10. Pickard, G. L.; Emery, W.J. *Descriptive Physical Oceanography*; Pergamon Press: Oxford, England, 5th Edition, 1990.

11. Curry, R.; Dickson, B.; Yashayaev, I. A change in the freshwater balance of the Atlantic Ocean over the past four decades. Nature **2003**, *426*, 826–829.

12. Durack, P.J.; Wijffels, S.E. Fifty-year trends in global ocean salinities and their relationship to broad-scale warming. J. Clim. **2010**, *23*, 4342–4362, doi: 10.1175/2010JCLI3377.1

13. Lagerloef, G.; Font, J. SMOS and Aquarius/SAC-D missions: The era of spaceborne salinity measurements is about to begin. Chapter 3 in *Oceanography from Space*; Barale, V., Gower, J.R.F., Alberotanza, L. Eds.; Springer: 2010.

14. Chen, J.L.; Wilson, C.R.; Chambers, D.P; Nerem, R.S. Seasonal global water mass balance and mean sea level variations. Geophys. Res. Lett. **1998**, *25*, 3555–3558.

15. Levitus, S.; Antonov, J.; Boyer, T. Warming of the world ocean, 1955–2003, Geophys. Res. Lett. **2005**, *32*, L02604, doi:10.1029/2004GL021592.

16. Minster, J.F.; Cazenave, A.; Rogel, P. Annual cycle in mean sea level from Topex-Poseidon and ERS-1: Inference on the global hydrological cycle. Global Planet. Change **1999**, *20*, 57–66.

17. Chambers, D.P.; Wahr, J.; Nerem, R.S. Preliminary observations of global ocean mass variations with GRACE, Geophys. Res. Lett. **2004**, *31*, L13310, doi:10.1029/2004GL020461.

18. Tapley, B.D.; Bettadpur, S.; Watkins, M.; Reigber, C. The gravity recovery and climate experiment: Mission overview and early results. Geophys. Res. Lett. **2004**, *31*, L09607, doi:10.1029/2004GL019920.

19. Chambers, D.P.; Schröter, J. Measuring ocean mass variability from satellite gravimetry. J. Geodyn. **2011**, *52* (5), 333–343.

20. Velicogna, I. Increasing rates of ice mass loss from the Greenland and Antarctic ice sheets revealed by GRACE. Geophys. Res. Lett. **2009**, *36*, L19503, doi:10.1029/2009GL040222.

21. Willis, J.K.; Chambers, D.P.; Nerem, R.S. Assessing the globally averaged sea level budget on seasonal to interannual time scales. J. Geophys. Res. **2008**, *113,* C06015, doi:10.1029/2007JC004517.

22. Leuliette, E.W; Willis, J.K. Balancing the sea level budget. Oceanography **2011**, *24* (2),122–129, doi:10.5670/oceanog.2011.32.

23. Nerem, R.S.; Chambers, D.P.; Leuliette, E.; Mitchum, G.T.; Giese, B.S. Variations in global mean sea level during the 1997-98 ENSO event, Geophys. Res. Lett. **1999**, *26*, 3005–3008.

24. Chambers, D.P.; Mehlhaff, C.A.; Urban, T.J.; Fujii, D; Nerem, R.S. Low-frequency variations in global mean sea level: 1950–2000, J. Geophys. Res., **2002**, *107* (C4), 3026, doi:10.1029/2001JC001089.

25. Ngo-Duc, T.; Laval, K.; Polcher, J.; Cazenave, A. Contribution of continental water to sea level variations during the 1997–1998 El Niño–Southern Oscillation event: Comparison between Atmospheric Model Intercomparison Project simulations and TOPEX/Poseidon satellite data. J. Geophys. Res. **2005**, *110*, D09103, doi:10.1029/2004JD004940.

Riparian—Wetlands

Water Quality and Monitoring

John C. Clausen
Department of Natural Resources and the Environment, University of Connecticut, Storrs, Connecticut, U.S.A.

Abstract

Water quality monitoring is conducted for various reasons. In order to conduct water quality monitoring in an appropriate and efficient manner, a design approach should be used. Proper water quality designs consider defining the water quality problem, forming monitoring objectives, developing a study or statistical design, determining the scale of the study, selecting variables to monitor and the type of sample to collect, sampling location, the frequency and duration of sampling, methods for collecting the sample and analysis of its contents, and land use and management monitoring.

INTRODUCTION

Water quality monitoring is the collection of information on the physical, chemical, biological, sometimes radiological characteristics of water.[1] Water quality monitoring is conducted for several reasons. Monitoring is used to determine baseline values or examine trends in the data. Best-known examples of this approach includes individual state monitoring in response to Section 305(b) of the Clean Water Act (PL92-500), and monitoring conducted by the U.S. Geological Survey at its National Stream Quality Accounting Network stations.[2] Monitoring is used to determine the fate and transport of pollutants. For example, how far has a contaminant moved in a ground water aquifer? Monitoring has been used to define critical areas needing treatments, especially in agricultural watersheds. The results of such monitoring can then be used to target practices. Compliance with water quality plans and permit provisions is a common objective of monitoring. For example, wastewater treatment plants often monitor discharge as a requirement of their NPDES permit. Measuring effectiveness of best management practices, either on an individual basis or on a programmatic watershed basis is often a goal of monitoring. Wasteload allocations from point sources and load allocations from nonpoint sources are requirements of total maximum daily loads (TMDLs). Determining these loads and the maximum loading capacity of the receiving water body require monitoring. When the source of a problem is unknown, water quality monitoring is used to determine the cause of impairment. Monitoring is used to validate and calibrate models as well as to provide input data for models used to predict transport rates and concentrations regionally. One such model in common use is Spatially Referenced Regressions on Watershed attributes (SPARROW).[2] Researchers also conduct monitoring to address specific research questions and test hypotheses.

Monitoring water quality is expensive and time consuming because water is constantly flowing and changing. Care is needed in interpreting the results of water quality monitoring because the samples collected may not represent the full range of values that may occur. Weather conditions have an important influence on water quality. The important components to consider in water quality monitoring are outlined in this entry.[3]

MONITORING NETWORK DESIGN

Much has been written regarding how to design and conduct water quality monitoring networks.[4] The Intergovernmental Task Force on Monitoring Water Quality[5] recommended a strategy for nationwide water quality monitoring that focused primarily on understanding the condition of the Nation's waters to support sound water quality decision making. The National Water Quality Monitoring Council was formed subsequent to the intergovernmental task force to continue guidance and coordination of water quality monitoring. Examples of monitoring networks include major programs by the U.S. EPA and the U.S. Geological Survey. EPA conducts National Aquatic Resource Surveys of lakes and streams that are probability-based on a multiyear revolving approach. Lakes are selected by ecoregion using stratified random sampling based on the density of lakes in the region. Streams are stratified by Strahler order. A proportion of sites are resampled during each rotation. The U.S. Geological Survey monitors nearly 400 National Stream Quality Accounting Network stations throughout the United States.[2] Several approaches have been used to design monitoring networks other than the fixed station and randomized methods. Optimization approaches have been developed to design networks based on economic and logistic considerations, hydrodynamics,

Encyclopedia of Natural Resources DOI: 10.1081/E-ENRW-120047558

Riparian—
Wetlands

and proximity to sources.[6] A critical sampling points methodology[7] has been developed to locate sampling sites in watersheds that is based on GIS, fuzzy logic, and a model that considers loading from the land and upstream areas. The remainder of this entry describes components to consider in water quality monitoring for all purposes and designs.

WATER QUALITY PROBLEM

When formulating a water quality problem statement, it is important to distinguish between a problem and a symptom. A water quality problem is a water quality issue requiring a solution, often stated in the form of a question. A symptom is a characteristic or condition of a water body indicating a problem. For example, algal blooms might be symptomatic of excessive nutrient loadings. For every water quality problem, there are typically several symptoms. Table 1 summarizes some typical symptoms and problems. A complete water quality problem statement should include information about the specific water body, the cause of the problem, and how the problem manifests itself, and the source of the causal agents. A good example of a definition of a water quality problem is: *The problem is eutrophication of St. Albans Bay due to excessive phosphorus loading from agricultural sources.*

OBJECTIVES

The formulation of monitoring objectives is a "decision-making" process, beginning with a problem definition and ending with the action to begin water quality monitoring. It is important that the objectives address the identified water quality problem. Formulating objectives can be viewed as a series of three steps that involve 1) identifying the objectives, 2) developing an objective hierarchy, and 3) specifying attributes to measure the level of achievement of these objectives.

An objective describes the answer to the following question: "What must be done?" By definition, an objective includes an object as part of the statement, an infinitive, and sometimes a constraint to limit all opportunities.

Table 1 Water quality symptoms and problems[3]

Symptom	Problem
Color	Algae, sediment, organic acids
Excess algae	Nutrients
Excess macrophytes	Nutrients, light
Hypoxia	Nutrients
Low biotic diversity	Toxics, nutrients
Taste	Salinity, algae, metals
Turbidity	Algae, sediment

For example, a complete objective, with a '+' to indicated parts of the statement, could be: *To determine + the effect of implementing conservation practices + on fecal coliform levels in Long Lake.*

STATISTICAL DESIGNS

Major water quality study designs include the plot, single watershed, above and below, two watersheds, paired watershed, multiple watershed, and trend station (Table 2). Plots are generally small areas (fractions of an acre) that are replicated on the landscape. All plots are treated alike except for the factor(s) under study. A major advantage of the plot design is that it has a control. A control is a plot that is monitored like all others but does not receive the treatment. The primary statistical approach used is the analysis of variance of a randomized complete block design. A single watershed has sometimes been used to evaluate the water quality changes both before and after a treatment is applied. This design should not be used because the effect due to the treatment cannot be separated from other confounding effects such as year-to-year differences in climate. Two watersheds, one with a treatment and one without, have been incorrectly used to evaluate the effects of a practice on water quality. This design should always be avoided because differences might be inherent in the two watersheds. The above-and-below watershed approach is sometimes viewed as a single watershed with monitoring above and below a practice. In actuality, there are two watersheds being monitored, one nested within the other. If the above-and-below approach is applied both before and after the practice is installed, this approach can be analyzed as a paired watershed design as described later. The above-and-below design is analyzed as a *t*-test of the differences between paired observations at the above and below stations.

The paired watershed approach requires a minimum of two watersheds—control and treatment—and two periods of study—calibration and treatment.[8] The control watershed serves as a check over year-to-year climate variations, and receives no changes during the study. During the calibration period, the two watersheds are treated identically and paired water quality data are collected. During the treatment period, one randomly selected watershed is treated, while the control watershed remains in the original management. The primary statistical approach is analysis of covariance.[9]

The multiple watershed approach involves several watersheds[10,11] that are selected with the treatments already in place from across the geographic region of interest. The greatest advantage of the multiple watershed approach is that the results are transferable to the region included in the monitoring. The basic statistical approach is the comparison of the means of two populations using the *t*-test.

Table 2 Statistical designs for water quality monitoring

Design	Characteristic	Advantage	Disadvantage	Statistics
Plot	small replicated	uses control	not transferable	ANOVA
Single	before and after	simple	temporal confounding	t-test
Above–below	nested watersheds	no climate effects	may not be independent	t-test
Two watersheds				
Paired watersheds	two treatment periods	variation controlled	no replication	ANCOVA
Multiple watersheds	> two watersheds	results transferable	large area	unpaired t-test
Trend stations	single watershed	large area	climate confounding	many

A trend is a persistent change in the water quality variable(s) of interest over time. Trend stations are single independent watersheds where a group of treatments might be implemented gradually over time or where the response might take a long time. A control watershed for trend detection is needed to distinguish climate trends. It is generally recommended to perform several different techniques before reaching a conclusion,[12] such as 1) a time plot, 2) a least-square fit regression, 3) comparison of annual means, 4) comparing cumulative distribution curves, 5) comparing quartiles (Q–Q plot), 6) double mass analysis, 7) paired regressions, 8) time series analysis, and 9) the seasonal Kendall test.[13]

SCALE OF STUDY

Three scale categories are commonly recognized: plot, field, and watershed; although, all three scales are in reality watersheds. Plots are appropriate monitoring units when the objective is to replicate several treatments. Limnocorrals and artificial stream channels are plots of a mesocosm scale[14] for lakes and streams. Monitoring on a field scale implies a larger area than an individual plot. Watershed scale monitoring is desired if the water resource system of concern is groundwater, a stream, or a lake/estuary.

VARIABLE SELECTION

Several methods for selecting variables include variable matrices, variable cross-correlations, and the probability of exceeding standards. Certain water quality characteristics are highly related, but the analysis cost of one is much cheaper than the other. Correlations between these variables and the probability of exceeding a standard can be used to reduce the variable list.[15]

SAMPLE TYPE

There are four types of water quality samples that can be collected: grab, composite, integrated, or continuous. The variable to sample influences the sample type. For

example, bacteria samples must be taken as grab samples with sterilized bottles. Other variables change dramatically during storage and therefore are inappropriate for compositing. These include all dissolved gases, chlorine, pH, temperature, and sulfide.[16]

A grab sample is a discrete sample that is taken at a specific point and time, usually by hand.[16,17] A series of grab samples, usually collected at different times or at flow intervals and lumped together, are considered composite samples, usually taken with an automatic sampler. A depth-integrated sample accounts for velocity- or stratification-induced changes with depth, especially for sediment sampling or stratified lakes.[18] Near continuous monitoring can be used for any water quality variable that is measured using electrometric methods, such as conductivity or dissolved oxygen.

SAMPLING LOCATION

Sampling locations should be stable and representative of the system being monitored. Stream sampling locations must account for vertical, horizontal, and longitudinal differences in water quality due in part to velocity differences. Lake systems also are heterogeneous due to vertical stratification, longitudinal gradients, and currents caused by winds and density differences. In order to describe lake water quality at a particular point, samples will be needed from each stratified layer in the system. Samples may have to be located within each bay or section of the lake. Groundwater monitoring must also consider vertical, horizontal, and longitudinal water quality differences.

Site selection criteria for a stream include accessibility, power available, ownership, streambed stability, permeability and gradient, uniform and straight section, control at all stages, and within a confined channel. Groundwater sampling sites consider the location of the water table divide, barrier locations (stream, strata), direction of flow, stratified concentrations, and depth to any confining layer. Lake sites are also concerned with stratified layers, longitudinal gradients, bays and beaches, and water circulation patterns.

SAMPLING FREQUENCY AND DURATION

One of the most frequently asked questions is "How many samples and for how long?" Unfortunately, the correct response is as follows: "It depends." The sample size can be calculated from the relationship:

$$n = \frac{t^2 s^2}{d^2} \tag{1}$$

where n is the calculated sample size, t is Student's 't' at $n-1$ degrees of freedom and a selected confidence level, s is the estimate of the population standard deviation, and d is the allowable difference from the mean.[19]

SAMPLE COLLECTION AND ANALYSIS

Sample Collection

A sample should be representative of the water it came from and not contaminated before analysis. Bottles are often rinsed with the sample water two or three times before filling.[16] If samples are collected from pipes under pressure, the system should be flushed for a sufficient period to guarantee that new water is being sampled. Bacteria samples are collected in sterilized bottles. Stream and lake grab samples should be collected just below the surface of the water to avoid the surface film.

Collection of samples from wells might involve purging or sampling at the rate of well recovery. Sampling of volatile substances requires specialized sampling equipment in wells.

Samples should be collected so that the bottle is completely full to reduce volatilization losses. An exception to this would be if the sample was to be frozen, in which case room for expansion upon freezing should be left in the container. Sampling of toxic substances requires extra precautions, including personal protection.

Sample Preservation and Transport

Collected samples require preservation to slow rates of changes in composition through chemical, physical, and biological processes; however, these changes may not be preventable and rapid analysis is recommended.[20] Examples of physical changes include settling of solids, adsorption of certain cations on container walls, and loss of dissolved gases. Chemical changes could include precipitation, dissolution from sediments, compexation with other ions, and changes in valence state. Biological reactions could result in both the uptake and/or release of certain constituents. Predation within the sample may occur or the specimens may degrade. Microbial activity may change the species of nitrogen present.[16] Primary preservation methods are acidification, refrigeration, filtration, and

preventing light from reaching the sample.[16,20] Glass containers may leach sodium and silica; plastic containers may sorb organics.[16] Transportation to the laboratory should be direct and use some method of preservation.

Quality Assurance

Quality assurance (QA) is "the total integrated program for assuring the reliability of monitoring and measurement data."[21] QA is composed of quality control and quality assessment. Quality control (QC) refers to activities conducted to provide high-quality data.[22] Quality assessment refers to techniques used to evaluate the effectiveness of the program.[23] Table 3 summarizes the major components of quality control and quality assessment programs. Indicators of data quality include: a) precision, b) accuracy, c) representativeness, d) comparability, and e) completeness. Transfer of sample custody takes place upon delivery of samples to the laboratory by signing over custody to a laboratory person receiving the samples. Performance audits use EPA-approved unknown ampules. Field equipment also requires calibration and is subject to sample blanks.

LAND USE AND MANAGEMENT MONITORING

An essential element of water quality monitoring is the tracking of land use and management activities in the watershed being monitored. Land use and management data are needed to explain any water quality changes that may occur. It is important to be able to attribute the water quality changes to the management practice and not other confounding influences such as climate or a point source.

Table 3 Components of a quality control and quality assessment program[3]

Quality control	Quality assessment
Good laboratory practices (GLPs)	**Internal**
Standard operating procedures (SOPs)	Duplicate samples
	Standard additions (spikes)
Education/training	Tests of sampling frequency
Sample custody procedures	Tests of reason with comparable data
Calibration and maintenance	
Supervision	Missing analysis records
	Standard curves
	Internal audit
	External
	Exchange sample with other lab
	External known materials
	External audit

There are four basic approaches for monitoring land treatment data: personal observations, field logs, personal interviews, and remote sensing.

CONCLUSION

Water quality monitoring is an important element of local, state, and federal programs as well as for research and education. It is expensive and therefore should be conducted in an appropriate manner to maximize utility.

REFERENCES

1. Sanders, T.G.; Ward, R.C.; Loftis, J.C.; Steele, T.D.; Adrian, D.D.; Yevjevich, V. *Design of Networks for Monitoring Water Quality*; Water Resources Publications: Littleton, CO, 1983; 336 p.
2. Smith, R.A.; Schwarz, G.E.; Alexander, R.B. Regional interpretation of water-quality monitoring data. Water. Res. Res. **1997**, *33* (12), 2781–2798.
3. USDA, Natural Resources Conservation Service. Water quality monitoring system design, Part 614, *National Water Quality Handbook*; U.S. Department of Agriculture; Washington, DC, 2003.
4. Strobl, R.O.; Robillard, P.D. Network design for water quality monitoring of surface freshwaters: A review. J. Environ. Mgmt. **2008**, *87*, 639–648.
5. Intergovernmental Task Force on Monitoring Water Quality. *The Strategy for Improving Water-Quality Monitoring in the United States. Final Report of the Intergovernmental Task Force on Monitoring Water Quality*; U.S. Geological Survey, Office of Water Data Coordination: Reston, VA. 1995.
6. Telci, I.T.; Nam, K.; Guan, J.; Aral, M.M. Optimal water quality monitoring network design for river systems. J. Environ. Mgmt. **2009**, *90*, 2987–2998.
7. Strobl, R.O.; Robillard, P.D.; Debels, P. Critical sampling point methodology: case studies of geographically diverse watersheds. Environ. Monit. Assess. **2007**, *129*, 115–131.
8. Wilm, H.G. How long should experimental watersheds be calibrated? Amer. Geophys. Union Trans. Part II. **1949**, *30* (2), 618–622.
9. Reinhart, K.G. Watershed calibration methods. In *Proceeding Internship Symposium on Forest Hydrology*; Sopper, W.E., Lull, H.W., Eds.; Pergamon Press: Oxford, 1967; 715–723.
10. Striffler, W.D. *The Selection of Experimental Watersheds and Methods in Disturbed Forest Areas*. Publisher No. 66. I.A.S.H. Symposium of Budapest: 1965; 464-473.
11. Wicht, C.L. The validity of conclusions from South African multiple watershed experiments. In *Proceeding International Symposium on Forest Hydrology*; Sopper, W.E., Lull, H.W., Eds.; Pergamon Press: Oxford, 1967; 749–760.
12. World Health Organization. *Water Quality Surveys - A Guide for the Collection and Interpretation of Water Quality Data*; IHD-WHO Working Group on Quality of Water. UNESCO/WHO: Geneva, Switzerland, 1978.
13. Hirsch, R.M.; Slack, J.R.; Smith, R.A. Techniques of trend analysis for monthly water quality data. Water Res. Res. **1982**, *18* (1), 107–121.
14. Odum, E.P. The mesocosm. BioScience **1984**, *34* (9), 558–562.
15. Moser, J. H.; Huibregtse, K.R. *Handbook for Sampling and Sample Preservation of Water and Wastewater*; U.S. Environment Protection Agency: 600/4-76-049, 1976.
16. Clesceri, L.S.; Greenberg, A.E.; Eaton, A.D. *Standard Methods for the Examination of Water and Wastewater* 17th Ed.; American Public Health Association: Washington, D.C., 1989.
17. Ponce, S.L. *Water Quality Monitoring Programs*; WSDG Technical Paper WSDG-TP-0002; Watershed Systems Development Group, USDA Forest Service: Fort Collins, CO, 1980.
18. U.S. Geological Survey. Chemical and Physical Quality of Water and Sediment. Chap. 5 In *National Handbook of Recommended Methods for Water-Data Acquisition*; Office of Water Data Coordination, Geological Survey: Reston, VA, 1977.
19. Snedecor, G.W.; Cochran, W.G. *Statistical Methods*, 7th Ed.; The Iowa State University Press: Ames, IO, 1980.
20. U.S. Environmental Protection Agency. *Methods for Chemical Analysis of Water and Wastes*; EPA-600/4-79-020. Office of Research and Development: Cincinnati, OH, 1983.
21. U.S. Environmental Protection Agency. *Guide for the Preparation of Quality Assurance Project Plans for the National Estuary Program*; EPA 556/2-88-001. Office of Marine and Estuarine Protection: Washington, D.C., 1988.
22. Lawrence J.; Chau, A.S.Y. *Quality Assurance for Environmental Monitoring*; Symposium on monitoring, modeling, and mediating water quality. American Water Resources Association: 1987; 165–175.
23. Taylor. J.E. Essential features of a laboratory quality assurance program. In *Statistics in the Environmental Sciences*; Gertz, S.M.; London, M.D., Eds.; American Society for Testing and Materials: 1984; 66–73.

BIBLIOGRAPHY

1. USDI, Geological Survey. National field manual for the collection of water-quality data. *Techniques of Water-Resources Investigations*; Book 9. Handbooks for Water-Resources Investigations. http://pubs.water.usgs.gov/twri9A/.
2. Ward, R.C.; Loftis, J.C.; McBride, G.B. *Design of Water Quality Monitoring Systems*; Van Nostrand Reinhold: New York, NY. 1990.

Riparian— Wetlands

Watershed Hydrology and Land-Use and Land-Cover Change (LULCC)

Yuyu Zhou

Joint Global Change Research Institute, Pacific Northwest National Laboratory, College Park, Maryland, U.S.A.

Abstract

Land-use and land-cover change (LULCC), a process of human modification of Earth's terrestrial surface, affects biodiversity, climate change, watershed hydrology, and other processes on the Earth largely in the forms of urbanization, agriculture, and deforestation. Urban and suburban development, primarily increasing impervious surface area, can dramatically influence watershed hydrology by reducing infiltration and increasing runoff, and hence raising the possibility of increased storm flow and flood frequency. Hydrologic models provide the capability to comprehend the complex hydrologic processes and have been widely and effectively used to understand the hydrologic impact of LULCC. Despite much progress made in the study of the hydrologic impact of LULCC, there are still numerous uncertainties and even controversies. Interdisciplinary and multidisciplinary collaborations can contribute to a comprehensive and improved understanding of the hydrologic impact of LULCC, and therefore help decision makers in watershed management practices.

INTRODUCTION

Carbon dioxide concentration increase and land-use and land-cover change (LULCC) are two major factors of human-induced change in our global environment.[1] LULCC, mainly urbanization, agriculture practice, and deforestation, has a profound impact on natural resources and the environment.[2,3] Change in watershed hydrology is one of the major impacts from LULCC, and it causes a series of environmental problems such as soil erosion and nonpoint source pollution. The hydrologic impact of LULCC varies with climate, location, scale, and other related factors. This entry reviews human-induced LULCC, its hydrologic impact, and methods used to study the hydrologic impact of LULCC.

LULCC IMPACT

Land cover refers to the physical material on the Earth's surface, such as water and vegetation, whereas land use is a more complex term that describes how people use the land.[4] LULCC is a process of human modification of earth's terrestrial surface.[5] LULCC is driven by factors such as population growth and migration, economic development, culture, regional and local policies, environmental conditions, their interactions, and people's responses to economic opportunities.[6] Remote sensing was identified as a major source of consistent and continuous data and has been used to derive information about LULCC across a variety of temporal and spatial scales.[7–9]

LULCC plays an important role in many processes such as biodiversity loss, climate change, and water and carbon cycle. LULCC is one of the most important factors contributing to biodiversity loss via processes such as forest fragmentation induced by human activities. For example, it was found that richness of plant species was significantly negatively related to nitrogen input, an indicator of land-use intensity.[10] LULCC impacts climate significantly across a variety of spatial and temporal scales in mainly two ways. First, LULCC-induced greenhouse gas emissions increase the concentration of carbon dioxide, an important driver of climate change, in the atmosphere. Second, LULCC-related albedo change can alter the surface energy balance. Pielke et al.[11] found that land-use change influenced regional and global climate through surface energy budget, as well as carbon cycle. Change in land-use and land-cover is also an important factor impacting watershed hydrology and associated environmental problems, such as soil erosion and water pollution. The impact of LULCC on watershed hydrology, the focus of this entry, is discussed in the following sections.

HYDROLOGIC IMPACT OF LULCC

Human-induced LULCC affects hydrologic processes such as evapotranspiration, interception, infiltration, surface

Encyclopedia of Natural Resources DOI: 10.1081/E-ENRW-120048174

Riparian—
Wetlands

flow, ground flow and stream flow, response time, and watershed annual and seasonal water yield. Urban and suburban development, agricultural practice, and deforestation and reforestation are typical forms of human-induced LULCC, and they influence watershed hydrology in different ways owing to their special characteristics. Due to the variability of agricultural practice, the results about its hydrologic impact are very different across numerous studies.[12] In this section, the focus is on the LULCC of urbanization and deforestation for which some consensus has been reached.

Impervious surface area (ISA), a key indicator of urban and suburban development, is often used to study the environmental impact of urbanization.[13,14] Increasing ISA influences a series of hydrologic processes (Fig. 1), and it reduces infiltration, raises runoff volumes and velocities, increases peak flow and duration, decreases baseflow, and increases sediment loading. As a result, these processes bring environmental issues such as flooding, erosion, sedimentation, increased temperatures, habitat changes, and loss of fish populations. Although the magnitude of hydrologic change due to urbanization varies significantly in numerous studies, a consensus has been reached that increasing peak flow is one of the major hydrologic impacts of urbanization.

A number of studies have reported that reforestation and deforestation impact various aspects of the hydrologic cycle. Reforestation and deforestation change the hydrologic processes such as interception of precipitation, evapotranspiration, and soil moisture, and the impact varies with spatial and temporal scales and local conditions. Inconsistent conclusions have been drawn about how reforestation and deforestation impact the watershed hydrology in previous studies. According to Siriwardena

et al.,[15] deforestation increases the runoff, and thus the annual water yield. A study by Lin and Wei[16] indicated that forest harvesting increased the mean and peak flows in annual and spring periods but the mean and peak flows in summer and winter periods were not significantly affected.

HYDROLOGIC MODELING

Although there is no doubt that human-induced LULCC influences watershed hydrology, the magnitude and mechanism of the impact are not clear for most LULCC types. Numerous methods have been developed to understand the hydrologic consequences of LULCC. Hydrologic modeling is the most widely used method for this purpose.

Numerous hydrologic models have been used effectively to investigate the hydrologic impact of LULCC with great flexibility. For example, the Variable Infiltration Capacity (VIC) model has been employed to study the hydrologic effects of long-term land and water management in North America and Asia over the past 300 years.[17] The hydrologic effects of LULCC in four catchments within the Columbia River basin have been studied with the topographically explicit distributed hydrology–soil–vegetation model (DHSVM) and a comparison between the results using this method and those from the experiment using the VIC model have also been conducted.[18] In particular, because distributed hydrologic models can take spatial heterogeneities into account and provide detailed descriptions of the hydrologic processes in a watershed, the impact of LULCC and its spatial pattern on hydrology can be investigated through these models in a more spatially explicit way. For example, a distributed rainfall–runoff simulation model was developed to evaluate the impact of urbanization on runoff using high spatial resolution ISA data.[19] With the capacity of hydrologic models, the hydrologic impact from LULCC can be studied with the consideration of other exogenous variables such as climate in the evaluation, and therefore understanding of the hydrologic impact of LULCC can be improved. It is worth noting that the calibration of hydrologic model is important and it may introduce uncertainties in the evaluation.

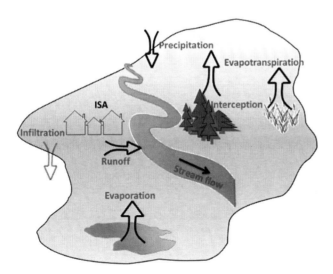

Fig. 1 Typical hydrologic processes in the urbanizing environment.

CONCLUSIONS

Due to people's different responses to economic opportunities, a variety of LULCC types such as urbanization, agriculture practice, and deforestation and reforestation occur across spatial and temporal scales on the Earth's surface. LULCC with spatial and temporal characteristics has a profound impact on the Earth's environment and can be effectively studied with the help of remote sensing data and technology.

LULCC affects biodiversity, climate change, watershed hydrology, and other processes on the Earth. The change in watershed hydrology induced by LULCC not only has direct influence on the water system, but also brings about a series of associated environmental issues such as water pollution. Watershed hydrology is complex and impacted by a number of factors, such as climate and soil condition, as well as LULCC. Hydrologic models provide the capability to comprehend the complex hydrologic processes and have been widely and effectively used to understand the hydrologic impact of LULCC.

Although much progress has been made in the study of hydrologic impact of LULCC, there are still numerous uncertainties and even controversies. There are uncertainties from data sources, LULCC quantification, and hydrologic modeling in the evaluation. The hydrologic impact of LULCC may vary with study area, climate condition, geography, and spatial scale, and it is also dependent on the temporal scale analyzed. All of these factors make the evaluation of the hydrologic impact of LULCC more challenging.

Although a great deal of effort has focused on the improvement of LULCC quantification using remote sensing across spatial and temporal scales and distributed and process-based hydrologic modeling, interdisciplinary and multidisciplinary collaborations are still necessary to comprehend the response of watersheds with different size, topography, soil moisture, and local condition to LULCC, thus contributing to an improved understanding of the hydrologic impact of LULCC.

REFERENCES

1. Bronstert, A. Rainfall-runoff modelling for assessing impacts of climate and land-use change. Hydrol. Process. **2004**, *18* (3), 567–570, doi:10.1002/hyp.5500.

2. Foley, J.A.; DeFries, R.; Asner, G.P.; Barford, C.; Bonan, G.; Carpenter, S.R.; Chapin, F.S.; Coe, M.T.; Daily, G. C.; Gibbs, H.K.; Helkowski, J.H.; Holloway, T.; Howard, E.A.; Kucharik, C.J.; Monfreda, C.; Patz, J.A.; Prentice, I.C.; Ramankutty, N.; Snyder, P.K. Global consequences of land use. Science **2005**, *309* (5734), 570–574, doi:10.1126/science.1111772.

3. DeFries, R.; Eshleman, K.N. Land-use change and hydrologic processes: A major focus for the future. Hydrol. Proces. **2004**, *18*, 2183–2186, doi: 10.1002/hyp.5584.

4. Fisher, P.; Comber, A.J.; Wadsworth, R. Land use and land cover: Contradiction or complement. In *Re-Presenting GIS.*; Fisher, P.; Unwin, D., Eds.; Wiley: London, UK, 2005; 85–98.

5. Ellis, E. Land-use and land-cover change. In *Encyclopedia of Earth. Environmental Information Coalition*, Cleveland, C.J., Ed.; National Council for Science and the Environment: Washington, D.C., 2012.

6. Lambin, E.F.; Turner, B.L.; Geist, H.J.; Agbola, S.B.; Angelsen, A.; Bruce, J.W.; Coomes, O.T.; Dirzo, R.; Fischer, G.; Folke, C.; George, P.S.; Homewood, K.; Imbernon, J.; Leemans, R.; Li, X.; Moran, E.F.; Mortimore, M.; Ramakrishnan, P.S.; Richards, J.F.; Skånes, H.; Steffen, W.; Stone, G.D.; Svedin, U.; Veldkamp, T.A.; Vogel, C.; Xu, J. The causes of land-use and land-cover change: moving beyond the myths. Global Env. Change **2011**, *11* (4), 261–269, doi:10.1016/s0959-3780(01)00007-3.

7. Singh, A. Digital change detection techniques using remotely-sensed data. Int. J. Remote Sensing **1989**, *10* (6), 989–1003.

8. Brink, A.B.; Eva, H.D. Monitoring 25 years of land cover change dynamics in Africa: A sample based remote sensing approach. Appl. Geogr. **2009**, *29* (4), 501–512. doi:10.1016/j.apgeog.2008.10.004.

9. Xiao, J.; Shen, Y.; Ge, J.; Tateishi, R.; Tang, C.; Liang, Y.; Huang, Z. Evaluating urban expansion and land use change in Shijiazhuang, China, by using GIS and remote sensing. Landscape Urban Plan. **2006**, *75* (1–2), 69–80. doi:10.1016/j.landurbplan.2004.12.005.

10. Kleijn, D.; Kohler, F.; Báldi, A.; Batáry, P.; Concepción, E.D.; Clough, Y.; Díaz, M.; Gabriel, D.; Holzschuh, A.; Knop, E.; Kovács, A.; Marshall, E.J.P.; Tscharntke, T.; Verhulst, J. On the relationship between farmland biodiversity and land-use intensity in Europe. Pro. R. Soc. B: Biol. Sci. **2009**, *276* (1658), 903–909, doi:10.1098/rspb.2008.1509.

11. Pielke, R.A.; Marland, G.; Betts, R.A.; Chase, T.N.; Eastman, J.L.; Niles, J.O.; Niyogi, DdS; Running, S.W. The influence of land-use change and landscape dynamics on the climate system: relevance to climate-change policy beyond the radiative effect of greenhouse gases. Philos. Transact. A Math. Phys. Eng. Sci. **2002**, *360* (1797), 1705–1719, doi:10.1098/rsta.2002.1027.

12. Li, L.; Jiang, D.; Li, J.; Liang, L.; Zhang, L. Advances in hydrological response to land use/land cover change. J. Nat. Resour. **2007**, *22* (2), 211–224.

13. Shuster, W.; Bonta, J.; Thurston, H.; Warnemuende, E.; Smith, D. Impacts of impervious surface on watershed hydrology: A review. Urban Water J. **2005**, *2* (4), 263–275, doi:10.1080/15730620500386529.

14. Zhou, Y.; Wang, Y.Q. Extraction of impervious surface areas from high spatial resolution imagery by multiple agent segmentation and classification. Photogramm. Eng. Rem. S. **2008**, *74* (7), 857–868.

15. Siriwardena, L.; Finlayson, B.L.; McMahon, T.A. The impact of land use change on catchment hydrology in large catchments: The Comet River, Central Queensland, Australia. J. Hydrol. **2006**, *326* (1–4), 199–214.

16. Lin, Y.; Wei, X. The impact of large-scale forest harvesting on hydrology in the Willow watershed of central British Columbia. J. Hydrol. **2008**, *359* (1–2), 141–149, doi:10.1016/j.jhydrol.2008.06.023.

17. Haddeland, I.; Skaugen, T.; Lettenmaier, D.P. Hydrologic effects of land and water management in North America and Asia: 1700–1992. Hydrol. Earth Syst. Sci. **2007**, *11* (2), 1035–1045, doi:10.5194/hess-11-1035-2007.

18. VanShaar, J.R.; Haddeland, I.; Lettenmaier, D.P. Effects of land-cover changes on the hydrological response of interior Columbia River basin forested catchments. Hydrol. Process. **2002**, *16* (13), 2499–2520, doi:10.1002/hyp.1017.

19. Zhou, Y.; Wang, Y.; Gold A.; August, P. Modeling watershed rainfall–runoff relations using impervious surface-area data with high spatial resolution. Hydrogeol. J. **2010**, *18* (6), 1413–1423, doi:10.1007/s10040-010-0618-9.

Riparian—
Wetlands

Wetlands: Indices of Biotic Integrity

Donald G. Uzarski
Neil T. Schock
Ryan L. Wheeler
Institute for Great Lakes Research, CMU Biological Station, and Department of Biology, Central Michigan University, Mt. Pleasant, Michigan, U.S.A.

Abstract

To make well-informed decisions, managers, policy-makers, and stakeholders have explored ways to assess habitat quality and the impacts of human activity on ecosystems of concern. Wetland assessment and monitoring have received considerable attention due to the recognition of the importance of wetland systems in maintaining the overall health of surrounding ecosystems and their benefit to humans. To effectively assess and monitor these habitats, methods relating biological communities to the effects of anthropogenic disturbance have been developed. One method, called an index of biotic integrity (IBI), establishes a measure of relative health for wetlands and has proven to be a valuable tool for conservation efforts. In this entry, we discuss the process of creating an IBI for wetland habitats and overview situations in which they may be implemented. Our intended goal in this entry is to give a basic understanding of how IBIs can be created along with a theoretical framework from which future developments may be possible.

INTRODUCTION

The assessment of habitat quality has been recognized as an important part of habitat and wildlife management. The goal of this entry is to introduce the framework of indices of biotic integrity (IBIs) and describe how they are constructed, validated, and applied as tools to evaluate habitat quality using biological indicators. We hope to clearly outline the value of IBIs as tools for habitat quality assessments and explain the advantages of this process when compared to previously developed approaches to habitat quality assessment.

This entry details the process of building metrics based on the biological community characteristics that make up the main components of an IBI and describes the implementation of these tools as they pertain to wetland habitats. These tools and concepts have proven successful in aquatic habitats, and it is our belief that these same concepts can be applied to numerous biological communities that populate various types of habitats and ecosystems.

THE IMPORTANCE OF WETLANDS

Wetlands maintain biological diversity, are very productive, and provide extensive functions and values that translate to ecological services.[1,2] Wetlands are unique in that they are transitional zones between the upland proper and true aquatic habitats (Fig. 1). Wetland types are also diverse and are structured by hydrology, geography, and geology.

The high variability in wetland habitat results in many very specialized organisms that are unique to wetlands.[1] Only an estimated 3.5% of the United States is covered by wetlands; yet they support over 50% of the species on the federal endangered species list.[3] High concentrations of nutrients and moisture make wetlands highly diverse and productive systems.[4]

Wetlands provide many ecosystem services.[5] Water flowing from terrestrial landscapes is filtered as it passes through wetlands into aquatic environments. Wetland plant communities filter out nutrients by assimilation into biomass, while trapping toxicants and sediments before they enter true aquatic habitats such as oceans or lakes.[6] In this way, wetlands play an important role in carbon and nutrient cycles.[7,8] The water-cleansing and nutrient-sequestering functions of wetlands are only a small portion of the services that wetlands provide. Other important services include habitat for wildlife, sources of food and medicine, floodwater storage, groundwater recharge, carbon storage, climate stabilization, erosion control, tourism and recreation, building materials, fertilizer, and cultural heritage.[2]

THE NEED FOR IBIs

Expanding human populations has given rise to the alteration of millions of acres of land, generally resulting in agricultural and urbanized areas. This often results in the removal or degradation of wetlands. In the United States, as much as 50% of wetland habitats have been drained or

Encyclopedia of Natural Resources DOI: 10.1081/E-ENRW-120051310

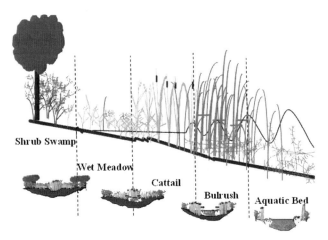

Fig. 1 Example of wetland zones along the transition from terrestrial to aquatic habitats. The zonation of lacustrine wetlands is the spatial equivalent of succession through time in palustrine wetlands.
Source: Taken from Uzarski.[9]

developed since European settlement.[10] Additionally, it is suspected that all wetlands have been negatively affected by humans in some way. Due to the essential services and habitat that wetlands provide, it is important that we have some way of evaluating and monitoring the health of our remaining wetlands to focus restoration and preservation efforts. Catching degradation before it is irreversible is essential. This can prove to be a difficult task due to the highly variable nature of wetlands. Traditional approaches to detect degradation relied heavily on chemistry. Unfortunately, this approach failed to account for human-induced habitat alteration, introduction of exotic species, and episodic events such as spills and effluent discharge. Biota are integrators of overall habitat and water quality and reveal both episodic and cumulative disturbance. Therefore, in the 1980s, the concept of IBIs was introduced. The majority of the initial work was done in streams and rivers. Initially, Karr[11] used fish community characteristics as indicators of environmental health, taking a more holistic approach toward assessing aquatic environmental quality. In general, biological communities provide information that a water sample cannot. Episodic insults to aquatic ecosystems may quickly become non-detectable as pollutants are diluted or washed away with currents. However, biological communities will maintain a "scar" from the event.[12] IBIs attempt to detect these "scars." A properly developed IBI will use these community "scars" as individual metrics to point toward specific insults that may have occurred. Each metric that makes up an IBI is a biological attribute related to specific types of human disturbance. Because of this, IBIs are sometimes referred to as multimetric IBIs. Since the 1980s, IBIs have been expanded to many different habitat types using a variety of biological communities. Currently, IBIs include the use of fish, macroinvertebrates, birds, reptiles, amphibians, zooplankton, algae, and vegetation.[13–16] The development of wetland IBIs has been a challenge in particular. Early

stream and river versions of IBIs relied heavily on detecting community changes as a result of altered temperature and dissolved oxygen regimes.[17] This approach was not applicable to wetlands because substantial temperature and dissolved oxygen fluctuations occur naturally in wetlands regardless of anthropogenic disturbance. Many of these hurdles have been breached, and wetland indices have proven to be cost-effective and accurate tools for assessing wetland condition. Wetland IBIs are currently being used by many government agencies and private organizations to make regulatory and conservation decisions.

IBI DEVELOPMENT

The process of developing indices of biological integrity requires the categorization of wetlands based on habitat type and natural disturbance to isolate variables directly associated with human disturbance. The delineation of these wetland categories is based upon hydrology and vegetation. Hydrology dictates the amount of natural disturbance shaping communities, and vegetation represents structure for other organisms that also naturally shape communities. The goal in development is to control for this natural variation and isolate community responses only due to human degradation. An example of wetland categories may include riverine, palustrine, and lacustrine wetlands or those associated with rivers, depressions in the landscape, or lakes, respectively. These categories are then further delineated based on hydrology, geography, and geology because this natural variation can create substantial variation in community composition regardless of human disturbance. The process of grouping sites with consistent natural conditions allows researchers to more confidently attribute observed variation among sites to human influences. While building IBIs for specific wetland categories, researchers use known levels of human disturbance as a baseline for building IBI metrics.[18]

IBI Development: Establishment of a Disturbance Gradient

Building IBI metrics involves identifying sites experiencing a range of known human disturbance. Reference sites receiving the least amount of human disturbance represent intact community composition. Those sites experiencing a range of stressors will then be used to determine variation in community composition, and this variation will be used to establish metrics. Ideally, a range of stressor types is also included to establish stressor–response relationships so that individual metrics will indicate specific ecosystem insults. Metrics can then be combined into an overall multimetric IBI. Known stressors at sites should be linked to chemical–physical characteristics and land use–land cover information.[19] Chemical–physical parameters and land use–land cover data

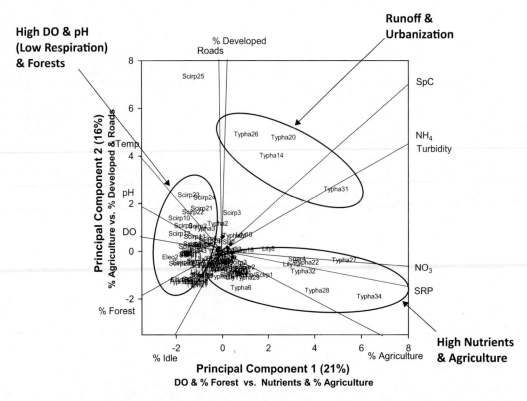

Fig. 2 Principal components analysis showing the distribution of sites based on habitat conditions and land use variables. Abbreviations: % Forest, forested land use; % Idle(?), inundated land cover; % Agriculture, Agricultural land use; SRP, Soluble reactive phosphorus; NO₃, Nitrate; NH₄, Ammonium; SpC, Specific conductance; % Developed, developed land use; DO, dissolved oxygen.

specific to each wetland can be combined using principal components analysis (PCA; Fig. 2). Sites with known human disturbance are used to identify related variables. Principal components that ordinate sites from the most human impact to the least human impact, or vice versa, can be used to identify chemical–physical and land use/cover variables associated with known human disturbances. These should then be related to variation in community composition. To capture local conditions unique to specific wetland categories, *in situ* chemical–physical variables, as well as adjacent land use and cover, are included in the disturbance gradient.

IBI Development: Metric Creation

Biological community data from the same sites used in PCA can be combined using ordination methods such as nonmetric multidimensional scaling (NMDS) or correspondence analysis (CA). Synthetic community variables or dimensions of these analyses can then be used in Pearson correlation analysis to relate biotic communities to principal components representing chemical–physical and land use and cover characteristics along a gradient of human disturbance. NMDS and CA dimensions can then be decomposed using species weights or factor loadings to point toward taxa responding to human disturbance. These taxa can then be used to generate metrics. Other commonly used metrics such as diversity, richness, and evenness can also be

explored and related to disturbance gradients. Any measures of the biotic communities related to measures of known stress are candidates for metrics to be included in an IBI.

IBI Development: Validation and Application

Developing an IBI is an iterative process, which involves a great deal of exploratory data analysis. Index scoring systems have been designed to use biological community characteristics such as taxa richness and diversity along with the relative abundance of tolerant and sensitive organisms to evaluate wetland health (Table 1).[14,15] Once the metrics have been calculated, the performance of these indices is tested on a group of sites that span a known disturbance gradient.[14] Index scores from sites that are known to be impacted by human activities are compared to scores from sites that are relatively pristine. A confounding factor for IBI use is the interpretation of IBI scores when more than one community index is used to assess the condition of a single wetland. For instance, a fish-based IBI may give a wetland a relatively high score, but a plant-based IBI used to assess the same wetland may result in a relatively low score. The observed differences in assessment scores may be associated with the nature of human influences on the wetland. In the previous example, the plant-based IBI may be low due to the influence of invasive plant species located too far from water depths where fish could be influenced.

Table 1 This is a basic example of the types of information used to calculate an IBI score for a wetland. These types of categorical calculations are typically performed for specific zones or areas within the wetland depending on the complexity of the habitat. Individual metric scores are added together and category scores are assigned based on the total score's magnitude

Example of a simplified index of biotic integrity worksheet

Individual metrics

1 Taxa richness from a specific family (species)

0 score = 1	>0 to 3 score = 3	>3 score = 5

2 Total genera richness

<10 score = 1	10 to 18 score = 3	>18 score = 5

3 Relative abundance of a specific family (%)

0 to 1 score = 1	>1 to 25 score = 3	>25 score = 5

4 Combined taxa richness of two specific families (genera)

0 to 2 score = 1	>2 to 4 score = 3	>4 to 6 score = 5	>6 score = 7

5 Shannon diversity index

0 to 0.4 score = 1	>0.4 to 0.9 score = 3	>0.9 score = 5

Category scores

0% to 15% of possible score

> *Degraded*: In comparison to other wetlands of the same type, this wetland is among the most impacted.

>15% to 50% of possible score

> *Moderately degraded*: The wetland shows obvious signs of anthropogenic disturbance.

>50% to 85% of possible score

> *Mildly impacted*: The wetland is beginning to show signs of anthropogenic disturbance.

>85% to 100% of possible score

> *Reference conditions*: The wetland is among the most pristine of wetlands of a similar type.

Riparian— Wetlands

An opposite result could be seen if the near-open water portion of the wetland was receiving pollutants from a nearby boat channel or drainage ditch. In this case, the fish-based IBI may indicate a low score, while the plant IBI may be high due to lack of influence from the pollutants. Taxa-based IBIs may also detect human disturbance at different scales. The distribution, colonization rates, and mobility of each indicator organism determine the scale of detection. For example, the relatively slow life cycle and distribution rates of plant species suggest that they could be indicating larger-scale influences, whereas macroinvertebrates, with their relatively short life cycles and limited range of mobility, might indicate finer scales of influence. As such, all wetlands within a particular region may yield a high quality rating using a plant-based IBI, but a macroinvertebrate-based IBI may yield a wide range of scores for the region and serve as a fine adjustment within the "high quality"

category. These possibilities are not absolute, but concepts such as these should be considered when implementing information gathered by IBIs into management decisions.

CONCLUSION

IBIs are valuable tools that researchers and managers can use to assess aquatic habitat conditions using biological community characteristics. When developed and validated thoroughly, these tools effectively assess the status and trends of wetland habitats and are at the forefront of aquatic habitat assessment techniques. The concepts discussed in this entry can and should be applied toward the development of new IBIs for other ecosystems. The information obtained from these methods will allow for more efficient usage of conservation efforts and funding. Future developments should focus on improving these tools to increase accuracy and ease of use and further decrease the costs of conservation efforts.

ACKNOWLEDGMENTS

We would like to dedicate this entry to Dr. Thomas M. Burton, a pioneer in wetland bioassessment. We thank Tom Clement, Dave Schuberg, Molly Gordon, Carli Gurholt, Lee Schoen, Tom Langer, Alexandra Bozimowski, Nicole Schmidt, Wen-jing Yang, Jenna MacDonald, and Mailen Staulter for editorial assistance. This is Contribution Number 37 of the Central Michigan University Institute for Great Lakes Research.

REFERENCES

1. Leveque, C.; Balian, E.V.; Martens, K. An assessment of animal species diversity in continental waters. Hydrobiologia **2005**, *542*, 39–67.
2. Woodward, R.T.; Wui, Y. The economic value of wetland services: A meta-analysis. Ecol. Econ. **2001**, *37*, 257–270.
3. Mitsch, W.A.; Gosselink, J.G. *Wetlands*, 3rd Ed.; John Wiley and Sons, Inc.: New York, 2000; 920.
4. Wassen, A.J.; Verkroost, A.W.M.; DeRuiter, P.C. Species richness-productivity patterns differ between N-, P-, and K-limited wetlands. Ecology **2003**, *84* (8), 2191–2199.
5. Mitsch, W.J.; Gosselink, J.G. The value of wetlands: Importance of scale and landscape setting. Ecol. Econ. **2000**, *35* (1), 25–33.
6. Chambers, R.M.; Meyerson, L.A.; Saltonstall, K. Expansion of *Phragmites australis* into tidal wetlands of North America. J. Aquat. Bot. **1999**, *64* (3–4), 261–273.
7. Comin, F.A.; Romero, J.A.; Astorga, V.; Garcia, C. Nitrogen removal and cycling in restored wetlands used as filters of nutrients for aricultural run-off. Water Sci. Technol. **1997**, *35* (5), 255–261.
8. Ottovia, V.; Balcarova, J.; Vymazal, J. Microbial characteristics of constructed wetlands. Water Sci. Technol. **1997**, *35* (5), 117–123.
9. Uzarski, D.G. *Wetlands of Large Lakes. Encyclopedia of Inland Waters*; Oxford: Elsevier, 2009; 599–606.
10. Krieger, K.M.; Kalarer, D.M.; Heath, R.T.; Herdendorf, C.E. Coastal wetlands of the Laurentian Great Lakes: Current knowledge and research needs. J. Great Lakes Res. **1992**, *18*, 525–528.
11. Karr, J.R. Assessment of biotic integrity using fish communities. Fisheries **1981**, *6*, 21–27.
12. Adams, S.M.; Greeley, M.S. Ecotoxicological indicators of water quality: Using multi-response indicators to assess the health of aquatic ecosystems. Water Air Soil Pollut. **2000**, *123*, 103–115.
13. Bhagat, Y.J.; Ciborowski, J.H.; Johnson, L.B.; Burton, T.M.; Uzarski, D.G.; Timmermans, S.A.; Cooper, M. Testing a fish index of biotic integrity for Great Lakes coastal wetlands: Stratification by plant zones. J. Great Lakes Res. **2007**, *33* (3), 224–235.
14. Uzarski, D.G.; Burton, T.M.; Genet, J.A. Validation and performance of an invertebrate index of biotic integrity for Lakes Huron and Michigan fringing wetlands during a period of lake level decline. Aquat. Health Manag. **2004**, *7*, 269–288.
15. Uzarski, D.G.; Burton, T.M.; Cooper, M.J.; Ingram, J.W.; Timmermans, S. Fish habitat use within and across wetland classes in coastal wetlands of the five Great Lakes: Development of a fish-based index of biotic integrity. J. Great Lakes Res. **2005**, *31* (1), 171–187.
16. Lougheed, V.L.; Chow-fraser, P. Development and use of a zooplankton index of wetland quality in the Laurentian Great Lakes basin. Ecol. Appl. **2002**, *12* (2), 474–486.
17. Wilhm, J.L.; Dorris, T.C. Biological parameters for water quality criteria. Bioscience **1968**, *18*, 477–481.
18. Burton, T.M.; Uzarski, D.G.; Gathman, J.P.; Genet, J.A.; Keas, B.E.; Stricker, C.A. Development of a preliminary invertebrate index of biotic integrity for Lake Huron coastal wetlands. Wetlands **1999**, *19* (4), 869–882.
19. Houlahan, J.E.; Keddy, P.A.; Makkay, K.; Findlay, C.S. The effects of adjacent land use on wetland species richness and community composition. Wetlands **2006**, *6* (1), 79–96.

Riparian— Wetlands

Acid Rain and Nitrogen Deposition

George F. Vance
Department of Ecosystem Sciences and Management, University of Wyoming, Laramie, Wyoming, U.S.A.

Abstract

One of the more highly publicized and controversial aspects of atmospheric pollution is that of acidic deposition. Although air pollution has occurred naturally since the formation of the Earth's atmosphere, the industrial era has resulted in human activities greatly contributing to global atmospheric pollution. Acidic materials can be transported long distances, some as much as hundreds of kilometers. This entry reviews the breadth of its impact, and how different regions are developing solutions to this global problem.

INTRODUCTION

Air pollution has occurred naturally since the formation of the Earth's atmosphere; however, the industrial era has resulted in human activities greatly contributing to global atmospheric pollution.[1,2] One of the more highly publicized and controversial aspects of atmospheric pollution is that of acidic deposition. Acidic deposition includes rainfall, acidic fogs, mists, snowmelt, gases, and dry particulate matter.[3] The primary origin of acidic deposition is the emission of sulfur dioxide (SO_2) and nitrogen oxides (NO_x) from fossil fuel combustion; electric power generating plants contribute approximately two-thirds of the SO_2 emissions and one-third of the NO_x emissions.[4]

Acidic materials can be transported long distances, some as much as hundreds of kilometers. For example, 30–40% of the S deposition in the northeastern U.S. originates in industrial midwestern U.S. states.[5] After years of debate, U.S. and Canada have agreed to develop strategies that reduce acidic compounds originating from their countries.[5,6] In Europe, the small size of many countries means that emissions in one industrialized area can readily affect forests, lakes, and cities in another country. for example, approximately 17% of the acidic deposition falling on Norway originated in Britain and 20% in Sweden came from eastern Europe.[5]

The U.S. EPA National Acid Precipitation Assessment Program (NAPAP) conducted intensive research during the 1980s and 1990s that resulted in the "Acidic Deposition: State of the Science and Technology" that was mandated by the Acid Precipitation Act of 1980.[6] NAPAP Reports to Congress have been developed in accordance with the 1990 amendment to the 1970 Clean Air Act and present the expected benefits of the Acid Deposition Control Program,[6,7] http://www.nnic.noaa.gov/CENR/NAPAP/. Mandates include an annual 10 million ton or approximately 40% reduction in point-source SO_2 emissions below 1980 levels, with national emissions limit caps of 8.95 million

tons from electric utility and 5.6 million tons from point-source industrial emissions. A reduction in NO_x of about 2 million tons from 1980 levels has also been set as a goal; however, while NO_x has been on the decline since 1980, projections estimate a rise in NO_x emissions after the year 2000. In 1980, the U.S. levels of SO_2 and NO_x emissions were 25.7 and 23.0 million tons, respectively.

Acidic deposition can impact buildings, sculptures, and monuments that are constructed using weatherable materials like limestone, marble, bronze, and galvanized steel,[7,8] http://www.nnic.noaa.gov/CENR/NAPAP/. While acid soil conditions are known to influence the growth of plants, agricultural impacts related to acidic deposition are of less concern due to the buffering capacity of these types of ecosystems.[2,5] When acidic substances are deposited in natural ecosystems, a number of adverse environmental effects are believed to occur, including damage to vegetation, particularly forests, and changes in soil and surface water chemistry.[9,10]

SOURCES AND DISTRIBUTION

Typical sources of acidic deposition include coal- and oil-burning electric power plants, automobiles, and large industrial operations (e.g., smelters). Once S and N gases enter the Earth's atmosphere they react very rapidly with moisture in the air to form sulfuric (H_2SO_4) and nitric (HNO_3) acids.[2,3] The pH of natural rainfall in equilibrium with atmospheric CO_2 is about 5.6; however, the pH of rainfall is less than 4.5 in many industrialized areas. The nature of acidic deposition is controlled largely by the geographic distribution of the sources of SO_2 and NO_x (Fig. 1). In the midwestern and northeastern U.S., H_2SO_4 is the main source of acidity in precipitation because of the coal-burning electric utilities.[2] In the western U.S., HNO_3 is of more concern because utilities and industry burn coal with low S contents and populated areas are high sources of NO_x.[2]

Encyclopedia of Natural Resources DOI: 10.1081/E-ENRA-120001785

Acid—Atmospheric

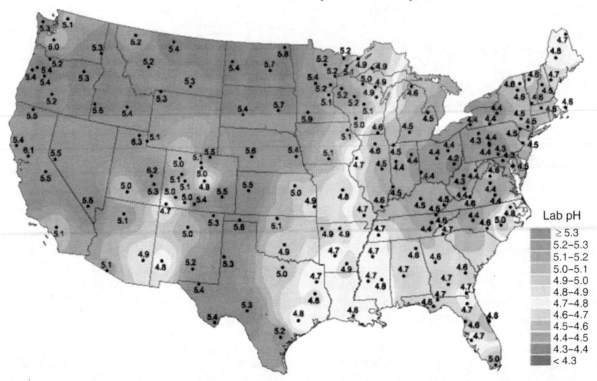

Hydrogen ion concentration as pH from measurements
made at the Central Analytical Laboratory, 1999

National Atmospheric Deposition Program/National Trends Network
http://nadp.sws.uluc.edu

Fig. 1 Acidic deposition across the U.S. during 1999.

Acid—
Atmospheric

Emissions of SO_2 and NO_x increased in the 20th century due to the accelerated industrialization in developed countries and antiquated processing practices in some undeveloped countries. However, there is some uncertainty as to the actual means by which acidic deposition affects our environment,[11,12] http://nadp.sws.uiuc.edu/isopleths/maps1999/. Chemical and biological evidence, however, indicates that atmospheric deposition of H_2SO_4 caused some New England lakes to decrease in alkalinity.[13,14] Many scientists are reluctant to over-generalize cause and effect relationships in an extremely complex environmental problem. Although, the National Acid Deposition Assessment Program has concluded there were definite consequences due to acidic deposition that warrant remediation[6,7] http://www.nnic.noaa.gov/CENR/NAPAP/. Since 1995, when the 1990 Clean Air Act Amendment's Title IV reduction in acidic deposition was implemented, SO_2 and NO_x emissions have, respectively, decreased and remained constant during the late 1990s.[4]

Both H_2SO_4 and HNO_3 are important components of acidic deposition, with volatile organic compounds and inorganic carbon also components of acidic deposition-related emissions. Pure water has a pH of 7.0, natural rainfall about 5.6, and severely acidic deposition less than 4.0. Uncontaminated rainwater should be pH 5.6 due to CO_2

chemistry and the formation of carbonic acid. The pH of most soils ranges from 3.0 to 8.0.[2] When acids are added to soils or waters, the decrease in pH that occurs depends greatly on the system's buffering capacity, the ability of a system to maintain its present pH by neutralizing added acidity. Clays, organic matter, oxides of Al and Fe, and Ca and Mg carbonates (limestones) are the components responsible for pH buffering in most soils. Acidic deposition, therefore, will have a greater impact on sandy, low organic matter soils than those higher in clay, organic matter, and carbonates. In fresh waters, the primary buffering mechanism is the reaction of dissolved bicarbonate ions with H^+ according to the following equation:

$$H^+ + HCO_3^- = H_2O + CO_2 \qquad (1)$$

HUMAN HEALTH EFFECTS

Few direct human health problems have been attributed to acidic deposition. Long-term exposure to acidic deposition precursor pollutants such as ozone (O_3) and NO_x, which are respiratory irritants, can cause pulmonary edema.[5,6] Sulfur dioxide (SO_2) is also a known respiratory irritant, but is generally absorbed high in the respiratory tract.

Indirect human health effects due to acidic deposition are more important. Concerns center around contaminated drinking water supplies and consumption of fish that contain potential toxic metal levels. With increasing acidity (e.g., lower pH levels), metals such as mercury, aluminum, cadmium, lead, zinc, and copper become more bioavailable.[2] The greatest human health impact is due to the consumption of fish that bioaccumulate mercury; freshwater pike and trout have been shown to contain the highest average concentrations of mercury.[5,15] Therefore, the most susceptible individuals are those who live in an industrial area, have respiratory problems, drink water from a cistern, and consume a significant amount of freshwater fish.

A long-term urban concern is the possible impact of acidic deposition on surface-derived drinking water. Many municipalities make extensive use of lead and copper piping, which raises the question concerning human health effects related to the slow dissolution of some metals (lead, copper, zinc) from older plumbing materials when exposed to more acidic waters. Although metal toxicities due to acidic deposition impacts on drinking waters are rare, reductions in S and N fine particles based on Clean Air Act Amendments will result in annual public health benefits valued at \$50 billion with reduced mortality, hospital admissions and emergency room visits.[16]

STRUCTURAL IMPACTS

Different types of materials and cultural resources can be impacted by air pollutants. Although the actual corrosion rates for most metals have decreased since the 1930s, data from three U.S. sites indicate that acidic deposition may account for 31–78% of the dissolution of galvanized steel and copper,[7,8] http://www.nnic.noaa.gov/CENR/NAPAP/. In urban or industrial settings, increases in atmospheric acidity can dissolve carbonates (e.g., limestone, marble) in buildings and other structures. Deterioration of stone products by acidic deposition is caused by: 1) erosion and dissolution of materials and surface details; 2) alterations (blackening of stone surfaces); and 3) spalling (cracking and spalling of stone surfaces due to accumulations of alternation crusts).[8] Painted surfaces can be discolored or etched, and there may also be degradation of organic binders in paints.[8]

ECOSYSTEM IMPACTS

It is important to examine the nature of acidity in soil, vegetation, and aquatic environments. Damage from acidification is often not directly due to the presence of excessive H^+, but is caused by changes in other elements. Examples include increased solubilization of metal ions such as Al^{3+} and some trace elements (e.g., Mn^{2+}, Pb^{2+}) that can be toxic to plants and animals, more rapid losses of basic cations (e.g., Ca^{2+}, Mg^{2+}), and the creation of unfavorable soil and aquatic environments for different fauna and flora.

Soils

Soil acidification is a natural process that occurs when precipitation exceeds evapotranspiration.[2] "Natural" rainfall is acidic (pH of ~5.6) and continuously adds a weak acid (H_2CO_3) to soils. This acidification results in a gradual leaching of basic cations (Ca^{2+} and Mg^{2+}) from the uppermost soil horizons, leaving Al^{3+} as the dominant cation that can react with water to produce H^+. Most of the acidity in soils between pH 4.0 and 7.5 is due to the hydrolysis of Al^{3+},[17,18] http://www.epa.gov/airmarkets/acidrain/effects/index.html. Other acidifying processes include plant and microbial respiration that produces CO_2, mineralization and nitrification of organic N, and the oxidation of FeS_2 in soils disturbed by mining or drainage.[2] In extremely acidic soils (pH < 4.0), strong acids such as H_2SO_4 are a major component.

The degree of accelerated acidification depends both upon the buffering capacity of the soil and the use of the soil. Many of the areas subjected to the greatest amount of acidic deposition are also areas where considerable natural acidification occurs.[19] Forested soils in the northeastern U.S. are developed on highly acidic, sandy parent materials that have undergone tremendous changes in land use in the past 200 years. However, clear-cutting and burning by the first European settlers have been almost completely reversed and many areas are now totally reforested.[5] Soil organic matter that accumulated over time represents a natural source of acidity and buffering. Similarly, greater leaching or depletion of basic cations by plant uptake in increasingly reforested areas balances the significant inputs of these same cations in precipitation.[20,21] Acidic deposition affects forest soils more than agricultural or urban soils because the latter are routinely limed to neutralize acidity. Although it is possible to lime forest soils, which is done frequently in some European countries, the logistics and cost often preclude this except in areas severely impacted by acidic deposition.[5]

Excessively acidic soils are undesirable for several reasons. Direct phytotoxicity from soluble Al^{3+} or Mn^{2+} can occur and seriously injure plant roots, reduce plant growth, and increase plant susceptibility to pathogens.[21] The relationship between Al^{3+} toxicity and soil pH is complicated by the fact that in certain situations organic matter can form complexes with Al^{3+} that reduce its harmful effects on plants.[18] Acid soils are usually less fertile because of a lack of important basic cations such as K^+, Ca^{2+}, and Mg^{2+}. Leguminous plants may fix less N_2 under very acidic conditions due to reduced rhizobial activity and greater soil adsorption of Mo by clays and Al and Fe oxides.[2] Mineralization of N, P, and S can also be reduced because of the lower metabolic activity of bacteria. Many plants and microorganisms have adapted to very acidic conditions (e.g., pH < 5.0). Examples include ornamentals such as azaleas and rhododendrons and food crops such as cassava, tea, blueberries, and potatoes.[5,22] In fact, considerable efforts in plant breeding and biotechnology are directed towards developing Al- and Mn-tolerant plants that can survive in highly acidic soils.

Agricultural Ecosystems

Acidic deposition contains N and S that are important plant nutrients. Therefore, foliar applications of acidic deposition at critical growth stages can be beneficial to plant development and reproduction. Generally, controlled experiments require the simulated acid rain to be pH 3.5 or less in order to produce injury to certain plants.[22] The amount of acidity needed to damage some plants is 100 times greater than natural rainfall. Crops that respond negatively in simulated acid rain studies include garden beets, broccoli, carrots, mustard greens, radishes, and pinto beans, with different effects for some cultivars. Positive responses to acid rain have been identified with alfalfa, tomato, green pepper, strawberry, corn, lettuce, and some pasture grass crops.

Agricultural lands are maintained at pH levels that are optimal for crop production. In most cases the ideal pH is around pH 6.0–7.0; however, pH levels of organic soils are usually maintained at closer to pH 5.0. Because agricultural soils are generally well buffered, the amount of acidity derived from atmospheric inputs is not sufficient to significantly alter the overall soil pH.[2] Nitrogen and S soil inputs from acidic deposition are beneficial, and with the reduction in S atmospheric levels mandated by 1990 amendments to the Clean Air Act, the S fertilizer market has grown. The amount of N added to agricultural ecosystems as acidic deposition is rather insignificant in relation to the 100–300 kg N/ha/yr required of most agricultural crops.

Forest Ecosystems

Perhaps the most publicized issue related to acidic deposition has been widespread forest decline. For example, in Europe estimates suggest that as much as 35% of all forests have been affected.[23] Similarly, in the U.S. many important forest ranges such as the Adirondacks of New York, the Green Mountains of Vermont, and the Great Smoky Mountains in North Carolina have experienced sustained decreases in tree growth for several decades.[6] Conclusive evidence that forest decline or dieback is caused solely be acidic deposition is lacking and complicated by interactions with other environmental or biotic factors. However, NAPAP research[6] has confirmed that acidic deposition has contributed to a decline in high-elevation red spruce in the northeastern U.S. In addition, nitrogen saturation of forest ecosystems from atmospheric N deposition is believed to result in increased plant growth, which in turn increases water and nutrient use followed by deficiencies that can cause chlorosis and premature needle-drop as well as increased leaching of base cations from the soil.[24]

Acidic deposition on leaves may enter directly through plant stomates.[1,22] If the deposition is sufficiently acidic (pH ~ 3.0), damage can also occur to the waxy cuticle, increasing the potential for direct injury of exposed leaf mesophyll cells. Foliar lesions are one of the most common symptoms. Gaseous compounds such as SO_2 and SO_3 present in acidic mists or fogs can also enter leaves through the stomates, form H_2SO_4 upon reaction with H_2O in the cytoplasm, and disrupt many metabolic processes. Leaf and needle necrosis occurs when plants are exposed to high levels of SO_2 gas, possibly due to collapsed epidermal cells, eroded cuticles, loss of chloroplast integrity and decreased chlorophyll content, loosening of fibers in cell walls and reduced cell membrane integrity, and changes in osmotic potential that cause a decrease in cell turgor.

Root diseases may also increase in excessively acidic soils. In addition to the damages caused by exposure to H_2SO_4 and HNO_3, roots can be directly injured or their growth rates impaired by increased concentrations of soluble Al^{3+} and Mn^{2+} in the rhizosphere,[2,25] http://nadp.sws.uiuc.edu. Changes in the amount and composition of these exudates can then alter the activity and population diversity of soil-borne pathogens. The general tendency associated with increased root exudation is an enhancement in microbial populations due to an additional supply of carbon (energy). Chronic acidification can also alter nutrient availability and uptake patterns.[8,22]

Long-term studies in New England suggest acidic deposition has caused significant plant and soil leaching of base cations,[1,21] resulting in decreased growth of red spruce trees in the White Mountains.[6] With reduction in about 80% of the airborne base cations, mainly Ca^{2+} but also Mg^{2+}, from 1950 levels, researchers suggest forest growth has slowed because soils are not capable of weathering at a rate that can replenish essential nutrients. In Germany, acidic deposition was implicated in the loss of soil Mg^{2+} as an accompanying cation associated with the downward leaching of SO_4^{2-}, which ultimately resulted in forest decline.[2] Several European countries have used helicopters to fertilize and lime forests.

Aquatic Ecosystems

Ecological damage to aquatic systems has occurred from acidic deposition. As with forests, a number of interrelated factors associated with acidic deposition are responsible for undesirable changes. Acidification of aquatic ecosystems is not new. Studies of lake sediments suggest that increased acidification began in the mid-1800s, although the process has clearly accelerated since the 1940s.[15] Current studies indicate there is significant S mineralization in forest soils impacted by acidic deposition and that the SO_4^{2-} levels in adjacent streams remain high, even though there has been a decrease in the amount of atmospheric-S deposition.[24]

Geology, soil properties, and land use are the main determinants of the effect of acidic deposition on aquatic chemistry and biota. Lakes and streams located in areas with calcareous geology resist acidification more than those in granitic and gneiss materials.[16] Soils developed from calcareous parent materials are generally deeper and more buffered than thin, acidic soils common to granitic areas.[2] Land management decisions also affect freshwater acidity. Forested watersheds tend to contribute more acidity than those dominated by meadows, pastures, and agronomic ecosystems.[8,14,20] Trees and other vegetation in forests are known to "scavenge" acidic compounds in fogs, mists, and atmospheric particulates. These acidic compounds are later deposited in forest soils when rainfall leaches forest vegetation surfaces. Rainfall below forest canopies (e.g., throughfall) is usually more acidic than ambient precipitation. Silvicultural operations that disturb soils in forests can increase acidity by stimulating the oxidization of organic N and S, and reduced S compounds such as FeS_2.[2]

A number of ecological problems arise when aquatic ecosystems are acidified below pH 5.0, and particularly below pH 4.0. Decreases in biodiversity and primary productivity of phytoplankton, zooplankton, and benthic invertebrates commonly occur.[15,16] Decreased rates of biological decomposition of organic matter have occasionally been reported, which can then lead to a reduced supply of nutrients.[20] Microbial communities may also change, with fungi predominating over bacteria. Proposed mechanisms to explain these ecological changes center around physiological stresses caused by exposure of biota to higher concentrations of Al^{3+}, Mn^{2+}, and H^+ and lower amounts of available Ca^{2+}.[15] One specific mechanism suggested involves the disruption of ion uptake and the ability of aquatic plants to regulate Na^+, K^+, and Ca^{2+} export and import from cells.

Acidic deposition is associated with declining aquatic vertebrate populations in acidified lakes and, under conditions of extreme acidity, of fish kills. In general, if the water pH remains above 5.0, few problems are observed; from pH 4.0 to 5.0 many fish are affected, and below pH 3.5 few fish can survive.[23] The major cause of fish kill is due to the direct toxic effect of Al^{3+}, which interferes with the role Ca^{2+} plays in maintaining gill permeability and respiration. Calcium has been shown to mitigate the effects of Al^{3+}, but in many acidic lakes the Ca^{2+} levels are inadequate to overcome Al^{3+} toxicity. Low pH values also disrupt the Na^+ status of blood plasma in fish. Under very acidic conditions, H^+ influx into gill membrane cells both stimulates excessive efflux of Na^+ and reduces influx of Na^+ into the cells. Excessive loss of Na^+ can cause mortality. Other indirect effects include reduced rates of reproduction, high rates of mortality early in life or in reproductive phases of adults, and migration of adults away from acidic areas.[16] Amphibians are affected in much the same manner as fish, although they are somewhat less sensitive to Al^{3+} toxicity. Birds and small mammals often have lower populations and lower reproductive rates in areas adjacent to acidified aquatic ecosystems. This may be due to a shortage of food due to smaller fish and insect populations or to physiological stresses caused by consuming organisms with high Al^{3+} concentrations.

REDUCING ACIDIC DEPOSITION EFFECTS

Damage caused by acidic deposition will be difficult and extremely expensive to correct, which will depend on our ability to reduce S and N emissions. For example, society may have to burn less fossil fuel, use cleaner energy sources and/or design more efficient "scrubbers" to reduce S and N gas entering our atmosphere. Despite the firm conviction of most nations to reduce acidic deposition, it appears that the staggering costs of such actions will delay implementation of this approach for many years. The 1990 amendments to the Clean Air Act are expected to reduce acid-producing air pollutants from electric power plants. The 1990 amendments established emission allowances based on a utilities' historical fuel use and SO_2 emissions, with each allowance representing 1 ton of SO_2 that canbought, sold or banked for future use,[4,6,7] http://www.nnic.noaa.gov/CENR/NAPAP/. Short-term remedial actions for acidic deposition are available and have been successful in some ecosystems. Liming of lakes and some forests (also fertilization with trace elements and Mg^{2+}) has been practiced in European counties for over 50 years.[16,23] Hundreds of Swedish and Norwegian lakes have been successfully limed in the past 25 years. Lakes with short mean residence times for water retention may need annual or biannual liming; others may need to be limed every 5–10 years. Because vegetation in some forested ecosystems has adapted to acidic soils, liming (or over-liming) may result in an unpredictable and undesirable redistribution of plant species.

REFERENCES

1. Smith, W.H. Acid rain *The Wiley Encyclopedia of Environmental Pollution and Cleanup*; Meyers, R.A., Dittrick, D.K. Eds.;, Wiley: New York, 1999; 9–15.
2. Pierzynski, G.M.; Sims, J.T.; Vance, G.F. *Soils and Environmental Quality;* CRC Press: Boca Raton, FL, 2000; 459 pp.
3. Wolff, G.T. Air pollution. *The Wiley Encyclopedia of Environmental Pollution and Cleanup* Meyers, R.A., Dittrick, D.K. Eds.; Wiley: New York 1999; 48–65.
4. U.S. Environmental protection agency *Progress Report on the EPA Acid Rain Program* EPA-430-R-99-011 U.S. Government Printing Office: Washington, DC.
5. Forster, B.A. *The Acid Rain Debate: Science and Special Interests in Policy Formation*; Iowa State University Press: Ames, IA, 1993.

6. *National Acid Precipitation Assessment Program Task Force Report*, National Acid Precipitation Assessment Program 1992 Report to Congress; U.S. Government Printing Office: Pittsburgh, PA, 1992; 130 pp.

7. *National Science and Technology Council*, National Acid Precipitation Assessment Program Biennial Report to Congress: An Integrated Assessment; 1998 (accessed July 2001).

8. Charles, D.F., The acidic deposition phenomenon and its effects: critical assessment review papers. *Effects Sciences*; EPA-600/8-83-016B; U.S. Environmental Protection Agency: Washington, DC, 1984; Vol. 2.

9. McKinney, M.L.; Schoch, R.M. *Environmental Science: Systems and Solutions*; Jones and Bartlett Publishers: Sudbury, MA, 1998.

10. United Nations, World band and World Resources Institute-World Resources: *People and Ecosystems—The Fraying Web of Life*; Elsevier: New York, 2000.

11. *National Atmospheric Deposition Program (NRSP-3)/ National Trends Network. Isopleth Maps*. NADP Program Office, Illinois State Water Survey, Champaign, IL, 2000 (accessed July 2001).

12. Council on Environmental Quality. *Environmental Quality, 18th and 19th Annual Reports*; U.S. Government Printing Office: Washington, DC, 1989.

13. Charles, D.F., *Acid Rain Research: Do We Have Enough Answers?* Proceedings of a Speciality Conference. Studies in Environmental Science #64, Elsevier: New York, 1995.

14. Kamari, J. *Impact Models to Assess Regional Acidification*; Kluwer Academic Publishers: London, 1990.

15. Charles, D.F. *Acidic Deposition and Aquatic Ecosystems*; Springer-Verlag: New York, 1991.

16. Mason, B.J. *Acid Rain: Its Causes and Effects on Inland Waters*; Oxford University Press: New York, 1992.

17. U.S. environmental protection agency. Effects of acid rain: human health. EPA Environmental Issues Website. Update June 26, 2001 (accessed July 2001).

18. Marion G.M.; Hendricks, D.M.; Dutt, G.R.; Fuller, W.H. Aluminum and silica solubility in soils. *Soil Science* **1976**, *121*, 76–82.

19. Kennedy, I.R. *Acid Soil and Acid Rain*; Wiley: New York, 1992.

20. Reuss J.O.; Johnson, D.W. *Acid Deposition and the Acidification of Soils and Waters*; Springer-Verlag: New York, 1986.

21. Likens G.E.; Driscoll, C.T.; Buso, D.C. Long-term effects of acid rain: response and recovery of a forest ecosystem. *Science* **1996**, *272*, 244–246.

22. Linthurst R.A. *Direct and Indirect Effects of Acidic Deposition on Vegetation*; Butterworth Publishers; Stoneham, MA, 1984.

23. Bush, M.B. *Ecology of a Changing Planet*; Prentice Hall: Upper Saddle River, NJ, 1997.

24. Alawell, C.; Mitchell, M.J.; Likens, G.E.; Krouse, H.R. Sources of stream sulfate at the Hubbard Brook experimental forest: long-term analyses using stable isotopes. Biogeochemistry **1999**, *44*, 281–299.

25. National Atmospheric Deposition Program. *Nitrogen in the Nation's Rain*. NADP Brochure 2000–2001a (accessed July 2001).

Acid—
Atmospheric

Acid Rain and Precipitation Chemistry

Pao K. Wang
Atmospheric and Oceanic Sciences, University of Wisconsin, Madison, Wisconsin, U.S.A.

Abstract

Acid rain is a phenomenon of serious environmental concern. By definition, acid rain refers to rainwater that is acidic. But in reality, it is more accurate to use the term *acid deposition* since not only rain but also snow, sleet, hail, and even fog can become acidic. In addition to the process where acids become associated with precipitation (called *wet deposition*), acid gases and particles can also be deposited on the Earth's surface directly (called *dry deposition*). However, since the name "acid rain" has become a household term and its formation is better understood than other types of acid deposition, the following discussions will focus on acid rain.

INTRODUCTION

Whereas an aqueous solution is acidic if its pH value is less than 7.0, acid rain refers to rainwater with pH less than 5.6. This is because, even without the presence of man-made pollutants, natural rainwater is already acidic as CO_2 in the atmosphere reacts with water to produce carbonic acid:

$$CO_2 + H_2O(l) \Leftrightarrow H_2CO_3(l) \tag{1}$$

The pH value of this solution is around 5.6. Even though the carbonic acid in rain is fairly dilute, it is sufficient to dissolve minerals in the Earth's crust, making them available to plant and animal life, yet not acidic enough to cause damage. Other atmospheric substances from volcanic eruptions, forest fires, and similar natural phenomena also contribute to the natural sources of acidity in rain. Still, even with the enormous amounts of acids created annually by nature, normal rainfall is able to assimilate them to the point where they cause little, if any, known damage.

However, large-scale human industrial activities have the potential of throwing off this acid balance, and converting natural and mildly acidic rain into precipitation with stronger acidity and far-reaching environmental effects. This is the root of the acid rain problem, which is not only of national but also international concern. This problem may have existed for more than 300 years starting at the time when the industrial revolution demanded a large scale burning of coal in which sulfur was a natural contaminant. Several English scholars, such as Robert Boyle in the 17th century and Robert A. Smith of the 19th century, wrote about the acids in air and rain; though there was a lack of appreciation of the magnitude of the problem at that time.

Individual studies of the acid rain phenomenon in North America started in the 1920s, but true appreciation of the problem came only in the 1970s.

To address this problem, the U.S. Congress established the National Acid Precipitation Assessment Program (NAPAP) to study the causes and impacts of the acid deposition. This research established that acid rain does cause broad environmental and health effects; the pollution causing acid deposition can travel hundreds of miles, and electric power generation is mainly responsible for SO_2 (~65%) and NO_x emissions (~30%). Subsequently, Congress created the Acid Rain Program under Title IV (Acid Deposition Control) of the 1990 Clean Air Act Amendments. Electric utilities are required to reduce their emissions of SO_2 and NO_x significantly. It was expected that by 2010 the would lower their emissions by 8.5 million tons compared to their 1980 levels. They also need to reduce their NO_x emissions by 2 million tons each year compared to the levels before the Clean Air Act Amendments.

However, it may not be adequate to solve the acid emission merely at the national level. With increasing industrialization of the Third World countries in the twenty-first century, one can expect great increase of the atmospheric loading of SO_2 and NO_x because many of these countries will burn fossil fuels to satisfy their energy needs. Clearly, some form of international agreements need to be forged to prevent serious environmental degradation due to acid rain.

THE CHEMISTRY OF ACID RAIN

Sulfuric acid (H_2SO_4) and nitric acid (HNO_3) are the two main acid species in the rain. The partitioning of acids in

Encyclopedia of Natural Resources DOI: 10.1081/E-ENRA-120010318

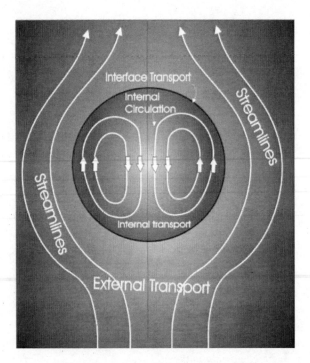

Fig. 1 A schematic of the acid rain formation process.

rain may be different in different places. In the United States, the partitioning is H_2SO_4 (~65%), HNO_3 (~30%), and others (~5%). While there are many possible chemicals that may serve as the precursors of acid rain, the two main substances are SO_2 and NO_x (and NO_x consists of NO and NO_2), and both are released to the atmosphere via the industrial combustion process. While power generation is the predominant source of these precursors, industrial boilers and automobiles also contribute substantially. When these precursors enter the cloud and precipitation systems, acid rain occurs. Fig. 1 shows a schematic of the acid rain formation process.

Once airborne, these chemicals can be involved in milliards of chemicals reactions. The main paths that lead to acid rain formation are described as follows.

Sulfuric Acid

SO_2 is believed to be the main precursor for the formation of sulfuric acid drops. Its main source in the atmosphere is the combustion of fossil fuels. This is because sulfur is a natural contaminant in coal (especially the low grade ones) and oil. The following reactions are thought to occur when SO_2 is absorbed by a water drop (see e.g., Pruppacher and Klett,[1] Seinfeld and Pandis[2]):

$$SO_2(g) + H_2O(l) \Leftrightarrow SO_2 \cdot H_2O(l) \qquad (2)$$

$$SO_2 \cdot H_2O \Leftrightarrow H^+ + HSO_3^- \qquad (3)$$

$$HSO_3^- \Leftrightarrow H^+ + SO_3^{-2} \qquad (4)$$

$$SO_3^{-2} \xrightarrow{\text{oxidation}} SO_4^{-2} \qquad (5)$$

The oxidant of the last step can be H_2O_2, O_3, OH, and others. There are still controversies about the identity of the oxidants.

Note that the equilibrium of the above reaction system is controlled by the pH values of the drop, and the presence of ammonia is often considered together with these reactions since it affects the pH of the drop. A detailed discussion of these reactions and their rates is given in Chapter 17 of Pruppacher and Klett.[1]

Nitric Acid

The main ingredients for the formation of nitric acid are NO and NO_2 (and are often combined into one category, NO_x). It is commonly thought that the main path of nitric acid found in clouds and raindrops is the formation of gas phase, HNO_3, followed by its uptake by liquid water. Although there are reactions of NO_x with liquid water that can lead to nitric acid, they are thought to be unimportant due to their slow reaction rates.

The main reaction for HNO_3 formation is

$$NO_2 + OH + M \rightarrow HNO_3 + M \qquad (6)$$

where M can be any neutral molecule. NO can be converted to NO_2 by the following reaction:

$$2NO + O_2 \rightarrow 2NO_2 \qquad (7)$$

DROP-SCALE TRANSPORT PROCESSES OF ACID RAIN

The chemical reactions described earlier must be considered together with the transport processes to obtain a quantitative picture of acid rain formation. This is especially true for SO_2 because absorption and reactions occur simultaneously. The convective transport influences the concentrations of different species and hence the reaction rates. Fig. 2 illustrates these processes schematically. These include the following.

External Transport

This refers to the transport of SO_2 gas toward the surface of the drop. It is a convective diffusion process (both convective transport and diffusional transport occur) and is influenced by the flow fields created by the falling drop and atmospheric conditions (pressure and temperature).

Fig. 2 A schematic of the drop-scale transport process of sulfur species involved in the acid rain.

Interfacial Transport

Once SO_2 is adsorbed on the surface of the drop, it must be transferred into the interior for further reactions to occur. The time for establishing phase equilibrium is controlled by Henry's law constant and mass accommodation coefficient of SO_2.

Internal Transport

In the interior of the drop, reactions[2-5] occur. At the same time, these species are transported by both diffusion and internal circulation. The latter is caused by the motion of the liquid drop in a viscous medium and can influence the production rates of these species (see Pruppacher and Klett[11]).

ENVIRONMENTAL FACTORS INFLUENCING THE ACID RAIN FORMATION AND IMPACTS

Like many environmental hazards, the acid rain process is not driven by a few well-controlled physical and chemical processes, but involves complicated interactions between the chemicals and the environments they exist in. While the main ingredients of acid rain come from industrial activities, many other factors may influence the formation of acid rain and its impacts. The following are some of the most important.

Meteorological Factors

Acid rain occurs in the atmosphere and hence is greatly influenced by meteorological factors such as wind direction and speed, amount and frequency of precipitation, pressure patterns, and temperature. For example, in drier climates, such as the western United States, wind blown alkaline dust is abundant and tends to neutralize the acidity in the rain. This is the *buffering effect* of the dust. In humid climates, like the Eastern Seaboard, less dust is in the air, and precipitation tends to be more acidic.

Seasonality may also influence acid precipitation. For example, while it is true that rain may be more acidic in summer (because of higher demands for energy and hence more fossil fuel used), the snow in winter can also pick up a substantial amount of acids. These snow-borne acids can accumulate throughout winter (if the weather is cold enough) and then are released in large doses during the spring thaw. These large doses of acid may have more significant effects during fish spawning or seed germination than the same doses at some less critical time.

Topography and Geology

The topography and geology of an area have marked influence on acid rain effects. Research from the U.S. EPA pointed out that areas most sensitive to acid precipitation are those with hard, crystalline bedrock and very thin surface soils. Here, in the absence of buffering properties of soil, acid rains will have direct access to surface waters and their delicate ecosystem. Areas with steep topography, such as mountainous areas, generally have thin surface soils and hence are very vulnerable to acid rain. In contrast, a thick soil mantle or one with high buffering capacity, such as most flatlands, helps keep acid rain damage down.

The location of water bodies is also important. Headwater lakes and streams are especially vulnerable to acidification. Lake depth, the ratio of water-shed area to lake area, and the residence time in lakes all play a part in determining the consequent threat posed by acids. The transport mode of the acid (rains or runoff) also influences the effects.

Biota

Acid rain may fall on trees causing damages. The kinds of trees and plants in an area, their heights, and whether they are deciduous or evergreen may all play a part in the potential effects of acid rain. Without a dense leaf canopy, more acid may reach the Earth to impact on soil and water chemistries. Stresses on the plants will also affect the balance of local ecosystem. Additionally, the rate at which different types of plants carry on their normal life processes influences an area's ratio of precipitation to evaporation. In locales with high evaporation rates, acids will concentrate on leaf surfaces. Another factor is that leaf litter decomposition may add to the acidity of the soil due to normal biological actions.

REFERENCES

1. Pruppacher, H.R.; Klett, J.D. *Microphysics of Clouds and Precipitation*; Kluwer Academic Publishers: 1997; 954 pp.
2. Seinfeld, J.H.; Pandis, S.N. *Atmospheric Chemistry and Physics*; John Wiley and Sons: 1998.

Acid–
Atmospheric

Agroclimatology

Dev Niyogi
Department of Agronomy and Department of Earth, Atmospheric, and Planetary Sciences, Purdue University, West Lafayette, Indiana, U.S.A.

Olivia Kellner
Department of Earth, Atmospheric, and Planetary Sciences, Purdue University, West Lafayette, Indiana, U.S.A.

Abstract

This entry provides a broad introduction to agroclimatology. Agricultural productivity is intricately related to the climate and a region's soil types, hydrologic cycle, and meteorology. Because of this, agronomy and climate are disciplines that are linked together in the science of agroclimatology: the applied science of water, soil, and crop management incorporating the knowledge of regional weather and climate information such as precipitation and temperature with an aim to improve crop yields. The interlinked roles of the disciplines of soil science, surface hydrology and the hydrologic cycle, agronomy (i.e., crop management and biomass), and meteorology and climatology on agroclimatology are introduced. A brief history of agroclimatology, discussion of agroclimatology and technologies associated with it today, the importance of agroclimatology, along with the projected path of agroclimatology into the future are discussed.

INTRODUCTION

Agroclimatology can be considered the study of local climate (determined by water, soil, and radiation in a given area, along with biomass and daily weather) and that local climate's interaction with agriculture and crop production for food, fiber, fuel, and availability of feed for livestock. It can help answer questions such as: how do temperature and precipitation patterns affect agricultural productivity? Agroclimatology can have many foci: climatic influences on crop production, availability of feed for livestock, crop modification to help plants withstand climate extremes, resilience of pests, sustaining yields in the face of an increasing world population, the influence of the microscale environment on crop yield, and agroclimatic modeling.[1,2] An emerging feature of the agroclimatic sciences has seen studies completed to understand the effect of climate change on agriculture, as well as the impact of agriculture on regional climates.[3–6]

Agroclimatology in the Beginning

The relation between world agricultural regions and global climatic patterns is well appreciated. It is also linked to a region's culture with seasonal festivals and events that celebrate the beginning and end of growing seasons. Agroclimatology is a relatively young science rooted in systematic weather observations of temperature and precipitation undertaken by many regional and national weather agencies such as, but not limited to, the U.S.A.'s National Weather Service, the Met Office of the United Kingdom, and the India Meteorological Department. Large-scale weather and climate events such as the Dust Bowl (~1920s, 1930s, and 1950s) that struck the U.S.A. have also contributed to the growth and evolution of agroclimatology as a practical science.[1,2,7] Agroclimatology can, however, be traced back to the domestication of native plants in societies that began thousands of years ago with a series of trial and error with local flora. This primitive science of plant domestication sometimes led to the collapse of societies due to weather and climate shifts that led to prolonged drought and crop failures (i.e., Anasazi of North America, the collapse of Polynesian society on Easter Island, and Mesopotamia that spanned modern day Turkey, Syria, Iran, and Iraq)[8] showing the need for a detailed understanding of weather, climate, and food production. Successes in crop domestication through the recognition of weather and climate patterns led to the establishment of distinctive crop-growing regions such as the Fertile Crescent in the Mediterranean region due to its mild climate and the U.S. Corn Belt due to its suitable glacial till soils and adequate rains. These two locations are examples of the many suitable crop-growing regions around the world where agricultural economies are well rooted.

Regular weather observations began with the development of the telegraph and railroad networks around the world during the Industrial Revolution (mid to late 1800s), which eventually led to the development of the national weather observation organizations across developed countries that kept record of daily weather data across the country. Daily weather observations were then shared via telegraph allowing for the development of daily weather

Encyclopedia of Natural Resources DOI: 10.1081/E-ENRA-120047622

Acid—
Atmospheric

maps that have grown through the century into the modern weather maps that meteorologists and climatologists use today.[9] As meteorology grew in the late 1800s, the development of regional weather climatologies from archived weather data of observed temperature and precipitation around the world began as well.[2,10] Building from these climatic data sets, climate systems were developed as a practical guide for sustaining flora and fauna. The Köppen climate system (1918) and the Holdridge life zones system (1947) are examples of two climate classification systems that characterize specified regions/climate zones based on temperature and precipitation (Köppen) or on precipitation, potential evapotranspiration, and humidity (Holdridge). These climate systems provide a quick overview of the region's agricultural potential.

Agroclimatology Today

In the early 1920s, the initial step of implementing agroclimatology as a practicing science was made by the United Kingdom Royal Meteorological Society with an expression of the need to study the effect of weather phenomena on crop growth and total yield.[7] Research investigating the impacts between climatic variables and crop yields began with simple regression analysis. These analyses eventually expanded over subsequent decades to simple statistical and empirical crop models by the 1970s and 1980s.[1,2,7] The development of remote-sensing technologies after World War II and into the 1970s and 1980s resulted in research programs and instruments capable of direct measurements of soil and moisture fluxes in the field (in situ) without human interference (remotely sensed data). Microsensors, automated weather stations, and remote-sensing technologies opened up a new dimension of detail in agroclimatology. The collection of meteorological, biophysical, and biogeochemical variables is now available in an unprecedented manner and this data is used to study and develop plant–climatic relationships. Some examples include assessing the effect of nitrogen levels on plant productivity or the effect of leaf area index on evapotranspiration and water demand in local areas, each of which is used to understand crop sustenance. This has allowed for much more detailed research and discoveries between crops and locally observed weather/climate patterns to take place.[1,2,11] Through the 1980s, crop modification, decision support systems and tools, and more detailed research have energized the development of crop models. Availability of observed weather patterns and observed surface data at micro to county-level scales has led to a broader understanding of hydrologic cycles within crop-climate analysis. Today's crop production models can simulate crop physiology and productivity reasonably well and have advanced to incorporate climate change scenarios providing insight into the possible impacts of varying weather and climate conditions on crop yields. In other words, while simple and statistical relations between climatic patterns and crop response continue to be important, there is a growing ability to develop more sophisticated, predictive tools that appear to adequately mimic complex agroclimatic interactions at multiple scales.

Current efforts to improve crop production amidst shifting climate patterns and dynamic economies include the efforts of international research scientists and crop models working in sync on projects such as the Agricultural Model Intercomparison and Improvement Project (AgMIP) and the Decision Support System for Agrotechnology Transfer (DSSAT).[12,13] Community efforts such as AgMIP consist of specialized teams comprising experts in agronomy, economics, and climate, devoted to the detailed study of specific crops, and collaborating and comparing more than multiple crop-specific (such as maize, wheat, rice, and sugarcane) models for crop response to carbon dioxide (CO_2), temperature, and other environmental factors. The research framework model for such collaborative assessments is to conduct studies at different locations to represent variation in crop production around the world and seek to find answers to, e.g., the following questions: Are models similar when responding to climate forcing parameters such as increased CO_2? Is the accuracy of ensemble model prediction better than individual model prediction? Does the detail of input data into the model affect how the model responds? The end goals of such efforts are to increase capacity through new adaptation strategies and methods for major agricultural regions in the developing and developed world. A number of crop models such as DSSAT, which is a compilation of crop simulation models founded on soil–plant–atmosphere dynamics for more than 28 crops, are now routinely available. The models simulate crop growth when given soil and meteorological input parameters to help assess the impacts of climate variability and change from farm-level to regional agricultural areas (Fig. 1).

Importance of Agroclimatology

Agroclimatology is important for numerous reasons. Understanding weather and climatic impacts to crops and soils is important in order to feed increasing world populations. It is also important in turn to economic/commodity markets, livestock production (a large majority of livestock feed is developed from corn), and climatic risk assessment. Agroclimatology also becomes important in cost-benefit analysis when considering irrigation, fertilizers, tile drainage systems, cropping systems, and yield.[14]

The U.S.A. is the world's largest producer of corn. Extreme climatic events such as the 2012 drought can greatly influence crop production/yield, drastically impact commodity prices, and impact the agricultural sector of the economy. The 2012 drought across a large portion of the U.S.A. is a recent example of an extreme climatic event

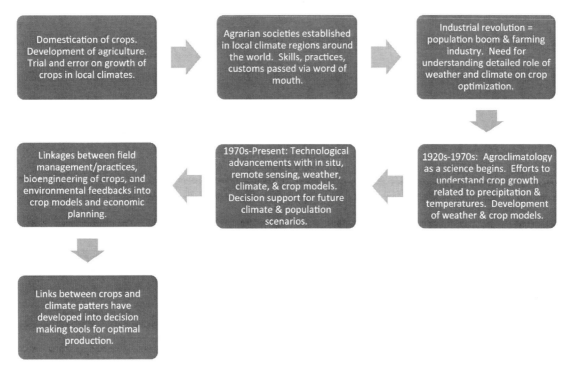

Fig. 1 General narrative chart of the growth of agroclimatology.

that dropped yield estimates of corn to 123.4 bushels per acre, the lowest since 1995. While not immediately felt in food prices by consumers, the effect of low yields can take a year to trickle through production and increase the cost of consumables. Feedlots paid lower prices for feed cattle in 2013 as a result of the higher cost of feed in 2013 caused by reduced availability of pasture and decreased yield from the 2012 drought. The decrease in available pasture results in cattle being fed over a longer term with feed, but at lower weights because of the higher cost of feed in 2013. It is speculated that this will lead to greater production declines by 2014, which will increase cattle prices for almost 2 years after the drought occurred.[15]

Climatic risk assessment determines the risk (i.e., potential) of crop failure or success due to weather and climate variability over a growing season or over several growing seasons. Risk is often determined from the analysis of past growing seasons and potential future weather and/or climate scenarios as computed by crop models and expert projections. Usually, once a climatic risk assessment has been completed, a cost-benefit analysis is undertaken by producers based on the assessment of global vulnerabilities. Decisions include choice of when to plant amidst the balance of late frost risk at the beginning of the season, pest issues during the growing season, and having enough radiation/temperature (degree days) to complete maturity through harvest. Decisions related to crop insurance, fertilizers, and pesticide purchases are agronomically as well as economically important, and weather plays a dominant role in ensuring profitability. Thus, agroclimatology plays an important role in helping manage weather and climate

risks for producers and stakeholders, and in turn better mitigating and adapting to climate change and variability.

AGROCLIMATOLOGY TODAY: SOME IMPORTANT VARIABLES

Agroclimatology in the Twenty-First Century

Agroclimatology today focuses more readily on providing guidance in the form of weather and climate products, aiming to develop resilience in crops to climatic extremes, pests, diseases, and weeds to improve crop growth and production at local and regional scales.[1] Although agroclimatology is specifically focused on studying the impacts of weather/climate patterns on crop production and soils, it is a multidisciplinary field comprised of, but not limited to, knowledge from agronomists, soil scientists, biologists, hydrologists, meteorologists, climatologists, physical geographers, human geographers, sociologists, and economists.[2] Embedded within the agroclimatic notion is the societal need to provide decision support for current agriculture practices and weather catastrophes such as droughts or late frosts.[16]

Agroclimatology is connected to several sub-sciences: soils, hydrology, weather and climate, and agronomy. Different soil types possess distinctive hydraulic and thermodynamic properties that influence the growth rate of plants and the movement of water and nutrients through the soil and on the land surface. Surface and subsurface hydrology are affected not only by the type of soil but the type and amount of vegetation present over that soil. The density of a plant biomass on the surface influences the amount of rainfall reaching soils and the amount of radiation reaching the land surface. Plant roots can also affect infiltration and runoff rates. Weather and climate affect agronomic productivity mainly through precipitation, temperature, and radiation for plant photosynthesis and growth.

Multi-scale coupled soil–vegetation–atmosphere transfer (SVAT) processes regulate the hydrologic, energy, and nutrient transfer balance in agricultural landscapes (Fig. 2). These transfer processes cascade across different scales that ultimately impact crop yield and profitability. Crop transpiration and photosynthesis regulate the nutrients used by the plant via the gross primary production (GPP) and the net primary productivity (NPP) of the agroclimatic system. The NPP, or yield, is linked with transpiration through the plant canopy and evaporation from the soil surface. The soil surface is the fundamental level where the carbon/nutrient and water link is established, but can be scaled beyond the leaf, plant, and to larger regional scales of total biomass/vegetation in a field. Thus, a detailed understanding of SVAT processes often becomes essential in providing agroclimatic guidance (Table 1).[17]

Changes in regional agricultural crop cover and greenness/phenology lead to changes in regional dew points and temperatures from the evapotranspiration from plants and/or irrigation. The spatial extent and greenness of crop cover can result in increased moisture in the atmosphere and lowering of the surface air temperature.[18,19] This in turn can lead to changes in regional convective potential, cloud cover, and in some cases rainfall (Fig. 3). Understanding these linkages is difficult, yet this is where agroclimatological assessments tend to help by providing a framework and understanding how a particular region would be expected to behave in a statistical/climatological sense. The fundamentals of such statistical/climatological relationships lie in the understanding of variables that are important for agricultural principles. Some of the climatic variables that are analyzed include temperature (maximum, minimum, average), rainfall characteristics (distribution, intensity), and solar radiation. Additional variables such as sunshine hours, humidity, winds, soil temperature, soil moisture, and evaporation are also needed but are generally difficult to measure with high spatiotemporal resolution or fidelity and are estimated from different models.[20,21]

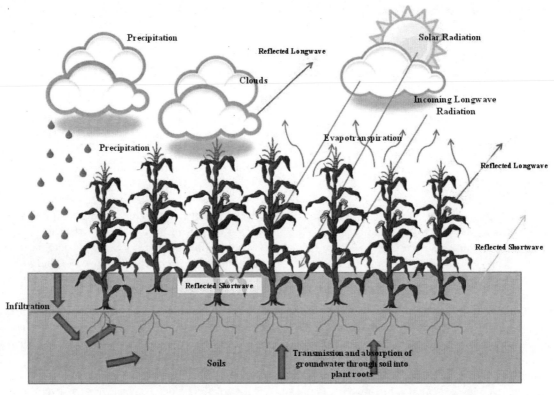

Fig. 2 The soil–vegetation–atmosphere–transfer model (SVAT) represents the continuous feedback of radiative energy between the atmosphere, biosphere (crop), lithosphere (soil/land surface), and water in all its physical states: solid–ice crystals of clouds; liquid–raindrops; and gas–water vapor/evapotranspiration. Primary components and associated parameters include the following: 1) soil—water retention, soil hydraulic conductivity, bulk density, water content, water potential, temperatures, infiltration, and evaporation; 2) canopy (vegetation/biomass)—leaf area index (LAI), physiology of the plant, leaf and plant density, seasonality, and evapotranspiration; 3) atmosphere—air temperature, humidity, wind speed, and solar radiation; and 4) hydrological cycle—evaporation, condensation, precipitation, infiltration, groundwater flow, surface runoff, and ponding.

Table 1 Common soil and vegetation parameters in land surface models embedded within some crop models and weather/climate models

Primary parameters: Dominant types of vegetation and/or land cover (USGS/NLCD classifications)	Secondary parameters (estimated or prescribed)
1) Vegetation type: cultivated or hay/pasture land (more detailed in crop models), forest type: deciduous, evergreen, mixed, shrubs	Saturated volumetric moisture content
2) Dominant type of soil texture (USDA textural classification):	Wilting point volumetric water content
Sand	Soil thermal coefficient at saturation
Loamy sand	Depth of the soil column
Sandy loam	Fraction of vegetation
Loam	Minimum surface resistance
Sandy clay loam	Leaf area index (LAI)
Silty clay loam	Roughness length
Clay loam	Albedo
Sandy clay	Emissivity
Silty clay	Soil thermal conductivity
Clay	Soil thermal diffusivity
3) Meteorological parameters:	Soil hydraulic conductivity
Daily T_{max} and T_{min}	Soil hydraulic diffusivity
Net solar radiation	Soil heat capacity
Precipitation	Soil bulk density
Relative humidity	Moisture flux resistance
Evapotranspiration	
Soil moisture	

Source: Adapted from Noilhan & Planton.[21]

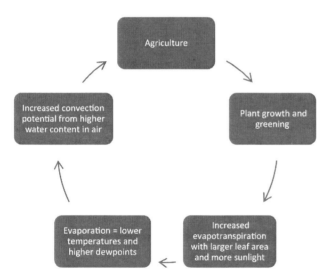

Fig. 3 Examples of agricultural impacts on physical climate.

Primary Agroclimatological Parameters

Weather and climate

Weather can be defined by daily precipitation, temperature, and other dynamic (wind) and thermodynamic (humidity) weather patterns. Weather characterization ranges from temporal and spatial scales of seconds to decades and millimeters to thousands of kilometers. The most common temporal and spatial scales for daily weather are microscale (areas less than 1 km, seconds to minutes), mesoscale (1–100 km, several hours to a day), synoptic scale (100–1000 s km, days to a week), and planetary scale (1000 s km and lasts up to several weeks).[22,23] The average weather and its variability over a period of time delineate a specific area's climate.[10] The climate of a region is typically defined by the average and variance of temperature and precipitation, which is influenced by topography, proximity to water bodies, size of the given landmass, or any geographical feature including urban areas. Mountain ranges can lead to regional climate zones with higher precipitation climates on the windward side, while more arid and temperate climates are found downwind.[10,24] Coastal climates are moderated daily and seasonally by ocean waters and currents due to the higher specific heat capacity of water and resulting land/sea breezes.[10,22] Urban climates impact local temperatures, wind patterns, and boundary layer depths because of differences in surface layer energy and heat balances. A lack of vegetation and prolific extent of impervious, highly reflective, and/or highly absorbent land-cover types generate this type of localized climate.[22,23,25]

The average (and variation) over 30 years of recorded weather patterns such as daily rainfall, high and low temperature, and mean temperature is generally used to define a region's (climate zone, state, or multi-state region) set of climate "normals."[26] These climate normals are used extensively in agriculture to determine the baseline and the ensuing anomaly for any given season. In the U.S.A., climate divisions (typically nine per state) are determined to help the broader applications community be aware of temperature and precipitation patterns/and shifts at regional and state scales. Sometimes, additional information on frost depth, soil temperatures, and average soil moisture are available through the 30-year climatological summaries (Fig. 4). For many regions within the U.S.A., the climate divisions are also aligned with the United States

Department of Agriculture's crop reporting districts and assist climate–crop yield assessments.

Soils

The basic soil classifications are based on particle size. Soils are typically classified as gravel (greater than 2.00 mm), sand (0.05–2.00 mm), silt (0.002–0.05 mm), or

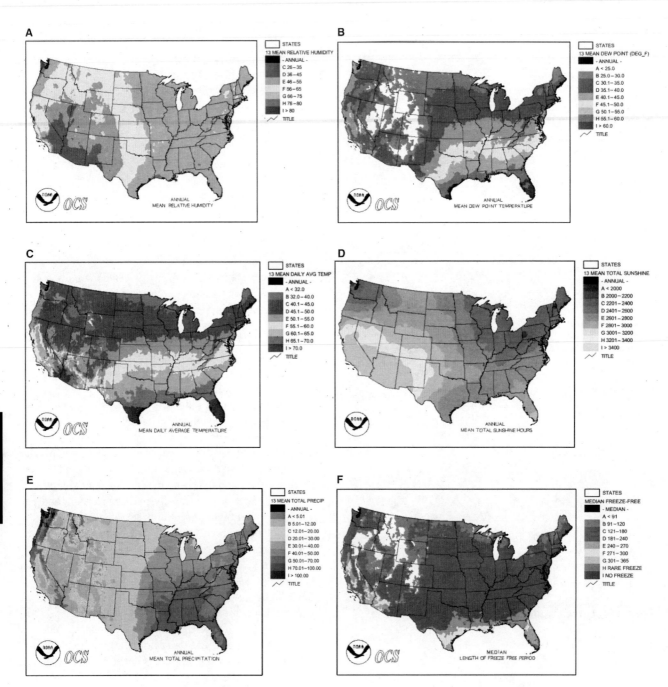

Fig. 4 Examples of climatology maps collected from the National Climatic Data Center's Climate Maps of the United States database. Maps are developed from the 1961 to 1990 period of record from official weather and climate station sites unless otherwise noted. (**A**) Annual mean relative humidity in percent, (**B**) annual mean dew point temperature in degrees Fahrenheit, (**C**) annual mean daily average temperature in degrees Fahrenheit, (**D**) annual mean total sunshine hours, (**E**) annual mean total precipitation in inches, and (**F**) median length of freeze-free period in days.

clay (less than 0.002 mm). The amount of quartz and carbon content of a given soil type is also included as a characteristic in some models to clarify the surface layer soil type from the underlying soil layers. Each basic soil type has specific thermal and hydraulic characteristics that influence the temperature and moisture content of that specific soil. Soil heat capacity (amount of heat required to produce a given change of the temperature of the body that is largely influenced by the presence of water), thermal conductivity (how fast heat is transferred through soil), and hydraulic conductivity (how fast water is transferred through soil) are just a few of the many parameters that govern how suitable a subsurface climate may be for a specific crop.[27]

The land-surface (soil and vegetative surface) response is governed by solar radiation. Shortwave (SW) radiation is absorbed and continually reemitted by the land surface, with peak absorption and emission during the afternoon, and a dominant outward flux of longwave (LW) radiation in the form of sensible and ground heat flux back to space in the night.[20,22,23] Land surface radiation absorption or reflection is characterized by a ratio of absorbed to emitted SW radiation called albedo (α). Typical values of albedo are 0.05–0.40 for dark and wet to light and dry soils, 0.18–0.25 for agricultural crops, 0.15–0.20 for deciduous forests, 0.05–0.15 for coniferous forests, 0.0–0.10 (small zenith angle) to 0.10–1.0 (large zenith angle) for water, and 0.40 (old) to 0.95 (fresh) for snow.[23]

The values of incident net radiation (NR) are quantified through the radiation balance equation, which is the sum of incoming shortwave radiation (SW↓), outgoing longwave radiation (LW↑), sensible heat (H) (thermal heat transfer from the ground to overlying air via conduction and convection), latent heat (E) (heat released into the air as water vapor condenses back into water), and ground heat (G) (thermal heat energy transferred from the ground).[20,22,23] All together, these variables help quantify the surface radiation budget in relation to surface albedo that is primarily governed by surface characteristics:

$$SW\downarrow (1-\alpha) + LW\uparrow = NR$$

$$H + E + G = NR$$

Data to understand energy fluxes is typically available from research field sites (e.g., Fluxnet, fluxnet.ornl.gov[28]). Data availability is limited because specialized measurements of surface sensible heat flux, latent heat flux, ground heat flux, carbon dioxide flux, moisture flux, evaporation/transpiration, surface temperature, soil moisture, and soil temperatures are challenging to obtain and require sophisticated equipment that needs ample care and maintenance. These data are typically used to develop and test newer plant-yield, environmental response relationships. Crop modelers develop empirical equations that can be applied in models and regional-scale analysis. In recent years, satellite platforms such as the NASA Moderate Resolution Imaging Spectroradiometer (MODIS) aboard Terra and Aqua, Atmospheric Infrared Sounder (AIRS), and the Advanced Microwave Scanning Radiometer-Earth Observing System (AMSR-E) have also greatly contributed to the understanding of the Earth's surface and atmospheric radiation balances.[29–31] Near real-time analysis of leaf area index (LAI), normalized difference vegetation index (NDVI), enhanced vegetation index, evapotranspiration, surface temperature, and moisture stress index are some of the measurements these satellite platforms provide for research and application into crop models, along with near real-time information pertinent to irrigation and harvest.[32] Surface flux data along with satellite data has been assimilated into weather forecast, climate, and crop growth models, achieving a higher level of understanding of weather/climate and crop production forecasts.[33,34]

Water

Agroclimatic decisions are acutely linked to hydroclimatology. Surface and ground water are readily coupled with the Earth's surface because they impact crop growth and evapotranspiration into the overlying atmosphere (which determines surface moisture flux). The moisture flux of the land surface is controlled by the saturation and temperature of the overlying air, wind speeds, turbulent eddies, and intensity of sunlight reaching the surface in the form of SW radiation.[8,11,20–22,35] Radiation reaching the Earth's soil surface contributes to the evaporation rate of water from the soil and is further impacted by vegetation. Plant physiology plays a large role in surface moisture flux: stomata (microscopic pores or openings on the leaves of plants) opens and closes based on environmental conditions such as carbon dioxide availability, soil moisture, atmospheric saturation, wind, and sunlight, releasing moisture into the atmosphere along with water vapor and oxygen exchange.[20,22,35] Moisture flux is a quantitative measure of soil moisture that moves up through the soil, through the plant, and evaporates out of the soil and transpires from the plant.[22,23,27,35] Soil moisture excess or deficits in the form of floods or droughts (along with temperature) have a direct impact on plant health and yield. The timing of water stress is also important in determining the crop response to the impact of stress. Crops 2–3 weeks after planting are typically more vulnerable than 2 months into the growing season.

Crop management (vegetation/biomass)

Climate variables such as base temperature, growing degree days (GDDs), average frost days, first frost day, and last frost day help determine crop progress. Temperature becomes an important driver of crop development and is typically used as a resource available to the plant. This is quantified as GDDs, which are used to determine phenology such as when the seeds of row crops will germinate and grow in conjunction with determining pest development rates within the soil:

$$GDD = ((T_{max} - T_{min})/2) - T_{base}$$

The base temperature for corn is commonly 10°C, and typically after 10 consecutive GDDs (base time for corn), seeds germinate and grow.[20,35] Base temperature will vary by crop.

The last spring frost consideration is also crucial to seed germination and plant growth. If a seed is planted before the last frost, farmers risk plant loss because the seed will fail to germinate. As a result, climatologies of important crop management variables have been developed: 1) the first and last frost dates (e.g., in the Midwest U.S.A. in states such as Indiana, temperatures below 32°F and 28°F are used as baseline temperatures); 2) snowfall amounts; 3) temperatures above 95°F; 4) rainfall; 5) evapotranspiration loss; etc. Climatologies also exist for high-impact weather events such as high winds that can impact spray operations, hail that can cause crop losses, and heavy rain that can result in erosion and/or flood and surface ponding in fields. Developing a climatology of a specific variable includes taking the actual values and anomalies and averaging them with quality control procedures such as duration of measurement, time of day the measurement is observed, and site-specific location information such as proximity to trees, buildings, concrete, and elevation to develop the desired climatology. These climatologies are taken either at a station level (like the U.S. Cooperative Observer Network) if observations exist, or at climate division levels over a 30-year period (e.g., 1981–2010). Additional weather networks exist across different spatial and temporal scales that contribute to the development of climatologies and provide real-time weather data to producers and forecasters. Many states have what are called "mesonetworks" that comprise weather observation stations in counties across a given state. These stations can provide hourly readings of temperature, precipitation, winds, evapotranspiration estimates, and pressure in addition to daily maximum and minimum temperatures, and 24-hour rainfall/precipitation totals. In the U.S.A., these data are typically available from respective state climate offices.

Extreme climatic events such as floods or droughts during the 1930s, 1950s, 1980s, and most recently 2012 contribute greatly to reduced yield, crop profitability, and world market stability. Forecasting climate extremes such as floods and drought still remains highly uncertain despite advancements in weather and climate modeling. Global features such as the El Niño Southern Oscillation (ENSO) are a large determining factor in North American weather regimes.[10] When compounded across seasons or even years, ENSO-related weather patterns can manifest and lead to anomalous events such as the 2012 drought.[36]

Current agroclimatic research is aligned toward characterizing droughts and their impacts on crop yields, including linking the interplay between agriculture–droughts–economic decisions and impacts. Drought is more than a lack of rainfall, and a number of indices have been applied to assess it. Drought impacts are characterized on the basis of timing (e.g., middle of the growing season well after planting), intensity, duration, spatial extent, and location (e.g., urban versus rural regions). Drought indices include basic analysis such as percentage of normal and quantitative meteorological measures, e.g., the Standardized Precipitation Index (SPI) and Palmer Drought Severity Index (PDSI).[37]

Development of climatologies include weather and climate variables such as evapotranspiration, annual precipitation, total sunshine hours, plant hardiness zones, and mean freeze-free period for agricultural applications (Fig. 4). These climatologies provide a baseline value to compare to current weather and climate patterns, letting agricultural producers and stakeholders know if seasonal weather patterns are trending above or below normal. This could allow producers to better prepare for a change in their projected yield for the season, and hopefully mitigate loss if sufficient time is still present to act accordingly. For example, the Useful to Usable (U2U) climate information project sponsored by the U.S. Department of Agriculture (USDA) is developing an agricultural climatology for the U.S. Corn Belt region that includes the ENSO phase at the climate division level for cereal crop producers.[38] This climatology can provide guidance to farmers related to irrigation, fertilizer application, planting, and harvest-reducing crop vulnerability to changing weather patterns.[14]

Agroclimatology in the Future

An important question going forward is how can societies, economies, and countries sustain and continue to expand food supplies for a continually growing world population. This problem is further compounded with shifts in climate patterns, particularly temperature and precipitation across the world.[1,2,39,40] Agronomic decisions and productivity is thus linked to environmental and economic sustainability. Climate projections indicate a high probability of the subtropics becoming drier while the midlatitudes will have continued shifts in temperature, rainfall, cloud cover, and related climatic patterns.[41] The agricultural community has been working with adaptation approaches, which at a macroscale will need to address the current distribution of arable land for agriculture and the increased demand for more hybrid crops to withstand wetter/drier climates. New soil diseases can pose a threat in regions experiencing changes in seasonal temperatures as growing seasons become longer and the time a field lays fallow shortens.[42,43] Understanding these climate–pathogen relationships in a changing climate regime will likely become increasingly important.

In addition to the demand for hardier crops to surpass the challenges of a changing climate, the demand for biofuels will likely continue to rise as governments continue to mandate the reduction of greenhouse gas emissions. Current research in this area includes determining the best crop rotation cycles to protect soil nutrients while also maximizing the use of crops and crop waste such as corn and corn stalks, oil seeds, crop residue, and woody biomass for the production of bioenergy and ethanol. However, growing crops for bioenergy and ethanol instead of food for populations leads to societal questions that are also at the forefront of the climate change debate. In essence, future challenges facing agroclimatology include better decision-making tools, data acquisition, availability and uniformity (spatial and temporal), links between economic and regional decision makers, the development of more accurate and detailed crop and forecast models, and a more detailed understanding of climate change. These challenges require collaboration between different disciplines.

CONCLUSION

Agroclimatology is at an interesting juncture, becoming central to today's most challenging questions in a world of continued population growth, increasing food demand, and climate change. With technological advancements such as multiscale remote-sensing data, vegetation and moisture stress mapping, land data assimilation systems linked with crop models, and high resolution crop, weather and climate datasets, models will likely play an important role in the understanding and evolution of agroclimatology. Weather and climate data collection, assessment and mitigation of extremes such as floods and droughts, and climate and crop modeling will continue into the future as agroclimatology grows as a science. In addition, the feedback of agricultural climates at field and regional scales needs to be further researched to grasp the full understanding of heat and moisture fluxes between soils, the overlying boundary layer, vegetation, and the feedback into crop yields.

Despite large technological advancements in data collection and analysis, and a more detailed understanding of the agroclimatic system than ever before, questions regarding the future vulnerability of agro-climate systems still remain unanswered. Collaborative research and efforts to mitigate the effects of climate change need to be aggressively supported and addressed to allow for continued food production, development of bioenergies, and human survival in a world currently experiencing climate variability and change.[2] Changes in temperature and precipitation as projected using climate change studies will impact crops and non-crop species. As summarized by a recent USDA synthesis report,[43] the projected variability in precipitation and location shifts will require changes in water management practices (which will further feedback into the local climate system). Projected changes in temperature indicate the northward advance of frost-free days, opening the doors to additional regions for crop growth but also creating an environment that would be potentially conducive for invasive weeds and pests.[43]

There are at least two enduring challenges that impact current monitoring and modeling efforts in agroclimate. The first is due to scale disparity: 1) field measurements are often "point" data while effects are often regional scale in nature and not well captured; and 2) if remotely sensed/satellite data are used for assessing the regional view, the dominant impact of local-scale decisions and micro features that can influence crop yields are not captured. Combination approaches involving the assimilation of multiscale products into a gridded assessment are currently underway and may likely alleviate the uncertainty due to this problem.[44,45] The second challenge can be linked to capturing the diversity in agronomic practices. For example, most studies consider the relationship between climate and crop patterns for a "typical" crop and the variations between hybrids are poorly assessed. Similarly, variability in farm-scale practices such as planting date, presence of tilling, crop cover, distances between crop rows, fertilizer use, and pest risk are also poorly captured and cause uncertainty in current assessments. Additional challenges due to socioeconomic choices and economic tradeoffs are also difficult to capture.

GLOSSARY

Anomaly: The deviation of a measurable unit, (e.g., temperature or precipitation) in a given region over a specified period from the long-term average, often the 30-year mean, for the same region.[46]

Base temperature: The temperature below which plant growth is zero. The base temperature varies by crop.[47]

Biochemistry (biochemical): The study of chemical processes within (or pertaining to the physical processes of) and related to living organisms that explain the processes of life such as cell biology and signaling, development and disease, energy and metabolism, genetics, molecular biology, and plant biology.[48]

Bioenergy: Renewable energy developed from biological sources such as plant material to be used for heat, electricity, or fuel for vehicles.[49]

Biogeochemistry (biogeochemical): The study of the physical (or pertaining to the physical processes of), chemical, biological, and geological processes and reactions that govern the composition of and changes to the natural environment. It mainly consists on studying the cycles and interaction of elements carbon, nitrogen, sulfur, and phosphorous as they move through the atmosphere, hydrosphere, and lithosphere.[50]

Boundary layer: In general, a layer of air adjacent to a bounding surface. Specifically, the term most often refers to the **planetary boundary layer**, which is the layer within which the effects of friction are significant. For the Earth, this layer is considered to be roughly the lowest one or two kilometers of the atmosphere. It is within this layer that atmospheric turbulence dominates the exchanges, and temperatures are most strongly affected by daytime insolation and nighttime radiational cooling, and winds are affected by friction with the Earth's surface. The effects of friction die out gradually with increasing height, so the "top" of this layer cannot be exactly defined.[51]

Climate: The average weather over a 30-year period.[46]

Climate change: A non-random change in climate that is measured over several decades or longer. The change may be due to natural or human-induced causes.[46]

Climate system: The system consisting of the atmosphere (gases), hydrosphere (water), lithosphere (solid rocky part of the Earth), and biosphere (living) that determine the Earth's climate.[46]

Climate variability: Variation in a climatic parameter such as temperature or precipitation that varies from its long-term mean over a shorter time scale than climate change. Variability occurs on times scales of daily, seasonal, annual, inter-annual, to several years and is inclusive of changes observed with teleconnections such as the El Niño Southern Oscillation.[52]

Climatologies: A quantitative description of climate showing the characteristic values of climate variables over a specific region.[46]

Climatology: The description and scientific study of climate.[46]

Condensation: The physical process by which water vapor in the atmosphere changes to liquid in the form of dew, fog, or cloud; the opposite of evaporation.[46]

Convection: Transfer of heat by fluid motion between two areas with different temperatures. In meteorology, the rising and descending air motion is caused by differential heat or density/pressure. Atmospheric convection is almost always turbulent and is the dominant vertical transport process over tropical oceans and during sunny days over continents. The terms "convection" and "thunderstorms" are often used interchangeably, although thunderstorms are only one form of convection. In the ocean, convection is prominent in regions of high heat loss to the atmosphere and is the main mechanism for deep water formation.[46]

Degree day: For any individual day, degree days indicate how far that day's average temperature departed from 65°F. Heating degree days (HDD) measure heating energy demand. It is a measure to indicate how far the average temperature fell below 65°F. Similarly, cooling degree days (CDD), which measure cooling energy demand, indicate how far the temperature averaged above 65°F. In both cases, smaller values represent less fuel demand, but values below 0 are set equal to 0, because energy demand cannot be negative. Furthermore, since energy demand is cumulative, degree day totals for periods exceeding 1 day are simply the sum of each individual day's degree day total.

For example, if some location had a mean temperature of 60°F on day 1 and 80°F on day 2, there would be 5 HDDs for day 1 (65 minus 60) and 0 for day 2 (65 minus 80, set to 0). For the day 1 + day 2 period, the HDD total would be 5 + 0 = 5. In contrast, there would be 0 CDDs for day 1 (60 minus 65, reset to 0), 15 CDDs for day 2 (80 minus 65), resulting in a 2-day CDD total of 0 + 15 = 15.[46]

Dew point: The point at which the air at a certain temperature contains all the moisture possible without precipitation occurring. When the dew point is 65°F, one typically begins to feel the humidity. The higher the temperature associated with the dew point, the more uncomfortable one feels.[46]

Drought: Drought is a deficiency of moisture that results in adverse impacts on people, animals, or vegetation over a sizeable area. NOAA and USDA together with its partners provide short- and long-term drought assessments through the U.S. Drought Monitor.[46]

Eddy: Swirling currents of air at variance with the main current.[51]

Enhanced vegetation index (EVI): A satellite product designed to monitor the health of vegetation. It is considered to be an improvement upon NDVI (see below) correcting for distortions in reflected light caused by particulate matter and corrects for ground cover below vegetation. EVI is calculated in a similar manner to NDVI.[53]

Ensemble forecast: Multiple predictions from an ensemble of slightly different initial conditions and/or various versions of models. The objective is to improve the accuracy of the forecast through averaging the various forecasts, which eliminates non-predictable components, and to provide reliable information on forecast uncertainties from the diversity amongst ensemble members. Forecasters use this tool to measure the fidelity of a forecast.[46]

ENSO (El Niño-Southern Oscillation): Originally, ENSO referred to El Niño/Southern Oscillation, or the combined atmosphere/ocean system during an El Niño warm event. The ENSO cycle includes La Niña and El Niño phases as well as neutral phases, or ENSO cycle, of the coupled atmosphere/ocean system though sometimes it is still used as originally defined. The Southern Oscillation is quantified by the Southern Oscillation Index (SOI).[46]

Evaporation: The physical process by which a liquid or solid is changed to a gas; the opposite of condensation.[46]

Evapotranspiration: Combination of evaporation from free water surfaces and transpiration of water from plant surfaces to the atmosphere.[51]

Extratropical: In meteorology, the area north of the Tropic of Cancer and the area south of the Tropic of Capricorn. In other words, the area outside the tropics.[46]

Flux: The rate of transfer of fluids, particles, or energy per unit area across a given surface (amount of flow per unit of time).[51]

Gridded assessment: The process of spatially interpolating observed weather and/or climate data across spatial grids or a specified resolution for the development of a continuous data set for analysis.

Gross primary production: The total amount of energy or nutrients created by an ecological unit/organism in a given length of time. Some of this energy is used for cellular respiration and maintenance of the organism tissue. The remaining energy not used for this process is **net primary production**.[54]

Gross primary productivity: The rate at which photosynthesis or chemosynthesis (gross primary production) occurs in an organism/ecosystem.[54]

Growing degree days: A heat index that relates the development of plants, insects, and disease organisms to environmental air temperature. The index varies depending on whether it is a cool, warm, or very warm season plant. For example, a corn growing degree day (GDD) is an index used to express crop maturity. The index is computed by subtracting a base temperature of 50°F from the average of the maximum and minimum temperatures for the day. Minimum temperatures less than 50°F are set to 50, and maximum temperatures greater than 86°F are set to 86. These substitutions indicate that no appreciable growth is detected with temperatures lower than 50 or greater than 86. If the maximum and minimum temperatures were 85 and 52, you would calculate the GDD by $((85 + 52/2) - 50) = 18.5$ GDD.[46]

Hardiness zone: Eleven zones in the U.S.A. based on average annual minimum temperatures. Zones are used by horticulturalists and nurseries to characterize plants by their hardiness; the hardiness zone maps can then be used to determine the likely survivability of particular plant species and varieties according to one's local growing area.[47]

Holdridge life zone system: A classification of ecosystems/climate zones developed by American botanist, Leslie Holdridge. The life zone system is based on the principle understanding that climate and plants have an intertwined relationship. This model relates climates and plants based on three properties: temperature, humidity, and precipitation. The classification of a life zone is determined though a triangular diagram that quantifies annual precipitation on the right side, potential evapotranspiration ratios on the left side, and humidity provinces along the base of the triangle. Tiers of the triangle designate latitudinal regions and altitudinal belts.[56]

Hybrid: A heterozygous (different alleles that code for the same gene or trait) offspring of two genetically distinct parents. An example would be a mule, offspring of a female horse and a male donkey. Sweet corn varieties are hybrids selectively bred for color(s) and sweetness.[55]

Insolation hours measure the duration of direct solar radiation for a given location on the Earth.[60]

Invasive species: Non-native or exotic species of plants, animals, and pests whose introduction to an ecosystem causes or is likely to cause economic or environmental harm or harm to human health.[47]

Köppen climate system: A widely used climate classification system developed by Vladimir Köppen. The Köppen classification system is based on a subdivision of terrestrial climates into five major types with four climates defined by temperature criteria and one type defined by dryness/aridity. Although Köppen's classification did not consider highland climate regions, the highland climate category, or H climate, is sometimes added to climate classification systems to account for elevations above 1,500 m (about 4,900 ft).[57]

Leaf area index (LAI): A dimensionless value that is used to characterize plant canopies and is defined as the one-sided green leaf area per unit ground surface area. It ranges from 0 to 10, with lower values representing less dense plant canopies and higher values more dense canopies. It is used as an eco-physiological measure of the photosynthetic and transpirational surface within a canopy, and/or as a remote-sensing measure of the leaf reflective surface within a canopy.[58]

Moisture/water stress: Most commonly expressed via discoloration and wilting of the plant due to the plant having to use more energy to remove water from the soil. The amount of force needed for a plant to remove water from the soil is known as the matric potential. When soil moisture is low, plants have to use more energy to remove water from the soil, thus the matric potential is greater. When the soil is dry and the matric potential is strong, plants show symptoms of stress.[61]

Normalized Difference Vegetation Index (NDVI): A satellite product that shows the health of vegetation. It is a ratio of the difference between near-infrared radiation and visible radiation detected by the satellite sensor to the sum of near-infrared radiation and visible radiation detected by the satellite sensor. Healthy vegetation absorbs most of the visible spectrum and reflects a large portion of the near-infrared spectrum. Unhealthy or sparse vegetation reflects more visible radiation and less near-infrared radiation. NDVI values range from −1 to 1. A zero means no vegetation and a positive value close to 1 indicates a high density of green leaves/healthy vegetation.[53]

Palmer Drought Severity Index (PDSI): An index that compares the actual amount of precipitation received in an area during a specified period with the normal or average amount expected during that same period. It was developed to measure the lack of moisture over a relatively long period of time and is based on the supply and demand concept of a water balance equation. Included in the equation are evaporation, soil recharge, runoff, temperature, and precipitation data.[46]

Phenotype: The outward appearance of the individual or organism. The phenotype results from interactions between genes, and between the genotype and the environment.[55]

Relative humidity: An estimate of the amount of moisture in the air relative to the amount of moisture that the air can hold at a specific temperature. For example, if it is 70°F near dawn on a foggy summer morning, the relative humidity is near 100%. In the afternoon, the temperature soars to 95°F and the fog disappears. The moisture in the atmosphere has not changed appreciably, but the relative humidity drops to 44% because the air has the capacity to hold much more moisture at a temperature of 95°F than it does at 70°F. But even when the relative humidity is "low" at 44%, it is a very humid day when the temperature is 95°F. For this reason, a better measure of comfort is dew point.[46]

Remote sensing: The science/process of collecting data through the detection of energy by sensors that is reflected from earth. Sensors are commonly onboard satellites or attached to aircraft. The sensors are not in the immediate environment where data is being collected. Sensors are either active or passive. Passive sensors detect only reflected radiation. Active sensors emit energy that is reflected back to the sensor to detect features on the Earth.[59]

Spectral instrument: An in situ (within the immediate environment) or remotely placed sensor that detects and records electromagnetic radiation from the sun that is reflected, transmitted, or absorbed (and subsequently emitted) from a surface or element. All surfaces or elements (such as trees, crops, soils, etc.) have a distinct spectral signature based on wavelength(s).

Standardized Precipitation Index (SPI): A probability-based way of measuring drought as a measure of precipitation across different time scales. This index is negative for drought and positive for wet conditions.[62]

Subtropical: A climate zone adjacent to the tropics with warm temperatures and little rainfall.[46]

Sunshine hours: A climatological indicator of cloudiness—measures the duration of sunshine in a given period (commonly a day or year) for a given location on the Earth.

Transpiration: Water discharged into the atmosphere from plant surfaces.[51]

Tropics: Areas of the Earth within 20° north and south of the equator.[51]

ACKNOWLEDGMENTS

NIFA/USDA—Useful to Usable (U2U): Transforming Climate Variability and Change Information for Crop Producers: Agriculture and Food Research Initiative Competitive Grant no. 2011-68002-30220, NSF INTEROP dri-NET, USDA NIFA Drought Trigger projects at Purdue through Texas A&M, and NSF COSIEN project at Purdue through UC Berkeley.

REFERENCES

1. Decker, W.L. Developments in agricultural meteorology as a guide to its potential for the twenty-first century. Agric. Forest Meteorol. **1994**, *69* (1), 9–25.

2. Steiner, J.L.; Hatfield, J.L. Winds of change: A century of agroclimate research. Agron. J. **2008**, *100* (Suppl. 3), S-132–S-152. doi: 10.2134/agronj2006.0372c.

3. Pielke, R.A., Sr.; Pitman, A.; Niyogi, D.; Mahmood, R.; McAlpine, C.; Hossain, F.; Goldewijk, K.K.; Nair, U.; Betts, R.; Fall, S.; Reichstein, M.; Kabat, P.; de Noblet-Ducoudré, N. Land use/land cover changes and climate: Modeling analysis and observational evidence. WIREs Clim. Change **2011**, *2* (6), 828–850. doi: 10.1002/wcc.144.

4. Dirmeyer, P.A.; Niyogi, D.; de Noblet-Ducoudré, N.; Dickinson, R.E.; Snyder, P.K. Impacts of land use change on climate. Int. J. Climatol. **2010**, *30* (13), 1905–1907. doi: 10.1002/joc.2157.

5. Niyogi, D.; Kishtawal, C.; Tripathi, S.; Govindaraju, R.S. Observational evidence that agricultural intensification and land use change may be reducing the Indian summer monsoon rainfall. Water Resour. Res. **2010**, *46* (3), W03533. doi: 10.1029/2008WR007082.

6. Mishra, V.; Cherkauer, K.A.; Niyogi, D.; Lei, M.; Pijanowski, B.C.; Ray, D.K.; Bowling, L.C.; Yang, G. A regional scale assessment of land use/land cover and climatic changes on water and energy cycle in the upper midwest United States. Int. J. Climatol. **2010**, *30* (13), 2025–2044. doi: 10.1002/joc.2095.

7. Monteith, J.L. Agricultural meteorology: Evolution and application. Agric. Forest Meteorol. **2000**, *103* (1), 5–9.

8. Diamond, J.M. *Collapse: How Societies Choose to Fail or Succeed*; The Viking Press (Penguin Group USA): New York, 2004; 592 pp.

9. Cox, J.D. *Storm Watchers: The Turbulent History of Weather Prediction from Franklin's Kite to El Nino*; Wiley & Sons: Hoboken, 2002; 252 pp.

10. Robinson, P.J.; Henderson-Sellers, A. *Contemporary Climatology*; Pearson Prentice Hall: Essex, 1999; 317 pp.

11. Lenschow, D.H. Boundary layer processes and flux measurements. In *Handbook of Weather, Climate, and Water: Atmospheric Chemistry, Hydrology, and Societal Impacts*; John Wiley & Sons, Inc.: Hoboken, 2003; 966 pp.

12. Rosenzweig, C.; Jones, J.W.; Hatfield, J.L.; Mutter, C.Z.; Adiku, S.G.K.; Ahmad, A.; Beletse, Y.; Gangwar, B.; Guntuku, D.; Kihara, J.; Masikati, P.; Paramasivan, P.; Rao, K.P.C.; Zubair, L. The Agricultural Model Intercomparison and Improvement Project (AgMIP): Integrated regional assessment projects. In *Handbook of Climate Change and Agroecosystems: Global and Regional Aspects and Implications*; Hillel, D.; Rosenzweig, C., Ed.; Imperial College Press, 2012; Vol. 2, 263–280.

13. Jones, J.W.; Hoogenboom, G.; Porter, C.H.; Boote, K.J.; Batchelor, W.D.; Hunt, L.A.; Wilkens, P.W.; Singh, U.; Gijsman, A.J.; Ritchie, J.T. DSSAT cropping system model. Eur. J. Agron. **2003**, *18* (3), 235–265.

14. Takle, E.S.; Anderson, C.J.; Anderson, J.; Angel, J.; Elmore, R.; Gramig, B.M.; Guinan, P.; Hilberg, S.; Kluck, D.; Massey, R.; Niyogi, D.; Schneider, J.; Shulski, M.; Todey, D.; Widhalm, M. Climate forecasts for corn producer decision-making. Earth Interact. **2013**, doi: 10.1175/2013EI000541.1, *in press*.

15. Crutchfield, S. U.S. Drought 2012: Farm and Food Impacts. United States Department of Agriculture, Economic Research Service, http://www.ers.usda.gov/topics/in-the-news/us-drought-2012-farm-and-food-impacts.aspx#.UnEd8HDqhyI (Last updated on 26 July, 2013).

16. Prokopy, L.S.; Haigh, T.; Mase, A.S.; Angel, J.; Hart, C.; Knutson, C.; Lemus, M.C.; Lo, Y.; McGuire, J.; Morton, L.W.; Perron, J.; Todey, D.; Widhalm, M. Agricultural advisors: A receptive audience for weather and climate information? Weather Clim. Soc. **2013**, *5* (2), 162–167.

17. Campbell, G.S.; Norman, J.M. *An Introduction to Environmental Biophysics*, 2ⁿᵈ Ed.; Springer Science + Business Media: New York, 1998; 286 pp.

18. Fall, S.; Niyogi, D.; Gluhovsky, A.; Pielke Sr., R.A.; Kalnay, E.; Rochon, G. Impacts of land use land cover on temperature trends over the continental United States: Assessment using the North American regional reanalysis. Int. J. Climatol. **2010**, *30* (13), 1980–1993. doi: 10.1002/joc.1996.

19. Niyogi, D.; Mahmood, R.; Adegoke, J.O. Land-use/land-cover change and its impacts on weather and climate. Boundary Layer Meteorol. **2009**, *133* (3), 297–298. doi: 10.1007/s10546-009-9437-8 (Editorial).

20. Griffiths, J.F., Ed. *Handbook of Agricultural Meteorology*; Oxford University Press: New York, 1994; 320 pp.

21. Noilhan, J.; Planton, S. A simple parameterization of land surface processes for meteorological models. Mon. Weather Rev. **1989**, *117* (3), 536–549.

22. Stull, R.B. *An Introduction to Boundary Layer Meteorology*; Springer LLC: New York, 1988; 683 pp.

23. Oke, T.R. *Boundary Layer Climates*, 2ⁿᵈ Ed.; Routledge: New York, 1988; 464 pp.

24. Holton, J.R. *An Introduction to Dynamic Meteorology*, 4ᵗʰ Ed.; Elsevier Academic Press: Waltham, 2004; 535 pp.

25. Landsberg, H.E. *The Urban Climate*; Academic Press: New York, 1981; 275 pp.

26. Arguez, A. NOAA's 1980–2010 Climate Normals. 14 February 2013, http://lwf.ncdc.noaa.gov/oa/climate/normals/usnormals.html, 2013.

27. Houser, P.R. Infiltration and soil moisture processes. In *Handbook of Weather, Climate, and Water: Atmospheric Chemistry, Hydrology, and Societal Impacts*; John Wiley & Sons, Inc.: Hoboken, 2003; 966 pp.

28. Baldocchi, D.; Falge, E.; Gu, L.; Olson, R.; Hollinger, D.; Running, S.; Anthoni, P.; Bernhofer, C.; Davis, K.; Evans, R.; Fuentes, J.; Goldstein, A.; Katul, G.; Law, B.; Lee, X.; Malhi, Y.; Meyers, T.; Munger, W.; Oechel, W.; Paw, K.T.; Pilegaard, K.; Schmid, H.P.; Valentini, R.; Verma, S.; Vesala, T.; Wilson, K.; Wofsy, S. FLUXNET: A new tool to study the temporal and spatial variability of ecosystem-scale carbon dioxide, water vapor, and energy flux densities. Bull. Am. Meteorol. Soc. **2001**, *82* (11), 2415–2434.

29. Maccherone, B.; Frazier, S. About MODIS, http://modis.gsfc.nasa.gov/about/, 2013.

30. Graham, S.; Parkinson, C. AMSR-E, http://aqua.nasa.gov/about/instrument_amsr.php, 2011.

31. Ray, S. How AIRS Works, http://airs.jpl.nasa.gov/instrument/how_AIRS_works/, 2013.

32. Frolking, S.; Xiao, X.; Zhuang, Y.; Salas, W.; Li, C. Agricultural land-use in China. A comparison of area estimates from ground-based census and satellite-borne remote sensing. Global Ecol. Biogeogr. **1999**, *8* (5), 407–416.

33. Olioso, A.; Chauki, H.; Courault, K.; Wingeron, J. Estimation of evapotranspiration and photosynthesis by assimilation of remote sensing data into SVAT models. Remote Sens. Environ. **1999**, *68* (3), 341–356.

34. Kumar, A.; Niyogi, D.; Chen, F.; Barlage, M.; Ek, M.B.; Peters-Lidard, C.D. Assessing impacts of integrating MODIS vegetation data in the Weather Research Forecasting (WRF) model coupled to two different canopy-resistance approaches. Conference papers and proceedings, 90th American Meteorological Society Annual Meeting, Atlanta, Georgia, Jan 17–21, 2010.

35. Monteith, J.L. Climatic variation and the growth of crops. Q. J. R. Meteorol. Soc. **1981**, *107* (454), 749–774.

36. Hoerling, M.; Schumert, S.; Mo, K. *An Interpretation of the Origins of the 2012 Central Great Plains Drought*, NOAA Drought Task Force, Narrative Team, Silver Spring, Maryland, 2013; 1–50, http://www.drought.gov/media/pgfiles/2012-Drought-Interpretation-final.web-041013_V4.0.pdf

37. Guttman, N.B. Comparing the Palmer Drought Index and the Standardized Precipitation Index. J. Am. Water Resour. Assoc. **1998**, *34* (1), 113–121.

38. Prokopy, L. *Useful to Useable: Transforming Climate Variability and Change Information for Cereal Crop Producers*. Agriculture and Food Research Initiative Competitive Grant no. 2011-68002-30220, https://drinet.hubzero.org/groups/u2u, 2013.

39. Downing, T.E.; Stowell, Y. Household food security and coping with climatic variability in developing countries. In *Handbook of Weather, Climate, and Water: Atmospheric Chemistry, Hydrology, and Societal Impacts*; John Wiley & Sons, Inc.: Hoboken, 2003; 966 pp.

40. Glantz, M.H. *Currents of Change El Nino Impact on Climate and Society*; Cambridge University Press: Cambridge, 1996, 194 pp.

41. Bates, B.C.; Kundzewicz, Z.W.; Wu, S.; Palutikof, J.P., Eds. *Climate Change and Water*. IPCC Secretariat: Geneva, 2008, 210 pp.

42. Southworth, J.; Pfeifer, R.A.; Habeck, M.; Randolph, J.C.; Doering, O.C.; Gangadhar Rao, D. Sensitivity of winter wheat yields in the midwestern United States to future changes in climate, climate variability and CO_2 fertilization. Clim. Res. **2002**, *22* (1), 73–86.

43. Walthall, C.L.; Hatfield, J.; Backlund, P.; Lengnick, L; Marshall, E.; Walsh, M.; Adkins, S.; Aillery, M.; Ainsworth, E.A.; Ammann, C.; Anderson, C.J.; Bartomeus, I.; Baumgard, L.H.; Booker, F.; Bradley, B.; Blumenthal, D.M.; Bunce, J.; Burkey, K.; Dabney, S.M.; Delgado, J.A.; Dukes, J.; Funk, A.; Garrett, K.; Glenn, M.; Grantz, D.A.; Goodrich, D.; Hu, S.; Izaurralde, R.C.; Jones, R.A.C.; Kim, S.-H.; Leaky, A.D.B.; Lewers, K.; Mader, T.L.; McClung, A.; Morgan, J.; Muth, D.J.; Nearing, M.; Oosterhuis, D.M.; Ort, D.; Parmesan, C.; Pettigrew, W.T.; Polley, W.; Rader, R.; Rice, C.; Rivington, M.; Rosskopf, E.; Salas, W.A.; Sollenberger, L.E.; Srygley, R.; Stöckle, C.; Takle, E.S.; Timlin, D.; White, J.W.; Winfree, R.; Wright-Morton, L.; Ziska, L.H. Climate change and agriculture in the United States: Effects and adaptation. In *USDA Technical Bulletin*; U.S. Department of Agriculture: Washington, DC, 2013, http://www.usda.gov/oce/climate_change/effects_2012/CC%20and%20Agriculture%20Report%20(02-04-2013)b.pdf

44. Liu, X.; Niyogi, D.; Charusombat, U. Estimating Corn Yields Regionally across Midwest Using the Hybrid Maize Model with a Land Data Assimilation System, Abstract

Acid—
Atmospheric

GC21H-07 of Presentation, American Geophysical Union Fall Meeting, San Francisco, California, Dec 3–7, 2012, http://198.61.161.98/abstracts/meetings/2012/FM/sections/GC/sessions/GC21H/abstracts/GC21H-07.html

45. Niyogi, D.; Liu, X. Adaptability of the Hybrid-Maize Model and the Development of a Gridded Crop Modeling System for the Midwest U.S. Presentation, ASA, CSSA, and SSSA International Annual Meetings, Cincinnati, Ohio, Oct 23, 2012. https://scisoc.confex.com/crops/2012am/webprogram/Paper70435.html

46. Climate Prediction Center Online Climate Glossary, http://www.cpc.ncep.noaa.gov/products/outreach/glossary.shtml#C, 2013.

47. Womach, J. *Agriculture: A Glossary of Terms, Programs, and Laws, 2005 Ed.*, CRS Report for Congress; The Library of Congress: Washington D.C., http://www.cnie.org/NLE/CRSreports/05jun/ 97-905.pdf, 2005. (29 October 2013).

48. What is biochemistry? Biochemical Society, http://www.biochemistry.org/?TabId=456 (accessed November 2013).

49. Bioenergy. United States Department of Agriculture, Economic Research Service, http://www.ers.usda.gov/topics/farm-economy/bioenergy.aspx#.Um_cl3DqhyI (accessed November 2013).

50. Biogeochemistry. Woods Hole Oceanographic Institution, https://www.whoi.edu/main/topic/biogeochemistry (accessed November 2013).

51. National Weather Service Online Glossary, http://w1.weather.gov/glossary/ (accessed November 2013).

52. Ramamasy, S.; Baas, S. *Climate Variability and Change: Adaptation to Drought in Bangladesh*; Asian Disaster Preparedness Center, Food and Agriculture Organization of the United Nations, ftp://ftp.fao.org/docrep/fao/010/a1247e/a1247e00.pdf, 2007. (29 October 2013).

53. Weier, J.; Herring, D. Measuring Vegetation (NDVI and EVI). NASA Earth Observatory, http://earthobservatory.nasa.gov/Features/MeasuringVegetation/measuring_vegetation_1.php, 2000. (29 October 2013).

54. Amthor, J.S.; Baldocchi, D.D. Terrestrial higher plant respiration and net primary production. In *Terrestrial Global Productivity*; Academic Press: San Diego, CA, 2001; 33–59.

55. Agricultural Thesaurus and Glossary. United States Department of Agriculture, National Agricultural Library, http://agclass.nal.usda.gov/glossary.shtml (accessed November 2013).

56. Meier, B.L.; Osborn, J.; Knight, C. Life Zones Reflect Climate: Climate Change Demands Future Planning. NOAA Earth System Research Laboratory, Global Systems Division, http://www.esrl.noaa.gov/gsd/outreach/education/poet/LifeZones.pdf, 2013.

57. Arnfield, A.J. Köppen Climate Classification. *Encyclopedia Britannica*, http://www.britannica.com/EBchecked/topic/322068/Koppen-climate-classification (accessed October 2013).

58. Chen, J.M.; Rich, P.M.; Gower, S.T.; Norman, J.M.; Plummer, S. Leaf area index of Boreal forests: Theory, techniques, and measurements. J. Geophys. Res. **1997**, *102* (D24), 29429–29443.

59. Remote Sensing. National Oceanic and Atmospheric Administration http://oceanservice.noaa.gov/facts/remote-sensing.html (accessed October 2013).

60. Jarraud, M. *Guide to Meteorological Instruments and Methods of Observation*, WMO-No. 8, 7th Ed.; World Meteorological Organization: Geneva, Switzerland, 2008, ISBN 978-92-63-10008-5.

61. Pearson, K.; Bauder, J. How and When Does Water Stress Impact Plant Growth and Development? Presentation, American Society of Agronomy, Crop Science Society of America, and Soil Science Society of America 2003 Annual Meeting, Denver, Colorado, http://waterquality.montana.edu/docs/irrigation/a9_bauder.shtml, 2005.

62. Heim, R. Climate of 2013 – April: U.S. Standardized Precipitation Index. National Climatic Data Center, http://www.ncdc.noaa.gov/oa/climate/research/prelim/drought/spi.html (accessed November 2013).

Air Pollutants: Elevated Carbon Dioxide

Hans-Joachim Weigel
Jürgen Bender
Federal Research Institute for Rural Areas, Forestry and Fisheries, Thünen Institute of Biodiversity, Braunschweig, Germany

Abstract

The assessment of the potential combined effects of air pollutants and elevated atmospheric concentrations of carbon dioxide (CO_2) on vegetation is of critical importance during the next decades. The interactive effects of these atmospheric compounds on crops, trees, and other types of vegetation have been shown. Existing evidence on such interactions is almost entirely restricted to CO_2 and ozone (O_3), the concentrations of which are increasing globally. Results from a number of studies indicate that elevated CO_2 may reduce the adverse effects of O_3 on plant growth and productivity, but the available information is inconsistent as several studies show that elevated CO_2 did not ameliorate the negative effects of O_3. The future interactions of elevated CO_2 and enhanced atmospheric nitrogen (N) deposition are of concern in many ecosystem types with respect to carbon sequestration and biodiversity. The overall impact of climate change including elevated CO_2 on future air pollutant effects is difficult to predict because of the largely uncertain influence and feedback of other growth variables such as plant genotype, soil water deficit, nutrient availability, or temperature.

INTRODUCTION

The concentrations of various compounds in the atmosphere have changed during the last century, and they continue to change. Most of these compounds interact with the terrestrial biosphere as they are part of the overall biogeochemical cycling of, e.g., carbon, oxygen, nitrogen, and sulfur.[1] For example, depending on their concentrations, gaseous compounds [sulfur dioxide (SO_2), nitrogen monoxide and dioxide (NO_2/NO)] may be beneficial to terrestrial ecosystems or remain inert (O_3) at low concentrations, whereas at higher levels, they may act as air pollutants affecting these systems in an adverse manner. Although atmospheric CO_2 is the basic plant resource for photosynthesis, its current concentration is still limiting C_3 plant growth. The rapid increase of the global atmospheric CO_2 concentration $[CO_2]$, along with the overall changes in climate and atmospheric chemistry, require an assessment of the potential future interactive effects of air pollutants and elevated $[CO_2]$ on terrestrial ecosystems.

ATMOSPHERIC CHANGE: CONCENTRATIONS AND TRENDS

On a global scale, the concentrations of a variety of gaseous and particulate compounds in the atmosphere, including CO_2, NO/NO_2, SO_2, O_3, ammonia (NH_3), heavy metals, and volatile organic compounds (VOC) have undergone temporal and spatial changes during the last century.[2] After peak emissions in the 1960s to the 1980s in industrialized countries particularly, the concentrations of SO_2, and to a smaller extent of NO_x (NO/NO_2), VOCs, and particulate matter, have declined during the past decades in Europe and North America. NH_3, which is the most important reduced N species, is of importance as a direct air pollutant in the vicinity of local emitters. However, wet and dry N deposition from oxidized and reduced N species are predicted to increase in other regions of the world.[3] The occurrence and distribution of airborne VOCs are difficult to assess because there are both anthropogenic and biogenic sources. With respect to heavy metals such as lead (Pb), cadmium (Cd), nickel (Ni), mercury (Hg), and zinc (Zn), a decline in emission and subsequent deposition was observed in most of Europe since the late 1980s. Unlike the development in Europe and North America, emissions and consequently atmospheric concentrations of many of the above-mentioned compounds have been increasing over the last two decades particularly in the rapidly growing regions of Asia, Africa, and Latin America.[4] For example, China and India are now the leading emitters of SO_2 in the world. Also, the predicted further increase in global nitrogen oxide (NO_x) emissions may be attributed largely to these countries. On the other hand, concentrations of ground-level O_3 and atmospheric CO_2 have increased and continue to increase on a global scale. In most industrialized countries, O_3 concentration $[O_3]$ has nearly doubled during the last 100 years. Current background $[O_3]$ in the Northern hemisphere is within the range of 23–34 ppbv (parts-per-billion by volume). Although at least in most parts of Western Europe there is a clear trend of decreasing O_3 peak values ("photosmog episodes"), models predict that background $[O_3]$ will continue to increase at a rate of

Encyclopedia of Natural Resources DOI: 10.1081/E-ENRA-120049199

Acid—
Atmospheric

0.5% to 2% per year in the Northern Hemisphere during the next several decades, and that global surface $[O_3]$ is expected to be in the range of 42–84 ppbv by 2100.[5] O_3 pollution has also become a major environmental problem in many of the countries with rapidly developing population and related economic growth, respectively. O_3 is currently considered the most important atmospheric pollutant that has direct negative effects on vegetation worldwide. Its concentrations vary considerably in time and space and show distinct annual and diurnal patterns.

Since the beginning of the 19th century, the CO_2 in the atmosphere has increased globally from approximately 280 ppmv (parts-per-million by volume) to current values of about 395 ppmv. It is expected that CO_2 will continue to increase even more rapidly and may reach about 550–650 ppmv between 2050 and 2070.[6] CO_2 is the substrate for plant photosynthesis, and its current atmospheric concentration is limiting for photosynthesis and growth of C_3 plants. It is expected that the increase in CO_2 will have far-reaching consequences for most types of vegetation.

EFFECTS OF O_3 AND CO_2 ALONE

Due to their global importance and their contrasting effects on vegetation, plant growth responses to either O_3 or CO_2 alone are briefly described. Primary O_3 effects include subtle biochemical and ultrastructural changes in the plant cell, which may result in impaired photosynthesis, alterations of carbon allocation patterns, symptoms of visible injury, enhanced senescence, reduced growth and economic yield, altered resistance to other abiotic and biotic stresses, or reduced flowering and seed production at the whole-plant level.[7,8] At the ecosystem level, this may result in a loss of competitive abilities of plant species in communities along with shifts in biodiversity and impaired ecosystem functions and services like reduced carbon sequestration and altered hydrology.[9] For example, current ambient (O_3) in many industrialized areas has been shown to suppress crop yields of sensitive species and to retard growth and development of trees and other plant species of the non-woody (semi)natural vegetation. Overall quantification of O_3 effects on vegetation is complicated by large inter- and intra-specific variability in the O_3 susceptibility of plants.[10]

By contrast, plants of the C_3 type most frequently respond to elevated CO_2 with a stimulation of photosynthesis accompanied by a reduced stomatal conductance and transpiration rate, an enhanced concentrations of soluble carbohydrates, and a stimulation of biomass production and economic yield.[11] Similarly, in C_4 plants, higher CO_2 concentrations reduce stomatal conductance and transpiration, i.e., both C_3 and C_4 plants may benefit from elevated CO_2 by improved water-use efficiency and a reduced demand for water. Under well-watered conditions, no significant growth stimulation has been found so far in C_4 plants, because C_4 photosynthesis is saturated under ambient CO_2.[12,13] Growth and yield enhancements of up to 25–35% as compared to ambient CO_2 have been observed when crop plants were exposed to 550–750 ppmv CO_2. Experiments with tree species ranging from short-term studies with seedlings to long-term whole-stand manipulations have also shown that elevated CO_2 stimulated net photosynthesis and resulted in enhanced tree growth in almost all cases.[14,15] As with O_3, plant species differ widely in their response to high CO_2, which makes an overall assessment of its potential effects on vegetation difficult.

INTERACTIVE EFFECTS OF AIR POLLUTANTS AND CO_2

Along with the ongoing and predicted further changes of global climate and atmospheric chemistry, there is considerable interest in how terrestrial ecosystems will respond to these multiple environmental changes and particularly how the individual changes in atmospheric constituents may interact with each other when they impact vegetation. The majority of studies dealing with this issue have addressed two-way interactions of O_3 and elevated CO_2, although there is much less information on how other air pollutants interact with high CO_2. There are no studies describing three-way interactions, i.e., in which two air pollution components and elevated CO_2 are combined together. In a biological sense, the combined action of multiple factors in comparison to single-factor effects can be described as additive (effect directly predictable from single-factor treatment) or as interactive. Interactive effects can be synergistic (effect > than expected from single-factor treatment) or antagonistic (effect < than expected from single-factor treatment).[16]

CO_2 and O_3

A great number of previous and more recent studies using different experimental approaches ranging from controlled environment to free-air O_3- and CO_2-enrichment systems have been carried out on the combined effects of the two gases. The bulk of these studies has shown that high CO_2 in the range of 200–400 ppmv above current ambient CO_2 levels either partially or totally compensates for adverse O_3 effects, whether these effects have been addressed at the biochemical and physiological level or at the whole-plant level including growth and yield. This has been demonstrated for crop (e.g., wheat, soybean, potato, rice) as well as for tree (e.g., trembling aspen, paper birch, sugar maple) species, although little information is available for grassland species.[17–20] For example, elevated CO_2 reduces O_3 effects, such as a loss in root and main stem biomass, a decrease in leaf area and mass, general foliar damage, lower growth and yield, lower starch levels, and an altered carbon balance. Results from recent free-air concentration enrichment (FACE) studies, however, have indicated that the mitigating effect of elevated CO_2 against O_3 damage might be less than predicted from earlier chamber studies.[12]

The proposed mechanisms to explain the protective effect of elevated CO_2 against the phytotoxic effects of O_3 include the following: i) reduced uptake or flux of O_3 through the stomata due to a CO_2-induced stomatal closure, ii) improved supply of carbon skeletons supporting the synthesis of antioxidants involved in the scavenging of O_3 and its toxic products, iii) protection of the RuBisCo protein from O_3-induced degradation, and iv) CO_2-induced changes in the cell surface/volume ratio.[9,21,22] However, it has been shown that in spite of decreased stomatal conductance under elevated CO_2, adverse effects of O_3 may still occur.[8,19] Additionally, elevated O_3 has been found to impair stomatal responsiveness to CO_2, i.e., O_3 causes less-sensitive ("sluggish") stomatal responses to elevated CO_2.[23] As CO_2 effects on stomatal conductance may be species specific, it is not yet possible to support a general concept of a CO_2-induced reduction in the flux of O_3 into the plant. Nevertheless, a reduction in stomatal conductance and thus in the O_3 uptake may increase atmospheric O_3 in the boundary layer.[24] Moreover, in a given plant species, protection by high CO_2 from a particular adverse effect is not necessarily associated with the protection against another adverse effect. For instance, in wheat plants, elevated CO_2 provided full protection from effects of O_3 on total plant biomass, but not on grain yield. From the available database of studies that have examined the interactive effects of O_3 and CO_2, the information is not entirely consistent, as several studies revealed that elevated CO_2 may not always protect plants from the adverse effects of O_3 (Table 1).

CO_2 and Other Air Pollutants

Very few studies have addressed the combined effects of elevated CO_2 and of other air pollutants. SO_2 has long been known to adversely affect agricultural crops and forest plants above a certain threshold concentration.[25] Reduced photosynthesis, altered water relations, growth retardations, yield losses, and altered susceptibilities to other stresses are common plant responses observed under SO_2 stress. Due to the diminishing importance of SO_2 as a widespread air pollutant, few studies have been conducted on the combined action of SO_2 and elevated CO_2. In earlier studies, it was shown for some crop species that elevated CO_2 reduced the sensitivity of the plants to SO_2 injury or protected them from the negative effects of SO_2 on growth and yield.[26] With the combined exposure of crop species to both gases, the yield increments were sometimes even

Table 1 Selected examples of the effects of elevated O_3 and CO_2, alone or in combination, on plant responses (examples with significant adverse effects of O_3 on visible injury, photosynthesis, growth, and yield)

Species	O_3 effect	CO_2 effect	O_3/CO_2 effect
Potato	Decreased chlorophyll content; visible foliar leaf injury	n.e.	Adverse effect of O_3 on chlorophyll content unchanged; reduced degree of visible O_3-induced leaf injury
Wheat	Visible leaf injury; reduced photosynthesis; reduced growth; reduced yield	Increased photosynthesis; increased growth; increased yield	Reduced degree of visible O_3-induced leaf injury; amelioration of negative O_3 effects on photosynthesis, growth, and yield
Soybean	Reduced photosynthesis; reduced growth; reduced seed yield	Increased photosynthesis; increased growth; insignificant increase of seed yield	O_3 impact on photosynthesis lessened; amelioration of negative O_3 effects on photosynthesis, growth, and yield
Cotton	Reduced leaf area per mass; reduced starch contents	Increased leaf area per mass and starch contents	Prevention of adverse effects of O_3 by CO_2
Norway spruce	Visible leaf injury (chlorotic mottling)	n.e.	No effect of CO_2 on the degree of O_3-induced leaf injury
Trembling aspen (different O_3-sensitive and -tolerant clones)	Reduced tree growth parameters (height, diameter, volume)	Enhancement of growth parameters	No effect of CO_2 on the degree of O_3-induced growth reductions
Paper birch	Reduced photosynthesis; decreased dry matter production	Increased photosynthesis; increased dry matter production	Decrease in photosynthesis and dry matter production similar to O_3 alone
White clover (sensitive clone)	Visible leaf injury	n.e.	Little effect on the degree of O_3-induced foliar injury

Source: Adapted from Karnosky et al.,[19] Vandermeiren et al.,[20] Polle & Pell,[21] and Runeckles.[26]
Abbreviations: CO_2, carbon dioxide; n.e., No effect; O_3, Ozone.

larger when compared to the stimulation observed with exposure to elevated CO_2 alone, suggesting that the plants were able to use the airborne sulfur more effectively under the conditions of enhanced carbon availability. Low-to-moderate SO_2 concentrations may confer a nutritional benefit to plants, particularly under conditions of low sulfur availability in the soil.

Studies on the interactive effects of elevated CO_2 and nitrogen oxides (NO and NO_2) are confined to commercial greenhouses under conditions of horticultural crop production under very high CO_2 and are not considered here. However, it has been shown repeatedly that positive plant growth responses to elevated CO_2 are smaller at low relative to high soil N supply. This is related to the question on the role of future atmospheric N deposition and "aerial carbon fertilization" by elevated CO_2 in shaping the size of the terrestrial carbon sink and how plant biodiversity might be affected by these inputs. Assuming that aerial N supply via enhanced N deposition causes similar effects as soil N fertilization, a few experimental and modeling studies addressed the question of how elevated CO_2 interacts with N deposition. For example, it has been shown that N addition enhanced the CO_2 stimulation of plant productivity in the first phase of a multiyear CO_2–N manipulation study with a herbaceous wetland plant community. But in the longer term, the observed N-induced shift in the plant community composition suppressed the CO_2 stimulation of plant productivity, indicating that plant community shifts can act as a feedback effect that alters ecosystem responses to elevated CO_2.[27] In a long-term study with simulated grassland systems with 16 species, high N supply reduced species richness by 16% under ambient CO_2 but only by 8% under elevated CO_2, i.e., high CO_2 ameliorated negative N effects.[28,29] Elevated CO_2 and N addition have been found to affect above and belowground C allocation in temperate forest trees in an opposite way, i.e., elevated CO_2 increases belowground allocation, whereas N increases aboveground allocation; however, the ratio of above vs. belowground C flow does not change in the combination of both treatments.[30]

CONCLUSION

The assessment of the potential combined effects of air pollutants and elevated atmospheric concentrations of CO_2 on vegetation is of critical importance during the next decades. Interactive effects of these atmospheric compounds on crops, trees, and other types of vegetation have been shown. Existing evidence on such interactions is almost entirely restricted to CO_2 and O_3, the concentrations of which are increasing globally. Although rising CO_2 will be mostly beneficial to plants, current ambient O_3 are high enough to impair plants in many regions of the world. There is prevailing information that elevated CO_2 may protect plants from adverse effects of O_3, but this has not been demonstrated unequivocally. There is also some information that rising CO_2 may protect plants against

phytotoxic SO_2 concentrations. The future interactions of elevated CO_2 and enhanced atmospheric N deposition are of concern in many ecosystem types with respect to carbon sequestration and biodiversity. Overall, our understanding has to be improved about how other growth variables, such as plant genotype, soil water deficit, nutrient availability, or temperature, may modify the interaction between air pollutants and elevated CO_2.

REFERENCES

1. Dämmgen, U.; Weigel, H.J. Trends in atmospheric composition (nutrients and pollutants) and their interaction with agroecosystems. In *Sustainable Agriculture for Food, Energy and Industry*; El Bassam, N., Behl, R.K., Prochnow, B., Eds.; James & James (Science Publishers) Ltd.: London, 1998; 85–93.
2. Bender, J.; Weigel, H.J. Changes in atmospheric chemistry and crop health: A review. Agron. Sustain. Dev. **2011**, *31*, 81–89.
3. Grübler, A. Trends in global emissions: Carbon, sulfur, and nitrogen. In *Encyclopedia of Global Environmental Change. Causes and Consequences of Global Environmental Change*; Douglas, I., Ed.; John Wiley & Sons: Chichester, 2003; Vol. 3, 35–53.
4. Emberson, L.D., Ashmore, M.R., Murray, F., Eds.; *Air Pollution Impacts on Crops and Forests: a Global Assessment*; Imperial College Press: London, 2003.
5. Vingarzan R. A review of surface ozone background levels and trends, Atmos. Environ. **2004**, *38*, 3431–3442.
6. IPCC. *The 4th Assessment Report, Working Group I Report: The Physical Scientific Basis*, Report of the Intergovernmental Panel on Climate Change; USA Cambridge University Press: Cambridge, UK and New York, 2007.
7. Ashmore, M.R. Assessing the future global impacts of ozone on vegetation. Plant Cell Environ. **2005**, *28*, 949–964.
8. Fiscus, E.L.; Booker, F.L.; Burkey, K.O. Crop responses to ozone: uptake, modes of action, carbon assimilation and partitioning. *Plant Cell Environ.* **2005**, *28*, 997–1011.
9. Fuhrer, J. Ozone risk for crops and pastures in present and future climates. Naturwissenschaften **2009**, *96*, 173–194.
10. Bender, J.; Weigel, H.-J. Ozone stress impacts on plant life. In *Modern Trends in Applied Terrestrial Ecology*; Ambasht, R.S., Ambasht, N.K., Eds.; Kluwer Academic/Plenum Publishers: New York, 2003; 165–182.
11. Ainsworth, E.A.; McGrath, J.M. Direct effects of rising atmospheric carbon dioxide and ozone on crop yields. In *Climate Change and Food Security, Advances in Global Change Research 37*; Lobell, D., Burke, M., Eds.; Springer Science + Business Media: Dordrecht, 2010; 109–130.
12. Long, S.P.; Ainsworth, E.A.; Leakey, A.D.B.; Morgan, P.B. Global food insecurity. Treatment of major food crops with elevated carbon dioxide or ozone under large-scale fully open-air conditions suggests recent models may have overestimated future yields. Phil. Trans. Royal Soc. **2005**, *B 360*, 2011–2020.
13. Leakey, A.D.B. Rising atmospheric carbon dioxide concentration and the future of C_4 crops for food and fuel. Proc. R. Soc. Lond. B **2009**, *276*, 2333–2343.

14. Norby, R.J.; DeLucia, E.H.; Gielen, B.; Calfapietra, C.; Giardina, C.P.; King, J.S.; Ledford, J.; McCarthy, H.R.; Moore, D.J.P.; Ceulemans, R.; De Angelis, P.; Finzi, A.C.; Karnosky, D.F.; Kubiske, M. E.; Lukac, M.; Pregitzer, K.S.; Scarascia-Mugnozza, G.E.; Schlesinger, W.H.; Oren, R. Forest response to elevated CO_2 is conserved across a broad range of productivity. PNAS **2005**, *102*, 18052–18056

15. Wittig, V.E.; Ainsworth, E.A.; Naidu, S.L.; Karnosky, D.F.; Long, S.P. Quantifying the impact of current and future tropospheric ozone on tree biomass, growth, physiology and biochemistry: a quantitative meta-analysis. Glob. Change Biol. **2009**, *15*, 396–424.

16. Fangmeier, A.; Bender, J.; Weigel, H.J.; Jäger, H.J. Effects of pollutant mixtures. In *Air Pollution and Plant Life*; Bell, J.N.B., Treshow, M., Eds.; John Wiley & Sons: Chichester, 2002; 251–272.

17. Olszyk, D.M.; Tingey, D.T.; Watrud, R.; Seidler, R.; Andersen, C. Interactive effects of O_3 and CO_2: implications for terrestrial ecosystems. In *Trace Gas Emissions and Plants*; Singh, S.N., Ed.; Kluwer Academic Publishers: Netherlands, 2000; 97–136.

18. Feng, Z.; Kobayashi, K. Assessing the impacts of current and future concentrations of surface ozone on crop yield with meta-analysis. Atmos. Environ. **2009**, *43*, 1510–1519.

19. Karnosky, D.F.; Pregitzer, K.S.; Zak, D.R.; Kubiske, M.E.; Hendrey, G.R.; Weinstein, D.; Nosal, M.; Percy, K.E. Scaling ozone responses of forest trees to the ecosystem level in a changing climate. Plant Cell Environ. **2005**, *28*, 965–981.

20. Vandermeiren, K.; Harmens, H.; Mills, G.; De Temmerman, L. Impacts of ground level ozone on crop production in a changing climate. In *Climate Change and Crops, Environmental Science and Engineering*; Singh, S.N., Ed.; Springer: Berlin, 2009; 213–243.

21. Polle, A.; Pell, E.J. The role of carbon dioxide in modifying the plant response to ozone. In *Carbon Dioxide and Environmental Stress*; Luo, Y.; Mooney, H.A., Eds.; Academic Press: San Diego, 1999; 193–213.

22. Tausz, M.; Grulke, N.E.; Wieser, G. Defense and avoidance of ozone under global change. Environ. Pollut. **2007**, *147*, 525–531.

23. 23 Onandia, G.; Olsson, A.-K.; Barth, S.; King, J.S.; Uddling, J. Exposure to moderate concentrations of tropospheric ozone impairs tree stomatal response to carbon dioxide. Environ. Pollut. **2011**, *159* 2350–2354.

24. Sanderson, M.G.; Collins, W.J.; Hemming, D.L.; Betts, R.A. Stomatal conductance changes due to increasing carbon dioxide levels: projected impacts on surface ozone levels. Tellus **2007**, *59b*, 404–411.

25. Legge, A.H., Krupa, S.V. Effects of sulphur dioxide. In *Air Pollution and Plant Life*; Bell, J.N.B., Treshow, M., Eds.; John Wiley & Sons: Chichester, 2002; 135–162.

26. Runeckles, V. Air pollution and climate change. In *Air Pollution and Plant Life*; Bell, J.N.B., Treshow, M., Eds.; John Wiley & Sons Ltd.: Chichester, 2002; 431–454.

27. Langley, J.A.; Megonigal, J.P. Ecosystem response to elevated CO_2 levels limited by nitrogen-induced plant species shift. Nature **2010**, *466*, 96–99.

28. Reich, P.B., Hungate, B.A., Luo, Y. Carbon–nitrogen interactions in terrestrial ecosystems in response to rising atmospheric carbon dioxide. Annu. Rev. Ecol. Evol. Syst. **2006**, *37*, 611–36.

29. Reich, P.B. Elevated CO_2 reduces losses of plant diversity by nitrogen deposition. Science **2009**, *326*, 1399–1402.

30. Drake, J.E.; Oishi, A.C.; Giasson, M.-A.; Oren, R.; Finzi, A.C. Trenching reduces soil heterotrophic activity in a loblolly pine (*Pinus taeda*) forest exposed to elevated atmospheric CO_2 and N fertilization. Agric. Forest Meteorol. **2012**, *165*, 43–52.

Acid— Atmospheric

Asian Monsoon

Congbin Fu
Institute of Atmospheric Physics, Chinese Academy of Sciences, Beijing, China

Abstract

The entry starts with the original meaning of the word "monsoon" and the major factors of Asian monsoon formation. It then describes the major features of Asian monsoon climate and its two subsystems: South Asian monsoon and East Asian monsoon. The variability of monsoon systems depends on different time scales, including intraseasonal, interannual, and decadal, and their relationships with large-scale atmospheric–ocean interaction, such as El Niño/Southern Oscillation (ENSO), Indian Ocean Dipole (IOD), and Pacific Decadal Oscillation (PDO). Variation on the geological scale and its link with uplifting of the Tibetan Plateau are also introduced. The potential influence of global warming and other anthropogenic factors on Asian monsoons are mentioned. The significant impacts of monsoon climate and its variability on society and economics, particularly the climate-related disasters, have been summarized. The concept of "global monsoon" is introduced at the end.

INTRODUCTION

Asian monsoon is a component of the global climate system. The name "monsoon" is derived from the Arabic word "*mausim*" meaning "season," large-scale seasonal reversals of dominant wind direction and distinct rainy and dry seasons of climate over the Indian subcontinent, which was named "Indian monsoon." Later on, it became clear that the monsoon is a much larger and complex system that affects the weather and climate over a large domain of tropical and subtropical Asia. A more general term "Asian monsoon" is widely used in climatology.

Classical monsoon theory believes that the seasonal change of land–ocean thermal contrast forced by the seasonal change of solar radiation is the major cause of the formation of a monsoon system. In summer, land is warmer than the ocean and a thermal depression forms over land. The airflow, called summer monsoon, brings the moist air from the ocean into the continent where there are more clouds, higher humidity, and more rainfall; while in winter, land is cooler than the ocean and a high-pressure system forms over land. The airflow, called winter monsoon, brings cold and dry continent air-mass from higher latitudes to the tropical and subtropical continent where there are relatively fewer clouds, low humidity, and less rainfall. Research also points out the role of dynamic and thermal dynamic effects of the Tibetan Plateau in the formation of Asian monsoon. However, the term "monsoon" is very often used to refer to its rainy phase only, or the summer monsoon.

MAJOR FEATURES OF ASIAN MONSOON CLIMATE AND ITS TWO SUB-SYSTEMS: SOUTH ASIAN MONSOON AND EAST ASIAN MONSOON

The Asian monsoon has been classified into two subsystems: the South Asian monsoon and the East Asian monsoon. These two subsystems are linked to each other, but with different features. The South Asian monsoon, or Indian monsoon, is a tropical system affecting the Indian subcontinent and surrounding regions. In summer, the southwest monsoon comes from the Indian Ocean and Arabian Sea. The onset of southwest monsoon usually occurs at the end of May or early June in southern India and then advances rapidly northward. In mid-July, it spreads over the whole Indian Peninsula. In early September, the southwest monsoon begins to retreat southeastward. The dominant period of southwest monsoon, from June through September, is the rainy season of India when the rainfall is 75% of the annual total. Around September, with the sun retreating south, the northern land mass of the Indian subcontinent begins to cool off rapidly. When a high-pressure system begins to build over northern India, the Indian Ocean and its surrounding region remain warm. This causes the cold wind to sweep down from the Himalayas and the Indo-Gangetic Plain toward the vast expanse of the Indian Ocean and south of the Deccan Peninsula. This is known as the northeast monsoon. From November to April, the India subcontinent enters into a dry season when the northeast monsoon dominates.

The East Asian monsoon is a tropical and subtropical system that affects large parts of Indo-China, the Philippines,

Encyclopedia of Natural Resources DOI: 10.1081/E-ENRA-120047623

Acid—
Atmospheric

China, Korea, and Japan. The East Asian summer monsoon comes from tropical and subtropical oceans. The onset of the East Asian summer monsoon has been generally recognized to start in mid-May over the South China Sea and then moves northward and eastward. It reaches the Yangtze River basin in mid-June and further across the Yellow River basin in late July to early August when the summer monsoon enters into its peak period. The East Asian summer monsoon begins to retreat in September. The beginning and ending of the rainy season in East Asia are closely linked with the advance and retreat of the summer monsoon.

The rain occurs in a concentrated belt that stretches east–west over China and tilts east–northeast over Korea and Japan. Such seasonal rain is known as *Meiyu* in China, *Changma* in Korea, and *Bai-u* in Japan. Different from the South Asia monsoon, the East Asian winter monsoon is stronger than the summer monsoon. It originates from the cold and dry continent air-mass of the Siberian high pressure system with dominant northwest or northeast flows. The East Asian winter monsoon is often accompanied by cold ocean waves, snow, and strong wind. Figure 1 presents the long-term mean precipitation (mm/day)

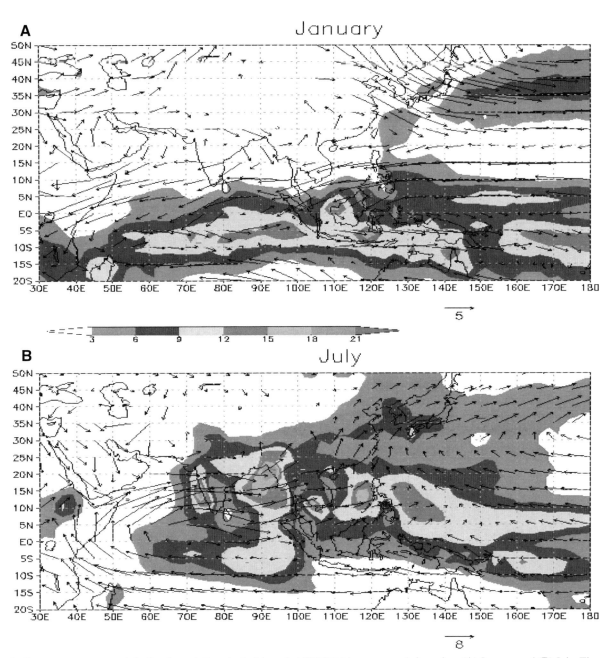

Fig. 1 Long-term mean precipitation (mm d-1) and wind (m s-1 at 850 hPa) in monsoon Asia region, (**A**) January and (**B**) July. The scale for precipitation is in shadow. The scale for wind is indicated by the arrow below the figure.
Source: Adapted from CMAP,[1] NCEP.[2]

Acid—
Atmospheric

and wind (m/s at 850 hPa) in monsoon Asia region in January and July.[1,2]

VARIABILITY OF ASIAN MONSOONS

Asian monsoon is characterized by high variability at various time scales, from synoptic, intraseasonal, interannual, decadal, and interdecadal, centennial to geological scales. On a synoptic scale, the advance of summer monsoon normally has several phases: onset, burst, break, surge, and retreat, and so forth. Recent research concluded that the onset of the Asian summer monsoon is composed of three sequential stages, that is, the Bay of Bengal, South China Sea, and Indian monsoon onsets. On an interannual scale, there is a close link between Asian monsoon and ENSO, the strongest signal of a tropical atmospheric–ocean system. Some analysis and numerical modeling studies indicate that during the warm phase of ENSO, the South Asian monsoon tends to be weaker than during the cold phase of ENSO, but the East Asian monsoon–ENSO relationship is somewhat different from that of the South Asian monsoon. Recent research also indicates the weakening of such a relationship in the last several decades. The newly discovered IOD, a mode of interannual variability of sea surface temperature of the Indian Ocean, may contribute to the weakening or modifying of the ENSO–Asian monsoon relationship. On an interdecadal scale, the so-called PDO, the out of phase oscillation between the central to western North Pacific and the central to eastern tropical Pacific, a prominent feature of North Pacific sea surface temperature variability is believed to contribute to the Asian monsoon variation and to modify the ENSO–monsoon relationship. On the geological scale, the strengthening of the Asian monsoon has been linked to the uplift of the Tibetan Plateau after the collision of the Indian subcontinent and Asia around 50 million years ago. Based on records from the Arabian Sea and the record of wind-blown dust in the Loess Plateau of China, many geologists believe the Asian monsoon first enhanced around 8 million years ago. More recently, plant fossils in China and new long-duration sediment records from the South China Sea led to timing of the Asian monsoon starting 15–20 million years ago and linked to early Tibetan uplift.

HUMAN–MONSOON INTERACTION

The behaviors of Asian monsoons under global warming have been examined by both observation and climate modeling. It is likely that both Asian summer and winter monsoons have become weaker as the Northern Hemisphere warms. Recent studies also suggest that anthropogenic aerosols related to air pollution in Asia may also contribute to the monsoon variation and changing rainfall patterns.

The region affected by the monsoon system, the so-called monsoon Asia, supports the largest population on Earth, being home to 3.6 billion people. Almost all aspects of societal and economic activities in monsoon Asia are critically dependent on the monsoon climate and its variability. It has direct impacts on water resources, air quality, and occurrence of climate-related disasters and indirectly on agriculture, industry, health, urban life, and ecosystem services. Monsoon rainfall provides the major water resources for the region to support human beings and ecosystems, especially in the development of agriculture. Related to high variability of climate, the monsoon Asia region is characterized by high frequency and intensive climate-related disasters, such as floods and drought, which have profound impacts on the food security, water security, and daily life of the people as well as the sustainable development of the region. For example, during the period from 1986 to 2006, the fatalities of flood-related disasters in Asia, including floods, landslides, windstorm, and wave/surge, are 83.7% of global total (Fig. 2). Prediction of Asian monsoon activities and their impacts on human society has long been an important task for the meteorological agencies of the countries in this region.

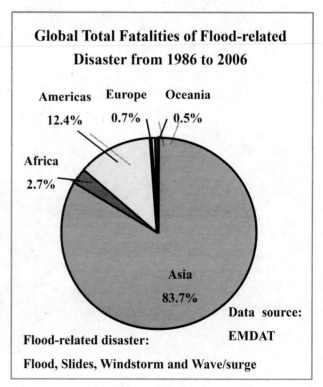

Fig. 2 Total global fatalities from flood-related disasters from 1986 to 2006.
Source: EMDAT, The international Disaster Database, Center for Research on the Epidemiology of Disaster (CRED) supported by USAID.

GLOBAL MONSOON

In the past several decades, it has been more widely accepted that monsoonal climate appears not only in Asia but also in Australia, Africa, South and North America, and the western North Pacific. A new concept of global monsoon was proposed to represent a response of the coupled atmosphere–land–ocean–creosphere–biosphere system to annual variation of solar radiative forcing.

REFERENCES

1. Xie, P.; Arkin, P.A. Global precipitation: A 17-year monthly analysis based on gauge observations, satellite estimates, and numerical model outputs. Bull. Am. Meteor. Soc. **1997**, *78* (11), 2539–2558.
2. Kalnay, E.; Kanamitsu, M.; Kistler, R.; Collins, W.; Deaven, D.; Gandin, L.; Iredell, M.; Saha, S.; White, G.; Woollen, J.; Zhu, Y.; Leetmaa, A.; Reynolds, R.; Chelliah, M.; Ebisuzaki, W.; Higgins, W.; Janowiak, J.; Mo, K.C.; Ropelewski, C.; Wang, J.; Jenne, R.; Joseph, D. The NCEP/NCAR 40-year reanalysis project, Bull. Am. Meteor. Soc. **1996**, *77* (3), 437–471.

BIBLIOGRAPHY

1. Chang, C.-P ; Ding, Y.; Lau, N.-C.; Johnson, R.H.; Wang, B.; Yasunari, T. *The Global Monsoon System: Research and Forecast*; World Scientific Publishing Co. Pte. Ltd.: Singapore, 2011; 594 pp.
2. Gao, Y.X. *Several Issues of East Asian Monsoon*; Science Press of China: Beijing, 1962; 278 pp.
3. Ramage, C.S. *Monsoon Meteorology*; Academic Press, 1971; 296 pp.
4. Wang, B. Ed; *The Asia Monsoon*; Springer-Praxis Books: Berlin, 2006; 787 pp.

**Acid—
Atmospheric**

Atmospheric Acid Deposition

Gene E. Likens
*Cary Institute of Ecosystem Studies, Millbrook, New York, and
University of Connecticut, Storrs, Connecticut, U.S.A.*

Thomas J. Butler
Cary Institute of Ecosystem Studies, Millbrook, and Cornell University, Ithaca, New York, U.S.A.

Abstract

Acid deposition, or more commonly called acid rain, is the wet and dry deposition of acidic substances from the atmosphere, usually derived from the burning of fossil fuels and the gaseous emission of sulfur dioxide (SO_2) and nitrogen oxides (NO_x). Acid rain (largely sulfuric and nitric acid) was detected on a region-wide basis in northwestern Europe in the late 1960s, and in eastern North America in the early 1970s, and was linked to the decline of fish populations in remote lakes far from pollution sources. Acid deposition also causes acidification in acid-sensitive landscapes where soil waters have poor acid-neutralizing capacity (ANC), or little ability to neutralize acid inputs. Government regulation of SO_2 and NO_x emissions has reduced acid deposition considerably in Europe and North America, but soils in acid-sensitive regions that have received acid deposition over many decades are now even more sensitive to further deterioration due to loss of base cations and continuing abundance of hydrogen and aluminum ions. Acid deposition has become a growing problem in developing areas, such as China and India, where there have been large increases in fossil fuel use, and limited controls on SO_2 and NO_x emissions. Further reductions in these emissions will be necessary to protect acid-sensitive areas and reduce other air pollution impacts, such as human health effects (e.g., ozone formation, fine particulate pollutants) related to SO_2 and/or NO_x emissions.

INTRODUCTION

"Acid rain" is the popular term for atmospheric acid deposition, which includes both wet and dry deposition. Wet deposition includes all forms of precipitation: rain, snow, fog, and cloud water. Dry deposition, which can generally range from 20% to 70% of total acid deposition, includes acidifying particles and gases which adsorb to surfaces, or are sometimes taken up by foliage. Some forms of acid deposition are natural, resulting from volcanic emissions (e.g., (hydrochloric and sulfuric acids), forest fires (nitric and organic acids), and lightning (nitric acid)). Acid deposition is of concern when additional acidity is added to the atmosphere from human-made (or anthropogenic) emissions of sulfur dioxide (SO_2) and nitrogen oxides (NO_x = NO + NO_2), largely from the burning of fossil fuels for power generation, industrial processes, transportation, heating, and cooking. These sources far exceed natural sources of acidity in many industrialized countries.

Anthropogenic acid deposition was first investigated in 1852 by Robert Angus Smith, an English chemist who studied the chemistry of rain in and around Manchester, and other industrialized cities in England and Scotland. Rain in these industrialized cities was discovered to contain sulfuric acid.[1] However, it was not until the 1960s that acid rain became an environmental issue first in northern Europe[2] and later in North America,[3,4] and was recognized as a widespread regional problem, rather than just a local concern.

Sulfuric acid is formed in the atmosphere by

$$SO_2 + H_2O \rightarrow H_2SO_4 \longleftrightarrow H^+ + HSO_4^-$$
$$\longleftrightarrow 2H^+ + SO_4^=, \text{ or}$$

$$2SO_2 + O_2 + 2H_2O \rightarrow 2H_2SO_4 \text{ etc.}$$

and nitric acid by

$$NO_2 + H_2O \rightarrow HNO_3 \longleftrightarrow H^+ + NO_3^- \text{ or}$$
$$NO_2 + OH \rightarrow HNO_3 \text{ etc.}$$

HNO_3 can also form an aerosol that can be dry deposited.

Acidity in precipitation is measured as pH, which is the negative log of the hydrogen ion (H^+) concentration. For every 1-unit change in pH, the acidity, or hydrogen ion concentration changes 10-fold. For example, a pH of 5.0 = 10 µeq/l (microequivalents per liter of solution) of H^+, a pH of 4.0 = 100 µeq H^+/l etc. Thus, acidity increases as pH decreases. Natural background levels of precipitation tend

Encyclopedia of Natural Resources DOI: 10.1081/E-ENRA-120047613

to be around pH 5.1 to 5.2 (8 to 6 μeq H⁺/l), but during the mid-1970s, pH yearly averages were on the order of 4.1 (80 μeq H⁺/l) to 4.3 (50 μeq H⁺/l) in northwestern Europe, and the northeastern United States and southeastern Canada, respectively.[5] Thus acidity in precipitation was about 8–10 times background levels. Individual events as low as pH 2.4 were recorded during this period.

MAJOR EMISSION SOURCES AND TRENDS

The major anthropogenic sources of SO_2 emissions are coal and oil-fired electric generating plants, metal smelters, and industrial fuel combustion. Major anthropogenic sources of NO_x are vehicles, electric generating facilities, and industrial fuel combustion. Trends for the United States are shown in Fig. 1A and 1B. European and North American emissions of SO_2 and NO_x have declined sharply over the past 40 years. However, emissions of SO_2 and NO_x have increased in eastern Asia (e.g., China and India), where industrial expansion and electric generation have expanded rapidly in recent decades. Emission controls in this region were limited through 2000 (Fig. 2A and 2B). In 2005, total global emissions of SO_2 were ~115 million metric tons/yr, with China accounting for 28% of the total, 5% for India, 15% for North America and 10% for Central and Western Europe.[7] Total global anthropogenic NO_x emissions for 2010 were estimated at ~77 million metric tons with China, India, North America, and Central and Western Europe accounting for 17%, 8%, 25%, and 12%, respectively.[8] Eastern North America, Europe, and eastern Asia are areas that have been impacted by

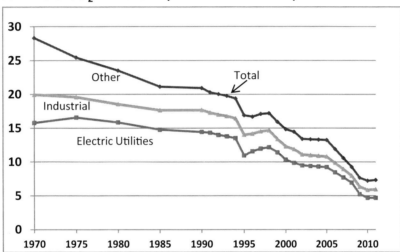

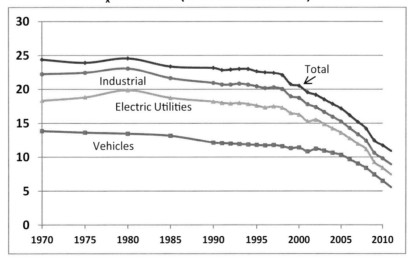

Fig. 1 Annual (**A**) SO_2 and (**B**) NO_x emissions in the United States.
Source: Adapted from EPA National Emissions Inventory.[6]

A   SO$_2$ Emissions (million metric tons)

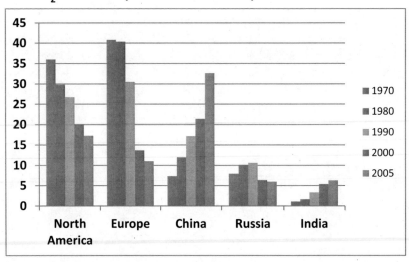

B   NO$_x$ Emissions (million metric tons)

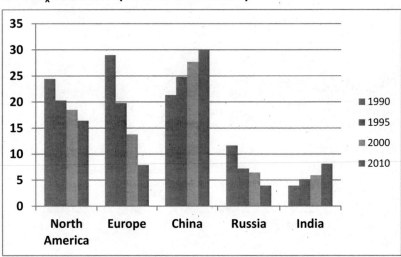

Fig. 2   Trends in (**A**) SO$_2$ emissions, and (**B**) in NO$_x$ emissions for selected regions based on model results. Note years are different (1990 to 2010) than years for SO$_2$ trends (1970 to 2005).
Source: Adapted from Smith et al.,[7] Cofala et al.[8]

acid deposition due to several factors such as deposition history and the presence of weathering-resistant bedrock and soils that heighten sensitivity in these regions (see Fig. 3).

TRENDS IN ACID DEPOSITION

Changes in emissions of SO$_2$ and NO$_x$, in most instances, are accompanied by comparable changes in acid deposition in regions downwind of the emission sources. For example, declines in emissions from the eastern and midwestern United States in recent decades are reflected in significant declines in H$^+$ concentration. One of the best examples of this change is from Hubbard Brook Experimental Forest (HBEF) located in

the White Mountains, New Hampshire, and far from large emission sources. The HBEF has the longest record of precipitation and stream water chemistry in the United States, with records that extend back to 1964.[12] This long-term record shows a 70% decline in H$^+$ concentration, from a 5-year average of 77 μeq H$^+$/l (pH = 4.11) in 1966 to 1970, to a 5-year average of 23 μeq H$^+$/l (pH 4.64) from 2005 to 2009 (Fig. 4).

The National Atmospheric Deposition Program tracks changes in precipitation chemistry in the United States through the NTN and AIRMoN networks.[13] The trend in acidity of precipitation in the United States is clearly shown by comparing Fig. 5A with Fig. 5B.

The USA Clean Air Amendments of 1990 included provisions to reduce acid deposition by decreasing SO$_2$ and NO$_x$

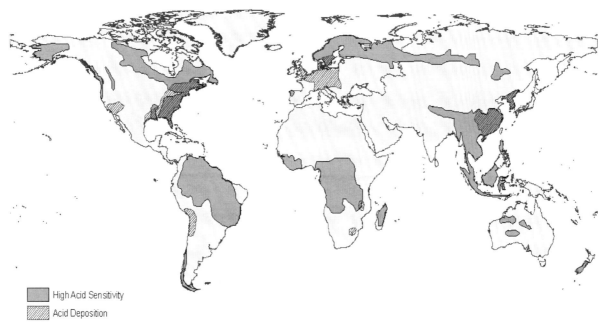

Fig. 3 A global perspective of the extent of acid deposition (hatched areas) and areas most likely to be impacted because of the sensitivity of the landscape to acid inputs (shaded areas).
Source: Adapted from Likens et al.,[9] Rhode et al.,[10] Kuylenstierna et al.[11]

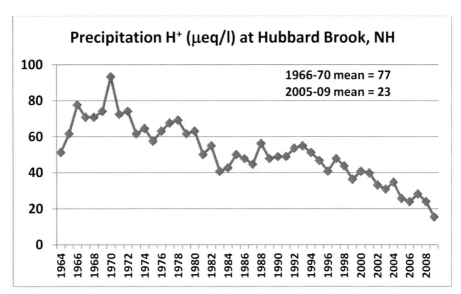

Fig. 4 Long-term record of annual H+ concentration in precipitation at Hubbard Brook Experimental Forest.

emissions, mainly from power plants. These emission reductions were first phased in after 1994. The effectiveness of these emission reductions is well documented. From 1994 to 2009, pH in precipitation increased (acidity decreased) from about 4.2 to 4.7 in the Northeast, from 4.7 to 5.0 in the southeast, and from 4.4 to 4.9 in the Midwest. These represent declines in H+ concentrations of 43, 10 and 27 μeq H+/l, respectively. Reduced emissions also simultaneously reduced dry deposition of acid generating species.[14,15] More recent air pollution rules, including mandated reductions in NO$_x$ from vehicles, have continued to reduce emissions and

further decrease acid deposition in the United States.[16] On a similar time scale, reductions in emissions from clean air legislation and resulting reduction in acid deposition has also occurred in Canada and Europe (Fig. 2) and further emissions reductions are anticipated.[17,18]

ENVIRONMENTAL EFFECTS OF ACID DEPOSITION

Acid deposition became an environmental issue when it was first linked to the loss of fish populations in the 1960s in

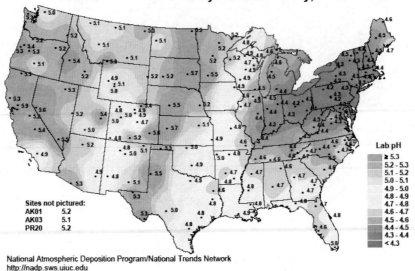

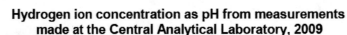

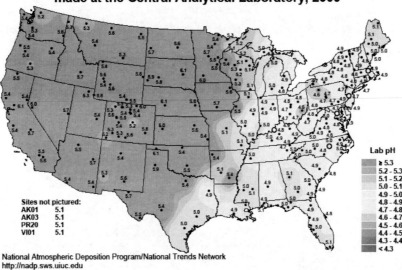

Fig. 5 National levels of pH in precipitation for 1994 and 2009.
Source: Adapted from NADP.[13]

remote areas of Northern Europe where no local sources of pollution were evident.[2] As further research into acid deposition occurred, clear impacts became evident that extended to terrestrial ecosystems, human-made structures, and indirectly (largely through inhalation of particles) to human health. Ecosystem impacts occur where the landscape is acid sensitive, which results from a combination of factors that includes precipitation amount, vegetation type and density, the resistance of bedrock to weathering, and the acidity of soils. Geology and soils in these sensitive areas are resistant to weathering, and therefore do not have enough readily

available base cations, such as calcium (Ca^{2+}) and magnesium (Mg^{2+}), to neutralize soil water acidity. Other sensitive areas include highly weathered tropical soils under the influence of high rainfall rates.[11] The overlap between areas receiving acid deposition and those that are acid sensitive is illustrated in Fig. 3.

Aquatic Effects

In the 1960s, Svante Odén, a Swedish soil scientist, linked the widespread acidification of Swedish lakes and rivers to

SO_2 emissions originating in Great Britain and Central Europe. Acid rain became an issue in North America in the early 1970s when the phenomenon was documented in the northeastern United States by Gene Likens et al.[3] and was linked to loss of fish populations in remote acid-sensitive lakes in eastern Canada[19] and the Adirondack Mountains of New York.[20]

Many fish species prefer waters that are above pH 6, although there is a range of acid tolerance among different fish species. There are typically 5–6 fish species in Adirondack lakes with pH values of 6 and above, but only 1 or 2 fish species when pH values are <5.[21] This pattern of lower species diversity with increasing acidity (and lower pH) is observed throughout the food web including plants, insects, and other invertebrates that are food sources for fish.[22,23] Lakes with a low pH, and low ANC (e.g., low-calcium ion (Ca^{2+}) concentration), often have high dissolved aluminum (Al^{3+}) concentrations, which are toxic to both plants and animals. Aluminum is generally bound and immobilized in nonacidic soils. However, aluminum becomes 1000 times more soluble as pH declines from 5.6 to 4.6.[24] Aluminum toxicity leads to fish mortality from deterioration of fish gills and impairment of respiration.

Another complication for aquatic ecosystems, especially in cool temperate climates where snowfall is common, is episodic acidification. During winter, snowpack accumulates; then with the first major thaws, much of the acidity stored in the snowpack leaches out preferentially in concentrated form into streams and lakes. This acid shock can result in fish kills and mortality to other aquatic species. Increased release of dissolved aluminum from soils can further increase the toxicity of the waters entering streams and lakes. Episodic acidification can also occur in non-snowy environments, where large rain storms lead to rapid runoff and little time for acid neutralization of these runoff waters.

Terrestrial Effects

In the 1970s and 1980s, forest health was in decline in parts of Europe and eastern North America, especially in acid-sensitive areas. Acid deposition was considered a factor, but linking acid deposition to forest decline was complicated by many other environmental factors. Tree vigor can be compromised by drought, insect infestation and disease, and other forms of air pollution such as high levels of atmospheric ozone, and excess fertilization from nitrogen deposition. The latter stressors (ozone and excess nitrogen) are also related to acid deposition since NO_x is a precursor to ozone formation and nitrogen deposition, as well as acid deposition.

There is substantial evidence supporting the hypothesis that acid deposition is at least one major factor in forest decline, both directly and indirectly. Red Spruce, common at high elevations, is directly damaged by leaching of calcium from cell membranes in the tree needles, due to

acid deposition. This impact has been linked to loss of cold hardiness and freezing damage. High-elevation forests are also exposed to high levels of acidity from clouds and fog, which often have acidity levels many times higher than precipitation in the same area.

Acid deposition on poorly buffered soils removes secondary plant nutrients such as calcium and magnesium, which can already be in short supply. The inability of base-poor soils to neutralize acid deposition can lead to increased levels of Al^{3+} in soil waters, which can be toxic to plant roots and aquatic organisms (as mentioned earlier). High aluminum concentrations and low calcium concentrations in soil waters have been correlated with sugar maple decline in Canada and the northeastern USA.

While ammonia (NH_3) and ammonium (NH_4^+) are considered basic substances, they can also lead to soil acidification when microbial nitrification occurs. Nitrification is the conversion of NH_3 or NH_4^+ to NO_3^- (nitrate). The simplest form of this conversion is

$$2NH_4^+ + 3O_2 \rightarrow 4H^+ + 2H_2O + 2NO_2^-$$
(The H^+ increases acidity)

$$2\,NO_2^- + O_2 \rightarrow 2NO_3^-.$$

NH_3 and NH_4^+ are largely derived from agricultural activity, including fertilizer application, but mainly from production of livestock waste. Ammonia, unlike SO_2 and NO_x emissions, has not shown large regional declines, and in many agricultural areas NH_3 emissions are increasing.[25]

CONCLUSION: ACID DEPOSITION TODAY AND IN THE FUTURE

Legislation has led to large declines in emissions of SO_2 and NO_x in Europe and North America, which has resulted in a substantial reduction in acid deposition. However, reduction in acid deposition has not always produced the expected recovery in ecosystems. Many decades of acid deposition have depleted acid-sensitive watersheds and their soils of the acid-buffering capacity necessary to reduce acidifying impacts on water and forest ecosystems. [26] While some recovery has occurred (e.g., rising pH levels in formerly acidified lakes and streams in Europe and North America),[27] these regions over time have become more acid sensitive and therefore continue to be impacted, even though the amount of acid deposition has declined. Recovery in these ecosystems will be a slow process. Decades of acid deposition may require decades of significantly reduced deposition for full recovery to occur.[28]

While acid deposition has declined in both Europe and North America, emissions of SO_2 and NO_x (and NH_3) have been increasing in rapidly developing areas such as Southeast Asia. China, India (see Fig. 2a and 2b), and other southeastern Asian countries are experiencing rapid growth in fossil fuel use for electric generation, industrial development and transportation, but have limited controls on the associated SO_2 and NO_x emissions (although since 2007, China has begun reductions in SO_2 from power plants).[29] Acidification from SO_2 and NO_x emissions is partially being neutralized by large amounts of alkaline particulates which are also being released from some of these same facilities.[11] However, because of the health-related problems associated with these multiple forms of air pollution, it will be important to control particulate,[28] as well as SO_2 and NO_x emissions. If only particulates emissions were reduced, there would be less atmospheric ANC and levels of acid deposition would rise rapidly in this area of the world. In summary, continued reductions in emissions of SO_2 and NO_x will be necessary in areas that have experienced significant acid deposition in the past, and in areas where the potential exists for increasing acid deposition in the future.

REFERENCES

1. Smith, R.A. *Air and Rain, the Beginning of a Chemical Climatology*; Longmans Green: London, 1872.
2. Odén, S. *The Acidification of Air Precipitation and its Consequences in the Natural Environment*; Bulletin of the Ecological Research Communications NFR. Translation Consultants Ltd.: Arlington VA, 1968.
3. Likens, G.E.; Bormann, F.H.; Johnson, N.M. Acid rain. Environment **1972**, *14*, 33–40.
4. Likens, G.E.; Bormann, F.H. Acid rain: a serious regional environmental problem. Science **1974**, *184* (4142), 1176–1179.
5. Likens, G.E.; Wright, R.F.; Galloway, J.N.; Butler, T.J. Acid rain. Sci. Am. **1979**, *241* (4), 43–51.
6. EPA National Emissions Inventory, http://www.epa.gov/ttnchie1/trends/ (accessed April 2012).
7. Smith S.J.; van Aardenne, J.; klimont, Z.; Andres, R.J.; Volke, A.; Delgado, A. Anthropogenic sulfur dioxide emissions: 1850–2005. Atmos. Chem. Phys. **2011**, *11*, 1101–1116 doi: 10:5194/acp-11-1101-2011.
8. Cofala, J.; Amann, M.; Mechler, R. Scenarios of world anthropogenic emissions of air pollutants and methane up to 2030. *Regional Air Pollution Information and Simulation*; Atmospheric Pollution and Economic Development, 2006. http://www.iiasa (accessed April 2012)
9. Likens, G.E.; Butler, T.J; Rury, M. Acid Rain. In *Encyclopedia of Global Studies*; Anheier, H; Juergensmeyer, M., Eds.; Sage Publications, Inc.: 2009.
10. Rhode, H.; Dentener, F.; Schulz, M. The global distribution of acidifying wet deposition. Environ. Sci. Technol. **2002**, *36*, 4382–4388.
11. Kuylenstierna, J.C.I.; Rhodhe, H.F.; Cinderby, S; Hicks, K. Acidification in developing countries: Ecosystem Sensitivity and the Critical Load Approach on a Global Scale. *Ambio* **2001**, *30*, 20–28.

12. Likens, G.E.; Buso, D.C. Dilution and the elusive baseline. Environ. Sci. Tech. **2012**, *46* (8), 4382–4387, DOI: 10.1021/es3000189.
13. NADP. National Atmospheric Deposition Program (NRSP-3). 2012, NADP Program Office, Illinois State Water Survey, 2204 Griffith Dr., Champaign, IL 61820, http://nadp.sws.uiuc.edu/ (accessed March 2012).
14. Burns, D.A. The National Acid Precipitation Assessment Program Report to Congress 2011: An Integrated Assessment. 2011, http://ny.water.usgs.gov/projects/NAPAP/
15. Butler, T.J.; Likens, G.E.; Vermeylen, F.M.; Stunder, B.J. The impact of changing nitrogen oxide emissions on wet and dry nitrogen deposition in the northeastern USA. Atmos. Environ. **2005**, *39*, 4851–4862.
16. USEPA. 2010 Progress Report - Environmental and Health Results. 2012, http://www.epa.gov/airmarkets/progress/ARPCAIR10_02.html#litigation (accessed June 2012).
17. Baseline Scenerios for the Clean Air for Europe (CAFÉ) Programme. Amann, M.; Bertok, I.; Cofala, J.; Gyarfas, F.; Heyes, C.; Klimont, Z.; Schöpp, W.; Winiwarter, W. submitted to the European Commission Directorate General for Environment 2005, http://www.behouddeparel.nl/zandverwerkingsinstallatie/Fijnstof%20cafe_lot11.pdf
18. Canada-United States Air Quality Agreement Progress Report 2010 http://www.ec.gc.ca/Publications/4B98B185-7523-4CFF-90F2-5688EBA89E4A%5CCanadaUnitedStatesAirQualityAgreementProgressReport2010.pdf
19. Beamish, R.J.; Harvey, H.H. Acidification of the La Cloche Mountain Lakes, Ontario, and resulting fish mortalities. J. Fish. Res. Bd. Canada **1972**, *29* (8), 1131–1143.
20. Schofield, C.L. Acid precipitation: effects on fish. Ambio **1976**, *5*, 228–230.
21. Driscoll, C.T.; Lawrence, G.B.; Bulger, A.J.; Butler, T.J.; Cronan, C.S.; Eager, C.; Lambert, K.F.; Likens, G.E.; Stoddard J.L.; Weathers, K.C. Acid deposition in the northeastern United States: sources and inputs, ecosystem effects, and management strategies. BioScience **2001**, *51* (3), 180–198.
22. Schindler, D.W.; Mills, K.H.; Malley, D.F.; Findlay, D.L.; Shearer, J.A.; Davies, I.J.; Turner, M.A.; Linsey, G.A.; Cruikshank, D.R. Long-term ecosystem stress: the effects of years of experimental acidification on a small lake. Science **1985**, *228* (4706), 1395–1401.
23. Confer, J.L.; Kaaret, T.; Likens, G.E. Zooplankton diversity and biomass in recently acidified lakes. Can. J. Fish. Aquat. Sci. **1983**, *40* (1), 36–42.
24. Johnson, N.M.; Driscoll, C.T.; Eaton, J.S.; Likens, G.E.; McDowell, W.H. Acid rain, dissolved aluminum and chemical weathering at the Hubbard Brook Experimental Forest New Hampshire, USA. Geochim. Cosmochim. Acta **1981**, *45*, 1421–1438.
25. Galloway, J.N.; Townsend, A.R.; Erisman, J.W.; Bekunda, M.; Cai, Z.; Freney, J.R.; Martinelli, L.A.; Seitzinger, S.P.; Sutton, M.A. Transformation of the nitrogen cycle: recent trends, questions, and potential solutions. Science **2008**, *320*, 889–892.
26. Likens, G.E.; Driscoll, C.T.; Buso, D.C. Long-term effects of acid rain: response and recovery of a forest ecosystem. Science **1996**, *272*, 244–246.

27. Stoddard, J.L.; Jeffries, D.S.; Lükewille, A.; Clair, T.A.; Dillon, P.J.; Driscoll, C.T.; Forsius, M.; Johannessen, M.; Kahl, J.S.; Kellogg, J.H.; Kemp, A.; Mannio, J.; Monteith. D.T.; Murdoch, P.S.; Patrick, S.; Rebsdorf, A.; Skjeikvåle, B.L; Stainton, M.P.; Traaen, T.; van Dam, H.; Webster, K.E.; Wieting, J.; Wilander, A. Regional trends in aquatic recovery from acidification in North America and Europe. Nature **1999**, *401*, 575–578.

28. Likens, G.E.; Buso, D.C. Dilution and the elusive baseline. Environ. Sci. Tech.**2012**, *46* (8), 4382–4387 DOI: 10.1021/es3000189.

29. Li, C.; Zhang, Q.; Krotkov, N.A.; Streets, D.G.; He, K.; Tsay, S.-C; Gleason, J.F. Recent large reduction in sulfur dioxide emissions from Chinese power plants observed by the Ozone Monitoring Instrument. J. Geophys. Res. Lett. **2010**, *37*, L08807.

Atmospheric Circulation: General

Thomas N. Chase
David Noone
Department of Civil, Environmental and Architectural Engineering and Cooperative Institute for Research in the Environmental Sciences (CIRES), University of Colorado at Boulder, Boulder, Colorado, U.S.A.

Abstract

We discuss the general motion of the Earth's atmosphere starting with basic energy transformations that drive the circulation, pressure relations, and the form the circulation takes in observations. We then discuss large-scale eddies that form on the zonally averaged flow. We emphasize the fundamental physical processes of energy conversion and transport that define the general circulation and how they relate to the hydrological cycle. Finally, we examine important regional asymmetries and human effects imposed on the basic flow.

INTRODUCTION

The Earth as a whole is in approximate radiative equilibrium. The amount of energy from the sun is balanced by the emission of infrared radiation from the Earth. However, there is an uneven distribution of energy. Because of the spherical shape of the Earth, sunlight directly illuminates the tropical regions, while close to the poles solar energy strikes the Earth at an oblique angle reducing the amount of energy received per unit area. In the extreme case, the winter pole receives no sunlight at all. The reflectivity of Earth depends on the cloud cover, other atmospheric constituents, and variations in the characteristics of the surface (snow, land, vegetation, and liquid ocean), further influencing the total amount of solar energy input and also further modifying the spatial distribution of energy. On the other hand, the amount of infrared radiation emitted by the Earth is more uniform and depends on only the average temperature of the atmosphere, which varies by about 15% from equator to pole (290 K in the tropics to 250 K at the pole). These two features taken together describe a net surplus of energy in the tropics (where energy gained from the sun dominates) and a net energy deficit near the polar regions (when loss of energy by infrared emission dominates). For energy balance, there is therefore a requirement for energy transport from the tropics to the polar regions. The winds that provide this required energy transport in the atmosphere are the atmospheric general circulation. A brief, historical overview of the concept of the general circulation is given by Lorenz,[1] and a more technical account of atmospheric motions is given by Schneider.[2]

Figure 1 shows the long-term mean sea level pressure and surface wind for July and January. For many centuries, it has been known that surface winds are predominantly easterly (wind coming from the east) along the equator and westerly (coming from the west) in the midlatitudes (between about 35° and 55° N and S). The subtropics, centered around 30° N and S, are characterized by low wind speeds (known as the doldrums due to the slow travel for sail-powered ships) and are usually free from poor weather and clouds. The very consistent easterlies in the tropics provide reliable travel for trade and are known as the trade winds. Alongside the wind patterns, atmospheric pressure at sea level is generally low along the equator, high in the subtropics, and low again near the midlatitudes. These features reflect and characterize many aspects of the average atmospheric general circulation and are explained by considering the combination of overturning of the atmosphere with conservation of momentum.

OVERTURNING CIRCULATION

Pressure measures the atmospheric mass and, on average, is the force exerted by the mass of the atmosphere per unit area. The global average pressure at the Earth's surface is 986 hPa. Pressure at sea level is typically around 1000 hPa, while the pressure at higher altitudes is lower. Pressure decreases with altitude because the fraction of the atmospheric mass above a point at high altitude is smaller than the fraction of the mass above a point at low altitude. Air can be treated as an ideal gas and so there are relationships among pressure, temperature, and volume. When air is heated at constant pressure, the volume increases (i.e., for a constant mass, density decreases). In the tropics, heating of the atmosphere is associated with the radiative imbalance (and is caused by latent heat realized during precipitation, which is associated with the energy excess in the

Encyclopedia of Natural Resources DOI: 10.1081/E-ENRA-120047618

A January

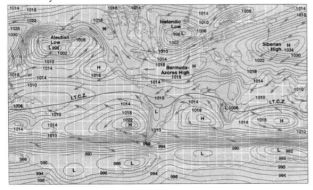

B July

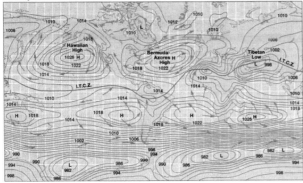

Fig. 1 Long-term average sea-level pressure and wind vectors for: (**A**) January, (**B**) July.
Source: From Aguado & Burt.[3]

tropics that had caused previous evaporation at the Earth's surface). The expansion of the column of air in the tropics is synonymous with ascending motion. With expansion, the pressure half way up the column will be higher than a shorter column away from the equator, and the difference in pressure from south to north will force the air at the top of the column poleward. Air moving out of the tropical column is a loss of mass, and so the surface pressure is lower in the tropics than outside the tropics. Air near the surface will be forced to move from high to low pressure and converge toward the equatorial zone from both hemispheres. This region of convergence is also the region of most intense tropical thunderstorm convection and is identified as the Intertropical Convergence Zone (ITCZ). The ITCZ encompasses land masses in the summer months in both hemispheres due to the greater heating of land relative to oceans, which causes an onshore flow moving the ITCZ inland. These continental scale onshore flows are referred to as monsoon circulations and are associated with regions of intense seasonal precipitation and moist land ecosystems such as rainforests.

Poleward flow aloft and equatorward flow at the surface describes a thermally direct overturning circulation cell (the Hadley Cell). Since the poleward moving air has been heated, it transports energy poleward. Radiative cooling

(net emission of radiation to space) at higher latitudes ensures that the equatorward moving air has less energy. The sum of these two components provides a net transport of energy out of the tropics, partially meeting the energy transport requirement for radiative balance.

The winds can be described by combining the motion of the overturning with consideration of conservation of momentum on the rotating Earth. Specifically, if one views the Earth rotating from space, a small volume of air that is at rest near the equator is moving at a speed given by the tangent speed of the Earth in solid body rotation. If this air mass is moved poleward at constant altitude, this velocity is conserved. Because the Earth is spherical, the distance between the Earth's surface and the rotating axis is smaller and the tangential speed of the Earth's surface will be smaller than the speed of the air mass. In which case, an observer at the surface will see the air moving eastward (termed westerly). Therefore, air in the upper troposphere (about 10–15 km above the surface) in the subtropics will be westerly because of the upper poleward branch of the overturning cell. This westerly wind region is the subtropical jet stream. The same argument can be applied to an air mass initially in the subtropics that moves to the west as it converges toward the equator (i.e., easterly trade winds). This phenomenon was described succinctly by George Hadley in 1735,[4] and the apparent force that gives rise to the east-west flow was named after Gaspard-Gustave Coriolis in the nineteenth century—the Coriolis force.[5]

While the tropical overturning circulation transports energy from the equator to the subtropics initially, the relationship between surface pressure and wind reveals a conundrum in the need for energy export from the subtropics to further poleward—as is required. For the Earth's atmosphere outside of the tropics, the Coriolis force tends to become equal and opposite to the pressure force and a balance is approximately reached (geostrophic balance). The conservation of momentum requires that the Coriolis force acts perpendicular to the direction of air motion (to the right in the Northern Hemisphere and to the left in the Southern Hemisphere). At the poleward edge of the subtropics, the decrease in pressure toward the pole provides a force that is balanced by a Coriolis force directed to the equator, which in turn requires the wind to be westerly. Indeed, westerly winds outside the tropics are seen in Fig. 1 to be approximately parallel to the isobars as predicted by geostrophic balance. The component of the wind vectors not parallel to the isobars is due to the slowing of wind speeds affected by surface friction. Upper level winds are more nearly geostrophic.

If this were the entirety of the process, the energy transported poleward from the equator would accumulate in the subtropics and the winds would be purely westerly, preventing further transport of energy toward the poles because the westerly flow would simply transport the energy eastward. This is not the case, however.

As energy continues to accumulate, the continual increase in the pressure difference between the subtropics and higher latitudes eventually becomes sufficiently great that instability in the westerly flow causes the development of continental-size waves embedded in the westerly flow which are associated with frontal weather systems that are composed of low pressure centers near the Earth's surface. Air circulates around the low pressure center, again with the flow almost parallel to the lines of constant pressure when friction is negligible (generally about 1–2 km above the Earth's surface). In the region of the low pressure system, warm moist air is moved poleward on the eastern side and colder and drier polar air travels equatorward of the western side, the sum of the two processes providing a net transport of energy poleward. As such the transport of energy required to meet the demands of radiative balance is met in the midlatitudes by the sequence of high- and low-pressure systems continually being generated and dissipated in the region of large temperature gradients and locally strong jet streams.

POLEWARD ENERGY TRANSPORT BY THE GENERAL CIRCULATION

Atmospheric motions are fundamentally caused by an excess of energy near the equator and an energy deficit near the poles. This gradient in energy is reflected in temperature and moisture (warm and moist tropics, cold and dry polar regions). The atmospheric circulation exists physically as a mechanism to limit these gradients in spite of continual forcing. Figure 2 shows the zonally averaged northward energy transport for the globe. The maximum energy transport poleward in both hemispheres by the

atmosphere occurs in midlatitudes at about 40°. This reflects the mixing action of midlatitude cyclonic (frontal) systems discussed in the previous section. Ocean transport of energy is of smaller magnitude with maxima in the subtropics.[6] A basic property of the overall atmospheric flow is that it is "chaotic" and therefore unpredictable beyond a week or two.[7]

ROLE OF THE HYDROLOGICAL CYCLE IN THE GENERAL CIRCULATION

The fundamental role of the water molecule in the general circulation cannot be overstated. While H_2O in all its phases is universally recognized as fundamental for the existence of life in general and human activities such as agriculture in particular, from the perspective of the atmospheric circulation water substance plays a central role in major energy transport mechanisms. This is expressed by the latent heating due to phase changes. Energy is needed to evaporate water from a wet land surface or ocean, and this energy is later released in the upper atmosphere if enough cooling occurs to cause condensation, resulting in a net upward energy transport. The radiative properties of the atmosphere dictate that the atmosphere is mostly heated from below with energy transfer from the Earth's surface to the lower atmosphere. Approximately three quarters of the net radiation arriving at the Earth's surface (Fig. 3) is transferred back to the atmosphere as latent heat (energy transfer due to the phase changes of H_2O that would dominant over ocean areas) as opposed to sensible

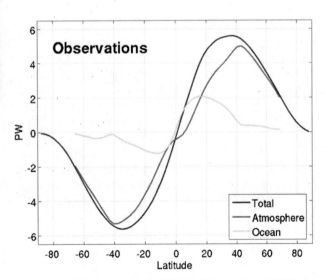

Fig. 2   Poleward energy transport by the general circulation by atmosphere, ocean, and the sum of the two.
Source: Modified from Trenberth & Caron.[6]

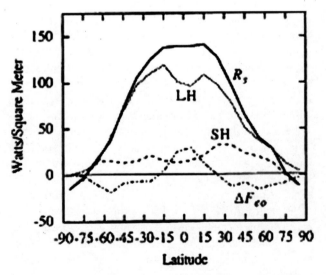

Fig. 3   Annual average surface energy balance over the lobe as a whole indication the proportion of net radiation (R_s) returned to the atmosphere as sensible heating (SH), latent heating (LH) by latitude. ΔF_{eo} represents energy transferred horizontally out of the land ocean below surface column.
Source: From Hartmann.[8]

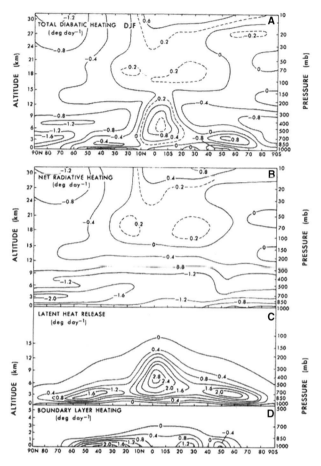

Fig. 4 Comparison by altitude and latitude of: (**A**) total diabatic heating due to radiation + sensible + latent heating, (**B**) net radiative heating, (**C**) latent heat, and (**D**) sensible heat.
Source: From Peixoto & Oort.[9]

heat (direct heating of the lower atmosphere) and infrared radiation.

Figure 4 shows the partitioning of atmospheric heating by process, altitude, and latitude and indicates that latent heating (third panel from top) is the major process supplying energy to the atmosphere. Compared to sensible heating (fourth panel from top and labeled boundary layer heating), latent heating is of considerably larger magnitude and is realized through a much greater depth of the atmosphere. Net radiative heating indicates mostly a cooling effect due to radiative losses to space. The top panel, total diabatic heating, represents the sum of the other three panels.

The role of the hydrological cycle in the general circulation of the future is of immense interest because three of the largest positive feedback mechanisms (processes that add to the magnitude of the original forcing) to warming due to increased CO_2 involve the H_2O molecule, one feedback for each of its three phases. The water vapor feedback (increased atmospheric water vapor content due to increased temperature and therefore increased saturation

point) is responsible for the largest part of the simulated global warming signal.[10,11] Increased cloudiness due to the increased water vapor is also thought to amplify global warming, though the precise magnitude remains poorly known because the net response is a function of the exact composition and altitude of changing clouds. The ice albedo feedback, where melting ice leads to decreased reflection of solar energy and therefore increased energy absorption, is also recognized as a positive feedback to global warming.[12]

ASYMMETRIC FLOW: TELECONNECTION PATTERNS, MONSOONS

We have previously discussed the symmetric overturning circulation and the asymmetric waves superimposed upon it that take the form of frontal systems at the Earth's surface. Other asymmetric parts of the general circulation that are important in both medium- to long-range weather forecasting and for understanding changes in the regional climate on time scale from decades to many thousands of years are atmospheric teleconnection patterns. These are defined as quasi-periodic fluctuations in part of the atmospheric–ocean system that then excite waves which propagate large distances away from the initial disturbance. El Nino-Southern Oscillation (ENSO) is one such phenomenon that involves teleconnections and that begins with a disturbance in the equatorial Pacific but subsequently has remote impacts on temperature and precipitation elsewhere on the globe. A number of other teleconnection patterns have been identified[13,14] that change the general atmospheric circulation over periods of years to decades. A discussion of more recently discovered teleconnection patterns is given in Thompson and Wallace.[15]

A second major class of asymmetric motions are monsoon circulations that are defined classically as a reversal of winds from onshore to offshore following the heating and cooling of land surfaces with season (i.e., an annual oscillation). Tropical monsoon systems are linked to the ITCZ in the sense that the heating of land surfaces in the summer draws the ITCZ over the land surface and so shifts the latitude of the rising branch of the tropical overturning circulation.

Another class of atmospheric circulation related to monsoons by the same physical process of differential heating of land and adjacent ocean is the regional sea breeze. Such locally confined circulations are vital, e.g., for the initiation of convection over the numerous islands in the tropical oceans that feed large parts of the ITCZ.

HUMAN EFFECTS: CO₂, AEROSOLS, LAND COVER CHANGE

Human activity alters the general circulation in a variety of ways. The increase in atmospheric CO_2 leads to a general atmospheric warming and consequently changes the global hydrological cycle while also reducing the speed of the tropical atmospheric circulations. These effects will necessarily involve changes in atmospheric circulations globally.[16]

Aerosols, particles resulting from combustion and residing in the atmosphere and that tend to reflect solar energy therefore disrupting the surface energy balance, are also implicated in atmospheric circulation changes.[17]

There is evidence that human-induced land-cover changes alter the general circulation.[18,19] For example, tropical deforestation tends to reduce evapotranspiration and increase albedo creating links to both the distribution of net surface radiation and the hydrological cycle both of which affect pressure patterns and winds. New diagnostic techniques, such as the use of isotope ratios of water as tracers, are proving useful in exploring linkages between the water cycle and both ambient general circulation and perturbations due to humans.[20]

CONCLUSION

The general motion of the Earth's atmosphere arises from differential energy production with excess energy in the tropics and near the earth's surface. These energy gradients are reflected in pressure relations that drive a circulation which transports energy upward and poleward on average. Deviations from symmetry include baroclinic waves (frontal systems), teleconnection patterns, and monsoon systems. We have emphasized the fundamental processes of energy conversion and transport that define the general circulation and how these relate to the hydrological cycle. Specifically, the atmospheric and oceanic general circulation can be well described by basic laws of physics including Newton's laws of motion applied to air on a rotating spherical Earth, the definition of air as an ideal gas, conservation of energy, and conservation of mass keeping in mind that this is a nonlinear system. We also described some ways in which human activity affects the general circulation through changes that are imposed on the distribution of energy associated with the radiative properties of the atmosphere (including changes to the concentration of gases like carbon dioxide and methane and aerosol particulates), and changes to the properties of the land surface (reflectivity of different types of surfaces and transfer of water from the landscape to the atmosphere).

The features of the general circulation described above are well understood in a basic sense. However, how the atmospheric circulation might change with future perturbations and the precise magnitude of the many feedbacks within the climate system remain without complete description.

REFERENCES

1. Lorenz, E.N. A history of prevailing ideas about the general circulation of the atmosphere. Bull. Am. Meteorol. Soc. **1983**, *64* (7), 730–733.
2. Schneider, T. The general circulation of the atmosphere. Annu. Rev. Earth Planet. Sci. **2006**, *34*, 655–688, DOI: 10.1146/annrev.earth.34.031405.125144.
3. Aguado, E.; Burt, J.E. *Understanding Weather and Climate*, 6th Ed.; Prentice Hall: New Jersey, 2013; 576 pp. ISBN-13: 9780321769633©2013.
4. Hadley, G. Concerning the cause of the general trade winds. Phil. Trans. Royal Soc. **1735**, *39*, 58–62.
5. Coriolis, C.G. Sur les equations du movement relative des systemes de corps. J. De L'ecole Royale Polytechnique **1835**, *15*, 144–156.
6. Trenberth, K.E.; Caron, J.M. Estimates of meridional atmosphere and ocean heat transports. J. Climate **2001**, *14* (16), 3433–3443. doi: http://dx.doi.org/10.1175/1520-0442(2001)014<3433:EOMAAO>2.0.CO;2
7. Lorenz, E.N. Deterministic non-periodic flow. J. Atmos. Sci. **1963**, *20* (2), 130–141.
8. Hartmann, D.L. *Global Physical Climatology*; Academic Press: San Diego, 1994; 411 pp.
9. Peixoto, J.P.; Oort, A. *Physics of Climate*; American Institute of Physics: New York, USA, 1992; 520 pp.
10. Arrhenius, S. On the influence of carbonic acid in the air upon the temperature of the ground. Lond. Edinb. Dublin Philos. Mag. J. Sci. (Fifth Series) **1896**, *41* (251), 237–275.
11. Held, I.M.; Soden, B.J. Water vapor feedback and global warming. Annu. Rev. Energy Environ. **2000**, *25* (1), 441–475.
12. Solomon, S.; Qin, D.; Manning, M.; Chen, Z.; Marquis, M.; Averyt, K.B.; Tignor, M.; Miller, H.L., Eds. Contribution of working group I to the fourth assessment report of the intergovernmental panel on Climate Change. In *Climate Change 2007: The Physical Science Basis*; Cambridge University Press: Cambridge, UK and New York, USA, 2007; 996 pp.
13. Namias, J. Thirty-day forecasting: A review of a ten-year experiment. Meteorol. Monogr. (American Meteorological Society) **1953**, *2* (6), 24–25.
14. Barnston, A.G.; Livzey, R.E. Classification, seasonality and persistence of low-frequency atmospheric circulation patterns. Mon. Weather Rev. **1987**, *115* (6), 1083–1126.
15. Thompson, D.W.J.; Wallace, J.M. Annular modes in extratropical circulation. Part I: Month to month variability. J. Climate **2000**, *13* (5), 1000–1016.
16. Held, I.M.; Soden, B.J. Robust responses of the hydrological cycle to global warming. J. Climate **2006**, *19* (21), 5686–5699.
17. Ramanathan, V.; Chung, C.; Kim, D.; Bettge, T.; Buja, L.; Kiehl, J.T.; Washington, W.M.; Fu, Q.; Sikka, D.R.; Wild, M. Atmospheric brown clouds: Impacts on South Asian climate and hydrologic cycle. Proc. Natl. Acad. Sci. **2005**, *102* (15), 5326–5333.
18. Eltahir, E.A.B. Role of vegetation in sustaining large-scale atmospheric circulations in the tropics. J. Geophys. Res. **1996**, *101* (D2), 4255–4268.

19. Lee, E.; Chase, T.N.; Rajagopalan, B. Highly improved predictive skill in the forecasting of the North East Asian summer monsoon. Water Resour. Res. **2008**, *44* (10), W10422. doi: 10.1029/2007WR006514.

20. Noone, D.; Galewsky, J.; Sharp, Z.; Worden, J.; Barnes, J.; Baer, D.; Bailey, A.; Brown, D.; Christensen, L.; Crosson, E.; Dong, F.; Hurley, J.; Johnson, L.; Strong, M.; Toohey, D.; Van Pelt, A.; Wright, J. Properties of air mass mixing and humidity in the subtropics from measurements of the D/H isotope ratio of water vapor at the Mauna Loa Observatory. J. Geophys. Res.-Atmos. **2011**, *116* (D22), 113. doi: 10.1029/2011JD015773.

Acid—
Atmospheric

Circulation Patterns: Atmospheric and Oceanic

Marcia Glaze Wyatt
Department of Geology, University of Colorado at Boulder, Boulder, Colorado, U.S.A.

Abstract

Fueled by the Sun, intrinsic dynamics of Earth systems drive global climate. Multiple interactions among climate-system components—ocean, atmosphere, land, and ice—are influenced by geography. They are further complicated by external forcings, both natural and anthropogenic. Dissimilar response times of system components to perturbations, combined with non-linear reactions among the components, have the potential to generate oscillatory signals on a variety of timescales. Positive and negative feedbacks further complicate these manifold interactions. Complex interplay among these various processes results in the redistribution of planetary heat, basically from where it is plentiful to where it is not. Through process-related changes in wind and precipitation regimes; atmospheric chemistry; the biosphere; inventories of clouds, sea ice, snow cover, and the like, Earth's radiative energy balance—the difference between incoming and outgoing energy—is modulated. These climate-driving dynamics can be distilled into a global collection of individual oceanic and atmospheric circulation patterns. In the short term, these patterns manipulate local and remote weather. Cumulatively and collectively, in the long term, they influence climate, and their combined effect on global surface temperature is complex.

INTRODUCTION

Climate: The Big Picture

Weather is a cooling process. The Sun heats Earth's surface; weather removes the surplus. Weather, averaged over time, is climate. Climate is what happens between delivery of energy and exit of its excess, the net result of which modulates Earth's average surface temperature.

The details are less simple. Interactions among myriad internal and external boundary conditions result in the absorption and partitioning of heat among various reservoirs—ocean, ice, land, and atmosphere—and at various levels within each. On different timescales, from seasonal to millennial, stored heat is moved in an assortment of ways via processes, many of which are coupled, to new latitudes, longitudes, depths, altitudes, and even to different reservoirs, with the end result being a global relocation of heat, from where there is an abundance to where there is less. Along this journey of heat redistribution, some heat escapes, emitted to space. How the incoming absorbed (solar) energy matches up to the outgoing emitted infrared (heat) energy determines Earth's radiative energy balance. Climate processes collectively affect, over time, the uptake, redistribution, and exit of planetary heat. Radiative energy imbalances are thereby minimized, and their influence on surface temperature is complex.

Climate's Task Masters: Atmospheric and Oceanic Circulation Patterns

Processes whose collective outcome modulates Earth's radiative energy imbalances can be traced to a global collection of individual oceanic and atmospheric circulation patterns. Locally and regionally, each circulation pattern impacts short-term weather. Interactions between and among them manipulate long-term climate. Each pattern possesses distinct traits.

All patterns fluctuate, most with preferred timescales of variability. A pattern's timescale of variability need not be governed by a similarly cadenced external source. The source of variability can be intrinsic to the system. Different response times of components to a forcing or disturbance, and non-linear interactions (where cause-and-effect relationships are not strictly proportional) among those components, leave the subsystem in a state of disequilibrium. Hence, the components never achieve collective stability. Therefore, they are varying constantly. The complexity of the subsystem, with extreme sensitivity to details regarding perturbations and boundary conditions, ensures that the resulting oscillatory behavior is not rigidly periodic. Feedback responses add another layer of complexity. Due to dissimilar response times of subsystem components to disturbances, it is not unusual for the initial system response to involve an immediate positive feedback (enhancing the initial perturbation), followed by a delayed negative

Encyclopedia of Natural Resources DOI: 10.1081/E-ENRA-120047620

Circulation—
Crops

feedback (damping or reversing the initial response), thereby modifying the timescale of variability through this means as well. And because the fluctuations are not temporally regular, the oscillatory character is best described as quasi-periodic. Most circulation patterns vary at more than one "quasi-regular" frequency, often including a low-frequency component. Preferred frequencies can be internally generated or externally forced, or internally generated with excitation from or entrainment by an external source.

Characteristic processes and traits tend to repeat for a given pattern; yet, again due to the non-linear nature of the subsystems, no two snapshots in time will be identical. In addition, teleconnections—climate features or responses highly correlated to, and distantly forced by, an associated regional circulation pattern—may co-exist with a given pattern for years and then wane or disappear, only to return to its former relationship at a later date—a further reminder of climate's complexity.

Debate exists over whether or not climate, with its numerous interacting circulation patterns, is itself, chaotic. In the mathematical sense, deterministic chaos refers to behavior that is acutely sensitive to initial conditions. Slight differences in initial conditions result in vastly divergent outcomes in chaotic systems. Were one to know with certainty each element of initial conditions, outcome could be predicted. Although this knowledge is unattainable, a chaotic system is largely, but not entirely, unpredictable. Nor is the system without order. Although a complex, non-linear system can be chaotic, and often is, it can also merely reflect behavior that appears chaotic—hence the debate. Weather tends to be chaotic. Climate, on the other hand, with its longer timescale, global energy-balance modulation over time, and negative feedbacks buffering the system from extreme disequilibrium, makes the situation less clear. Chaotic or not, it is clear that the climate system is complex and non-linear. It involves components that behave chaotically, but it is not without order.

On multidecadal timescales, individual expression of regional patterns may yield to collective interaction (network behavior), as suggested by recent research.[1] According to this hypothesis, synchronization and local coupling give rise to a low-frequency signal that propagates across the Northern Hemisphere through a sequence of atmospheric and lagged oceanic circulations (Fig. 1). In turn, this collective, slowly varying signature furnishes the modulating low-frequency background upon which higher-frequency behavior of regional circulations is superimposed.

Over 30 large-scale modes, or leading patterns, of climate variability have been identified across the globe. Most modes are spatially local or regional; yet with many of these, their impacts are hemispheric to global. Some patterns are dominated by atmospheric processes, some by oceanic ones. Some involve sea ice. Coupling between systems (ocean, ice, and atmosphere) and between levels within a system is common, particularly for low-frequency

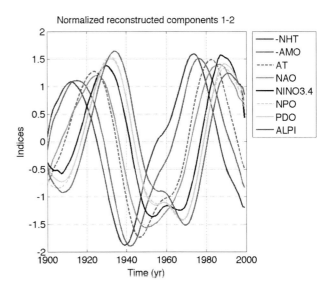

Fig. 1 A low-frequency signal propagating through a network of synchronized climate indices is reflected in this graph. The propagation pattern is termed the "stadium-wave" signal by its original authors—a reference to its communication through a sequence of indices. The plot shows normalized reconstructed components (RCs) of an eight-index climate network. Indices include anomalies of the averaged Northern Hemisphere surface temperature (NHT), the Atlantic Multidecadal Oscillation (AMO), the Atmospheric Mass Transfer index (AT: refers to zonal component of basin-scale wind direction), the North Atlantic Oscillation (NAO), (NINO 3.4) – an index for El Nino, the North Pacific Oscillation (NPO), the Pacific Decadal Oscillation (PDO), and the Aleutian Low index (AL). The RC time series shown have been normalized to have a unit variance. The indices are synchronized at non-zero lags (except for NPO and PDO, whose rescaled RCs are virtually identical).
Note: The reconstructions of NHT and AMO are of negative polarity.
Source: Reproduced with permission: Wyatt et al.[1]

oscillations. Dizzying detail of the numerous circulation modes renders their complete description unfeasible in limited space. Thus, the goal here is to weave together a brief overview of select individual regional circulation patterns. Discussion touches on their geographically governed dynamics, their individual behaviors, and their interconnectivity. The abridged inventory of patterns described here is divided into two broad categories: circulation patterns of the tropics and those of the extratropics. Links to websites featuring graphics related to many of the described patterns can be found after the bibliography at the end of this contribution.

CIRCULATION PATTERNS OF THE TROPICS

Geographically Governed Dynamics

Three tropical-specific traits play large roles in interrupting zonal symmetry imposed along the equator by the annual mean delivery of solar radiation: 1) At the equator, the Coriolis effect (the apparent deflection of moving objects due to planetary rotation—a function of latitude) is zero. On either side of the equator, directions of deflection are opposite. 2) Large-scale winds from each hemisphere converge near, but not necessarily aligned with, the geographical equator. Convergence is marked by a region of intense atmospheric convection known as the Intertropical Convergence Zone (ITCZ); and 3) planetary-scale subsurface ocean waves (eastward-traveling Kelvin and westward-propagating Rossby), when excited by anomalous winds, transmit information regarding overlying atmospheric disturbances zonally across the equatorial basin. Within months (the exact time involved dependent upon basin width), the equatorial subsurface adjusts to overlying atmospheric shifts. In turn, modified subsurface dynamics affect surface conditions. Ocean adjustment time—the ocean's "dynamic memory"—is shorter in the tropics than elsewhere on the planet. Two aspects, unique to the tropics, make possible this rapid oceanic adjustment to overlying atmospheric anomalies: One, latitudinally confined, eastward-directed Kelvin waves travel swiftly. They exist only along the equator. And two, although westward-propagating Rossby waves appear at all latitudes, their velocity is fastest in the near-equatorial zone.

Together, this collection of tropical-specific properties generates a mean state of zonal asymmetry in sea-surface temperatures (SSTs), subsurface thermocline depths, and atmospheric patterns, and from which zonal and meridional deviations occur with relative rapidity.

Zonal (East–West) Modes of Coupled Ocean Atmosphere Variability in the Tropics

El Nino–Southern oscillation

The mean state of the tropical Pacific is characterized by warm SSTs, a deep thermocline, and strong overlying atmospheric convection in the west; and a cool tongue of SSTs and a shallow thermocline in the east. Strong easterly winds co-occur with this zonally non-uniform profile. El Nino–Southern Oscillation (ENSO), a coupled ocean-atmosphere system, fluctuates interannually from this mean state. Zonal asymmetry is either magnified (La Nina) or diminished (El Nino).

At the helm of the ocean-atmosphere coupled system is the sea-level pressure (SLP) configuration of an equatorially circumnavigating phenomenon, the Southern Oscillation (SO). Zonal exchange of atmospheric mass along the equator manifests as fluctuations in the normalized difference in SLP anomalies between high- and low-pressure centers of the SO. SLP differences between a low-pressure region over north central Australia [near Darwin, Australia (12°28′S; 130°51′E)] and a high-pressure zone over the central South Pacific [Tahiti (17°40′S; 149°25′W)] are reflected in the SO index (SOI) and is one metric used to

define ENSO phase and intensity. A negative SOI describes the ENSO warm phase (El Nino), a positive SOI, a cool one (La Nina). Surface waters slosh east–west in concert with the SOI, influencing subsurface and overlying features in their wake. Adjustments of the subsurface ocean and overlying atmosphere to the equatorial zone's SLP redistribution occur regionally. Their impact is global.

Positive and negative departures from the tropical Pacific mean state are initially magnified through positive feedback responses (Bjerknes mechanism).[2] Interplay of subsurface Kelvin and Rossby ocean waves initially amplifies, then ultimately damps, or even reverses the sign of, the excursions. Together, the Bjerknes mechanism and the lagged Kelvin–Rossby interaction describe a leading model for the ENSO pattern, a delayed oscillator system.[3–5]

Although El Nino events rarely conform to a typical profile, a "standard" or "canonical" El Nino can be described. Hints of a building "canonical" El Nino can be seen in decreased SLP over Tahiti and increased SLP over Darwin. With the corresponding drop in SLP difference between the two pressure centers, southeasterly winds over the south and central tropical Pacific weaken. The atmospheric "push" on warm waters pooled in the west relaxes. Warm waters spread eastward. Westerly wind anomalies further advance the collapse of zonal asymmetry along the equatorial zone. Initial signs of an El Nino event typically manifest in late fall, particularly around the end of December [thus its name "El Nino" (Christ child) in reference to the time of year]. Once started, an incipient El Nino may reverse course and disappear. But if it survives seasonally competing processes imposed on its development during the boreal spring (e.g., seasonally related strengthened southeasterlies), it generally matures. El Nino "onset" typically takes hold around March, often persisting 18 to 24 months thereafter.

With El Nino's continued evolution, a weakened atmospheric zonal component (Walker circulation) shifts eastward with the eastward spread of warm SSTs and precipitation. Dry conditions supplant the moisture displaced from the western tropical Pacific. Upper-level subtropical jets on either side of the equator strengthen and shift equatorward, in particular, the jet within the winter hemisphere. Jet flow is extended eastward. Weather systems are redirected accordingly. Regions may experience warmth where usually it is cool; moisture where usually it is dry.

Teleconnection features associated with a circulation pattern can persist for decades and then wane or disappear. An example lies in the strong correlation between El Nino events and the suppression of North Atlantic hurricanes. Such was the norm for much of the 1950s into the 1980s. In the mid-1980s, and throughout the 1990s, this relationship began to falter. Spatial distribution and location of El Nino-related SST warming play a role. Although warm SST anomalies in the eastern basin characterized El Nino events throughout the mid-20th century, centrally located anomalies ("El Nino Modoki") occurred more frequently in the century's latter decades. Shifts in tropical Pacific anomalies govern shifts in the Walker circulation, thereby scripting different outcomes for remotely affected regions.[6] Similar inconsistency is seen in the often negatively correlated relationship between El Nino and the Indian summer monsoon, with the El Modoki "flavor" tending to be more consistently associated with suppression of Indian summer rains than its more easterly positioned earlier-century counterpart.[7]

Simultaneously occurring with the mean state's collapsed zonal asymmetry is a strengthened meridional component (Hadley circulation) that redirects atmospheric heat flow away from the tropics. This atmospheric reorganization is communicated to planetary-scale subsurface ocean waves (Kelvin and Rossby). They, in turn, restructure subsurface flow of heat and ocean mass. The result of El Nino's massive ocean-atmosphere reorganization is a decrease in influx of ocean heat and ocean mass to the tropics. During a warm ENSO event, net flow of heat is from the tropics poleward. In contrast, net heat flow is equatorward during a La Nina. During a cool ENSO event, warm surface water and the overlying Walker (zonal) circulation shift far westward; Hadley (meridional atmospheric) circulation weakens; and cold waters dominate the eastern basin, absorbing atmospheric heat and carrying it into the equatorial ocean subsurface. Warm subsurface waters from the subtropics flow toward the equator. Envisioned by some as a heat pump,[8–10] ENSO collects heat in its neutral and cool polarity phases and expels it poleward via ocean and atmosphere during its warm polarity phase, with local, remote, short-term, and long-term impacts on weather and climate dynamics. Multidecadally, the frequency and intensity of events are modulated, with frequency of El Nino events (or La Nina events) being greater in some decades than in others,[11] with indications for heat redistribution on decadal timescales.

The question of why easterlies occasionally subside or strengthen, and why the interannual (two-to-seven year) oscillation period is irregular, stirs debate. Source water from the subtropics,[9] and/or velocity of subtropical-water transport,[12,13] could play roles in subsurface water temperature feeding the cold tongue, altering the zonal SST gradient, thereby influencing easterly wind strength. In general, it is suggested that whatever the mechanism, it likely involves the decreased/increased temperature gradient between SSTs of the western warm pool and the subsurface waters feeding the cold tongue in the east, a condition whose causes could be local or global, external or internal, short term or long.

Atlantic Nino and Benguela Nino

Mean-state traits of the tropical Atlantic are similar to those in the Pacific: a coupled ocean-atmosphere system with prevailing easterly winds, warm water pooled in the west, and a cold tongue of upwelled deep water in the east. As with the ENSO system, overlying SLP variations and interplay in the Atlantic between perturbation-generated subsurface planetary-scale waves—Kelvin and Rossby waves—result in growth, then demise, of a warm event. In the Atlantic, there are two types of warm events—the Atlantic Nino, associated with northeasterly winds, and the Benguela Nino, associated with southeasterly winds. With either type, when a warm tropical Atlantic event occurs, easterlies subside; SSTs increase in the east. The zonal atmospheric circulation—the Walker circulation—shifts eastward, weakens, and is supplanted by a strengthened meridional atmospheric component—the Hadley circulation. An intensified Hadley cell carries heat poleward and strengthens the Atlantic sector's subtropical jet. Due to the impact of a narrower basin width on interacting subsurface waves, events in the Atlantic are weaker, shorter, and more frequent than in the Pacific. Despite this, end results of an El Nino and an Atlantic event are similar: redistribution of heat from the tropics poleward through oceanic and atmospheric circulation changes, with regional and remote impacts.

There are times when dynamics of the Atlantic shift in apparent response to a tropical Pacific-centered El Nino, suggesting interbasin communication. Within several months of an El Nino, an Atlantic Nino sometimes follows—1997/1998, for example. Yet, at other times, no relationship is apparent—1982/1983 was such an instance. Reasons for the inconsistent response to similarly large El Nino events are unclear. Hypotheses include destructive interference,[14] influence from a sometimes El Nino-modified feature—the Pacific North American (PNA) pattern,[15] or an atmospheric bridge.[16]

Indian ocean dipole or Indian Nino

Equatorial zonal interannual variability of a coupled ocean-atmosphere system also exists in the Indian Ocean—the Indian Ocean Dipole (IOD),[17–19] but its character is not fully analogous to its Atlantic and Pacific counterparts. Although the IOD is comparable in fundamental ways to tropical zonal variability in the Atlantic and Pacific Oceans, the geography surrounding the Indian Ocean imposes significant differences—a discontinuous eastern boundary and the vast landmass of India to the north. The former allows direct oceanic communication with the Pacific basin to its east. The latter—a southern extension of the Indian subcontinent—blocks east–west oceanic flow within the Indian Ocean basin north of 25°N. The imposing influence of this landmass scripts prevailing wind regimes of monsoonal flow: northeasterly in the boreal winter to southwesterly in the boreal summer. These seasonally reversing

Circulation—Crops

winds within the northern Indian Ocean tend to mask the equatorial zonal mode.

Equatorial Indian Ocean conditions differ from those in the monsoon-dominated regions northward. A weak zonal wind component lingers over the equatorial Indian Ocean when monsoons are strong. Yet during transitions from one monsoonal flow to the other—i.e., during boreal spring and fall—equatorial westerlies intensify. The end result of these diverse wind patterns across the Indian Ocean is a mean equatorial state of warm water pooled in the eastern basin, below which the thermocline is deep and above which, the zonal convection of a Walker cell is strengthened. These mean conditions—a reversed profile of the Pacific's—supply abundant moisture to Indonesia and little to East Africa Interannual fluctuations occur between amplified and weakened expressions of this equatorial mean state.

The first signs of a developing event emerge characteristically in the late boreal spring or early summer. If the seasonally weak equatorial westerly winds intensify, the equatorial mean state is amplified. A negative (cool) IOD has potential to develop, delivering more moisture to Indonesia, less to East Africa. On the other hand, if the equatorial westerly winds fail, and are supplanted by equatorial anomalous easterlies and southeasterlies, a positive (warm) IOD may take hold. Zonal asymmetry collapses. The SST anomaly gradient reverses. Atmospheric convection shifts west. Strong rains pelt East Africa. Indonesia and Australia are unusually dry. An "event" typically persists until the following winter.

Teleconnections reach beyond the surrounding region, some apparently decadally modulated or, in some cases, either enhanced or damped by interactions with teleconnections from ENSO or with processes in the tropical Atlantic.[20] Although the IOD is thought to be an independent intrinsic coupled mode of variability of the Indian Ocean,[17–18] warm ENSO events often co-occur with positive IOD events and cool ENSO with negative IOD. ENSO has been posited as both being a potential trigger for,[21–23] and potentially triggered by the IOD.[24]

Meridional (North–South) Modes of Coupled Ocean–Atmosphere Variability in the Tropics

Atlantic meridional mode and Pacific meridional mode

Meridional modes of the tropics involve latitudinal migration of the ITCZ in relatively swift response to winds and associated SST anomalies. Meridional modes, which are dominantly interannual, yet multidecadally modulated by oceanic influence,[25] have been identified in both the Atlantic [Atlantic Meridional Mode (AMM)] and the

Pacific [Pacific Meridional Mode (PMM)].[26] They are variable coupled ocean–atmosphere phenomena in the tropics that are excited by boreal winter extratropical atmospheric variability (see section "Annular modes and related extratropical circulation") over their northern basins. Wind-induced latent heat flux leads to a summer-centered coupled response in tropical SST anomalies. A meridional SST gradient across the mean latitude of the ITCZ develops. Strongly sensitive to small meridional SST gradients, the ITCZ migrates toward the warmer hemisphere, accompanied by cross-equatorial surface winds flowing in the direction of that hemisphere. Remotely forced by extratropical atmospheric circulation in their respective basins, meridional modes impact regional temperature, precipitation, and hurricane[25–27] regimes, with potential impact extending to zonal tropical modes, e.g., ENSO.[28]

CIRCULATION PATTERNS OF THE EXTRATROPICS

Geographically Governed Dynamics

The Coriolis effect is non-zero away from the equator and throughout the extratropics. Its parameter, which is same-signed throughout a given hemisphere, increases with latitude. In part because of Coriolis-related peculiarities, the travel of Kelvin waves in the extratropics is restricted to poleward (equatorward) flow within a narrow zone along the western (eastern) coastlines of continents or cyclonically around a closed boundary. Kelvin waves cannot travel in the open ocean along lines of latitude in the extratropics. Only westward-propagating Rossby waves can, with slower velocities at higher latitudes. Consequence of these dynamics is twofold: 1) communication of overlying atmospheric changes to the subsurface ocean is propagated only toward the west; and 2) time required for the Rossby waves to travel the cross-basin takes years to decades in the mid-to-high latitudes. This renders adjustment of the subsurface ocean to overlying large-scale wind field slow in the extratropics. Extratropical ocean "dynamic memory" is long; persistence of a climate signal at mid-to-high latitudes is thereby extended.

In addition to the extratropical traits responsible for modifying climate signals and their transmission, equator-to-pole atmospheric transport of heat is similarly amended. It is broken into three large-scale convection cells—this due to the Coriolis effect. Related rising, sinking, diverging, and converging basin-scale air flow within these cells establish semipermanent surface high- and low-pressure regions and associated belts of prevailing winds, thus governing fundamental aspects of regional circulation patterns.

Dominantly Atmospheric Circulations

Annular modes and related extratropical circulation

Atmospheric variability at mid-to-high latitudes in each hemisphere is strongly influenced by annular modes: the Northern Annular Mode (NAM) in the Northern Hemisphere and the Southern Annular Mode (SAM) in the Southern Hemisphere.[29–31] Shifts of atmospheric mass between high latitudes poleward of ~60° and a ring, or annulus, around the midlatitudes (~45°) characterize these patterns with no preferred timescale of variability, manifesting as large-scale variability of angular momentum in their respective hemispheres. Lower-than-normal SLPss over the polar regions, higher-than-normal pressures in mid-latitudes, and positive westerly wind anomalies at about 55° to 60° characterize the high-index (positive) polarity. The reverse is true for the low-index (negative) polarity.

During the boreal winter for the NAM and the austral spring for the SAM, these tropospheric patterns merge with stratospheric dynamics. Although often used interchangeably with NAM and SAM, the Arctic Oscillation (AO) and Antarctic Oscillation (AAO), respectively, are terms that apply to the tropospheric expression of the seasonal stratosphere–troposphere coupling. This stratosphere–troposphere coupling involves a polar vortex of westerly winds that extends seasonally from the stratosphere to the surface, thereby enabling stratospheric perturbations to influence activity in the troposphere, from the poles to the equator. In turn, when the two atmospheric levels are coupled, tropospheric processes impact the stratosphere. This vertical two-way coupling imparts a low-frequency component to the annular mode's time scale of variability. Associated with this lower-frequency component is a recently proposed relationship between ocean heat flux from western boundary currents (WBCs) and their extensions (WBCEs), and sudden stratospheric warmings (SSWs);[32] the latter, in turn, force the troposphere. Both phenomena—WBCE ocean heat flux and SSWs—exhibit decadal to multidecadal variability. The WBCE-related ocean heat flux out of the ocean is positively correlated to lateral advection of upper-ocean heat to WBCs. This lateral advection of ocean heat, itself, covaries at a one-year lag with westerly wind strength associated with a multidecadal component of the NAM.[33,34]

The North Atlantic Oscillation (NAO)[35] is a North Atlantic–centered manifestation of the NAM. Both NAM and NAO characterize similar temporal and spatial leading modes of Northern Hemispheric variability. Whereas NAM is hemispheric, NAO is confined to the North Atlantic region. Atmospheric mass redistribution between subpolar and subtropical latitudes manifests as changes in the normalized SLP anomaly difference between the North Atlantic atmospheric centers of action (COA): the Icelandic Low and the Azores High. A positive SLP anomaly difference represents the high index, or positive polarity, of NAO and a negative SLP anomaly difference indicates the negative polarity, or low index, of the NAO. Low-frequency modulation of interannual-to-interdecadal variability is apparent. A characteristic tripolar pattern of wind-generated SST anomalies within the North Atlantic is generated by higher-frequency fluctuations of the NAO;[36,37] whereas a more uniform, basin-wide SST anomaly pattern, likely involving large-scale ocean dynamics,[37–40] distinguishes the low-frequency component, with significant teleconnection impact.[39,40]

The NAM also extends its influence into the North Pacific, although not as conspicuously as in the North Atlantic. In fact, SLP fluctuations over the North Pacific and North Atlantic exhibit no significant correlation—an observation seemingly inconsistent with the paradigm of a hemispheric NAM. But reconciliation might be found in an argument that has been made for the indirect correlation between the Atlantic and Pacific ocean basins. Although the Atlantic and Pacific centers may not correlate positively with one another, both correlate negatively with the Arctic COA. Through that indirect association, it is proposed that observed SLP behavior in the North Pacific is compatible with the hemispheric paradigm of polar–subtropical see-saw exchange of atmospheric mass.[41] Obscuring that otherwise potentially obvious NAM-North Pacific relationship may be the coexistence of another strong mode of climate variability. That mode may be found in the PNA pattern.[15]

PNA is a prominent upper-atmospheric mode of low-frequency variability involving anomalous atmospheric pressure. Located at midtropospheric heights over the extratropics, influence on its behavior is not confined to the mid-to-high latitudes. Strongly associated with the Aleutian Low (AL) in the North Pacific, the PNA also is modified by tropical processes. Associated with jet-stream tracks, modified by underlying topography and constraints of vorticity conservation, the PNA is linked to the flow of storm systems (wave trains) across the Northern Hemisphere. Intensity and location of weather systems, particularly those affecting North America in the boreal winter, are most impacted by the PNA. Liu et al.[42] attempted to tease apart behavior of the AL and AO (NAM). Teleconnected influence of each individual COA differed from the teleconnections resulting from the combination of their influences. This observation is consistent with the idea of coexistence of patterns and the consequent possible masking of the NAM signature in the North Pacific.

Although the PNA is strongly associated with the North Pacific via the AL, its scope is hemispheric. More regional in nature is the North Pacific's analogue to the NAO, namely, the North Pacific Oscillation (NPO).[43] In the North Pacific area, including the most northwestern sector of North America and the eastern coastal areas of Eurasia, the NPO is the most important teleconnection mode. As with the NAO, NPO can be described as a north–south

Circulation—Crops

seesaw of atmospheric mass between subpolar and subtropical regions, with the normalized SLP anomaly difference between the regions measuring the polarity and strength of the pattern's interannual-to-interdecadal fluctuations.

Originally described in 1932, NPO was said to result primarily from geographical shifts of the mean position of the AL.[43] Indeed, it is observed that AL migrates west and north when weak (increased SLP) and to the east and south when strong (decreased SLP), the most extreme shifts occurring on multidecadal timescales.[44] Spatial scale of NPO influence fluctuates similarly.[45] During decadal-plus intervals, the footprint of the NPO appears confined to the North Pacific basin. The AL is weak and skewed west of its mean position. During these "regional" time spans, ENSO plays a strong role in the NPO behavior, with pronounced influence on AL. In contrast, during "hemispheric" intervals, the reach of NPO influence spreads far beyond the Pacific basin. The AL is strong and shifts southeast of its mean position. ENSO forcing on AL is minimal. Instead, extratropical forcing of the low-pressure system is dominant. With an intensified AL, the hemispheric PNA pattern is more pronounced.

NAO, too, shows indications of vacillating on low-frequency timescales between more regionally confined intervals and more hemispherically spanning ones.[46] Not unexpectedly, details of NAO/NPO high-frequency variability and of associated teleconnected impacts related to precipitation, surface temperature, ocean gyre activity, and sea-ice patterns, vary accordingly at the low-frequency tempo. Geographical shifts of the Arctic, subpolar, and subtropical atmospheric COAs on decadal-plus timescales can explain much of the observed behavior. Evidence suggests indirect oceanic influence on atmospheric COAs.[47–51] Low-frequency solar variability, too, may play a role.[44,52]

The SAM, many of its features analogous to the NAM, overlies relatively simple geography. Absence of landmass in the midlatitudes around 60°S conveys unparalleled strength to the SAM-associated mid-high-latitude westerly winds. These winds drive the Antarctic Circumpolar Current (ACC) in the Southern Ocean—a globally girdling ocean current that connects all three ocean basins and is fundamental to global intercommunication of climate signals. Through upper-atmospheric communication [via the Pacific South American Pattern (PSA)], this southernmost atmospheric pattern, the SAM, whose influence strongly impacts Antarctic temperatures, sea-ice dynamics, and associated deep-water formation, appears to be linked to tropical dynamics, particularly ENSO in the Pacific sector.[53–55]

Dominantly Oceanic Circulations

The atmospheric circulations described earlier are dominated by stochastic, high-frequency behavior. In contrast, dominantly oceanic circulations operate at low-frequency

timescales. Water's high heat capacity, in conjunction with long time spans required for both extratropical subsurface Rossby wave travel and vertical mixing, contribute to the ponderous pace of oceanic processes. At these low-frequency timescales, the ocean can exert direct and indirect influence on the atmosphere, thereby adding persistence to overlying atmospheric circulation. This influence is reflected in a low-frequency modulation of that atmospheric circulation's frequency and intensity at interdecadal-to-multidecadal timescales. In addition, the ocean modifies hemispheric and global teleconnections associated with the atmospheric patterns.

Atlantic multidecadal oscillation

Ocean heat transport is northward at all latitudes of the Atlantic Ocean. This cross-equatorial ocean heat transport is unique to the Atlantic, generating a dipole of SST anomalies between the North and South Atlantic basins. This SST distribution is not static. Proxy and instrumental records support the existence of an intrinsic low-frequency mode of variability, characterized by a multidecadally (50–80 years) varying monopolar pattern of sea-surface temperature anomalies averaged across 0° to 60°N and 75° to 7.5°W. It is known as the Atlantic Multidecadal Oscillation (AMO).[56–60] Climate impacts of the AMO-related SST anomalies are regional—North America and Eurasia are particularly impacted. AMO's reach can be hemispheric, as well, extending to the North Pacific.[57,58,61,62]

Recent studies introduce the notion that AMO operates at a second timescale of variability, a 20–30-year quasi-regular period.[63–66] Both the higher-frequency mode and the more well-known low-frequency mode elude full description, although both appear related to fluctuations in strength of the Atlantic sector's Meridional Overturning Circulation (AMOC),[65–69] as is suggested by model studies incorporating an interactive ocean[67,70] and by variations in instrumental,[71] and in proxy[72] records of the interhemispheric SST dipole pattern. Mechanistic details for the AMO–AMOC connection remain under debate.[67,73–76]

Westward propagation of SST anomalies is associated with the hypothesized higher-frequency AMO mode.[75] Also associated are sea-level variations[64] along the North American and European coast lines. One idea is that the signal is generated from an internal ocean mode within mid-to-high latitudes of the North Atlantic, perhaps excited by low-frequency atmospheric noise.[75] From the North Atlantic, the thermal anomalies propagate into the Arctic.[63]

In contrast, studies support a strong active role for Arctic involvement in the lower-frequency mode.[66,75,76] This more-slowly paced pattern might involve a salinity oscillation between the Arctic and high latitudes of the North Atlantic.[66,76,77] According to modeled results, this proposed oscillation is associated with saline Rossby modes in the Arctic, which perhaps are excited by variability in processes that influence freshwater balance in the region (e.g., Atlantic inflow, sea-ice changes, or variability in river runoff).[66] The approximate 60-year rhythm found in 20th-century AMO SST anomalies and Arctic/Atlantic salinity oscillations can be found in numerous patterns: the i) Atlantic core water temperature;[78,79] ii) the Arctic sea-ice extent in the Western Eurasian Arctic Shelf Seas (Greenland, Barents, Kara Seas);[80–82] iii) basin-scale meridional temperature gradient (MTG);[83] iv) basin-scale westerly wind strength;[80–84] and v) surface temperature over Eurasia.[83]

Potentially related to the salinity oscillations and the above-mentioned set of processes is evidence for an approximate 60-year variability in the geographical positioning of the axis between SLP COAs (i.e., Icelandic Low and Azores High) overlying the AMO-related SST anomaly pattern. This axis realignment, with its low-frequency component projected onto the NAO, is hypothesized to influence the North Atlantic's high-latitude salinity concentration, and by extension, the multidecadal character of the AMOC strength and AMO polarity.[47,58]

Pacific decadal oscillation

The Pacific Decadal Oscillation [PDO (Fig. 2)] is an oceanic circulation pattern characterized by a boreal-winter, basin-wide, quasi-periodically varying SST pattern north of 20°N in the North Pacific.[85,86] A positive PDO hosts a pattern of cool SST anomalies in the vast central/western sector of the basin and warm El Nino-like anomalies in the tropical sector and along the west coast of North America. A negative PDO pattern is reversed. A related pattern is the Interdecadal Pacific Oscillation (IPO),[87] which includes both the North and South Pacific Oceans

Pacific Decadal Oscillation

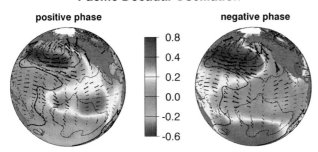

Fig. 2 Anomalous climate conditions associated with warm (positive) and cold (negative) phases of the Pacific Decadal Oscillation (PDO). Values are shown for degrees Celsius for sea-surface temperature (SST), millibars for sea-level pressure (SLP), and direction and intensity of surface wind stress (arrow width proportional to wind strength). The longest wind vectors represent pseudostress of 10m^2/s^2. Anomalies would need to be multiplied by associated index value (not given here) in the range between -3 to +3.
Source: Reproduced with permission: Steven Hare; adapted and updated from Mantua et al.[85]

and of which the PDO is likely a part. Both the PDO and IPO have been shown to display two dominant quasi-periodicities, one of 15 to 30 years and another of 50 to 70 years (Fig. 1). Both are intimately coupled with the overlying atmospheric circulation, which involve the AL and the Hawaiian (or Pacific) High (HH).

Longitudinal/latitudinal migrations of these COAs covary with phases of the PDO, the lower frequency shifts being more extreme. With a positive PDO, an intensified AL shifts east and south; the strengthened HH moves west and south.[52] Movements are opposite for a negative PDO. Repositioned COAs can yield unexpected results. Consider the behavior of El Nino-associated phenomena. They can differ according to AL position, and by extension, PDO phase. An example can be found in fluctuating sea-ice growth in the Bering Sea. When PDO is in its negative phase, El Nino events correlate with decreased ice extent. In contrast, when PDO is in its positive phase, Bering Sea ice extent increases during El Nino events. This is due to the repositioning of AL and the consequent placement of accompanying winds.[88]

Repositioned COAs can also modify the extent and the consequences of PDO-related teleconnections, many of which involve ENSO. The multidecadal component of PDO modulates the frequency and intensity of ENSO. During a positive PDO, warm ENSO events increase in both intensity and frequency. During a negative PDO, cool ENSO events are favored.[1,10]

With PDO modulation of ENSO due to COA migrations, remote teleconnections of ENSO can build in impact due to cumulative effects. Some PDO-modulated ENSO teleconnections impact multidecadal variations in the freshwater balance of the North Atlantic. Impacts occur on differing timescales. They include the i) alteration of hemispheric-scale precipitation patterns (increased precipitation in the tropical Atlantic with La Nina[89] and decreased with El Nino);[90] ii) the basin-scale flow of relatively fresh water out of the Pacific (increased fresh Pacific water to the Arctic[88,91] and decreased to the Indian Ocean during El Nino); and iii) a possible increase in occurrence of positive IOD events with El Nino. With them comes a decadal-scale-lagged suppression of salt delivery to the Atlantic from the Indian Ocean (Agulhas leakage).[92,93] The opposite occurs with La Nina.

PDO low-frequency control of ENSO teleconnections also influences sea-ice formation regions off the Antarctic coasts [more in the Weddell Sea (Atlantic sector) during El Nino and more in Ross Sea (Pacific sector) during La Nina], with delayed potential impact on strength of the AMOC. Delivery of upper-ocean heat by the North Pacific subtropical gyre to the WBCE is similarly paced.[94]

From the examples outlined earlier, it can be seen that the Atlantic and Pacific multidecadally varying ocean modes, each through indirectly teleconnected influences, appear to modify the other at decadal-plus timescales, influencing their temporal relationships. That temporal relationship during the 20th century shows that the 50–80-year mode of PDO has occurred in quadrature with that of the AMO.[1,95] In other words, their phases show an approximate offset by a quarter of a period (Fig. 1). Their relationship is correlated with drought patterns in the contiguous United States. Observation indicates that drought has been most extreme during positive phases of the AMO, whereas the accompanying phase—positive or negative—of PDO has scripted the distribution of drought-impacted regions (Fig. 3).[96] The record of the Colorado River basin flow over the last century appears consistent with this observation.[97] In addition, 500-year-long proxy records reflect a similar cadence in the Colorado River basin stream flow, suggesting a strong and persistent natural component for this observation.[98]

Antarctic circumpolar wave

Identified in 1996,[99] the Antarctic Circumpolar Wave (ACW), a product of polar–subtropical temperature contrasts, consists of SLP and SST anomalies in a wave number-two structure that propagates in the Southern Ocean around Antarctica. Its interannual quasi-periodicity of four to five years is modulated on decadal, interdecadal, and multidecadal timescales. The ACW propagates westward in the vicinity of the eastward flowing, SAM-related, hemispherically encircling Antarctic Circumpolar Current

(ACC). Influenced by the fast eastwardly flowing ACC, net motion of the westward ("upstream")—propagating ACW is eastward, yet at a slower rate than that of the ACC. The ACW makes a full trip around the globe at southern latitudes (~65°S to 30°S) in eight to nine years. Weather patterns in southern regions of Australia, South America, and Africa are influenced by the ACW.

Influence of and on the ACW extends to the tropics, where it appears to engage in two-way communication with ENSO. The atmosphere transmits tropical information to high southern latitudes, imprinting an ENSO footprint upon the ACW. Piggybacked upon the ACW, the original ENSO footprint is slowly modified as it circumnavigates the Antarctic continent. ACW, through various bifurcations equatorward, returns the amended ENSO signature to the tropics, and the return routing of the signal is guided by the phase of PDO.[100,101]

CONCLUSION

Earth's regional climate patterns work on a variety of timescales to globally redistribute heat from where there is an excess to where there is a deficit. All Earth system components—ocean, atmosphere, land, and sea ice—participate in tailoring characteristics of individual circulation patterns. Impact of individual circulation patterns on weather

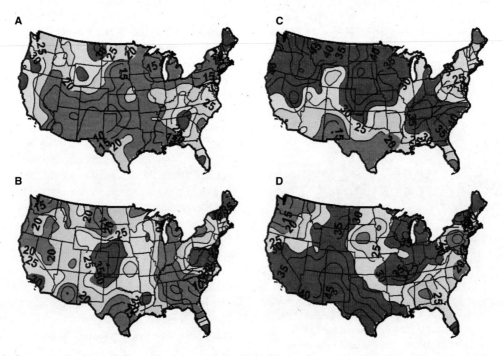

Fig. 3   Drought frequency (in percent of years) in the contiguous United States for warm and cool (positive and negative) low-frequency regimes of the Atlantic Multidecadal Oscillation (AMO) and the Pacific Decadal Oscillation (PDO). A) +PDO and –AMO; B) –PDO and –AMO; C) +PDO and +AMO; and D) –PDO and +AMO. Note that drought is most widespread when AMO is in its warm (positive) state (Figs. 3C, 3D). The distribution of drought is modulated with phase of PDO (more in northwest and north central and southeast with positive PDO and positive AMO; more southwest and midwest with negative PDO and positive AMO). The "stadium wave" in Fig. 1 gives insight into when the various AMO and PDO phases coincide.
Source: Reproduced with permission: Greg McCabe et al.[96]

and climate, on local, regional, hemispheric, and global spatial scales, is modified according to interaction among patterns and according to timescales. Complexity of local and regional detail tends to yield to hemispheric order in the long term. This may be attributable, at least in part, to synchronized network behavior of climate indices at low-frequency timescales, as recently hypothesized by Wyatt et al.,[1] Wyatt,[81] and Wyatt and Curry.[82] Interaction of regional circulation patterns, especially at multidecadal tempos, is often accompanied by characteristic phenomena. Interconnectedness of Earth's climate network is profound, the foundation of the planet's heat engine.

ACKNOWLEDGMENTS

Thanks go to Roger Pielke, Sr., professor emeritus at the University of Colorado, Boulder. He has been instrumental in motivating work related to this project. Thanks also go to an anonymous reviewer for helpful suggestions on manuscript content and style.

REFERENCES

1. Wyatt, M.G.; Kravtsov, S.; Tsonis, A.A. Atlantic multidecadal oscillation and Northern Hemisphere's climate variability. Clim. Dyn. **2012**, *38* (5/6), 929–949.
2. Bjerknes, J. Atmospheric teleconnections from the equatorial pacific. Month. Weather Rev. **1969**, *97* (3), 163–172.
3. Schopf, P.S.; Suarez, M.J. Vacillations in a coupled ocean-atmosphere model. J. Atmospheric Sci. **1988**, *45*, 549–566.
4. Suarez, M.J.; Schopf, P.S. A delayed action oscillator for ENSO. J. Atmospheric Sci. **1988**, *45* (21), 3283–3287.
5. Battisti, D.S.; Hirst, A.C. Interannual variability in a tropical atmosphere-ocean model: influence of the basic state, ocean geometry, and nonlinearity. J. Atmospheric Sci. **1989**, *46* (12), 1687–1712.
6. Kim, H.-M.; Webster, P.J.; Curry, J.A. Impact of shifting patterns of Pacific Ocean warming on north atlantic tropical cyclones. Science **2009**, *325* (5936), 77–80.
7. Kumar, K.K.; Rajagopalan, B.; Hoerling, M.; Bates, G.; Cane, M. Unraveling the mystery of Indian monsoon failure during El Nino. Science **2006**, *314*, 115–119.
8. Sun, D.-Z. A possible effect of an increase in the warm-pool SST on the magnitude of El Nino warming. J. Climate **2003**, *16* (2), 185–205.
9. Sun, D.-Z.; Zhang, T. A regulatory effect of ENSO on the time-mean thermal stratification of the equatorial upper ocean. Geophysical Res. Lett. **2006**, *33*, L07710.
10. Sun, D.-Z. The control of meridional differential heating over the level of ENSO activity: *A Heat-Pump Hypothesis. Earth's Climate: The Ocean-Atmosphere Interaction; Geophysical Monograph Series*; Wang, C., Xie, S.-P.,Carton, J.A., Eds.; American Geophysical Union: Washington, D.C., **2004**, *147*, 71–83.
11. Federov, A.V.; Philander, S.G. Is El Nino changing? Science **2000**, *288* (5473), 1997–2002.
12. McPhaden, M.J.; Zhang, D. Slowdown of the meridional overturning circulation in the upper Pacific Ocean. Nature **2002**, *415*, 603–608.
13. McPhaden, M.J.; Zhang, D. Pacific Ocean circulation rebounds. Geophysical Res. Lett. **2004**, *31*, L18301.
14. Chang, P.; Fang, Y.; Saravanan, R.; Link, J.; Seide, H. The cause of the fragile relationship between the Pacific El Nino and the Atlantic Nino. Nature **2006**, *443*, 324–328.
15. Wallace, J.M.; Gutzler, D.S. Teleconnections in the geopotential height field during the Northern Hemisphere winter. *Monthly Weather Rev.* **1981**, *109*, 784–812.
16. Wang, C. An overlooked feature of tropical climate: Inter-Pacific–Atlantic variability. Geophysical Res. Lett. **2006**, *33*, L12702.
17. Saji N.H.; Goswami, B.N.; Vinayachandran, P.N.; Yamagata, T.A. Dipole mode in the tropical Indian Ocean. Nature **1999**, *401*, 360–363.
18. Webster, P.J.; Moore, A.M.; Loschnigg, J.P.; Leben, R.R. Coupled ocean-atmosphere dynamics in the Indian Ocean during 1997–1998. Nature **1999**, *401*, 356–360.
19. Yamagata, T.; Behera, S.K.; Luo, J.-J.; Masson, S.; Jury, M.R.; Rao, S.A.. Coupled ocean-atmosphere variability in the tropical Indian Ocean, earth's climate. In *The Ocean-Atmosphere Interaction; Geophysical Monograph Series*; Wang, C., Xie, S.-P., Carton, J.A., Eds.; American Geophysical Union: Washington, D.C., **2004**, *147*, 189–211.
20. Saji, N.H.; Yamagata, T. Interference of teleconnection patterns generated from the tropical Indian and Pacific Oceans. Climate Res. **2003**, *25*, 151–169.
21. Masumoto Y.; Meyers, G. Forced Rossby waves in the southern tropical Indian Ocean. J. Geophysical Res. **1998**, *103*, 27589–27602.
22. Annamalai, H.; Xie, S.-P.; McCreary, J.-P.; Murtugudde, R. Impact of Indian Ocean sea surface temperature on developing El Nino. J. Climate **2005**, *18*, 302–319.
23. Yu, J.-Y.; Lau, K.M. Contrasting Indian Ocean SST variability with and without ENSO influence: A coupled atmosphere-ocean GCM study. Meteorol. Atmospheric Phys. **2005**, *90* (3–4), 179–191.
24. Izumo, T.; Vialard, J.; Lengaigne, M.; de Boyer Montegut, C.; Behera, S.K.; Luo, J.-J.; Cravatte, S.; Masson, S.; Yamagata, T. Influence of the state of the Indian Ocean dipole on the following year's El Nino. Nat. Geosci. **2010**, *3*, 168–172.
25. Vimont, D.J.; Kossin, J.P. The Atlantic meridional mode and hurricane activity. Geophysical Res. Lett. **2007**, *34*, L07709.
26. Chiang, J.C.H.; Vimont, D.J. Analogous Pacific and Atlantic meridional modes of tropical atmosphere-ocean variability. J. Climate **2004**, *17*, 4143–4158.
27. Smirnov, D.; Vimont, D.J. Variability of the Atlantic meridional mode during the atlantic hurricane season. J. Climate **2011**, *24*, 1409–1424.
28. Chang, P.; Zhang, L.; Saravanan, R.; Vimont, D.J.; Chiang, J.C.H.; Ji, L.; Seidel, H., Tippett, M.K. Pacific meridional mode and El Nino-Southern Oscillation. Geophysical Res. Lett. **2007**, *34*, L16608.
29. Thompson, D.W.J.; Wallace, J.M. The arctic oscillation signature in the wintertime geopotential height and temperature fields. Geophysical Res. Lett. **1998**, *25* (9), 1297–1300.
30. Thompson, D.W.J.; Wallace, J.M. Annular modes in the extratropical circulation, Part I: month-to-month variability. J. Climate **2000**, *13*, 1000–1016.

31. Thompson, D.W.J.; Wallace, J.M.; Hegerl G.C. Annular modes in the extratropical circulation. Part II: trends. J. Climate **2000**, *13*, 1018–1036.

32. Schimanke, S.; Korper, J.; Spangehl, T.; Cubasch, U. Multi-decadal variability of sudden stratospheric warmings in an AOGCM; Geophysical Res. Lett. **2011**, *38*, L01801. doi:10.1029/2010GL045756.

33. Kelly, K.A.; Dong, S. The relationship of western boundary current heat transport and storage to midlatitude ocean-atmosphere interaction. In *Earth's Climate: The Ocean-Atmosphere Interaction; Geophysical Monograph Series*; Wang, C., Xie, S.-P., Carton, J.A., Eds.; American Geophysical Union: Washington, D.C., **2004**, *147*, 347–363.

34. Xie, S.-Ping. Satellite observations of cool ocean-atmosphere interaction. Bull. Am. Meteorological Soc. **2004**, *85*, 195–208.

35. Hurrell, J.W. Decadal trends in the North Atlantic oscillation: regional temperatures and precipitation. Science **1995**, *269* (5224), 676–679.

36. Deser, C.; Blackmon, M.L. Surface climate variations over the North Atlantic Ocean during winter: 1900–1993. J. Climate **1993**, *6*, 1743–1753.

37. Eden, C.; Jung, T. North Atlantic interdecadal variability: Oceanic response to the North Atlantic oscillation. J. Climate **2001**, *14*, 676–691.

38. Bjerknes, J. *Atlantic Air-Sea Interaction*. Adv. Geophysics;. Academic Press, 1964; 1–82.

39. Kushnir, Y.; Interdecadal variations in North Atlantic sea-surface-temperature and associated atmospheric conditions. J. Climate **1994**, *7*, 142–157.

40. Visbeck, M.; Chassignet, E.P.; Curry, R.G.; Delworth, T.L.; Dickson, R.R.; Krahmann, G, The North Atlantic oscillation: climatic significance and environmental impact. American Geophysical Union. Geophysical Monogr. **2003**, *134*, 279.

41. Wallace J.M.; Thompson, D.W.J. The Pacific center of action of the Northern hemisphere annular mode: real or artifact? J. Climate **2002**, *15*, 1987–1991.

42. Liu, J.; Curry, J.A.; Dai, Y.; Horton, R. Causes of the northern high-latitude land surface winter climate change. Geophysical Res. Lett. **2007**, *34*, L14702.

43. Walker, G.T.; Bliss, E.W. World weather V. Memoirs R. Meteorological Soc. **1932**, *4* (36), 53–84.

44. Kirov, B.; Georgieva, K. Long-term variations and inter-relations of ENSO, NAO, and solar activity. Phys. Chem. Earth **2002**, *27*, 441–448.

45. Wang, L.; Chen, W.; Huang, R. Changes in the variability of North Pacific oscillation around 1975/1976 and its relationship with East Asian winter climate. J. Geophysical Res. **2007**, *112*, D11110.

46. Walter, K.; Graf, H.F. On the changing nature of the regional connection between the North Atlantic oscillation and sea surface temperature. J. Geophysical Res. **2002**, *107* (D17), 4388.

47. Polonsky, A.B.; Basharin, D.V.; Voskresenskaya, E.N.; Worley, S.J.; Yurovsky, A.V. Relationship between the North Atlantic oscillation, Euro-Asian climate anomalies and Pacific variability. Marine Meteorology. Pac. Oceanography **2004**, *2* (1–2), 52–66.

48. Grosfeld, K.; Lohmann, G.; Rimbu, N.; Fraedrich, K.; Lunkeit, F. Atmospheric multidecadal variations in the North Atlantic realm: proxy data, observations, and atmospheric circulation model studies. Climate Past **2007**, *3*, 39–50.

49. Msadek, R.; Frankignoul, C., Li, L.Z.X. Mechanisms of the atmospheric response to North Atlantic multidecadal variability: a model study. Climate Dynamics **2011**, *36*, 1255–1276.

50. Sugimoto, S.; Hanawa, K. Decadal and interdecadal variations of the Aleutian low activity and their relation to upper oceanic variations over the North Pacific. J. Meteorological Soc. Jap. **2009**, *87* (4), 601–614.

51. Frankignoul, C.; Sennechael, N.; Kwon, Y. -Oh; Alexander, M.A. Influence of the meridional shifts of the Kuroshio and the Oyashio extensions on the atmospheric circulation. J. Climate **2011**, *24*, 762–777.

52. Georgieva, K.; Kirov, B.; Tonev, P.; Guineva, V.; Atanasov, D. Long-term variations in the correlation between NAO and solar activity: the importance of north-south solar activity asymmetry for atmospheric circulation. Adv. Space Res. **2007**, *40*, 1152–1166.

53. Rind, D.; Chandler, M.; Lerner, J.; Martinson, D.G.; Yuan, X. Climate response to basin-specific changes in latitudinal temperature gradient and implications for sea-ice variability. J. Geophysical Res. **2001**, *106*, 20161–20173.

54. Ding, Q.; Steig, E.J.; Battisti, D.S.; Wallace, J.M. Influence of the tropics on the Southern Annular Mode. J. Climate **2012**, *25* (18), 6330–6348.

55. Okumura, Y.M.; Schneider, D.; Deser, C.; Wilson, R. Decadal-interdecadal climate variability over antarctica and linkages to the tropics: analysis of ice core, instrumental, and tropical proxy data. J. Climate **2012**, *25*, 7421–7441.

56. Kerr, R.A. A North Atlantic climate pacemaker for the centuries. Science **2000**, *288* (5473), 1984–1985.

57. Enfield, D.B.; Mestas-Nuñez, A.M.; Trimble, P.J. The Atlantic multidecadal oscillation and its relation to rainfall and river flows in the continental U. S. Geophysical Res. Lett. **2001**, *28*, 277–280.

58. Dima, M.; Lohmann, G. A hemispheric mechanism for the Atlantic multidecadal oscillation. J. Climate **2007**, *20*, 2706–2719.

59. Sutton, R.T.; Hodson, D.L.R.. Influence of the ocean on North Atlantic climate variability 1871–1999. J. Climate **2003**, *16*, 3296–3313.

60. Delworth, T.L.; Mann, M.E. Observed and simulated multidecadal variability in the Northern Hemisphere. Climate Dynamics **2000**, *16*, 661–676.

61. Sutton, R.T.; Hodson, D.L.R. Atlantic Ocean forcing of North American and European summer climate. Science **2005**, *309*, 115–118.

62. Knight, J.R.; Folland, C.K.; Scaife, A.A. Climate impacts of the Atlantic multidecadal oscillation. Geophysical Res. Lett. **2006**, *33*, L17706.

63. Frankcombe, L.M.; Dijkstra, H.A.; von der Heydt, A. Sub-surface signatures of the Atlantic multidecadal oscillation. Geophysical Res. Lett. **2008**, *35*, L19602.

64. Frankcombe, L.M.; Dijkstra, H.A. Coherent multidecadal variability in North Atlantic sea level. Geophysical Res. Lett. **2009**, *36*, L15604.

65. Chylek, P.; Folland, C.K.; Dijkstra, H.A.; Lesins, G.; Dubey, M. Ice-core data evidence for a prominent near 20 year time-scale of the Atlantic mulitdecadal oscillation. Geophysical Res. Lett. **2011**, *38*, L13704.

Circulation—
Crops

66. Frankcombe, L.M.; Dijkstra, H.A. The role of Atlantic-Arctic exchange in North Atlantic multidecadal climate variability. Geophysical Res. Lett. **2011**, *38*, L16603.

67. Knight, J.R.; Allan, R.J.; Folland, C.K.; Vellinga, M.; Mann, M.E. A signature of persistent natural thermohaline circulation cycles in observed climate. Geophysical Res. Lett. **2005**, *32*, L20708.

68. Latif, M.; Böning, C.; Willebrand, J.; Biastoch, A.; Dengg, J.; Keenlyside, N.; Schweckendiek, U.; Madec, G.. Is the thermohaline circulation changing? J. Climate **2006**, *19* (18), 4631–4637.

69. Ottera, O.H.; Bentsen, M.; Drange, H.; Suo, L. External forcing as a metronome for Atlantic multidecadal variability. Nat. Geosci. **2010**, *3*, 688–694.

70. Msadek, R.; Frankignoul, C.; Li, L.Z.X. Mechanisms of the atmospheric response to North Atlantic multidecadal variability: a model study. Climate Dynamics **2010**, 36 (7 8), 1255–1276.

71. Keenlyside, N.S.; Latif, M.; Jungclaus, J.; Kornblueh, L.; Roeckner, E. Advancing decadal-scale climate prediction in the North Atlantic sector. Nature **2008**, *453* (7191), 84–88.

72. Black, D.; Peterson, L.C.; Overpeck, J.T.; Kaplan, A.; Evans, M.N.; Kashgarian, M. Eight centuries of North Atlantic Ocean Atmosphere Variability. Science **1999**, *286*, 1709–1713.

73. Vellinga, M.; Wu, P. Low-latitude freshwater influence on centennial variability of the Atlantic thermohaline circulation. J. Climate **2004**, *17*, 4498–4511.

74. Timmermann, A.; Latif, M.; Voss, R.; Grotzner, A. Northern Hemisphere interdecadal variability: a coupled air-sea mode. J. Climate **1998**, *11*, 1906–1931.

75. Frankcombe, L.M.; von der Heydt, A.; Dijkstra, H.A. North atlantic multidecadal climate variability: an investigation of dominant time scales and processes. J. Climate **2010**, *23*, 3626–3638.

76. Jungclaus, J.H.; Haak, H.; Latif, M.; Mikolajewicz, U. Arctic-North Atlantic interactions and multidecadal variability of the meridional overturning circulation. J. Climate **2005**, *18*, 4013–4031.

77. Delworth, T.L.; Manabe, S.; Stouffer, R.J. Multideceadal climate variability in the Greenland Sea and surrounding regions: a coupled model simulation. Geophysics Res. Lett. **1997**, *24* (3), 257–260.

78. Polyakov, I.V.; Alekseev, G.V.; Timokhov, L.A.; Bhatt, U.S.; Colony, R.L.; Simmons, H.L.; Walsh, D.; Walsh, J.E.; Zakharov, V.F. Variability of the intermediate Atlantic water of the Arctic Ocean over the last 100 years. J. Climate **2004**, *17* (23), 4485–4497.

79. Polyakov, I.V.; Bhatt, U.S.; Simmons, H.L.; Walsh, D.; Walsh, J.E.; Zhang, X. Multidecadal variability of North Atlantic temperature and salinity during the twentieth century. J. Climate **2005**, *18*, 4562–4581.

80. Frolov, I.E.; Gudkovich, A.M.; Karklin, B.P.; Kvalev, E.G.; Smolyanitsky, V.M., Eds.; Arctic and Antarctic Research Institute (AARI), St. Petersburg, Russia. In *Climate Change in Eurasian Arctic Shelf Seas: Centennial Ice Cover Observations*; Springer-Praxis Books in Geophysical Sciences, Praxis Publishing: Chichester, UK, 2009; 1–165.

81. Wyatt, M.G. *A Multidecadal Climate Signal Propagating Across the Northern Hemisphere through Indices of A Synchronized Network.* Ph.D. Dissertation. University of Colorado: Boulder, CO, 2012; 201.

82. Wyatt, M.G.; Curry, J.A. Role for Eurasian Arctic shelf sea ice in a secularly varying hemispheric climate signal during the 20th century. Clim. Dyn. **2013**, DOI: 10.1007/s00382-013-1950-2 (in press).

83. Outten, S.D.; Esau, I. A link between Arctic sea ice and recent cooling trends over Eurasia. Climatic Change **2012**, *110* (3–4), 1069–1075.

84. Girs A.A. Multiyear oscillations of atmospheric circulation and long-term meteorological forecasts. L. Gidrometeroizdat 1971 480 p. (in Russian)

85. Mantua, N.J.; Hare, S.R.; Zhang, Y.; Wallace, J.M.; Francis, R.C. A Pacific interdecadal climate oscillation with impacts on salmon production. Bull. Am. Meteorological Soc. **1997**, *78*, 1069–1079.

86. Minobe, S. A 50–70-year climatic oscillation over the North Pacific and North America. Geophysical Res. Lett. **1997**, *24*, 683–686.

87. Power, S.; Casey, T.; Folland, C.K.; Colman, A.; Mehta, V. Inter-decadal modulation of the impact of ENSO on Australia. Climate Dynamics **1999**, *15*, 319–323.

88. Niebauer, H. Variability in Bering Sea ice cover as affected by a regime shift in the North Pacific in the period 1947–1996. J. Geophysical Res. **1998**, *103* (C12), 27717–27737.

89. Schmittner, A.; Appenzeller, C.; Stocker, T.F Enhanced Atlantic freshwater export during El Nino. Geophysical Res. Lett. **2000**, *27 (8)*, 1163–1166.

90. Latif, M.; Roeckner, E.; Mikolajewicz, U.; Voss, R. Tropical stabilization of the thermohaline circulation in a greenhouse warming simulation. J. Climate **2000**, *13*, 1809–1813.

91. Gordon, A.L. Interocean Exchange. In *Ocean Circulation and Climate: Observing and Modelling the Global Ocean, International Geophysics Series*; Siedler, G., Church, J., Gould, J., Eds.; Academic Press: San Diego, CA, **2001**; *77*, 306.

92. Gordon, A.L.; Weiss, R.F.; Smethie, W.M.; Warner, M.J. Thermocline and intermediate water communication between the South Atlantic and Indian Oceans. J. Geophysical Res. **1992**, *97*, 7223–7240.

93. Schouten, M.W.; de Ruijter, W.P.M.; van Leeuwen, P.J.; Dijkstra, H.A.; An oceanic teleconnection between the equatorial and southern Indian Ocean. Geophysical Res. Lett. **2002**, *29* (16), 1812.

94. Hasegawa, T.; Yasuda, T.; Hanawa K. Multidecadal variability of the upper ocean heat content anomaly field in the North Pacific and its relationship to the aleutian low ans the kuroshio transport. Papers Meteorol. Geophysics **2007**, *58*, 155–166.

95. Zhang R.; Delworth T.L. Impact of the Atlantic multidecadal oscillation on North Pacific climate variability. Geophysical Res. Lett. **2007**, *34*, L23708.

96. McCabe, G.J.; Palecki, M.A.; Betancourt, J.L. Pacific and Atlantic ocean influences on multidecadal drought frequency in the United States. PNAS **2004**, *101* (12), 4136–4141.

97. Nowak, K.; Hoerling, M.; Rajagopalan B.; Zagona, E. Colorado River Basin hydroclimatic variability. J. Climate **2012**, *25* (2), 4389–4403.

98. Woodhouse, C.A; Gray, S.T; Meko, D.M. Updated streamflow reconstruction for the Upper Colorado River Basin. Water Resources Res. **2006**, *42*, W05415.

99. White, W.B.; Peterson, R.G. An antarctic circumpolar wave in surface pressure, wind, temperature, and sea-ice extent. Nature **1996**, *380* (6576), 699–702.

100. White, W.B.; Annis, J. Influence of the Antarctic circumpolar wave on El Nino and its multidecadal changes from 1950 to 2001. J. Geophysical Res. **2004**, *109*, C06019.

101. White, W.B.; Chen, S.-Chin; Allan, R.J.; Stone, R.C. Positive feedbacks between the Antarctic circumpolar wave and the global El Nino-Southern oscillation wave. J. Geophysical Res. **2002**, *107* (C10), 3165–3162.

BIBLIOGRAPHY

1. Frolov, I.E.; Gudkovich, A.M.; Karklin, B.P.; Kvalev, E.G.; Smolyanitsky, V.M.; Arctic and Antarctic Research Institute (AARI), St. Petersburg, Russia. In *Climate Change in Eurasian Arctic Shelf Seas: Centennial Ice Cover Observa-tion*s; Philppe Blondel, C. Geology, F.G.S., Eds.; Springer-Praxis Books in Geophysical Sciences, Praxis Publishing: Chichester, UK, 2009; 1–165.

2. Klyashtorin, L.B.; Lyubushin, A.A. *Government of the Russian Federation; State Committee for Fisheries of the Russian Federation; Federal State Unitary Enterprise; Russian Federal Research Institute of Fisheries and Oceanography; Moscow; Cyclic Climate Changes and Fish Productivity;* Dr. Gary D. Sharp Editor of English version of book; Center for Climate/Ocean Resources Study; Salinas, CA, USA; VNIRO Publishing: Moscow; 2007; 1–223.

3. Mann, K.H.; Lazier, J.R.N. *Department of Fisheries and Oceans; Bedford Institute of Oceanography; Dartmouth, Nova Scotia; Canada. Dynamics of Marine Ecosystems; Biological-Physical Interactions in the Oceans,* 3rd Ed.; Blackwell Publishing; Malden, MA, Oxford, UK, and Carlton, Victoria, Australia, 2006; 1–495.

4. Burroughs, W.J. *Weather Cycles: Real or Imaginary,* 2nd Ed.; Cambridge University Press: U.K., 2003; ISBN 0 521 52822 4; 1–317.

Climate Change

Robert L. Wilby
Department of Geography, University of Loughborough, Loughborough, U.K.

Abstract

Evidence of the causes and consequences of climate change originates from many sources. Signals from past variations in solar output and the Earth's orbit are found in isotopes extracted from sediment and ice cores. Detailed meteorological observations following explosive volcanic eruptions show the extent to which aerosols blasted into the stratosphere can scatter sunlight and cause global cooling. Laboratory experiments show how greenhouse gases (GHGs; principally carbon dioxide and water vapor) absorb radiation and thereby warm the Earth's surface. Climate models help to organize knowledge of the forces that drive climate change (solar variation, volcanic aerosols, GHG emissions, and other human influences), alongside indicators of change (from long-term meteorological, ice, and ocean observations) and understanding of Earth systems (including ocean–atmosphere, carbon, and hydrological cycles). Considerable uncertainty surrounds future forcing and process–responses but models help to detect and attribute the likelihood of changing patterns of global temperature and rainfall as well as the frequency and intensity of extreme weather events. Climate models also inform policy-making on stabilization targets for future GHG emissions.

INTRODUCTION

Climate change is an abstract concept in the sense that no one can feel a 30-year average rainfall total or global mean temperature. Yet these types of indexes are widely used by scientists to describe the state of the climate system (Fig. 1). However, climate averages are accumulated from day-to-day weather, which is perceptible, as are meteorological variations from pole to the equator or from sea level to summit. Retreating ice sheets, increasing ocean heat content, and rising sea levels are further manifestations of a warming planet.

Climate change is also a plastic concept, because the term means very different things to different people.[1] More than a century ago, atmospheric warming linked to "carbonic acid" (now termed carbon dioxide or CO_2) was regarded by Arrhenius as beneficial by delaying the onset of the next ice age.[2] In contrast, many modern-day governments and their citizens view climate change as a present and/or imminent threat to be managed. Others do not see the same urgency or question whether current (policy and economic) responses are proportionate to the established risk.

This entry begins with an overview of the natural and human drivers of climate variability and change over decades to millennia. Global climate models and their projections are then described. The final section considers the technical challenges surrounding formal detection and attribution of climate change. These are all areas of rapid scientific development and vigorous debate.

NATURAL CAUSES OF CLIMATE CHANGE

Natural causes of climate change are conventionally described as either external or internal to the climate system. External forces affect the amount of solar radiation entering the upper atmosphere; internal forces determine the amount of radiation that actually reaches the surface of the Earth. This is distinct from internal natural climate variability, which arises from the interplay between ocean and atmosphere processes in response to radiative forcing.

Even minor changes in solar output and receipt of radiation by the Earth are sufficient to trigger large responses in the planet's ocean–atmosphere system. Variations in solar radiation output are linked to levels of sunspot activity. The number of sunspots varies, on an average, over 11-year, 22-year, 80–90-year, and longer timescales. However, the length of individual sunspot cycles varies between 10 and 12 years. During periods of high sunspot activity and rapid cycles (such as the 1950s), solar output increases by about one-tenth of a percent relative to the long-term average of 1370 Wm^{-2}. Conversely, periods with fewer sunspots and slower cycles have coincided with notable cool interludes, such as the Little Ice Age, which was harshest during the Maunder minimum of 1645 to 1715.

Solar activity is not the only determinant of the Earth's solar energy balance. By the mid-19th century, Agassiz recognized that Alpine glaciers had once been much more extensive; Croll's astronomical theory of Ice Ages provided an explanation. Calculations by Milanković subsequently showed how variations in the eccentricity (stretch of the

Encyclopedia of Natural Resources DOI: 10.1081/E-ENRA-120047634

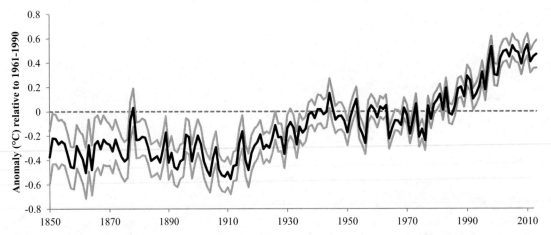

Fig. 1 Global annual mean temperature anomalies (black line) based on analyses of surface data (HadCRUT4) showing lower and upper bounds of the 95% confidence interval of the combined effects of measurement and sampling, bias, and coverage uncertainties (gray lines). Annual anomalies are calculated with respect to the 1961–1990 mean.
Source: UK Met Office, http://www.metoffice.gov.uk/hadobs/hadcrut4/.

Earth's elliptical orbit), obliquity (tilt of the Earth's axis), and precession of the equinoxes (wobble of the North Pole orientation) affect receipt of solar radiation by the upper atmosphere. The rhythm of these cycles has since been corroborated by oxygen isotope analysis of deep sediment cores as 23,000 years for precession, 41,000 years for obliquity, and 96,000 years for eccentricity. Rapid uplift of mountain chains and/or changes in ocean currents due to continental drift can trigger glacial epochs over even longer periods.

The amount of solar radiation received at the Earth's surface depends partly on the composition of the atmosphere. Violent volcanic eruptions like Tambora (1815), Krakatoa (1883), Santa Maria (1902), Novarupta (1912), Mount St. Helens (1980), and Pinatubo (1991) inject huge quantities of sulfur dioxide into the stratosphere. Atmospheric processes convert the gas into aerosols that can linger and reflect radiation for several years. For example, the eruption of Mount Pinatubo added an estimated 10 teragrams of sulfur (TgS) to the stratosphere and caused a global cooling of ~0.5°C up to 18 months later.[3] Volcanic eruptions also provide opportunities to observe the effect of changing atmospheric temperatures on water vapor content—a potent climate feedback mechanism represented in "climate models." Moreover, data from eruptions have been used to show the potential efficacy and side effects of solar radiation management via artificial seeding of the atmosphere with sulfate aerosols (a controversial geoengineering option).

Internal climate variability is generated by major transfers of energy, mass, and momentum within the climate system. Some exchanges exhibit organization at the hemispheric or even planetary scales. For example, the El Niño is characterized by warm waters and high rainfall extending westward across the Pacific from Peru and Equador. Quasiperiodic oscillations between warm and cool (La Niña) phases occur every 4 to 7 years. The influence of El Niño is expressed most forcefully by drought in northeast Brazil, eastern Australia, Indonesia, and India, but greater-than-average rainfall over southwest and southeast United States. The North Atlantic Oscillation (NAO) and Pacific Decadal Oscillation (PDO) are further ocean–atmosphere variations that modulate regional rainfall anomalies on interannual to interdecadal timescales. Positive NAO is generally associated with strong westerly airflows, warmer- and wetter-than-average winter rainfall, and flood-rich episodes across northwest Europe. When a positive PDO and El Niño coincide, the Pacific coast dipole (dry northwest–wet southwest) is amplified.

Other harmonics of natural climate variability such as the Atlantic Multidecadal Oscillation (AMO) occur over even longer timescales. Long instrumental and paleoclimate records show that sea surface temperatures fluctuate through anomalously warm and cool phases, each lasting several decades. The recent warming of the North Atlantic to a condition not seen since the 1950s is thought to explain the spate of unusually wet summers in northern Europe, and hot, dry summers in southern Europe since the 1990s.[4] Similarly, very long reconstructions of rainfall and river flow indicate decades—even centuries—of drought in North America. For example, drought-sensitive tree-ring chronologies suggest four "mega droughts" in the Great Plains between AD 1 and 1200. These events have no modern counterpart, and their origins are poorly understood.[5] Some explanations refer to changes in the strength and location of the North Atlantic thermohaline circulation; others theories suggest that El Niño/La Niña effects may have been amplified by external climate forcing.

HUMAN CAUSES OF CLIMATE CHANGE

Anthropogenic climate change arises from changes in the composition of the atmosphere and/or modifications to surface energy, moisture, carbon fluxes, and stores.

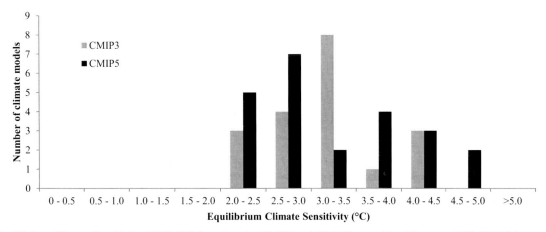

Fig. 2 Equilibrium Climate Sensitivity (ECS) (°C) based on the CMIP3 and CMIP5 ensembles. The mean ECS (3.2°C) is approximately the same for CMIP3 and CMIP5.
Source: IPCC (2007) Fourth Assessment Report and IPCC (2013) Fifth Assessment Report.

By the 1850s and 1860s, Tyndall had determined that CO_2 (and other greenhouse gases [GHGs] such as nitrous oxide, methane, and water vapor) selectively absorbs solar radiation in the infrared band. The atmospheric concentration of CO_2 was about 280 ppm prior to the Industrial Revolution; by 2013, the concentration had exceeded 400 ppm for the first time at the Mauna Loa Observatory in Hawaii. Emissions from fossil fuel combustion and land use changes are adding to the CO_2 concentration and thereby strengthening radiative forcing. What remains less clear is how laboratory measurements scale up to a global warming response, once feedbacks are taken into account (e.g., from changing cloud, snow, or ice cover). Climate sensitivity is typically defined as the expected global warming by doubling the effective CO_2 concentration. The term "effective" is used because the sensitivity implicitly captures forcing by other GHGs. The value of the sensitivity parameter is important because it strongly determines the amount of climate change projected by models and is a point of contention.

Arrhenius calculated that either halving or doubling the concentration of CO_2 would change global mean surface temperatures by 4°C to 5°C. Over a century later, the Intergovernmental Panel on Climate Change (IPCC) Fifth Assessment Report (AR5) stated that the equilibrium climate sensitivity is likely in the range 1.5°C to 4.5°C with high confidence, extremely unlikely less than 1°C (high confidence), and very unlikely greater than 6°C (medium confidence) (Fig. 2). As climate sensitivity cannot be measured directly, the range reflects different interpretations of transient changes in surface, upper air, and ocean temperature measurements, estimated radiative forcing, satellite data, and environmental reconstructions for the last millennium. Upper bound sensitivities are difficult to constrain because of non-linear feedbacks in the climate system and large uncertainty about aerosols and ocean heat uptake. One study suggests that the reduced rate of global mean warming observed over the past decade supports the case for low-end climate sensitivity: somewhere in the region of 1.2°C to 3.9°C.[6]

Human modifications of land cover contribute to climate change in three main ways. First, through forest clearance, crop cultivation, and soil degradation, carbon stores are converted to carbon sources. During the period 1990 to 2005, net emissions of carbon from land use change were 1.5 ± 0.7 Pg Cyr^{-1} (compared with 8.7 ± 0.7 Pg Cyr^{-1} from fossil fuel combustion in 2008).[7] Replacement of natural vegetation by pasture or arable landscapes may also contribute emissions of methane from cattle or submerged fields and nitrous oxide from fertilizers. Permafrost thawing and forest dieback through drought and fire driven by human-induced regional rainfall and temperature changes could further enhance GHG emissions.

Second, destruction or degradation of terrestrial and marine environments affects their capacity to act as carbon sinks. It is estimated that nearly half of all CO_2 emitted from burning fossil fuels and land use change between 1959 and 2008 has remained in the atmosphere.[7] This fraction exhibits much year-to-year variability and is thought to be increasing as the effectiveness of natural carbon sinks declines. However, considerable uncertainty surrounds the future behavior of the carbon cycle under a changed climate. For example, changes in ocean currents or acidity could affect the rate of carbon sequestration by the marine ecosystem.

Third, changes in land use affect regional energy and moisture budgets, an effect that is most pronounced in built environments. It has long been recognized that urban centers can be several degrees warmer than the surrounding countryside.[8] Compared with vegetated surfaces, building materials retain more solar energy during the day and have lower rates of radiant cooling during the night. Urban areas also have lower wind speeds, less convective heat losses, and less evapotranspiration, yielding more energy for surface warming. Artificial space heating, air conditioning, transportation, cooking, and industrial processes further add to the heat load. Other regional climate changes have been reported following reservoir creation or widespread irrigation: the former stimulating more intense precipitation events and the latter local cooling.

CLIMATE MODELS

The climate system is represented in models by equations for radiation transfer, momentum, mass, and water vapor, solved at regular grid points on the Earth's surface, heights in the atmosphere, and depths in the ocean. The sophistication and number of climate model experiments have been rising since the first double CO_2 experiments in the 1970s. Early models had simplified "slab" oceans, fixed cloudiness, and no heat transport by oceans, idealized topography, and coarse grid resolution. State-of-the-art global climate models now incorporate many more physical elements including atmospheric chemistry, cloud convection, snow and ice stores, carbon cycling, rudimentary biomes, and hydrology, ocean salinity, and circulation (Fig. 3). They also include important mechanisms that either amplify or dampen radiative forcing via positive (e.g., water vapor, ice-albedo, carbon and methane release, ocean acidity) or negative (e.g., low cloud, lapse rate) feedbacks. The number of calculations involved is so immense that climate model simulations can only be performed on super-computers or via massively distributed networks of computers (such as the public participation ClimatePrediction.net experiment).

Despite rapid developments in computing hardware, global climate models still operate at comparatively coarse spatial resolutions, relative to some of the phenomena that they seek to represent. The Japanese Earth Simulator runs at 20 km globally but most global climate models have resolutions of 100 km. This means that important features of the climate system such as clouds and convective storms must be described statistically (parameterized). Notwithstanding these constraints, climate models yield convincing representations of zonal average temperature trends;[9] they are less skillful at replicating regional rainfall patterns such as the South Asia monsoon.[10] Climate models are also known to produce precipitation more often and more lightly than is observed and to exaggerate soil moisture feedbacks in convection schemes. Although other techniques can be used to "downscale" to finer spatial scales, these are entirely dependent on the quality of climate information transferred from the coarser resolution global model.[11]

Future rates and patterns of climate change are uncertain because of natural forcing (by variations in solar output and volcanic aerosols), natural climate variability (due to ocean-atmosphere patterns such as El Niño), and socioeconomic and demographic trends (translated into GHG emissions, then atmospheric concentrations). Further uncertainty

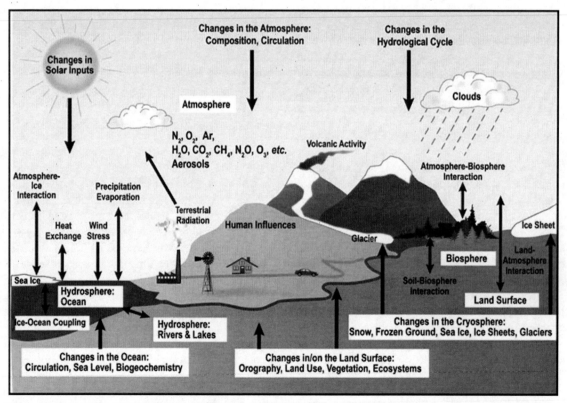

Fig. 3   The major constituents of a global climate model.
Source: Intergovernmental Panel on Climate Change (2007).

arises from parameter and process representations within climate models themselves. At global scales, natural climate variability and climate model uncertainty dominate the total uncertainty for the next few decades; emission scenario uncertainty increases in importance with time and is the major contributor to total uncertainty beyond the mid-21st century (Fig. 4). Uncertainty in aerosol emissions is important much sooner but mainly on regional scales.

These elements of uncertainty can be quantified by running large numbers of models (ensembles) with different initial conditions for oceans and atmosphere and different emission scenarios. For example, the climate model ensemble used in the IPCC Fifth Assessment Report projects a likely global mean temperature change of 0.3°C to 4.8°C by 2081–2100 relative to 1986–2005 (Table 1). Some climate models have been run beyond year 2100 to

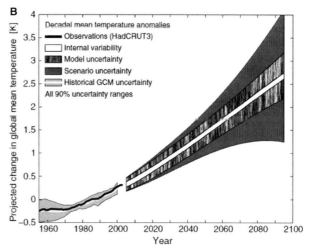

Fig. 4 The total uncertainty in an ensemble (CMIP3) of global mean, decadal mean, and model projections for the 21st-century, separated into three components: internal variability (thick white line), model uncertainty (shadow gray), and scenario uncertainty (dark gray). The gray regions show the uncertainty in the 20th century integrations of the same climate models, with the mean in white. The black lines show an estimate of the observed historical changes. (**A**) Precipitation, with observations (GPCP). (**B**) Temperature, with observations (HadCRUT3). All anomalies are calculated relative to the 1971–2000 mean, except for the precipitation observations, for which a 1979–2000 mean is used.
Source: Hawkins and Sutton (2011).

Table 1 Projected global mean surface temperature change and global mean sea level rise by 2046–2065 and 2081–2100. The AR5 applied four new scenarios, termed Representative Concentration Pathways (RCPs), identified by their approximate total radiative forcing in year 2100 relative to 1750. These are 2.6 Wm[-2] (RCP2.6: a mitigation scenario leading to very low forcing); 4.5 Wm[-2] (RCP4.5: a lower stabilization scenario); 6.0 Wm[-2] (RCP6.0: a higher stabilization scenario); and 8.5 Wm[-2] (RCP8.5: a very high emissions scenario)

Variable	Scenario	2046–2065		2081–2100	
		Mean	Likely range[c]	Mean	Likely range[c]
Global Mean Surface Temperature Change (°C)[a]	RCP2.6	1.0	0.4 to 1.6	1.0	0.3 to 1.7
	RCP4.5	1.4	0.9 to 2.0	1.8	1.1 to 2.6
	RCP6.0	1.3	0.8 to 1.8	2.2	1.4 to 3.1
	RCP8.5	2.0	1.4 to 2.6	3.7	2.6 to 4.8
		Mean	Likely range[d]	Mean	Likely range[d]
Global Mean Sea Level Rise (m)[b]	RCP2.6	0.24	0.17 to 0.32	0.40	0.26 to 0.55
	RCP4.5	0.26	0.19 to 0.33	0.47	0.32 to 0.63
	RCP6.0	0.25	0.18 to 0.32	0.48	0.33 to 0.63
	RCP8.5	0.30	0.22 to 0.38	0.63	0.45 to 0.82

[a]Based on the CMIP5 ensemble; anomalies calculated with respect to 1986–2005.
[b]Based on 21 CMIP5 models; anomalies calculated with respect to 1986–2005.
[c]Calculated from projections as 5–95% model ranges.
[d]Calculated from projections as 5–95% model ranges.
Source: IPCC (2013) Fifth Assessment Report Summary for Policymakers.

quantify the multicentury commitment, even irreversible, global warming, and sea level rise. One study found thermal expansion of the ocean (an important component of sea level rise) of up to 1.9 m if 21st century CO_2 concentrations exceed 1000 ppmv.[12]

The above climate model results are not predictions. This is because other, first-order climate forcings beyond CO_2 (e.g., aerosol effects on clouds, black carbon deposition, reactive nitrogen, and changes in land use/cover) are not always included, yet are known to be significant over multidecadal timescales.[13] Nonetheless, models were originally conceived to aid policy decisions about the costs and benefits of cutting GHG emissions; increasingly, model outputs are applied to impact and adaptation studies (Table 2). Climate models are also used to explore potential tipping points in the climate system; to set stabilization targets for emissions to avoid dangerous climate change; or to evaluate the consequences of more radical interventions

such as geoengineering. For example, one climate model ensemble suggests that cumulative carbon emissions should not exceed 1 trillion tonnes if a peak warming of 2°C above preindustrial temperatures is to be avoided.[14] Other studies examine not *if* but *when* global warming could reach 4°C, and one found the best estimate to be some time in the 2070s.[15]

DETECTING AND ATTRIBUTING CLIMATE CHANGE

Climate change detection and attribution is only possible when three ingredients are present: a coherent theory of causation; a physically plausible model of the climate system; and long, high-quality records of environmental change (e.g., air or ocean temperatures, atmospheric pressure, rainfall patterns).

Table 2   Examples of national climate change risk assessments and adaptation programs

Country	Report	Year	Lead agency
Australia	*Climate Change Risks to Australia's Coast: A First Pass National Assessment*	2009	Department of Climate Change
Belgium	*Belgian National Climate Change Adaptation Strategy*	2010	Flemish Environment Nature and Energy Department
Canada	*From Impacts to Adaptation: Canada in a Changing Climate 2007*	2008	Natural Resources Canada
Denmark	*Danish strategy for Adaptation to a changing climate*	2008	Danish Energy Agency
Finland	*Evaluation of the Implementation of Finland's National Strategy for Adaptation to Climate Change 2009*	2009	Ministry of Agriculture and Forestry
France	*French National Climate Change Impact Adaptation Plan 2011–2015*	2011	Ministry of Ecology, Sustainable Development, Transport and Housing
Germany	*German Strategy for Adaptation to Climate Change*	2008	Federal Ministry for the Environment, Nature Conservation and Nuclear Safety
Iceland	*Iceland's Climate Change Strategy*	2007	Ministry for the Environment
Ireland	*Ireland National Climate Change Strategy 2007–2012*	2007	Department of the Environment, Heritage and Local Government
Japan	*Wise Adaptation to Climate Change*	2008	Ministry of Environment
Netherlands	*Working on the Delta: Acting Today, Preparing for Tomorrow*	2012	Ministry of Infrastructure and the Environment; Ministry of Economic Affairs, Agriculture and Innovation
Spain	*Evaluación Preliminar de los Impactos en España por Efecto del Cambio Climático*	2005	Ministerio de Medio Ambiente
Sweden	*Sweden Facing Climate Change – Threats and Opportunities*	2007	Swedish Commission on Climate and Vulnerability
Switzerland	*Stratégie Suisse d'adaptation aux changements climatiques: Rapport intermédiaire au Conseil federal*	2010	Département fédéral de l'Environnement, des Transports, de l'Energie et de la Communication
UK	*The UK Climate Change Risk Assessment 2012 Evidence Report*	2012	Department for Environment, Food and Rural Affairs
USA	*Scientific Assessment of the Effects of Global Change on the United States*	2008	U.S. Global Change Research Program; Committee on Environment and Natural Resources, National Science and Technology Council

Circulation—
Crops

Climate change *detection* involves demonstrating that environmental data have changed in some defined statistical sense, without having to offer any reason(s) for that change. Detection occurs when an event or trend lies beyond the range expected to occur by chance. Change need not occur in a linear or in a monotonic way. There are plenty of cases where a record has changed abruptly, as with rainfall in southeast Australia in the 1980s[16] or Sahel in the late 1960s (Fig. 5). Detection may be confounded if an underlying trend is weak compared with the "noise" of natural climate variability; on the other hand, there is always a chance of spurious detection even in random data. False trends can also emerge if records are not homogeneous or have been influenced by non-climatic influences such as expansion of urban areas, changes in observer practices, meteorological equipment, station location and surroundings, or monitoring network density. Assessing trends in climate change *impacts* can be just as problematic. For example, much debate has surrounded trends detected in U.S. hurricane damages, a rise that is thought to be explained by increased prosperity and exposure of the population.[17]

Climate change *attribution* involves establishing the probable cause(s) of detected changes at a given level of statistical confidence. The difficulty for climate change studies is that there are no observational controls, as in epidemiological studies. Attribution, therefore, depends on comparing observations against climate models run with and without anthropogenic GHG emissions. In such experiments, attribution of observed climate change to human influence is accepted if there is consistency between the modeled response (to combined human-plus-natural forcing) and observed behavior. Alternatively, models can be used to compute the change in risk of a certain climate event with and without human influences. For instance, the chance of temperature anomalies like the European heat wave of the summer of 2003 is believed to have doubled due to historic emissions.[18]

Callendar was the first to connect rising atmospheric concentrations of CO_2 with evidence of increasing global mean temperatures—the basis for climate change detection and attribution. Using a network of less than 150 stations and records from the 1890s to 1930s, he calculated that "excess" CO_2 was increasing world temperatures at a rate of 0.3°C per century.[19] Seventy-five years later, the IPCC declared that "it is extremely likely that human influence has been the dominant cause of the observed warming since the mid-20th century." However, it is now recognized that climate models reproduce global mean temperatures in different ways due to compensating effects between aerosols and climate sensitivity and that the trade-off between these parameters affects their respective future warming estimates.[20]

Evidence of change detection and attribution at global, hemispheric, and zonal scales has accumulated over the last two decades, but detection and attribution of changes in the intensity and frequency of extremes or attribution of individual extreme events are more demanding propositions.[21] Nonetheless, there are good reasons to expect robust detection of certain extreme events. For instance, regional climate models suggest that heavy rainfall could increase disproportionately more than average events under a warmer atmosphere with greater moisture content. However, precipitation observations show mixed results: evidence of increases in North America and Europe but no change for the Asia-Pacific or central African regions. This may be explained by the inability of climate models to adequately resolve extreme precipitation at local scales, the mismatch between point (observations) and grid (modeled) rainfall, natural climate variability, and by definition, the rarity of extreme events. Others focus on explaining the causes of specific extreme events. For example, analysis of the Thailand floods, East African and Texan droughts, and UK winter coldness in 2011 suggests that the human influence is making certain types of events more likely.[22] But these studies also demonstrate the importance of understanding the role of natural climate variability and non-climatic factors.

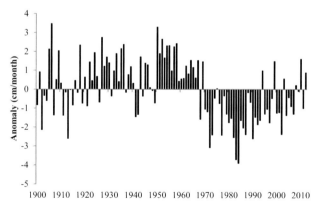

Fig. 5 Sahel precipitation anomalies for the rainy season (June to October) over the domain 20–10°N, 20°W–10°E. Explanations for the abrupt transformation include the following: (1) changes in meteorological networks; (2) agricultural extensification and vegetation removal; (3) natural forcing by sea surface temperatures in the Atlantic, tropical Pacific, Indian Ocean, and Mediterranean; or (4) forcing by a combination of natural SST variability, anthropogenic greenhouse gases, and aerosols. Climate model experiments enable testing of the last three hypotheses.
Source: http://jisao.washington.edu/data/sahel/

CONCLUSION

Climate change arises from three sets of forces: external variations in solar radiation receipt caused by periodicities in the earth's orbit and changes in solar output; internal variations in aerosol concentrations; and human-induced changes in atmospheric GHG concentrations due to fossil

Circulation—Crops

fuel combustion and modifications to land surface properties. Observed variations in global climate since the Industrial Revolution are best explained by the interplay of natural and human forces, along with natural variability in ocean–atmosphere systems. However, the climate sensitivity to rising concentrations of GHGs remains a highly contested figure that partly shapes attitudes to both the urgency and necessity of decarbonizing economies. Decadal climate projections also show divergent behavior because of large uncertainty in future socioeconomic drivers of GHG emissions and the nascent ability of climate models to represent aerosols, clouds, ice sheet dynamics, and ocean heat uptake.

REFERENCES

1. Hulme, M. *Why We Disagree about Climate Change: Understanding Controversy, Inaction and Opportunity*; Cambridge University Press: 2009; 393 p.
2. Crawford, E. Arrhenius' 1896 model of the greenhouse effect in context. Ambio **1997**, *26*, 6–11.
3. Soden, B.J.; Wetherald, R.T.; Stenchikov, G.L.; Robock, A. Global cooling after the eruption of Mount Pinatubo: A test of climate feedback by water vapor. Science **2002**, *296*, 727–730.
4. Sutton, R.T.; Dong, B. Atlantic Ocean influence on a shift in European climate in the 1990s. Nat. Geosci. **2012**, *5*, 788–792.
5. Cook, E.R.; Seager, R.; Heim, R.R.; Vose, R.S.; Herweijer, C.; Woodhouse, C. Megadroughts in North America: Placing IPCC projections of hydroclimatic change in a long-term palaeoclimate context. J. Q. Sci. **2009**, *25*, 48–61.
6. Otto, A.; Otto, F.E.L.; Boucher, O.; Chruch, J.; Hegerl, G.; Forster, P.M.; Gillett, N.P.; Gregory, J.; Johnson, G.C.; Knutti, R.; Lewis, N.; Lohmann, U.; Marcotzke, J.; Myhre, G.; Shindell, D.; Stevens, B.; Allen, M.R. Energy budget constraints on climate response. Nat. Geosci. **2013**, *6*, 415–416.
7. Le Quéré, C.; Raupach, M.R.; Canadell, J.G.; Marland, G.; et al. Trends in the sources and sinks of carbon dioxide. Nat. Geosci. **2009**, *2*, 831–836.
8. Wilby, R.L.; Jones, P.D.; Lister, D. Decadal variations in the nocturnal heat island of London. Weather **2011**, *66*, 59–64.
9. Sakaguchi, K.; Zeng, X.B.; Brunke, M.A. The hindcast skill of the CMIP ensembles for the surface air temperature trend. J. Geophys. Resear. Atmospheres **2012**, *117*, D16113.
10. Annamalai, H.; Hamilton, K.; Sperber, K.R. The South Asian summer monsoon and its relationship with ENSO in the IPCC AR4 simulations. J. Clim. **2007**, *20*, 1071–1092.
11. Wilby, R.L.; Wigley, T.M.L. Downscaling General Circulation Model output: A review of methods and limitations. Prog. Phys. Geogr. **1997**, *21*, 530–548.
12. Solomon, S.; Plattner, G-K.; Knutti, R.; Friendlingstein, P. Irreversible climate change due to carbon dioxide emissions. Proc. Nat. Acad. Sci. **2009**, *106*, 1704–1709.
13. Pielke Sr., R.; Beven, K.; Brasseur, G.; Calvert, J.; Chahine, Climate change: The need to consider human forcings besides greenhouse gases. Eos **2009**, *90*, 413–414.
14. Allen, M.R.; Frame, D.J.; Huntingford, C.; Jones, C.D.; Lowe, J.A.; Meinshausen, M.; Meinshausen, N. Warming caused by cumulative carbon emissions towards the trillionth tonne. Nature **2009**, *458*, 1163–1166.
15. Betts, R.A.; Collins, M.; Hemming, D.L.; Jones, C.D.; Lowe, J.A.; Sanderson, M.G. When could global warming reach 4 degrees C? Philos. Trans. R. Soc. A **2011**, *369*, 67–84.
16. Cai, W.; Cowan, T. Southeast Australia rainfall reduction: A climate-change-induced poleward shift of ocean-atmosphere circulation. J. Clim. **2013**, *26*, 189–205.
17. Pielke Jr., R.A.; Gratz, J.; Landsea, C.W.; Collins, D.; Saunders, M.A.; Musulin, R. Normalized hurricane damage in the United States: 1900–2005. Nat. Hazards Rev. **2008**, *13*, 29–41.
18. Stott, P.A.; Stone, D.A.; Allen, M.R. Human contribution to the European heatwave of 2003. Nature **2003**, *432*, 610–614.
19. Callendar, G.S. 1938. The artificial production of carbon dioxide and its influence on temperature. Q. J. R. Meteorol. Soc. **1938**, *64*, 223–240.
20. Knutti, R. Why are climate models reproducing the observed global surface warming so well? Geophys. Resear. Lett. **2008**, *35*, L18704.
21. Hegerl, G.C.; Karl, T.R.; Allen, M.; Bindoff, N.L.; Gillett, N.; Karoly, D.; Zhang, X.B.; Zwiers, F. Climate change detection and attribution: Beyond mean temperature signals. J. Clim. **2006**, *19*, 5058–5077.
22. Peterson, T.C.; Stott, P.A.; Herring, S. Explaining extreme events of 2011 from a climate perspective. Bull. Am. Meteorol. Soc. **2012**, *93*, 1041–1067.

BIBLIOGRAPHY

1. Hawkins, E.; Sutton, R. The potential to narrow uncertainty in projections of regional precipitation change. Clim. Dyn. **2011**, *37*, 407–418.
2. Intergovernmental Panel on Climate Change (IPCC) Fifth Assessment Report, http://www.ipcc.ch/.
3. Mauna Loa Observatory, Earth System Research Laboratory, NOAA Global Monitoring Division, http://www.esrl.noaa.gov/gmd/obop/mlo/.
4. O'Hare, G.; Sweeney, J.; Wilby, R.L. *Weather, Climate and Climate Change: Human Perspectives*; Pearson Education Ltd: Harlow, 2005; 403.
5. Oppenheimer, M.; Petsonk, A. Article 2 of the UNFCCC: Historical origins, recent interpretations. Clim. Change **2005**, *73*, 195–226.
6. The Earth Simulator Center, Japanese Agency for Marine-Earth and Technology http://www.jamstec.go.jp/esc/index.en.html.

Circulation—Crops

Climate Change: Coastal Marine Ecosystems

Jennifer P. Jorve
Rebecca L. Kordas
Kathryn M. Anderson
Jocelyn C. Nelson
Manon Picard
Christopher D. G. Harley
Department of Zoology, University of British Columbia, British Columbia, Vancouver, Canada

Abstract

Coastal marine ecosystems contain a staggering amount of useable resources: an estimated US$14 trillion annually in economic and ecological goods and services. However, the health of these ecosystems (from tropical reefs to polar waters) is vulnerable to the effects of climate change. Increasing temperature and oceanic CO_2 concentrations combined with other changing environmental parameters are predicted to cause changes in species reproduction and distribution, and affect biodiversity worldwide. Together, climate change stressors, overfishing, and species invasions continuously put pressure on these nearshore communities. The vast majority of scientific research indicates that many species, and overall biodiversity, will continue to decline this century. Curbing the effects of climate change on coastal marine ecosystems should begin with reductions in greenhouse gas emissions, and implementation of effective management strategies to allow these ecosystems to recover and persist.

INTRODUCTION

Coastal marine ecosystems extend from the high tide mark to the edge of the continental shelf. They contain a staggering amount of life and range in form from tropical coral reefs to temperate-zone kelp forests to high-latitude sea ice habitats. Seventy-five percent of all countries are bordered by oceans or seas, and almost half of the world's population lives within 100 km of coasts. Coastal marine ecosystems supply an estimated 43% of the world's ecosystem goods (e.g., food and raw materials) and services (e.g., disturbance regulation and nutrient cycling), a contribution valued at over US$14 trillion per year.[1] However, the health of coastal ecosystems and the benefits associated with them are vulnerable to the effects of climate change.

CLIMATE CHANGE IN THE OCEAN

Climate change is caused by anthropogenic emissions of greenhouse gases such as carbon dioxide (CO_2), which will soon reach the highest atmospheric concentration in tens of millions of years.[2] The rise in CO_2 concentrations since the industrial revolution are 100–1000 times faster than at any point in the past 420,000 years, and these rates of increase are expected to accelerate by a factor of 2–6 by the year 2100.[3] Although one of the most widely publicized repercussions of greenhouse gas emissions is the predicted 2–3°C increase in average air and ocean temperatures over the next 100 years,[2] several additional changes will occur in the ocean. For example, excess CO_2 reacts with seawater to increase the concentration of hydrogen ions (i.e., reduce pH) and reduce the concentration of carbonate ions, which are a key building block for calcified shells and skeletons. This suite of changes in ocean chemistry is known as "ocean acidification." The acidity of the surface ocean (the uppermost ocean layer where mixing due to a combination of temperature changes, waves, and currents acts to homogenize chemical conditions) has already increased by 30%, and may increase up to 150% by the end of this century.[2] Additional factors associated with climate change include sea level rise and changes in upwelling, storminess, and coastal salinity (see Solomon, Qin, et al.[2] for details). Sea level rise, caused by thermal expansion of water and melting of land-based ice sheets, is expected to occur at a magnitude of anywhere between 0.18 and 0.59 meters by the end of the twenty-first century. Uneven heating of the Earth's surface will affect wind patterns, which in turn determine storm frequency and the strength of upwelling along coastlines. Changes in precipitation will vary from place to place, resulting in increasing or decreasing river outflow and thus changes in coastal salinity and terrestrially derived nutrients and contaminants.

Encyclopedia of Natural Resources DOI: 10.1081/E-ENRA-120047626

Circulation—Crops

BIOLOGICAL RESPONSES TO CLIMATE CHANGE IN COASTAL MARINE ECOSYSTEMS

Climate change will have a wide range of biological consequences, including changes in the rates and seasonal timing of growth and reproduction, the distribution of species, and the biodiversity of entire ecosystems.[4] Although the specific ecological impacts of climate change depend upon the system of interest (see Table 1), there are some generalities that are widely applicable. In general, changes in water temperature and CO_2 concentration are predicted to have the largest impacts on marine life. Experimental studies indicate that warming negatively impacts individuals living near their upper thermal tolerance limits, such as many corals,[20] while ocean acidification has a broadly negative effect on calcifying organisms but a positive effect on seaweeds.[21] While each abiotic change (e.g., warming or ocean acidification) may cause specific changes in an individual's physiology or growth rate, such direct effects are frequently modified by interactions with other organisms, such as competitors or consumers. For example, a small amount of warming has been shown to benefit the Atlantic population of the habitat-forming seaweed *Sargassum muticum* in the absence of herbivores. However, the higher energy demands of herbivores in warmer water increased the grazing pressure on *Sargassum*, and therefore the overall effect of warming on the seaweed was negative.[22]

Table 1   Key impacts by habitat type. Note that all impacts may include important changes in community structure, defined as the species present in the system, their abundance, and how they interact with one another (e.g., in a food web). Omission of a particular stressor does not mean that it will not be important in some situations

Habitat	Key stressors	Impacts	References
Coral reefs	Warming Ocean acidification Sea level rise Changing storminess	Increased coral bleaching and mortality events Increased susceptibility to coral disease Reduced coral calcification and increased dissolution Reef drowning by sea level rise	[3]
Mangroves	Warming Sea level rise Changing storminess	Slowed juvenile mangrove growth and eventual mortality Habitat loss Reduced species richness Loss of connectivity between populations and systems	[3,5–7]
Sandy beaches	Sea level rise Changing storminess	Changes in beach composition and morphology Beach erosion and habitat loss	[8]
Estuaries	Warming Ocean acidification Sea level rise Altered run-off	Eutrophication and changes in water clarity Changes in phytoplankton growth rates Reduction of calcifying species (e.g., oysters)	[9,10]
Seagrass beds	Warming Sea level rise Altered run-off	Changes in sedimentation Reduced seagrass biomass and productivity Change in species distributions	[11]
Rocky shores	Warming Ocean acidification	Compressed vertical habitat space Localized species loss in "hot spots" Poleward range shifts Changes in species composition	[12]
Salt marshes	Sea level rise Altered run-off	Habitat loss via marsh drowning Reduced water filtration and carbon sequestration	[9,13,14]
Kelp forests	Warming Ocean acidification Changing storminess	Changes in kelp abundance, density, and productivity Community shift from kelp forests to algal turfs in some regions	[11,15,16]
Continental shelf habitats	Warming Ocean acidification Altered run-off Altered currents and upwelling	Eutrophication, low oxygen levels in bottom waters Reduction of demersal fish and benthic invertebrate populations Biogeographic range shifts	[17]
Open water (pelagic) habitats	Warming Ocean acidification	Shifts in latitudinal ranges and depth distributions Changes in primary and secondary productivity Changes in biogeochemical cycling	[6,11]
Sea ice habitats	Warming	Declines in phytoplankton body size and population size Poleward range shifts Habitat loss	[6,18,19]

Circulation–Crops

The sum of all impacts of environmental change (from both direct and indirect effects) can create large changes in the abundance of species.[12] When the affected species are particularly ecologically important—e.g., habitat-forming species such as kelps and corals, and top predators such as sea stars and sharks—dramatic shifts in entire ecosystems may result. In coral reef ecosystems, warming and ocean acidification cause coral bleaching and mortality, promoting a transition to a system dominated by fleshy algae.[3] Following this shift, coral-associated fish and other reef residents are lost from the system, resulting in a decline in biodiversity. Dramatic ecological changes such as shifts from coral reefs to seaweed beds can occur rapidly when systems are pushed beyond some threshold or "tipping point."[23] Once an ecosystem has shifted into a new phase, it may take a long time to recover even if environmental conditions become more favorable. Because of these time lags, and because ecosystems respond to multiple climate stressors and other human pressures simultaneously, predicting the future state of marine ecosystems is no easy task.

ECONOMIC IMPLICATIONS

Fisheries and Aquaculture

Marine fisheries provide protein for approximately 1.5 billion people worldwide, and the demand for seafood continues to rise.[24] As with all marine species, the distribution and abundance of fished species are linked to changes in climate. Warming promotes poleward shifts of fish and plankton and may positively affect fisheries in higher latitudes as species migrate to track optimal environmental conditions, although this may occur at the expense of native, cold-water fish stocks.[25] The spread of disease also tracks temperature changes: historically limited by lower temperatures, diseases are spreading poleward,[26] and warming increases disease prevalence by increasing susceptibility to pathogens.[27] Although increasing CO_2 boosts growth rates for few commercially important species such as seaweeds, lobsters, crabs, and shrimp,[28] the net effect of ocean acidification on most commercially important calcifying species, like oysters, clams, sea urchins, and abalone, is negative.[21] Aquaculture can provide an alternative to reliance on wild stocks, but the ocean is the primary source of water used in aquaculture facilities and farmed animals will face many of the same physiological problems with warming and acidification as wild stocks.[29] As with land crops, many of the areas currently used for farming marine animals will no longer be suitable as optimal thermal conditions shift poleward and local environments change.

Species Invasions

Along with climate change, species invasions are considered one of the greatest overall threats to marine biodiversity.[30]

Species introductions can carry disease, increase competition for resources, reduce native species abundance, or result in the homogenization of ecosystems.[31] In 2005, damage and management due to species invasions cost nearly $120 billion in the USA alone.[32] Climate change, which disrupts local ecosystems and improves environmental conditions for many invaders, is predicted to increase the frequency and severity of invasions.[33] In general, invasive species tolerate significantly higher temperatures than native species, and as temperature regimes change, experimental evidence shows that the abundance of invasive species increases while native species abundance decreases.[34] Additionally, some invasive species recruit and grow earlier in the year than native species, which in turn shifts community dominance to invasive species.[30] Overall increases in invasive species numbers in the past decade, as in Southern California and Mexico, have been linked to climactic shifts promoting the spread of new invaders arriving via shipping, aquaculture, and ballast water vectors.[35]

MANAGING FOR RISK

As ecological shifts in coastal marine systems become more apparent, it may become advantageous to implement management strategies focused on mitigating climate change impacts. For example, managers can control several stressors at local scales and in the short term by reducing nutrient pollution in run-off and over-fishing.[11] Additionally, the control of fishing through management tools such as Marine Protected Areas (MPAs; see Caselle entry on "Marine Protected Areas") may enable ecosystems to retain enough resilience to weather some of the impacts of climate stressors such as warming and acidification. Although climate change will certainly impact the effectiveness with which MPAs meet specific conservation goals, it should be emphasized that the overall importance of MPAs—via their role in enhancing resilience—increases with climate change. Therefore, research is needed on assessment methods of ecosystem health and function, but also on the identification of key indicators of change in communities (e.g., the "canary in the coal mine"). Because many ecological responses to climate change only become apparent when they are combined with other stressors, such as over-fishing or pollution,[36] MPAs can act to reduce the total stress on ecosystems by managing those aspects that are under short-term regulatory control.

CONCLUSIONS

Nearshore marine systems supply economically and ecologically valuable goods and services to human communities worldwide.[1] As the global population size increases, the demand for these provisions will only increase.[24] Sadly, the recurring theme in scientific research addressing the future state of nearshore marine communities indicates that ecosystem function and health are in decline. Climate change stressors, such as warming and ocean acidification, combined with over-fishing and species invasions, continuously put pressure on the current state and shape of marine communities. Although it is difficult to precisely predict how whole ecosystems will respond to simultaneous climate stressors, the sum of negative impacts on key species implies that the overall abundance of many species and biodiversity as a whole will likely decline during the current century.[4] If we are going to mitigate the impacts of climate change on nearshore marine systems, we need to curb greenhouse gas emissions and implement effective management strategies that set realistic goals (e.g., fishing levels) that would allow species and ecosystems to recover and persist.

ACKNOWLEDGMENTS

We thank Rebecca Gooding and Norah Brown for valuable contributions to the text, the Hakai Network for Coastal People, Ecosystems, and Management, and the Killam Trusts for support during the writing of this entry.

REFERENCES

1. Costanza, R.; d'Arge, R.; de Groot, R.; Farber, S.; Brasso, M.; Hannon, B.; Limburg, K.; Naeem, S.; O'Neill, R.V.; Paruelo, J.; Raskin, R.G.; Sutton, P.; van den Belt, M. The value of the world's ecosystem services and natural capital. Nature **1997**, *387* (6630), 253–260.
2. Solomon, S.; Qin, D.; Manning, M.; Chen, Z.; Marquis, M.; Averyt, K.B.; Tignor M.; Miller, H.L., Eds.; IPCC. *Climate Change 2007: The Physical Science Basis, Contribution of Working Group I to the Fourth Assessment Report of the Intergovernmental Panel on Climate Change*; Cambridge University Press: Cambridge, UK, and New York, NY, USA, 2007.
3. Hoegh-Guldberg, O.; Mumby, P.J.; Hooten, A.J.; Steneck, R.S.; Greenfield, P.; Gomez, E.; Harvell, C.D.; Sale, P.F.; Edwards, A.J.; Caldeira, K.; Knowlton, N.; Eakin, C.M.; Iglesias-Prieto, R.; Muthiga, N.; Bradbury, R.H.; Dubi, A.; Hatziolos, M.E. Coral reefs under rapid climate change and ocean acidification. Science **2007**, *318* (5857), 1737–1742.
4. Munday, P.L.; Jones, G.P.; Pratchett, M.S.; Williams, A.J. Climate change and the future for coral reef fishes. Fish Fish. **2008**, *9* (3), 261–285.
5. Alongi, D. Mangrove forests: Resilience, protection from tsunamis, and responses to global climate change. Est. Coast. Shelf Sci. **2008**, *76* (1), 1–13.
6. Doney, S.C.; Ruckelshaus, M.; Duffy, J.E.; Barry, J.P.; Chan, F.; English, C.A.; Galindo, H.M.; Grebmeier, J.M.; Hollowed, A.B.; Knowlton, N.; Polovina, J.; Rabalais, N.N.; Sydeman, W.J.; Talley, J.D. Climate change impacts on marine ecosystems. Annu. Rev. Mar. Sci. **2012**, *4*, 11–37.
7. Gilman, E.L.; Ellison, J.; Duke, N.C.; Field, C. Threats to mangroves from climate change and adaptation options: A review. Aquat. Bot. **2008**, *89* (2), 237–250.
8. Brown, A.C.; McLachlan, A. Sandy shore ecosystems and the threats facing them: Some predictions for the year 2025. Environ. Conserv. **2002**, *29* (1), 62–77.
9. Scavia, D.; Field, J.C.; Boesch, D.F.; Buddemeier, R.W.; Burkett, V.; Cayan, D.R.; Fogarty, M.; Harwell, M.A.; Howarth, R.W.; Mason, C.; Reed, D.J.; Royer, T.C.; Sallenger, A.H.; Titus, J.G. Climate change impacts on U.S. coastal and marine ecosystems. Estuaries **2002**, *25* (2), 149–164.
10. Kennedy, V.S. Anticipated effects of climate change on estuarine and coastal fisheries. Fisheries **1990**, *15* (6), 16–24.
11. Hoegh-Guldberg, O.; Bruno, J.F. The impact of climate change on the world's marine ecosystems. Science **2010**, *328* (5985), 1523–1528.
12. Harley, C.D.G. Climate change, keystone predation, and biodiversity loss. Science **2011**, *334* (6059), 1124–1127.
13. Gedan, K.; Silliman, B.R.; Bertness, M.D. Centuries of human-driven change in salt marsh ecosystems. Annu. Rev. Mar. Sci. **2009**, *1*, 117–141.
14. Day, J.W.; Christian, R.R.; Boesch, D.M.; Arancibia, A.Y.; Morris, J.; Twilley, R.R.; Naylor, L.; Schaffner, L. Consequences of climate change on the ecogeomorphology of coastal wetlands. Est. Coast. **2008**, *31* (3), 477–491.
15. Harley, C.D.G.; Hughes, A.R.; Hultgren, K.M.; Miner, B.G.; Sorte, C.J.B.; Thornber, C.S.; Rodriguez, L.F.; Tomanek, L.; Williams, S.L. The impacts of climate change in coastal marine systems. Ecol. Lett. **2006**, *9* (2), 228–241.
16. Connell, S.D.; Russell, B.D. The direct effects of increasing CO_2 and temperature on non-calcifying organisms: Increasing the potential for phase shifts in kelp forests. Proc. R. Soc. B **2010**, *277* (1686), 1409–1415.
17. Hall, S.J. The continental shelf benthic ecosystem: Current status, agents for change and future prospects. Biol. Conserv. **2002**, *29* (3), 350–374.
18. Schofield, O.; Ducklow, H.W.; Martinson, D.G.; Meredith, M.P.; Moline, M.A.; Fraser, W.R. How do polar marine ecosystems respond to rapid climate change? Science **2010**, *328* (5985), 1520–1523.
19. Wassmann, P.; Duarte, C.M.; Agustí, S.; Sejr, M.K. Footprints of climate change in the Arctic marine ecosystem. Global Change Biol. **2011**, *17* (2), 1235–1249.
20. Hughes, T.P.; Baird, A.H.; Bellwood, D.R.; Card, M.; Connolly, S.R.; Folke, C.; Grosberg, R.; Hoegh-Guldberg, O.; Jackson, J.B.C.; Kleypas, J.; Lough, J.M.; Marshal, P.; Nystrom, M.; Palumbi, S.R.; Pandolfi, J.M.; Rosen, B.; Roughgarden, J. Climate change, human impacts, and the resilience of coral reefs. Science **2003**, *301* (5635), 929–933.
21. Kroeker, K.J.; Kordas, R.L.; Crim, R.N.; Singh, G.G. Meta-analysis reveals negative yet variable effects of ocean acidification on marine organisms. Ecol. Lett. **2010**, *13* (11), 1419–1434.
22. O'Connor, M.I. Warming strengthens an herbivore-plant interaction. Ecology **2009**, *90* (2), 388–398.
23. Mumby, P.J.; Hastings, A.; Edwards, H.J. Thresholds and the resilience of Caribbean coral reefs. Nature **2007**, *450* (7166), 98–101.
24. FAO. *The State of World Fisheries and Aquaculture 2008*; Food and Agriculture Organization of the United Nations: Rome, 2009.
25. Roessig, J.M.; Woodley, C.M.; Cech, J.J.; Hansen, L.J. Effects of global climate change on marine and estuarine

fishes and fisheries. Rev. Fish Biol. Fish. **2004**, *14* (2), 251–275.

26. Hofmann, E.; Ford, S.; Powell, E.; Klinck, J. Modeling studies of the effect of climate variability on MSX disease in eastern oyster (*Crassostrea virginica*) populations. Hydrobiologia **2001**, *460* (1–3), 195–212.

27. Harvell, C.D.; Mitchell, C.E.; Ward, J.R.; Altizer, S.; Dobson, A.P.; Ostfeld, R.S.; Samuel, M.D. Climate warming and disease risks for terrestrial and marine biota. Science **2002**, *296* (5576), 2158–2162.

28. Ries, J.B.; Cohen, A.L.; McCorkle, D.C. Marine calcifiers exhibit mixed responses to CO_2-induced ocean acidification. Geology **2009**, *37* (12), 1131–1134.

29. Cooley, S.; Doney, S. Anticipating ocean acidification's economic consequences for commercial fisheries. Environ. Res. Lett. **2009**, *4* (2), 8 pp.

30. Stachowicz, J.; Terwin, J.; Whitlatch, R.; Osman, R. Linking climate change and biological invasions: Ocean warming facilitates nonindigenous species invasions. Proc. Natl. Acad. Sci. **2002**, *99* (24), 15497–15500.

31. Crooks, J.A. Characterizing ecosystem-level consequences of biological invasions: The role of ecosystem engineers. Oikos **2002**, *97* (2), 153–166.

32. Larson, D.L.; Phillips-Mao, L.; Quiram, G.; Sharpe, L.; Stark, R.; Sugita, S.; Weiler, A. A framework for sustainable invasive species management: Environmental, social, and economic objectives. J. Environ. Manag. **2011**, *92* (1), 14–22.

33. Hellmann, J.J.; Byers, J.E.; Bierwagen, B.G.; Dukes, J.S. Five potential consequences of climate change for invasive species. Conserv. Biol. **2008**, *22* (3), 534–543.

34. Sorte, C.J.B; Williams, S.L.; Zerebecki, R.A. Ocean warming increases threat of invasive species in a marine fouling community. Ecology **2010**, *91* (8), 2198–2204.

35. Miller, K.A.; Aguilar-Rosas, L.E.; Pedroche, F.F. A review of non-native seaweeds from California, USA Baja California, Mexico. Hidrobiologica, **2011**, *21* (3), 365–379.

36. McLeod, E.; Salm, R.; Green, A.; Almany, J. Designing marine protected area networks to address the impacts of climate change. Front. Ecol. Environ. **2009**, *7* (7), 362–370.

Climate Change: Polar Regions

Roger G. Barry
Cooperative Institute for Research in Environmental Sciences, National Snow and Ice Data Center, Boulder, Colorado, U.S.A.

Abstract

The record of climatic changes in the Cenozoic era is briefly discussed. The Antarctic ice sheet formed approximately 33 million years ago (Ma), whereas perennial Arctic sea ice was not established until 14 Ma and the Greenland ice sheet formed approximately 3 Ma. Glacial/interglacial cycles from 2.6 to 0.43 Ma had a 41,000-year periodicity, which then switched to 100,000 years. Long-term changes in the climates of the polar regions were largely in phase. During the last glacial cycle, however, there was a bipolar seesaw on a millennial scale. The late-glacial Younger Dryas cold event was absent in the Antarctic region. During the last millennium the Medieval Warm Period and the Little Ice Age (LIA) were concentrated around the North Atlantic sector. In the twentieth century the Arctic warmed to the 1930s and then substantially more than the global average from 1980. Arctic sea ice declined dramatically in the 2000s. Arctic glaciers and ice caps have retreated and thinned with a recent acceleration in this trend. In the 2000s, Greenland outlet glaciers have retreated and thinned. Permafrost temperatures have risen recently. Temperatures have risen sharply over the last five decades in the Antarctic Peninsula and the ice shelves there are disintegrating. The Antarctic sea ice expanded slightly over the last three decades, although the mechanisms are uncertain.

INTRODUCTION

The two polar regions have undergone intervals in the past when their climatic conditions were broadly similar with extensive ice sheets and sea ice cover and other times, as today, when their climates are quite dissimilar. This entry begins by sketching the climatic history of past geological epochs and then treats in detail the climate of the two regions over the last century.

THE GEOLOGICAL RECORD OF THE CENOZOIC ERA AND THE QUATERNARY PERIOD

From approximately 65 to 50 million years ago (Ma) in the early Cenozoic (Greek for new life) era, global climatic conditions were warm, probably due to high levels of atmospheric carbon dioxide (CO_2) (up to 1600 ppmv). As a greenhouse gas, this traps infrared radiation emitted by the surface in the lower atmosphere raising the temperature. During this time there were no ice sheets, although the Antarctic continent was located in high southern latitudes. Now we discuss how world climate evolved during the Tertiary period (the start of the Cenozoic era; Cenozoic is Greek for new life) that began approximately 65 Ma, after the age of dinosaurs. Information about climatic conditions during this remote past is derived from sedimentary rocks that contain plant and animal fossils, and especially limestones that yield data on past CO_2 levels in the atmosphere. Conditions remained warm from 65 to 50 Ma with no ice sheets, perhaps due to high CO_2 concentrations. Between 55.5 and 52 Ma there were several abrupt, extreme global warming events. These have been linked to intervals with high eccentricity and high obliquity in the Earth's orbit that triggered massive releases of soil organic carbon that had been locked up in permafrost in the Arctic and Antarctic regions.[1]

Then, temperatures especially in the deep ocean began a steady decline and approximately 33 Ma the climate of Antarctica changed from mild and temperate to glacial conditions, with the accumulation of a thick ice sheet. The cooling is primarily attributed to a decrease in the atmospheric CO_2, but it is also possible that the opening of Drake Passage south of Tierra del Fuego allowed the formation of the Antarctic Circumpolar Current that excluded warm water from the seas around the continent. Since that time the climate of Antarctica has been essentially the same, although the extent of ice on the continental shelves fluctuated, particularly over the last five million years or so. The West Antarctic Ice Sheet developed approximately 6 Ma, but subsequently disappeared at times of warming and then reformed.

The Arctic Ocean was formed in the Middle Eocene approximately 45 Ma. Climatic conditions in the Arctic were quite different from those in the Antarctic; the climate

Encyclopedia of Natural Resources DOI: 10.1081/E-ENRA-120047632

Circulation—
Crops

remained temperate until near the end of the Cenozoic era. Hence, the Earth had unipolar (Southern Hemisphere) glaciation from approximately 33 Ma until perhaps 14 Ma when the Arctic Ocean formed a perennial sea ice cover. However, the Greenland ice sheet did not form until the Pliocene period approximately 3 Ma when CO_2 levels had decreased sufficiently. The closure of the Panama Isthmus approximately 2.9 Ma that affected ocean circulation in the North Atlantic may have been a contributory factor. Ice accumulated in northern Eurasia and North America approximately 2.7 Ma, but Greenland was probably ice free during a warm interval approximately 2.4 Ma.

The start of the Quaternary era that was marked by repeated glacial and interglacial intervals is dated to 2.6 Ma. Until 0.8 Ma, global climate was characterized by cycles of 41,000-year duration that were driven by the axial tilt of the Earth (the obliquity) varying between 22.1° and 24.5°. The obliquity effect on solar radiation receipts is most marked in high latitudes giving a 10% range between obliquity extremes at latitudes 80–90°. Then, for poorly understood reasons, the cyclicity switched to ~100,000 years. This periodicity roughly coincides with the timing of the eccentricity of the Earth's orbit around the Sun, but the effect of this on solar radiation is too small by itself to cause glacial/interglacial cycles. Other processes, particularly ice-albedo feedback, must amplify the small fluctuations in solar energy. This positive feedback involves the cooling effect of an ice sheet, as a result of its high reflectivity (albedo), leading to further ice growth and temperature decrease.

Between 0.8 and 0.43 Ma, interglacial episodes have been shown by the isotopic record in the European Project for Ice Coring in Antarctica (EPICA) ice core from Dome C in Antarctica to have been longer and cooler than subsequent interglacials. The interglacial from 428 ka to 397 ka known as Marine Isotope Stage (MIS)-11 was the longest and gave rise to the disappearance of the Greenland ice sheet and the West Antarctic ice sheet. Subsequent interglacials each lasted only approximately 10% of the 100,000 years occupied by the glacial cycles. These glacial cycles, lasting between 10,000 and 100,000 years, were basically in phase in the two polar regions. Ablation on the Northern Hemisphere ice sheets responded mainly to solar variations in summer and ice-albedo feedback effects as a result of the main ice sheet margins being on land. In Antarctica, in contrast, the ice terminates in the ocean and most ice loss is by calving. Hence, the main control on temperature here is the length of the summer season.

The penultimate interglacial interval, known as the Eemian, peaked at approximately 125,000 years ago (125 ka), with temperatures approximately 1–2°C above those in the twentieth century. Trees grew as far north as northern Norway in what is now arctic tundra. However, the Greenland ice sheet persisted, although it shrank slightly.

During the last glacial cycle (115 to 15 ka), the climates of the two polar regions were slightly out of phase. Antarctic temperatures led Arctic temperatures by a quarter of a period, leading to a bipolar seesaw on a millennial timescale. The Antarctic warms while Greenland is cold and then the Antarctic cools as Greenland warms rapidly. This linkage of the two polar regions is attributed to the interhemispheric ocean circulation—the Meridional Overturning Circulation. Greenland ice cores reveal that there were 25 major climatic oscillations during the last glacial cycle (and these continued into the postglacial Holocene period). The oscillations were characterized by rapid warming followed by a gradual cooling with each event over the last 50,000 years averaging approximately 1470 years. They were probably caused by influxes of fresh water into the North Atlantic related to "binge–purge" cycles in the mass of the Laurentide ice sheet over North America.

During the Last Glacial Maximum (LGM) approximately 25–18 ka vast ice sheets covered much of North America and Fennoscandia, extending to the British Isles. In North America, the Canadian Arctic Archipelago was occupied by the Innuitian ice sheet, the western cordilleras by the Cordilleran ice sheet, and most of central and eastern Canada and the northern United States by the 3 km thick Laurentide ice sheet. The locking up of water in the ice sheets led to a drop in the sea level of approximately 130 m. Consequently, continental shelves were exposed, particularly in the Arctic, and a Bering land bridge formed, enabling humans to enter Alaska from eastern Siberia. Permafrost developed in the sediments of the exposed shelves and this persists extensively in the Laptev and East Siberia seas at present. Global mean temperature during the LGM was approximately 6°C below that in the twentieth century. In Beringia, July temperatures were approximately 4°C lower than late twentieth century.

The ice began to retreat approximately 16 ka, but then between 12,800 and 11,500 years Before Present (BP) there was a pronounced cool event known as the Younger Dryas when postglacial warming was replaced by a return to glacial conditions in high latitudes of the Northern Hemisphere. Temperatures in Greenland and Europe dropped by 15°C. Accumulation on the Greenland ice sheet was halved and there were high dust loadings in the atmosphere.

There is no evidence of the event in Antarctica, where a thousand years earlier an Antarctic Cold Reversal has been documented. The mechanism of the Younger Dryas is attributed to a surge of melt water from Glacial Lake Agassiz in North America into the Arctic Ocean and North Atlantic. This influx of freshwater shut down the subtropical ocean conveyor belt by putting a lid of low-density water at the surface. The onset and termination of the Younger Dryas each took place in only about a decade.

Circulation—Crops

HOLOCENE

The start of the post-glacial Holocene period is dated from 11,500 years ago when conditions approached those of today. In the Arctic region, the warmest conditions appear to have occurred in the early Holocene, coincident with the precessional timing of the boreal summer solstice, occurring when the Earth was nearest to the Sun (perihelion), making northern summers relatively warmer. The precession of the equinoxes has a period of approximately 23,000 years; currently, summer solstice occurs at aphelion, when the Earth is most distant from the Sun, making summers relatively cooler. In the early Holocene the distribution of bowhead whale bones indicates at least periodically ice-free summers along the length of the Northwest Passage and the same pattern is repeated much later from approximately A.D. 500–1250.

In the Northern Hemisphere the massive ice sheets delayed the Holocene warming. The Fennoscandian ice sheet disappeared approximately 9 ka and the Laurentide finally approximately 6.5 ka in northern Labrador-Ungava. Its last surviving remnant is the present day Barnes ice cap in Baffin Island. In northern mid-latitudes the Holocene thermal maximum occurred from approximately 8–5 ka, since which time there has been a slight cooling trend. This is driven by precessional effects decreasing solar radiation, which in summer at 65°N has declined by approximately 6 W/m² over the last two millennia This has been punctuated by several neoglacial events each lasting a few centuries. The best known and most studied of these is the Little Ice Age (LIA), dated around the North Atlantic to A.D. 1550–1850. Recent work suggests that it was triggered by four volcanic events in the late thirteenth century that each led to two to three cold summers as a result of sulfur dioxide aerosols injected into the stratosphere.[2] The effect of these was amplified by ice-albedo feedback resulting from sea ice expansion in the Arctic region, which initiated the onset of the LIA in Iceland and Baffin Island. Further volcanism in the mid-fifteenth century gave this cooling a boost and later, during the Maunder Minimum in sunspot activity from A.D. 1645 to 1715, a decrease of approximately 1% in solar radiation added to the cooling. The presence of sunspots paradoxically gives rise to increased solar radiation because, although the sunspot itself is cooler, the surrounding faculae have higher levels of emission.

The LIA followed an interval known as the Medieval Warm Period, approximately A.D. 950–1250. Temperatures were approximately 0.5–1.4°C above those of 1961–1990 for the northwestern and central North Atlantic. During this interval the Vikings sailed the North Atlantic and built settlements in southwest Greenland that later succumbed to the deteriorating climate.

THE ANTHROPOCENE

The Anthropocene is a term that has recently been proposed for the period when human influences on global climate have become widely apparent. The start date of this interval is uncertain, but it coincides approximately with the beginning of the Industrial Revolution approximately A.D. 1800. At that time atmospheric concentrations of CO_2, methane, and pollutants began to rise. From A.D. 1800, CO_2 concentrations increased by almost 40% from approximately 280 ppm to 390 ppm today due to fossil fuel burning and changes in land use, whereas methane concentrations rose 2.5 times from approximately 700 ppb to 1750 ppb due to agriculture (paddy rice and cattle), termites, and release from wetlands and landfill sites. Methane is a more potent greenhouse gas than CO_2. It has a lifetime in the atmosphere of only 12 years, but compared with CO_2 (value 1), methane has a greenhouse warming potential of 25 over a century. For comparative purposes, we may note that Antarctic ice cores show that, in response to fluctuations of CO_2 concentration from 180 to 280 ppm during glacial/interglacial cycles, mean air temperature fluctuated by approximately 8°C.

Temperatures in the twentieth century rose to an initial peak in the 1930s and 1940s, then declined or leveled off, and have risen continuously since the mid-1980s. The early to mid-century warming, which was most marked in the northern North Atlantic, is thought to be attributable to natural variability. Cool season temperatures on Svalbard rose by 2–4°C from 1912 to the 1930s associated with changes in air circulation and sea-surface temperatures.

The decline in Northern Hemisphere temperatures in the 1960s–1970s is attributed to higher levels of sulfate and soot (black carbon) aerosols in the atmosphere due to industrial activity. Over the period 1950s–1980s, there was a "global dimming," with a reduction in direct solar radiation by 4–6% attributed to this effect. Surface solar radiation decreased by 3–7 W/m² per decade in major land areas. A "global brightening" (+2–8 W/m² per decade) succeeded this during the 1980s to 2000, and the trend has continued in the United States and Europe as a result of controls on air pollution. The brightening has contributed to the post-1980s global warming.[3]

THE ARCTIC

Since approximately 1980, the mean global temperatures have risen due to increasing greenhouse gas concentrations by 0.5°C, with the three warmest years on record (since 1850) being 2010, 2005, and 1998. The pattern differs from that of the high northern latitude warming in the early twentieth century in that it was a global phenomenon. In the Arctic, temperatures have increased at almost twice the global average rate in the past 100 years as a result of "polar amplification" involving the effects of the ice-albedo positive feedback of sea ice and snow cover. This effect, together with the advection (horizontal transport) of warm water and warm air from middle latitudes, has given rise to the amplified Arctic warming. In the central Arctic Ocean temperatures from 1979 to 1995 rose by 0.5°C or more in April–June and January and fell slightly during October–December. For locations north of 64°N there have been Arctic-wide warm conditions since the 1990s in spring and, less strongly, in summer that are unique in the instrumental record. Before this there was generally interdecadal negative covariability between northern Europe and Baffin Bay in winter. This pattern of temperature anomalies is associated with the North Atlantic Oscillation (NAO) in mean sea-level pressure between the Azores anticyclone and the Icelandic low. The Arctic-wide warming pattern of the 1990s indicates a change in atmospheric circulation.

The Arctic Ocean sea ice cover has undergone major shrinkage and thinning since the beginning of all-weather satellite mapping with passive microwave sensors in 1979. The trend in ice area from 2001 to 2012 is approximately −3% per decade in the winter months and −12.4% per decade in September when the ice reaches its annual minimum. This can be compared with a decrease of only 2.2% per decade through 1996. The annual minimum area has decreased from approximately 6.5 million km^2 in 1979 to 3.4 million km^2 in 2012.[4] The five years since 2007 have shown the lowest ice areas in the 33-year record. The ice thickness has also decreased based on submarine sonar measurements. The mean ice draft decreased from 3.1 m in 1958–1976 to 1.8 m in the 1990s. Further analysis for 1988–2003 shows a thinning of 1.31 m or 43%. Between 2004 and 2008, satellite LiDAR data show a thinning of 67 cm and a 42% decrease in the area of multiyear ice. These changes are attributed to a complex set of factors. Since 2000, air temperatures over the Arctic Ocean and absorbed solar radiation at the surface have increased, and the Pacific surface water entering via Bering Strait has warmed. Earlier, changes in air circulation led to increased export of multiyear ice via Fram Strait associated with a cyclonic circulation. This last factor was due to the positive phase of the Arctic Oscillation (low atmospheric pressure over the Arctic Ocean and high pressure in mid-latitudes) during winters of the late 1980s to early 1990s.

The shrinkage and thinning of sea ice have resulted in the Northwest Passage through the Canadian Arctic Archipelago becoming open for shipping for the first time in summer 2007 and this situation has been repeated in successive summers. The Northern Sea Route north of Siberia also became open to shipping in summer 2008 and subsequent summers. This is a major commercial benefit of global warming.

Increased winter precipitation has characterized the Arctic for the past 40–50 years. However, snow cover in northern high latitudes has generally declined.[5] Visible satellite imagery shows that Arctic snow cover extent in May–June decreased by an average of 18% over the 1966–2008 period of record. The largest and most rapid decreases in snow water equivalent (SWE) and snow cover duration have been observed in the maritime regions of the Arctic, which also have the highest precipitation amounts. The North American sector has exhibited decreases in snow cover duration and depths from approximately 1950, whereas widespread decreases in snow cover were not apparent over Eurasia until after approximately 1980. However, there have been long-term increases in winter snow depth over northern Scandinavia and Eurasia.

Lakes in the high latitudes of Canada are responding to the amplified polar warming by increasingly earlier breakup and later freezeup.[6] The respective rates increased from −0.18 d/yr and +0.123 d/yr for 1950–2004 to −0.23 d/yr and +0.16 d/yr, respectively, for 1970–2004.

Glaciers and ice caps in the Arctic region have been retreating like those in most mountain ranges of the world since the 1960s. Icelandic glaciers and ice caps were in retreat during 1930–1960 and then accelerated retreat and thinning have been observed since 1990. Glaciers and ice caps in the Canadian Arctic region retreated and thinned since records began approximately in 1960. The Ward Hunt ice shelf in the northern Ellesmere Island has lost 90% of its area since the beginning of the twentieth century and has largely disintegrated. Glaciers and ice caps on the Arctic islands north of Eurasia have also retreated and thinned since the 1950s–1960s, with an acceleration of this trend since 1990.

The area of summer melt on the Greenland ice sheet, mapped from satellite passive microwave data, has shown a steady increase, with interannual fluctuations, since 1978. The large outlet glaciers in Greenland have shown enhanced calving and thinning in the 2000s as a result of bottom melting from incursions of warm ocean water. This followed a long period of terminus stability. The mass balance of the ice sheet changed from more or less in balance in the 1970s–1980s to a loss of approximately 250 Gt per year during 2005–2011.

Arctic permafrost is warming in response to higher air temperatures and a shortening of the snow cover season. On the North Slope of Alaska this warming was pronounced in the 1980s and 1990s. New record high temperatures at 20 m depth were recorded in 2011 at all permafrost observatories on the North Slope of Alaska, where measurements began in the late 1970s.[7] In northern Russia, permafrost temperatures rose by 1–2°C over the last three decades. During the last 15 years, active-layer thickness has increased in the northern European Russia, northern East Siberia, Chukotka, Svalbard, and Greenland. However, active-layer thickness on the Alaskan North Slope and in the western Canadian Arctic was relatively stable during 1995–2008.

Circulation— Crops

ANTARCTICA

Ice-core reconstructions indicate that Antarctica has warmed approximately 0.2°C since the late nineteenth century.[8] Temperatures were broadly in phase with those in the Southern Hemisphere, but temperatures on the continent and in the Antarctic Peninsula were basically anti-phase.

Temperatures over most of Antarctica showed slight changes in the twentieth century. The exception is the Antarctic Peninsula where a 2.5°C warming has occurred since the 1950s, mainly in winter and spring, when observational records began. Analysis of records, including those from automatic weather stations, since 1957 show warming of 0.10°C per decade in East Antarctica and 0.17°C per decade in West Antarctica. A postulated bipolar seesaw in Arctic and Antarctic air temperatures in the twentieth century has been shown to be absent using ice core data and a reconstruction of the Southern Annular Mode index.[9] Neither of these indices is correlated with climate records from the Arctic or North Atlantic, nor with the Atlantic Multidecadal Oscillation. Instead, these indices are correlated with tropical Pacific sea surface temperatures. However, they point out the complexity of Antarctic climate variability.

Major changes in ice shelves along the Antarctic Peninsula have been observed beginning in the 1960s. Most losses occurred from the 1990s onward. Larsen-B on the northeast coast largely disintegrated in 2002. The Wordie and Wilkins ice shelves on the west coast underwent several large calving events in the late 1990s and 2000s. The loss of ice shelf buttressing of glaciers entering the Larsen-B ice shelf led to their acceleration and rapid thinning.

Whaling ship observations and early ice atlases indicate that Antarctic sea ice in austral summer declined from the 1930s–1950s to the satellite record of the 1970s–1980s. Over the last three decades there has been a slight increase of approximately 3% in Antarctic sea ice area overall, with a maximum in March, and with large regional variations. This increase has been attributed to a slight cooling in summer and autumn.[10] Another suggested mechanism involves increased snowfall raising the surface albedo, but further work is needed.

Records of snow accumulation from ice cores, snow pits, stake networks, and weather stations for both West and East Antarctica indicate that there have been no changes in annual mean values since the late 1950s.

In Antarctica, there is rapid thinning of glaciers in the Amundsen Sea sector, especially on Pine Island Glacier where it is up to 9 m per year. However, it has proved very difficult to estimate the overall mass balance of the ice sheet. Satellite radar altimetry covering 72% of the grounded ice sheet during 1992–2003 indicated a small mass growth. West Antarctica appears to be losing mass and East Antarctica gaining slightly.[11] Permafrost in the Antarctic Peninsula shows warming over the last few decades but there are few data points.

CONCLUDING REMARKS

On geological time scales the two polar regions have shown generally synchronous changes, as well as on the 10,000–100,000 year scale. The major exception was from 33 Ma to 14 Ma when the Antarctic had a major ice sheet and the Arctic did not. The last glacial cycle saw major Northern Hemisphere ice sheets in North America and Fennoscandia that only disappeared in the Holocene. The post-glacial warming was interrupted by the 1300-year-long Younger Dryas cold event, particularly around the North Atlantic sector. In the last millennium the same sector experienced the LIA. In the early twentieth century the northern North Atlantic warmed strongly. From the 1980s, there is a marked global warming signature, strongly amplified in the Arctic. Arctic glaciers, the Greenland ice sheet, and particularly Arctic sea ice all are showing a marked response to this warming. In contrast, in southern high latitudes only the Antarctic Peninsula has displayed substantial warming over the last five decades and this is reflected in the retreat or collapse of ice shelves. Antarctic sea ice has shown minor changes.

REFERENCES

1. DeConto, R.M.; Galeotti, S.; Pagani. M.; Tracy, D.; Schaefer, K.; Zhang, T.; Pollard, D.; Beering, D.J. Past extreme warming events linked to massive carbon release from thawing permafrost. Nature **2012**, *484*, 87–91.
2. Miller, G.H.; Geirsdóttir, A.; Zhong, Y.; Larsen, D.J.; Otto-Bliesner, B.L.; Holland, M.M.; Bailey, D.A.; Refsnider, K.A.; Lehman, S.J.; Southon, J.R.; Anderson, C.; Björnsson, H.; Thordarson, T. Abrupt onset of the Little Ice Age triggered by volcanism and sustained by sea-ice/ocean feedbacks. Geophys. Res. Lett. **2012**, *39*, L02708
3. Wild, M. Enlightening global dimming and brightening. Bull. Am. Met. Soc. **2012**, *93*, 27–37.
4. Stroeve, J.C.; Serreze, M.C.; Holland, M.P.; Kay, J.E.; Malanik, J.; Barrett, A.P. The Arctic's rapidly shrinking sea ice cover: a research synthesis. Clim. change **2012**, *110*, 105–127.
5. Callaghan, T.V.; Johansson, M.; Brown, R.D.; Groisman, P.Y.; Labba, N.; Radionov, V.; Barry, R.G.; Bulygina, O.N.; Essery, R.L.H.; Frolov, D.M.; Golubev, V.N.; Grenfell, T.C.; Petrushina, M.N.; Razuvaev, V.N.; Robinson, D.A.; Romanov, P.; Shinedell, D.; Shmakin, A.B.; Sokratov, S.A.; Warren, S.; Yang, D. The changing face of Arctic snow cover: a synthesis of observed and projected change. Ambio **2012**, *40*, 17–31.
6. Latifovic, R.; Pouliot, D. Analysis of climate change impacts on lake ice phenology in Canada using the historical satellite data record. Remote Sens. Environ. **2007**, *106*, 492–507.

7. Romanovsky, V.E.; Gruber, S.; Instanes, A.; Jin, H.; March-enko, S.S.; Smith, S.L.; Trombotto, D.; Walter, K.M. Frozen Ground, Chapter 7, In *Global Outlook for Ice and Snow*; Earthprint, UNEP/GRID: Arendal, Norway, 2007; 181–200.

8. Schneider, D.P: Steig, E.J; van Ommen, T.D.; Dixon, D.A.; Mayewski, P.A.; Jones, J.M.; Bitz, C.M. Antarctic tempera-tures over the past two centuries from ice cores. Geophys. Res. Lett. **2006**, *33*, L16707.

9. Schneider, D.P; Noone, D.C. Is a bipolar seesaw consistent with observed Antarctic climate variability and trends? Geophys. Res. Lett. **2012**, *39*, L06704.

10. Shu, Q.; Qiao, F.; Song, Z.; Wang, C. Sea ice trends in the Antarctic and their relationship to surface air temperature during 1979–2009. Clim. Dynam. **2012**, *38*, 2355–2563.

11. Wingham, D.J; Shepherd, A.; Muir, A.; Marshall, G.J. Mass balance of the Antarctic ice sheet. Phil. Trans. Roy. Soc. **2006**, *364*, 1627–1635.

BIBLIOGRAPHY

1. Barry, R.G; Gan, T.Y. *The Global Cryosphere: Past, Present and Future*; Cambridge University Press: Cambridge, 2011; 472.

2. Serreze, M.C; Barry, R.G. *The Arctic Climate System*; Cambridge University Press: Cambridge, 2005; 385.

3. Turner, J; Marshall, G.J. *Climate Change in the Polar Regions*; Cambridge University Press: Cambridge, 2011; 434.

Circulation—
Crops

Climate: Classification

Long S. Chiu

Department of Atmospheric, Oceanic and Earth Sciences, George Mason University, Fairfax, Virginia, U.S.A.

Abstract

Climate is the average weather condition over a region, and climate classification is the grouping of regions of similar climate. Approaches to climate classifications include the genetic, empirical, and application schemes. Examples of each approach, including the Bergeron air mass classification, the Köppen–Geiger empirical classification, and the Thornthwaite soil-water approach are described. Advances in measurement, collection, and analysis techniques will allow refinement and further development of climate classifications applicable to climate assessment, monitoring, change detection, and sustainable exploitation of the Earth's natural resources.

INTRODUCTION

Weather is a snapshot of the state of the atmosphere. The state variables include air pressure, temperature, moisture, cloud conditions, wind, and other environmental parameters. In a broad sense, climate is a "long-term" average of the weather.[1] According to the Intergovernmental Panel on Climate Change (IPCC), "Climate [is the] statistical description in terms of the mean and variability of relevant quantities over a period ranging from months to thousands or millions of years. The period is usually taken to be 30 years, as defined by the World Meteorological Organization (WMO)" while other periods are also used, depending on the available data and specific application.[2] Climate classification is the division of the Earth's climate into contiguous regions of relatively homogeneous statistics of climate elements or according to the dominant physical processes at work. This entry describes three basic approaches to climate classification with an example each, discusses state-of-the-art observations, and argues for integrating the observations for refinement and further development of climate classes to facilitate the assessment, monitoring, change detection of the Earth's climate, and sustainable exploitation of the Earth's natural resources.

CLIMATE CLASSIFICATION SCHEMES

Brief History

The earliest climate classification was by the Greeks who divided the Earth according to the location relative to the Sun's inclination. Contemporary works in climate are mostly shaped by German meteorologists. For example,

in 1817, Alexander von Humboldt (1769–1859) drew annual mean temperatures on a world map. Seasonal temperature ranges with improvements were included by Wladimir Köppen (1846–1940) in 1884. He later collaborated with Rudolf Geiger (1894–1981) in establishing his system of climate classification.[3,4] Tor Bergeron (1891–1971), the Germany-educated Swedish meteorologist, developed a classification scheme in terms of causes of climate in 1928. Alisov[5] classified climate regimes according to the physical causes such as circulation type, air mass formation, and frontal systems.[1] American climatologist and geographer, C. W. Thornthwaite (1899–1963) developed a hierarchical classification in 1931, essentially in terms of the annual pattern of soil-moisture conditions. His scheme was made more explicit with the introduction of potential evaporation in 1948, which, however, was superseded by the physical-based formulation of evaporation by Penman.[6]

Broadly, we can define three approaches to climate classification: genetic, empirical, and applied classification schemes. The genetic approach focuses on the cause of climate. An example is the air mass classification developed by Bergeron that later developed into the spatial synoptic classification (SSC) system often employed in synoptic climatology. An empirical approach is data driven and is focused on the effect of climate on a specific region. An example is the Köppen–Geiger classification scheme, which utilizes frequently observed atmospheric variables such as temperature and precipitation that are related to the land biomes. An application classification is created for, or is an outgrowth of, specific applications. An example is the Thornthwaite classification, which considers the role of vegetation on the soil-water availability.

Encyclopedia of Natural Resources DOI: 10.1081/E-ENRA-120047628

Circulation—
Crops

Mathematical Climate Classification

Climate comes from the Greek *klima* which means inclination. The earliest Greek climate classification divides the hemispheres into three zones based on their latitudes: the "summerless," "intermediate," and "winterless" zones, labeled the torrid, temperate, and frigid zones, which are bounded by the Arctic Circle, Antarctic Circle, Tropic of Cancer, and Tropic of Capricorn, in that order.[1] Major improvements by Supan[7] take account of the actual observations of temperature, precipitation, wind, and orography. The Earth is divided into a hot belt, two temperate belts, and two cold caps, and regions are classified as polar, temperate, tropical, continental, marine, and mountain climates, and other types, with variations. He further classified the Earth into 34 climate provinces. However, no attempts were made to relate climate types to different regions.[1]

Bergeron and Spatial Synoptic Classification

The Bergeron classification is the most widely accepted form of air mass classification, focused primarily on the origin of the air masses. Three letters in the air mass classification specify the air mass types. The first letter describes its moisture properties; c is used for continental air masses (dry) and m for maritime air masses (moist). The second letter describes the thermal characteristic of its source region: T for tropical, P for polar, A for Arctic or Antarctic, M for monsoon, E for equatorial, and S for subsidence air (dry air formed by significant downward motion in the atmosphere). The third letter designates the stability of the atmosphere. If the air mass is colder (warmer) than the ground below it, it is labeled k (w).[1] During the 1950s, air mass identification was used in weather forecasting that later developed into synoptic climatology.[8]

Based upon the Bergeron classification scheme, the SSC includes six categories of air masses: Dry Polar (similar to continental polar), Dry Moderate (similar to maritime superior), Dry Tropical (similar to continental tropical), Moist Polar (similar to maritime polar), Moist Moderate (a hybrid between maritime polar and maritime tropical), and Moist Tropical (similar to maritime tropical, maritime monsoon, or maritime equatorial).

Figure 1 shows the world distribution of air masses according to the SSC system. Note that the SSC classifies air masses over land as well as over ocean. More refined techniques such as cluster analyses have been used for SSC based on weather observations in the United States.[9]

Köppen–Geiger Classification

The objective of Köppen's classification is the association of climate elements that determine the land biomes. His scheme is based on annual and seasonal temperature and precipitation values and is represented by three letters. The first letter recognizes the five primary types and are labeled A through E, namely A for tropical; B, dry; C, mild midlatitude; D, cold midlatitude; and E, polar. The second letter shows the seasonal precipitation pattern, with w, s, and f, where w represents a dry winter—the driest winter month has at the most 1/10th of the precipitation found in the wettest summer month; s is a dry summer—the driest summer month has at the most 30 mm (1.18 in.) of rainfall and has at the most one-third the precipitation of the wettest winter month, and f, when the condition of neither s nor w is satisfied. The third letter represents the seasonal temperature conditions and is designated a when the warmest month averages above 22°C, and b, when conditions do not meet the requirements for a, but there still are at least four months above 10°C.

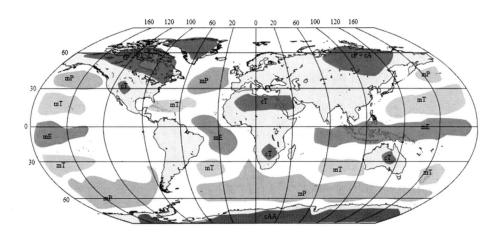

Fig. 1 World distribution of air mass classification; m and c denote maritime and continental, respectively, and P, T, E, M, A, and S denote polar, tropical, equatorial, monsoon, arctic (or Antarctic), and subsiding air masses, in that order. A third letter k or w indicates whether the air is colder (k) or warmer (w) than the underlying surfaces to indicate the stability of the atmosphere.
Source: http://en.wikipedia.org/wiki/Air_mass, cited March 16, 2012.

The five primary classifications can be further divided into secondary classifications such as rain forest, monsoon, tropical savanna, humid subtropical, humid continental, oceanic, Mediterranean, steppe, subarctic, tundra, polar ice cap, and desert climates.

Figure 2 shows a distribution of the climate classification according to the Köppen's scheme developed by the United Nations Food and Agricultural Organization (FAO).[10,11] The temperature data are based on the Climate Research Unit (CRU)[12] analysis and the rainfall data are from the Global Precipitation Climatology Centre (GPCC)[13] data. The climate classes are described as follows.

Rain forests are characterized by heavy rainfall, with minimum normal annual rainfall between 1750 and 2000 mm. Mean monthly temperatures exceed 18°C during all months of the year.

A monsoon is characterized by a seasonal prevailing wind, which lasts for several months before the rainy season in the region. Regions within North America, South America, sub-Saharan Africa, Australia, and East Asia are monsoon regimes.

A tropical savanna is a grassland biome located in semi-arid to semihumid climate regions of subtropical and tropical latitudes, with average temperatures remaining at or above 18°C year round and rainfall between 750 and 1270 mm a year. They are widespread in Africa, and are found in India, the northern parts of South America, Malaysia, and Australia.

In the humid subtropical climate zone, winter rainfall (and sometimes snowfall) is associated with large storms that are steered by westerlies (winds from west to east). Most summer rainfall is accompanied by thunderstorms and occasionally associated with tropical cyclones. Humid subtropical climates lie on the east side continents, roughly between latitudes 20° and 40° away from the equator.

A humid continental climate is marked by variable weather patterns and a large seasonal temperature variance. Places with more than 3 months of average daily temperatures above 10°C, a coldest month temperature of below −3°C, but which do not meet the criteria for an arid or semiarid climate, are classified as continental.

An oceanic climate is typically found along the west coasts at the middle latitudes of all the world's continents, and in southeastern Australia, and is accompanied by plentiful precipitation year round. It is characterized by warm, but not hot summers and cool, but not cold winters, and a relatively narrow annual temperature range and an adequate and evenly distributed precipitation throughout the year.

The Mediterranean climate regime resembles the climate of the lands in the Mediterranean Basin, parts of western North America, parts of West and South Australia, southwestern South Africa and parts of central Chile. The climate is characterized by hot, dry summers and cool, wet winters.

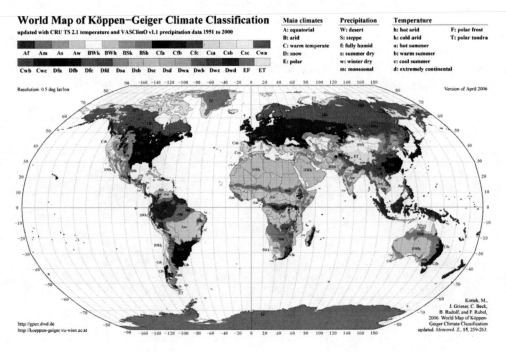

Fig. 2 Climate classification according to the Koppen Scheme Map is developed by the United Nations Food and Agricultural Organization (FAO). Temperature and rainfall data are based on the Climate Research Unit (CRU) and the Global Precipitation Climatology Centre (GPCC) data, respectively.
Source: Grieser et al.,[11] Mitchell et al.,[12] and Beck et al.[13]

A steppe is a dry grassland with an annual temperature range in the summer of up to 40°C and during the winter down to −40°C.

A subarctic climate has little precipitation, with monthly temperatures above 10°C for 1 to 3 months of the year, and permafrost in large parts of the area due to the cold winters. Winters within subarctic climates usually include up to 6 months of average below-freezing temperatures (0°C).

Tundra occurs in the far Northern Hemisphere, north of the Taiga Belt, including vast areas of northern Russia and Canada, characterized by a temperature range of 10°C to 0°C for the warm months, annual rainfall 150–250 mm, and absence of trees.

A polar ice cap, or polar ice sheet, is a high-latitude region that is covered in ice. Subfreezing temperatures occur most of the year due to a net radiative deficit of incoming solar and outgoing longwave radiation.

A desert is a landscape form or region that receives very little precipitation. Deserts usually have a large diurnal and seasonal temperature range. Depending on location, daytime high temperatures can reach up to 45°C in summer, and low night-time temperatures in winter can go down to below 0°C due to extremely low humidity. Many deserts are formed in rain shadows, as mountains block the moisture on the windward side and leave the leeward side as rain shadows.

The Thornthwaite Scheme

The Thornthwaite [14] scheme is a hierarchical classification scheme, explained essentially in terms of the annual pattern of soil-moisture conditions. The soil-moisture conditions are determined by the monthly rainfall as input and evaporation as an output. Evaporation is indicated by temperature conditions. Thornthwaite [15] quantified his earlier scheme and introduced potential evapotranspiration (PE) to describe thermal efficiency (TE). PE is the maximum evaporation available, which is not limited by the available moisture of the underlying surface. The TE is compared to precipitation to form a moisture index (Im) to show amounts and periods of soil-water surplus (deficit) when precipitation is larger (smaller) than PE. Definite break points in the moisture regimes are used to define climatic boundaries. Table 1 shows the moisture provinces corresponding to the Im computed from monthly PE and precipitation. The thermal provinces are defined by the annual PE.

The moisture descriptors are hyperhumid (A), humid (B1,...,B4), moist and dry subhumid (C_1 and C_2), semiarid (D), and arid (E) conditions. The moist conditions (A, B, and C_1) are further subdivided into the following subclasses: r showing little or no water deficiency, s: moderate summer water deficiency, w: moderate winter water deficiency, s_2: large summer water deficiency, and w_2: large winter water deficiency. Similarly, dry conditions (C_2, D, and E) have the following subclasses; d shows little or no water surplus, s: moderate winter water surplus, w. moderate summer water surplus, s_2: large winter water surplus, and w_2: large summer water surplus.

Thermal descriptors represents the water need and include frost (denoted E′), tundra (D′), microthermal (C′$_1$ and C′$_2$), mesothermal (B′$_1$..B′$_4$), and megathermal (A′) regimes. A microthermal climate is one of low annual mean temperatures, generally between 0°C and 14°C, which experiences short summers and has a potential evaporation between 14 and 43 cm. A mesothermal climate lacks persistent heat or persistent cold, with potential evaporation between 57 and 114 cm. A megathermal climate is one with persistent high temperatures and abundant rainfall, with potential annual evaporation in excess of 114 cm. The seasonality is represented as a′ for regions with summer precipitation contributing to less than 48% of annual total, b′ (between 48% and 68%), c′$_1$ and c′$_2$ between 68% and 88%, and d′ higher than 88%.

To illustrate this scheme, San Francisco, California, is classified as C_1B′$_1$s$_2$a′, i.e., dry subhumid, first mesothermal, with large winter water surplus, and a temperature efficiency regime of normal to megathermal.

Table 1 Moisture and thermal indices used in the Thornthwaite Scheme classification

	Moisture province	Moisture index (Im)	Thermal province	Annual PE (cm)
A	Perhumid	100 and above	Megathermal	>114.0
B4	Humid	80 to 99.9	Mesothermal	99.8 to 114.0
B3	Humid	60 to 79 9	Mesothermal	85.6 to 99.7
B2	Humid	40 to 59.9	Mesothermal	71.3 to 85.5
B1	Humid	20 to 39.9	Mesothermal	57.1 to 71.2
C2	Moist subhumid	0 to 19.9	Mesothermal	42.8 to 57.0
C1	Dry subhumid	−19.9 to 0	Mesothermal	28.6 to 42.7
D	Semiarid	−39.9 to −20	Tundra	14.3 to 28.5
E	Arid	−60 to −40	Frost	0 to 14.2

More Recent Developments

With advances in measurement techniques, collection methodology, analysis procedures, and improved understanding of the working of the Earth systems, more sophisticated classification schemes are available. For example, better quality-controlled rainfall data are collected by the Global Precipitation Climatology Center as part of the WMO Global Precipitation Climatology Project, which are used in Fig. 2. Budyko's climate classification is based on the soil-energy requirement. Population growth and man's activities on the Earth system prompted Brooks's[16] classification scheme. The emergence of megacities with their own climate leads to the development of urban climate by Landsberg.[17] The Thornthwaite scheme is modified and is the basis of the Palmer Drought Indices used in monitoring drought and crop conditions.

Climate data were collected on the Earth's surface with ground-based instruments. The advance of space technology and information acquired through remote sensing has increased our knowledge of the Earth systems. Atmospheric variables, such as air pressure, wind, temperature and humidity profile, precipitation, soil moisture, evapotranspiration, atmospheric composition, chemical constituents, and vegetation, are now available from remote-sensing techniques.[18] For example, improved precipitation data are collected by the international Tropical Rainfall Measuring Mission (TRMM) and its follow-on Global Precipitation Mission (GPM). International collaborations have also resulted in co-ordinated measurements of atmospheric trace gases, biogeochemical substances, and flux measurements. These observations cover the large expense of the earth's surface that has not been surveyed before. Advances in information technology facilitate data integration and data mining. More sophisticated statistical procedures, such as eigenvector analysis and cluster analysis, have been used to group the climates of places and to define areas of "similar" climates. Climate classifications are suited for monitoring climate change and impact on society. Generally, they provide more robust change detection criteria for monitoring climate change due to human activities.

CONCLUSION

The climate at each location on the Earth is dependent on the interactions of the components of the Earth systems: the atmosphere, hydrosphere, land surface, and biosphere and are shaped by human activities (or the anthrosphere). The progress on climate classification is dependent on the available data and their targeted applications. We described three approaches to climate classification, the genetic, empirical, and application classifications with examples.

The Köppen–Geiger classification scheme describes the integrated effect of climate on vegetation, which is related to the biomes. The Thornthwaite scheme includes the effect of vegetation on regulating climate through evapotranspiration. As photosynthetic process ultimately determines the habitability of the Earth, land-based classification schemes can be extended to include the ocean biota and primary productivity. This will include oceanic variables, such as sea-surface temperature, salinity, and biogeochemical variables not used before in climate classification. The integration of terrestrial and aquatic biomes will allow a better assessment, monitoring, and sustainable exploitation of the natural resources of the Earth on which our lives depend.

ACKNOWLEDGMENTS

The author acknowledges support for his research on Earth system science by the U.S. National Aeronautics and Space Administration and National Oceanic and Atmospheric Administration.

REFERENCES

1. American Meteorological Society (AMS). *Glossary of Meteorology*, 2nd Ed.; AMS, 45 Beacon Street: Boston, MA 02108–3693, 2000. http://glossary.ametsoc.org/
2. Solomon, S.; Qin, D.; Manning, M.; Chen, Z.; Marquis, M.; Averyt, K.B; Tignor, M.; Miller, H.L., Eds.; *Contribution of Working Group I to the Fourth Assessment Report of the Intergovernmental Panel on Climate Change*; Cambridge University Press: Cambridge, United Kingdom and New York, NY, USA, 2007. IPCC; http://www.ipcc.ch/.
3. Köppen, W.P. *Grundriss der Klimakunde*, 2d Ed.; Walter de Gruyter: Berlin, 1931.
4. Köppen, W.P.; Geiger, R. *Handbuch der Klimatologie*; Gebruder Borntraeger: Berlin, 1930–1939; 6 Vol.
5. Alisov, B.P. *Die Klimate der Erde*; Deutscher Verlag der Wissenschaften: Berlin, 1954; 277 pp.
6. Penman, H.L. Natural evaporation from open water, bare soil and grass. Proc. Roy. Soc. Lond. **1948**, A (194), S. 120–145.
7. Supan, A. Die Temperaturzonen der Erde. Petermanns Geog. Mitt. **1879**, *25*, 349–358.
8. Schwartz, M.D. Detecting structural climate change: an air mass-based approach in the North Central United States, 1958–1992. Ann. Assoc. Am. Geographers **1995**, *85* (3), 553–568.
9. Sheridan, S.C. The redevelopment of a weather-type classification scheme for North America. Intl. J. Climatol. **2002**, *22*, 51–68. (http://sheridan.geog.kent.edu/ssc.html, cited May 31, 2012)
10. Kottek. M.; Geisier, J.; Beck, C.; Rudolf, B; Rubel, F. Map of Koppen-Geiger climate classification update. Meteorol. Z. **2006**, *15*, 259–263;http://koeppen-geiger.vu-wien.ac.at/pdf/kottek_et_al_2006_A4.pdf. Retrieved January 20, 2012.
11. Jürgen G.; René, G.; Stephen, C.; Michele, B. New gridded map of the Koeppen's Climate classification, 2006. http://

www.fao.org/nr/climpag/globgrids/KC_classification_
en.asp (cited May, 2012).

12. Mitchell, T.L Jones, P. An improved method of constructing a database of monthly climate observations and associated high-resolution grids. Int. J. Climatol. **2005**, *25* 693–712. http://www.cru.uea.ac.uk/.

13. Beck, C.; Grieser J.; Rudolf, B. A new monthly precipitation climatology for the global land areas for the period 1951 to 2000, Klimastatusbericht. 2004; 181–190, DWD.

14. Thornthwaite, C. The climates of North America according to a new classification. Geogr. Rev. **1931**, *21*, 633–655.

15. Thornthwaite, C.W. An approach toward a rational classification of climate. Geogr. Rev. **1948**, *38*, 55–94.

16. Brooks, C.E.P. *Climate in Everyday Life;* Greenwood Press, London, 1950; 314 pp.

17. Landsberg, H.; *Urban Climate*; International Geophysics Series. Academic Press: New York, 1981; Vol. 28, 277 p.

18. Chiu, L.S. Earth observations. In *Advanced Geoinformation Science*; Yang, C., Miao, Q., Wong, D.,Yang, R., Eds.; CRC Press: Taylor and Francis, 2011; 486.

BIBLIOGRAPHY

1. Pidwirny, M. Climate Classification and Climatic Regions of the World. *Fundamentals of Physical Geography, 2nd Ed.* 2006, http://www.physicalgeography.net/fundamentals/7v. html, Viewed January 20, 2012.

2. Ritter, Michael E. *The Physical Environment: An Introduction to Physical Geography.* 2006. http://www.uwsp.edu/geo/faculty/ritter/geog101/textbook/title_page.html (cited May 31, 2–12, 2012).

3. Rohli, R.V.; Vega, A.J. *Climatology*, 2nd Ed; Jones and Barlett Learning, 2010; 426 pp.

4. Trewartha, G.T. *An Introduction to Climate*, 3d Ed.; Appendix A, 1954; 223–238.

Climate: Extreme Events

Philip Sura
Department of Earth, Ocean and Atmospheric Science and Center for Ocean-Atmospheric Prediction Studies,
Florida State University, Tallahassee, Florida, U.S.A.

Abstract
One very important topic in climatology, meteorology, and related fields is the understanding of extremes in a changing climate. There is broad consensus that some of the most hazardous effects of climate change are due to a potential increase (in frequency and/or intensity) of extreme weather and climate events. This entry reviews the basic statistical definitions of weather and climate extremes, discusses how they are sampled and commonly used, and provides a brief overview on how extremes are expected to change in a warming climate.

INTRODUCTION

Extreme events in weather and climate are by definition scarce, but they can have a significant physical and socio-economic impact on people and countries in the affected regions. Alongside the intuitive knowledge that hurricanes, tornados, severe storms and rain, droughts, floods, and extreme temperatures might qualify, how can we define extreme events more quantitatively? There are several approaches used in climate research. One often-used [e.g., by the Intergovernmental Panel on Climate Change (IPCC)] definition of an extreme event of the variable under consideration is based on the tails of its climatological (i.e., reference) probability density function (PDF) at a particular geographical location. Extreme events would normally be as rare as or rarer than the, for example, 5th and 95th percentiles. The specific percentile values are not rigorously defined, so often other, more or less stringent, ranges are used (e.g., the 1st and 99th or 10th and 90th percentiles), depending on the particular application. Another routinely used definition of extreme events is that of block maxima or minima. Block maxima/minima are the highest/lowest values attained at a specified location during a given time interval (e.g., daily, monthly, seasonal, or annual). If the interval is the whole period for which observations are available, a block maximum/minimum is called the "absolute extreme." Note that both definitions do not depend on the particular shape (e.g., Gaussian or non-Gaussian) of the PDF. Yet the Gaussian distribution is very often used to estimate the odds of extreme events, neglecting the non-Gaussianity of real world observations. Therefore, an extreme event can also be defined as the non-Gaussian tail of the data's PDF. This definition implies that a high amplitude event does not qualify as extreme if it is described by Gaussian statistics. In a nutshell, because there are different definitions, it is important to be aware of the one being used in a particular application.

Whichever definition is used, understanding extremes has become an important objective in weather and climate research because weather and climate risk assessment depends on knowing and understanding the statistics of the tails of PDFs. At this point it is essential to define weather and climate. As described in almost all fundamental meteorological textbooks,[1,2] weather is varying on timescales of hours, days, to a few weeks, whereas climate varies on longer timescales of months, years, and decades. There is, of course, a certain overlap and the terms weather and climate are typically used in a loose way, specifying the timescales as needed for particular applications. It should be noted that nonlinear multi-scale interactions render a strict separation of timescales unfeasible, making it all put impossible to attribute an individual extreme event to a changing climate.

There is broad consensus that some of the most hazardous effects of climate change are related to a potential increase (in frequency and/or intensity) of extreme weather and climate events. (A notable exception is gradual sea level rise, which will result in the inundation of large densely populated regions; by our definition this is not an extreme event, but it is surely catastrophic.) The overarching goal of studying extremes is, therefore, to understand and then manage the risks of extreme events and related disasters to advance strategies for efficient climate change adaption. Although numerous important studies have focused on changes in mean values under global warming, such as mean global temperature (one of the key variables in almost every discussion of climate change; see, for example, reports from the IPCC available at http://www.ipcc.ch), the interest in how extreme values are altered by a changing climate is a relatively new topic in climate research. The reasons for that are primarily twofold. First, for a comprehensive statistical analysis of extreme events, high-quality and high-resolution (in space and time) observational data sets are needed. It is only recently that global

Encyclopedia of Natural Resources DOI: 10.1081/E-ENRA-120047635

high-quality daily observations became available to the international research community. Second, we need extensive simulations of high-resolution climate models to (hopefully) simulate realistic climate variability. Again, only recently long enough high-resolution numerical simulations of climate variability became feasible to study extreme events in some detail.

SAMPLING

The general problem of understanding extremes is, of course, their scarcity: it is very hard to obtain reliable (if any) statistics of those events from a finite observational record. Therefore, the general task is to somehow extrapolate from the well-sampled center of a PDF to the scarcely or unsampled tails. The extrapolation into the more or less uncharted tails of a distribution can be roughly divided into four major, by no means mutually exclusive categories. In fact, the study of extreme events in weather and climate is most often done by combining the strategies of the following methods.[3,4]

The *statistical approach* (*extreme value theory*) is solely based on mathematical arguments.[3,5–8] It provides methods to extrapolate from the well-sampled center to the scarcely or unsampled tails of a PDF using mathematical tools. The key point of the statistical approach is that, in place of an empirical or physical basis, asymptotic arguments are used to justify the extreme value model. In particular, the generalized extreme value (GEV) distribution is a family of PDFs for the block maxima (or minima) of a large sample of independent random variables drawn from the same arbitrary distribution. Although the statistical approach is based on sound mathematical arguments, it does not provide much insight into the physics of extreme events. Extreme value theory is, however, widely used to explore climate extremes.[9–11] In fact, the foundation of extreme value theory is very closely related to the study of extreme values in meteorological data.[7,8] Nowadays this is very often done in conjunction with the numerical modelling approach discussed below. That is, model output is analyzed using extreme value theory to see if statistics are altered in a changing climate.

The *empirical–physical approach* uses empirical knowledge and/or physical reasoning to provide a basis for an extreme value model. The key point here is that, in contrast to the purely statistical method that primarily uses asymptotic mathematical arguments to model only the tail of a PDF, empirical and/or physical reasoning is employed to model the entire PDF. The empirical–physical method can itself be further split into either empirical or physical strategies (or both), focusing on the empirical or physical aspects of the problem, respectively. For example, a purely empirical approach would simply fit a suitable PDF to given data, whereas a more physical ansatz would also determine the physical plausibility of a specific PDF. However, often a clear distinction between the physical and empirical aspects is impossible. The empirical–physical method lacks the mathematical rigor of the statistical method, but it provides

valuable physical insight into relevant real world problems. An example for an empirical–physical application is the Gamma distribution, which is often used to describe atmospheric variables that are markedly asymmetric and skewed to the right. Often skewness occurs when there is a physical limit that is near the range of data.[6] Well-known applications are precipitation and wind speed, which are physically constrained to be non-negative. It should be noted that the empirical–physical approach can be, in principle, put on a more rigorous foundation using the principle of maximum entropy.[12–14] That is, given some physical information (i.e., constraints) of a process, the PDF that maximizes the information entropy under the imposed constraints is the one most likely found in nature (i.e., the least biased given the constraints). However, not all PDFs commonly used to explain weather and climate data can be justified this way. For example, the Gamma and Weibull distributions can be obtained by the principle of maximum entropy, but the constraints are not necessarily physically meaningful.[15]

The *numerical modelling* approach aims to estimate the statistics of extreme events by integrating a general circulation model (GCM) for a very long period.[16–18] That is, this approach tries to effectively lengthen the limited observational record with proxy data from a GCM, filling the unsampled tails of the observed PDF with probabilities from model data. Numerical modelling allows for a detailed analysis of the physics (at least model-physics) of extreme events. In addition, the statistical and empirical–physical methods can also be applied to model data, validating (or invalidating) the quality of the model. It is obvious that the efforts by the IPCC to understand and forecast the statistics of extreme weather and climate events in a changing climate fall into this category. Because very long model runs are needed to sample the tails of a PDF, global GCMs used for that purpose are currently run at a relatively coarse spatial resolution and are, therefore, unable to resolve important sub-grid scale features such as clouds, tornadoes, and local topography. Because of that, GCMs cannot be used for very localized studies of extreme events. To overcome this problem downscaling methods are commonly used, for which the prediction of extremes is not based on direct GCM output but on subsequent statistical or dynamical models to link the coarse GCM output to local events.[19]

The *non-Gaussian stochastic approach* makes use of stochastic theory to evaluate extreme events and the physics that governs these events.[4] Assuming that weather and climate dynamics are split into a slow (i.e., slowly decorrelating) and a fast (i.e., rapidly decorrelating) contribution, weather and climate variability can be approximated by a stochastic system with a predictable deterministic component and an unpredictable noise component. In general, the deterministic part is non-linear and the stochastic part is state dependent. The stochastic approach takes advantage of the non-Gaussian structure of the PDF by linking a stochastic model to the observed non-Gaussianity. This can be done in two conceptually different ways. On the one hand, if the deterministic component is non-linear and the stochastic component is state independent, the non-Gaussianity is due to the non-linear deterministic part. On the other hand, if the deterministic component is linear and the stochastic component is state dependent, the non-Gaussianity is due to the state dependent noise. Of course, any combination of the two mechanisms is also possible. Although the non-linear approach with state-independent noise captures some types of non-Gaussian climate variability well,[20,21] it recently became clear that state-dependent (or multiplicative) noise plays a major role in describing weather and climate extremes.[4] The physical significance of multiplicative noise is that it has the potential to produce non-Gaussian statistics in linear systems. In particular, Sura and Sardeshmukh,[22] Sardeshmukh and Sura,[23] and Sura[4] attribute extreme anomalies to stochastically forced linear dynamics, where the strength of the stochastic forcing depends linearly on the flow itself (i.e., linear multiplicative noise). Most important, because the theory makes clear and testable predictions about non-Gaussian variability, it can be verified by analyzing the detailed non-Gaussian statistics of oceanic and atmospheric variability.

What do these approaches have in common? Every approach effectively extrapolates from the known to the scarcely known (or unknown) using certain assumptions and, therefore, requires a leap of faith. For the statistical approach the assumptions are purely mathematical. For example, the assumption of classical extreme value theory, that the extreme events are independent and drawn from the same distribution, and that sufficient data are available for convergence to a limiting distribution (the GEV distribution) may not be met.[5,6] The potential drawback of the empirical–physical approach is its lack of mathematical rigor (with the exception of the principle of maximum entropy); it primarily depends on empirical knowledge and physical arguments. The weakness of numerical modelling (including downscaling) lies in the largely unknown ability of a model to reproduce the correct statistics of extreme events. Currently, GCMs are calibrated to reproduce the observed first and second moments (mean and variance) of the general circulation of the ocean and atmosphere. Very little is known about the credibility of GCMs to reproduce the statistics of extreme events. Likewise, the non-Gaussian stochastic approach relies on the assumption that weather and climate variability can be modelled by a stochastic process. It has to be concluded that the common methods to study extreme events have some limitations and that the study of extreme weather and climate events is largely empirical. In particular, there exists no closed quantitative theory on how the statistics of weather and climate extremes might change in a warming climate.

CLIMATE CHANGE

What can be said about likely changes in frequency or intensity of climate extremes in a warming climate? Overall, not very much. In fact, there are only few processes we understand well enough to have some confidence in projected changes.[24] At the top of the list is an increase in the number of extremely warm days and heat waves. In fact, many land areas are already experiencing significant increases in maximum temperatures. This is also consistent with projected temperature changes obtained from GCM projections. Because warm air can hold more water, there is also an agreement that the intensity (mean and variability) of the hydrological cycle increases with increasing temperatures.[25] Given a more intense hydrological cycle, larger amounts of rainfall will come from heavy showers and more intense thunderstorms. Of course, more intense precipitation increases the likelihood of severe floods. Somewhat counter intuitively, the likelihood of severe drought will also increase in a warmer climate because, with a more intense hydrological cycle, a larger proportion of rain will fall in the more extreme events. In addition, higher temperatures will result in increased evaporation reducing the amount of moisture available at the surface. Note that the global spatial distribution of temperature and precipitation extremes (and the probability of droughts) is highly variable.[24,26] Of course, many people are mostly interested in the projected change of severe winds. Unfortunately, we have very little definite knowledge about how the strength and frequency of hurricanes and severe mid-latitude storms might change under global warming. The reason for that is, that the genesis of tropical and mid-altitude storms is controlled by many physical processes, including the large-scale atmospheric flow, whose interactions we currently do not fully understand. Also, there is a large uncertainty with regard to how our global climate models are capable of simulating the plethora of small-scale processes. That is also the reason why we cannot make a definitive projection for small-scale events such as tornadoes, hail, and thunderstorms.

CONCLUSION

Knowing the tails of weather and climate PDFs is an important goal in the atmospheric and ocean sciences because weather and climate risk assessment depends on understanding extremes. Although the commonly used definitions of extremes, and their conceptual implementation in a meteorological framework, are straightforward, it is very hard to obtain statistically significant information of extreme weather and climate events from scarce data. In particular, we only have a very limited physical and statistical understanding of how extremes are going to be altered in a changing climate. Many climate projections just look into the change of the mean and the variance, that is, assuming Gaussian statistics. However, the non-Gaussian statistics (the shape of the distribution) will most likely also be altered in a changing climate. More research (more observations, better theoretical and numerical models) is needed to improve our understanding of how a PDF might change in the future.

REFERENCES

1. Wallace, J.M.; Hobbs, P.V. *Atmospheric Science: An Introductory Survey (Second Edition)*; Academic Press: 2006; 504 pp.
2. Hartmann, D.L. *Global Physical Climatology;* Academic Press: 1994; 411 pp.
3. Garrett, C.; Müller, P. Extreme events. Bull. Amer. Meteor. Soc. **2008**, *89*, ES45–ES56.
4. Sura, P. A general perspective of extreme events in weather and climate. Atmos. Res. **2011**, *101*, 1–21.
5. Coles, S. *An Introduction to Statistical Modeling of Extreme Values*; Springer-Verlag: 2001; 208 pp.
6. Wilks, D.S. *Statistical Methods in the Atmospheric Sciences*; Second Edition. Academic Press: 2006; 627 pp.
7. Gumbel, E.J. On the frequency distribution of extreme values in meteorological data. Bull. Amer. Meteor. Soc. **1942**, *23*, 95–105.
8. Gumbel, E.J. *Statistics of Extremes*; Columbia University Press: 1958; 375 pp.
9. Katz, R.W.; Parlange, M.B.; Naveau, P. Statistics of extremes in hydrology. Adv. Water Resour. **2002**, *25*, 1287–1304.
10. Katz, R.W.; Naveau, P. Editorial: Special issue on statistics of extremes in weather and climate. Extremes 2010, *13*, DOI 10.1007/s10 687–010–0111–9.
11. Smith, R.L. Extreme value statistics in meteorology and the environment. Environ. Stat. **2001**, *8*, 300–357.
12. Jaynes, E.T. Information theory and statistical mechanics. Phys. Rev. **1957**, *106*, 620–630.
13. Jaynes, E. T. Information theory and statistical mechanics. II. Phys. Rev. **1957**, *108*, 171–190.
14. Jaynes, E.T. *Probability Theory: The Logic of Science*; Cambridge University Press: 2003; 758 pp.
15. Lisman, J.H.C.; van Zuylen, M.C.A. Note on the generation of most probable frequency distributions. Stat. Neerlandica **1972**, *26*, 19–23.
16. Easterling, D.R., Meehl, G.A.; Parmesan, C.; Changnon, S.A.; Karl, T.R.; Mearns, L.O. Climate extremes: Observations, modeling, and impacts. Science **2000**, *289*, 2068–2074.
17. Kharin, V. V.; Zwiers, F.W. Estimating extremes in transient climate change simulations. J. Clim. **2005**, *18*, 1156–1173.
18. Kharin, V.V.; Zwiers, F.W.; Zhang, X.; Hegerl, G.C. Changes in temperature and precipitation extremes in the IPCC ensemble of global coupled model simulations. J. Clim. **2007**, *20*, 1419–1444.
19. Wilby, R.L.; Wigley, T.M.L. Downscaling general circulation model output: a review of methods and limitations. Prog. Phys. Geogr. **1997**, *21*, 530–548.
20. Kravtsov, S.; Kondrashov, D.; Ghil, M. Multi-level regression modeling of nonlinear processes: derivation and applications to climate variability. J. Clim. **2005**, *18*, 4404–4424.
21. Kravtsov, S.; Kondrashov, D.; Ghil, M. Empirical model reduction and the modelling hierarchy in climate dynamics and the geosciences. In *Stochastic Physics and Climate Modelling*; Palmer, T.; Williams, P. Eds.; Cambridge University Press: 2010; 35–72.
22. Sura, P.; Sardeshmukh, P.D. A global view of non-Gaussian SST variability. J. Phys. Oceanogr. **2008**, *38*, 639–647.
23. Sardeshmukh, P.D.; Sura, P. Reconciling non-Gaussian climate statistics with linear dynamics. J. Clim. **2009**, *22*, 1193–1207.
24. Houghton, J. *Global Warming - The Complete Briefing. Fourth Edition*; Cambridge University Press: 2009; 438 pp.
25. Allen, M.R., Ingram, W.J. Constraints on future changes in climate and the hydrological cycle. Nature **2002**, *419*, 224–232.
26. Christensen, J.H.; Hewitson, B.; Busuioc, A.; Chen, A.; Gao, X.; Held, I.; Jones, R.; Kolli, R.K.; Kwon, W.-T.; Laprise, R.; Magaña Rueda, V.T.; Mearns, L.; Menéndez, C.G.; Räisänen, J.; Rinke, A.; Sarr, A.; Whetton, P. Regional Climate Projections. In *Climate Change 2007: The Physical Science Basis. Contribution of Working Group I to the Fourth Assessment Report of the Intergovernmental Panel on Climate Change* Solomon, S.; Qin, D.; Manning, M.; Chen, Z.; Marquis, M.; Averyt, K.B.; Tignor, M.; Miller, H.L. Eds.; Cambridge University Press: Cambridge, United Kingdom and New York, NY, USA, 2007.

Climatology

Jill S. M. Coleman
Department of Geography, Ball State University, Muncie, Indiana, U.S.A.

Abstract

Climate is the prevailing state of the atmosphere that characterizes a particular region. The study of climate and the climate system or climatology examines long-term weather conditions and the physical processes that generate or alter those conditions. Although climate is often described in terms of weather variable averages (e.g., mean annual temperature), climate includes the extremes, frequencies, trends, and other statistical properties of the atmosphere. The climate of a location is determined by several factors such as latitude, water proximity, global semipermanent pressure systems, ocean currents, topography, and land cover type. Climate classification systems, such as the Köppen or Thornthwaite schemes, highlight geographic regions with similar atmospheric conditions. Climate is not static, but a very dynamic part of the Earth–atmosphere system that responds to natural and anthropogenic influences.

INTRODUCTION

Atmospheric science examines the physical processes of the atmosphere, the influence of the atmosphere on other systems, and the impacts these other systems have on the atmosphere at a variety of spatial and temporal scales. Traditionally, the study of the atmosphere is subdivided into two main areas: meteorology and climatology. Meteorology emphasizes weather and weather forecasting, the short-term variation of the atmosphere on the order of hours, days, or a few weeks. Climatology or the study of climate focuses on the long-term or average weather conditions over an extended period of time, usually over multiple decades, and the processes that create those conditions. Specialization areas within climatology include paleoclimatology, the study of past climates using techniques (e.g., ice cores) prior to instrumental data; historical climatology, the study of earlier climates within the timeframe of the written record (i.e., the past few thousand years); hydroclimatology, the study of the interaction of the climate system and the hydrological cycle; bioclimatology, the study of the interaction of the climate system and living organisms; and climatology specializations based on a region (e.g., tropical climatology), atmospheric system size (e.g., synoptic climatology), unique environment (e.g., urban climatology) or application (e.g., tourism).

Climate describes the general state of the atmosphere or expected weather conditions. Whereas weather describes the immediate or near-future state of the atmosphere for given time and place through atmospheric variables such as temperature, pressure, humidity, wind speed and direction, and cloud cover, climate relays the statistical properties and persistent behavior of the atmosphere. Climate statistics are often expressed as normals, extremes, and frequencies.[1] Climatic normals refer to the mean weather conditions of a location, usually averaged over a 30-year period; however, shorter or longer averaging periods are also used depending on data availability and time scale of comparison. Extremes show the range, variability, and anomalies in atmospheric conditions, such as the record low and high temperature. Frequencies give information on the probability or likelihood of a particular weather event (e.g., tornado) occurring. Climate is more than a descriptive science detailing statistics on typical or abnormal atmospheric conditions. Moreover, climate represents the coupling between the atmosphere and the surfaces of the Earth.

CLIMATE SYSTEM CONTROLS

The dynamic linkages of the Earth–atmosphere system dictate the general atmospheric conditions or climate over a region. Climate is a direct response to the complex interactions of energy, mass, and momentum between the atmosphere and the other major Earth subsystems, the hydrosphere (water), biosphere (living organisms), and the lithosphere (land).[2] Several physical and geographical components of these subsystems combine to control the climate system. Global climate types and patterns are governed by features such as latitude, semipermanent pressure systems, water proximity, oceanic circulation, and local topography.

Latitude

Seasonal changes in the orientation of the Earth with respect to the Sun determine the amount and intensity

Encyclopedia of Natural Resources DOI: 10.1081/E-ENRA-120047625

Circulation— Crops

of solar radiation different locations receive. Solar radiation or insolation receipt not only directly impacts diurnal temperature changes but atmospheric circulation, including pressure distribution, wind flow, precipitation, cloud formation, and other features of weather and climate. Sun angle and day length regulate the insolation amount and are largely a function of latitude, the geographic coordinates that specify the north–south position of a place relative to the equator.

Latitude is the angle between the equatorial plane and a location on the surface of the Earth with values ranging from a minimum of 0° at the equator to a maximum of 90° at the poles. As the Earth revolves around the Sun, the axial tilt of the Earth results in the Northern (Southern) Hemisphere directed toward the Sun for half of the year and away from the Sun the other half of the year. Equatorial locations receive on average around 12 hours of daylight per day throughout the year, whereas polar locations fluctuate between periods of 24 hours of daylight or darkness.

While the amount of daylight hours is an important factor in insolation amount, equally as important is the intensity of the solar radiation. The higher the Sun remains above the horizon, the more intense the radiation. When the Sun is high in the sky, solar radiation beams are highly concentrated over a small area and are more effective at warming a surface than when the sun is low in the horizon and the energy from the beams is distributed over a greater area. In addition, low sun angles require insolation beams to pass through a thicker atmosphere, resulting in greater beam depletion from the scattering, reflection, and absorption by the atmosphere and less receipt then by the surface. On average, low latitudes of the Tropics and Subtropics receive the most insolation and experience higher temperatures over the course of a year and high latitudes of the Arctic and Antarctic receive the least; hence, an inverse relationship generally exists between latitude and annual solar radiation receipt.

Global Atmospheric Circulation: Semipermanent Pressure Systems

Latitude and surface-type heating differences combined with the rotational speed of the Earth produce a well-defined pattern of atmospheric pressure systems and wind flow, often explained using the three-cell model. An idealized depiction of mean global atmospheric circulation, the three-cell model refers to the number of distinct circulation cells present in each hemisphere: the Hadley cell in the low latitudes; the Ferrell cell in the midlatitudes; and the Polar cell in the high latitudes. These cells redistribute energy between the warmer regions that are energy-rich and colder regions that are energy-poor, thereby preventing the Tropics from becoming increasingly hot or the Poles increasingly cold. The rising and sinking motions associated with each cell produce latitudinal zones dominated by either

low (cyclonic) or high (anticyclonic) atmospheric pressure, varying in intensity and position with the seasons. Locations dominated by anticyclonic flow, such as the subtropics, typically have minimal cloud cover, lower precipitation frequency and higher surface insolation receipt whereas low pressure regions of the Tropics and high midlatitudes are often areas of rising motion and extensive cloud cover.

Water: Oceanic Circulation and Land Proximity

Similar to global wind and pressure patters, oceanic circulation cells redistribute surplus energy from lower latitudes to high latitudes with an annual energy deficit. Each oceanic basin contains a system of circulation cells or gyres with warm and cold currents whose large-scale movements are dictated by the strength and position of the semipermanent subtropical anticyclones.[1] The eastern oceanic basins are dominated by cold ocean currents, such as the California current off the western North American coast, that promote cooler temperatures and stable (nonprecipitation forming) atmospheric conditions. In contrast, the western oceanic basins have warm ocean currents that produce warm, humid air masses prone to cloud formation and precipitation. Oceanic circulation patterns (and the prevailing climate) may also shift from interannual variations in salinity, surface water movement, and energy storage, such that occurs during El Niño or La Niña events of the equatorial Pacific.

Water also regulates other aspects of climate, particularly temperature. In comparison to most land surfaces, water has a high specific heat capacity or the amount of energy required to raise the temperature of a substance 1°C. The specific heat capacity of water is approximately five times greater than land, meaning much more energy is required to heat and cool water than land.[3] Consequently, coastal locations usually have narrower temperature ranges and less temperature extremes than locations further inland, an effect described as continentality. The degree of continentality is also moderated by other factors such as the prevailing wind and local topography. For instance, in the midlatitudes where the prevailing wind is from the west, east coast locations will have a diminished oceanic influence and enhanced continentality than similar west coast locations.

Local Features

Regional topography and other site-specific factors can influence local climate characteristics, producing unique temperature and precipitation patterns within prevailing climate zones. The height above mean sea level or altitude of a location directly impacts temperature, with temperature decreasing by an average rate of 6.5°C per kilometer (or 3.6°F per 1000 feet) in the lower troposphere.[3] For instance, despite being on the equator, Quito, Ecuador has

Circulation—Crops

a relatively low mean annual maximum temperature (22°C or 72°F) primarily due to its high-altitude location at 2879 m (9446 ft).[4] High-altitude environments also have rapid evaporation rates and weaker radiation retention which produces large diurnal temperature ranges.

Topography moreover influences insolation, precipitation patterns, and local winds. The orientation and steepness of mountain slopes determine direct daylight energy receipt, with south-facing slopes in the Northern Hemisphere receiving longer daylight hours and more direct sun angles than any other direction. Steep slopes can also generate katabatic or mountain winds, high velocity winds driven by localized pressure differences between high- and low-elevation areas and intensified by nighttime cooling; Greenland and Antarctica are well-known for this type of extreme wind. Locations oriented toward the prevailing wind flow generally receive much higher cloud cover, precipitation, and cooler temperatures than areas on the leeward side of a topographic barrier. Even in relatively flat regions, the prevailing wind flow may influence local moisture transport. For example, winds across the North American Great Lakes are predominantly from the west or northwest thereby producing lake-effect enhanced snowfall for areas on the eastern sides of the water bodies.

CLIMATE CLASSIFICATION

The Köppen classification system is perhaps the most widely known technique for grouping climates with similar thermal, moisture, and natural vegetation characteristics. Devised by Vladimir Köppen in early 20th century and subsequently modified, the multitiered method utilizes mean monthly temperature and precipitation to delineate climate regions into one of five major types (designated by a capital letter): tropical (A); dry (B); midlatitude-mild (C); midlatitude-cold (D); and polar (E).[3] A sixth major type, highland (H), is also commonly used for areas with large climate variation over short distances primarily due to altitude. Climate zones A, C, D, and E are differentiated according to latitude-based temperature characteristics, whereas climate zone B contains arid and semiarid areas with annual precipitation deficits. These broad categories are further subdivided (designated by lower case letters) using criteria such as dry season duration and timing and seasonal temperature extremes. For example, the Mediterranean climate type common to southern Europe and the southwestern coasts of South America, Australia and Africa is *Csb* in the Köppen system, indicating a midlatitude location with mild winters and warm, dry summers. Glen Trewartha and others have subsequently modified the Köppen system to allow for greater within type variability and more consideration of natural vegetation types.

Based on a water balance approach, the Thornthwaite climate classification scheme examines the seasonal changes in the relationship between moisture availability

and energy. Categories are based on several thermal and moisture indices that utilize the concept of potential evapotranspiration (PE) or the maximum amount of moisture that could be evaporated to the atmosphere from the surface (including vegetation) if moisture was not a limiting factor.[1,5] Humid conditions occur when precipitation (and soil water storage) exceeds the amount necessary for PE, whereas arid conditions arise when water sources are insufficient to meet PE requirements. Other indices generate divisions according to annual and seasonal thermal efficiency, with higher (lower) values indicating a greater (diminished) capacity for evapotranspiration and generally warmer (colder) temperatures. Analogous to the Köppen classification system, individual climate types are designated by a multilevel letter combination generated from the indices. The Thornthwaite classification system contains nine major divisions based on moisture availability and an additional nine based on thermal efficiency of which both groups are further subdivided according to seasonal variation.

While the Köppen and Thornthwaite systems are the most extensively used global climate classification systems, other methods of climate delineation have gained popularity for their emphasis on local and regional climate system factors, dominant forcing mechanisms, and/or application. Hubert Lamb developed a manual climate classification scheme for the British Isles based on the midlatitude cyclone model whereby relative position and the relative frequency of the synoptic-scale atmospheric circulation pattern determines climate parameters such as temperature, precipitation and wind direction.[6,7] The Lamb catalogue consists of 26 unique weather types with regional atmospheric situations dominated by either anticyclonic, cyclonic, or directional flow. Similar regional weather typing classification schemes have been developed for the United States,[8–10] including the newer spatial synoptic classification (SSC) scheme.[11] The SSC identifies North American (and more recently Western Europe) locations with similar weather and air mass characteristics into one of seven major types: dry polar (DP); dry moderate (DM); dry tropical (DT); moist polar (MP); moist moderate (MM); moist tropical (MT); and transitional (TR). While regional climates can be classified according to annual and seasonal weather type frequency, the SSC also has the ability for operational (real-time) meteorology usage.[12]

CONCLUSION

Climate represents not only the long-term average weather conditions of a location or the climatic normal but the extremes and frequencies of those conditions. As a discipline, climatology has evolved from a predominantly descriptive science detailing the properties and geographic distribution of climate types to determining the physical processes that govern and modify the climate system.

Circulation—Crops

Consequently, several climatology subdisciplines have arisen to address the different interactions between the atmosphere and other Earth systems (i.e., hydrosphere, biosphere, and lithosphere) that create climate. While climate classification systems, such as the Köppen scheme, retain widespread implementation, other classification methods have gained popularity for their ability to incorporate region-specific attributes or provide explanatory power. Regardless of the system, a common theme is integration of the major climate system controls such as latitude, water proximity, oceanic circulation, semipermanent pressure systems, and topography.

Investigation into the origins and modifications of past, current, and future climates will continue to be a major focus of climatology, particularly the impact of human activities on the climate system. Although the climate system has varied considerably throughout the geologic past from natural forcing mechanisms (e.g., variations in solar output), modern climate shifts have become readily apparent since the late 18th century when the industrialization era began to impact atmospheric composition and the climate system balance. As data resolution and technology capabilities continue to progress, scientists are increasingly more adept at modeling the impacts of increased greenhouse gas emissions (e.g., carbon dioxide and methane) and atmospheric pollutants on the Earth–atmosphere system. Understanding these potential impacts will be an important factor in environmental sustainability policy and practices.

REFERENCES

1. Rohli, R.V.; Vega, A.J. *Climatology*; 2nd Edition; Jones and Bartlett: Sudbury, MA, 2012.
2. Shelton, M.L. *Hydroclimatology: Perspectives and Applications*; Cambridge University Press: Cambridge, U.K., 2009.
3. Aguado, E.; Burt, J.E. *Understanding Weather and Climate*; 6th Edition; Prentice Hall: Upper Saddle River, NJ, 2012.
4. Pearce, E.A.; Smith, G. *World Weather Guide*; Random House: New York, NY, 1990.
5. Thornthwaite, C.W. An approach toward a rational classification of climate. Geog. Rev. **1948**, *38* (1), 55–94.
6. Lamb, H.H. *British Isles Weather Types and a Register of Daily Sequence of Circulation Patterns,* 1861–1971; Geophysical Memoir, Vol. 116, HMSO: London, U.K., 1972.
7. Briffa, K.R. The simulation of weather types in GCMs: A regional approach to control-run validation. In *Analysis of Climate Variability: Applications of Statistical Techniques,* 2nd Edition; von Storch, H., Navarra, A., Eds.; Springer-Verlag: Heidelberg, New York, 1999; 119–138.
8. Muller, R.A.A synoptic climatology for environmental baseline analysis: New Orleans. J Appl. Meteorol. **1977**, *16* (1) 20–33.
9. Coleman, J.S.M.; Rogers, J.C. A synoptic climatology of the central United States and associations with Pacific teleconnection pattern frequency. J. Clim. **2007**, *20* (14) 3485–3497.
10. Keim, B.D.; Meeker, L.D.; Slater, J.F. Manual synoptic classification for the East Coast of New England (USA) with an application to $PM_{2.5}$ concentration. Clim. Res. **2005**, *28* (2) 143–153.
11. Sheridan, S.C. The redevelopment of a weather-type classification scheme for North America. Int. J. Climatol. **2002**, *22* (1) 51–68.
12. Sheridan, S.C. Spatial synoptic classification, http://sheridan.geog.kent.edu/ssc.html (accessed June 2012).

BIBLIOGRAPHY

1. Bonan, G.B. *Ecological Climatology: Concepts and Applications*; Cambridge University Press: Cambridge, UK, 2008.
2. Bridgeman, H.A.; Oliver, J.E. *The Global Climate System: Patterns, Processes and Teleconnections*; Cambridge University Press: Cambridge, UK, 2006.
3. Hidore, J.J.; Oliver, J.E.; Snow, M.; Snow, R. *Climatology: An Atmospheric Science,* 3rd Edition; Prentice Hall: Upper Saddle River, NJ, 2009.

Circulation—
Crops

Climatology: Moist Enthalpy and Long-Term Anomaly Trends

Souleymane Fall
College of Agriculture Environment and Nutrition Science, and College of Engineering, Tuskegee University, Tuskegee, Alabama, U.S.A.

Roger A. Pielke, Sr.
Cooperative Institute for Research in Environmental Sciences (CIRES), University of Colorado at Boulder, Boulder, Colorado, U.S.A.

Dev Niyogi
Department of Agronomy, Crops and Earth System Sciences, and Department of Earth and Atmospheric Sciences, Purdue University, West Lafayette, Indiana, U.S.A.

Gilbert L. Rochon
Tuskegee University, Tuskegee, Alabama, U.S.A.

Abstract

Moist enthalpy hereafter referred to as equivalent temperature (TE), expresses the atmospheric heat content by combining into a single variable air temperature (T) and atmospheric moisture. As a result, TE, rather than T alone, is an alternative metric for assessing atmospheric warming, which depicts heat content. Over the mid-latitudes, TE and T generally present similar magnitudes during winter and early spring, in contrast with large differences observed during the growing season in conjunction with increases in summer humidity. TE has generally increased during the recent decades, especially during summer months. Large trend differences between T and TE occur at the surface and lower troposphere, decrease with altitude, and then fade in the upper troposphere. TE is linked to the large scale climate variability and helps better understand the general circulation of the atmosphere and the differences between surface and upper air thermal discrepancies. Moreover, when compared with T alone, TE is larger in areas with higher physical evaporation and transpiration rates and is more correlated to biomass seasonal variability.

INTRODUCTION

Presently, air temperature is the key metric for assessing climate change and more specifically global warming over land. A huge body of studies dealing with surface air temperature trends suggest that at global scale, an increase took place over the last century.[1–5] This widely scrutinized warming observed at the surface and in the troposphere is associated with anthropogenic greenhouse forcing of the climate system,[6–8] although natural effects have also been suggested as being important.[9]

However, warming is related to atmospheric energy content and temperature is only one of its components, as emphasized in some studies.[10–13] Another component which plays an important role in the warming process is atmospheric moisture content, which has been reported to have increased during the past decades,[14–20] although recent studies suggest that this increase has stopped[21] and even reversed.[22] A broader assessment of warming could, therefore, take into consideration moist enthalpy, which includes both temperature and moisture and denotes the heat content of air.

Although this study is limited to the atmospheric heat content, it is worth mentioning that at global scale, ocean heat content 1) is the major contributor of the increase of heat content of the whole Earth system; 2) is found to be well correlated with the global net radiation flux; and 3) is the main driver of the variability of the Earth's climate system.[23–28]

So far, few studies have simultaneously quantified temperature and moisture by combining them into a single variable. Steadman[29,30] utilized a scale of apparent temperature (A), which expresses levels of human comfort. His method has been employed by the National Oceanic and Atmospheric Administration (NOAA) as a heat stress index. To investigate summertime heat stress over the United States, Gaffen and Ross[31] used a simplified version of Steadman's A and derived thresholds defined by the 85th percentile values of July and August daily temperature and A averages. They found that the annual frequency of days exceeding the thresholds as well as the number of high heat-stress nights did increase at most stations. Using observed temperature and humidity datasets from 188 first-order weather stations for the period 1961–1990, Gaffen and Ross[32] found upward trends in A over the United States, in accordance with upward temperature and humidity trends. More recent studies focus on moist enthalpy, which combines both air temperature and humidity in a

Encyclopedia of Natural Resources DOI: 10.1081/E-ENRA-120047641

single variable, to assess surface heating trends. At a global scale, Ribera et al.[33] used the NCEP/NCAR re-analysis temperature to study the relationships between equivalent temperature (*TE*) and modes of climate variability. Although an increase of the globally averaged *TE* was found, significant differences were observed between oceanic and continental areas. Pielke et al.[10] compared values of year 2002 daily temperature and moist enthalpy for the city of Fort Collins (Colorado) and the Central Plains Experimental Range of the U.S. Department of Agriculture's Agricultural Research Service located 60 km northeast of the city. Their results show that temperature and moist enthalpy are nearly equal when absolute humidity is low, but as humidity increases during the growing season, moist enthalpy values become much larger. Davey et al.[11] examined 1982–1997 temperature and moist enthalpy trend differences for surface sites in the eastern United States. They found that moist enthalpy trends are warmer than temperature trends during the winter season, but relatively cooler in the fall. Rogers et al.[34] analyzed 124-year records of summer moist enthalpy for Columbus (Ohio) and found that the highest values of moist enthalpy occurred during the summer of 1995 when both temperature and moisture were very high, in contrast with the hot summers of 1930–1936 which, despite high temperatures, experienced lower moist enthalpy because of relatively low or negative anomalies of absolute humidity. More recently, Fall et al.[12] used National Centers for Environmental Prediction North American Regional Reanalysis data to investigate temperature and moist enthalpy at near-surface and various upper-air standard levels. They noted that the moisture component induces larger trends and variability of moist enthalpy relative to temperature. Their results indicated that, while moist enthalpy values and trend were much larger than temperature ones at near-surface level, there was almost no difference at 300–200 mb. Peterson et al.[13] examined the energy content of the surface atmosphere, which is composed of temperature (enthalpy), kinetic energy, and latent heat. They found that the global surface atmospheric energy has increased since the 1970s, mainly because of increases in both enthalpy and latent heat, which equally contribute to the global increases in heat content. However, at regional scale, the two components were in some cases found to be of opposite signs: for example, Peterson et al.[13] observed that in Australia, despite the increase in the surface temperature, heat content was found to be decreasing.

DEFINITION OF MOIST ENTHALPY

Moist enthalpy, also referred to as moist static energy[10] or equivalent temperature (*TE*),[11,12,33,34] expresses the surface air heat content (*H*) by taking into account air temperature and moisture as single variable. A previous variant of moist enthalpy, namely *A*, was developed by Steadman[29,30] by combining four variables: summer temperature, humidity (vapor pressure), wind speed, and extra-radiation. Gaffen and Ross[31,32] simplified Steadman's *A* by ignoring the effects of wind and radiation and computing it from ambient temperature (*T*, in °C) and water vapor pressure (*e*, in kPa)

$$A = -1.3 + 0.92\,T + 2.2\,e \tag{1}$$

Recent studies have mainly focused on moist enthalpy,[10] which is written as:

$$H = CpT + Lvq \tag{2}$$

where *Cp* is the specific heat of air at constant pressure, *T* is the air temperature, Lv is the latent heat of vaporization, and *q* is the specific humidity. Thus, it can be seen that, as described by Peterson et al.,[13] moist enthalpy is the sum of two terms: enthalpy (calculated from *T*) and latent heat.

The approximate value of Lv (30°C) has been used in most of the studies. Fall et al.,[12] following the Priestley–Taylor method, estimated Lv (J/kg) with the temperature function:

$$Lv = 2.5 - 0.0022 \times T \tag{3}$$

Such an estimate accounts for the variation of Lv with temperature.

The specific humidity *q* can be computed from the dewpoint temperature and surface pressure using Bolton's empirical relationship:[35]

$$q = \frac{0.62197e}{p - 0.37803e}, \text{ where } e = 6.112\,\exp\left(\frac{17.67Td}{Td + 243.5}\right) \tag{4}$$

where *e* is the saturated vapor pressure in hPa, *p* is the surface pressure in hPa, and *Td* is the dewpoint temperature in °C.[34,36]

H is the heat in units of Joule and must be scaled into degree units in order to obtain *TE* for easy comparison to air temperature.

$$TE = H/Cp \tag{5}$$

Equation 5 can be also written as

$$TE = T + (Lvq/Cp) \tag{6}$$

where Lv is in units of Joules per kilogram and *Cp* is in units of Joules per kilogram per degree K. As *q* is dimensionless (i.e., kg per kg), the ratio has units of degree K.

The above equations show that both sensible and latent heat contribute to the magnitude of *TE*. Pielke et al.[37] have shown that, for example, at 1000 mb, an increase of

2.5°C in air temperature will produce the same change in *TE* as a 1°C increase in dew point temperature. *TE* becomes larger as both *T* and *q* increase. Conversely, when *T* is abnormally high but out of phase with a low *q*, *TE* exhibits modest–to-low values.[34] However, in the long term, the combination of the two terms seems to result in larger trend and variability of *TE*, regardless of the magnitude of *q*. For example, in their comparison between *T* and *TE* over the United States (1979–2005), Fall et al.[12] found larger trends and variability of *TE* (relative to *T*), even though in terms of contribution to the magnitude of *TE*, the moisture component (Lv*q*) was much smaller than the enthalpy (or sensible heat: *CpT*).

CLIMATOLOGY

In general, moist enthalpy increases progressively from winter to summer and then decreases, thus exhibiting patterns that are similar to surface temperature annual cycle. A comparison between monthly mean *T* and *TE* over the United States shows that in winter and early spring there is almost no difference between the two variables. However, with increasing humidity from late spring to early fall, *TE* increases much more than *T*, in particular during summer (Fig. 1). Therefore, large differences are observed during the growing season (up to 22.74°C in July according to results from Fall et al.).[12] The same patterns are noted at daily time scale and for both maximum and minimum *T* and *TE*.[10]

MOIST ENTHALPY VARIABILITY AND ANOMALY TRENDS

In general, at global scale, there is scientific consensus on an increase of the heat content of the ocean, especially during the latter half of the past century.[25] Findings from Levitus et al.[26] show that from 1955 to 1998, approximately 85% of this warming occurred in the world's oceans. With the ocean, only the temperature is required to diagnose moist enthalpy. However, in the atmosphere, the humidity also must be included.

As shown in recent studies,[13,33] over the past decades moist enthalpy anomalies at global scale have generally exhibited warmer trends. Ribera et al.[33] found *TE* increments between +0.05 and +0.29°K/decade during the 1958–1998 period, with most of this increase occurring during the 1958–1978 period over densely populated coastal areas (e.g., Eastern Asia, Northern Europe, southeastern North America, and South Africa) and oceanic zones, in contrast with dry continental areas, which generally are characterized by negative trends (e.g., Sahara).

Positive *TE* trends are also found at regional and local scales, although Peterson et al.[13] indicate that the two terms can be of opposite signs in some regions (e.g., Southern Hemisphere sub-tropics, as shown in Fig. 2). Results from Gaffen and Ross studies[31,32] indicate that the occurrence of hot and humid days increased in the United States during the past decades (1961–1995). As a result, extreme heat-stress events became more frequent, and because of a pronounced increase in atmospheric moisture content,

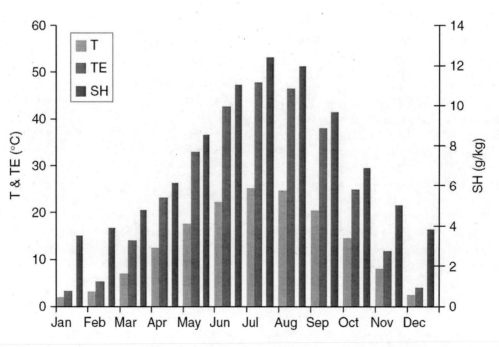

Fig. 1 Monthly climatology of temperature (*T*), equivalent temperature (*TE*), and specific humidity (SH) at 2 m (average 1979–2005) over the United States. The ordinate scale on the right pertains to values of specific humidity.
Source: Reprinted with permission from Fall et al.[12]

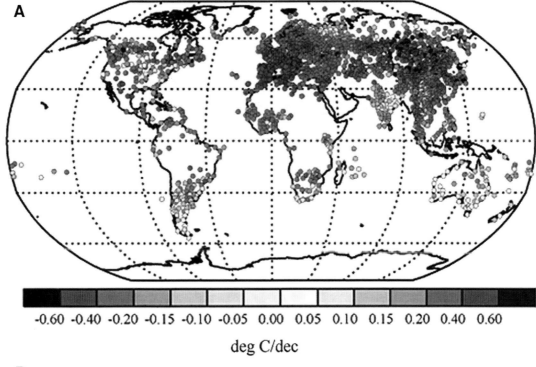

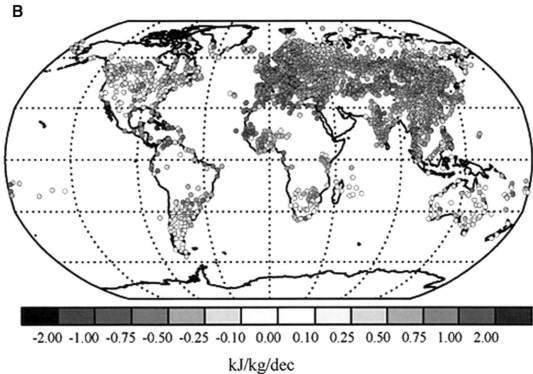

Fig. 2 Decadal trends (1973–2003) calculated for HadCRUH stations using pentad anomaly-specific humidity and temperature and pentad climatologies (1974–2003) from Lott et al. (2008); (**A**) surface temperature trends (°C/decade); (**B**) heat content trends (kJ/kg/decade).

Source: Reprinted with permission from Peterson et al.[13]

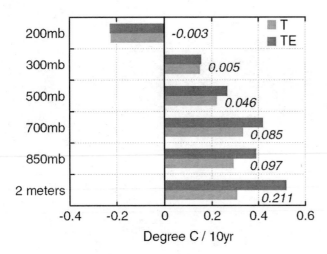

Fig. 3 Decadal anomaly trends computed from monthly T and TE at different pressure levels (1979–2005; units: °C/10 yr). Italicized values denote the differences $TE - T$. All trends are significant at the 5% level (P value < 0.05).
Source: Reprinted with permission from Fall et al.[12]

summertime TE trends were positive and higher than summertime T ones. In more recent studies, comparisons between near-surface T and TE have confirmed that TE trends are generally warmer and significantly different from T.[11,12] However, differences between T and TE trends decrease gradually with altitude and almost disappear at 300–200 mb [about 9,000–12,000 m]. For example, results from Fall et al.[12] who computed trend differences (TE minus T) at various levels for the United States during the period 1979–2005 (Fig. 3) show not only the decreasing trend differences with altitude (e.g., 0.211°C/decade at near-surface and 0.005°C/decade at 300 mb), but also a shift to an opposite trend sign (−0.003°C/decade for both T and TE at 200 mb).

Seasonally, at global scale during the 1958–1998 period, Ribera et al.[33] found that TE trends remained positive for all seasons and most of the increase generally took place during the summer season; the largest trends were found over the continental southern hemisphere and oceanic areas. Over the United States, surface TE trends are found to be warmer in winter and cooler in fall and summer.[11,12] These seasonal patterns persist up to 700 mb and above that level, most of TE increase occurs during the fall season; at 200 mb, TE trends are negative, with the most substantial cooling taking place in winter.[12]

CONCLUSION: ON THE IMPORTANCE OF HEAT CONTENT

The significance of considering heat content for a broader understanding of the Earth system climate variability has been demonstrated by various studies. The assessment of warming rates at global and regional scales requires

considering the variability of ocean, atmosphere, and cryosphere heat content.[25–27] In particular, ocean heat content varies in conjunction with the annual cycle of the Earth net radiation balance and has been found to be a major source of variability of the global heat balance.[23,25,27,28,38] The observed ocean heat content variations also appear to be correlated with simulated variations in greenhouse gases, sulfate aerosols, solar irradiance, and volcanic aerosols.[27]

Several studies have highlighted the importance of atmospheric heat content (alternatively moist enthalpy or TE) in climate variability and change assessments. The relevance of TE includes, but is not limited to the following:

- Moist enthalpy, which expresses the combined effects of temperature and humidity, represents an important forcing of the general circulation of the atmosphere through different transport mechanisms and provides a better understanding of the large-scale dynamics in the atmospheric heat balance.[39,40] Ribera et al.[33] found a close relationship between TE and large scale modes of climate variability (the North Atlantic Oscillation and the Arctic Oscillation).

- Apparent temperature (A), a variant of TE, is closely related to human comfort. The A climatology developed by Gaffen and Ross[31,32] has been extended and adopted by NOAA as a useful index for heat stress and is periodically updated.

- Pielke et al.[10] stated that… *"The difference in temporal trends in surface and tropospheric temperatures [National Research Council,2000], which has not yet been explained, could be due to the incomplete analysis of the surface and troposphere for temperature, and not the more appropriate metric of heat content."* In a subsequent study, Fall et al.[12] compared TE and T at different atmospheric levels (up to 200 mb) and investigated the vertical structure of the combined effects of temperature and moisture. They analyzed the climatology, time series, and decadal trends of the two variables and found a contrast between 1) pronounced temporal and spatial differences at the near-surface level, and 2) almost no difference at 300–200 mb. More specifically, the thermal discrepancies between surface and upper air are much larger when TE is used instead of T alone. Fall et al.[12] concluded that the use of TE to assess tropospheric heating trends may *"help obtain an improved estimate of the impacts of surface properties on heating trends."*

- Observation- and reanalysis-based studies have shown that moist enthalpy (TE) is more sensitive to surface vegetation properties than is air temperature (T). Davey et al.[11] found that over the eastern United States from 1982 to 1997, TE trends were similar or slightly cooler than T trends at predominantly forested and agricultural sites, and significantly warmer at predominantly grassland and shrubland sites. Results from Fall et al.[12] indicate that TE 1) is larger than T in areas with higher physical evaporation and transpiration rates (e.g., deciduous broadleaf forests and croplands) and

2) shows a stronger relationship than T to vegetation cover, especially during the growing season (biomass increase). These moist enthalpy-related studies confirm previous results showing that changes in vegetation cover, surface moisture, and energy fluxes generally lead to significant climatic changes[41–43] and responses, which can be of a similar magnitude to that projected for future greenhouse gas concentrations.[44,45] Therefore, it is not surprising that TE, which includes both sensible and latent heat, more accurately depicts surface and near-surface heating trends than T does.

In general, studies dealing with heat content have suggested that despite its undeniable relevance and popularity, temperature needs to be supplemented with additional metrics for assessing global warming. For this purpose, moist enthalpy, which includes both temperature and atmospheric moisture content, is a useful variable.[10–12] Using the Bowen ratio (ratio of sensible heat to latent heat) is another efficient way for analyzing the combined effects of temperature and moisture and their co-variation.[13,46]

Overall, a large majority of studies still use temperature as the key metric for assessing global warming. Global temperature is a convenient variable because of its higher availability and greater spatial/temporal coverage. However, several recent studies have addressed moist enthalpy (which is expressed in Joules and actually represents a robust measure of heat) and recommended both temperature and atmospheric moisture be considered in climate change assessments.

ACKNOWLEDGMENTS

We thank Dallas Staley for her outstanding contribution in editing and finalizing the entry. The study benefited from the DOE ARM Program (08ER64674; Dr. Rick Petty and Dr. Kiran Alapaty), and in parts from NASA Terrestrial Hydrology Program (Dr. Jared Entin), and NSF CAREER-0847472 (Liming Zhou and Jay Fein). Roger Pielke, Sr. acknowledges support from NSF Grant 0831331 and received support from the University of Colorado at Boulder (CIRES/ATOC).

REFERENCES

1. Crowley, T.J.; Lowery, T. How warm was the medieval warm period? Ambio **2000**, *29*, 51–54.
2. Mann, M.E.; Jones, P.D. Global surface temperatures over the past two millennia. Geophysical Res. Lett. **2003**, *30* (15), 1–4.
3. Soon, W-H.; Legates, D.R.; Baliunas, S. Estimation and representation of long-term (>40 year) trends of Northern Hemisphere-gridded surface temperature: a note of caution. Geophysical Res. Lett. **2004**, *31*, L03209, doi: 1029/2003GRL019141.
4. Moberg, A.; Sonechkin, D.M.; Holmgren, K.; Datsenko, N.M.; Karlen, W. Highly variable Northern Hemisphere temperatures reconstructed from low- and high-resolution proxy data. Nature **2005**, *433*, 613–618.
5. Intergovernmental Panel on Climate Change (IPCC). Summary for policymakers. In *Climate Change: The Physical Science Basis. Contribution of Working Group I to the Fourth Assessment Report of the Intergovernmental Panel on Climate Change*; Solomon, S., Qin, D., Manning, M., Chen, Z., Marquis, M., Averyt, K.B., Tignor, M., Miller, H.L., Eds.; Cambridge University Press: Cambridge, New York, 2007.
6. Mears, C.A.; Wentz, F.J. The effect of diurnal correction on satellite-derived lower tropospheric temperature. Science **2005**, *309*, 1548–1551.
7. Sherwood, S.C.; Lanzante, J.R.; Meyer, C.L. Radiosonde daytime biases and late 20th century warming. Science **2005**, *309*, 1556–1559.
8. Santer, B.D.; Thorne, P.W.; Haimberger, L.; Taylor, L.K.E.; Wigley, T.M.L.; Lanzante, J.R.; Solomon, S.; Free, M.; Gleckler, P.J.; Jones, P.D.; Karl, T.R.; Klein, S.A.; Mears, C.; Nychka, D.; Schmidt, G.A.; Sherwood, S.C.; Wentz, F.J. Consistency of modelled and observed temperature trends in the tropical troposphere. Int. J. Climatol. **2008**, *28*, 1703–1722.
9. Spencer, R.W.; Braswell, W.D. On the misdiagnosis of surface temperature feedbacks from variations in earth's radiant energy balance. *Remote Sensing* **2011**, *3*, 1603–1613.
10. Pielke, R.A., Sr.; Davey, C.; Morgan, J. Assessing "global warming" with surface heat content. Eos Trans. **2004**, *85* (21), 210–211.
11. Davey, C.A.; Pielke R.A., Sr.; Gallo, K.P. Differences between near surface equivalent temperature and temperature trends for the eastern United States: equivalent temperature as an alternative measure of heat content. Global Planetary Change **2006**, *54*, 19–32.
12. Fall, S.; Diffenbaugh, N.; Niyogi, D.; Pielke, R.A., Sr.; Rochon, G. Temperature and equivalent temperature over the United States (1979–2005). Int. J. Climatol. **2010**, *30*, 2045–2054, doi: 10.1002/joc.2094.
13. Peterson, T.C.; Willett, K.M.; Thorne, P.W. Observed changes in surface atmospheric energy over land. Geophysical Res. Lett. **2011**, *38*, L16707, doi: 10.1029/2011GL048442.
14. Wentz, F.J.; Schabel, M.C. Precise climate monitoring using complementary satellite data sets. Nature **2000**, *403*, 414–416.
15. Trenberth, K.E.; Fasullo, J.; Smith, L. Trends and variability in column-integrated atmospheric water vapor. Climate Dynamics **2005**, *24*, 741–758.
16. Held, I.M.; Soden, B.J. Robust responses of the hydrological cycle to global warming. J. Climate **2006**, *19* (21), 5686–5699.
17. Santer, B.D.; Mears, C.; Wentz, F.J.; Taylor, K.E.; Gleckler, P.J.; Wigley, T.M.L.; Barnett, T.P.; Boyle, J.S.; Brueggemann, W.; Gillett, N.P.; Klein, S.A.; Meehl, G.A.; Nozawa, T.; Pierce, D.W.; Stott, P.A.; Washington, W.M.; Wehner, M.F. Identification of human-induced changes in atmospheric moisture content. Proc. Natl. Acad. Sci. **2007**, *104*, 15248–15253, doi: 10.1073/pnas.0702872104.
18. Wentz, F.J.; Ricciardulli, L.; Hilburn, K.; Mears C. How much more rain will global warming bring? Science **2007**, *317*, 233–235.

19. Willett, K.M.; Jones, P.D.; Gillett, N.P.; Thorne, P.W. Recent changes in surface humidity: development of the HadCRUH dataset. J. Climate 2008, 21, 5364–5383.

20. Willett, K.M.; Jones, P.D.; Thorne, P.W.; Gillett, N.P. A comparison of large scale changes in surface humidity over land in observations and CMIP3 GCMs. Environ. Res. Lett. 2010, 5, 025210 doi:10.1088/1748–9326/5/2/025210.

21. Wang, J.-W., Wang, K., Pielke, R.A., Sr; Lin, J.C., Matsui, T. Towards a robust test on North America warming trend and precipitable water content increase. Geophysical Res. Lett. 2008, 35, L18804, doi: 10.1029/2008GL034564.

22. Solomon, S.; Rosenlof, K.; Portmann, R.; Daniel, J.; Davis, S.; Sanford, T; Plattne, G-K. Contributions of stratospheric water vapor to decadal changes in the rate of global warming. Science 2010, 327 (5970), 1219–1223, doi: 10.1126/science.1182488.

23. Ellis, J.S.; Vonder Haar, T.H.; Levitus, S.; Oort, A.H. The annual variation in the global heat balance of the Earth. J. Geophysical Res. 1978, 83, 1958–1962.

24. Piexoto, J.P.; Oort, A.H. Physics of climate. Am. Inst. Phys. 1992, 520 pp.

25. Levitus, S.; Antonov, J.I.; Boyer, T.P.; Stephens, C. Warming of the World Ocean. Science 2000, 287, 2225–2229.

26. Levitus, S.; Antonov, J.; Boyer, T. Warming of the world ocean, 1955–2003. Geophysical Res. Lett. 2005, 32, L02604, doi: 10.1029/2004GL021592.

27. Levitus, S.; Antonov, J.; Wang, J.; Delworth, T.L.; Dixon, K.W.; Broccoli, A.J. Anthropogenic warming of Earth's climate system. Sciences 2001, 292, 267–270.

28. Pielke, R.A., Sr. Heat storage within the Earth system. Bull. Am. Meteorol. Soc. 2003, 84, 331–335.

29. Steadman, R.G. The assessment of sultriness, part 2: effects of wind, extra radiation and barometric pressure on apparent temperature. J. Appl. Meteorol. 1979, 18, 874–885.

30. Steadman, R.G. A universal scale of apparent temperature. J. Appl. Meteorol. Climatol. 1984, 23, 1674–1687.

31. Gaffen, D.J.; Ross, R.J. Increased summertime heat stress in the U.S. Nature 1998, 396, 529–530.

32. Gaffen, D.J.; Ross, R.J. Climatology and trends in U.S. surface humidity and temperature. J. Climate 1999, 12, 811–828.

33. Ribera, P.; Gallego, D.; Gimeno, L.; Perez, J.F.; García, R.; Hernandez, E.; de la Torre, L.; Nieto, R.; Calvo, N. The use of equivalent temperature to analyze climate variability. Studia Geophysica et Geodaetica 2004, 48, 459–468.

34. Rogers, J.C.; Wang, S.H.; Coleman, J.S.M. Evaluation of a long-term (1882–2005) equivalent temperature time series. J. Climate 2007, 20, 4476–4485.

35. Bolton, D. The computation of potential equivalent temperature. Monthly Weather Rev. 1980, 108, 1046–1053.

36. Pielke, R.A., Sr.; Wolter; K.; Bliss, O.; Doesken, N.; McNoldy, B. The July 2005 Denver heat wave: How unusual was it? Natl. Weather Digest. 2006, 31, 24–35.

37. Pielke, R.A., Sr. Influence of the spatial distribution of vegetation and soils on the prediction of cumulus convective rainfall. Rev. Geophys. 2001, 39, 151–177.

38. Rossby, C. Current problems in meteorology. In The Atmosphere and Sea in Motion. Rockefeller Inst. Press: New York 1959; 9–50.

39. Riehl, H.; Malkus, J. On the heat balance in the equatorial trough zone. Geophysica 1958, 6, 3–4.

40. Tian, B.; Zhang, G.; Ramanathan, V. Heat balance in the Pacific warm pool atmosphere during TOGA COARE and CEPEX. J. Climate 2001, 14, 1881–1893.

41. Bonan, G.B. Effects of land use on the climate of the United States. Climatic Change 1997, 37, 449–486.

42. Bounoua, L.; Collatz, G.J.; Los, S.; Sellers, P.J.; Dazlich, D.A.; Tucker, C.J.; Randall, D.A. Sensitivity of climate to changes in NDVI. J. Climate 2000, 13, 2277–2292.

43. Niyogi, D; Kishtawal, C.M.; Tripathi, S.; Govindaraju, R.S. Observational evidence that agricultural intensification and land use change is reducing the Indian summer monsoon rainfall. Water Resources Res. 2010, 46, W03533, 17 pp., doi: 10.1029/2008WR007082.

44. Feddema, J.J.; Oleson, K.W.; Bonan, G.B.; Mearns, L.O.; Buja, L.E.; Meehl, G.A.; Washington, W.M. The importance of land-cover change in simulating future climates. Science 2005, 310, 1674–1678.

45. Diffenbaugh, N.S. Atmosphere-land cover feedbacks alter the response of surface temperature to CO_2 forcing in the western United States. Climate Dynamics 2005, 24, 237–251, doi: 10.1007/s00382-004-0503-0.

46. Pielke, R.A., Sr; Davey, C.; Niyogi, D.; Fall, S.; Hubbard, K.; Lin, X.; Cai, M.; Lim, Y-K.; Li, H.; Nielsen-Gammon, J.; Gallo, K.; Hale, R.; Angel, J.; Mahmood, R.; Foster, S.; Steinweg-Woods, J.; Boyles, R.; McNider, R.T.; Blanken, P. Unresolved issues with the assessment of multi-decadal global land surface temperature trends. J. Geophysical Res. 2007, 112, D24S08, doi: 10.1029/2006JD008229.

Circulation—
Crops

Crops and the Atmosphere: Trace Gas Exchanges

Jürgen Kreuzwieser
Heinz Rennenberg
Intitute of Forest Botany and Tree Physiology, University of Freiburg, Freiburg, Germany

Abstract

In its natural atmospheric environment, vegetation is exposed to a plethora of trace gases that interact with plants in numerous different ways. The impact of such gases can be critical for natural ecosystems and crops as they potentially affect plant health, growth, and yield. On the contrary, under specific conditions, some of these gases can even be of nutritional use for the plants. Most trace gases can be exchanged between foliage and the atmosphere, i.e., taken up or emitted usually by the plants' foliage. The present entry focuses on the plant physiological and physicochemical background determining the exchange of trace gases between leaves and the atmosphere.

INTRODUCTION

The atmosphere mainly consists of nitrogen (78% by volume), oxygen (21% by volume) and the noble gas argon (0.93% by volume) together making up >99.9% of the atmosphere's composition; the remainder is known as trace gases. Trace gases are typically present in the range of parts per trillion by volume (pptv) to parts per million by volume (ppmv); they include the greenhouse gases carbon dioxide (CO_2), methane (CH_4), nitrous oxide (N_2O), ethane, water vapor, and ozone (O_3); the air pollutants sulfur dioxide (SO_2), ammonia (NH_3), nitric oxide (NO), nitrogen dioxide (NO_2), peroxyacylnitrates (PAN), nitric acid (HNO_3), and carbon monoxide (CO); and a number of volatile organic compounds (VOCs) (Table 1). Among VOCs are isoprenoids (mainly isoprene and monoterpenes) and many oxygenated species such as alcohols, aldehydes, and organic acids.[2] Because of their reactivity, VOCs strongly affect the oxidation capacity of the troposphere and influence the concentration and distribution of several other trace gases, including CH_4 or CO.[3] On a regional scale, VOCs significantly contribute to the formation of tropospheric O_3.[4] The main source of VOCs (~90%) is natural emission by vegetation.[2]

IMPACTS OF TRACE GASES ON CROPS

The effects of atmospheric trace gases on crops are quite diverse and depend on the type of gas, its concentration, the duration of exposure, and the amount taken up, as well as a range of plant internal factors. Direct phytotoxic effects due to exposure to high concentrations of pollutants such as O_3 and SO_2 on crop plants include, among others, visible leaf injury; changes in chloroplast structure and cell membranes; disturbances of stomatal regulation, respiration, and photosynthesis; and reductions in growth and yield.[5] However, because sulfur (S) is an essential nutrient for plants, SO_2 absorbed by foliage may also be used as an additional source of sulfur in polluted areas, in addition to sulfate from the soil.[6] The same principle applies to nitrogen (N). Thus, effects of trace gases can be divided into phytotoxic effects caused by protons, organic compounds, SO_2, NO_2, NH_3, and O_3; and nutritional effects caused by S- and N-containing gases and CO_2.[7]

FACTORS CONTROLLING TRACE GAS EXCHANGE

Trace gases can be exchanged between the atmosphere and aboveground plant parts by 1) dry deposition as gases or aerosols; 2) wet deposition as dissolved compounds in rainwater or snow; or 3) interception of compounds dissolved in mist or cloud water[7] (Fig. 1). The direction of the exchange (i.e., emission vs. deposition) and its velocity (i.e., the exchange rates) is controlled by the physicochemical conditions and by internal plant factors.

The gradient in the gas concentration between substomatal cavities and the atmosphere is the driving force for gas exchange. The gas flux is a diffusive (passive) process and can be described by Fick's law of diffusion. Accordingly, the net flux of a trace gas is zero when the substomatal concentration is equal to the concentrations in the surrounding ambient air. This concentration is referred to as the compensation point for the particular gas. When trace gas concentrations outside the leaves are higher than those in the substomatal cavities, a net flux

Encyclopedia of Natural Resources DOI: 10.1081/E-ENRA-120049201

Circulation—Crops

Table 1 Range of ambient concentrations of different trace gases in the atmosphere

Trace gas	Ambient concentrations
Ar	9,340 ppm
CO_2	379 ppm (2005)
CH_4	1774 ppb (2005)
N_2O	319 ppb (2005)
Isoprene	ppt–several ppb
Monoterpenes	ppt–several ppb
Alcohols	1–30 ppb
Carbonyls (aldehydes, ketones)	1–30 ppb
Alkanes	1–3 ppb
Alkenes	1–3 ppb
Esters	ppt–several ppb
Carbonic acids	0.1–16 ppb
SO_2	0.5–20 ppb
H_2S	0–0.2 ppb
NO_2	0.5–15 ppb
O_3	20–80 ppb

Source: Data from Forster, et al.,[1] Guenther, et al.,[2] Kesselmeier & Staudt,[3] Bender & Weigel,[5] and Herschbach, et al.[6]

into the leaves will take place (deposition), and vice versa. Therefore, crops may act both as a source (if ambient concentrations are lower than substomatal concentrations) or as a sink of a specific gas (Table 2). This dual behavior has been observed for a variety of gases (SO_2, H_2S, NO_2, NH_3, organic acids, and aldehydes). Compensation points for pollutants such as NH_3, e.g., range between 1 and 6 parts per billion by volume (ppbv) (see Husted et al.[9] and references therein). They depend mainly on the plant species or cultivar, development stage, temperature, and status of N nutrition of the plants. Generally, compensation points increase with increasing availability of nutrients in the soil, which suggests that this is one mechanism by which plants cope with excess nutrient supply.[8]

PLANT PHYSIOLOGICAL CONTROLS

The existence of a compensation point depends on the capacity of a plant to produce the trace gas to be exchanged, or to consume it. Therefore, compensation points do not exist for compounds that cannot be produced (e.g., O_3) or consumed (e.g., isoprene). It is evident that for each of the many gases exchanged between crop plants and the atmosphere, specific metabolic pathways exist, not all of which are understood yet. As an example, some details on the exchange of nitrogen compounds are presented (Fig. 2).

Both NO_2 and NH_3 can be taken up by aboveground parts of plants, mainly via the stomata of leaves.[7] In the aqueous phase of the apoplast, NO_2 is either disproportionated yielding equal amounts of NO_2^- and NO_3^-, or it reacts with apoplastic ascorbate.[10] Because disproportionation of NO_2 in water is slow at atmospheric NO_2 concentrations, the reaction with ascorbate may be of more importance. Upon conversion to either NO_3^- or NO_2^-, these anions are transported to the cytoplasm, where they are reduced by the assimilatory nitrate reduction pathway yielding NO_4^+ and the amino acid glutamine. Atmospheric NH_3 dissolves in the aqueous phase of the apoplastic space to yield NO_4^+, which is then taken up into the cytoplasm.

Both NH_3 and NO_2 can also be emitted by plants. NH_3 may be released from cellular NO_4^+ pools when plants are exposed to excess nitrogen in the soil. In addition, it can be released from drying water films at the leaf surface. During this process, the remaining NH_4^+ concentrations on the surface will increase. By contrast, the chemical source of NO_2 emitted by the leaves is largely unknown; it has been proposed that nitrate reductase may be involved in the reduction of NO_2^- to NO_2.[8]

The rate of trace gas emission does not necessarily depend on the actual rate of production. Some volatile compounds are produced and then stored in particular pools. For example, some plant reservoirs contain high amounts of monoterpenes, which can be emitted throughout the day and night independent of light-dependent biosynthesis.[3]

Through their effect on biochemical pathways, biotic and abiotic factors (e.g., stress factors) and the developmental stage of plants influence the rate of trace gas emission. For example, stress caused by wounding, chilling, iron deficiency, O_2 deficiency, or induction of oxidative stress caused by exposure to O_3 or SO_2 can lead to the production of the VOCs hexenal, hexanal, formaldehyde, formate, ethene, ethane, ethanol, and acetaldehyde.[3]

INTERNAL TRANSFER RESISTANCES

The transfer of gases in and out of plants is often described by a resistance analogy. Along the path from the sites of production (or consumption) to the atmosphere, a series exists of mainly internal plant resistances. Because biosynthesis and consumption of volatile compounds usually take place in the cytoplasm or in other compartments of the cell, the gas must pass across the bordering membranes. The diffusive flux through these membranes is determined by the molecular size and the lipophilic character of the particular compound. Polar molecules such as organic and inorganic acids are not likely to be dissolved in the lipophilic membranes; therefore, the diffusive flux should be

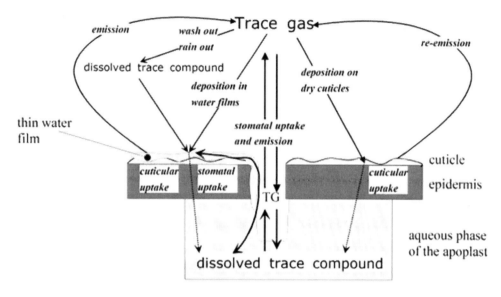

Fig. 1 Main access routes for trace gases to enter a leaf: Uptake via the stomata and cuticular uptake.
Source: Redrawn from Rennenberg and Geßler.[8]

Table 2 Typical range of rates of exchange for different trace gases between crops and the atmosphere

Trace gas	Range of exchange[a] $[\mu g\ g^{-1}$ (leaf d.wt.) $h^{-1}]$
SO_2	−300 to 0
H_2S	−2 to 2
O_3	−300 to 0
NO_2	−3 to 1
NH_3	−3 to 3
Isoprene	0 to 0.1
Monoterpenes	0 to 20
Oxygenated VOCs	−50 to 50

[a]Negative values indicate deposition and positive values indicate emission of respective trace gas.

slow. Carrier proteins can facilitate the transport of these polar compounds across membranes from the cytoplasm into the apoplastic space, similar to the transport of organic acids.[11]

The compounds are transferred from the liquid phase in the apoplastic space to the gaseous phase in the substomatal cavity, or vice versa. The volatilization into the internal leaf air space depends on 1) chemical properties of both the aqueous phase and the individual compound to be emitted (e.g., its solubility in the apoplastic solution); and 2) physical factors such as the ambient concentration of the gas, its vapor pressure, and temperature.[11] A reduction of the apoplastic pH reduces the resistance for inorganic and organic acids, because these compounds become protonated and thereby more volatile. When present in the gaseous phase of the apoplastic space, trace gases can escape from the

leaves through either the cuticle or the stomata. However, because of their polarity, the lipophilic cuticle constitutes a strong barrier, and therefore diffusion through the stomata is the main pathway. However, nonstomatal emission and deposition have also been observed in air pollutants such as NO, NO_2, and SO_2, but at much lower rates than stomatal exchange. For this reason, factors that influence the stomatal aperture exert a strong influence on the rate of gas exchange between plants and the atmosphere. The control by the stomata of emission and deposition of NO_2, SO_2, O_3, PAN, and other trace gases was observed in many studies, including in investigations of crop plants. Thus, concentration and time of exposure are not the only factors determining the effect of an air pollutant on vegetation. Plants usually close their stomata during hot, dry conditions when, for instance, O_3 levels are high. This may provide some protection for the plants from O_3 injury. Alternatively, in northern Europe where O_3 concentrations are lower than in southern and central Europe, the potential O_3 uptake at a given O_3 concentration may be higher because of higher air humidity, leading to high rates of stomatal O_3 uptake.

CONCLUSION

Trace gases influence not only natural ecosystems but also agricultural crops. Future studies should focus on the impact of different combinations of air pollutants (e.g., increased nitrogen input combined with elevated O_3 concentrations) on plants and should include aspects of global climate change (e.g., higher temperatures, droughts, and increased frequency of heavy rainfalls and droughts in combination with elevated CO_2 concentrations). Future

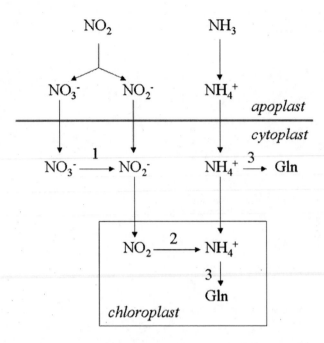

Fig. 2 Main processes involved in the assimilation of atmospheric NH$_3$ and NO$_2$ taken up by the leaves. 1: nitrate reductase, 2: nitrite reductase, 3: glutamine synthetase (cytoplasmatic and chloroplastic isoforms).
Source: Redrawn from Rennenberg, and Geßler.[8]

management practices should allow for optimum plant growth, while simultaneously reducing the loss of pollutants from the plant–soil system.[12]

REFERENCES

1. Forster, P.; Ramaswamy, V.; Artaxo, P.; Berntsen, T.; Betts, R.; Fahey, D.W.; Haywood, J.; Lean, J.; Lowe, D.C.; Myhre, G.; Nganga, J.; Prinn, R.; Raga, G.; Schulz, M.; Van Dorland, R. Changes in Atmospheric Constituents and in Radiative Forcing. In: *Climate Change 2007: The Physical Science Basis. Contribution of Working Group I to the Fourth Assessment Report of the Intergovernmental Panel on Climate Change.* Solomon, S.; Qin, D.; Manning, M.; Chen, Z.; Marquis, M.; Averyt, K.B.; Tignor, M.; Miller, H.L. Eds.; Cambridge University Press: Cambridge and New York, 2007; 137–153 pp.

2. Guenther, A.B.; Hewitt, C.N.; Erickson, D.; Fall, R.; Geron, C.; Graedel, T.; Harley, P.; Klinger, L.; Lerdau, M.; McKay, W.A.; Pierce, T.; Scholes, B.; Steinbrecher, R.; Tallamraju, R.; Taylor, J.; Zimmerman, P.A. Global model of natural volatile organic compound emissions. J. Geophys. Res. **1995**, *100*, 8873–8892.

3. Kesselmeier, J.; Staudt, M. Biogenic volatile organic compounds (VOC): An overview on emission, physiology and ecology. J. Atmos. Chem. **1999**, *33*, 23–88.

4. Ashworth, K.; Wild, O.; Hewitt, C.N. Impacts of biofuel cultivation on mortality and crop yields. Nat. Clim. Change **2013**, *3*, 492–496.

5. Bender, J.; Weigel, H.-J. Changes in atmospheric chemistry and crop health: A review. Agron. Sustain. Dev. **2011**, *31*, 81–89.

6. Herschbach, C.; De Kok, L.J.; Rennenberg, H. Net uptake of sulfate and its transport to the shoot in tobacco plants fumigated with H$_2$S or SO$_2$. Plant Soil. **1995**, *175*, 75–84.

7. Wellburn, A.R. Why are atmospheric oxides of nitrogen usually phytotoxic and not alternative fertilizers? New Phytol. **1990**, *115*, 395–429.

8. Rennenberg, H.; Geßler, A. Consequences of N Deposition to Forest Ecosystems—Recent Results and Future Research Needs. In *Forest Growth Responses to the Pollution Climate of the 21st Century*; Sheppard, L.J.; Neil Cape, J., Eds.; Kluwer Academic Publication: Dordrecht, the Netherlands, 1999; 47–64.

9. Husted, S.; Schjoerring, J.K.; Nielsen, K.H.; Nemitz, E.; Sutton, M.A. Stomatal compensation points for ammonia in oilseed rape plants under field conditions. Agr. Forest Meteorol. **2000**, 105, 371–383.

10. Ramge, P.; Badeck, F.W.; Ploechl, M.; Kohlmaier, G.H. Apoplastic antioxidants as decisive elimination factors within the uptake of nitrogen dioxide into leaf tissues. New Phytol. **1993**, *125*, 771–785.

11. Gabriel, R.; Schäfer, L.; Gerlach, C.; Rausch, T.; Kesselmeier, J. Factors controlling the emissions of volatile organic acids from leaves of *Quercus ilex* L. (Holm oak). Atmos. Environ. **1999**, *33*, 1347–1355.

12. Rogasik, J.; Schroetter, S.; Schnug, E. Impact of air pollutants on agriculture. Phyton **2002**, *42*, 171–182.

Dew Point Temperature

John S. Roberts
New Technology Department, Research and Development, Rich Products Corporation, Buffalo, New York, U.S.A.

Abstract

Dew point temperature is an important parameter in areas of processing involving interaction of water vapor in air, such as dehydration systems, control of heating, ventilating, and air conditioning (HVAC) systems including humidification/dehumidification of air, optimization of fuel cell operation, and plant sanitation, to name a few. Dew point temperature is defined, and how it relates to other properties of moist air, such as dry bulb and wet bulb temperatures, humidity ratio, water vapor pressure, and relative humidity is described. Methods used to determine and calculate dew point temperature are presented, such as through charts, tables, and equations, and results from these various methods are compared. Finally, the most common instruments used to measure dew point temperature are also described, such as capacitive sensors, saturated salt sensors, and optical surface condensation hygrometers, along with their advantages and disadvantages.

INTRODUCTION

Dew point temperature is an aspect of moist air below which condensation of water vapor present in the air begins to occur. At the dew point temperature, the relative humidity of moist air is 100%, and further cooling of the air initiates water vapor condensing due to losing the energy it gained during evaporation. Some common examples of dew formation as a result of moist air being cooled below its dew point temperature are observed on windowpanes, on grass in the evening, on a glass containing a cold beverage, on cooling pipes, and morning fog. When the dew point is below freezing temperature, 0°C, sublimation occurs where the water vapor converts directly to frost. When this condition applies, the term frost point temperature is used. The objectives of this entry are to further explain dew point temperature and related concepts and to provide and compare several methods in which dew point temperature is determined.

DEFINITION AND CONCEPTS

Dew point temperature is the temperature at which moist air must be cooled at constant pressure and constant humidity ratio for the partial pressure of water vapor in the air to equal the equilibrium water vapor pressure at that same temperature and pressure.[1,2] Some definitions state that it is the temperature where moist air must be cooled under constant pressure and humidity ratio to reach saturation.[3,4] The term saturation can be misleading because it indicates that the air cannot accept more water vapor molecules physically, when in fact it can if temperature increases. Several published texts have stated concerns in using saturation in terms of moist air.[1,5,6] This misunderstanding of saturation of air has led to the misconception of how much water vapor air can "hold." The limitation of water vapor in air is not due to a lack of space in the air or due to any binding of water vapor to oxygen or nitrogen gas molecules, which does not occur. Rather, the limitation of water vapor in air is solely due to the amount of available thermal energy to maintain water in the vapor state in the air.[1] This is shown in the relationship of the equilibrium (saturation) water vapor pressure with temperature. As temperature increases, the equilibrium water vapor pressure increases. In maintaining the use of saturation with respect to air, several texts have defined saturated air as an equilibrium between water vapor in the air and the condensed water phase at the existing temperature and pressure.[3,5] This concept of saturation of air is with respect to equilibrium and should be kept in mind when studying water vapor in air. In addition, moist air is a term typically used when describing the amount of water vapor in air between complete dryness and complete humidification.[3]

The composition of moist air can be described by the humidity ratio, or specific humidity, and the relative humidity. The specific humidity, W, is the ratio of the mass of water vapor, m_v, to the mass of the dry air, m_a:

$$W = \frac{m_v}{m_a} \tag{1}$$

The humidity ratio can also be expressed in terms of partial pressures:

$$W = \frac{m_v}{m_a} = \frac{M_v p_v V / RT}{M_a p_a V / RT} = \frac{M_v p_v}{M_a p_a} \tag{2}$$

Encyclopedia of Natural Resources DOI: 10.1081/E-ENRA-120047643

Dew—Wind

where M is molecular weight, V is volume, p is partial pressure, R is gas constant, T is temperature, subscript v represents water vapor, and subscript a represents dry air. The ratio of the molecular weight of water vapor to the molecular weight of dry air is approximately 0.62198. The vapor pressure of air, p_a, can be expressed as $p - p_v$ from the relationship $p = p_a + p_v$, where p is the total barometric pressure of the moist air. Thus, the equation for humidity ratio can be reduced to

$$W = 0.62198 \frac{p_v}{p - p_v} \qquad (3)$$

Dew point temperature, T_d, is defined mathematically as the solution $T_d(p, W)$ of the equation:

$$W_s(p, T_d) = W \qquad (4)$$

W_s is the humidity ratio of air at saturation (equilibrium) and can be expressed similar to Eq. 1:

$$W_s = 0.62198 \frac{p_{vs}}{p - p_{vs}} \qquad (5)$$

where p_{vs} is the saturation vapor pressure at the dew point temperature. Equation 5 reduces to

$$p_{vs}(T_d) = p_v = \frac{pW}{0.62198 + W} \qquad (6)$$

An important aspect either in determining the dew point or as moist air approaches the dew point during cooling is that both the pressure and the composition of moist air (humidity ratio or absolute humidity) remain constant while the relative humidity increases. The relative humidity is defined as the mole fraction of water vapor in moist air, n_v, to the mole fraction of equilibrium water vapor in air, n_{vs}, at the same temperature and pressure. As partial vapor pressure, p_v, equals $n_v p$ and saturation vapor pressure, p_s, equals $n_{vs} p$, the relative humidity can be expressed in terms of pressure:

$$\varphi = \left. \frac{p_v}{p_s} \right)_{T,p} \qquad (7)$$

Expressed in words, the relative humidity is the ratio of the amount of moisture in the air to the equilibrium moisture in air at the same temperature and pressure. When the pressure and humidity ratio remain constant, the relative humidity increases as temperature decreases. So as the air cools, the partial pressure of the water vapor remains constant (constant moisture content in the air) and the equilibrium vapor pressure of water decreases as the temperature decreases. Eventually, the air temperature would have decreased such that the equilibrium vapor pressure of water equals the partial vapor pressure. Thus, as the air cools, the

numerator remains constant while the denominator decreases until unity is reached in Eq. 7. In the case where pressure remains constant, the relative humidity would be a function of both temperature and humidity ratio. As the dew point temperature is solely a function of the humidity ratio (at constant pressure), the dew point temperature is a better indicator of the amount of moisture in the air than relative humidity.[1,5,6]

DETERMINATION OF DEW POINT TEMPERATURE

Dew point temperature can be determined by the following means when other properties of moist air are known: from a psychrometric chart, a table on thermodynamic properties of water at saturation, equations, or physical measurements using hygrometers. This section describes how each method is used to determine dew point temperature.

Psychrometric Chart

The easiest means to determine dew point temperature is from a psychrometric chart. Figure 1 is a reduced version of the psychrometric chart showing only the dry bulb temperature, humidity ratio, and the saturation (equilibrium) curve. The condition of the moist air can be located on the chart by knowing any two of the following properties: dry bulb temperature, humidity ratio or relative humidity, and wet bulb temperature. Once the state of the moist air has been established on the chart, the dew point temperature of this moist air can be determined by following along constant humidity ratio to the saturation curve. The temperature at this intersection is the dew point temperature.

Tables and Equations

When a psychrometric chart is not available, the dew point temperature can be determined by tables or by equations when the dry bulb temperature is given along with either the relative humidity or the humidity ratio. The state of moist air at the dew point temperature is saturated, 100% relative humidity; thus the partial vapor pressure equals the saturation vapor pressure. The corresponding temperature is the dew point temperature from the saturation vapor pressure–temperature relationship, which can be determined either from tables or from equations. If humidity ratio is given, as expressed in Eq. 3, the partial pressure of water vapor is equal to the saturation vapor pressure of the moist air and can be directly calculated from Eq. 6. This direct determination is valid because the humidity ratio remains constant when the dew point temperature is determined, as illustrated in Fig. 1. When relative humidity is given and is below 100%, the process of determining the partial pressure of the water is more involved because the relative humidity of the moist air is

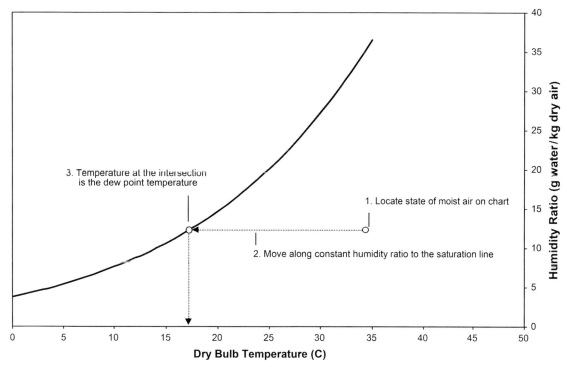

Fig. 1 Dew point temperature determination using a psychrometric chart.

less than the relative humidity of the moist air at the dew point temperature. Once the partial vapor pressure is determined, it can be set equal to the saturation vapor pressure because that is the condition of the moist air at the dew point temperature.

As shown in the expression for relative humidity in Eq. 7, the saturation vapor pressure–temperature relationship must be known to determine the partial pressure of water vapor. This relationship is also needed to calculate the dew point temperature, as shown in Eq. 4. The saturation vapor pressure can be found in a table on thermodynamic properties of water at saturation or can be calculated from equations. A table on thermodynamic properties of water at saturation consists of temperature and corresponding saturation water vapor pressure, specific volume, enthalpy, and entropy.[3] Several equations are widely used to express the saturation vapor pressure of water with respect to temperature. One set of equations widely used is given below.[3]

For the temperature range $-100°C$–$0°C$:

$$\ln(p_{vs}) = (-5674.5329/T) + (6.3925247) \\ + (-0.009677843)T + (6.22115701 \times 10^{-7})T^2 \quad (8) \\ + (2.0747825 \times 10^{-9})T^3 + (-9.484204 \times 10^{-13})T^4 \\ + (4.1635019)\ln(T)$$

For the temperature range 0–$200°C$:

$$\ln(p_{vs}) = (-5.8002206/T) + (1.3914993) \\ + (-0.048640239)T + (4.1764768 \times 10^{-5})\,T^2 \quad (9) \\ + (1.4452093 \times 10^{-8})T^3 + (6.5459673)\ln(T)$$

where p_{vs} is the saturation vapor pressure, Pa; and T is absolute temperature, K. Another widely used equation to calculate saturation vapor pressure as a function of temperature is the Magnus–Tetens equation:[7]

$$p_{vs} = 0.6105exp\left[\frac{aT}{b+T}\right] \quad (10)$$

where $a = 17.27$, $b = 237.7$, and T is in °C. This equation is an approximation of the variation of saturation vapor pressure with temperature, but its advantage is in its simplicity. The saturation vapor pressure of water can then be used to determine the corresponding dew point temperature either by tables or through equations.

The following equations are expressed to solve for dew point temperature. Equations 11 and 12 have been commonly used to calculate dew point temperature from saturated vapor pressure.[3]

For the frost point temperatures below 0°C:

$$T_d = 6.09 + (12.608)\ln\left(p_v\right) + 0.4959\left(\ln\left(p_v\right)\right)^2 \quad (11)$$

and for dew point temperature in the range 0°C to 93°C:

$$T_d = 6.54 + (14.526)\ln\left(p_v\right) + 0.7389\left(\ln\left(p_v\right)\right)^2 \\ + 0.09486\left(\ln\left(p_v\right)\right)^3 + 0.4569\left(p_v\right)^{0.1984} \quad (12)$$

where T_d is in °C and p_v is in kPa.

Dew—Wind

The dew point temperature is often calculated from the Magnus–Tetens equation, Eq. 10, which can be expressed to solve for dew point temperature as shown below:

$$T_d = \frac{\ln\left(\dfrac{p_v}{0.6105}\right) \cdot b}{a - \ln\left(\dfrac{p_v}{0.6105}\right)} \tag{13}$$

A good overview of dew point temperature as it relates to relative humidity is presented along with various equations to illustrate the relationship, such as one based on the Clausius–Clapeyron equation[8]

$$\frac{d\rho_{vs}}{dT} = \frac{L\rho_{vs}}{R_v T^2} \tag{14}$$

which can be simplified to the following expression to calculate dew point temperature:

$$T_d = T_{db}\left(1 - \frac{T_{db}\ln(\phi)}{L / R_v}\right) \tag{15}$$

where temperature is in Kelvin, L is the enthalpy of vaporization (2.501×10^6 J/kg at $T_{db} = 273.15$ K and 2.257×10^6 J/kg at $T_{db} = 373.15$ K) and R_v is the gas constant for water vapor (461.5 J/K kg).

A new set of empirical equations were introduced to calculate dew point.[9]

For $-20 < T < 10°C$:

$$T_d = 193.03 + 28.633 p_v^{0.1609}$$

For $10 < T < 180°C$: $\qquad\qquad$ (16)

$$T_d = \frac{3816.44}{23.197 - \ln(p_v)} + 46.13$$

The dew point temperature calculated in Eqs. 15 and 16 are in Kelvin.

Table 1 shows the comparison of dew point temperatures determined from the different methods described in the preceding text with dry bulb temperature and relative humidity known. Equations 12–14 produced very similar dew point temperatures and were close to the dew point temperature determined from a table on thermodynamic properties of water at saturation.[3] The difficulty in determining dew point temperature from the table is that the saturation vapor pressures are given in whole-number dry bulb temperatures, and therefore interpolation is needed when the saturation vapor pressure lies between two dry bulb temperature entries. The psychrometric chart is useful in assessing approximate dew point temperatures, usually in whole numbers, with reasonable accuracy. But as illustrated in Fig. 1, the chart's advantage is in the ease of determining the dew point temperature. The results of using Eq. 16, which was recently offered as an alternative empirical equation in calculating dew point temperature,[9] is also shown in Table 1. The dew point temperatures were predicted slightly higher in all calculations than the other methods, more at lower dry bulb temperature conditions.

Methods of Measurement

The gravimetric hygrometer is regarded as a primary standard of measurement of moisture in air because it determines the humidity ratio directly.[3,4] As shown by the psychrometric chart in Fig. 1, the humidity ratio directly correlates with the dew point temperature. This standard method of measurement consists of passing moist air through a series of three U-tubes filled with desiccant. The amount of water vapor absorbed and the volume of dry gas can be measured precisely under controlled temperature and pressure conditions. This instrument is not used in industrial applications because the equipment is cumbersome, requires special care, and measurement times are long (up to 30 hours for low dew point temperature measurements).[3,4] There are several more practical devices to measure dew point temperature, and the three most commonly used in industry and research are capacitive

Table 1 Comparison of dew point temperatures, T_d, given dry bulb temperature, T_{db}, and relative humidity, φ, using a table, a psychrometric chart, and various equations

T_{db} (°C)	φ (×100) (%)	p_{vs} (table) (kPa)	p_v ($\varphi = p_{vs}$) (kPa)	T_d (table) (°C)	T_d (psych. chart) (°C)	T_d (Eq. 12) (°C)	T_d (>Eq. 13) (°C)	T_d (Eq. 14) (°C)	T_d (Eq. 16) (°C)
9	70	1.1481	0.80367	3.82	4	3.84	3.85	3.86	4.17
29	20	4.0083	0.80166	3.79	4	3.80	3.81	3.46	4.13
19	40	2.1978	0.87912	5.11	5	5.14	5.13	5.00	5.43
19.5	70	2.2683	1.5878	14.43	14	13.93	13.93	13.87	14.12
34	30	5.3239	1.5972	14.48	14	14.02	14.02	13.75	14.21
29.5	50	4.1272	2.0636	17.99	18	18.02	18.04	17.90	18.12
46	29	10.098	2.9283	23.68	24	23.68	23.74	23.37	23.82
35	80	5.6278	4.5022	31.02	31	31.01	31.10	31.00	31.12

(aluminum oxide) sensors, saturated salt (lithium chloride) sensors, and optical surface condensation (chilled mirror [CM]) hygrometers.[10,11] These instruments vary in the method of measurement, range of measurement, level of accuracy, and cost. The choice of measurement depends on the environment it will be used and the level of accuracy required. The following sections describe the operation of these devices along with their advantages and disadvantages.

Capacitive (aluminum oxide) sensor

Aluminum oxide sensors are the most widely used, especially in heating, ventilating, and air conditioning (HVAC) applications, where cost is more critical than accuracy. A capacitor consists of two electrodes with a dielectric in between. These capacitive sensors consist of an aluminum substrate, followed by a layer of a hygroscopic material, aluminum oxide, and then followed by a coating of gold film.[10] The aluminum substrate and gold film are the electrodes of the capacitor, and the aluminum oxide acts as the dielectric. Some recent capacitive sensors have a polymer as their dielectric. Water vapor from moist air absorbs into aluminum oxide, and the amount of absorbed water vapor directly correlates with the capacitance of the sensor. The instrument determines the dew point based on a series of correlations, first on the capacitance–water vapor relationship and then on the temperature–vapor pressure relationship.

Capacitive sensors have the advantage of measuring dew points in extreme temperature conditions. These sensors can measure low frost points, down to −100°C and are able to withstand very high temperatures as found in pressurized heating applications. Capacitive sensors have good sensitivity at low humidity levels; so these sensors are frequently used in petrochemical and power industries.[11] Capacitive sensors are small and portable; so they are often mounted in ducts or walls in industrial processing stream applications. These sensors also have a fast response time, thus making them suitable in applications where a quick read is critical in saving time and product, such as spot testing of gas cylinders.[12] However, capacitive sensors can become saturated in environments where the relative humidity is above 85%.[11] Also, these sensors are based on a secondary measurement, capacitance; so periodic recalibration is necessary.

Saturated salt (lithium chloride) sensor

Lithium chloride-saturated salt sensors consist of a thin wall metal tube covered with a glass sleeve impregnated with saturated lithium chloride solution. Lithium chloride salt solution is the most desired because at saturation it has a low equilibrium relative humidity of 11% and maintains this humidity over a wide range of temperatures, 11.2% at 0°C and 10.8% at 70°C.[4] This sensor bobbin is wrapped spirally with two wire electrodes, or bifilars, which are connected to an alternating current voltage source.[3] The ionic solution completes the electrical circuit between the wires. There is a direct relationship between the amount of condensation and the amount of current drawn between the two electrodes.

The operation of this sensor is best described by explaining how the dew point is determined from the following two conditions: initial moist air with a partial vapor pressure greater than the saturated salt, and initial moist air with a partial vapor pressure less than the saturated salt. In the first condition, when the initial moist air has a partial vapor pressure greater than that of the saturated salt solution, water vapor from the air will condense onto the sensor due to the vapor pressure gradient. As more water vapor condenses, the salt solution becomes more conductive. The current creates resistive heating, which increases the vapor pressure of the salt solution. The rate of condensation and rate of heating decrease until equilibrium are reached between the partial vapor pressure of the solution and the surrounding ambient moist air. A temperature probe is embedded into the sensor to measure the solution temperature needed to achieve this equilibrium vapor pressure. At this increased sensor temperature, the increased partial vapor pressure of the solution is proportional to the saturated vapor pressure of water at that same temperature to maintain the 11% equilibrium relative humidity of the lithium chloride-saturated solution. So, with the temperature reading, the saturated water vapor pressure can be determined. The partial vapor pressure can then be determined from this saturated water vapor pressure using the correlation shown in Eq. 7 and based on 11% equilibrium relative humidity of the saturated salt solution. This partial pressure of the solution is the same partial vapor pressure of the ambient moist air. As dew point temperature directly correlates with vapor pressure, the dew point can be determined based on the temperature–vapor pressure relationship. The hygrometer internally calculates all of these relationships to determine dew point. If the initial moist air has a partial vapor pressure less than the saturated salt solution, evaporation would occur until the sensor is at equilibrium with the air. As the solution loses some water to its environment, less current is drawn, and the temperature sensor would record the evaporative cooling.

The limits of this sensor are that moist air conditions below 11% relative humidity cannot be measured because this is the equilibrium vapor pressure of saturated lithium chloride, and dew point above 70°C cannot be measured. As equilibrium has to be reached, the measurement times are longer than with most sensors. Also, exposure to liquid water can wash away the salt crystals in the sensor. These low-cost sensors are typically used in applications where an inexpensive, slow, and moderate accuracy is needed, such as in refrigeration controls, dryers, dehumidifiers, airline monitoring, and pill coaters.[11]

Dew—Wind

Optical surface condensation (CM) hygrometer

The most direct method to determine dew point temperature is the CM hygrometer and is the most widely used for precise dew points.[4,10] This method of measurement involves a metallic mirrored surface, a mechanism to cool this mirrored surface, a light-emitting device (LED), and a photodetector. The principle of measurement is based on the reflectance of light from the LED by the mirror. The gas (i.e., moist air) is passed over the mirror, and when the temperature of the mirror is above the dew point of the moist air, the photodetector reads the direct reflectance of the light. However, when the mirror is cooled to the dew point, water vapor begins to condense on the mirror surface. The light is then scattered, and the decrease of reflected light intensity is picked up by the detector. The key factors in accurate measurement of the CM hygrometers are controlling the rate of cooling of the mirror, thus controlling the amount of condensing water vapor on the mirror surface, preventing contaminants on the mirror surface, and precise reading of the temperature of the mirror.

The mirrored surface is made of a good conductor, normally of copper or silver, and coated with an inert metal to prevent tarnishing and oxidation. As water vapor can condense at or below the dew point temperature, precise control of the mirror temperature is important not to allow significant dew formation. If significant dew has formed, the humidity ratio will reduce slightly, and the reading will not be the precise dew point temperature of the initial air condition. Modern CM hygrometers cool the mirror using a solid-state heating element, such as a Peltier device, to control cooling of the mirror. High-quality CM hygrometers have a feedback control mechanism where the photodetector communicates with the Peltier device to maintain the mirror temperature such that the dew drop level is at a minimum and remains constant; thus, the rates of evaporation and condensation are equal.

Contaminants, either insoluble or water soluble, on the mirror surface can scatter light and give false readings of dew formation. When insoluble contamination occurs, the mirror surface has to be cleaned manually. Water-soluble contamination is normally in the form of dissolved salts, and there are several methods to correct this type of contamination. One method developed by General Eastern is known as the Programmable Automatic Contaminant Error Reduction (PACER) system.[10–12] This method chills the mirror for an extended time, so excessive amounts of dew form to dissolve the salts on the surface. Then the system rapidly heats the mirror to evaporate the water. This results in localized regions of salt crystals on the mirror surface while the majority of the surface is clean. The other method that optical CM hygrometers use to correct for contamination is the cycling chilled mirror (CCM) hygrometer. In this method of measurement, the mirror is chilled to the dew point for a short time (5% of the measurement time) and then heated above the dew point.[10] This process is repeated to give a reproducible reading of dew point. Having the mirror at dew point for a limited time reduces the chance of contamination. In both methods, eventually the surface will have to be manually cleaned.

Moisture on the mirror is at equilibrium with the water vapor in the air at one unique temperature; so the measurement of the mirror temperature is a fundamental and primary measurement of dew point temperature.[10] A precise temperature sensor, such as a platinum resistance thermometer, measures the mirror temperature at the dew point. CM hygrometers have a dew point measurement accuracy of ±0.2°C.

The advantages of CM hygrometers are in their accuracy, reliability, fast response time, ability for longer continuous and unattended operation due to its self-correcting capabilities, and ability to measure dew points ranging from −70°C to 95°C. However, these instruments can be expensive. Typical applications of the CM hygrometers are for humidity calibration standards, for critical environmental monitoring (i.e., clean rooms, pharmaceutical laboratories, areas where sensitive computers and electronics are located, and environmental chambers), and for monitoring extreme environmental conditions (i.e., heat-treating furnaces, engine test beds).[11]

REFERENCES

1. Bohren, C.F.; Albrecht, B.A. *Atmospheric Thermodynamics*; Oxford University Press, Inc.: New York, 1998.
2. Mohen, M.J.; Shapiro, H.N. *Fundamentals of Engineering Thermodynamics*, 4th Ed.; John Wiley and Sons, Inc.: New York, 2000.
3. *ASHRAE Handbook Fundamentals,* SI Ed.; American Society of Heating, Refrigeration, and Air Conditioning Engineers, Inc.: Atlanta, GA, 1989.
4. Kuehn, T.H.; Ramsey, J.W.; Threlkeld, J.L. *Thermal Environmental Engineering*, 3rd Ed.; Prentice Hall: Saddle River, NJ, 1998.
5. Williams, J. *The Weather Book*, 2nd Ed.; Vintage Books: New York, 1997.
6. http://www.shorstmeyer.com/wxfaqs/humidity/humidity.html (accessed October 2009).
7. http://www.paroscientific.com/dewpoint.htm (accessed October 2009).
8. Lawrence, M.G. The relationship between relative humidity and the dew point temperature in moist air. Bull. Am. Meteorol. Soc. **2005**, *86* (2), 225–233.
9. Li, S.Q.; Gong, Z.X. Calculations of the state variables of moist air. In *Drying'92*; Mujumdar, A.J., Ed.; Elsevier Science Publishers B.V.: Amsterdam, the Netherlands, 1992.
10. Wiederhold, P.R. *Water Vapor Measurement. Methods and Instrumentation*; Marcel Dekker, Inc.: New York, 1997.
11. http://iceweb.com.au/Analyzer/humidity_sensors.html (accessed October 2009).
12. Larson, K. Humidity measurement update. Control **1990**, June, 49–59.

Drought: Management

Donald A. Wilhite
Michael J. Hayes
National Drought Mitigation Center, Lincoln, Nebraska, U.S.A.

Abstract

Droughts are a normal part of climate for virtually all areas of the world, and droughts affect more people worldwide than any other natural hazard. Yet officials from both developing and developed nations struggle to deal with the wide range of economic, environmental, and social impacts related to droughts. There are important management actions that officials at local, regional, and national levels can take to reduce the impacts from droughts. The long-term goal is to reduce the impacts of drought through the adoption of drought preparedness plans.

INTRODUCTION

Some would argue that drought cannot be "managed." Yes, it is true droughts are a normal part of climate for virtually all areas of the world (e.g., Fig. 1), and that droughts affect more people worldwide than any other natural hazard.[1] It is also true that officials from both developing and developed nations struggle to deal with the wide range of economic, environmental, and social impacts related to droughts. However, these officials are not powerless to reduce the impacts of drought. Rather, there are important management actions that officials at local, regional, and national levels can take to reduce the impacts from droughts. The approach taken to address drought impacts and reduce their effects is called drought management, or perhaps more appropriately, drought risk management. The long-term goal is to reduce the impacts of drought through the adoption of drought preparedness plans.

SHIFTING THE EMPHASIS FROM CRISIS TO DROUGHT RISK MANAGEMENT

Traditionally, droughts have been viewed as unusual occurrences that creep up on officials who are typically unprepared to deal with the impacts droughts create. This is why drought has been called the "creeping phenomenon."[1] In reality, drought is a normal feature for virtually all climates. Officials often react to the occurrence of drought through "crisis management." After a drought is over, officials turn their attention to the next crisis, and any lessons learned about responding to the drought are most likely lost and forgotten. This crisis management approach is illustrated in the "Hydro-Illogical Cycle" (Fig. 2). Crisis management approaches to dealing with droughts are reactive, poorly coordinated and targeted, untimely, and generally too late. As a result, they are largely ineffective.

In order to break the Hydro-Illogical Cycle, officials around the world at local, regional, and national scales need to adopt a drought risk management approach. Drought risk management involves taking actions before droughts occur in order to reduce the drought impacts. It has three main components: 1) a comprehensive drought monitoring and early warning system; 2) planning and building the institutional capacity to respond to droughts; and 3) identification and implementation of mitigation actions and policies that can be taken before the next drought. These components will be discussed in greater detail.

A comprehensive drought monitoring and early warning system is a critical component of drought risk management because effective, timely decisions related to droughts can only be made if officials have an accurate assessment of the potential or developing drought event. This early warning system must incorporate all of the critical components of the hydrologic system (e.g., precipitation, streamflow, groundwater, snowpack, soil moisture, and reservoir and lake levels) because drought severity cannot be defined by precipitation deficiencies alone. A comprehensive system will assist officials by providing appropriate "triggers" for actions that the officials need to take, or by identifying when particular impacts are going to occur. An effective drought monitoring and early warning system requires synthesis and analysis of timely data and an efficient dissemination system to communicate this information (e.g., the media, extension services, or the World Wide Web).

Drought planning is a very important component of drought risk management because it establishes and

Encyclopedia of Natural Resources DOI: 10.1081/E-ENRA-120010110

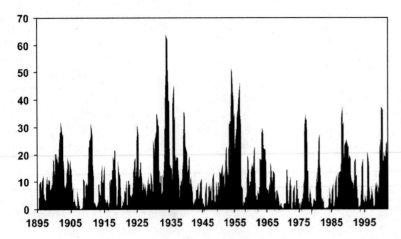

Fig. 1 The percent area of the United States in severe to extreme drought by month from 1895 through March 2002. Similar periodic patterns appear on graphs depicting regional hydrological basins in the United States, and would likely appear for most regions in the world.
Source: Adapted from National Climatic Data Center, Asheville, North Carolina, U.S.A.

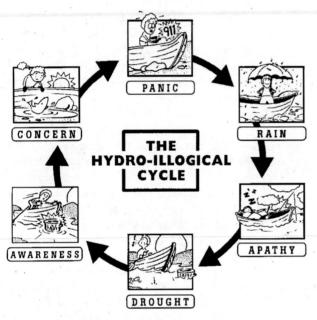

Fig. 2 The Hydro-Illogical Cycle.
Source: National Drought Mitigation Center, University of Nebraska, Lincoln, Nebraska, U.S.A.

Several methodologies exist for assisting officials with the development of drought plans. One of these methodologies, described by Wilhite et al.,[2] is a 10-step drought planning process that targets drought planners in the United States and elsewhere (http://drought.unl.edu/center/pdfpubs/10step.pdf) (accessed April 2002). The process was designed to be generic and adaptable because it is important for planners to develop a plan appropriate for their regional and governmental structures. These plans must be dynamic, reflecting changing government policies, technologies, personnel, and natural resources management practices.

The third important component of drought risk management is mitigation. Mitigation is defined as the policies and actions taken before a drought that will reduce drought impacts. Sometimes, if officials are alert enough and can see the development of drought in its early stages, mitigation can take place during the drought's early stages and may be very effective in reducing impacts as the drought becomes more severe. Otherwise, actions taken during a drought are generally responses directly related to the drought's severity and impacts. These responses are important, of course, and need to be well documented ahead of time within a drought plan. But it is important to keep in mind that mitigation is most effective if it takes place during times when drought is not occurring and officials are not responding to drought during a crisis. Mitigation actions should address vulnerabilities associated with drought with the goal of reducing impacts in future events.

What are some examples of drought mitigation? Certainly the development of a comprehensive drought monitoring and early warning system and the development of a drought plan, as described above, are two examples of mitigation. Both of these actions should be taken before a region is experiencing drought. Other broad categories for potential drought mitigation actions include revising or developing legislation or public policies related to drought and water supplies; water supply augmentation and the

preserves the institutional capacity with which officials can respond to droughts and reduce drought impacts. There are many benefits of a drought plan. A drought plan serves as the organizational framework for dealing with droughts and improving the coordination between and within levels of government. In addition, drought plans enable proactive mitigation and response to droughts; enhance early warning through integrated monitoring efforts; involve stakeholders, which are necessary for successful programs; identify areas, groups, and sectors particularly at risk; improve information dissemination by outlining the information delivery systems and strategies; and build public awareness of the need for improved drought and water management.

development of new supplies; demand reduction and the development of water conservation programs; public education and awareness programs; specific priorities for water allocations; and water use conflict resolution.[3]

As with drought planning, there are several methodologies for identifying the appropriate mitigation actions to take in a region. In 1998, as part of the activities of the Western Drought Coordination Council (WDCC), a methodology was developed to look at drought risk. An important part of this methodology was the identification of mitigation actions and how these actions would be implemented.[4] The methodology also involves identifying and understanding the people and sectors that are vulnerable to droughts and why, allowing officials to target their mitigation efforts more effectively.

DROUGHT RISK CHALLENGES AND OPPORTUNITIES

Serious challenges still remain in drought risk management. One of these challenges is the acceptance of drought as a natural hazard, and a hazard that needs to be prepared for. Fragmented resource management and numerous federal programs present challenges, as do the declining financial and human resources. Confusion over the difference between mitigation and response is a challenge, and many times officials have a difficult time identifying innovative mitigation actions and implementing new policies and programs. Stakeholder involvement and acceptance still needs improvement. Perhaps one of the biggest challenges is to maintain the momentum for risk management in a changing political climate.

Some progress toward drought risk management is being made around the world. Australia and New Zealand, for example, have had national drought policies and strategies to reduce drought impacts.[5,6] Other nations are looking at establishing national drought policies. A global drought preparedness network is in the development stages; this network would assist nations by promoting drought risk management and sharing lessons learned about drought monitoring, planning, and mitigation. The network, based at the National Drought Mitigation Center/International Drought Information Center at the University of Nebraska, would be made up of regional networks coordinated by institutions around the world. Collectively this network of regional networks may enhance the drought management capability of many nations.

In the United States, three states had drought plans in 1982. By 2002, 33 states had drought plans, and 6 of those states incorporate mitigation actions into their plans. In 1998, New Mexico became the first state to develop drought plan that emphasizes mitigation. Five states are currently in the process of developing drought plans, and it is hoped that mitigation will be a major component of each

of these new plans. A number of Native American nations in the southwestern United States have developed drought mitigation plans as well. In addition, improved coordination has occurred within federal agencies and between federal and state governments.

New drought monitoring efforts and products have been developed. One of the best examples of progress in this area is the Drought Monitor product, developed to assess current drought conditions in the United States. The first Drought Monitor map was issued in August 1999, and a weekly update is posted every Thursday morning (http://drought.unl.edu/dm/) (accessed April 2002). The unique feature of this product is that four agencies rotate creating the map: the National Drought Mitigation Center, the United States Department of Agriculture, the Climate Prediction Center, and National Climatic Data Center of the National Oceanic and Atmospheric Administration. In addition, a feedback network of more than 160 local experts provides input about the map's portrayal of drought conditions before the map is released each week.

CONCLUSION

Clearly, there is reason for optimism about drought risk management and reducing drought impacts in the future. But it is also clear that officials around the world need to take proactive steps to develop comprehensive and integrated drought monitoring and early warning systems, determine who and what is at risk to droughts and why, and create drought mitigation plans with specific actions that address these risks with the goal of reducing the impacts of future drought events. There is a growing recognition that drought risk management is a critical ingredient of sustainable development planning and must be addressed systematically through risk-based policies and plans.

REFERENCES

1. Wilhite, D.A. Drought as a natural hazard: concepts and definitions. In *Drought: A Global Assessment*; Wilhite, D.A., Ed.; Routledge: New York, 2000; Vol. 1, 1–18.
2. Wilhite, D.A.; Hayes, M.J.; Knutson, C.; Smith, K.H. Planning for drought: moving from crisis to risk management. J. Am. Water Res. Assoc. **2000**, *36* (4), 697–710.
3. Wilhite, D.A. State actions to mitigate drought: lessons learned. In *Drought: A Global Assessment*; Wilhite, D.A., Ed.; Routledge: New York, 2000; Vol. 2, 149–157.
4. Knutson, C.; Hayes, M.J.; Phillips, T. How to Reduce Drought Risk, Preparedness and Mitigation Working Group of the Western Drought Coordination Council: Lincoln, NE; 1998; 1–43, http://drought.unl.edu/handbook/risk.pdf (accessed April 2002).

Dew—Wind

5. O'Meagher, B.; Smith, M.S.; White, D.H. Approaches to integrated drought risk management: Australia's national drought policy. In *Drought*: *A Global Assessment*; Wilhite, D.A., Ed.; Routledge: New York, 2000; Vol. 2, 115–128.

6. Haylock, H.J.K.; Ericksen, N.J. From state dependency to self-reliance: agricultural drought policies and practices in New Zealand. In *Drought: A Global Assessment*; Wilhite, D.A., Ed.; Routledge: New York, 2000; Vol. 2, 105–114.

Drought: Precipitation, Evapotranspiration, and Soil Moisture

Joshua B. Fisher
Konstantinos M. Andreadis
Jet Propulsion Laboratory, California Institute of Technology, Pasadena, California, U.S.A.

Abstract

Drought is the major hydrological disaster for human society, and encompasses multiple hydrological components. There are numerous definitions of drought, many of which do not include transpiration or evaporation (evapotranspiration, ET). We describe here the role of ET in drought, and why ET is important to consider when assessing drought. We emphasize the dynamic roles of ET, precipitation, and soil moisture, with a focus on the vegetation response during water stress. Finally, we conclude with a case study example of drought in Amazonia.

INTRODUCTION

Drought ranks as one of the most expensive natural disasters in terms of human welfare and food security. For example, in the United States the annual cost of drought relief measures has been estimated between US$6 and US$8 billion. Droughts can cover very large areas and last for several years with so-called megadroughts documented from medieval times.[1] In general terms, drought is caused by extremes within the natural variability of climate, but can be exacerbated by human activity (e.g., deforestation). The literature on drought is extensive, with definitions categorically ranging from meteorological (or, climatological, atmospheric), agricultural, hydrologic, and socio-economic (e.g., management based),[2–5] but we focus here on the vegetation transpiration and evaporation (or, actual evapotranspiration, ET) component of drought.[6] From a vegetation perspective in general, physical drought is the drying of soil such that the overlying vegetation experiences physiological water stress manifested in a reduction of productivity, loss of leaves/needles, and, ultimately, mortality. As such, a given soil moisture content (SMC) would correspond to different classes of drought depending on the ability of vegetation to adapt to decreased soil moisture. For instance, some species or stages of succession may have plants with deep roots that can tap deep sources of soil moisture, even the groundwater table, so the ability of these plants to withstand what would otherwise be considered a drought may, in fact, not be considered a drought for some time.[7] Other species, however, may be poorly adapted to low levels of soil moisture through sparse root distribution, low water use efficiency, or high temperature sensitivity, and thus may enter into a drought much more quickly than other, better adapted, species.

SMC is inherently coupled directly to precipitation (PPT) and ET, with PPT as the moisture input, and ET as the moisture withdrawal; soil water holding capacity acts as the intermediate "bucket" size if considering a bucket model of SMC change. Among those three variables (i.e., SMC, PPT, ET), the end members of SMC can be outlined: 1) for two areas with equivalent PPT and ET, SMC may be different because of different SMC retention properties (e.g., sandy soils may hold less water than soil with more clay); 2) for two areas with equivalent soils and PPT, SMC may be different because of different ET; and 3) for two areas with equivalent soils and ET, SMC may be different because of different PPT. The same exercise can be applied for areas to have similar SMC (e.g., two areas with equivalent soils, one with high PPT and high ET, and the other with low PPT and low ET). By this definition, and with reference to the title of this entry, ET can vary under drought or non-drought situations, although ET will go to zero under persistent and intense drought. Between the ET components, transpiration will go to zero as drought persists (assuming no deep water sources), because stomata close to avoid water loss (assuming no leaky stomata),[8] and plants will maintain respiration through carbon stores; evaporation from the soil surface will also go to zero as the soil dries out (assuming no hydraulic lift/redistribution from deep water sources).[9] The role of ET in drought is particularly pertinent in already water-limited environments where increasing temperatures over time accelerate ET, which leads to greater drought severities.[10]

Plants are typically able to withstand relatively short periods of SMC decline, so a given day with no PPT and high ET may not be considered a drought. Although there are different metrics of drought that take ET into account such as the widely used Palmer Drought Severity Index,[2]

Dew—Wind

which uses the potential ET, it is the *cumulative water deficit* (CWD) that plants respond to – that is, the summation of days in which the soil water deficit (SWD) is below a critical water stress threshold. SWD may be calculated as follows:

IF PPT − ET > 0, THEN $SWD_1 = 0$,
ELSE $SWD_1 = P − ET + SWD_0$

where the subscripts indicate adjacent time steps. The maximum CWD (MCWD) reached during the time period of interest relative to the time-averaged climatological MCWD may be considered "drought."[11] Both the length of the CWD and the MCWD in a given period must surpass the long-term means of the two for that period to be considered a drought, although a corresponding vegetation water stress response is also necessary.

An example of MCWD drought and vegetation response is shown in the *Science* paper by Phillips et al.,[11] "Drought sensitivity of the Amazon rainforest." Here, we describe how measurements of anomalously low PPT indicated the possibility of an intense drought over Amazonia – the lowest PPT at the time, in fact, in the past 100 years. With few soil moisture measurements available, SWD was constructed from measured PPT, estimated ET using meteorological measurements,[12,13] and measurements of the soil water holding capacity. We calculated MCWD at 136 sites where we also observed vegetation response, and determined that sites experiencing the greatest hydrologic drought as defined by MCWD

also had the greatest vegetation response, specifically mortality and biomass loss. It can be seen at one of the sites, for example (Fig. 1), that PPT varies seasonally, as does CWD, but in 2005 (also 1997 and 2001) CWD spikes well beyond the mean CWD for the 10-year record. In this analysis, the length and peak (MCWD) of the CWD spike vary by site and are proportional to the vegetation response (e.g., mortality).

CONCLUSION

Evaporation and transpiration are critical components to drought, although many traditional definitions of drought ignore ET. Vegetative drought inherently implies a response from vegetation, and this response must be calculated from a cumulative water deficit, as ET > P adds up over time, drying out the soil. In the absence of soil moisture measurements, soil moisture may be calculated from precipitation and ET. Even with soil moisture measurements, the understanding of the bioclimatic variables controlling ET helps elucidate and predict how drought will change given changes in the controlling factors.

ACKNOWLEDGMENTS

This entry was written by the Jet Propulsion Laboratory, California Institute of Technology, under a contract with the National Aeronautics and Space Administration.

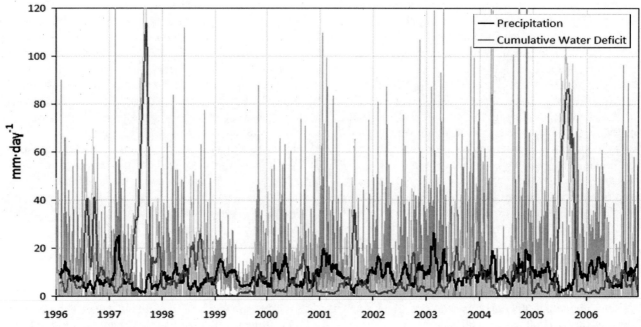

Fig. 1 Precipitation (black) and cumulative water deficit (gray) over 10 years for a site in Amazonia, described in Phillips et al.[11] The light colors are the daily values, and the dark colors are the 30-day moving averages.

Dew—Wind

REFERENCES

1. Cook, E.R.; Woodhouse, C.A.; Eakin, C.M.; Meko, D.; Stahle, D.W. Long-term aridity changes in the western United States, Science **2004**, *306* (5698), 1015–1018.

2. Palmer, W.C. Meteorological drought Rep., U.S. Department of Commerce, 1965; 58.

3. Dracup, J.A.; Lee, K.S.; Paulson, Jr., E.G. On the definition of droughts. Water Resour. Res. **1980**, *16* (2), 297–302.

4. Wilhite, D.A.; Glantz, M.H. Understanding the drought phenomenon: the role of definitions. Water Int. **1985**, *10*, 111–120.

5. McKee, T.B.; Doesken, N.J.; Kleist, J. The relationship of drought frequency and duration to time scales, in *Eight Conference on Applied Climatology*; American Meteorological Society: Anaheim, CA, 1993; 179–184 pp.

6. Fisher, J.B.; Whittaker, R.H.; Malhi, Y. ET Come Home: A critical evaluation of the use of evapotranspiration in geographical ecology, Global Ecol. Biogeogr. **2012**, *20*, 1–18.

7. Fisher, R.A.; Williams, M.; Da Costa, A.L.; Malhi, Y.; Da Costa, R.F.; Almeida, S.; Meir, P. The response of an Eastern Amazonian rain forest to drought stress: results and modelling analyses from a throughfall exclusion experiment, Global Change Biol. **2007b**, *13* (11), 2361–2378.

8. Fisher, J.B.; Baldocchi, D.D.; Misson, L.; Dawson, T.E.; Goldstein, A.H. What the towers don't see at night: Nocturnal sap flow in trees and shrubs at two AmeriFlux sites in California., Tree Physiol. **2007a**, *27* (4), 597–610.

9. Dawson, T.E. Hydraulic lift and water use by plants: implications for water balance, performance and plant-plant interactions, Oecologia **1993**, *95* (4), 565–574.

10. Andreadis, K.M.; Lettenmaier, D.P. Trends in 20th century drought over the continental United States., Geophys. Res. Lett. **2006**, *33* (L10403), doi:10.1029/2006GL025711.

11. Phillips, O.L.; Aragão, L.E.O.C.; Lewis, S.L.; Fisher, J.B.; Lloyd, J.; López-González, G.; Malhi, Y.; Monteagudo, A.; Peacock, J.; Quesada, C.; van der Heijden, G.; Almeida, S.; Amaral, I.; Arroyo, L.; Aymard, G.; Baker, T.R.; Bánki, O.; Blanc, L.; Bonal, D.; Brando, P.; Chave, J.; Oliveira, A.C.A.; Cardozo, N.D.; Czimczik, C.I.; Espejo, J.; Feldpausch, T.; Freitas, M.A.; Higuchi, N.; Jiménez, E.; Lloyd, G.; Meir, P.; Mendoza, C.; Morel, A.; Neill, D.; Nepstad, D.; Patiño, S.; Peñuela, M.C.; Prieto, A.; Ramírez, F.; Schwarz, M.; Silveira, M.; Thomas, A.S.; ter Steege, H.; Stropp, J.; Vásquez, R.; Zelazowski, P.; Dáavila, E.A.; Andelman, S.; Andrade, A.; Chao, K.-J.; Erwin, T.; Di Fiore, A.; Honorio, E.; Keeling, H.; Killeen, T.; Laurance, W.; Cruz, A.P.; Pitman, N.; Vargas, P.N.; Ramírez-Angulo, H.; Rudas, A.; Salamão, R.; Silva, N.; Terborgh, J.; Torres-Lezama, A. Drought sensitivity of the Amazon rainforest., Science **2009**, *323* (5919), 1344–1347.

12. Fisher, J.B.; Tu, K.; Baldocchi, D.D. Global estimates of the land-atmosphere water flux based on monthly AVHRR and ISLSCP-II data, validated at 16 FLUXNET sites., Remote Sensing Env. **2008**, *112* (3), 901–919.

13. Fisher, J.B.; Malhi, Y.; de Araújo, A.C.; Bonal, D.; Gamo, M.; Goulden, M.L.; Hirano, T.; Huete, A.; Kondo, H.; Kumagai, T.; Loescher, H.W.; Miller, S.; Nobre, A.D.; Nouvellon, Y.; Oberbauer, S.F.; Panuthai, S.; von Randow, C.; da Rocha, H.R.; Roupsard, O.; Saleska, S.; Tanaka, K.; Tanaka, N.; Tu, K.P. The land-atmosphere water flux in the tropics, Global Change Biol. **2009**, *15*, 2694–2714.

Dew—Wind

Drought: Resistance

Graeme C. Wright
Department of Primary Industries, Queensland Department of Primary Industries, Kingaroy, Queensland, Australia

Nageswararao C. Rachaputi
Queensland Department of Primary Industries, Kingaroy, Queensland, Australia

Abstract

Drought can be considered as a set of climate pressures that can result from a combination of heat, aerial, or soil water deficits, as well as salinity. The diversity of drought created from these phenomena has led to the selection of numerous types of resistance mechanisms that operate at different levels of life organization. This entry reviews drought in the context of crop production, defining circumstances in which growth or yield of the crop is reduced because of insufficient water supply to meet the crop's water demand.

INTRODUCTION

Drought can be considered as a set of climate pressures that can result from a combination of heat, aerial, or soil water deficits, as well as salinity. The diversity of drought created from these phenomena has led to the selection of numerous types of resistance mechanisms that operate at different levels of life organization (molecule, cell, organ, plant, and crop). Decades of research have been dedicated to the understanding of these mechanisms, with a premise that the improved understanding would contribute to the long-term improvement of plant and crop production under drought conditions.[1]

DISCUSSION

This entry will concentrate on crop production, and in that context drought is a term used to define circumstances in which growth or yield of the crop is reduced because of insufficient water supply to meet the crop's water demand. During the 1960s to 1980s most of the drought research was dedicated to understanding the mechanisms of survival and growth under drought conditions. It is only in fairly recent times that attention has been given to recognizing the complex nature of drought and to separating the *productivity* of crop plants under drought from *survival* mechanisms. Drought resistance in modern agriculture requires sustainable and economically viable crop production, despite stress. However, plant survival can be a critical factor in subsistence agriculture, where the ability of a crop to survive drought and produce some yield is of critical importance.

Hence, in the context of agricultural production, drought resistance in a crop can be best defined in terms of the optimization of crop yield in relation to a limiting water supply.[2] Multitudes of options exist for farmers to alleviate the effects of drought on crop yield, depending on the probability of drought. These can be categorized into management and genetic options, although they can be integrated into a package to manage drought in the target environment. The basis of most management technologies adopted by farmers revolves around optimizing water conservation and its subsequent utilization by the crop. Examples include the use of deep tillage to increase rainfall infiltration, stubble retention to minimize soil evaporation,[3] and intercropping.[4] Genetic options include the use of the best locally adapted varieties or landraces, as well as relaying and intercropping varieties with varying phenology to exploit differences in timing and severity of drought patterns.[5]

Future advances in crop drought resistance and associated improvements in productivity under drought are most likely to come from genetic improvement programs that can apply the wealth of knowledge created over the past century. As Richards[6] states, however, it will never be possible to *overcome* the effects of drought, any progress is likely to be slow, and the gains will only be small. The following sections will therefore concentrate on opportunities and emerging technologies for the improvement of drought resistance in crop plants, using genetic enhancement.

DROUGHT RESISTANCE TRAITS

Levitt[7] has proposed a terminology for drought resistance and its subdivision into different categories based on different mechanisms. These three categories of drought resistance have been widely accepted, and they continue today in a slightly modified form to provide a

Encyclopedia of Natural Resources DOI: 10.1081/E-ENRA-120010407

Dew—Wind

framework for evaluating potential traits for use in crop breeding programs.[8,10] They are drought escape, dehydration postponement, and dehydration tolerance.

Drought Escape

Matching the phenology to the expected water supply in a given target environment has been an important strategy for improving productivity in water-limited environments.[11] In most crop species there is large genetic variability in phenological traits, and these traits are highly heritable and amenable to selection in large-scale breeding programs.[12] Matching phenology has proven to be a highly successful approach in environments that have a high probability of end-of-season drought stress pattern.[13]

Dehydration Postponement

Crops use a variety of mechanisms to maintain turgor in leaves and reproductive structures despite declining water availability. They can effectively regulate water loss from leaves via stomatal control, with large varietal differences in stomatal conductance in response to leaf water potential recorded in cereals[14] and grain legumes.[15]

Production of abscisic acid (ABA) has been implicated as a mechanism behind stomatal control.[16] Other benefits of ABA, including maintenance of turgor in wheat spikelets and subsequent grain set, have also been reported.[17] Subsequent stimulation of research activity into the use of the ABA trait in crop breeding programs has followed. However, Blum[18] recently concluded that while ABA is undoubtedly involved in plant response to drought stress and even perhaps in desiccation tolerance, its value in the context of drought resistance breeding is still questionable.

Osmotic adjustment (OA) has been reported as an important drought-adaptive mechanism in crop plants where solutes accumulate in response to increasing water deficits, thereby maintaining tissue turgor despite decreases in plant water potential.[19] OA has been shown to maintain stomatal conductance, photosynthesis, and leaf expansion at low water potential,[20] as well as reducing flower abortion[21] and improving soil-water extraction despite declining water availability.[22,23] Research has confirmed that OA is directly associated with grain yield in a number of crops, including wheat,[24,25] sorghum,[26] and chickpea.[27]

Dehydration Tolerance

The ability of cells to continue metabolism at low-leaf water potential is known as dehydration tolerance. Membrane stability and the associated leakage of solutes from the cell[28] provide one measure of the ability of crop/genotypes to withstand dehydration. Sinclair and Ludlow[29] have suggested that the lethal water potential is a key measure

of dehydration tolerance, with significant variation among crops and cultivars observed. Some stages of the plant's life cycle are less susceptible to large reductions in water content. For instance, prior to establishment of a large root system, seedlings may often survive large reductions in water content.[30] Protection against lethal damage in seedlings and seeds is correlated with the accumulation of sugars and proteins.[31] Proline has also been implicated in cellular survival of water deficits, and has also been involved in osmotic adjustment. Its role as a selection trait for enhanced drought resistance has been questioned.[32] Although the work on drought resistance mechanisms has produced a few promising leads, their application in practical breeding programs has been limited.

HEIRARCHY OF DROUGHT RESISTANCE TRAITS

Richards[6] suggests there are two major principles to consider when identifying critical traits to use in breeding programs aimed at improving productivity under drought, namely, the influence of the trait in relation to time scale and to level of organization.

Time Scale

Traits that influence drought resistance in crops can span a wide range of time scales. Short-term responses to water deficit, for example, include many of the processes covered earlier (heat-shock proteins, stomatal closure, OA, ABA), which Passioura[2] suggests are often primarily concerned with "metabolic housekeeping." Although these processes are important, they tend to be associated with crop survival rather than with events that influence crop productivity. At the other end of the time scale are longer-acting processes such as control of leaf area development, which can be modulated by the crop to adjust water supply to prevailing demand.[33] It is not always clear which processes are operating to control these balances, but presumably hormonal signals are involved. Passioura[2] argues that researchers need to distinguish between traits linked to short-term responses that might be important for overall drought resistance and those that are unimportant when integrated over longer time scales.

Level of Organization

The capacity of the trait to influence yield is related to the level of organization (molecule–cell–organ–plant–crop) in which the trait is likely to be expressed.[34] Richards[6] cites an example that despite the doubling of crop yields since 1900, the rate of leaf photosynthesis, which is expressed at the cellular-organ level, has remained the same or decreased. Increases in leaf area during this period have been largely responsible for the yield increases. It is concluded that the closer the trait is to the

Dew—Wind

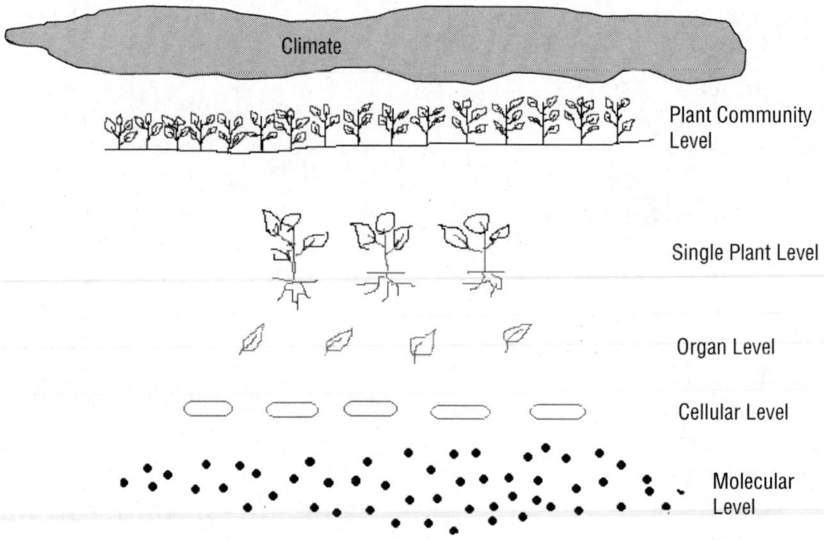

Fig. 1 The hierarchy of processes leading to crop yield.

level of organization of the crop the more influence it will have on productivity (Fig. 1).

DROUGHT RESISTANCE TRAITS IN TERMS OF THE PASSIOURA WATER MODEL

Passioura[2,34,35] argues that there are no traits that confer global drought resistance. Also, given the earlier discussion which clearly suggests that short term responses to drought stress operating at the cellular level may have no bearing on the yield of water-limited crops, crop productivity is best analyzed from the top down. Here the thinking has shifted from understanding defense mechanisms for survival to applying this knowledge to optimize economic productivity under a given water-limited condition.

Passioura proposed that when water is the major limit, grain yield (GY) is a function of the amount of water transpired by the crop (T), the efficiency of use of this water in biomass production (WUE), and the proportion of biomass that is partitioned into grain, or harvest index (HI), thus:

$$GY = T \times WUE \times HI \tag{1}$$

It was argued that individual traits could be assessed in terms of their contribution to each of these functional yield components and thereby increase yield under drought. This identification has provided a framework to more critically identify and evaluate important drought resistance traits. The following examples demonstrate the utility of this approach.

In a number of crops, improvement in the WUE trait has been achieved via selection for carbon isotope discrimination,[36] or correlated surrogate measures.[37–39] Readers are referred to reviews for more details on this subject.[6,40,41]

In maize, increased partitioning to the grain (or higher HI) has been brought about by using an ideotype selection index that focused on a reduction in anthesis to silking interval.[42] Grain yield of the final selections increased by 108 kg/ha/cycle and came about by an increase in HI, with no change in final biomass relative to the parents.

Wright et al.[43] proposed that estimates of each of the water model components for a peanut crop could be derived from simple and low-cost measurements of total and pod biomass at harvest, from estimates of WUE from correlated measures of specific leaf area, and by reverse-engineering the TDM component of the water model, such that T=TDM/TE. A selection study is being conducted on four peanut populations, in which a selection index approach utilizing T, WUE, and HI is assessing the value of these traits in a large-scale breeding program in India and Australia.[44]

CONCLUSION

Although drought is commonly referred to as "prolonged water deficit" periods during a crop's life, it is indeed a complex syndrome with various climate pressures operating together in infinite combinations, resulting in significant reductions in crop performance. Historically, options for managing agricultural drought have revolved around management techniques such as deep tillage, mulching, intercropping, etc. However, future options for drought management will increasingly be based on the genetic improvement of crops for targeted environments. With genetic options, matching phenology to water availability in target environments has proven to be highly a successful approach. Although our understanding of drought resistance mechanisms has improved significantly, it is essential

Dew—Wind

to distinguish the traits linked to short-term responses from those which are important when integrated over longer time scales.

REFERENCES

1. Schulze, E.D. Adaptation Mechanisms of Non-Cultivated Arid-Zone Plants: Useful Lessons for Agriculture. In *Drought Research Priorities for the Dryland Tropics*; Bidinger, F.R., Johansen, C., Eds.; ICIRSAT: Patancheru, A.P. 502324, India, 1988, 159–177.

2. Passioura, J.B. Drought and Drought Tolerance. In *Drought Tolerance in Higher Plants: Genetical, Physiological, and Molecular Biological Analysis*; Belhassen, E., Ed.; Kluwer Academic Publishers: the Netherlands, 1996; 1–6.

3. Unger, W.P.; Jones, O.R.; Steiner, J.L. Principles of Crop and Soil Management Procedures for Maximising Production per Unit Rainfall. In *Drought Research Priorities for the Dry Land Tropics*; Bidinger, F.R., Johansen, C., Eds.; ICIRSAT: Patancheru, A.P. 502324, India, 1988; 97–112.

4. Natarajan, M.; Willey, R.W. The effects of water stress on yield advantages of intercropping systems. Field Crops Res. **1986**, *13*, 117–131.

5. Nageswara Rao, R.C.; Wadia, K.D.R.; Williams, J.H. Intercropping of short and long duration groundnut genotypes to increase productivity in environments prone to end-of-season drought. Exp. Agric. **1990**, *26*, 63–72.

6. Richards, R.A. Defining Selection Criteria to Improve Yield Under Drought. In *Drought Tolerance in Higher Plants: Genetical, Physiological, and Molecular Biological Analysis*; Belhassen, E., Ed.; Kluwer Academic Publishers: the Netherlands, 1996; 71–78.

7. Levitt, J. *Response of Plants to Environmental Stresses*; Academic Press: New York, 1972.

8. Turner, N.C. Crop water deficits: A decade of progress. Adv. Agron. **1986**, *39*, 1–51.

9. Turner, N.C.; Wright, G.C.; Siddique, K.H.M. Adapatation of grain legumes (pulses) to water-limited environments. Adv. Agron. **2001**, *71*, 193–231.

10. Turner, N.C. Drought Resistance: A Comparison of Two Frameworks. In *Management of Agricultural Drought: Agronomic and Genetic Options*; Saxena, N.P., Johansen, C., Chauhan, Y.S., Rao, R.C.N. Eds.; Oxford and IBH: New Delhi, 2000.

11. Subbarao, G.V.; Johansen, C.; Slinkard, A.E.; Rao, R.C.N.; Saxena, N.P.; Chauhan, Y.S. Strategies for improving drought resistance in grain legumes. Crit. Rev. Plant Sci. **1995**, *14*, 469–523.

12. Jackson, P.; Robertson, M.; Cooper, M.; Hammer, G. The role of physiological understanding in plant breeding; From a breeding perspective. Field Crops Res. **1996**, *49*, 11–37.

13. Siddique, K.H.M.; Loss, S.P.; Regan, S.P.; Jettner, R. Adaptation of cool season grain legumes in Mediterranean type environments of south-western Australia. Aust. J. Agric. Resour. **1999**, *50*, 375–387.

14. Wright, G.C.; Smith, R.C.G.; Morgan, J.M. Differences between two grain sorghum genotypes in adaptation to drought stress. III. Physiological responses. Aust. J. Agric. Resour. **1983**, *34*, 637–651.

15. Flower, D.J.; Ludlow, M.M. Contribution of osmotic adjustment to the dehydration tolerance of water-stressed pigeonpea (*Cajanus cajan* (L.) millsp.) leaves. Plant Cell Environ. **1986**, *9*, 33–40.

16. Davies, W.J.; Tardieu, F.; Trejo, C.L. How do chemical signals work in plants that grow in drying soil? Plant Physiol. **1994**, *104*, 309–314.

17. Morgan, J.M. Possible role of abscisic acid in reducing seed set in water-stressed wheat plants. Nature (Lond.) **1980**, *285*, 655–657.

18. Blum, A. Crop Responses to Drought and Interpretation of Adaptation. In *Drought Tolerance in Higher Plants: Genetical, Physiological, and Molecular Biological Analysis*; Belhassen, E., Ed.; Kluwer Academic Publishers: the Netherlands, 1996; 57–70.

19. Morgan, J.M. Osmoregulation and water stress in higher plants. Annu. Rev. Plant Physiol. **1984**, *35*, 299–319.

20. Jones, M.M.; Rawson, H.M. Influence of the rate of development of leaf water deficits upon photosynthesis, leaf conductance, water use efficiency, and osmotic potential in sorghum. Physiol. Plant. **1979**, *45*, 103–111.

21. Morgan, J.M.; King, R.W. Association between loss of leaf turgor, abscisic acid levels and seed set in two wheat cultivars. Aust. J. Plant Physiol. **1984**, *11*, 143–150.

22. Morgan, J.M.; Condon, A.G. Water use, grain yield and osmoregulation in wheat. Aust. J. Plant Physiol. **1986**, *13*, 523–532.

23. Wright, G.C.; Smith, R.C.G. Differences between two grain sorghum genotypes in adaptation to drought stress. II. Root water uptake and water use. Aust J. Agric. Resour. **1983**, *34*, 627–636.

24. Morgan, J.M. Osmoregulation as a selection criterion for drought tolerance in wheat. Aust. J. Agric. Resour. **1983**, *34*, 607–614.

25. Blum, A.; Mayer, J.; Gozlan, G. Associations between plant production and some physiological components of drought resistance in wheat. Plant Cell Environ. **1983**, *6*, 219–225.

26. Santamaria, J.M.; Ludlow, M.M.; Fukai, S. Contributions of osmotic adjustment to grain yield in *Sorghum bicolor* (L.) Moench under water-limited conditions. I. Water stress before anthesis. Aust. J. Agric. Resour. **1990**, *41*, 51–65.

27. Morgan, J.M.; Rodriguez-Maribona, B.; Knights, E.J. Adaptation to water-deficit in chickpea breeding lines by osmoregulation: Relationship to grain-yields in the field. Field Crops Res. **1991**, *27*, 61–70.

28. Blum, A. *Plant Breeding for Stress Environments*; CRC Press: Boca Raton, FL, 1988.

29. Sinclair, T.R.; Ludlow, M.M. Influence of soil water supply on the plant water balance of four tropical grain legumes. Aust. J. Plant Physiol. **1986**, *13*, 329–341.

30. Chandler, P.M.; Munns, R.; Robertson, M. Regulation of Dehydrin Expression. In *Plant Responses to Cellular Dehydration During Environmental Stress*; Close, T.J., Bray, E.A., Eds.; American Society of Plant Physiologists: Rockville, MD, 1993; 159–166.

31. Close, T.J.; Fenton, R.D.; Yang, A.; Asghar, R.; DeMason, D.A.; Crone, D.E.; Meyer, N.C.; Moonan, F. Dehydrin: The Protein. In *Plant Responses to Cellular Dehydration During Environmental Stress*; Close, T.J., Bray, E.A., Eds.; Amer. Society of Plant Physiology, 1993; 104–118.

32. Hanson, A.D.; Hitz, W.D. Metabolic responses of mesophytes to plant water deficits. Annu. Rev. Plant Physiol. **1982**, *33*, 163–203.

33. Mathews, R.B.; Harris, D.; Williams, J.H.; Nageswara Rao, R.C. The physiological basis for yield differences between four groundnut genotypes in response to drought. II. Solar radiation interception and leaf movement. Exp. Agric. **1988**, *24*, 203–213.

34. Passioura, J.B. The Interaction between Physiology and the Breeding of Wheat. In *Wheat Science—Today and Tomorrow*; Evans, L.T., Peacock, W.J., Eds.; Cambridge University Press: Cambridge, 1981; 191–201.

35. Passioura, J.B. Grain yield, harvest index and water use of wheat. J. Aust. Inst. Agric. Sci. **1977**, *43*, 117–120.

36. Farquhar, G.D.; Richards, R.A. Isotopic composition of plant carbon correlates with water-use efficiency in wheat genotypes. Aust. J. Plant Physiol. **1984**, *11*, 539–552.

37. Nageswara Rao, R.C.; Wright, G.C. Stability of the relationship between specific leaf area and carbon isotope discrimination across environments in peanuts. Crop Sci. **1994**, *34*, 98–103.

38. Masle, J.; Farquhar, G.D.; Wong, S.C. Transpiration ratio and plant mineral content are related among genotypes of a range of species. Aust. J. Plant Physiol. **1992**, *19*, 709–721.

39. Clark, D.H.; Johnson, D.A.; Kephart, K.D.; Jackson, N.A. Near infrared reflectance spectroscopy estimation of 13C discrimination in forages. J. Range Manag. **1995**, *48*, 132–136.

40. Subbarao, G.V.; Johansen, C.; Rao, R.C.N.; Wright, G.C. Transpiration Efficiency: Avenues for Genetic Improvement. In *Handbook of Plant and Crop Physiology*; Pessarakli, M., Ed.; Marcel Dekker: New York, NY, 1994; 785–806.

41. Wright, G.C.; Rachaputi, N.C. Transpiration Efficiency. In *Water and Plants (Biology) for the Encyclopedia of Water Science*; Marcel Dekker, Inc., 2003; 982–988.

42. Fischer, K.S.; Johnson, E.C.; Edmeades, G.O. Breeding and Selection for Drought Resistance in Tropical Maize. In *Drought Resistance in Crops with Emphasis on Rice*; International Rice Research Institute: Los Banos, 1983; 377–399.

43. Wright, G.C.; Rao, R.C.N.; Basu, M.S. A Physiological Approach to the Understanding of Genotype by Environment Interactions—A Case Study on Improvement of Drought Adaptation in Groundnut. In *Plant Adaptation and Crop Improvement*; Cooper, M., Hammer, G.L., Eds.; CAB International: Wallingford, 1996; 365–381.

44. Nigam, S.N.; Nageswara Rao, N.C.; Wright, G.C. In *Breeding for Increased Water-Use Efficiency in Groundnut*, New Millenium International Groundnut Workshop, Shandong Peanut Research Institute, Qingdao, China, Sept. 4–7, 2001; 1–2.

El Niño, La Niña, and the Southern Oscillation

Felicity S. Graham
Wealth from Oceans National Research Flagship, CSIRO Marine and Atmospheric Research, and Institute for Marine and Antarctic Studies, University of Tasmania, Hobart, Tasmania, Australia

Jaclyn N. Brown
Wealth from Oceans National Research Flagship, CSIRO Marine and Atmospheric Research, Hobart, Tasmania, Australia

Abstract

El Niño-Southern Oscillation, ENSO, is a dominant mode of climate variability affecting the Earth. Occurring every 2–7 years, ENSO arises due to air–sea interactions in the tropical Pacific Ocean and is marked by anomalously warm or cool sea-surface temperatures in the central-to-eastern equatorial Pacific. Warm events in the eastern equatorial Pacific are called El Niño events, and cool events are called La Niña events. Climate and weather phenomena associated with ENSO include floods, droughts, heat waves, and tropical cyclones, which can lead to effects on fisheries, agriculture, health, and air quality.

INTRODUCTION

The equatorial Pacific Ocean and atmosphere are the stage for one of the most important phenomena affecting the world's climate: El Niño-Southern Oscillation (ENSO). The term El Niño refers to a period of prolonged warmer-than-usual sea-surface temperatures in the central-to-eastern equatorial Pacific Ocean, typically peaking around December. El Niño is strongly linked to an atmospheric phenomenon known as the Southern Oscillation, part of which is the Walker circulation, a zonal atmospheric cell characterized by ascending motion in the western Pacific and descending motion in the eastern Pacific, connected at the surface by the trade winds. The opposite phase of El Niño, denoted La Niña, is characterized by cooler-than-usual sea-surface temperatures in the central and eastern equatorial Pacific. The whole phenomenon—El Niño, La Niña, and the Southern Oscillation—is known as ENSO and affects the Pacific ocean–atmosphere system on interannual timescales. However, ENSO events have far broader-reaching effects on climate and weather patterns than just the Pacific Ocean. For instance, extreme weather associated with ENSO events range from droughts to increased rainfall and severe flooding.[1–4] Furthermore, ENSO is associated with changes in the incidence of tropical cyclones[5–7] and affects weather and climate in Pacific Island countries,[8] the United States,[9] South America,[10] Australia,[11] and even Europe.[12] The biogeochemical structure of the tropical ocean is altered by ENSO, which in turn affects the habitat availability for fish such as tropical tunas.[13] This entry summarizes the current understanding of ENSO in the context of the physical conditions of the tropical Pacific and its global significance.

WHAT IS ENSO?

The term "El Niño" was originally used to describe warming of a local current along the west coast of South America in December (El Niño in Spanish is "the Child Jesus"). However, it was not until the 1960s that scientists realized El Niño had a much broader impact on sea-surface temperatures than along the Peruvian coast.[10] During El Niño years, there tends to be broad warming of sea-surface temperatures from the International Date Line to the western coast of South America.[10] These warming events occur on an interannual timescale, approximately every 2–7 years, with the event lasting about 1 year. Cool ENSO events called La Niña (Spanish for "the girl child") events, are characterized by basin-scale cooling of sea-surface temperatures. There are many different definitions for what constitutes an ENSO event;[14] however, one common criterion used to define an El Niño is that the 5-month running mean of the sea-surface temperature anomaly in the eastern Pacific in the region bound by 120°W–170°W and 5°S–5°N (denoted Niño-3.4) is at least +0.5°C higher in that region for a period of 6 consecutive months.[15] The converse is true for La Niña. Using this definition, ENSO events can be identified in a time series of the sea-surface temperature averaged in the Niño-3.4 region over the last 60 years (Fig. 1A). Corresponding to the changes in the sea-surface temperature during ENSO events is a shift in atmospheric conditions. The Southern Oscillation is a see-saw oscillation of the surface air pressure between the western and eastern equatorial Pacific with two centers of action over Indonesia in the west and Tahiti in the east.[10] El Niño events and the Southern Oscillation are closely related,[16] with

Dew—Wind

Encyclopedia of Natural Resources DOI: 10.1081/E-ENRA-120047617

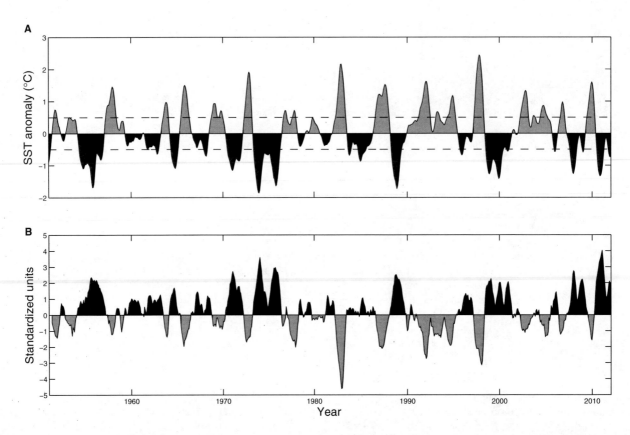

Fig. 1 (**A**) Time series of sea-surface temperature (SST) anomalies from 1951 to 2012 relative to the mean seasonal climatology over the whole period. Data is averaged over the Niño-3.4 region (120°W–170°W and 5°S–5°N). Events with sea-surface temperature anomalies equal to or greater (less) than 0.5°C (–0.5°C) for a period of 6 consecutive months are defined as El Niño (La Niña). (**B**) Time series of the Southern Oscillation Index, based on sea-surface pressure observations at Tahiti and Darwin from 1951 to 2012. The zero corresponds to the mean over the same period. El Niño events have negative units; La Niña events have positive units.
Source: Data provided by the NOAA/OAR/ESRL PSD, Boulder, Colorado, U.S.A, from their website at http://www.esrl.noaa.gov/psd/.

anomalously high (low) sea-level pressures in the west (east) during El Niño years. Increased sea-level pressures over Australia and southeast Asia during El Niño years result in drier-than-normal conditions, whereas in the low-pressure region over western parts of South America, prevailing conditions are wet (Fig. 2A). The reverse is true during La Niña years (Fig. 2B). The Southern Oscillation Index (SOI) is defined as the pressure anomaly at Tahiti minus that at Darwin, Australia (Fig. 1B), and is strongly anticorrelated with Niño-3.4. Understanding the underlying characteristics of the tropical Pacific ocean–atmosphere system is key to understanding why ENSO has such broad-reaching effects on weather and climate around the globe.

MEAN TROPICAL PACIFIC OCEAN–ATMOSPHERE INTERACTIONS

Sea-surface temperatures of approximately 28°C typically occur in the western tropical Pacific, in a body of warm water known as the "warm pool" (Fig. 3A). In the eastern Pacific, a "tongue" of cool water extends west along the equator

with a sea-surface temperature of approximately 22°C. Zonal variations in the depth of the thermocline (the depth of the sharpest change in temperature in the upper ocean) partly drive the differences between eastern and western Pacific sea-surface temperatures. In the western Pacific, the thermocline is relatively deep (on average, ~200 m below the surface), and prevents cooler subsurface waters from mixing with the warm waters above (Fig. 3D). On the other hand, a shallow thermocline in the eastern Pacific allows cooler subsurface waters to be transported upward ("upwelled") into the surface layer where they are readily mixed to bring about the cold "tongue." The atmospheric response to the contrast in sea-surface temperatures along the equatorial Pacific is for regions of low sea-level pressure to be located in the western equatorial Pacific above warm sea-surface temperatures and for regions of high sea-level pressure to be located in the east. The prevailing winds in the equatorial region—the equatorial trade winds—are influenced by the atmospheric pressure contrast ("gradient") and blow toward low-pressure regions where air converges and is convected. As a result, the western equatorial Pacific has a higher incidence of rainfall, thunderstorms, and tropical cyclones than the eastern equatorial Pacific. The trade winds reinforce the east–west

A

El Niño impacts (June - August)

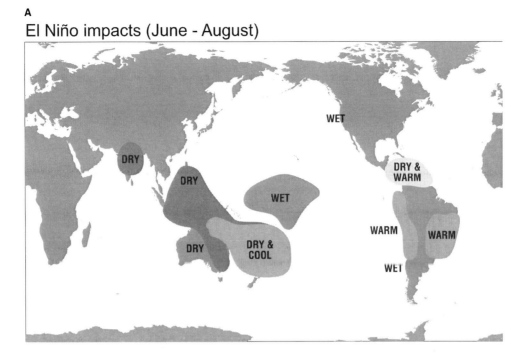

El Niño impacts (December - February)

Fig. 2 (**A**) El Niño regional impacts in June–August (upper panel) and December–February (lower panel). (*Continued*)

difference in sea-surface temperature by piling up warm water in the west and driving upwelling of cool waters in the east and along the equator. It is clear then that atmospheric responses to an anomalously warm oceanic state drive further sea-surface temperature increases in the ocean, through a positive feedback.

Much of our understanding of tropical Pacific ocean–atmosphere interactions has been through the success of the Tropical Ocean Global Atmosphere (TOGA) program

(see the special volume 103 of the *Journal of Geophysical Research*, June 1998, for a comprehensive review). TOGA was initiated in 1985 to study ENSO. The backbone of the program is an array of buoys known as the Tropical Atmosphere-Ocean (TAO) array. Each of the buoys measures temperature in the upper 500 m of the ocean, surface winds, sea-surface temperature, surface air temperature, and humidity, and relay hourly averaged data back to the United States National Oceanic and Atmospheric Administration's

Dew—Wind

B
La Niña impacts (June - August)

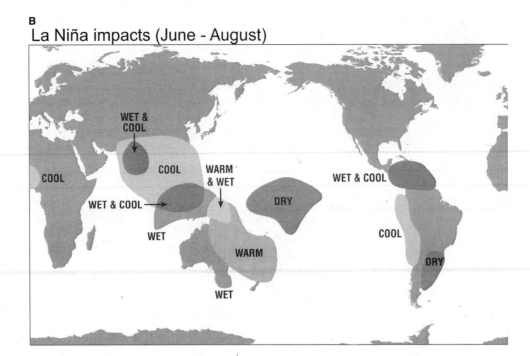

La Niña impacts (December - February)

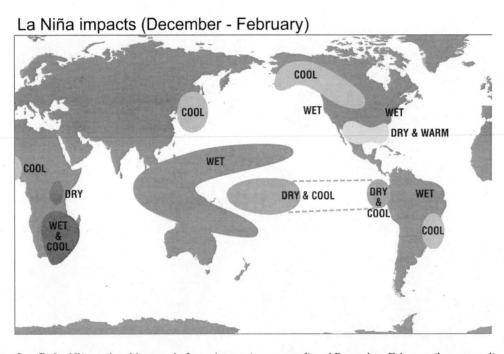

Fig. 2 (*Continued*)   (**B**) La Niña regional impacts in June–August (upper panel) and December–February (lower panel).
Source: Image modified from NOAA/OAR/ESRL PSD, Boulder, Colorado, U.S.A., from their website at http://www.ncdc.noaa.gov/paleo/ctl/clisci10.html, based on research by Halpert and Ropelewski[54] and Ropelewski and Halpert.[55]

(NOAA) Pacific Marine Environmental Laboratory (PMEL). The ENSO observing system also includes a Voluntary Observing Ship (VOS) program, an island tide-gauge network, surface drifters, and satellite data. The coupled nature of the tropical Pacific ocean–atmosphere described in the preceding text is important for the onset and development of an ENSO event, as well as its center of action and magnitude.

ENSO EVOLUTION

No two ENSO events are the same; rather, ENSO events vary considerably in how and where they evolve, their intensity, and the regions that they affect.[17] Furthermore, ENSO and its impacts vary from decade to decade and from generation to generation.[18–20] However, there are key ocean–atmosphere interactions involved in the onset,

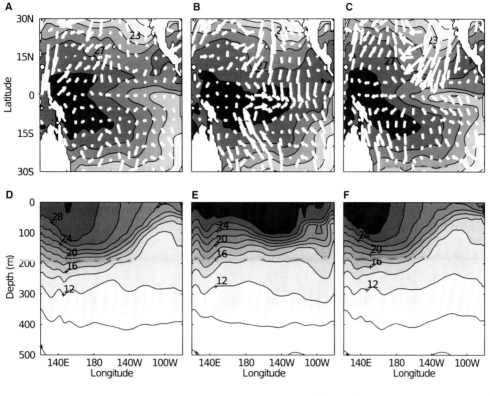

Fig. 3 The contours in the upper panels are of sea-surface temperature anomaly (SST) (in °C) with zonal wind stress (in N m⁻²) overplot-ted in each. The contours in the lower panels represent the subsurface temperature profile (in °C) along the equator. The left-most panels (**A** and **D**) are for December 1996, which was before the peak of the 1997–1998 El Niño event. The middle panels (**B** and **E**) show condi-tions during the peak of the 1997–1998 El Niño in December, and the right-most panels (**C** and **F**) correspond to the peak of the December 1998–1999 La Niña event.
Source: Data prepared by Neville Smith's group at the Australian Bureau of Meteorology Research Centre (BMRC).[56–60]

growth, and termination of an ENSO event. Here we describe these interactions for an El Niño event that evolves in the eastern equatorial Pacific.

Consider a slightly warmer sea-surface temperature in the central-to-eastern equatorial Pacific. This initial anomalous warming reduces the zonal gradient in sea-surface temperatures between the east and west. As a result, the air above the sea-surface also warms, reducing the overlying atmospheric pressure gradient and leading to a change in the atmospheric Walker circulation. The easterly trade winds reduce, moving the center of precipi-tation eastward and deepening the thermocline in the eastern Pacific. A deeper thermocline prevents cooler subsurface waters entering the mixed layer in the eastern Pacific. Consequently, the original warm sea-surface tem-perature anomaly is reinforced, and the associated responses in the atmosphere and ocean circulation are strengthened (Fig. 3B,E).[16] A similar but reversed mech-anism applies for La Niña (Fig. 3C,F), where a slightly cooler sea-surface temperature in the central-to-eastern equatorial Pacific strengthens the trade winds and raises the thermocline in the east. Increased upwelling in the east further cools the sea-surface temperature anomaly, growing the event.

Recently, a modified type of El Niño has been identi-fied, having its strongest signature in the central Pacific.[21] This type of El Niño is termed the central Pacific El Niño or El Niño Modoki (Japanese for "similar but different") and, as its name suggests, it has some dynamical similari-ties to the canonical, eastern Pacific El Niño.[22]

There are a number of mechanisms thought to be impor-tant in the termination of an ENSO event.[23–27] One impor-tant mechanism is the displacement of warm water (above the thermocline or 20°C isotherm).[28,29] During El Niño, the trade winds slacken, and the thermocline deepens in the east, causing the thermocline to flatten along the equator. A flattened thermocline changes the oceanic pressure gra-dient and causes the transport of warm water north and south from the equator. Such a discharge of warm water is sufficient to overcome the ocean–atmosphere interactions that grow an El Niño event, and so the El Niño event begins to decline.[29,30] The discharge mechanism works in reverse during a La Niña event; at the termination of the event, the equatorial Pacific is left in a "recharged" state. Other mechanisms that have been proposed include wave reflec-tion at the western boundary[23,24] and the zonal movement of mean and anomalous currents along with wave reflec-tion at the eastern boundary.[31] Some attempt has been made to unify these mechanisms in a simple model.[32] The characteristics of the onset, development, and termination of El Niño and La Niña events, including their quasi-periodicity, asymmetry, and decadal and generational variability, have been well documented in the literature.[33–36] Leading coupled models are now at a stage where they effectively simulate ENSO dynamics; however, they still suffer from a range of issues including overly cool water along the equator known as the cold tongue bias.[37]

PREDICTING EL NINO EVENTS

El Niño events tend to be followed by a La Niña event (Fig. 1A). The initiation of El Niño events, however, is still a subject of ongoing research. The ocean is capable of sitting in a weak La Niña or neutral ocean state for a number of years without an El Niño occurring.[38] In many situations, a westerly wind burst associated with a tropical cyclone or Madden–Julian Oscillation weather system can provide enough of a reduction in the easterly trades to trigger the positive feedback mechanism to begin generating an El Niño.[39–42]

Seasonal forecasting models are becoming more adept at forecasting El Nino events.[43] It still remains challenging to predict the generation of an El Niño through the March–April–May period known as the "Spring Predictability Barrier."[44] This period is key for making cropping decisions over the Austral winter period if an El Niño is to develop later in the year.[45]

The tropical Pacific Ocean is projected to warm under increasing greenhouse gas scenarios.[46] It is not clear how ENSO will change in response to this increased warming.[47–49] ENSO is expected to continue occurring; however, leading climate models disagree on whether it will be stronger or weaker in amplitude and more or less frequent.

CONCLUSION

ENSO events have broad-reaching impacts: they affect where rainfall occurs, determine the availability of food for fisheries in the eastern Pacific by suppressing or increasing the supply of nutrients, and set up conditions favorable for the development of tropical cyclones. ENSO can have devastating impacts on communities and the economy as well: the 1997–1998 extreme El Niño is estimated to have taken 32,000 lives, displaced 300 million people from their homes, and brought about total losses of US\$ 89 billion.[50] Scientists' understanding of ENSO over the past few decades has progressed significantly through the establishment of the TOGA observing system in the tropical Pacific Ocean, advances in theory,[23,24,29,51,52] and our ability to simulate ENSO,[53] all of which have also enabled early detection of the onset and development of an ENSO event. However, there are still many aspects of ENSO that are

poorly understood; for example, we have limited understanding of how the mechanisms of ENSO might be impacted by climate change. Given the potential impacts of ENSO on the environment, health, and the economy, better understanding and predicting ENSO should be a key driver for future research.

REFERENCES

1. McBride, J.L.; Nicholls, N. Seasonal relationships between Australian rainfall and the Southern Oscillation. Monthly Weather Rev. **1983**, *111*, 1998–2004.
2. Ropelewski, C.F.; Halpert, M.S. Global and regional scale precipitation patterns associated with the El Nino/Southern Oscillation. Monthly Weather Rev. **1987**, *115* (8), 1606–1626.
3. Zhang, X.-G.; Casey, T.M. Long-term variations in the Southern Oscillation and relationships with Australian rainfall. Aust. Meteorol. Mag. **1992**, *40*, 211–225.
4. Andrews, E.D.; Antweiler, R.C.; Neiman, P.J.; Ralph, F.M. Influence of ENSO on flood frequency along the California coast. J. Climate **2004**, *17* (2), 337–348.
5. Nicholls, N. Predictability of interannual variations of Australian seasonal tropical cyclone activity. Monthly Weather Rev. **1985**, *113*, 1144–1149.
6. Chen, G.; Tam, C.-Y. Different impacts of two kinds of Pacific Ocean warming on tropical cyclone frequency over the Western North Pacific. Geophysical Res. Lett. **2010**, *37*, 1–6.
7. Callaghan, J.; Power, S.B. Variability and decline in the number of severe tropical cyclones making land-fall over Eastern Australia since the late nineteenth century. Climate Dynamics **2011**, *37*, 647–662.
8. Australian Bureau of Meteorology and CSIRO. Climate change in the Pacific: Scientific Assessment and New Research. Volume 1: Regional Overview. Volume 2: Country Reports. 2011; http://www.cawcr.gov.au/projects/PCCSP/publications.html#1 (accessed December 2012).
9. Horel, J.D.; Wallace, J.M. Planetary-scale atmospheric phenomena associated with the Southern Oscillation. Monthly Weather Rev. **1981**, *109* (4), 813–829.
10. Philander, S.G. *El Niño, La Niña, and the Southern Oscillation*. Academic Press: San Diego, **1990**.
11. Nicholls, N. Historical El Niño/Southern Oscillation Variability in the Australian Region. In *El Niño, Historical and Paleoclimatic Aspects of the Southern Oscillation*; Diaz, H.F, Margrav, V., Eds.; Cambridge University Press: Cambridge, UK, 1992; 151–174.
12. Fraedrich, K. An ENSO Impact on Europe? Tellus. Series A Dynamic Meteorol. Oceanography **1994**, *46* (4), 541–552.
13. Ganachaud, A.; Sen Gupta, A.; Brown, J.N.; Evans, K.; Maes, C.; Muir, L.C.; Graham. F.S.; Projected changes in the tropical Pacific Ocean of importance to tuna fisheries. Climatic Change **2012**, *115*, 1–17.
14. Meyers, G.; McIntosh, P.; Pigot, L.; Pook, M. The years of El Niño, La Niña, and interactions with the tropical Indian Ocean. J. Climate **2007**, *20*, 2872–2880.
15. Trenberth, K.E. The definition of El Niño. Bull. Am. Meteorol. Soc. **1997**, *78* (12), 2771–2777.

16. Bjerknes, J. Atmospheric teleconnections from the equatorial Pacific. Monthly Weather Rev. **1969**, *97* (3), 163–172.

17. Brown, J.N.; McIntosh, P.C.; Pook, M.J.; Risbey, J.S. An investigation of the links between ENSO flavors and rainfall processes in Southeastern Australia. Monthly Weather Rev. **2009**, *137* (11), 3786–3795.

18. Power, S.; Folland, C.; Colman, A.; Mehta, V. Inter-decadal modulation of the impact of ENSO on Australia. Climate Dynamics **1999**, *15*, 319–324.

19. Power, S.B.; Smith, I.N. Weakening of the Walker Circulation and apparent dominance of El Niño both reach record levels, but has ENSO really changed? Geophysical Res. Lett. **2007**, *34*, GL30854.

20. Wang, B. Interdecadal changes in El Niño onset in the last four decades. J. Climate **1995**, *8*, 267–285.

21. Ashok, K.; Behera, S.K.; Rao, S.A.; Weng, H.; Yamagata, T. El Niño modoki and its possible teleconnection. J. Geophysical Res. **2007**, *112* (C11).

22. Kug, J.-S.; Jin, F.F.; An, S.-I. Two types of El Niño events: cold tongue El Niño and warm pool El Niño. J. Climate **2009**, *22* (6), 1499–1515.

23. Battisti, D.S.; Hirst, A.C. Interannual variability in a tropical atmosphere-ocean model: influence of the basic state, ocean geometry and nonlinearity. J. Atmospheric Sci. **1989**, *46* (12), 1687–1712.

24. Suarez, M.J.; Schopf, P.S. A delayed action oscillator for ENSO. J. Atmospheric Sci. **1988**, *45* (21), 3283–3287.

25. Boulanger, J.P.; Menkes, C.; Lengaigne, M. Role of high- and low-frequency winds and wave reflection in the onset, growth and termination of the 1997–1998 El Niño. Climate Dynamics **2004**, *22* (2–3), 267–280.

26. McPhaden, M.J.; Yu, X. Equatorial waves and the 1997–1998 El Niño. Geophysical Res. Lett. **1999**, *26* (19), 2961–2964.

27. Harrison, D.E.; Vecchi, G.A. On the termination of El Niño. Geophysical Res. Lett. **1999**, *26* (11), 1593–1596.

28. Wyrtki, K. Water displacements in the Pacific and the genesis of El Niño cycles. J. Geophysical Res. Biogeosci. **1985**, *90* (C4), 7129–7132.

29. Jin, F.-F. An equatorial ocean recharge paradigm for ENSO. Part I: conceptual model. J. Atmospheric Sci. **1997**, *54*, 19.

30. Meinen, C.S.; McPhaden, M.J. Observations of warm water volume changes in the equatorial pacific and their relationship to El Niño and La Niña. J. Climate **2000**, *13*, 9.

31. Picaut, J. An advective-reflective conceptual model for the oscillatory nature of the ENSO. Science **1997**, *277*, 663–666.

32. Wang, C. A unified oscillator model for the El Niño – Southern Oscillation. J. Climate **2001**, *14*, 98–115.

33. Rasmusson, E.M.; Carpenter, T.H. Variations in tropical sea surface temperature and surface wind fields associated with the Southern Oscillation/El Niño. Monthly Weather Rev. **1982**, *110*, 354–384.

34. Harrison, D.E.; Larkin, N.K. El Niño-Southern Oscillation sea surface temperature and wind anomalies, 1946–1993. Rev. Geophysics (1985) **1998**, *36* (3), 353–399.

35. Larkin, N.K.; Harrison, D.E. ENSO warm (El Niño) and cold (La Niña) event life cycles: ocean surface anomaly patterns, their symmetries, asymmetries, and implications. J. Climate **2002**, *15* (10), 1118–1140.

36. Dommenget, D.; Bayr, T.; Frauen, C. Analysis of the nonlinearity in the pattern and time evolution of El Niño-Southern Oscillation. Climate Dynamics **2012**, *14*, 1457.

37. Brown, J.N.; Sen Gupta, A.; Brown, J.R., Muir, L.C.; Risbet, J.S.; Whetton, P.; Zhang, X.; Ganachaud, A.; Murphy, B.; Wijffels. S.E. Implications of CMIP3 model biases and uncertainties for climate projections in the western tropical Pacific. Climatic Change **2012**, 1–15.

38. Kessler, W.S. Is ENSO a cycle or a series of events? Geophysical Res. Lett. **2002**, *29* (23).

39. Luther, D.S.; Harrison, D.E.; Knox, R.A. Zonal winds in the central equatorial Pacific and El Niño. Science **1983**, *222*, 327–330.

40. Lengaigne, M.; Guilyardi, E.; Boulanger, J.-P., Menkes, C.; Delecluse, P.; Innesse, P.; Cole, J.; Slingo, J. Triggering of El Niño by westerly wind events in a coupled general circulation model. Climate Dynamics **2004**, *23* (6), 601–620.

41. McPhaden, M.J.; Zhang, X.B.; Hendon, H.H.; Wheeler, M.C. Large scale dynamics and MJO forcing of ENSO variability. Geophysical Res. Lett. **2006**, *33* (16).

42. Latif, M.; Biercamp, J.; Von Storch, H. The response of a coupled ocean-atmosphere general circulation model to wind bursts. J. Atmospheric Sci. **1988**, *45* (6), 964–979.

43. Wang, W. Simulation of ENSO in the new NCEP coupled forecast system model (CFS03). Monthly Weather Rev. **2005**, *133* (6), 1574–1593.

44. Webster, P.J.; Yang, S. Monsoon and ENSO: selectively interactive systems. Q. J. R. Meteorol. Soc. **1992**, *118* (507), 877–926.

45. McIntosh, P.C.; Pook, M.J.; Risbey, J.S.; Lisson, S.N.; Rebbeck, M. Seasonal climate forecasts for agriculture: towards better understanding and value. Field Crops Res. **2007**, *104*, 130–138.

46. IPCC. Climate Change 2007: the physical science basis. In *Series Climate Change 2007: the Physical Science Basis*; Solomon, S., Qin, D., Manning, M., et al., Eds.; Cambridge University Press: Cambridge, United Kingdom and New York, NY, USA, 2007; http://www.ipcc.ch/publications_and_data/publications_ipcc_fourth_assessment_report_wg1_report_the_physical_science_basis.htm (December 2012).

47. Vecchi, G.A.; Wittenberg, A.T. El Niño and our future climate: where do we stand? Wiley Interdiscip. Rev. **2010**, *1* (2), 260–270.

48. Guilyardi, E.; Bellenger, H.; Collins, M.; Ferrett, S.; Cai, W, Wittenberg, A. A first look at ENSO in CMIP5. CLIVAR Exchanges No. 58 **2012**, *17* (1), 29–32.

49. Collins, M.; An, S.-I.; Cai, W.; Ganachaud, A.; Guilyardi, E.; Jin, F.-F.; Jochum, M.; Lengaigne, M.; Power, S.; Timmermann, A.; Vecchi, G.; Wittenberg, A. The impact of global warming on the tropical Pacific ocean and El Niño. Nat. Geosci. **2010**, *3*, 391–397.

50. Trenberth, K.E. The extreme weather events of 1997 and 1998. Consequences **1999**, *5* (1), 3–15.

51. Neelin, J.D.; Battisti, D.S.; Hirst, A.C.; Jin, F.-F.; Wakata, Y.; Yamagata, T.; Zebiak, S.E. ENSO Theory. J. Geophysical Res. **1998**, *103* (C7), 14261–14314.

52. Power, S.B. Simple analytic solutions of the linear delayed-action oscillator equation relevant to ENSO theory. Theor. Appl. Climatol. **2011**, *104* (1), 251–259.

Dew—Wind

53. Zebiak, S.E.; Cane, M.A. A model EI Nino-Southern Oscillation. Monthly Weather Rev. **1987**, *115* (10), 2262–2278.

54. Halpert, M.S.; Ropelewski, C.F. Surface temperature patterns associated with the Southern Oscillation. J. Climate **1992**, *5* (6), 577–593.

55. Ropelewski, C.F.; Halpert, M.S. Precipitation patterns associated with the high index phase of the Southern Oscillation. J. Climate **1989**, *2* (3), 268–284.

56. Smith, N.R. An improved system for tropical ocean subsurface temperature analyses. J. Atmospheric Oceanic Technol. **1995**, *12* (4), 850–870.

57. Smith, N.R. Objective quality controls and performance diagnostics of an oceanic subsurface thermal analysis scheme. J. Geophysical Res. Biogeosciences **1991**, *96*, 3279–3287.

58. Smith, N.R. The BMRC ocean thermal analysis system. Aust. Meteorol. Mag. **1995**, *44*, 93–110.

59. Smith, N.R.; Blomley, J.E.; Meyers, G.A Univariate statistical interpolation scheme for subsurface thermal analyses in the tropical oceans. Prog. Oceanography **1991**, *28* (Pergamon), 219–256.

60. Meyers, G.; Phillips, H.; Smith, N.R.; Sprintall, J. Space and time scales for optimum interpolation of temperature - tropical Pacific ocean. Prog. Oceanography **1991**, *28* (Pergamon), 189–218.

BIBLIOGRAPHY

1. Clarke, A.J. *An Introduction to the Dynamics of El Niño Southern Oscillation*. Academic Press. **2008**.

2. Fedorov, A.V.; Brown, J.N. Equatorial waves. In *Encyclopedia of Ocean Sciences, Second Edition*; Steele, J., Ed.; Academic Press, **2009**, 3679–3695.

3. McPhaden, M.J.; Busalacchi, A.J.; Cheney, R., Donguy, J.-R.; Gage, K.S.; Halpern, D.; Ji, P.; Meyers, G.; Mitchum, G.T.; Niiler, P.P.;Picaut, J.; Reynolds, R.W.; Smith, N.; Takeuchi, K. The tropical ocean-global atmosphere observing system: a decade of progress. J. Geophysical Res. **1998**, *103* (C7), 14169–14240.

4. Neelin, J.D.; Battisti, D.S.; Hirst, A.C.; Jin, F.-F.;Wakata, Y.; Yamagata, T.; Zebiak, S.E. ENSO theory. J. Geophysical Res. Biogeosci. **1998**, *103* (c7), 14261–14290.

5. Wang, C.; Picaut, J. Understanding ENSO physics—a review. *Earth's Climate: The Ocean-Atmosphere Interaction*; Wang, C., Xie, S.-P., Carton, J.A., Eds.; AGU: Washington, D.C., **2004**; 147, 21–48.

Floods: Riverine

William Saunders
Alison MacNeil
Edward Capone
Northeast River Forecast Center, National Oceanic and Atmospheric Administration (NOAA), Taunton, Massachusetts, U.S.A.

Abstract
Riverine flooding occurs whenever the flow of water in a river channel exceeds its carrying capacity. These floods are categorized as minor, moderate, or major, depending on their severity and impacts. Total damage from flooding in the United States averages over $8 billion annually. Surface runoff is the most common factor contributing to riverine floods and occurs whenever the amount or intensity of precipitation exceeds the soil's capacity to accommodate the runoff water. Snowmelt is another factor that typically leads to riverine flooding in the late winter or early spring. Failures of dams or other structures are far less common, but riverine flooding from these events can be significantly more devastating than floods from naturally occurring events. Flooding statistics, such as annual exceedance probability, are commonly used to assess the magnitude of any specific flood event. To provide advanced notice of potential riverine flooding in the United States, River Forecast Centers of the National Weather Service regularly execute watershed and river analysis computer models to establish predictive forecasts.

INTRODUCTION

Riverine flooding occurs whenever a river overtops its banks and water begins to spread laterally away from the river channel (Fig. 1). This happens when the flow within the river channel exceeds the channel's capacity to carry water downstream. Primary factors that result in riverine flooding include (a) an overabundance of rainfall and surface runoff in the drainage area (or "watershed") upstream of the flood location, (b) a quickly melting snowpack in the watershed, (c) failures of riverine structures, such as dams, resulting in a sudden release of stored water, and (d) any combination of the above.

Categories based on the severity of flooding in a particular area have been defined by the National Weather Service (NWS), in coordination with local emergency management officials nationwide. The flood categories are defined as follows:

- **Minor Flooding:** Minimal or no property damage, but possibly some public threat (e.g., minor inundation of roads).
- **Moderate Flooding:** Some inundation of structures and roads near stream. Some evacuations of people or transfer of property or both to higher elevations.
- **Major Flooding:** Life-threatening. Extensive inundation of structures and roads. Significant evacuations of people or transfer of property or both to higher elevations.

River flooding occurs at different rates for different locations. Slow and steady river rises to flood thresholds commonly occur on larger rivers that have more significant drainage areas and wider channels with greater assimilative capacity. Conversely, more rapid flooding events, or "flash floods," generally occur on smaller rivers that drain smaller watersheds. Other characteristics of locations that are prone to frequent flash flooding include (a) steeply sloping upstream river channels and (b) a high percentage of low-permeable soils in the contributing watershed. Flash floods are common in mountainous regions, where exposed rock formations inhibit the infiltration of rainfall into the soil and tributary streambeds are steep. Flash floods are also common in urban areas, where much of the land cover is developed with impermeable surfaces (e.g., roadways, parking lots, and roof structures).

According to the National Oceanic and Atmospheric Administration (NOAA), annual flood damages in the United States average over $8 billion (historical costs adjusted for 2012 inflation) and flood-related fatalities average over 90 persons per year.[1] Annual damages in any one year can be much higher, (e.g., flood damages over $52.5 billion in 2005). While these costs are impressive, it should be noted that floods from excessive rainfall and snowmelt are naturally occurring phenomena that can also bring tangible benefits, such as ground water recharge, soil fertilization, nutrient replenishment, and maintenance of biodiversity in riparian areas adjacent to river channels.

Dew—Wind

Encyclopedia of Natural Resources DOI: 10.1081/E-ENRA-120047656

Different rivers and watersheds typically demonstrate unique responses to the atmospheric and hydrologic stresses placed on them (although some regional commonalities can be observed). As such, understanding how and when riverine floods occur is critical to the accurate prediction of future floods, the mitigation of flood-related damages, and the efficient realization of flood-related benefits.

RIVERINE FLOODING DUE TO SURFACE RUNOFF

Surface water runoff is one step in the hydrologic cycle (Fig. 2) and is a leading cause of riverine flooding. During precipitation events, water will infiltrate the soil, evaporate, or flow directly overland to river channels. A number of

Fig. 1 Overtopping river with lateral flooding.
Source: Adapted from NWS Central Region, http://www.crh. noaa.gov/cys/floodweekWY.php?wfo=riw.

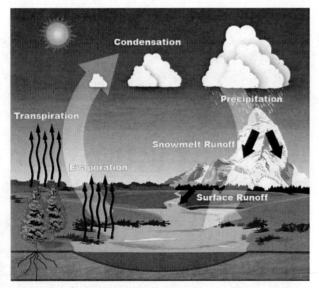

Fig. 2 Hydrologic cycle.
Source: Adapted from NWS Southern Region, http://www.srh. noaa.gov/jetstream/atmos/hydro.htm.

factors determine the amount of surface runoff generated, including rainfall intensity, duration, antecedent soil moisture conditions, topography, soil type, vegetation, and land use. Other factors, such as the presence of lakes and reservoirs, can help delay surface runoff from reaching the river channels and decrease flooding intensity.

Surface runoff occurs when the rainfall intensity is greater than the infiltration rate of the soil or when the amount of rainfall exceeds the storage capacity of the soil or both. Excess water from these conditions then flows over the land surface toward river channels (this process is called overland flow). The infiltration rate of water into the soil is dependent on the soil's porosity and permeability, gravity, and capillary action, with porosity and permeability being the most important. Highly porous soils, such as sand, allow water to infiltrate rapidly and surface runoff rarely occurs. Soils with low porosity and/or permeability, such as clay, have very slow infiltration rates and surface runoff can occur regularly.

Urbanization enhances flood potential by increasing the amount of surface runoff that occurs due to the impermeability of man-made surfaces, such as pavement, that inhibit water from infiltrating into the soil below. In colder regions, frozen soil can also increase the amount of surface runoff by preventing water from infiltrating into the soil column.

Antecedent soil moisture content is a measure of soil "wetness" prior to the onset of precipitation and is another factor that significantly impacts the amount and timing of surface runoff. How quickly the soil becomes saturated depends on the antecedent soil moisture content. The higher the moisture content at the start of a rainfall event, the faster the soil becomes saturated. Once the soil column is saturated, any additional rainfall results in surface runoff.

Not all surface runoff events result in riverine flooding. Flooding begins to occur once the channel capacity of a river is exceeded. The excess water then spills over the banks of the channel and spreads outward laterally into the floodplain. The amount of water in the river channel prior to a heavy rainfall event plays a significant role in whether flooding will occur. It takes more runoff at low flow or base flow conditions to cause flooding than for a river with elevated flow from recent rainfall.

SNOWMELT FLOODING

A riverine flood is deemed a snowmelt flood when the primary cause of the flood is from melting snow. Snowpacks have the potential to store large quantities of water for variable lengths of time, from days to months. Before a snowpack begins to warm up and "ripen," it can also act like a sponge and absorb large quantities of rainfall under certain conditions. Typically, snowmelt runoff occurs once the snowpack reaches 32°F and enough melt has occurred to saturate the snowpack. Once the snowpack becomes saturated, any additional melt is released and is then either absorbed by the soil column or becomes surface runoff, flowing downslope toward river and stream channels. Flooding then occurs as it does for surface runoff flooding, i.e., when the infiltration rate or storage capacity of the soil or both are exceeded, and river channels have exceeded channel capacity.

Six factors typically contribute to snowmelt flooding in winter and spring:[2]

- **High antecedent soil moisture conditions prior to snowmelt:** After the growing season has ended, heavy rains increase the potential for snowmelt flooding in the spring, because less evapotranspiration occurs, and there is less time for the soil to drain and dry before it freezes or a snowpack begins to form.
- **Ground frost or frozen soil:** A deep, hard frost layer will inhibit snowmelt from infiltrating into the soil, thus generating increased amounts of surface runoff during rainfall and snowmelt events. Frozen soil column conditions are aided by cold temperatures prior to a heavy snowfall and normal-to-above normal soil moisture. A heavy snowfall on top of a shallow ground frost will also insulate the ground and help prevent deep frosts from occurring.
- **Heavy winter snow cover:** Heavy snowfall naturally increases the amount of water typically stored and available for snowmelt. Widespread heavy snow cover also keeps air temperatures cooler and enhances nighttime radiational cooling, which extends cooler temperatures and delays spring warming. These conditions increase the potential for heavy rain events to coincide with snowmelt, leading to more rapid snowmelt in the later spring.
- **Widespread heavy rains during the snow season and melt period:** Heavy rain on a fresh, cold snowpack gets absorbed by the pack and increases the amount of water available during flooding. A warm, heavy rain also warms up cold snowpacks, causing them to begin melting earlier than normal. Heavy rain on a ripe snowpack enhances the immediate snowmelt rate. Some of the most devastating floods have occurred during "rain on ripe snowpack" events.
- **Rapid snowmelt:** Typically, the melting of large snowpacks occurs over an extended period of time, from days to weeks and even months. Snowmelt in the absence of rainfall follows a diurnal pattern along with air temperature, with snowmelt rates increasing during the day and then dropping off at night as temperatures cool. Rapidly melting snowpacks occur under certain conditions, such as during unusually warm periods with high dew point temperatures (humidity), increased wind speeds, and when nighttime temperatures remain above freezing.
- **Ice jams in rivers:** Ice jams can occur either during the early winter at the onset of ice formation (freeze-up jams) or during the melting period when ice begins to break up and move downstream. These ice jams act like temporary dams and restrict the flow through the river, thus causing the potential for flooding upstream as the water backs up behind the jam.

FLOODING DUE TO DAM/STRUCTURE FAILURE

The purpose of a dam is to retain or store water for any of several reasons, such as consumptive water supply, irrigation, flood control, energy generation, recreation, pollution and sediment control, and low flow augmentation. Many dams fulfill a combination of these functions.

In the United States, dams are classified by their hazard potential (i.e., high, significant, or low). The failure of a high hazard dam would probably result in loss of human life, economic and environmental losses, and disruption of basic lifeline facilities. A significant hazard dam failure would not be expected to cause loss of human life but would cause economic and environmental losses and disruption of lifeline facilities. Losses due to a low hazard dam failure would generally be limited to the owner of the dam/structure. The nation has approximately 84,000 dams with an overall average age of 52 years. Of these, nearly 14,000 dams are high hazard structures.[3]

Dams can fail with little warning (Fig. 3). While intense storms may produce flash flooding in a few hours, dam break flooding can occur within a half hour for downstream locations closest to the breach. Other catastrophic failures and breaches can take days or weeks to occur, as a result of debris jams, the accumulation of melting snow, or increasing water pressure on a damaged or weakened dam after days of heavy rain. Flooding can also occur when a dam operator releases excess water downstream to relieve pressure on the dam.

Some common causes of dam failure inducing significant riverine flooding include:

- Substandard construction materials/techniques (e.g., Gleno Dam, Pergamo, Italy, 1923)
- Spillway design errors (South Fork Dam, Johnstown, Pennsylvania, USA, 1889)

Fig. 3 Dam breach at Hadlock Pond Dam, Fort Ann, NY.
Source: Adapted from Hom, V. IWRSS Dam Failure Collaboration. National HIC Meeting, July 8–10, 2009. NWS Central Region, http://www.crh.noaa.gov/Image/lmrfc/about/HIC_july_2009/Hom_IWRSS_Dambreak.ppt.

- Geological instability (Malpasset, Cote d'Azur, France, 1959)
- Poor maintenance, especially of outlet pipes (Lawn Lake Dam, Estes Park, Colorado, USA, 1982)
- Extreme inflow (Shakidor Dam, Balochistan, Pakistan, 2005)
- Human, computer, or design error (Buffalo Creek Flood, Logan County, West Virginia, USA, 1972; Dale Dike Reservoir, South Yorkshire, England, 1864)
- Internal erosion, especially in earthen dams (Teton Dam, eastern Idaho, USA, 1976)
- Reservoir volumetric displacement (Vajont Dam, Monte Toc, Italy, 1963)
- Earthquakes

Failures of high hazard dam structures commonly result in floods of much greater magnitude than normal floods due to surface runoff or snowmelt. The area downstream of a dam that would flood in the event of a breach (or failure) is referred to as the "dam breach inundation zone" and is generally much larger than the area that would be flooded for a normal river or stream flood event. The dam breach inundation zones for high hazard dams are frequently larger than the flood zones that are depicted on Flood Insurance Rate Maps of the Federal Emergency Management Agency (FEMA). Some people who live in these zones may be completely unaware of the potential hazard lurking upstream.

COMMON RIVERINE FLOODING STATISTICS

As noted in the introduction, the NWS identifies minor, moderate, and major flood categories associated with discrete elevations at individual forecast sites. These flood categories are based on the specific impacts at the individual locations, (i.e., not United States Geological Survey [USGS] river gage data statistics or FEMA floodplain return period criteria). In the northeastern part of the nation, some forecast locations have seen an increase in the number of minor floods over the past 50 years, while the magnitude of major flooding in the region has also increased. Some studies[4] have indicated that a combination of basin urbanization/suburbanization and possibly climate change are partly responsible for the increase in the number of flood events.

The USGS, FEMA, and many other federal, state, local, and private agencies categorize the magnitude of a flood event according to its annual exceedance probability. The following is the present terminology used for flood magnitudes that replaces the common "nth" year return period terminology:

- 10% annual exceedance probability flood = 10-year flood
- 2% annual exceedance probability flood = 50-year flood
- 1% annual exceedance probability flood = 100-year flood
- 0.2% annual exceedance probability flood = 500-year flood

Accordingly, what was once commonly referred to as a "100-year flood" is now referred to as a 1% annual exceedance probability flood, because it has a 1% probability of occurring in any given year. This change in terminology helps to address a common misunderstanding that a 100-year flood can occur only once in a 100-year period. In fact, there is an approximate 63.4% chance of one or more 1% annual exceedance probability floods occurring in any 100-year period.[5]

The 1% annual exceedance probability flood can be expressed as a river-stage level or a flow rate. For a majority of their river gages, the USGS establishes a "rating curve," which provides a correlation of flow rate to river elevation at the gage location. Using this relationship, the 1% annual exceedance probability flood flow rate can also be expressed as an elevation, which can then be delineated on a detailed map (e.g., USGS quadrangle map) as an area of potential inundation. The resulting mapped floodplain is now referred to as the 1% annual exceedance probability floodplain, which figures prominently for building permits, environmental regulations, and flood insurance.

MONITORING AND FORECASTING OF RIVERINE FLOODS IN THE UNITED STATES

To monitor existing river conditions and deliver accurate predictions of future riverine flooding, the NWS employs watershed and river analysis computer models to help provide daily forecasts of water surface elevations and river flows at many locations across the United States. The models, which are used by hydrologists at the NWS River Forecast Centers (RFCs), incorporate observations and future predictions of precipitation and temperature, soil moisture and snowpack conditions, and routed flows from upstream locations and reservoir operations. The models are executed at least once per day and more frequently during high flow conditions.

The watershed and river analysis computer models are applied over a period of time that includes the recent past and the near future. Models are adjusted to ensure that simulated flows and elevations align with the recent hydrologic conditions and observations. By achieving this agreement between simulations and observations, modelers also establish better confidence that the near-future portion of the simulation provides an accurate forecast. The adjusted model results are then disseminated to Weather Forecast Offices (WFOs), which are, in turn, responsible for issuing flood watches and warnings. Other users of river forecast information include federal and state Emergency Management Agencies, municipal water providers, and the general public.

River forecasts are distributed in the form of hydrographs, which are graphical plots of elevation (or flow) vs. time. The NWS-generated hydrographs (e.g., Fig. 4) can be found at the Advanced Hydrologic Prediction Service (AHPS) website.[6]

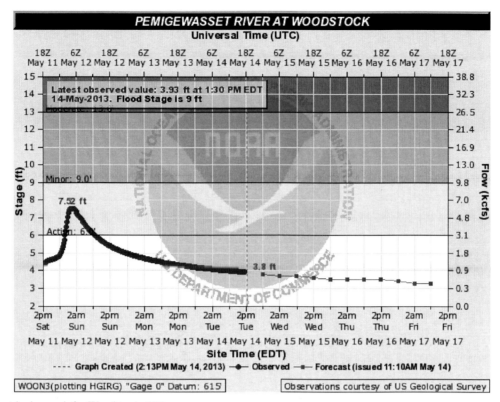

Fig. 4 Forecast hydrograph for Woodstock, NH.
Source: Adapted from NWS AHPS, http://water.weather.gov/ahps2/index.php?wfo=gyx.

FUTURE OF FLOOD FORECASTING IN THE UNITED STATES

Current RFC operational procedures require at least daily issuance of forecasts for most locations. Using the current suite of watershed and river analysis models, these forecasts are generally limited to discrete locations that demonstrate monodirectional flow. In the future, as computer processing times decrease, more complicated algorithms (such as those in hydrodynamic models) may be added to RFC operational modeling suites, allowing for the expansion of forecasting services to more points, particularly at tidally influenced locations. These advanced models will also allow for operational simulation of in-line and lateral structures (such as dams and levees), so that their potential failure consequences may be considered as part of a real-time forecast.

An NWS effort is currently underway to couple discrete in-stream forecast locations with detailed near-shore terrain models. Automated inundation maps are generated through this coupling, based on the forecast elevations from the model simulations. As this coupling gets established for more forecast locations nationwide, the inundation maps may become part of the routine forecasts.

Finally, the future of riverine forecasting within the NWS also includes the use of multiple atmospheric and hydrologic models that can produce a family of hydrologic simulations, rather than a single deterministic forecast. These simulation families will comprise an ensemble forecast, which will include an envelope of maximum and minimum elevation values, as well as higher probability simulations in between. This Hydrologic Ensemble Forecast System (HEFS) will also expand the forecast product to include probabilistic river forecasts over a time horizon of days to months to one year. These ensemble forecasts will help local emergency managers to plan for multiple scenarios of an upcoming event and for longer-term (e.g., seasonal) contingencies.

REFERENCES

1. National Oceanic and Atmospheric Administration (NOAA) National Weather Service (NWS). Hydrologic Information Center – Flood Loss Data. http://www.nws.noaa.gov/hic/ (accessed April 2013).
2. NOAA NWS. Snowmelt Flooding. http://www.floodsafety. noaa.gov/snowmelt.shtml (accessed April 2013).
3. American Society of Civil Engineers. 2013 Report Card for America's Infrastructure: Dams. http://www.infrastructurere-portcard.org/a/#p/dams/overview (accessed April 2013).
4. Vogel, R.M.; Yaindl, C.; Walter, M. Nonstationarity: Flood magnitude and recurrence reduction factors in the United States. J. Am. Water Resour. Assoc. **2011**, *47* (3), 464–474.
5. California Department of Water Resources. California Water Plan Update 2005, Bulletin 160-05, Volume 4; California Department of Water Resources, Sacramento, CA, 2005; 633–634.
6. NOAA NWS. Advanced Hydrologic Prediction Service. http:// water.weather.gov/ahps/forecasts.php (accessed April 2013).

Fronts

Jesse Norris
David M. Schultz
Centre for Atmospheric Science, School of Earth, Atmospheric and Environmental Sciences, University of Manchester, Manchester, U.K.

Abstract
A front is a boundary between different air masses in the atmosphere and, as such, constitutes a region of strong temperature gradient. Fronts may also be associated with changes in wind speed and direction, water vapor content, pressure, clouds, and precipitation. Fronts slope with height over the more stable air mass. Fronts form from large-scale wind patterns that increase the horizontal temperature gradient, resulting in rising warm air and sinking cold air. Most fronts are associated with low-pressure systems. There are three types of fronts that occur near the Earth's surface: cold, warm, and occluded fronts. In addition, upper-level fronts can be connected to the tropopause, associated with the strong winds of the jet stream. Smaller-scale fronts associated with sea breezes, land breezes, and convective storms also occur. The dryline is a front-like phenomenon unique to the southern and central United States. Fronts can also be associated with heavy precipitation and improving air quality.

INTRODUCTION

A front is a boundary that separates different air masses in the Earth's atmosphere. Frontal passages are usually characterized by a change in temperature, moisture content, and wind direction. Changes in cloud cover and type and precipitation occurrence may also be associated with fronts. The process by which a front forms is called frontogenesis and results in an increase in the horizontal temperature gradient across the front. Associated with frontogenesis, cross-frontal circulations result to maintain thermal wind balance, where the horizontal temperature gradient is proportional to the vertical shear of the horizontal wind parallel to the isotherms (contours of constant temperature). A variety of fronts are possible, including surface-based fronts in association with low-pressure systems (cold, warm, and occluded fronts), upper-level fronts associated with the tropopause, and smaller-scale fronts associated with land breezes, sea breezes, and convective storms.

DEFINITION

A front is a boundary separating different air masses in the Earth's atmosphere. An air mass is a volume of air near the surface of the Earth with relatively homogeneous air temperature, moisture content (e.g., relative humidity, mixing ratio, dewpoint temperature), and static stability (a measure of the atmosphere's resistance to vertical air movements). Air masses usually have a horizontal extent of thousands of kilometers and are defined primarily by their source regions, which are functions of latitude (tropical or polar) and underlying surface (continental or maritime). Thus, fronts are usually characterized by gradients in temperature, humidity, or both. Frontal passages are sometimes also indicated by changes in wind direction, minima in pressure, and changes in the sensible weather (cloud cover and type, precipitation occurrence and rate).

FORMATION

Frontogenesis is the process by which fronts form. It is defined mathematically as an increase in the horizontal temperature gradient over time.[1] The opposite is frontolysis, a decrease in the horizontal temperature gradient over time. A front can form from a broad region of initially weak temperature gradient that is increased as a result of the large-scale wind pattern. Two such wind patterns promote frontogenesis: confluence and horizontal shear (Fig. 1). Confluence is where two or more airstreams originating far from one another meet and merge into a single airstream (for example, airstreams from the southwest and southeast meeting along a north–south-oriented axis and flowing north; Fig. 1A), similar to the manner in which rivers merge at a confluence. If the two airstreams have different temperatures, then a front can form roughly along the axis of confluence.[2] In contrast, horizontal wind shear is where wind speed changes in a direction perpendicular to the flow (for example, wind speed changing from southerly to northerly from west to east). If the isotherms are oriented in a way to be rotated by the shear, then a front may form (Fig. 1B).[3]

Encyclopedia of Natural Resources DOI: 10.1081/E-ENRA-120047648

Dew—Wind

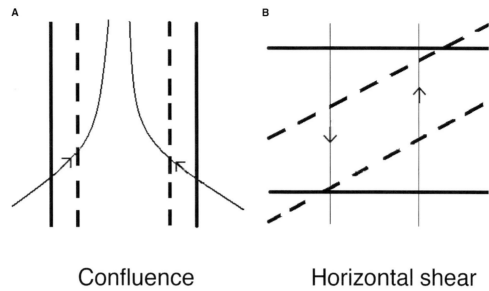

Confluence   Horizontal shear

Fig. 1 Wind patterns favorable for frontogenesis: (**A**) confluence and (**B**) horizontal shear. Thin solid lines represent the streamlines of the flow, thick solid lines represent the temperature contours at an initial time, and thick dashed lines represent the temperature contours at a later time.

CROSS-FRONTAL CIRCULATION

For weather phenomena the size of fronts and larger, the horizontal temperature gradient is proportional to the vertical shear of the horizontal wind parallel to the isotherms, a principle called thermal wind balance. Thus, fronts are accompanied by increasing wind speed with height, which is why surface fronts usually occur in conjunction with the polar-front jet stream, a region of strong winds 5–12 km above the earth's surface.

When frontogenesis occurs, the atmosphere responds in two ways to attempt to maintain thermal wind balance: by increasing the vertical wind shear and by decreasing the horizontal temperature gradient through ascent (and hence adiabatic cooling) of the warm air and descent (and hence adiabatic warming) of the cold air. The resulting horizontal and vertical motions are called the secondary circulation and can be expressed mathematically through the Sawyer–Eliassen equation.[4,5] At the surface, the weak vertical velocities inhibit the frontolytic effects of the adiabatic temperature changes that would weaken the horizontal temperature gradient.

Although most fronts occur near the surface of the Earth, sometimes fronts form aloft near the tropopause.[6] Aloft, the descent of the tropopause due to subsidence along the jet stream can lead to an intensifying horizontal temperature gradient and an upper-level front.

CYCLONES, JET STREAMS, AND FRONTS

Low-pressure systems known as extratropical cyclones develop along the polar-front jet stream in conjunction with meanders in the jet stream called Rossby waves. Around a low-pressure center, the air flows counterclockwise in the Northern Hemisphere and clockwise in the Southern Hemisphere. Confluence and horizontal shear induced by the wind flow around the cyclone cause the isotherms to rotate, contract, and form two surface fronts (stage I in Fig. 2), which are often linked to upper-level fronts. The cold front generally extends equatorward from the cyclone center and marks the boundary of the cold air that is transported around the cyclone from the polar region (stage II in Fig. 2). The warm front generally extends east from the cyclone center and marks the boundary of the warm air that is transported around the low center from the tropics. Fronts typically slope over the colder and more statically stable air mass.

As a low-pressure system deepens, the strengthening winds around the system rotate the cold and warm fronts around the low (Fig. 2). If the low becomes strong enough, these fronts are wrapped up, and the intervening warm air is lifted aloft, forming a boundary between two cold air masses called an occluded front (stages III and IV in Fig. 2). Because warm fronts tend to consist of relatively stable air compared to cold fronts, the cold front often advances over the warm front, forming a warm-type occluded front. Recent research has shown that the warm-type occluded front is the most common structure for occluded fronts.[7] Cyclones with occluded fronts are often intense, with strong surface winds and bands of heavy precipitation embedded within regions of light and moderate precipitation.

WEATHER ASSOCIATED WITH FRONTS

Each of these three principal types of fronts (cold, warm, occluded) is associated with a different sequence of weather. Traditional conceptual models of fronts, however, often oversimplify the variety of observed fronts. Consider cold fronts. By definition, cold fronts are associated with

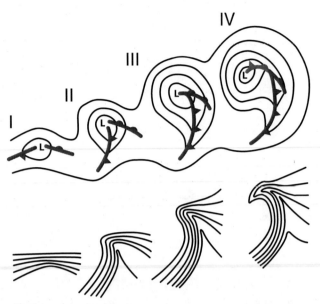

Fig. 2 Four stages in the evolution of an extratropical cyclone in the Northern Hemisphere. The stages are separated by approximately 6–24 hours. Top: thin solid lines represent isobars (contours of constant pressure) at mean sea level and thick solid lines represent fronts using conventional notation (line with triangles represents the cold front, line with semicircles represents the warm front, and line with alternating triangles and semicircles represents the occluded front). Bottom: solid lines represent isotherms (contours of constant temperature) in the lower troposphere near the Earth's surface with warm air to the south and cold air to the north.
Source: Copyright © 2011, American Meteorological Society.

near-surface temperatures decreasing as warm prefrontal air is replaced by advancing cold air behind the front. Yet, pools of cold, stable air in advance of the front may mask the temperature changes occurring above the surface or cause abrupt warming when replaced by the postfrontal air.[8,9] Conceptual models of cold fronts in the Northern Hemisphere often depict the wind shift from southerly to westerly or northerly coincident with the temperature decrease; yet, many cold fronts occur where the cold air arrives after the wind shift has occurred.[10] Finally, cold fronts are often depicted as having steeply sloping leading edges, forcing strong ascent and thunderstorms. In contrast, many cold fronts are not associated with such convective storms or even a change in cloud type or amount.[11,12] Thus, caution should be exercised in generalizing the specific characteristics of the fronts depicted in idealized conceptual models to real fronts.

A cross-section through a warm front, by comparison, is much more gently sloping than that of a cold front. Warm southerly air is generally depicted as gently ascending over cold easterly flow. Thus, a warm front in the Northern Hemisphere is usually characterized by increasing temperatures and a wind shift from easterly to southerly. This gentle ascent is consistent with the traditional conceptual model depicting low clouds and steady precipitation near the surface rising up to high cirrus clouds well in advance of the warm front. In reality, deep convective storms and

regions of heavier precipitation organized in bands may be present in the ascending air.

Occluded fronts are usually depicted as having weak, if any, temperature differences across them. Winds may shift at the frontal passage, but if the occluded front is part of an intense surface cyclone, the shift in wind direction may be minimal. Clouds and precipitation are often not collocated with the surface occluded front. Instead, the heaviest precipitation is associated with warm air aloft that is rising due to the secondary circulations aloft.[13]

OTHER FRONTS AND FRONT-LIKE PHENOMENA

In addition to the fronts associated with extratropical cyclones, other types of fronts occur on smaller scales. These fronts often occur near the boundaries between different land-surface characteristics. A sea-breeze front is one example. The sea-breeze front forms during the daytime in the warm season when cool moist air from over a water body moves onshore, replacing warm dry air inland. The sea breeze can be a focal point for convective storms. At night, the reverse land breeze occurs when cool air over the land moves offshore over the warmer water. The land-breeze front tends to be weaker because the temperature gradient between land and water is usually weaker at night than in the daytime. Other similar front-like behavior can be seen across gradients in vegetation types or soil moisture.

Convective storms also produce front-like phenomena. For example, the leading edge of cool outflow from convective storms can produce a gust front as it moves into the warmer humid air mass ahead of the storm.[14] The lifting along gust fronts can produce new convective cells that merge with the original storm, thus creating a longer-lasting storm.

The dryline of the southern and central United States is not a front, but sometimes has front-like properties. The dryline is a climatological boundary between the moist air originating over the Gulf of Mexico and the dry air originating over the southwestern United States.[15] As such, the dryline is primarily identifiable by the strong gradient in moisture across it. Front-like circulations have been observed across the dryline because the dry air tends to be warmer than the moist air late in the day and the moist air tends to be warmer than the dry air at night. The dryline also tends to be associated with stronger gradients when a synoptic-scale low-pressure system is present.[16] The dryline is often the focus for deep convective storms, which are often associated with hail, strong winds, and tornadoes.

IMPACTS OF FRONTS

The ascent along fronts can produce heavy and sustained precipitation. For example, the ascending warm air of an extratropical cyclone can have its origins deep from within the subtropics, bringing abundant moisture poleward. Long stretches of cloud can extend over thousands of kilometers ahead of the surface cold front. This flow is referred to as the warm conveyor belt and, when this moist flow rises over the warm front or over mountains, heavy precipitation can result. The mountains of the western United States receive much of their cool-season precipitation from such warm conveyor belts (also called the Pineapple Express or atmospheric rivers in some contexts).[17,18] As such, the precipitation associated with fronts can be integral to water management in some locations.

Fronts can also affect the air quality. Pollutants may be ventilated out of the boundary layer, and fresh clean air may replace the polluted air when fronts pass through a region. However, where a layer of cool stable air may persist near the surface, some fronts may be unable to penetrate this layer and freshen the air. Why some fronts can clear the boundary layer in such situations and others cannot is an active research topic.

CONCLUSION

Fronts commonly form in conjunction with the development of extratropical low-pressure systems. They are boundaries between air masses and are associated with changes in temperature, humidity, pressure, and clouds and precipitation. Despite their conceptual simplicity, many fronts do not fit this relatively simple framework. For example, wind shifts and temperature gradients may not be coincident, and mixing across the front means that these boundaries are not rigid surfaces.[10,19] Research to improve our understanding of frontal structure and evolution and their relationship to low-pressure systems is ongoing.

REFERENCES

1. Petterssen, S. Contribution to the theory of frontogenesis. Geofys. Publ. **1936**, *11* (6), 1–27.
2. Bergeron, T. Über die dreidimensional verknüpfende Wetteranalyse I. Geofys. Publ. **1928**, *5* (6), 1–111.
3. Williams, R.T. Atmospheric frontogenesis: A numerical experiment. J. Atmos. Sci. **1967**, *24*, 627–641.
4. Sawyer, J.S. The vertical circulation at meteorological fronts and its relation to frontogenesis. Proc. Roy. Soc. Lond. **1956**, *A234*, 346–362.
5. Eliassen, A. On the vertical circulation in frontal zones. Geofys. Publ. **1962**, *24* (4), 147–160.
6. Keyser, D.; Shapiro, M.A. A review of the structure and dynamics of upper-level frontal zones. Mon. Wea. Rev. **1986**, *114*, 452–499.
7. Schultz, D.M.; Vaughan, G. Occluded fronts and the occlusion process: A fresh look at conventional wisdom. Bull. Amer. Meteor. Soc. **2011**, *92*, 443–466, ES19–ES20.
8. Sanders, F.; Kessler, E. Frontal analysis in the light of abrupt temperature changes in a shallow valley. Mon. Wea. Rev. **1999**, *127*, 1125–1133.
9. Doswell, C.A., III.; Haugland, M.J. A comparison of two cold fronts: Effects of the planetary boundary layer on the mesoscale. Electronic J. Severe Storms Meteor. **2007**, *2*, 1–12.
10. Schultz, D.M. A review of cold fronts with prefrontal troughs and wind shifts. Mon. Wea. Rev. **2005**, *133*, 2449–2472.
11. Mass, C.F.; Schultz, D.M. The structure and evolution of a simulated midlatitude cyclone over land. Mon. Wea. Rev. **1993**, *121*, 889–917.
12. Schultz, D.M.; Roebber, P.J. The fiftieth anniversary of Sanders (1955): A mesoscale-model simulation of the cold front of 17–18 April 1953. *Synoptic–Dynamic Meteorology and Weather Analysis and Forecasting: A Tribute to Fred Sanders,* Meteor. Monogr; No. 55, Amer. Meteor. Soc. 2008, 126–143.
13. Novak, D.R.; Colle, B.A.; Aiyyer, A.R. Evolution of mesoscale precipitation band environments within the comma head of northeast U.S. cyclones. Mon. Wea. Rev. **2010**, *138*, 2354–2374.
14. Charba, J. Application of gravity current model to analysis of squall-line gust front. Mon. Wea. Rev. **1974**, *102*, 140–156.
15. Schaefer, J.T. The life cycle of the dryline. J. Appl. Meteor. **1974**, *13*, 444–449.
16. Schultz, D.M.; Weiss, C.C.; Hoffman, P.M. The synoptic regulation of dryline intensity. Mon. Wea. Rev. **2007**, *135*, 1699–1709.
17. Lackmann, G.M.; Gyakum, J.R. Heavy cold-season precipitation in the northwestern United States: Synoptic climatology and an analysis of the flood of 17–18 January 1986. Wea. Forecast. **1999**, *14*, 687–700.
18. Neiman, P.J.; Ralph, F.M.; Wick, G.A.; Lundquist, J.D.; Dettinger, M.D. Meteorological characteristics and overland precipitation impacts of atmospheric rivers affecting the west coast of North America based on eight years of SSM/I satellite observations. J. Hydrometeor **2008**, *9*, 22–47.
19. Schultz, D.M. Perspectives on Fred Sanders' research on cold fronts. *Synoptic-Dynamic Meteorology and Weather Analysis and Forecasting: A Tribute to Fred Sanders,* Meteor. Monogr.; No. 55, Amer. Meteor. Soc. 2008, 109–126.

BIBLIOGRAPHY

1. Browning, K.A. Conceptual models of precipitation systems. Wea. Forecasting **1986**, *1*, 23–41.
2. Hoskins, B.J. The mathematical theory of frontogenesis. Annu. Rev. Fluid Mech. **1982**, *14*, 131–151.
3. Hoskins, B.J.; Bretherton, F.P. Atmospheric frontogenesis models: Mathematical formulation and solution. J. Atmos. Sci. **1972**, *29*, 11–37.
4. Shapiro, M.A., Grønås, S. Eds.: *The Life Cycles of Extratropical Cyclones;* American Meteorological Society: Boston, 1999; 359.
5. Shapiro, M.A.; Keyser, D. Fronts, jet streams and the tropopause. In *Extratropical Cyclones, The Erik Palmén Memorial Volume;* Newton, C.W.; Holopainen, E.O., Eds.; Amer. Meteor. Soc. 1990, 167–191.

Land–Atmosphere Interactions

Somnath Baidya Roy
Deeksha Rastogi
Department of Atmospheric Sciences, University of Illinois, Urbana, Illinois, U.S.A.

Abstract

Land and atmosphere interactions are driven by continuous exchange of heat, moisture, momentum, and various gases. Land surface heterogeneity due to topography and land cover leads to heterogeneity in land–atmosphere exchanges resulting in various natural phenomena, such as sea breezes, valley winds, and monsoons, at wide ranges of spatial and temporal scales. Manmade land-use/land-cover change due to deforestation, agriculture, and urbanization modify these exchanges and thus affect weather, climate, water, and carbon cycle at multiple scales. Growing demands for food, energy, and natural resources imply that land-use/land-cover change will continue to intensify in the future. Understanding the underlying process of land–atmosphere interactions is critical for predicting the impact of future land-use/land-cover change on the natural environment.

INTRODUCTION

Land and atmosphere are integral components of our natural environment. Interactions between land and atmosphere occur through continuous exchange of heat, moisture, momentum, and various gases (Fig. 1). These exchanges result in various natural phenomena, such as sea breezes and monsoons, at wide ranges of spatial and temporal scales. Human activity such as deforestation, agriculture, and urbanization can modify these exchanges and thus affect weather, climate, water cycle, and other aspects of the natural environment. Hence, land–atmosphere interaction is a topic of interest for scientists and policymakers worldwide.

PROCESSES

The land surface is a major source of energy for the atmosphere. Even though all energy originally comes from the sun, this energy is in the form of shortwaves that cannot be readily absorbed by the atmospheric gases. Some of these solar shortwaves are reflected back but a major part is absorbed by the surface. Some of this energy is used by plants for photosynthesis but the rest heats up the surface. The surface then tries to cool down by transferring energy to the atmosphere by three processes: longwave radiation, conduction, and latent heating. Most of the energy transfer occurs through longwave radiation emitted by the surface. Unlike solar shortwaves, these longwaves are easily absorbed by the atmosphere, especially by clouds and water vapor. The second process is conduction where heat exchange occurs between the surface and the lowest layer of the atmosphere in direct contact with the surface. During the daytime heat flows from the warm surface into the lower

atmosphere and is then carried upward by convection. At night, the surface is cooler than the air above and consequently heat flows downward from the atmosphere into the ground. Finally, energy is also transferred in the form of latent heating by evaporation and transpiration from the ground and vegetation. During evaporation and transpiration, liquid water in the soil and vegetation absorbs heat and is released into the atmosphere in the form of water vapor. The heat carried by the vapors is released into the atmosphere when the vapors condense to form cloud droplets.

The Earth's surface acts as both a source and a sink for moisture in the atmosphere. Moisture is transported as water vapor from the land surface to the atmosphere by evaporation and transpiration. Most of this moisture is returned back to the surface in the form of rain, snow, hail, and freezing rain. A tiny amount of moisture is also transferred back to the surface when some of the water vapor in the lower atmosphere condenses directly onto the surface as dew and frost on cold nights. Some of this moisture seeps into the ground and becomes part of the groundwater system, whereas the rest flow downslope as surface runoff into nearby waterbodies. Even though land covers less than 30% of the Earth's surface, the continuous and vigorous back-and-forth exchange of moisture between land and the atmosphere plays a significant role in the overall water cycle.

The land surface is the primary sink of atmospheric momentum. Friction due to natural and manmade obstacles such as hills, trees, and buildings slow down the wind. During this process kinetic energy in the wind is converted to heat. Friction due to land surface is typically expressed as a function of the surface roughness. Surfaces with tall, rigid densely packed obstacles, such as tropical forests and urban areas, have higher roughness and hence generate stronger friction than relatively smooth surfaces such as sandy deserts.

Encyclopedia of Natural Resources DOI: 10.1081/E-ENRA-120047637

Dew—Wind

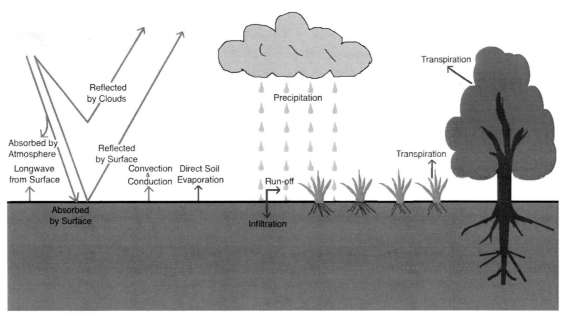

Fig. 1   Schematic diagram representing various processes of energy and water exchange between the land surface and the atmosphere.

The Earth's surface is also a source and a sink of atmospheric gases such as carbon dioxide, oxygen, and nitrogen. Carbon dioxide is absorbed by plants during photosynthesis and released from the surface back into the atmosphere during respiration by plants and animals. This natural carbon cycle has been significantly modified by humans since the Industrial Revolution due to the large amount of carbon dioxide released by fossil fuel burning without a corresponding natural sink. Oxygen is absorbed by plants and animals during respiration and released back into the atmosphere by plants as a byproduct of photosynthesis. Atmospheric nitrogen is absorbed by bacteria in the soil and nitrogen-fixing plants and released back into the atmosphere during decomposition of biomass.

NATURAL PHENOMENA

The land surface is spatially heterogeneous due to variability in topography and vegetation cover. Consequently, the magnitudes of the exchange processes also vary in space. For example, forests absorb more heat but also release more water during evapotranspiration than adjacent grasslands. Such spatial heterogeneity in land–atmosphere interactions result in various natural phenomena such as slope-valley winds, sea breezes, and monsoons.

Slope and Valley Winds

Slope and valley winds are localized wind circulations generated by topography (Fig. 2). They are part of the same circulations system but formed by distinctly different mechanisms. A fundamental difference is that slope winds flow across the valleys, whereas valley winds flow along the valleys. Slope winds are generated by temperature difference between mountain slopes and adjacent valleys. Air near the slope is warmer during the daytime compared with the air at the same altitude away from the slope. This causes air from the valley to move upward along the mountain slope creating the upslope or anabatic wind. The process is reversed in the night when winds flow downward leading to the downslope or katabatic winds.

Valley winds are formed due to temperature difference between narrow valleys and adjacent plains. Due to the topographic constraints, valleys contain a smaller volume of air than open plains. Consequently during the daytime a valley heats up faster leading to a localized low-pressure region over the valley floor. Upslope winds that develop in the morning transport air upward further reducing air pressure in the valley. By early afternoon this low-pressure region becomes strong enough so that air from the adjacent plains flows into the valley in response to the pressure differential. In the evening the valley cools fast and downslope winds bring cold air down creating a high pressure region over the valley floor. In response, valley winds start flowing out into the plains. Valley winds are typically stronger than slope winds and play an important role in the transport of heat, moisture, and pollutants in mountainous regions all over the world.

Sea Breezes

Sea breezes are circulations generated due to land–water temperature contrasts in coastal areas (Fig. 3). Because land has a lower heat capacity than water, the land surface warms up faster than the sea surface. After sunrise, as the land heats up, it heats the air above it. Since warm air is lighter, it rises up, creating a low-pressure region over the land. In response to

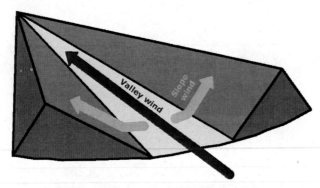

Fig. 2 Typical slope and valley winds during the afternoon.

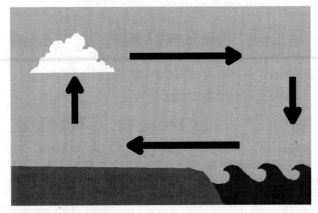

Fig. 3 Circulation patterns and clouds associated with a typical sea breeze.

this pressure gradient, cold air from the sea flows in towards the land resulting in a sea breeze. In the evening, the land cools faster than the sea causing the sea breezes to weaken and then finally die out. If temperature over the land falls below the sea surface temperature, a low pressure is created over the sea, which results in a land breeze that flows from land to sea. The strength of sea/land breeze depends on the temperature contrast between the land and the sea. Strong sea breezes can often generate thunderstorms, especially during the summer in tropical regions. Sea breezes are observed in coastal regions all over the world. An interesting example is the Florida peninsula where sea breezes develop both along the Gulf of Mexico coast in the west and the Atlantic coast in the east. Circulations forced by land–water temperature gradients similar to sea breezes are also observed at the boundaries of large lakes such as the Great lakes in North America and large rivers such as the Amazon where they are known as lake and river breezes, respectively.

Asian Monsoon

Monsoons are atmospheric processes characterized by seasonal reversal of winds. It is widely acknowledged that the Asian monsoon acts like a giant sea breeze. The Asian continental land mass quickly heats up in the summer while the ocean is still cold. This temperature difference causes moist winds to flow inland resulting in significant amount

of rainfall over south and east Asia. This rainfall is a significant source of freshwater for farmers in that region. The landmass starts to cool down in the Fall but the ocean is still warming up. Consequently, dry winds flow out into the ocean from the land. Scientists are currently investigating if similar land–atmosphere coupling exists in other recently discovered monsoon systems such as the North American Monsoon and the West African Monsoon.

MANMADE ACTIVITIES

Rapid population growth and increasing demand for food, housing, energy, and other natural resources is leading to an increase in tropical deforestation, agriculture, and urbanization. These land use changes are changing the magnitudes of land–atmosphere interactions, thereby significantly affecting regional climate, energy, and water resources.

Tropical Deforestation

Deforestation is an acute problem particularly for tropical regions where forests are being cleared to meet the growing demand for food, fuel, timber, and other agricultural products such as textiles and biofuels. By most estimates about half of the world's tropical rainforests have disappeared while the rest are under threat. About 80% of tropical deforestation is driven by commercial and subsistence agriculture with logging for timber also making a significant but smaller contribution. Large-scale conversion of natural forests into farmlands and pastures has severe impacts on the climate and hydrology at local, regional, and even global scales. Perhaps the most important impact of tropical deforestation is that on the global carbon cycle. Forests absorb carbon dioxide during photosynthesis and store it in their roots, stems, and leaves. Forest degradation releases this stored carbon back to the atmosphere thereby contributing to global climate change. Tropical deforestation released approximately 1 PgC/yr during 2000s, representing 6–17% of the total anthropogenic carbon emissions. Deforestation also leads to a suite of changes in land–atmosphere exchange of heat and moisture that act competitively to alter local and regional climate. Deforested lands have higher albedo or reflectivity and usually reflect more sunlight causing a cooling effect. On the other hand, deforestation reduces transpiration resulting in a strong warming that more than offsets the albedo-induced cooling effect. Deforestation also has a strong impact on the water cycle. Trees can access groundwater through their deep roots and release water into the atmosphere through transpiration. Deforestation cuts off this process thereby reducing atmospheric moisture content. Consequently, deforested regions tend to be warmer and drier than nearby forests. In spite of community and government efforts to control tropical deforestation, it is still continuing, albeit at a slower rate. Hence, the environmental impacts of deforestation are likely to persist in the near future.

Agriculture

Agriculture is the dominant driver of manmade land use/land cover change. As a result of continuous agricultural expansion throughout human history, farms have replaced almost 50% of global forests and grasslands. Conversion of forests to farmlands increases albedo leading to a cooling effect. This effect is masked in tropics and leads to a net warming as discussed in the previous section. However, clearing temperate forests for agriculture has a cooling effect because the albedo-induced cooling is much stronger in those regions. Converting grasslands to farmlands does not significantly affect the exchange of heat and momentum because crop albedo and roughness are similar to that of grasses. However, this conversion leads to a drying of the soil and more evapotranspiration due to increased water uptake by crops. Intensive agricultural practices also release stored soil organic carbon into the atmosphere thereby contributing to global warming. Increasing demand for food implies agricultural expansion is not going to slow down in the near future. However, improved agricultural practices and efficient irrigation techniques can reduce its impact on climate to a certain extent.

Urbanization

Driven by economic, political, and social factors, more and more people are moving from rural to urban areas. According to the United Nations, the proportion of human population living in cities has grown from 13% in 1900 to more than 50% today and is likely to go up to 70% by 2050. The expansion of cities and suburbs leads to the conversion of natural vegetation into residential, commercial, and industrial complexes. These changes affect the exchange of mass, momentum, and energy between the surface and the atmosphere. The most well-known impact of urbanization is the urban heat island (UHI) effect where urban areas are significantly warmer than the surrounding regions (Fig. 4). Brick, concrete, and other materials used in building construction and asphalt in the roads absorb more solar radiation than the natural vegetation. Additionally, human activity such as industries and vehicular transport generate more heat than natural ecosystems. The presence of buildings prevents this extra heat from getting radiated back into the atmosphere. Also, due to relative lack of vegetation there is less cooling

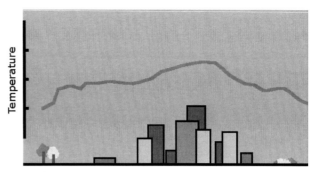

Fig. 4 Typical urban heat island temperature profile showing significant warming over urban areas compared with the surroundings.

due to evapotranspiration. All these processes cause urban areas to be warmer than their surroundings by several degrees. According to a recent NASA study, large cities in the northeastern United States are 13–16°F warmer than the surrounding area during the daytime in the summer due to the UHI effect. Urbanization also affects the surface water balance. Buildings, parking lots, and roads reduce the infiltration of rainwater into the ground resulting in increased surface runoff and reduced ground water recharge. Although urbanization significantly affects the environment in cities and suburbs, it tends to reduce the impact of human land use in rural areas that may revert back to their original natural state. The projected growth of urbanization in the future implies that environmental impacts will continue to intensify.

BIBLIOGRAPHY

1. Laurance, W.F. Reflections on the tropical deforestation crisis, Biol. Conserv. **1999**, *91*, 109–117.
2. Memon, R.A.; Leung, D.Y.C.; Chunho, L. A review on the generation, determination and mitigation of urban heat island, J. Environ. Sci. **2008**, *20*, 120–128.
3. Miller, S.T.K.; Keim, B.D.; Talbot, R.W.; Mao, H. Sea breeze: Structure, forecasting and impacts, Rev. Geophys. **2003**, *41*, 1011.
4. Moutinho, P.; Schwartzman, S. (Eds.), *Tropical Deforestation and Climate Change*; Amazon Institute for Environmental Research: Washington, DC, 2005; 132 pp.
5. Schmidli, J.; Rotunno, R. Mechanisms of along-valley winds and heat exchange over mountainous terrain, J. Atmospheric Sci. **2010**, *67*, 3033–3047.

Dew—Wind

Meteorology: Tropical

Zhuo Wang
Department of Atmospheric Sciences, University of Illinois at Urbana-Champaign, Urbana, Illinois, U.S.A.

Abstract
Tropical meteorology is the study of weather and climate systems in the tropics. The tropics are the source of energy and angular momentum for the global atmospheric circulation. Due to the weak Coriolis force and warm sea surface temperature, the atmospheric processes in the tropics differ from those in higher latitudes in various ways. The major tropical weather and climate systems, including El Niño-Southern Oscillation, monsoons, the Hadley circulation, the MJO and tropical cyclones, are briefly described here.

INTRODUCTION

Tropical meteorology is the study of weather and climate systems in the tropics. There are different definitions of the tropics. One may define the tropics as the region between 23.5°S and 23.5°N, or the region between 30°S and 30°N. In the former definition, the tropics are the region where the sun reaches the zenith twice a year, whereas the latter definition divides the globe into two equal halves, tropics and extratropics. A more meteorological definition of the "tropics" was provided by Riehl,[1] which defined the tropics as "that part of the world where atmospheric processes differ decidedly and sufficiently from those in higher latitudes."

Atmospheric processes in the tropics differ from those in higher latitudes in various ways. In the tropics, the Coriolis force is small (zero along the equator). The quasi-geostrophic theory, which provides a dynamic framework for understanding the large-scale phenomena in midlatitudes, is not readily applicable. Compared with the midlatitudes, the pressure gradient and temperature gradient are generally very weak in the tropics (except for tropical cyclones).[2] Also contrary to the midlatitudes, baroclinic instability does not play an important role in large-scale circulations in the tropics. On the other hand, the tropics are the source of angular momentum and energy for the global atmospheric system. Figure 1 shows the distributions of the zonally averaged absorbed solar flux and the emitted thermal infrared flux at the top of the atmosphere as a function of latitude. There is a surplus of radiative energy in the tropics and a net deficit in middle and high latitudes, which requires on average a poleward transport of energy. The tropics are also characterized by warm sea surface temperature (SST). Moist convection interacts with circulations of various spatial scales in the tropics and plays an important role in driving the global atmospheric circulation system.

Tropical meteorology studies atmospheric circulations of different spatial and temporal scales. Some of the major weather and climate systems in the tropics are described in the following sections.

MAJOR CLIMATE AND WEATHER SYSTEMS IN THE TROPICS

The Hadley Circulation and the ITCZ

The Hadley circulation is the mean meridional circulation in the tropics. It has two thermally direct circulation cells: air rises near the equator, goes poleward near the tropopause (~15–16 km) in both hemispheres, descends in the subtropics (~25–30°N/S), and then moves equatorward near the surface. Due to the Coriolis force, the poleward flow near the tropopause turns eastward and leads to the subtropical westerly jet in the upper troposphere. Similarly, the equatorward flow turns westward and results in the prevailing trade winds in the lower troposphere. The upper-level poleward flow transports energy from the tropics to higher latitudes (meridional energy transport is also carried out by eddy motions and oceanic currents),[4] whereas the trade winds act as accumulators of latent heat[5] and transport moisture equatorward, fueling convection in the rising branch of the Hadley circulation.

The convergence zone in the lower troposphere near the equator is the so-called intertropical convergence zone (ITCZ), otherwise known by sailors as the Doldrums. It is a band of heavy precipitation (Fig. 2), cyclonic relative vorticity, and relatively low surface pressure in the monthly or seasonal mean map. In a snap shot of an infrared image, the ITCZ is manifested as a zone of transient cloud clusters. In contrast, the subtropics are characterized by relatively high surface pressure and arid climates, where evaporation exceeds precipitation and most of the world's large deserts reside.

Encyclopedia of Natural Resources DOI: 10.1081/E-ENRA-120047655

Dew—Wind

The Walker Circulation

The Walker circulation is the east–west overturning circulation in the tropics. It was discovered by the British meteorologist Sir Gilbert Walker. The Walker circulation is driven by the zonal variations of SST and differential heating due to land distribution and topography. The ascending branches of the Walker circulation occur over the western North Pacific-Maritime continent, equatorial South America, and equatorial Africa, and the descending motions occur over the East Pacific, the East Atlantic, and the West Indian Ocean,

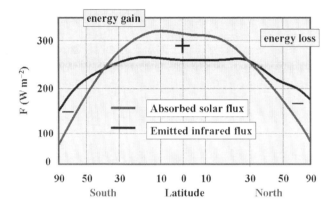

Fig. 1 Zonally averaged components of the absorbed solar flux and emitted thermal infrared flux at the top of the atmosphere. The plus and minus signs denote energy gain and loss, respectively.
Source: Adapted from Vonder Haar and Suomi,[3] courtesy of Roger Smith. (© American Meteorological Society. Used with permission).

where the SST is relatively cold. The most extensive Walker cell is over the Pacific basin, where SST is relatively warm in the west and cold in the east. The western North Pacific-Maritime continent region receives more precipitation than the East Pacific (Fig. 2). The large amount of latent heat release warms the atmospheric column over the western North Pacific and results in the eastward pressure gradient force in the upper troposphere and westward pressure gradient force in the lower troposphere, which lead to eastward winds in the upper troposphere and westward winds in the lower troposphere. The interannual variability of the Walker cell over the Pacific basin is closely related to the SST anomalies over the equatorial central and eastern Pacific, or the El Niño-Southern Oscillation (see the El Niño-Southern Oscillation section).

Monsoons

Monsoon is traditionally defined as the seasonal reversal of the prevailing surface wind over southern Asia and the Indian Ocean. The seasonal variations of the surface wind are accompanied by variations of precipitation: a rainy summer season with onshore flow and a dry winter season with offshore flow. A more general definition of monsoon has been adopted in recent years, which denotes the seasonal variations of wind and precipitation. Besides the Asian–Australian monsoon, the global monsoon system includes the African monsoon, the North American monsoon, and the South American monsoon.[6,7] Compared with the Asian–Australian monsoon, the other monsoon systems are less extensive and may not have a winter counterpart.

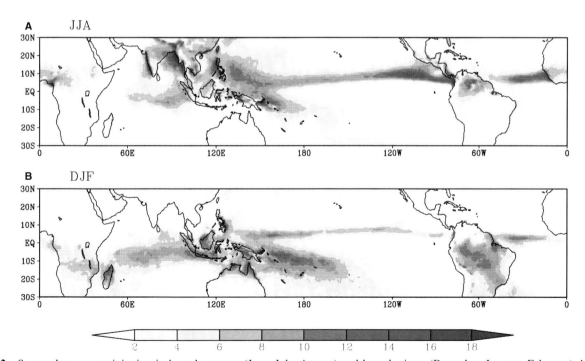

Fig. 2 Seasonal mean precipitation in boreal summer (June–July–August) and boreal winter (December–January–February) derived from TRMM 3B43 (averaged over 1998–2009; units: mm/day).

The large-scale monsoon circulation is driven by seasonally varying, continental-scale distributions of sensible and latent heating and cooling. In particular, the different heat capacity between land and ocean enhances the temperature gradient due to differential solar heating. The land–sea thermal contrast leads to horizontal pressure gradient and drives a regional circulation under the influence of the Coriolis force. Moist convection and cloud radiative processes also modulate the differential heating and provide feedback to the monsoon circulation.[8] At the regional scales, the monsoon circulation and precipitation are affected by local topography and SST gradient.[9,10]

Monsoon is important because it is one of the most energetic components of the global climate system, and also because it has profound social and economical impacts in many tropical regions. On the interannual time scales, the Asian–Australian monsoon is subject to the impacts of ENSO, the Indian dipole pattern, and the tropical tropospheric biennial oscillation (TBO); on the intraseasonal time scales, the Asian–Australian monsoon has active and break cycles in terms of rainfall and is influenced by the Madden–Julian oscillation.[11] Precipitation anomalies of the Asian summer monsoon also have remote impacts over North America.[12]

El Niño-Southern Oscillation

El Niño describes the unusual SST warming over the equatorial eastern and central Pacific. This phenomenon typically occurs in boreal winter around Christmas, and is hence named El Niño (Spanish for "the boy child"). La Niña is the counterpart to El Niño and is characterized by SST cooling across much of the equatorial eastern and central Pacific. The Southern Oscillation (SO) refers to an oscillation in sea-level pressure (SLP) between the tropical eastern and western Pacific, which is closely related to the variability of the Walker circulation. Bjerknes[13] pointed out that the El Niño and the SO are strongly coupled to each other: SST cooling over the equatorial eastern Pacific (i.e., La Niña) would enhance the zonal SST and SLP gradients across the Pacific basin, lead to a stronger Walker circulation and enhance the trade wind easterlies along the equator. The enhanced easterly wind can further amplify the east–west SST gradient through the effects of oceanic advection, upwelling, and thermocline displacement, leading to a positive feedback loop. Conversely, the oceanic warming over the tropical eastern Pacific during an El Niño event is associated with a weaker Walker circulation (the negative phase of the Southern Oscillation) and reduced surface easterly winds, which would in turn further amplify the anomalous warming over the equatorial eastern Pacific. The term El Niño-Southern Oscillation, or ENSO, refers to these coupled atmospheric and oceanic processes.

ENSO is the most important mode of interannual variability in the tropics. Although the strong SST anomalies are confined to the equatorial eastern and central Pacific, ENSO affects the Asian–Australian monsoon and tropical cyclones over both the Pacific and the Atlantic; through teleconnection patterns or modulations of the subtropical jet stream, it also has remote impacts in many extratropical regions.[14–16] As a low-frequency mode (its warming or cooling cycle occurs every 2–7 years), it is an important source of climate predictability in the tropics.

Madden–Julian Oscillation

The Madden–Julian oscillation (MJO), which was first documented by Madden and Julian,[17,18] is the dominant mode of intraseasonal variability in the tropics. It is a quasi-periodic oscillation, with the period varying between 30 and 60 days, and is characterized by the eastward progression of large areas of suppressed and enhanced precipitation (Fig. 3). The MJO originates over the western equatorial Indian Ocean. Over the Indian Ocean–western Pacific warm pool region, perturbations in the zonal wind and surface pressure fields are coupled with convection and propagate eastward with a phase speed 5–8 m/s.[19] East of the dateline, the convective signals of the MJO usually vanish, and the signals in the wind and surface pressure fields continue to propagate eastward at a much faster speed.

The MJO plays a significant role in the tropical weather and climate.[20,21] It has impacts on monsoons,[22,23] and modulates tropical convection and tropical cyclone activity over the East Pacific and the Atlantic.[24,25] Some studies suggest that it plays an active role in the onset and development of ENSO.[26] The MJO also has remote impacts on the midlatitude weather.[27]

Equatorial Waves

Equatorial waves are large-scale waves trapped in the equatorial region, which include equatorial Rossby waves, Kelvin waves, westward and eastward propagating inertiogravity waves, and mixed Rossby–gravity waves. Away from the equator the amplitude of these waves decays rapidly. The seminal paper by Matsuno[28] laid the theoretical foundation for understanding the dynamics of equatorial waves. Using a shallow-water model on an equatorial beta-plane, Matsuno[28] extracted different modes of waves, which were later confirmed by observational analysis. Gill[29] extended this work and interpreted the atmospheric response to steady heating in the tropics in terms of equatorial waves. Recent studies have shown that some types of equatorial waves are coupled with moist convection, which modifies the wave structure and tends to slow down the wave propagation.

Equatorial waves excited by convective heating can affect weather and climate in remote regions. Some low-frequency equatorial waves contribute to the intraseasonal variability in the tropics. Equatorial waves also modulate tropical cyclone activities or serve as synoptic-scale precursors for tropical cyclone formation.[30,31] Oceanic Kelvin waves are believed to play a role in the ENSO dynamics.[32]

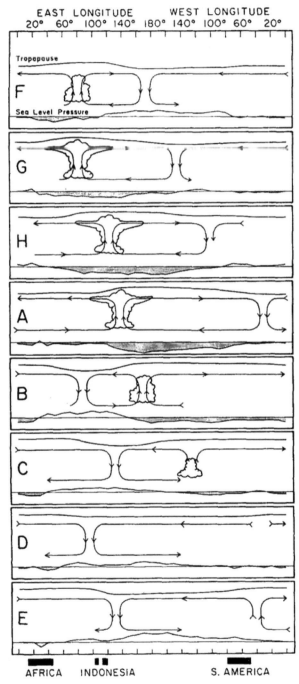

Fig. 3 Longitude-height schematic diagram of the MJO life cycle (from top to bottom). Arrows indicate the zonal circulation, and clouds represent regions of enhanced convection, and the curves below and above the circulation indicate the pressure perturbations in the upper troposphere and at the sea level.
Source: Adapted from Madden and Julian.[18] (© American Meteorological Society. Used with permission).

Tropical Cyclones

A tropical cyclone is a non-frontal, cyclonic circulation (counterclockwise swirling winds in the Northern Hemisphere and clockwise swirling winds in the Southern Hemisphere) in the tropics with organized convection. Based on the maximum sustained surface wind speed, tropical cyclones can be ranked into different categories (Table 1). A tropical cyclone with maximum sustained surface wind speed exceeding 33 m/s is called a hurricane over the Atlantic and the East Pacific, and a typhoon over the western North Pacific.

An intense tropical cyclone usually has an eye, which is a cloud-free region in satellite imagery. The eyewall of a tropical cyclone is a ring of deep convection and is characterized by heavy precipitation and strong surface wind. Outside the eyewall, deep convection tends to organize into long, narrow rainbands that are oriented in the same direction of the horizontal wind. Tropical cyclones have a warm-core vertical structure, and the cyclonic circulation weakens with height. Although the radius of maximum wind is of the order of 100 km, the low-level cyclonic wind extends to much larger radii (up to 1000 km).

Tropical cyclones form over warm oceans, which provide the energy source for the storms through the release of the latent heat of condensation. The formation of a tropical cyclone also requires a pre-existing synoptic-scale disturbance (such as tropical easterly waves), weak vertical wind shear and moist unstable air.[33] Tropical cyclones rarely form within 5° of the equator due to the weak Coriolis force (or the weak planetary vorticity) near the equator.[34] The track of a tropical cyclone is mainly determined by the steering flow and the beta-effect. Tropical cyclones usually move westward or northwestward (southwestward in the Southern Hemisphere) before being carried eastward by the extratropical westerly steering flow. Tropical cyclones

Table 1 Classification of tropical cyclones based on the maximum sustained surface wind speed

Category	Sustained winds
Tropical depression	Less than 38 mph (34 kt or 62 km/hr)
Tropical storm	39–73 mph 35–63 kt 63–118 km/hr
Category 1 hurricane	74–95 mph 64–82 kt 119–153 km/hr
Category 2 hurricane	96–110 mph 83–95 kt 154–177 km/hr
Category 3 hurricane	111–129 mph 96–112 kt 178–208 km/hr
Category 4 hurricane	130–156 mph 113–136 kt 209–251 km/hr
Category 5 hurricane	157 mph or higher 137 kt or higher 252 km/hr or higher

The hurricane categories follow the Saffir–Simpson Hurricane scale.

Dew—Wind

can induce coastal flooding due to storm surges and torrential rainfall, and they may have significant downstream impacts if undergoing extratropical transition.

Deep Cumulonimbus Clouds: Hot Towers

Riehl and Malkus[5] hypothesized that buoyant updrafts in deep cumulonimbus clouds carry energy-rich air from the boundary layer to the upper troposphere with little or no dilution and are essential to the vertical energy transport and the general circulation in the tropics. The updrafts in these convective clouds are referred to as "hot towers." Riehl and Malkus[5] estimated that between 1500 and 5000 hot towers were needed to balance the global heat budget. Recent studies confirmed the role of hot towers in vertical energy transport, but suggested that undiluted convective cores rarely exist in the tropics.[35,36] Although air parcels get diluted by mixing with the drier and low-energy environmental air, they become more buoyant in the upper troposphere due to latent heat release by ice microphysical processes.

Hot towers are also believed to play an important role in the formation and intensification of tropical cyclones. In a vorticity-rich environment, cumulonimbus clouds can generate small-scale, intense, cyclonic vortex tubes, the so-called vortical hot towers.[37] The merger of these vortices leads to the formation of a tropical cyclone proto vortex. The latent heat release associated with these cumulonimbus clouds collectively drives a transverse circulation with low-level inflow, which concentrates the low-level vorticity and intensifies the cyclonic wind.

CONCLUSION

Tropical meteorology is the study of tropical atmospheric processes of various spatial and temporal scales. The tropics are the source of angular momentum and energy for the global atmospheric circulation. Weather and climate systems in the tropics may have remote impacts over extratropical regions. Advances in tropical meteorology in recent decades, aided by satellite observations and modern numerical models, have significantly increased our knowledge of the tropical atmosphere and contributed to improved weather and climate prediction in both the tropics and extratropics.

REFERENCES

1. Riehl, H. *Climate and Weather in the Tropics.* Academic Press, 1979; 611 pp.
2. Sobel, A.H.; Nilsson, J.; Polvani, L.M. The weak temperature gradient approximation and balanced tropical moisture waves. J. Atmos. Sci. **2001**, *58*, 3650–3665.
3. Vonder H.; Thomas H.; Suomi, V.E. Measurements of the earth's radiation budget from satellites during a five-year period. Part I: extended time and space means. J. Atmos. Sci. **1971**, *28*, 305–314.
4. Fasullo, J.T.; Trenberth, K.E. The annual cycle of the energy budget: Meridional structures and poleward transports. J. Clim **2008**, *21*, 10, 2314–2326.
5. Riehl, H.; Malkus, J. On the heat balance in the equatorial trough zone. Geophysica **1958**, *6*, 503–538.
6. Chang, C.-P.; Wang, B.; Lau, N.-C. *The Global Monsoon System: Research and Forecast. Report of the International Committee of the Third International Workshop on Monsoons (IWM-III),* Vol. WMO/TD No. 1266 (TMRP Report No. 70); Secretariat of the World Meteorological Organization, 2005a; 542.
7. Chang, C.-P., Ding, Y.H., Johnson, R.H., Lau, G.N.-C., Wang, B., Yasunari, T., Eds.; *The Global Monsoon System: Research and Forecast,* 2nd Ed.; World Scientific, 2011; 608 pp.
8. Webster, P. The elementary monsoon. In *Monsoons;* Fein, J.S., Stephens, P.L., Eds.; John Wiley & Sons, 1987; 3–32 pp.
9. Chang, C.-P.; Wang, Z.; McBride, J.; Liu, C.H. Annual cycle of Southeast Asia-Maritime Continent rainfall and the asymmetric monsoon transition. J. Clim. **2005b**, *18*, 287–301.
10. Wang, Z.; Chang, C.P. A numerical study of the interaction between the large-scale monsoon circulation and orographic precipitation over South and Southeast Asia. J. Clim. **2012**, *25*, 2440–2455.
11. Pai, D.; Bhate, J.; Sreejith, O.; Hatwar, H. Impact of MJO on the intraseasonal variation of summer monsoon rainfall over India. Clim. Dyn. **2009**, doi:10.1007/s00382-009-0634-4.
12. Wang, Z. *Summertime Teleconnections Associated with U.S. Climate Anomalies and Their Maintenance.* Ph.D. Dissertation, Dept. of Meteorology, University of Hawaii: Honolulu, 2004; 195 pp.
13. Bjerknes, J. Atmospheric teleconnections from the equatorial Pacific. Mon. Wea. Rev. **1969**, *97*, 163–172.
14. Gray, W.M. Atlantic seasonal hurricane frequency. Part I: El Niño and the 30 mb quasi-biennial oscillation influences. Mon. Wea. Rev. **1984**, *112*, 1649–1668.
15. Lau, N.-C.; Wang, B. Interactions between Asian monsoon and the El Nino-Southern Oscillation. In *The Asian Monsoon*; Wang, B. Ed.; Springer/Praxis Publishing: New York, 2006; 478–512 pp.
16. Wang, Z.; Chang, C.P.; Wang, B. Impacts of El Niño and La Niña on the U.S. Climate during Northern Summer. J. Clim. **2007**, *20*, 2165–2177.
17. Madden, R.A.; Julian, P.R. Detection of a 40–50 day oscillation in the zonal wind in the Tropical Pacific. J. Atmos. Sci. **1971**, *28*, 702–708.
18. Madden, R.A.; Julian, P.R. Description of global-scale circulation cells in the tropics with a 40–50 day period. J. Atmos. Sci. **1972**, *29*, 1109–1123.
19. Zhang, C. Madden-Julian Oscillation. Rev. Geophys. **2005**, *43*, RG2003, doi:10.1029/2004RG000158.
20. Lau, K.-M., Waliser, D.E., Eds., *Intraseasonal Variability of the Atmosphere-Ocean Climate System*; Springer: Heidelberg, Germany, 2005; 474 pp.
21. Wang, B.; Ding, Y. An overview of the Madden-Julian oscillation and its relation to monsoon and mid-latitude circulation. Adv. Atmos. Sci. **1992**, *9*, 1, 93–111.
22. Yasunari, T. Cloudiness fluctuations associated with the Northern Hemisphere summer monsoon. J. Meteor. Soc. Jpn. **1979**, *57*, 227–242.

23. Lau, K.-M.; Chan, P.H. Aspects of the 40–50 day oscillation during the Northern summer as inferred from outgoing longwave radiation. Mon. Wea. Rev. **1986**, *114*, 1354–1367.

24. Chen, S.S.; Houze, R.A.; Mapes, B.E. Multiscale variability of deep convection in relation to large-scale circulation in TOGA COARE. J. Atmos. Sci. **1996**, *53*, 1380–1409.

25. Maloney, E.D.; Hartmann, D.L. Modulation of eastern north pacific hurricanes by the Madden–Julian oscillation. J. Clim. **2000**, *13*, 1451–1460.

26. Zhang, C.; Gottschalck, J. SST anomalies of ENSO and the Madden–Julian oscillation in the equatorial pacific. J. Clim. **2002**, *15*, 2429–2445.

27. Liebmann, B.; Hartmann, D.L. An observational study of tropical–midlatitude interaction on intraseasonal time scales during winter. J. Atmos. Sci. **1984**, *41*, 3333–3350.

28. Matsuno, T. Quasi-geostrophic motions in the equatorial area. J. Meteor. Soc. Jpn. **1966**, *44*, 25–43.

29. Gill, A.E. Some simple solutions for heat-induced tropical circulation. Quart. J. Roy. Meteor. Soc **1980**, *106*, 447–462.

30. Frank, W.M.; Roundy, P.E. The role of tropical waves in tropical cyclogenesis. Mon. Wea. Rev. **2006**, *134*, 2397–2417.

31. Dunkerton, T.J.; Montgomery, M.T.; Wang, Z. Tropical cyclogenesis in a tropical wave critical layer: easterly waves. Atmos. Chem. Phys. **2009**, *9*, 5587–5646.

32. Jin, F. An equatorial ocean recharge paradigm for ENSO. Part I: Conceptual model. J. Atmos. Sci. **1997**, *54*, 811–829.

33. Gray, W.M. Global view of the origin of tropical disturbances and storms. Mon. Wea. Rev. **1968**, *96*, 669–700.

34. Elsberry, R.L.; Foley, G.; Willoughby, H.; McBride, J.; Ginis, I.; Chen, L. *Global Perspectives on Tropical Cyclones. Tech. Doc.* No. 693; World Meteor. Organiz.: Geneva, Switzerland, 1995; 289 pp.

35. Zipser, E.J. Some views on "hot towers" after 50 years of tropical field programs and two years of TRMM data. Meteorol. Monogr. **2003**, *29*, 49–49.

36. Fierro, A.O.; Simpson, J.; LeMone, M.A.; Straka, J.M.; Smull, B.F. On how hot towers fuel the Hadley cell: an observational and modeling study of line-organized convection in the equatorial trough from TOGA COARE. J. Atmos. Sci. **2009**, *66*, 2730–2746.

37. Montgomery, M.T.; Nicholls, M.E.; Cram, T.A.; Saunders, A.B. A vortical hot tower route to tropical cyclogenesis. J. Atmos. Sci. **2006**, *63*, 355–386.

BIBLIOGRAPHY

1. Madden, R.A.; Julian, P.R. Observations of the 40–50 day tropical oscillation – A review. Mon. Wea. Rev. **1994**, *122*, 814–837.

2. Montgomery, M.T.; Smith, R.K. Paradigms for tropical-cyclone intensification. Q. J. R. Meteorol. Soc. **2011**, *137*, 1–31.

3. Philander, S.G.H. *El Niño, La Niña, and the Southern Oscillation.* Academic Press, San Diego, CA, 1990; 289 pp.

4. Trenberth, K.E.; Branstator, G.W.; Karoly, D.; Kumar, A.; Lau, N.; Ropelewski, C. Progress during TOGA in understanding and modeling global teleconnections associated with tropical sea surface temperatures. J. Geophys. Res. **1998**; *103*, 14291–14324.

5. Wang, B., Eds.; *The Asian Monsoon*; Springer/Praxis Publishing Co.: New York, 2006; 787 pp.

Ozone and Ozone Depletion

Jason Yang
Department of Geography, Ball State University, Muncie, Indiana, U.S.A.

Abstract

Ozone with three oxygen atoms is one of the vital atmospheric gases. It has either bad or good effects on the living system of Earth's surface depending on its location in the atmosphere. In the lower atmosphere, ozone is formed by chemical reaction involving nitrogen oxides and volatile organic compounds and is a major component of air pollutants, which has harmful effects to plants, animals, and human beings. Ozone in the upper atmosphere is formed naturally by absorbing the Sun's ultraviolet radiation and is beneficial to the living system, preventing these damaging radiations from reaching the Earth's surface. However, observations of the upper atmosphere showed that ozone depletion was indeed occurring since the 1980s while more and more human-produced chemicals such as chlorofluorocarbons were steadily increasing in the atmosphere. Therefore, efforts need to be made to regulate the ozone-forming substances to reduce the ozone increase in the lower atmosphere. Meanwhile, more efforts have to be made to reduce human-produced ozone-depleting substances to recover the ozone depletion in the upper atmosphere.

INTRODUCTION

Ozone is a colorless gas that is naturally present in the atmosphere. Each ozone molecule contains three atoms of oxygen and is denoted chemically as O_3. The word ozone is derived from the Greek word ózein (*ozein* in Latin), meaning "to smell."[1] Ozone mainly exists in two layers of the atmosphere. Approximately 10% of the total atmospheric ozone resides in the troposphere, the lowest layer of the atmosphere from ground up to 10 km altitude. The ozone in this layer is called tropospheric ozone or commonly known as ground-level ozone. The remaining 90% of ozone is in the stratosphere, a layer between 17 km and up to 50 km above the Earth's surface. The ozone in this region is called stratospheric ozone or commonly known as the ozone layer (Fig. 1).[2]

The ozone molecules in the troposphere and the stratosphere are chemically identical, but they play very different roles in the atmosphere and have very different effects on human beings and other living organisms. Ground-level ozone is a key component of photochemical smog, a familiar problem in the atmosphere of many cities around the world, and plays a destructive role on the environment and ecosystem. Many studies have documented the harmful effects of ground-level ozone on crop production, forest growth, and human health.[3–5] Stratospheric ozone, on the other hand, plays a beneficial role by absorbing most of the biologically damaging ultraviolet-B (UV-B) radiation ranging from 280 nm to 320 nm. Without the absorption by the ozone layer, more UV-B radiation would penetrate the atmosphere and reach the Earth's surface. Many

experimental studies of plants and animals and clinical studies of humans have shown the harmful effects of excessive exposure to UV-B radiation.[6,7] A simple statement summarizing the different effects of atmospheric ozone is that it is "good up high, bad nearby."[8]

The dual role of ozone links it to two separate environmental issues. One issue relates to the increases of "bad" ozone in the lower troposphere. Although ground-level ozone is not emitted directly from car engines or by industrial operations, motor vehicle exhaust, industrial emissions, and chemical solvents are the major anthropogenic sources of those chemicals to form ground-level ozone.[9] Many programs around the globe have already been developed to regulate the emission of ozone-forming substances that cause excess ozone near the Earth's surface by setting up air quality standards for ozone. Emission control programs, including removal of highway tolls to reduce traffic congestions, use of vapor recovery devices for gasoline pumps, and enforcement of vehicle emission inspections, have been implemented in the United States and some other countries since the 1980s to control the increase of ground-level ozone.[10]

The second environmental issue relates to the loss of "good" ozone in the upper stratosphere. Ground-based and satellite instruments have measured decreases in the amount of ozone in the stratosphere since the 1980s.[11] The extreme case occurs over some parts of Antarctica, where up to 60% of the total overhead amount of ozone disappears during some periods of the southern hemisphere spring (August–October). Similar processes occur in the Arctic polar region, which have also led to significant ozone depletion

Dew—Wind

Ozone in the Atmosphere

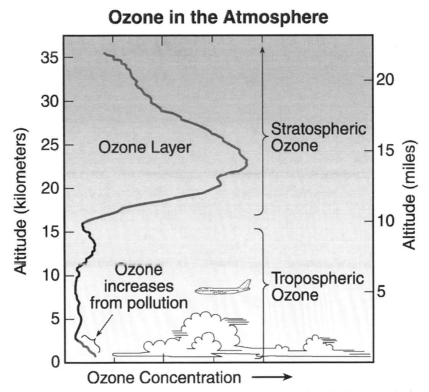

Fig. 1 Ozone is present throughout the troposphere and stratosphere. Most ozone resides in the stratospheric "ozone layer" above the Earth's surface.
Source: NOAA Earth System Research Laboratory, http://esrl.noaa.gov/csd/assessments/ozone/2006/chapterfigures.html.

during late winter and spring in many recent years. The primary cause of the ozone depletion in the stratosphere has been found to be the presence of chlorine- and bromine-containing gases.[12] These gases can dissociate and release chlorine atoms in the presence of UV-B radiation, which then go on to catalyze ozone destruction. Many countries have joined the effort to stop ozone depletion in the stratosphere by signing an international agreement known as the "Montreal Protocol" in 1987.[13] In this agreement, governments around the world have decided to eventually discontinue productions that emit ozone-depleting gases, and industries have developed more and more "ozone-friendly" substitutes. All other things being equal, and with adherence to the international agreements, the ozone layer is expected to recover over the next 50 years or so.

ATMOSPHERIC OZONE

Tropospheric Ozone

In the troposphere, ozone is formed by a chemical reaction involving nitrogen oxides (NO_x) and volatile organic compounds (VOCs) subjected to favorable weather conditions, such as low humidity, plenty of sunlight, and moderate to low wind speed.[14] Emissions from industrial facilities and electric utilities, motor vehicle exhaust, gasoline vapors, and chemical solvents are major anthropogenic sources for

NO_x and VOCs. As a major component of urban smog, ozone is harmful to human health because it can cause sore and watery eyes, soreness in the throat and sinuses, and difficulty in breathing.[5] A statistical study of 95 large urban communities in the United States showed a significant association between ground-level ozone levels and premature death.[15] It is estimated that a one-third reduction in urban ozone concentrations would save roughly 4000 lives per year. Ground-level ozone also damages vegetation, such as forests and crops. In the United States alone, ground-level ozone is responsible for an estimated $500 million in reduced crop production each year.[16] To address ozone pollution near the Earth's surface, the U.S. Environmental Protection Agency (EPA) has set up air quality standards for ozone since 1971 and kept revising it in 1979, 1997, 2008, and 2011, to provide requisite protection for public health and welfare.[17]

Ground-level ozone can be monitored, predicted, and declared in advance by the environmental protection agencies in many countries today. For example, U.S. EPA coordinates the collection of real-time ground-level ozone data from around the country and posts the information to the public through the AIRNow website.[18] Therefore, people can avoid unhealthy exposure to ozone on those days with higher ozone risk and take some simple steps in their daily life (e.g., limiting vehicle usage) to decrease the production of compounds that create bad ozone.

Dew—Wind

Stratospheric Ozone

In the stratosphere, ozone is formed and destructed naturally by chemical reactions involving oxygen molecules and UV-B radiation, which is illustrated in Fig. 2. In the first step, one oxygen molecule (O_2) is split into two oxygen atoms (O) by absorbing UV-B radiation. The free unstable oxygen atoms (O) then combine with other oxygen molecules (O_2) to form ozone molecules (O_3). The ozone molecule absorbs UV-B radiation and splits into oxygen molecule (O_2) and an oxygen atom (O) again. The oxygen atom then joins up with an oxygen molecule to regenerate ozone. This process is repeated over and over, thereby involving large amount of UV-B radiation that would otherwise reach the Earth's surface.

The stratospheric ozone can also be destroyed by emissions from natural processes and human activities, including large sources of chlorine- and bromine-containing gases that reach the stratosphere. First, the chlorine (Cl) in stratosphere can take away an oxygen atom (O) from ozone molecule (O_3) to form chlorine monoxide (ClO) and leave a normal oxygen molecule (O_2) (Eq. 1). The newly generated chlorine monoxide (ClO) can destroy a second ozone molecule (O_3) to yield another chlorine atom (Cl) (Eq. 2), which can repeat the first reaction and continue to destroy ozone molecules (Eq. 1).[19]

$$Cl + O_3 \rightarrow ClO + O_2 \qquad (1)$$

$$ClO + O_3 \rightarrow Cl + 2O_2 \qquad (2)$$

NO_x emitted from automobile and jet engine exhaust also have the ability to enter the stratosphere and destroy the ozone shield.

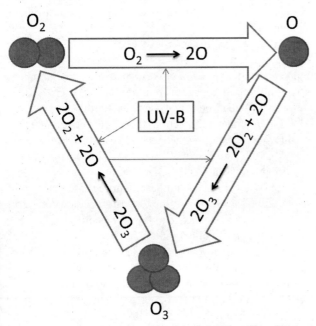

Fig. 2 Natural formation and destruction of oxygen atom (O), oxygen gas molecule (O_2), and ozone molecule (O_3) in stratosphere.

OZONE DEPLETION

The Ozone Hole

Ozone depletion in the stratosphere was first noticed based on the ground-based observations taken by the British Antarctica Survey. The finding was reported in the scientific journal *Nature* in 1985 that there was a substantial thinning of the ozone layer over Antarctica in early spring.[11] Soon after, measurements from satellite instruments confirmed the spring ozone depletion and further showed that in each early spring season, the ozone layer depletion extended over a large region centered near the South Pole.[20]

Ozone depletion in stratosphere varies strongly with latitude over the globe. There is no significant trend found in the Tropics, and only a very small amount of declines (3–6%) for middle latitudes. However, the ozone decrease can be up to 30% in the Arctic and even 60% over the Antarctica. Ozone depletion above Antarctic is also a seasonal phenomenon, occurring primarily in August–November with peak depletion in early October. This severe stratospheric ozone depletion is the so-called ozone hole, which is defined as an area of the Antarctic stratosphere in which the ozone levels have dropped to as low as 33% of their pre-1980 values. Therefore, the ozone hole is not technically a "hole" where no ozone presents, but a region of exceptionally depleted ozone in the stratosphere that happens at the beginning of Antarctic spring. The size of the ozone hole is usually determined by the geographic region contained within the 220-Dobson unit (DU) contour in the total ozone map. Based on the NASA Earth Observatory Image presented in Fig. 3, the maximum size of the ozone hole in most years after 1985 far exceeds the size of the Antarctic continent.

Reasons and Consequence of Ozone Depletion

It is now widely accepted that the stratospheric ozone is gradually being destroyed by man-made chemicals referred to as ODSs, including CFCs, hydrochlorofluorocarbons, halons, and other gases. These ODSs were formerly used and sometimes are still used in coolants, foaming agents, fire extinguishers, solvents, pesticides, and aerosol propellants. Once released into the atmosphere, these gases degrade very slowly until they reach the stratosphere where they are broken down by the intensive UV radiation and release chlorine- and bromine molecules, which destroy the ozone molecules there (Eqs. 1 and 2).

The role of sunlight in ozone depletion is the reason why the Antarctic ozone depletion is the greatest during early spring. Even though nacreous or polar stratospheric clouds (PSCs) are at their most abundant during winter, there is no light over the Antarctic to drive the chemical reactions. During the spring, however, the increasing solar radiation raises the stratospheric temperature and removes the PSCs via sublimation, thereby releasing the trapped compounds that can deplete ozone. At the end of spring, warming temperatures

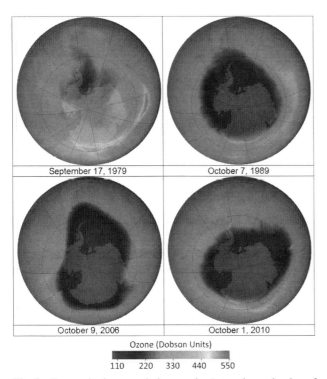

Fig. 3 Stratospheric ozone hole over the Antarctic on the day of its maximum depletion in four different years 1979, 1989, 2006, and 2010.
Source: NASA Earth Observatory Images, http://earthobservatory.nasa.gov/IOTD/view.php?id=49040.

can break up the vortex and ozone-rich air flows in from lower latitudes, so the PSCs are destroyed and the ozone depletion process shuts down, and the ozone hole closes.[21]

Ozone depletion is expected to increase the UV-B levels on the Earth's surface, which could lead to damage on the living systems, including crops and human health. Studies have shown that an increase of UV-B radiation would be expected to reduce crops yields for a number of economically important species, such as rice and soybeans.[7,22] Some scientists suggest that marine phytoplankton, which are the base of the ocean food chain, are already under stress from UV-B radiation. This could have adverse consequences for human food supplies from the oceans.[23] Another main public concern regarding the ozone hole has been the effects of increased UV-B radiation on human health. Research has shown that UV-B radiation is the main cause of sunburn and tanning, as well as the formation of vitamin D3 in the skin, which have negative influences on the immune system of human beings. UV-B radiation contributes significantly to the aging of the skin and eyes, and it is the UV-B range that is the most effective in causing skin cancer.[24]

Trends of Ozone Depletion

Research on ozone depletion advanced very rapidly in the 1970s and 1980s, leading to the identification of CFCs and other halocarbons as the cause. Since then, industries have developed more ozone-friendly substitutes for the CFCs and other ODSs. The 1987 international agreement,

"Montreal Protocol on Substances That Deplete the Ozone Layer," ratified now by over 190 nations, established legally binding controls on the production and consumption of ozone-depleting gases. As a result, the total abundance of ozone-depleting gases in the atmosphere has begun to decrease in recent years, which has been observed using instruments such as Total Ozone Mapping Spectrometer on NASA's satellites.[25] In addition, a report in 2007 showed that the ozone hole over the Antarctic region was closing and the smallest it had been for about a decade.[26] Another report in 2010 found that global ozone and ozone in the polar regions are no longer decreasing but are not yet increasing over the past decade, and the ozone layer outside the polar regions is projected to recover to its pre-1980 levels some time before the middle of this century.

However, significant depletion of ozone layer in the Arctic stratosphere has drawn more attention recently. A record ozone layer loss was observed on March 15, 2011, with about 50% of the ozone over the Arctic having been destroyed.[27] A study published in the journal *Nature* in October 2011 also stated that between December 2010 and March 2011 up to 80% of the ozone in the Arctic region at about 20 km above the surface was destroyed.[28] The level of ozone depletion above the Arctic region was severe enough that scientists opined that it could be compared to the "ozone hole" that forms over Antarctica every October. And for the first time, sufficient ozone loss occurred that can be reasonably described as an Arctic ozone hole.[29]

CONCLUSION

Ozone can be either harmful or beneficial to the environment depending on its location in the atmosphere. The tasks to reduce the tropospheric "bad" ozone and to recover the stratospheric "good" ozone depletion are both challenging. Great efforts have been made to reduce the emission of the ozone-forming substances such as NO_x and VOCs, which generate bad ozone on the Earth's surface. Efforts have also been made to reduce the emission of the ODSs such as CFCs and halons, which will eventually reach the stratosphere to destroy the good ozone. The interaction of science in identifying the problem, technology in developing alternatives, and governments in devising new policies is thus an environmental "success story in the making." Indeed, the Montreal Protocol and its subsequent Amendments and Adjustments have served as a model in international cooperation for other environmental issues now facing the global community. If the nations of the world continue to follow the provisions of the Montreal Protocol, the decrease of ODSs will continue throughout the twenty-first century and the stratospheric ozone is expected to restore to its pre-1980 level before the Antarctic "ozone hole" began to form.

REFERENCES

1. Rubin, M.B. The history of ozone. The Schönbein Period, 1839–1868. Bull. Hist. Chem. **2001**, *26* (1), 16.

2. WMO, Scientific Assessment of Ozone Depletion: 2002 in Global Ozone Research and Monitoring Project - Report No. 47. 2003, 498 p.

3. Fishman, J.; Creilson, J.K.; Parker, P.A.; Ainsworth, E.A.; Vining, G.G.; Szarka, J.; Booker, F.L.; Xu, X. An investigation of widespread ozone damage to the soybean crop in the upper Midwest determined from ground-based and satellite measurements. Atmos. Environ. **2010**, *44* (18), 2248–2256.

4. Emberson, L.D.; Büker, P.; Ashmore, M.R. Assessing the risk caused by ground level ozone to European forest trees: A case study in pine, beech and oak across different climate regions. Environ. Pollut. **2007**, *147* (3), 454–466.

5. Stedman, J.R.; Kent, A.J. An analysis of the spatial patterns of human health related surface ozone metrics across the UK in 1995, 2003 and 2005. Atmos. Environ. **2008**, *42* (8), 1702–1716.

6. Van Der Leun, J.C. Effects of Increased UV-B on Human Health, in *Studies in Environmental Science*; Lee, S.D.; Wolters, G.J.R.; Schneider, T.; Grant, L.D. Eds.; Elsevier: 1989: 803–812 p.

7. Kumagai, T.; Hidema, J.; Kang, H.-S.; Sota, T. Effects of supplemental UV-B radiation on the growth and yield of two cultivars of Japanese lowland rice (*Oryza sativa* L.) under the field in a cool rice-growing region of Japan. Agr. Ecosyst. Environ. **2001**, *83* (1–2), 201–208.

8. Ozone - Good Up High Bad Nearby. U.S. Environmental Protection Agency (EPA) July 21, 2011 Available from: http://www.epa.gov/oaqps001/gooduphigh/(accessed November 2011).

9. WMO, Scientific Assessment of Ozone Depletion: 2006, in Global Ozone Research and Monitoring Project - Report No. 50. 2007, 572 p.

10. Yang, J.; Miller, D.R. Trends and variability of ground-level ozone in Connecticut over the period 1981–1997. J. Air Waste Manage. Assoc. **2002**, *52*, 8.

11. Lubinska, A. Ozone depletion: Europe takes a cheerful view. Nature, **1985**. *313* (6005): p. 727–727.

12. Fahey, D.W.; Hegglin, M.I. (Coordinating Lead Authors), Twenty Questions and Answers about the Ozone Layer: 2010 Update, Scientific Assessment of Ozone Depletion: 2010, 72 pp. World Meteorological Organization, Geneva. 2011: Switzerland.

13. Morrisette, P.M. The evolution of policy responses to stratospheric ozone depletion. Nat. Resourc. J. **1989**, *29*, 27.

14. Brimblecombe, P. *Air Composition and Chemistry*. 2nd ed. 1996, New York: Cambridge University Press: 253.

15. Bell, M.L.; McDermott, A.; Zeger, S.L.; Samet, J.M.; Dominici, F. Ozone and short-term mortality in 95 U.S. urban communities, 1987–2000. J. Am. Med. Association., **2004**, *292*, p. 6.

16. Ground-level Ozone. U.S. Environmental Protection Agency (EPA) September 23, 2010 Available from: http://www.epa.gov/glo/basic.html. (accessed December 2011).

17. Ozone (O_3) Standards - Table of Historical Ozone NAAQS. U.S. Environmental Protection Agency (EPA) November 08, 2011; Available from: http://www.epa.gov/ttn/naaqs/standards/ozone/s_o3_history.html. (accessed 2011 November).

18. AIRNow. EPA/NOAA/NPS August 30, 2010; Available from: http://airnow.gov/ (accessed December 2011).

19. Gabler, R.E.; Petersen, J.F.; Trapasso, L.M.; Sack, D. The Atmosphere, Temperature, and the Heat Budget, in *Physical Geography*, Brooks/Cole, Cengage Learning: Belmont, CA, 2009, 89 p.

20. Garrett, O.; Carver, G. The Ozone Hole Tour, Part II. 1998; Available from: http://www.atm.ch.cam.ac.uk/tour/part2.html (accessed December 2011)

21. Ozone Facts: What is the Ozone Hole? U.S. National Aeronautics and Space Administration (NASA) November 18, 2009; Available from: http://ozonewatch.gsfc.nasa.gov/facts/hole.html (accessed January 2012)

22. Koti, S.; Reddy, K.R.; Kakani, V.G.; Zhao, D.; Gao. Effects of carbon dioxide, temperature and ultraviolet-B radiation and their interactions on soybean (*Glycine max* L.) growth and development. Environ. Exp. Bot. **2007**, *60* (1), 1–10 p.

23. Nielsen, T.; Ekelund, N.G.A. Influence of solar ultraviolet radiation on photosynthesis and motility of marine phytoplankton. FEMS Microbiol. Ecol. **1995**, *18* (4), 281–288 p.

24. Tevini, M. Editor, *UV-B Radiation and Ozone Depletion: Effects on Humans, Animals, Plants, Microorganisms, and Materials*; Lewis Publishers: Boca Raton, 1993.

25. Total Ozone Mapping Spectrometer (TOMS). Jan. 25, 2012; Available from: http://ozoneaq.gsfc.nasa.gov/instruments.md (accessed Febraury 2002)

26. Ozone hole closing up, research shows. ABC News November 16, 2007; Available from: http://www.abc.net.au/news/2007-11-16/ozone-hole-closing-up-research-shows/727460 (accessed December 2011).

27. Dell'Amore, C. First North Pole Ozone Hole Forming? National Geographic News 2011; Available from: http://news.nationalgeographic.com/news/2011/03/110321-ozone-layer-hole-arctic-north-pole-science-environment-uv-sunscreen/.

28. Manney, G.L.; Santee, M.L.; Rex, M.; Livesey, N.J.; Pitts, M.C.; Veefkind, P.; Nash, E.R.; Wohltmann, I.; Lehmann, R.; Froidevaux, L.; Poole, L.R.; Schoeberl, M.R.; Haffner, D.P.; Davies, J.; Dorokhov, V.; Gernandh, H.; Johnson, B.; Kivi, R.; Kyro, E.; Larsen, N.; Levelt, P.F.; Makshtas, A.; McElroy, C.T.; Nakajima, H.; Parrondo, M.C.; Tarasick, D.W.; Von der Gathen, P.; Walker, K.A.; Zinoviev, N.S. Unprecedented Arctic ozone loss in 2011. Nature **2011**, *478* (7370), 469–475

29. 29. Arctic ozone loss at record level. BBC News Online October 2, 2011; Available from: http://www.webcitation.org/6297yWsLm.

Transpiration and Physical Evaporation: United States Variability

Toshihisa Matsui
NASA Goddard Space Flight Center, National Aeronautics and Space Administration (NASA), Greenbelt, and ESSIC, University of Maryland, College Park, Maryland, U.S.A.

David M. Mocko
NASA Goddard Space Flight Center, National Aeronautics and Space Administration (NASA), Greenbelt, and SAIC, Beltsville, Maryland, U.S.A.

Abstract

Terrestrial transpiration and physical evaporation (*Et*) are important parameters for the surface energy and water budget, and have potential feedbacks to weather and climate. This entry introduces the energy and water balance equations, and examines seasonal and regional variability of *Et* associated with variability of land-cover, precipitation, and surface net solar radiation through state-of-the-art 30-year observations and simulations of the land-surface over the Conterminous United States. Results show that seasonal variability of *Et* is explained by the mean surface net solar radiation, whereas regional variability of *Et* is explained by the regional precipitation distribution.

INTRODUCTION

Terrestrial transpiration and physical evaporation are Earth's breath that transports water and heat from the soil or plants to the atmosphere. Transpiration is the water vapor exchange from plant leaves to the surrounding air through numerous tiny pores in the leaf, called *stomata*. The water that transpires from the leaves has its origin in the soil and travels from the plant's root structure through the stems to the leaves. Simultaneously, plants exchange other gas species (e.g., CO_2 and O_2) for photosynthesis, and uptake minerals from the soils. Thus, transpiration is the critical indicator for plant productivity. Physical evaporation is vaporization of the standing water on the soil or on plants to the overlying atmosphere, or the diffusion/turbulent transport of water vapor of soil moisture from the soil surface to the overlying air.

The seasonal and regional variability of transpiration and evaporation is important in the context of the surface water budget and energy budget. Over large areas and an annual time scale, the surface water budget equation is expressed as

$$P = Et + R + dW \qquad (1)$$

where *P* is surface precipitation; *Et* is transpiration and physical evaporation; *R* is surface runoff; and *dW* is the change in the surface moisture storage (such as soil moisture, snow cover, or reservoir). Note that ground-water movement is negligible in comparison with other terms for this time and spatial scale. Variability of water inputs from the atmosphere to the surface (*P*) will be balanced by the water outputs (*Et* + *R* + *dW*). On the same scale, the surface energy budget equation is expressed as

$$Q_{SW}(1-a) + (1-\varepsilon)\,Q_{LW} = Q_L + Q_H + Q_{LW}^{emit} \qquad (2)$$

where Q_{SW} is the incoming surface shortwave solar flux; *a* is the surface albedo; Q_{LW} is the downward atmospheric thermal flux; ε is the thermal emissivity; Q_L is the turbulent latent heat flux (i.e., physical evaporation and transpiration); Q_H is the turbulent sensible heat flux; and Q_{LW}^{emit} is the thermal flux emitted from the surface. Thus, variability of energy inputs from the atmosphere to the surface $[Q_{SW}(1-a) + (1-\varepsilon)\,Q_{LW}]$ will be balanced by the energy outputs from the surface to the atmosphere ($Q_L + Q_H + Q_{LW}^{emit}$).

Equations 1 and 2 are coupled through *Et* and Q_L. For example, if *Et* is in the unit of mm/day and Q_L is in the unit of W/m²,

$$Et = Q_L / (\rho_{liq}\,\lambda) = a\,Q_L$$

where ρ_{liq} is liquid water density (in kg/m³) and λ is the latent heat of vaporization (in J/kg); the density and λ vary slightly with temperature and pressure, but for typical conditions, $a = 0.0337$ (m²/W • mm/day). Plant transpiration is limited to the availability of root-zone soil moisture, and soil evaporation is limited to the top-soil moisture. Thus, Q_L is also coupled to *dW*. With respect to the energy and water balance equations, this entry examines seasonal and regional variability of *Et* through 30-year observations and simulations of a land-surface model (LSM).

Encyclopedia of Natural Resources DOI: 10.1081/E-ENRA-120047654

Dew—Wind

SEASONAL AND REGIONAL VARIABILITY OF *Et* OVER THE CONTERMINOUS UNITED STATES

According to Eqs. 1 and 2, variability of *Et* or Q_L is extracted from the seasonal and regional variability of input terms: P and $Q_{SW}(1 - a)$. In this section, we discuss the roles and variability of net solar radiation, precipitation, land-cover type, and *Et* over the Conterminous United States (CONUS) through reviewing state-of-the-art observational datasets and numerical model simulation results from the North American Land Data Assimilation System (NLDAS).[1] The NLDAS project constructs atmospheric forcing and LSM datasets from the best available observations and model outputs. Unlike in situ or satellite data, NLDAS can provide spatially and temporally complete datasets that satisfy energy and budget balances in Eqs. 1 and 2. The use of separate observational platforms for datasets of P, Et, and other terms can lead to inconsistencies and imbalances in the closure of the water and energy cycle equations. For this section, we have compiled a 30-year seasonal climatology (March/April/May (MAM); June/July/August (JJA); September/October/November (SON); December/January/February (DJF)) from NLDAS Phase 2[2] datasets using precipitation observations and output from the Mosaic LSM from 1979 to 2009. It is important to note that NLDAS-2 precipitation is a state-of-the-art dataset over CONUS for such a long period, based primarily on a 1/8th-degree daily gridded analysis of rain gauge observations. This precipitation (and other NLDAS-2 surface meteorology, such as Q_{SW}

and Q_{LW}) is used as input to the LSM. The Mosaic LSM (like all LSMs) contains several parameterizations and assumptions in its model formulations to calculate *Et*, *R*, *dW*, Q_L, and Q_H. In the following sub-sections, we discuss the seasonal and regional variability of the key water and energy terms.

Precipitation and Solar Radiation

Surface precipitation, P, varies seasonally and regionally, characterized primarily by synoptic weather (i.e., large-scale circulation), mesoscale circulation (local disturbance), and thermodynamics (daytime heating). When precipitation reaches the surface, some water may be trapped on the canopy, stored on the surface of the soil (including as snow), be infiltrated into the soil, or run off to a river. Infiltrated moisture to the soil becomes the storage term, *dW* (Eq. 1), that is the direct source of *Et*, through transpiration and/or soil evaporation.

Figure 1 shows a 30-year seasonal climatology of the precipitation from NLDAS-2. In the MAM, DJF, and SON periods, the coastal regions of the North West show a large amount of the precipitation generated by extratropical cyclones from the Pacific Ocean. The moisture flux from the north Pacific also generates snowfall or orographic rainfall over the mountains in the west. In the eastern domains, precipitation is peaked around the lower Mississippi Basin due to moisture flux from the Gulf of Mexico. Precipitation is higher along a narrow range of the northeast coastal region

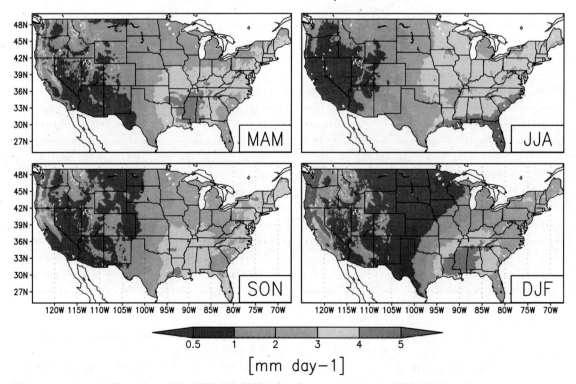

Fig. 1   Thirty-year seasonal climatology (Mar 1979–Feb 2009) of surface precipitation in NLDAS Phase 2.

in the MAM and the SON periods. In the JJA period, the largest precipitation appears in the Deep South due to effects of sea breezes. Excepting winter, the South West tends to be drier for all seasons. The Great Plains become very dry during winter, and have light amounts of precipitation during other seasons.

Surface incoming solar radiation (Q_{sw}) varies along with seasonal cycles. Its seasonal variability is, at the first order, governed by the mean solar zenith angle due to tilt of earth's rotation axis relative to the sun. Its regional variability is affected by the presence of clouds and aerosols. Surface albedo (a: defined as a fraction of the incoming shortwave flux that is reflected by the surface) also affects the regional variability of the surface net solar radiation. Albedo differs widely depending on the land-cover type and vegetation *senescence*, ranging from 10% with dense forest to 50% over desert, but it could be up to 95% with the presence of fresh snow.

Figure 2 shows a 30-year seasonal climatology of the surface net solar radiation, $Q_{sw}(1 - a)$, from NLDAS-2 Mosaic. It is obvious that the largest net solar radiation is in the JJA period, whereas the lowest is in the DJF period, and is generally stratified in the latitudinal zones. This confirms that the largest factor for determining the net surface solar radiation is the mean solar zenith angle. Even in the same latitudinal zone, net solar radiation shows regional (longitudinal) variability at some extent, especially highlighted in the MAM period. For example, net solar radiation is ~40 W/m² higher

in California than in Nevada. In Colorado, the Rocky Mountain region has spotty regions with low net solar radiation. These types of regional variability are attributed to differences in albedo associated with the land-cover type and the presence of cloud/aerosols and surface snow.

Natural vegetation competes and evolves by taking the best advantage of solar radiation and rainfall patterns. Figure 3 shows the satellite-derived land-cover map used in NLDAS. Forest types extend over much of the eastern domain of the CONUS, coastal regions of the North West, and spotted over the Mountain region, where enough orographic rainfall sustains their life. Trees with a large canopy require more soil moisture to maintain their photosynthesis, transpiration, and mineral uptake. Some tree species adapt to the region where mean temperature significantly changes seasonally by becoming dormant during the cold seasons (e.g., deciduous forest). For areas with moderate climate throughout the seasons, tree species tend to be evergreen (e.g., evergreen forest). Grass and shrub species have smaller bodies, which require less number of resources to maintain. They can be dormant for long periods during drought and/or cold. Thus, they are well adapted for the dry regions, such as in the Great Plains, South West, and some of the Mountain regions. Cropland exists elsewhere in the CONUS, especially extended over the Midwest, lower Mississippi river basin, and the Great Plains. Agricultural practices have modified crop species, soil, and moisture sources for their climate and economy through a long-time

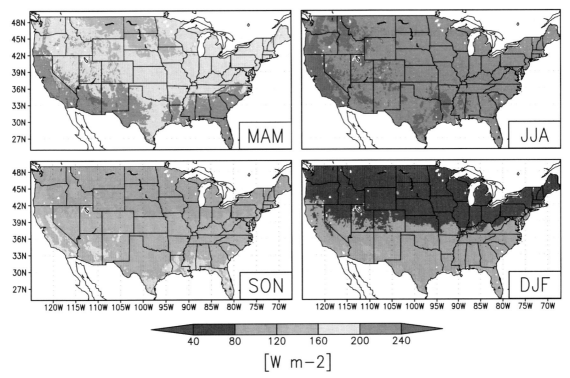

Fig. 2 Thirty-year seasonal climatology (Mar 1979–Feb 2009) of surface net solar radiation from the Mosaic LSM in NLDAS Phase 2.

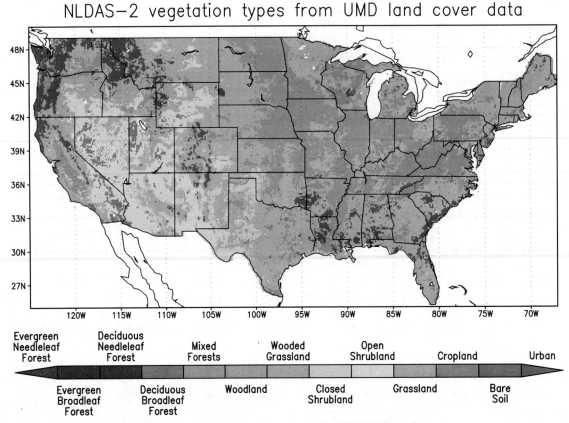

Fig. 3 University of Maryland (UMD) dominant land-cover type map used in NLDAS Phase 2.

history. Thus the distribution of cropland is not strictly regulated by the rainfall and net radiation patterns.

Transpiration and Physical Evaporation

For a given solar insolation and downwelling thermal radiation, surface total available energy is determined, balanced by energy release from the turbulent sensible heat flux (Q_H), the turbulent latent heat flux (Q_L), and the thermal flux emitted from the surface (Q_{LW}^{emit}). For further detail, these three terms in Eq. 2 can be expressed as follows. For a plant-leaf level,

$$Q_{net} = \rho_{atm} \lambda \, [q_{air} - q^{sat}(T_{leaf})]g_L + \rho_{atm} C_p \, (T_{air} - T_{leaf}) \, g_H \\ + \varepsilon \sigma T_{leaf}^4, \quad (3a)$$

and for soil level,

$$Q_{net} = \rho_{atm} \lambda \, (q_{air} - q_{soil})g_L + \rho_{atm} C_p (T_{air} - T_{soil}) \, g_H \\ + \varepsilon \sigma T_{soil}^4 \quad (3b)$$

where ρ_{atm} is dry air density; λ is latent heat of vaporization; q_{air} is water vapor mixing ratio of surrounding air; $q^{sat}(T_{leaf})$ is saturated water vapor mixing ratio at leaf temperature (T_{leaf}); g_L is water vapor conductance (stomatal and leaf aerodynamic conductance for plant leaf, soil aerodynamic conductance for soil); g_H is heat conductance (leaf or soil aerodynamic conductance); C_p is the specific heat of dry air, T_{air} is surrounding air temperature; ε is emissivity; σ is

Stefan–Boltzmann constant; q_{soil} is water vapor mixing ratio of soil surface; T_{soil} is soil surface temperature.

Et plays a primal role to determine the energy balance between Q_H, Q_L, and Q_{LW}^{emit}. For example, strong wind and high stomatal conductance provide the best environment for plants to transpire more water vapor from leaves to the surrounding air, and consequently less energy is available for sensible heat flux and thermal emission, which is strongly coupled with leaf temperature. Consequently, leaf temperature must be reduced until it balances the net radiation through Eq. 3a. Large *Et* indicates larger net solar radiation and available soil moisture from frequent precipitation, thus linking to larger exchange of CO_2 for photosynthesis (plant production) and larger uptake of root-zone soil moisture associated with various minerals required for their growth. Alternatively, a plant may become highly stressed due to lack of available soil moisture or a lack of wind, and transpiration and associated latent heat flux will be restricted, which will quickly increase leaf temperature, enhancing sensible heat flux and thermal emission. If such condition is extended for a long term, plants start wilting and become dormant to survive in a severe environment.

Soil evaporation performs a similar way to transpiration; it controls soil skin temperature and sensible heat flux (Eq. 3b). A significant difference between soil evaporation and plant transpiration is the depths of soil moisture. Soil evaporation is linked to top-soil moisture, whereas plant

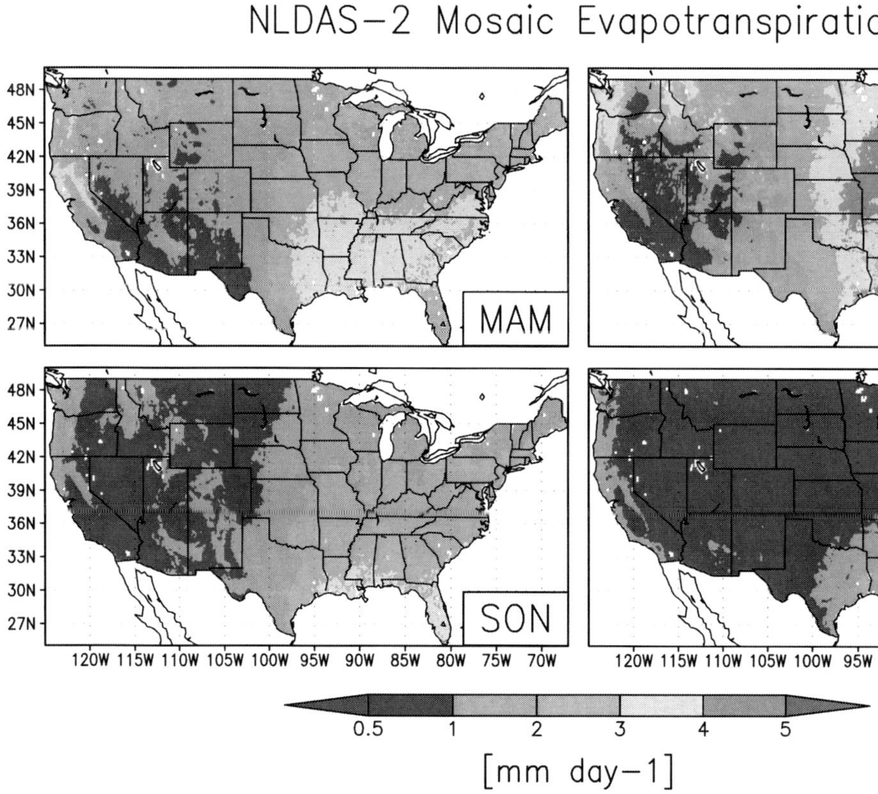

NLDAS-2 Mosaic Evapotranspiration

Fig. 4 Thirty-year seasonal climatology (Mar 1979–Feb 2009) of Et from the Mosaic LSM in NLDAS Phase 2.

transpiration is linked to root-zone soil moisture (up to several meters in depth). During short-term drought situation, deep-root tree species can maintain high Et and Q_L with deep soil moisture, even if physical evaporation is limited from the soil surface. It should be also noted that Eq. 3a will be scaled to the leaf area index. So Eq. 3a dominates the Eq. 3b for dense forests, and vice versa for a less vegetated area.

Figure 4 shows the seasonal Et from the 30-year NLDAS-2 Mosaic climatology. Overall patterns of seasonal and regional variability of Et can be extracted from those of precipitation (Fig. 1) and net solar radiation (Fig. 2). From MAM to DJF, the trend of CONUS-mean Et is well explained by that of net solar radiation; e.g., Et is largest in JJA, decreases from MAM to SON, and lowest in DJF. The same trend is shown in the surface net solar radiation. For a given net solar radiation, seasonal precipitation and land-cover type explains the regional variability. In MAM, light Et (~1–2 mm/day) is spread over the Great Plains and West in general. Moderate Et (2–4 mm/day) appears in eastern domains of the CONUS and the North West Coast. The similar regional pattern continues in the JJA period, but Et over the eastern domains is ~1 mm/day larger than the MAM period. Also, from MAM to JJA, the peak of Et shifts from the South to Midwest cropland regions. In the SON period, minimal Et (<0.5 mm/day) is shown across the South West, and Et decreases approximately 2 mm/day over eastern domains. In the DJF period, the minimal Et region is spread widely over the West and Midwest, and small Et (0.5–2 mm/ day) remains in the Southeast and the West Coast.

Figure 5 shows a scatter plot between NLDAS-2 precipitation and Mosaic Et. All values are a 30-year seasonal climatology, each mark represents a single grid-point within the 1/8th-degree NLDAS grid (52,476 marks for each season); it therefore correlates Figs. 1–4. For all seasons, precipitation and Et are well correlated with each other; it suggests that precipitation variability characterizes the regional Et variability. Slopes between Et and precipitation (defined as Et/precipitation) become steeper in the order from DJF (~0.25), SON (~0.6), MAM (~1.0), to JJA (~1.25), which is in the same order to the seasonal variability of surface net solar radiation over the CONUS (Fig. 2). This means that JJA (DJF) has the largest (smallest) energy that is used to recycle surface precipitation back to the atmosphere by Et. Also of note is the number of points where the Et is larger than precipitation, especially during JJA, as shown in Fig. 5 and comparing Figs. 1 and 4. This effect is caused by changes in the dW storage term, showing evidence of soil moisture and snow cover that fell as precipitation in previous seasons being converted to Et during a season when more energy is available. Overall, Fig. 5 suggests that, at first order, variability of mean surface net solar radiation determines the seasonal variability of Et, whereas precipitation variability determines the regional variability of Et. Variability of land-cover (vegetation) type (Fig. 3) and regional variability of net surface radiation (Fig. 2), as well as soil moisture and snow cover storage also affect the spread of the scatter plots.

Dew—Wind

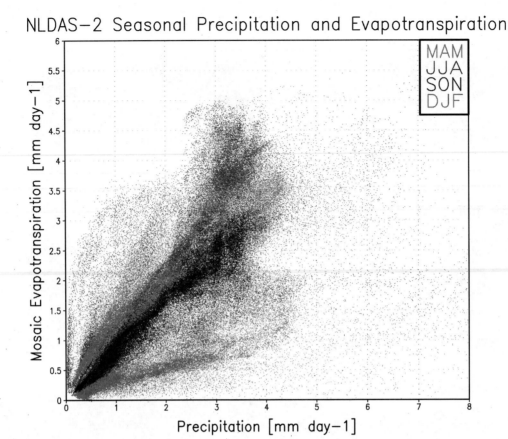

Fig. 5 Scatter plot between surface precipitation and Mosaic *Et* for the MAM, JJA, SON, and DJF periods in NLDAS Phase 2. Each scattered point represents a grid point over CONUS, and all values are based on a 30-year seasonal climatology (Mar 1979–Feb 2009).

CONCLUSION

Terrestrial transpiration and physical evaporation (*Et*) are important parameters for the surface energy and water budget equations. Variability of *Et* also feeds back to weather and climate through modulating planetary boundary layer, lower-troposphere thermodynamic profile, mesoscale circulation, and potentially general circulation.[3] It is critical to understand and monitor regional and seasonal variability of *Et* together with variability of vegetation senescence, land-cover, precipitation, and surface net radiation.

This entry examined *Et* variability from state-of-the-art NLDAS-2 data, and concluded that net surface radiation strongly controls seasonal variability of *Et*, whereas precipitation distribution strongly affects regional variability of *Et* over the CONUS. However, due to potential errors in observations and in numerical simulations, NLDAS (as well as other similar *Et* and precipitation datasets) and the conclusion could be biased to some extent. Currently, eddy turbulent covariance measurements at numerous locations globally have created an extensive surface network (FLUXNET).[4] It is the most direct and accurate *Et* in situ observational database, and in addition it is temporally continuous. Nevertheless, FLUXNET lacks spatial continuity, and its historical record of observations is somewhat limited.

Satellite- or aircraft-based visible–infrared imager sensors can measure various aspects of surface information, and can be translated into *Et* data. One of the notable measurements is the MOD16 *Et* product, which adapts the Penman–Monteith equation to the various surface information derived from the MODerate Resolution Imaging Spectroradiometer (MODIS) sensor in Terra and Aqua satellites.[5] MOD16 *Et* can estimate *Et* at high spatial resolution (1 km) over the globe, whereas satellite orbital patterns and algorithm limit the *Et* estimates at instantaneous (a.m. or p.m.) times to only once, possibly twice, a day. Technically, MOD16 *Et* product is available from an 8-day average; average *Et* of up to 8-day instantaneous *Et* estimate. It also involves uncertainties of *Et* retrieval with the Penman–Monteith algorithm.

An LSM, such as used within NLDAS, is composed of a set of surface boundary conditions and soil–vegetation–atmosphere–transfer parameterizations; thus, it can estimate *Et* spatially and temporally complete at satellite-resolvable resolution. Uncertainties of LSMs attribute to the boundary conditions, atmospheric forcing, and model physics. Thus none of the three is the best possible method to estimate *Et*; in other words, *Et* estimates from in situ, satellite, and LSMs are always necessarily to facilitate monitoring the regional and seasonal variability of *Et*. More comprehensive intercomparison and reviewing are available in Jiménez et al.[6] Continuous maintenance and development of an in situ *Et* network, satellite *Et* estimates, and LSMs will be required in the future, and simultaneously, a better assimilation method between the three datasets must be developed.

ACKNOWLEDGMENTS

This work was supported by the NASA Modeling Analysis and Prediction project. The authors are grateful to Dr. D. Considine at NASA HQ. The NLDAS project is supported by NOAA's Climate Prediction Program for the Americas (CPPA) and by the NOAA Climate Program Office's Modeling, Analysis, Predictions, and Projections (MAPP) program. The data used in this effort were acquired as part of the activities of NASA's Science Mission Directorate, and are archived and distributed by the Goddard Earth Sciences (GES) Data and Information Services Center (DISC).

REFERENCES

1. Mitchell, K.E.; Lohmann, D.; Houser, P.R.; Wood, E.F.; Schaake, J.C.; Robock, A.; Cosgrove, B.A.; Sheffield, J.; Duan, Q.; Luo, L.; Higgins, R.W.; Pinker, R.T.; Tarpley, J.D.; Lettenmaier, D.P.; Marshall, C.H.; Entin, J.K.; Pan, M.; Shi, W.; Koren, V.; Meng, J.; Ramsay, B.H.; Bailey, A.A. The multi-institution North American Land Data Assimilation System (NLDAS): Utilizing multiple GCIP products and partners in a continental distributed hydrological modeling system. J. Geophys. Res. **2004**, *109*, D07S90. doi:10.1029/2003JD003823.

2. Xia, Y.; Mitchell, K.; Ek, M.; Sheffield, J.; Cosgrove, B.; Wood, E.; Luo, L.; Alonge, C.; Wei, H.; Meng, J.; Livneh, B.; Lettenmaier, D.; Koren, V.; Duan, Q.; Mo, K.; Fan, Y.; Mocko, D. Continental-scale water and energy flux analysis and validation for the North American Land Data Assimilation System project phase 2 (NLDAS-2): 1. Intercomparison and application of model products. J. Geophys. Res. **2012**, *117*, D03109, 27 pp. doi:10.1029/2011JD016048.

3. Pielke Sr., R.A. Influence of the spatial distribution of vegetation and soils on the prediction of cumulus convective rainfall. Rev. Geophys. **2001**, *39*, 151–177.

4. Baldocchi, D.; Falge, E.; Gu, L.; Olson, R.; Hollinger, D.; Running, S.; Anthoni, P.; Bernhofer, Ch.; Davis, K.; Evans, R.; Fuentes, J.; Goldstein, A.; Katul, G.; Law, B.; Lee, X.; Malhi, Y.; Meyers, T.; Munger, W.; Oechel, W.; Paw, K.T.; Pilegaard, K.; Schmid, H.P.; Valentini, R.; Verma, S.; Vesala, T.; Wilson, K.; Wofsy, S. FLUXNET: A new tool to study the temporal and spatial variability of ecosystem-scale carbon dioxide, water vapor, and energy flux densities. Bull. Am. Meteorol. Soc. **2001**, *82* (11), 2415–2434.

5. Mu, Q.; Zhao, M.; Running, S.W. Improvements to a MODIS Global Terrestrial Evapotranspiration Algorithm. R. Sensing Environ. **2011**, *115*, 1781–1800.

6. Jiménez, C.; Prigent, C.; Mueller, B.; Seneviratne, S.I.; McCabe, M.F.; Wood, E.F.; Rossow, W.B.; Balsamo, G.; Betts, A.K.; Dirmeyer, P.A.; Fisher, J.B.; Jung, M.; Kanamitsu, M.; Reichle, R.H.; Reichstein, M.; Rodell, M.; Sheffield, J.; Tu, K.; Wang, K. Global intercomparison of 12 land surface heat flux estimates. *J. Geophys. Res.* **2011**, *116*, D02102. doi:10.1029/2010JD014545.

Tropical Cyclones

Patrick J. Fitzpatrick
Mississippi State University, Stennis Space Center, Mississippi, U.S.A.

Abstract

This entry provides an overview on tropical cyclones. A tropical cyclone is a rotating, organized system of clouds and thunderstorms that originates over tropical or subtropical waters and has a closed low-level cyclonic circulation. The life cycle from genesis to mature, intense storms is discussed, the physical causes of their development are described, and their wind/cloud structure is presented. The causes of storm surge are described in detail, along with a storm surge scale based on intensity, bathymetry, storm size, and storm speed. The mechanisms that damage property and vegetation from tropical cyclone wind, storm surge, and flooding are explained in detail. The Saffir–Simpson wind damage scale and the Australian wind damage scale are contrasted. A global context is provided for tropical cyclone climatology, classifications, and terminology. A table is provided of classification schemes—from genesis to intense categories—for the seven ocean regions where tropical cyclones occur. The process for assigning tropical cyclone names in each ocean basin is presented. A brief overview on forecasting procedures is discussed.

INTRODUCTION USING A WESTERN HEMISPHERE PERSPECTIVE

A tropical cyclone is a rotating, organized system of clouds and thunderstorms that originates over tropical or subtropical waters and has a closed low-level cyclonic circulation. These systems form and intensify from ocean heat extraction through complex processes, with a warm upper-level center which distinguishes it from other windy weather systems. Due to the Earth's rotation, these storms spin counterclockwise in the Northern Hemisphere, and clockwise in the Southern Hemisphere. Both hemispheric spins are referred to as *cyclonic rotation* because the sense of spin about a local vertical axis is the same as the Earth's rotation when viewed from above.

Classifications for tropical cyclones vary globally. In this initial discussion, we use Western Hemisphere terminology, which covers the North Atlantic, Northeast Pacific, and North Central Pacific basins. Classifications in the Eastern Hemisphere are different compared with the Western Hemisphere. Furthermore, classifications vary between the Northwest Pacific, the North Indian Ocean, the Southwest Indian Ocean, the Southeast Indian Ocean, and the Southwest Pacific Ocean. For example, the term *hurricane* in the Western Hemisphere is the equivalent to the term *typhoon* in the Northwest Pacific Ocean. Their origins are locally derived. Hurricane is derived from the Caribbean Indian word hurican meaning "evil spirit," who in turn translated it from the Mayan storm god Hurakan. Typhoon is derived probably from the Chinese words tung fung ("a terrible storm of east winds"), ta fung (a "great wind"), or t'ai fung (either "eminent wind" or "wind of Taiwan").

A Greek god is also called Typhon (a storm giant and father of all "monsters") but it is unknown if there is a connection to the word typhoon or China. The global classifications will be discussed in the last section.

Wind speed is the fundamental demarcation of tropical cyclone classifications. International units are in ms^{-1}, and most countries have adopted the metric standard. However, some regions, countries, and professions prefer other wind units. The United States uses mph, where $1\ ms^{-1} = 2.2$ mph. Other countries prefer kmh^{-1}, where $1\ ms^{-1} = 3.6\ kmh^{-1}$. Mariners prefer knots, where $1\ ms^{-1} = 1.9$ knots. This text uses the metric standard of ms^{-1} for wind speed. Because the fastest winds in tropical cyclones occur near the ground, a standard height of 10 m (33 ft) is used. Winds are also averaged by either 1 or 10 minutes (depending on global region), and referred to as *sustained winds*.

A tropical cyclone does not form instantaneously. Initially a tropical cyclone begins as a *tropical disturbance* when a mass of organized, oceanic thunderstorms persists for 24 hours. The tropical disturbance becomes a *tropical depression* when a closed circulation is first observed and all sustained winds are less than $17\ ms^{-1}$. When these sustained winds increase to $17\ ms^{-1}$ somewhere in the circulation, it is then classified as a *tropical storm*. A tropical cyclone becomes a *hurricane* when sustained surface winds are $33\ ms^{-1}$ or more somewhere in the storm.

To clarify the life cycle, this entry discusses tropical cyclone formation in phases. The first phase is the *genesis stage*, and includes tropical disturbances and tropical depressions. The second phase includes tropical storms and hurricanes, called the *mature stage*. These phases are separated because most disturbances and a few depressions never reach tropical storm intensity, eventually dissipating, whereas mature systems may intensify, remain steady-state, or weaken. About 25% of mature systems evolve into a non-tropical system as they move out of the tropics; these systems may even re-intensify, becoming a storm with sustained winds up to 38 ms^{-1} and impacting regions outside the tropics. The genesis and mature stages will now be described.

LIFE CYCLE

Tropical disturbances form where there is a net inflow of air at the surface, known as *convergence,* resulting in ascent to compensate. As air rises, it will often saturate and form a cloud base. Once the air is saturated, ascent may be enhanced where the atmosphere is in a state of *static instability.* In a *statically unstable* atmosphere, ascending saturated air is less dense than surrounding unsaturated air. As a result, it accelerates upward because the air is buoyant relative to its environment, forming a broad spectrum of vertically growing cumulus clouds.

However, thunderstorm formation is common in the tropics, and is only a prerequisite. Several conditions must simultaneously exist for a tropical disturbance to develop a closed circulation and become a tropical depression. First, the disturbance must be in a *trough,* defined as an elongated area of low *atmospheric pressure.* Atmospheric pressure is the weight of a column of air per area. All troughs away from the equator contain a weak, partial cyclonic rotation. These troughs can develop from a variety of mechanisms related to certain temperature and wind patterns beyond the scope of this entry. Some disturbances undergo the transition to tropical depression directly inside troughs. However, others experience this transition as *tropical waves* (called inappropriately *easterly waves* in the Western Hemisphere), which form when a trough "breaks down" into a cyclonic wave-like pattern in the wind field and travels westward away from the source region. About 60% of the Atlantic tropical cyclones actually originate from tropical waves, which form over Africa and propagate into the Atlantic. Tropical waves are fairly persistent features, and can propagate long distances. In the Atlantic about 55–75 tropical waves are observed, but only 10–25% of these develop into a tropical depression or beyond.

The second condition required for genesis is a water temperature of generally at least 27°C (80°F). Heat and water vapour transferred from the ocean to the air generates and sustains static instability (and therefore thunderstorms) in the disturbance. The third genesis condition is minimal *vertical wind shear,* defined as the difference between wind speed and direction generally at 12 km (40,000 ft) aloft and near the surface. In other words, for genesis to occur, the wind must be roughly the same speed and from the same direction above the surface at all height levels in the atmosphere. This allows thunderstorms and the wind structure to grow unimpeded. The three

conditions—warm water, surface trough, and weak vertical wind shear—are generally necessary but insufficient conditions to develop closed rotation (and, by definition, a tropical depression). Furthermore, a few exceptions exist with strong vertical wind shear or water temperature less than 27°C, but a depression still forms. Because of insufficient understanding on the genesis stage, much research is currently underway on this vexing forecast problem.

Once a tropical depression forms, the favorable conditions of wind shear, warm water, and complete cyclonic rotation provide the "ignition" process for further development. The ascending air, developing warm-core aloft, and decreasing surface pressure in the depression stimulates low-level inflow towards the center. The cyclonic circulation of the disturbance increases. Typically, the genesis timeframe of both disturbance and depression lasts for several days or longer. However, under ideal conditions, they can evolve much quicker. When the cyclonic sustained winds increase to 17 ms^{-1} somewhere in the depression, the system is upgraded to a tropical storm. At this point, the mature stage begins.

For tropical storm intensification into a hurricane, the same conditions that allowed its initial development (warm water, moist air, and weak wind shear) must continue. However, a system with closed rotation develops faster compared with the genesis stage because the fluxes of heat and moisture from the ocean become more efficient at faster winds, and because a larger percentage of the heat is retained in the storm center. The column of air begins to warm, which decreases atmospheric pressure. More air will flow toward the lower surface pressure, trying to redistribute the atmosphere's weight, resulting in faster winds. The faster cyclonic winds also enhance convergence. Both factors increase thunderstorm production and low-level inflow. A feedback mechanism now occurs in which faster cyclonic winds maintain thunderstorms by fluxes and convergence, dropping surface central surface pressure more, creating stronger inflow and faster cyclonic winds, and so on. When sustained winds reach 33 ms^{-1} somewhere in the storm, it is classified as a hurricane.

Water temperature is unquestionably linked to these storms' development. Tropical cyclones rarely form over water colder than 27°C. They also weaken dramatically if a mature system moves over water colder than 27°C, or if they make landfall, because their heat and moisture source has been removed. The warmer the water, the greater are the chances for genesis, the faster is the rate of development, and the stronger these storms can become. Because tropical cyclones mix the ocean column, it is also important the warm surface water is at least 60 m (200 ft) thick. Under conditions of prolonged weak shear and an ocean surface temperature greater than 29°C (85°F), sustained winds may reach 89 ms^{-1}.

Fortunately, few hurricanes reach their maximum potential because of some inhibiting factor. Conditions that stop intensification include wind shear, landfall, dry air intrusion, storm-induced ocean mixing or upwelling, and movement over colder water. Occurrence of any (or a combination) of these influences will stall development or cause weakening. Furthermore, even with no inhibiting factors, strong hurricanes rarely maintain their intensity. The internal physics of a hurricane preclude a strong steady-state storm for more than 1–2 days. Instead, strong wind conditions promote interior adjustments near the storm's center, weakening it (see discussion in next section on concentric eyewalls).

Persistent occurrence of any or several inhibiting factors will cause disintegration of a tropical cyclone. Of these possibilities, most dissipating cases occur due to landfall or movement over colder water. Tropical cyclones making landfall rapidly decay. If the storm remains over land, its maximum sustained winds will decrease on average 20 ms⁻¹/day and the rate of dissipation is even faster for initially strong storms. Thirty-six hours after landfall, inland tropical cyclones rarely contain winds above tropical depression strength.

TROPICAL CYCLONE STRUCTURE

The structure of a tropical cyclone is one of the most fascinating features in meteorology (Fig. 1). Distinct cloud patterns exist for each stage of a tropical cyclone's life cycle. These distinct patterns allow meteorology centres worldwide to classify these systems as a depression, tropical storm, hurricane, or major hurricane based on cloud organization. During genesis, typically a mass of thunderstorms with a weak rotation is first observed. Usually these thunderstorms will temporarily dissipate or weaken, leaving a residual circulation. Often, no further development occurs. But should thunderstorms return (sometimes within 12 hours, but more often taking several days), and a closed circulation develops, a depression has formed. A dominant band of clouds gradually takes on more curvature around a cloud minimum center. When the band curves at least one-half distance around the storm center, typically tropical storm intensity has been achieved. A cloud shield extends further out with squall lines of thunderstorms of less vertical growth than near the storm centre. These thunderstorm bands are known as *spiral bands*. Between the bands is light-to-moderate rain or areas of sinking air (downdrafts).

As the tropical storm strengthens, the dominant cloud band continues coiling around the center. When the band completely coils around the center, hurricane intensity usually has been reached. At this point, a clear region devoid of clouds forms in the center known as the *eye*, surrounded by a ring of thunderstorms known as the *eyewall*. The eyewall contains the fiercest winds and often the heaviest rainfall,

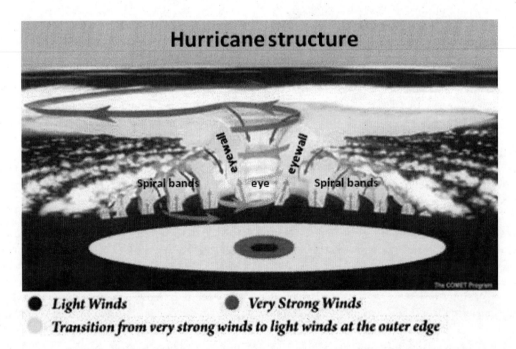

Fig. 1 Cross-section of a mature hurricane showing the eye in the center surrounded by the eyewall and spiral bands circling its environment. Arrows depict a surface cyclonic airflow, air rising in the eyewall and spiral bands, air descending between the spiral bands in the eye, and anticyclonic outflow aloft. The circular colors depict the horizontal surface wind structure, in which winds are relatively calm in the eye, fastest in the eyewall, and decrease away from the storm center. [The source of this material is the COMET® Website at http://meted.ucar.edu/ of the University Corporation for Atmospheric Research (UCAR), sponsored in part through cooperative agreement(s) with the National Oceanic and Atmospheric Administration (NOAA), U.S. Department of Commerce (DOC). © 1997–2013 University Corporation for Atmospheric Research. All Rights Reserved.]
Source: Adapted from COMET MetEd.[1]

making this feature the most dangerous part of a hurricane. The eyewall slants outward with height giving the eye a "coliseum" appearance, as if one is in a giant football stadium made of clouds. In the eye, winds become weak, even calm! This transition from hurricane force winds to calm is rather sudden (often within minutes). The average eye size diameter is 32–64 km (20–40 miles). Typically an eye starts at about 56 km wide during the transition from tropical storm to hurricane. As the hurricane intensifies, the eye usually contracts. Small eyes correlate to intense hurricanes, with diameters as little as 14 km (9 miles). However, intense hurricanes with large eyes also occur, and there is considerable variance with these numbers.

Sometimes a second eyewall forms outside the original eyewall about 64–96 km (40–60 miles) from the center. This outer eyewall "cuts-off" off the inflow to the inner eyewall, causing the inner one to weaken and dissipate. Because the eye is wider, temporary weakening occurs. The outer eyewall will begin to contract inward to replace the inner eyewall, and approximately 12–24 hours later, intensification resumes. This internal adjustment process, known as the *concentric eyewall cycle*, is one reason strong hurricanes experience intensity fluctuations in otherwise favourable conditions.

Outside the eyewall, weaker spiral bands accompanying the hurricane typically affect a large area. The average width of a hurricane's cloud shield is 800 km (500 miles), but varies tremendously. However, cloud size is not an accurate indication of strong wind coverage. One criterion for size is the radial extent of tropical storm-force winds ($17 \ ms^{-1}$). Mariners and navy fleets typically avoid winds stronger than $17 \ ms^{-1}$. Furthermore, many hurricane preparedness exercises require completion before tropical storm-force winds begin; for example, bridges are often closed if winds exceed $17 \ ms^{-1}$, impeding last-minute evacuees.

DAMAGE FROM TROPICAL CYCLONES

Coastal communities devastated by strong hurricanes usually take years to recover. Many forces of nature contribute to the destruction. Obviously, tropical cyclone winds are a source of structural damage. As winds increase, stress against objects increases at a disproportionate rate. For example, a $22\text{-}ms^{-1}$ wind causes a stress of $36 \ kg \ m^{-2}$ ($7.5 \ lb \ ft^{-2}$). In $45\text{-}ms^{-1}$ winds, stress becomes $147 \ kg \ m^{-2}$ ($30 \ lb \ ft^{-2}$). Although sustained wind is used as a reference standard, the actual wind will be 20–30% faster or slower than the sustained wind at any instantaneous period. These gusts often initiate structure damage, typically beginning at the roof. The removal of roof coverings often occurs with wind gusts at $36 \ ms^{-1}$, roof decks at $45 \ ms^{-1}$, and roof structure at $54 \ ms^{-1}$. Wind interacting with a building is deflected over and around it. Positive (inward) pressures are applied to the windward walls and try to "push" them down. Negative (outward) pressures are applied to the side and leeward walls. The resulting "suction" force can peel away any exterior covering (siding, shingles, and so on). Negative (uplift) pressures are applied to the roof, especially along windward eaves, roof corners, and leeward ridges, similar to the lifting aerodynamics on aircraft wings, and cause roof failure.

Building damage occurs in other ways. Wind damage begins with exterior items such as television antennas, satellite dishes, unanchored air conditioners, wooden fences, gutters, storage sheds, carports, and yard items. As the wind speed increases, cladding items on buildings become susceptible to wind damage including vinyl siding, gutters, roof coverings, windows, and doors. Debris is also propelled by strong winds, compounding the damage. The weak points to internal building exposure are usually through roof, window, door, or garage damage. Any cause which compromises the inside of a building usually results in extreme damage to that structure from rain and wind. Damage patterns tend to be uneven due to the streakiness of wind gusts; building construction quality; building age; wind interference, wind acceleration, or vortex shedding from other buildings; proximity to open fields, trees, or vegetation. In addition to structure damage, large tree branches will snap, trees will be uprooted, and power poles will be toppled (or power lines snapped), resulting in power outages which can last weeks.

Floods produced by the rainfall can also be quite destructive, and are a leading cause of tropical cyclone-related fatalities. A majority (57%) of the 600 U.S. tropical cyclone-related deaths between 1970 and 1999 were associated with inland flooding. Tropical cyclone rainfall averages 150–300 mm (6–12 in.) at landfall regions, but varies tremendously. Heavy rainfall is not just confined to the coast. The remnants of tropical cyclones can bring heavy rain far inland, and is particularly dangerous in hills and mountains where acute concentrations of rain turn tranquil streams into raging rivers in a matter of minutes or cause mudslides. In addition, mountains "lift" air in tropical cyclones, increasing cloud formation and rainfall. Rainfall rates of 300–600 mm (1–2 ft) per day are not uncommon in mountainous regions when tropical cyclones pass through.

Although all these elements (wind, rain, floods, and mudslides) are dangerous, historically most people have been killed in the *storm surge*, defined as an abnormal rise of the sea along the shore associated with cyclonic wind storms such as tropical cyclones. Death tolls in coastal regions can be terrible for those who do not evacuate inundation zones. Most storm surge fatalities are associated with structural collapse or drowning. The worst natural disaster in the United States history occurred in 1900 when a hurricane-related 5 m (16 ft) storm surge inundated Galveston Island, TX, U.S.A., and claimed over 6000 lives. In 1893, nearly 2000 were killed in Louisiana and 1000 in South Carolina by two separate hurricanes. Camille in 1969 (181 fatalities), Katrina in 2005 (1836 fatalities), and Ike in 2008 (20–40 fatalities) are the latest U.S. examples.

Dew—Wind

However, these U.S. statistics pale compared with the lives taken globally. Regions around the North Indian Ocean, Japan, and China have experienced fatalities by storm surge ranging from 10,000 to 300,000 in the last few centuries, with the latest tragedy of 138,000 killed in Myanmar by Cyclone Nargis in 2008. In contrast, the 1900 Galveston hurricane ranks 16th globally in terms of storm surge fatalities based on the latest research.

Storm surge is caused by several factors. The main contribution to the storm surge results from the interaction of coastal water with wind-driven water at landfall known as the *wind stress effect*. In deep offshore water, a vertical ocean circulation occurs, which compensates surface water wind transport, and no storm surge from wind forcing exists. But as a tropical cyclone approaches the shoreline and begins to interact with the ocean floor and land boundaries, this circulation is disrupted and water levels must rise to compensate. As a tropical cyclone approaches a region, coastal waters begin to rise gradually from the wind stress effect, then quickly at landfall.

A current also develops parallel to the shoreline ahead of the tropical cyclone, causing water to rise in response to the Earth's rotation, known as a *surge forerunner*. The forerunner effect is dangerous because it peaks before landfall, sometimes trapping residents before they complete evacuation. For example, many of the Hurricane Ike fatalities occurred when roads on the Bolivar Peninsula flooded one day before landfall, cutting off the only escape route.

The final factor to storm surge is a minor contribution associated with the low pressure of a hurricane, which causes a bulge of water known as the *inverse barometer effect*. For every 10 mb pressure drop, water rises 10 cm (3.9 in.). The contribution ranges from 0.3–1 m (1–3 ft) elevated water depending on intensity, peaking at landfall.

Other factors that determine a surge's height include coastal bathymetry, storm intensity, storm size, and storm translation speed (Fig. 2). All other factors being equal, the most intense tropical cyclones produce higher storm surge. However, for a given intensity, bathymetry causes tremendous variation in surge heights. A coastline with a shallow sloping ocean floor is prone to higher storm surge than a coastline with a steep ocean shelf nearby. For example, a Category 3 produces a surge of 1.5 m (5 ft) in very deep bathymetries but 6.7 m (22 ft) in very shallow, with most values in between for a given coastline. Low-lying regions adjacent to extended shallow seas, such as the Gulf of Mexico in the southern United States, and the Bay of Bengal bordering Bangladesh and India, are particularly vulnerable to the storm surge. In shallow bathymetries, large, slow-moving tropical cyclones produce more surge than average-sized, average speed hurricanes for the same intensity (Fig. 2). The storm surge is always highest on the side of the eye corresponding to onshore winds, which is usually the right side of the point of landfall in the Northern Hemisphere, called the *right front quadrant*.

Storm surge at high tide

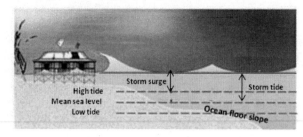

Storm surge for different bathymetries

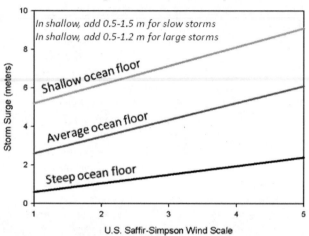

Fig. 2 (Top) Graphical portrayal of the storm surge at high tide impacting a structure along a shoreline. The definition of storm tide includes the storm surge plus water elevation departures from mean sea level due to the tide cycle. Waves are superimposed on the surge. (Bottom) Storm surge relationship to ocean floor slope and hurricane intensity using the U.S. wind scale. Note the large differences between shallow and steep bathymetry for a given intensity. Storm surge is 0.5–1.5 m higher for slow storms in shallow water, and is 0.5–1.2 m higher for large storms in shallow water. Storm size and speed only marginally modifies surge elevation in average and steep ocean floors, and is neglected.
Source: Top figure adapted from Australian Bureau of Meteorology.[2] Bottom figure adapted from Fitzpatrick et al.[3]

The total elevated water includes two additional components—the astronomical tide and ocean waves. The astronomical tide results from gravitational interactions between the Earth, moon, and sun, producing high and low tides every 12–24 hours. Should the storm surge coincide with high tide, additional inundation occurs. The total water elevation due to surge and tides is known as the *storm tide* (Fig. 2), but in practice it is difficult to distinguish from storm surge during post-storm inspections. Waves are superimposed on top of the storm surge, running up past inundation boundaries, spilling over structures, sloshing against and inside structures, and enhancing damage. Waves may also contribute to inundation through a process called *wave setup* in which incoming waves exceed retreating waves, and is currently the subject of considerable research.

Water is very heavy, weighing 1025 kg m^{-3} (64 lb ft^{-3}), and dense. Therefore, waves possess much greater force than wind, and can undermine foundations and destroy support walls. Furthermore, if a wave is breaking, the force is 4–7 times greater than moving water. In addition, waves also cause an "uplift" force if the waves are under a horizontal structure, displacing the object and even toppling it, including bridge decks. Even without wave forces, water is very damaging. Buoyancy causes a lifting force on submerged wood buildings and beam foundations, damaging structures. Storm surge currents contain forces equal to or slightly greater than wind. Destruction can be thorough, leaving only slabs behind near the coast. Erosion of beaches and islands also is severe, and coastal highways can be devastated. Further inland where currents are weaker with little wave activity, inundation is still damaging. Most possessions, floors, and drywall that are submerged will be ruined, plus mold will grow on the items when the water retreats.

The United States has developed the *Saffir–Simpson hurricane wind scale*, which describes the expected level of wind damage for a given hurricane intensity. This scale classifies hurricanes into five categories according to maximum sustained wind and escalating potential property damage. Categories 1–5 correspond to a lower threshold of 33, 43, 50, 58, and 70 ms^{-1}, respectively. Although all categories are dangerous, categories 3, 4, and 5 are considered *major hurricanes*, with the potential for widespread devastation and loss of life. Whereas only 24% of U.S. landfalling tropical cyclones are major hurricanes, they historically account for 85% of the damage. However, while low-intensity hurricanes cause less physical devastation, preventative measures are still required during a landfall threat. Low-intensity hurricanes also disrupt regional economic activity for several days to several weeks through business interruption and evacuation costs. Additionally, freshwater flood threat is present for any tropical storm or hurricane regardless of intensity.

It should be noted that storm surge ranges used to be included in the Saffir–Simpson scale, but were removed in 2010 since surge depends not only on wind but bathymetry, storm size, and translation speed. Central pressure was also removed because it does not correspond to eyewall winds but instead correlates to wind structure. The wind scale also does not include damage from floods, and small-scale intense features such as imbedded tornadoes. Because damage is dependent on so many factors, preparations for a tropical cyclone impact should be thorough.

GLOBAL TROPICAL CYCLONE PREDICTION, CLIMATOLOGY, AND TERMINOLOGY

Different naming nomenclatures related to intensity classifications are used worldwide, and can be confusing. A global context of tropical cyclones will now be discussed.

Tropical cyclones generally occur in every tropical ocean except the South Atlantic and eastern South Pacific (Fig. 3). Locations include the tropical North Atlantic Ocean (including the Caribbean Sea and Gulf of Mexico); the Northeast Pacific (off the west coast of Mexico); the Central North Pacific (near Hawaii); the Northwest Pacific (including the China Sea, Philippine Sea, and Sea of Japan); the North Indian Ocean (including the Bay of Bengal and the Arabian Sea); the Southwest Indian Ocean (off the coasts of Madagascar and extending almost to Australia); the Southeast Indian Ocean (off the northwest coast of Australia); and the Southwest Pacific Ocean (from the east coast of Australia to about 140°W). In addition, with improved satellite monitoring capabilities, several (mostly) weak, short-lived tropical cyclones in the South Atlantic Ocean have been unofficially identified the last two decades. Cyclone Catarina (2004) is the only official South Atlantic tropical cyclone with category 2 winds, named after the state in Brazil it made a landfall.

Regional Specialized Meteorological Centers (RSMCs) and Tropical Cyclone Warning Centers (TCWCs) monitor and forecast tropical cyclones used by each country, territory, and national meteorological service in their region (Fig. 3). RSMCs provide a spectrum of weather forecast products and responsibilities, which include hurricane forecasts, whereas TCWCs specialize in tropical cyclone services. Official warnings are the responsibility of each country/territory's national meteorological service. Monitoring is conducted with surface observations from land-based instruments; buoys and ships; upper-air balloon instruments, launched generally twice per day; satellite imagery; and satellite-derived diagnostics. Because these observations do not provide complete data coverage, the United States is the only nation to also use expensive reconnaissance aircraft.

Forecast agencies issue predictions for storm track, intensity, rainfall, storm surge, and other parameters. The priority is predicting tropical cyclone movement. To understand tropical cyclone motion, it is helpful to use the analogy of a wide river with a small eddy rotating in it. The river generally transports the eddy downstream, but the eddy will not move straight because the speed of the current varies horizontally, causing a slight left or right deviation, which also changes the eddy's speed of movement. Furthermore, this eddy's rotation may alter the current in its vicinity, which in turn will alter the motion of the eddy.

Likewise, one may think of a tropical cyclone as a vortex embedded in a river of air. The orientation and strength of large-scale pressure patterns generally dictate the storm's motion, except that the steering depends on both the horizontal and vertical distribution of the steering current. The tropical cyclone can also interact with steering currents, and even other nearby tropical cyclones, ultimately altering its own currents. Other factors impact track such as the Earth's rotation and storm asymmetry, and are beyond the scope of this entry. This is a difficult forecast problem, especially in data-void regions such as the ocean.

Dew—Wind

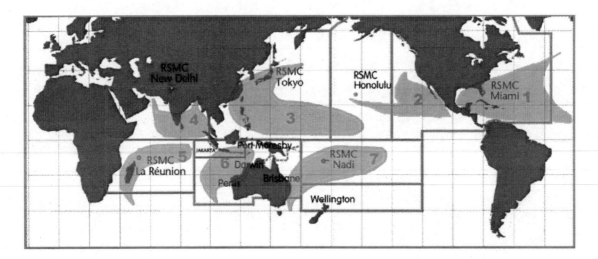

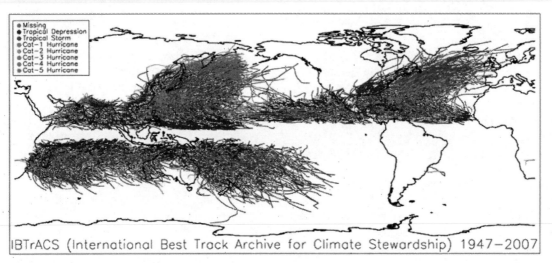

Fig. 3 (Top) Global genesis regions (in light gray), the Regional Specialized Meteorological Centers (RSMCs), and Tropical Cyclone Warning Centers (TCWCs; Wellington, Brisbane, Perth, Darwin, Port Moresby, and Jakarta). For example, the National Weather Service National Hurricane Center and the Japan Meteorological Agency Typhoon Center are RSMCs. (Bottom) Global tropical depression, tropical storm, and hurricane tracks by U.S. Saffir–Simpson wind categories for 1947–2007. "Missing" indicates a position was available but intensity data was missing.
Source: Top figure adapted from World Meteorological Organization.[4] Bottom figure courtesy of Knapp et al. adapted from.[5] Bottom figure is © Copyright March 2010 AMS.

The complexity of track—as well as intensity and storm surge—predictions requires the use of computer models. Computer models ingest current weather observations and approximate solutions to complicated equations for future atmospheric values such as wind, temperature, and moisture. Forecast agencies use a suite of models that differ in their mathematical assumptions and complexities in describing atmospheric processes. The most complex models must be run on the fastest computers in the world, known as supercomputers.

Many U.S. residents perceive the North Atlantic Ocean basin as a proliferate producer of hurricanes due to the publicity these storms generate. In reality, several oceans produce more hurricanes annually than the North Atlantic. For example, the most active ocean basin in the world—the Northwest Pacific—averages 17 hurricanes per year. The

second most active is the Northeast Pacific, which averages nine hurricanes. In contrast, the North Atlantic mean annual number of hurricanes is six. Table 1 summarizes each basin's average number of hurricanes and total tropical cyclones using U.S. definitions.

Tropical cyclone season is typically limited to the warm seasons. In the Atlantic, the official tropical cyclone season begins June 1 and ends November 30, although activity has been observed outside this timeframe. However, tropical cyclones are most numerous and strongest in late summer and early fall. Exceptions to this late summer/early fall peak in tropical cyclone activity occur in certain parts of the world such as India since their monsoon trough moves inland during the summer. In addition, while activity does peak in late summer, the Northwest Pacific tropical cyclone season lasts all year.

Table 1 The mean number of total tropical cyclones (hurricanes and tropical storms), hurricanes, and major hurricanes per year in all tropical ocean basins using U.S. definitions

Tropical ocean basin	Mean annual tropical storms and hurricanes	Mean annual hurricanes	Mean major hurricanes
Northwest Pacific	26	17	8
Central North Pacific and Northeast Pacific	17	9	5
East Coast Australia and Southwest Pacific	10	5	2
West Coast Australia and Southeast Indian	8	4	1
North Atlantic	12	6	2
Southwest Indian	9	5	2
North Indian	5	2	Between 0 and 1
South Atlantic	0	0	0
Southeast Pacific	0	0	0
Global	86	47	20

Improved satellite monitoring capabilities have also unofficially identified several (mostly) weak, short-lived tropical cyclones in the South Atlantic Ocean the last two decades. Cyclone Catarina (2004) is the only official South Atlantic tropical cyclone with category 2 winds, named after the state in Brazil it made landfall. The official average is zero in the South Atlantic, but apparently a tropical cyclone forms every few year.

Tropical cyclone classification schemes in different ocean basins

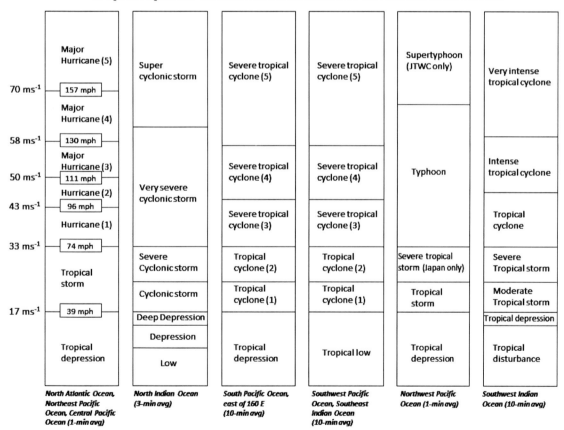

Fig. 4 Classification schemes for the six ocean regions where tropical cyclone terminology varies from genesis to intense categories. The U.S. definitions are used for reference in the leftmost column. Also shown are the wind damage scales used in the United States and in Australia. Wind averaging schemes are shown for each basin. One-minute averaging results in winds that are approximately 14% more than 10-minute average winds (1-minute winds = 1.14 times 10-minute winds). Wind units are provided in ms^{-1} and mph on the left side.

The classification schemes vary around the globe by assorted names, different time averaging for sustained winds, and diverse wind thresholds. Figure 4 summarizes global classifications. They all recognize classes from genesis to strong storms with consistent thresholds of 17 and 33 ms^{-1}, but use their own terms and wind thresholds for other delineations. For example, in the Northwest Pacific Ocean, the designation at 33 ms^{-1} is "typhoon," in India is "very severe cyclonic storm," and off Australia is "severe tropical cyclone." The rough equivalent term for major hurricanes in the North India Ocean is "super cyclonic storm," and in the Southwest Indian Ocean is "very intense tropical cyclone," although the wind thresholds are different. Finally, in the most active global region, the Northwest Pacific Ocean, storms with winds greater than 67 ms^{-1} are common enough that the Joint Typhoon Warning Center uses a special category exists for them—"supertyphoon."

Australia also uses a different wind damage scale than the United States (Fig. 4). Their category 1 begins at 17 ms^{-1} for minor damage, escalating to "some roof and structure damage" (Australian category 3), "significant damage" (Australian category 4), and "widespread damage" (Australian category 5). The U.S. Saffir–Simpson wind scale does not include tropical storms, the Australian wind scale of 3 is approximately the same as category 1 in America, Australian category 4 is approximately category 2–3 in America, and Australian category 5 is approximately category 4–5 in America.

When 17 ms^{-1} is reached, a name is assigned in most basins. However, the number of lists varies, as does whether a revolving cycle is used. For example, the Atlantic basin storms use a 6-year repeating list. However, the Northwest Pacific uses five lists, which do not rotate annually, but simply goes to the next list when the last name is reached. Whenever a storm has had a major impact in terms of damage and/or fatalities, any country affected by the storm can request the name be "retired" by agreement of the World Meteorological Organization.

All names are determined at international meetings, and reflect the regional culture. For example, Northeast Pacific storms tend to have Hispanic names, whereas central North Pacific storms have Hawaiian names. In regions with multiple countries, names will vary reflecting the different cultures. Southwest Indian tropical cyclone names are based on French, Madagascar, and many African countries. Northwest Pacific uses Asian nomenclature that do not reflect personal names but instead natural objects (such as trees, flowers, rivers, or animals), "stormy" adjectives (such as swift, strong, sharp, fast), and gods or goddesses.

In addition, the Philippines assign local names to tropical cyclones in their region, even though it will also have an "official" international name, resulting in two names for the same storm. The two purposes for names are to facilitate easy regional communication of a storm between forecasters and the general public, and for historical context.

BIBLIOGRAPHY

1. Fitzpatrick, P.J.; *Hurricanes – A Reference Handbook*; ABC-CLIO: Santa Barbara, CA, 2005; 412 pp.

Excellent background information and imbedded links on hurricanes are available at the following websites (accessed September 2, 2013):

1. The National Hurricane Center: http://www.nhc.noaa.gov
2. Australian Bureau of Meteorology: http://www.bom.gov.au/cyclone/
3. India Meteorological Department: http://www.imd.gov.in/section/nhac/dynamic/cyclone.htm
4. World Meteorological Organization Tropical Cyclone Programme: http://www.wmo.int/pages/prog/www/tcp/index_en.html
5. Frequently Asked Questions About Hurricanes: http://www.aoml.noaa.gov/hrd/tcfaq/tcfaqHED.html
6. International Best Track Archive for Climate Stewardship: http://www.ncdc.noaa.gov/oa/ibtracs/index.php
7. Historical Hurricane Tracks: http://www.csc.noaa.gov/hurricanes/

References used in figures (all websites accessed September 2, 2013)

1. COMET MetEd. Tropical cyclones. In *Introduction to Tropical Meteorology, 2nd edition*; Chapter 8, 2013. Available at https://www.meted.ucar.edu/training_module.php?id=868
2. Australian Bureau of Meteorology. Storm surge preparedness and safety, 2013. Available at http://www.bom.gov.au/cyclone/about/stormsurge.shtml
3. Fitzpatrick, P.J.; Lau, Y.; Tran, N.; Li, Y.; Hill, C.M. The role of bathymetry in local storm surge potential. Nat. Hazards Earth Syst. Sci. **2013**. In revision.
4. World Meteorological Organization. Organizational structure, 2013. Available at http://www.wmo.int/pages/prog/www/tcp/organization.html. Also see http://www.wmo.int/pages/prog/www/tcp/Advisories-RSMCs.html.
5. Knapp, K.R.; Kruk, M.C.; Levinson, D.H.; Diamond, H.J.; Neumann, C.J. The International Best Track Archive for Climate Stewardship (IBTrACS): Unifying tropical cyclone best track data. Bull. American Meteor. Soc. **2010**, *91*, 363–376.

Urban Heat Islands

James A. Voogt

Department of Geography, University of Western Ontario, London, Ontario, Canada

Abstract

The urban heat island (UHI) represents the relative warmth of urban areas compared to their nonurbanized surroundings. It can be observed in surface, air, and subsurface temperatures. The UHI arises because of changes to surface and atmospheric characteristics in cities that affect the surface energy budget. Urban surface cover, structure and material properties along with human energy and water use are important controls on the surface energy budget and therefore the UHI. The magnitude of the UHI is limited by weather conditions, especially wind and cloud cover as well as time of day and season. The geographical setting of the city also affects the temporal and spatial characteristics of the UHI. UHIs can impact human health and comfort, energy and water consumption, growing season length and biological processes, and can affect the weather over and downwind of cities. The UHI represents an unintentional change to climate at the scale of a city. Efforts to alter urban temperatures through application of "cool technologies" have shown promise and can help reduce the negative impacts of warmer urban temperatures associated with both the UHI and from large-scale climate change.

INTRODUCTION

The UHI represents the relative warmth of the air, surface, and subsurface in urban areas compared to their nonurbanized surroundings. It arises from the consequences of urban development on the surface and atmospheric characteristics in urban areas and represents an unintentional change to the climate of cities. The magnitude (or intensity) of an UHI is defined by the difference between urban and nonurban temperatures.

TYPES OF UHIS AND THEIR CHARACTERISTICS

The relative warmth of the urban atmosphere, surface, and substrate materials leads to several distinct heat islands (Fig. 1, Table 1).[1] The *urban canopy layer heat island* is the best-known heat island.[3] It describes the warmth of a near-surface air layer that extends from the ground to the mean height of the buildings and vegetation. It is the most commonly measured layer of the atmosphere in cities for air temperatures because it is easily accessed by ground-based instruments that are either fixed or mobile (e.g., on vehicles). Canopy layer heat islands are typically largest at night (Fig. 2) during weather conditions in which winds are calm and skies are clear. Heat island intensities under such conditions are typically a few degrees Celsius in large parts of most cities and may exceed 10°C for the most densely developed parts of a large city. They are smaller and sometimes even slightly negative (representing an

urban "cool island") during clear daytime conditions; together, these indicate a strong temporal variability of the heat island over the course of a day under calm, clear weather conditions.[1,3] When averaged over all weather conditions, the heat island magnitude in the urban canopy layer is typically 1–3°C.

The spatial structure of the heat island shows a pattern of isotherms—isolines of equal temperature—that follow the border of the city with a relatively tight spacing when winds are calm—hence, the topographic analogy with an island (Fig. 2). The spatial temperature gradient, the change of temperature over space, is typically large near the edge of the "island," forming the so-called "cliff" in response to the large relative change in surface characteristics in the region of rural to urban transition. Within the urban area, temperatures in the urban canopy layer are strongly controlled by the local characteristics of the urban surface. They may show substantial spatial variability (on the order of several °C) when weather conditions are favourable. The highest nighttime temperatures, the peak of the heat island, are usually associated with the area of most intense urban development and some warmer air is often transported horizontally downwind of the city.

Above the urban canopy layer, the urban *boundary layer heat island* represents an urban-scale warming through the depth of the urban boundary layer (up to 1–2 km during daytime with clear skies and a few 10s to 100s of meters at night) that has a smaller magnitude and is much less spatially and temporally variable than that of

Encyclopedia of Natural Resources DOI: 10.1081/E-ENRA-120047640

Dew—Wind

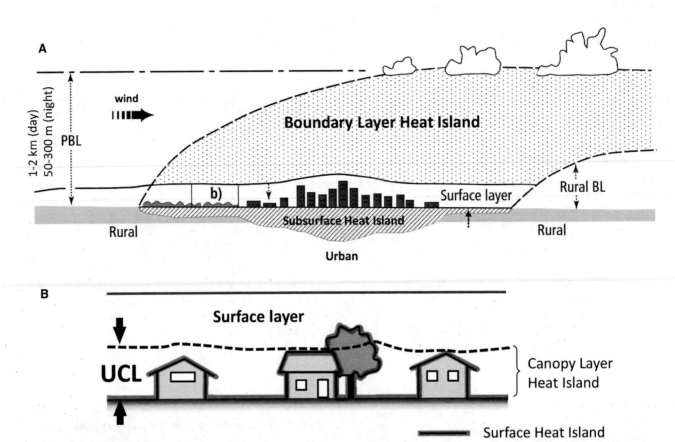

Fig. 1 Schematic diagram showing the different heat island types and their locations within an urban area. Shading represents the affected volume, but it does not show the expected spatial variations. (**A**) Under fair weather conditions, within the first 1–2 km of the atmosphere, known as the planetary boundary layer (PBL), a boundary layer heat island exists with a characteristic plume shape extending in the downwind direction. (**B**) Within the urban canopy layer (UCL), air volume a canopy layer heat island exists (warmer air temperatures). The surface heat island consists of all surfaces, but usually only a subset of these surfaces can be seen from any one observation point. Below the surface, a subsurface heat island exists.
Source: Adapted from Oke.[2]

the underlying urban canopy layer heat island. The heat island magnitude here is positive both day and night. The warmed boundary layer air above the city is often transported downwind by the mean wind leading to a plume of warmer air above downwind non-rural areas (Fig. 1). Measurements of the urban boundary layer heat island are relatively rare as access to this layer is difficult. Thermometers must be mounted on tall towers, balloons or aircraft; or remote-sensing techniques using ground-based instruments (Table 1).

The *surface UHI* represents the relative warmth of the surfaces in cities compared to those in rural areas. Surface temperatures can be observed using remote sensing techniques from satellite or aircraft-mounted instruments. These instruments provide a spatially continuous view of the urban area but tend to be biased toward viewing the highest upward facing unobstructed surfaces such as roofs, open spaces, and the tops of trees.[5,6] The implications of this method of observing the heat island must be borne in mind when interpreting remotely sensed images of the surface UHI. Under clear skies and light winds,

daytime surface temperatures seen from above are much warmer than air temperatures in the canopy layer and show more spatial variability (Fig. 2 and Table 1) due to the juxtaposition of surfaces with contrasting properties in urban areas (e.g., hot dry rooftops adjacent to an irrigated lawn). However, the variability is not always easily observed—for example when the sensors have spatial resolutions that are substantially larger than the scale of the surface structure that provides the temperature variability. This can be the case when using satellite observations of surface temperature. The daytime surface heat island is positive (Fig. 2) and is controlled by the amounts of impervious and vegetated surfaces as well as the surface characteristics of the rural reference. At night, the overall spatial structure and magnitude of the surface heat island is similar to that of the canopy layer heat island, reflecting the importance of surface controls (Table 2) on the near-surface air temperature. The spatially averaged surface heat island magnitude under favorable weather conditions is larger by day than at night,[7] making it the reverse of that in the canopy layer air. Use of simple

Table 1 Heat island types and their spatial and temporal characteristics

Heat island type	Variable affected	How measured	Spatial characteristics	Temporal characteristics
Canopy Layer Heat Island	Air temperature	Fixed or mobile (vehicle-based) measurements using thermometers.	Shows significant spatial variability associated with important elements of surface structure: building height to width ratio, amount of vegetation, topographic features.	Largest at night. Magnitude grows rapidly in late afternoon and early evening. May be negative (a cool island) during daytime. Highly sensitive to wind and cloud.
Boundary Layer Heat Island	Air temperature	Thermometers mounted on very tall towers, balloons, kites, or aircraft. Remote-sensing from ground-based instruments.	Exhibits a "domed" structure in near-calm conditions and a distinct downwind "plume" as winds increase. Boundary layer depth is 1–1.5 km by day but only 50–300 m at night. Heat island magnitude decreases with height in the boundary layer by night and is approximately constant by day.	Shows relatively small diurnal variation. Sensitive to wind and cloud.
Surface Heat Island	Surface temperature	Remote sensing from towers, aircraft, or satellites.	Significant spatial variability associated with variations in surface characteristics including: shading, surface orientation, moisture status, thermal properties, surface reflectivity and vegetation coverage.	Largest during daytime in summer season. Nighttime value is also positive and largest in summer. Highly sensitive to weather conditions.
Subsurface Heat Island	Ground temperature	Ground or borehole temperature measurements.	Spatial variability occurs due to variations in surface characteristics. Urban areas show a greater depth of temperature decrease from the surface before temperatures reverse to show the geothermal gradient.	Temporal response is increasingly lagged with depth so that subsurface patterns reflect past conditions at the surface. Temporal influences below the first meter are typically only seasonal or longer.

Dew—Wind

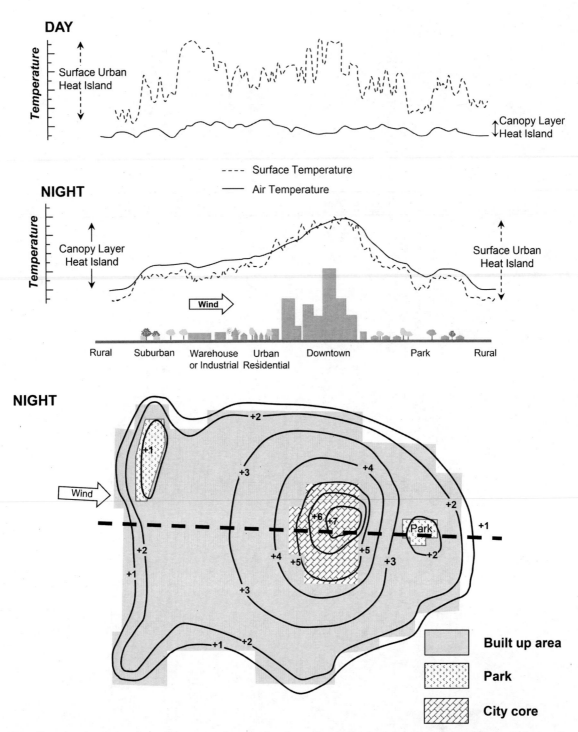

Fig. 2 Top: Conceptual cross-sections of the canopy layer and surface urban heat islands by day and night. Bottom: A plan view of the nighttime canopy layer heat island.
Source: Adapted from Voogt.[4]

correlations to relate surface and air temperatures and their heat islands is problematic without the use of fully coupled surface-atmosphere energy budget models that can represent the physical processes that govern the exchanges of energy between the surface and air.

The *subsurface UHI* is influenced by the relative warmth of the surface and atmosphere above it. It too shows spatial variability associated with the surface structure above, but this variability is much less than that of the surface and decreases with depth. It can be measured from temperature measurements deep in the soil or from boreholes or wells. Where substrate materials are water saturated, subsurface flows may also yield a warm plume associated with the subsurface heat island.

Table 2 Factors that lead to the formation of and/or influence the magnitude of urban heat islands

Heat island influence	Effect on UHI magnitude
Surface geometry	UHI magnitude increases as the ratio of building height to street width increases and view of the nighttime sky is obstructed.
Surface thermal properties	UHI magnitude increases with materials that make the city a better storer of heat; those with higher heat capacity and or thermal conductivity relative to rural materials.
Anthropogenic heat input	UHI magnitude increases as anthropogenic heat increases. This input can have large seasonal variations in some climates as well as intraurban spatial variability related to the density of development and magnitude of energy use.
City size	UHI magnitude tends to increase with city size up to a limiting amount.
Wind speed	UHI magnitude decreases rapidly as wind speed increases.
Cloud cover	UHI magnitude decreases as cloud cover increases.
Season	UHI magnitude is typically largest in the warm season in mid-latitudes. In high latitudes, the UHI is largest in the winter due to anthropogenic heat input. In tropical cities with distinct wet and dry seasons, the UHI is typically largest in the dry season.
Time of Day	The canopy layer UHI is largest at night (air temperatures). The surface urban heat island magnitude is larger during the day (clear, sunny conditions).

Source: Adapted from Arnfield.[10]

The exact spatial configuration of each heat island type in a given city depends on the layout of the city and its topographic setting; cities in coastal or mountainous areas will be influenced by sea- or mountain-valley breeze systems that can significantly impact the spatial pattern of the heat islands and their temporal development.[1]

CAUSES OF THE UHI: THE URBAN ENERGY BUDGET

The UHI arises from the fact that urban areas modify their surface and atmospheric characteristics relative to the surrounding rural regions. These alterations result in a modified *surface energy budget* in cities. The energy budget describes the exchanges of energy at the Earth's surface: how the radiation energy arriving at the city surface from the Sun and atmosphere is absorbed and partitioned into heat energy that is used to warm the air through mixing (convective sensible heat), heat energy used to evaporate water from the surface or to transpire water from vegetation (convective latent heat), and heat that is conducted and stored in the substrate materials. As a general rule, urban areas tend to increase both the relative amount of energy used to warm the air and that which is stored in the urban substrate materials, and to decrease the relative amount of energy directed toward evapotranspiration (the combined evaporation and transpiration of water) compared to the energy budget of surrounding nonurban areas. In addition, cities also directly add heat to their atmosphere through the use of energy. Building heating and cooling, electricity use and transportation are important contributors to this anthro-

pogenic energy component.[8] Therefore, we can say that altered energy budgets underlie UHI formation. To better understand the reasons for urban energy budget changes, we turn to an examination of the surface characteristics of urban areas.

CAUSES OF THE UHI: CHARACTERISTICS OF THE URBAN SURFACE

The surface characteristics of urban areas provide important controls on the surface energy budget and therefore on the UHI. Three groups of urban surface characteristics are important. First, the surface coverage (and relative fractions) of buildings, vegetation and impervious surfaces are important. Buildings and other impervious surfaces such as roads are designed to shed water so that energy absorbed by these surfaces is directed only towards heating the material and the air above it. As the relative fraction of buildings and impervious surfaces increases and vegetation decreases, greater fractions of energy will be directed toward warming the air and substrate because evapotranspiration from vegetation becomes more limited.

Second, the structure or form of the surface as determined by the dimensions and spacing of buildings and trees is important to UHI formation. By day, the urban surface is better at absorbing solar radiation than most rural surfaces. Absorption is enhanced by several factors: the reflectivity of the surface to solar radiation (known as the albedo) which may be less (darker) for some surfaces (such as asphalt roads and dark colored roofs) compared to the rural surroundings, and the three-dimensional structure of the urban surface which serves to increase the effective area of the surface that can absorb sunlight and which also serves to "trap" some radiation that is initially reflected off of individual surfaces. At night, the three-dimensional structure of the surface serves to effectively block the view of the sky for many parts of the urban surface, especially within the urban canopy layer (e.g., the roadway separating adjacent tall buildings) and thereby reduces the loss of radiative heat energy. The sky view factor that describes the relative obstruction of the sky to the surface has been shown to be strongly related to the UHI magnitude of the urban canopy layer air.[9] The structure of the urban surface is also responsible for altering winds near the surface; this provides both sheltering effects and increased turbulence (mixing). During the day, effective mixing of the atmosphere assisted by the rough surface helps to warm the urban boundary layer with heat from the surface. At night, the sheltering of air between buildings contributes to the increased warmth of the night time urban canopy layer heat island.

Third, the properties of materials that affect temperature, such as radiative, thermal, and moisture characteristics differ in urban areas from their rural surroundings. Many urban materials are drier, have higher heat capacities, and may be darker (lower albedo). This allows them to more readily absorb solar energy, to store it, and to have higher temperatures because no energy is used to evaporate water. Urban materials thus tend to act as a sink for heat during daytime and a source of heat at night. The material properties contribute to a temporal lag in the warming of urban areas in the morning and a similar lag in their cooling at night.

WEATHER AND CLIMATE INFLUENCES

Surface and atmospheric heat islands are significantly impacted by weather conditions. All heat islands are best expressed under clear and calm conditions when surface controls that affect microclimates have their maximum effect.[3] As winds increase, the heat island magnitude in the urban canopy layer decreases exponentially and as cloud cover increases, it decreases linearly. Seasonally, heat island magnitudes are typically largest in the summer or dry season when the input of solar energy to the surface is high.[10] In some special cases, such as in very high latitude settlements, the heat island may be largest in winter when there is a large anthropogenic heat input to the atmosphere from space heating requirements. Table 2 summarizes the factors that affect heat island magnitude. These apply most directly to the canopy layer heat island, but also affect the other heat island types.

HEAT ISLAND IMPACTS

UHIs have a range of impacts. The relative warmth of cities can increase the length of the growing season and alter the timing associated with the stages of plant development. At the same time, the length of the frost-season is reduced. Associated with the relative warmth is a decrease in the need for energy required for space heating of homes and businesses. In the summer season, heat islands may place additional demands for space cooling needs, driving a need for more energy use at a time when utilities may already be operating near peak load. Increasing local energy use at these times can increase the anthropogenic heat flux which further exacerbates the atmospheric heat islands. Daytime boundary layer heat islands can also impact air quality through an enhancement of smog (ozone) formation rates within the boundary layer. These formation rates are temperature dependent and are more efficient at higher temperatures. Higher temperatures also provide two positive feedbacks that can further exacerbate air quality. Warmer temperatures drive a greater demand for electricity to power cooling needs that can lead to both emissions of more pollutants and greenhouse gases from power plants and at the same time increase natural (biogenic) emissions of smog-forming pollutants. Summer heat islands can also exacerbate the demand for water use in cities.

The warmth of UHIs can reduce the incidence of cold exposure to urban inhabitants in winter, but is an important stressor during summertime. The urban canopy layer warmth is particularly of concern during hot summer nights, because it is then when the heat island effects are largest, and the physiological impacts on the body become important when it is unable to achieve a period of cooling at night during sleep. Prolonged exposure to such conditions can be a serious health threat and there have been several notable incidents of urban-enhanced mortality during heat waves, for example in Chicago in 1995 and Paris in the summer of 2003. Future climate change is expected to provide additional thermal stress to urban inhabitants because of the increased frequency of hot summer nights in urban areas relative to rural areas.[11]

The UHI plays a role in modifying the weather a city experiences. The relative warmth of the UHI can help induce a local circulation pattern, known as a city breeze, analogous to a sea breeze, in which cooler rural air is transported horizontally into the city. This wind system can have consequences for pollutant dispersion and urban design for pathways for the cleaner, cooler rural air. The relatively warm urban boundary layer air provides a deeper layer into which pollutants may be mixed and can also provide for a higher base to convective clouds that may form over the urban area. The warmer urban atmosphere can, in select cases, provide a greater likelihood of warm season convective cloud formation and possible increases in precipitation over and downwind of the city.[12] In the cold season, the heat island may influence the relative fraction of rain to snow events.

The presence of the UHI also has links to larger-scale environmental change. The UHI complicates the identification of long-term temperature trends from stations that may have become progressively more urbanized over time. These stations may incorporate effects of UHIs within their records, rendering their time series "contaminated" with respect to their ability to identify long-term, nonurban warming trends. To overcome this problem, distinctly nonurban stations are required, such as those that are part of the U.S. Climate Reference Network,[13] which must be maintained, along with distinctly urban stations accompanied by well-documented records regarding their surroundings and the development over time. The establishment of such rigorously maintained networks of stations would allow both the identification of nonurban changes in temperature due to climate forcings as well as changes in the urban environments.

In the future, increased conversion of land to urban use, intensification of the urban development and increases in urban energy use are factors that can promote an increase in UHI magnitude. Under global climate warming forced by increases in green house gas concentrations, the absolute value of urban (and rural) temperatures is expected to increase, but UHI magnitudes may not necessarily increase. This situation occurs because changes to the frequency of weather conditions that favor heat island formation and to the moisture conditions of surrounding rural areas may occur that can favor either increased or decreased heat island magnitude, depending on the location of the city in question and the regional impacts of large scale climate change. Urban scale modifications to the characteristics of the urban surface, that involve so-called "cool technologies" such as high reflectivity surfaces for reducing the absorption of the Sun's energy and increases in vegetative cover are increasingly being considered in many cities to ameliorate the negative impacts of warmer urban temperatures associated with both the UHI and from large scale climate change. A guide to such strategies is provided in Ref.[14]

REFERENCES

1. Yow, D.M. Urban heat islands: Observations, impacts and adaptation. Geogr. Compass **2007**, *1* (6), 1227–1251.
2. Oke, T.R. Urban Environments. In *The Surface Climates of Canada*; Bailey, W.G.; Oke, T.R.; Rouse, W.R., Eds.; McGill-Queen's University Press: Montreal, 1997; 303–327.
3. Oke, T.R. The energetic basis of the urban heat island. Q. J. R. Meteorol. Soc. **1982**, *108*, 1–24.
4. Voogt, J.A. Urban heat island. In *Encyclopedia of Global Environmental Change, Volume 3, Causes and Consequences of Global Environmental Change*; Douglas, I. Ed.; John Wiley & Sons: Chichester, 2002; 660–666.
5. Voogt, J.A.; Oke, T.R. Thermal remote sensing of urban climates. Rem. Sens. Environ. **2003**, *86*, 370–384.
6. Roth, M.; Oke, T.R.; Emery, W.J. Satellite-derived urban heat island from three coastal cities and the utilization of such data in urban climatology. Int. J. Rem. Sens. **1989**, *10*, 1699–1720.
7. Imhoff, M.L.; Zhang, P.; Wolfe, R.E.; Bounoua, L. Remote sensing of the urban heat island effect across biomes in the continental USA. Rem. Sens. Environ. **2010**, *114*, 504–513.
8. Sailor, D. A review of methods for estimating anthropogenic heat and moisture emissions in the urban environment. Int. J. Climatol. **2011**, *31*, 189–199.
9. Oke, T.R. *Boundary Layer Climates*; Routledge: 1987.
10. Arnfield, A.J. Two decades of urban climate research: A review of turbulence, exchanges of energy and water, and the urban heat island. Int. J. Climatol. **2003**, *23*, 1–26.
11. McCarthy, M.P.; Best, M.J.; Betts, R.A. Climate change in cities due to global warming and urban effects. Geophys. Res.Lett. **2010**, *37*, L09705, doi:10.1029/2010GL042845.
12. Shepherd, J.M.; Stallins, J.A.; Jin, M.L.; Mote, T.L. Urbanization: Impacts on clouds precipitation and lightning. In *Urban Ecosystem Ecology*; Aitkenhead-Peterson, J.; Volder, A. Eds.; Agron. Monogr. 55., ASA, CSSA, and SSA: Madison WI, 2010; 1–28.
13. U.S. Climate Reference Network, http://www.ncdc.noaa.gov/crn/ (accessed May 2012).
14. United States Environmental Protection Agency. Reducing Urban Heat Islands: Compendium of Strategies. http://www.epa.gov/hiri/resources/compendium.htm (accessed October 2011).

BIBLIOGRAPHY

1. Oke, T.R.; Mills, G.; Christen, A.; Voogt, J.A. *Urban Climates*; Cambridge University Press: Cambridge, Forthcoming.

Water Storage: Atmospheric

Forrest M. Mims III
Geronimo Creek Observatory, Seguin, Texas, U.S.A.

Abstract

Water vapor is the most variable of the major gases that form Earth's atmosphere. Despite its modest fraction of virtually zero to several percent of a given parcel of air, water vapor is a major factor in weather, climate, the hydrologic cycle, and agriculture. Water vapor is also the leading greenhouse gas and is responsible for maintaining Earth's global average temperature above the freezing point. Global monitoring by satellite instruments has vastly improved and expanded the knowledge of the regional and global distribution of water vapor.

INTRODUCTION

This entry examines the variability, global distribution, and abundance of water vapor stored in the atmosphere. The hydrologic cycle is summarized, and some of the environmental effects of water vapor are mentioned. Also discussed are various methods employed to measure water vapor, including ground-based Global Positioning System (GPS) receivers and satellite instruments that have provided detailed knowledge about the global distribution of water vapor. Water vapor is the principal greenhouse gas, and concerns are addressed about the need to improve the understanding of its role in the climate system and the spatial and temporal accuracy of global total water vapor measurements.

WATER VAPOR

The properties of water are unique among Earth's ingredients that comprise the climate system. Only water is present as a gas, liquid, and solid. Water vapor is the most variable of the major gases that form Earth's atmosphere. Its concentration in a given parcel of air, which can vary from virtually none to several percent, is significantly affected by temperature, geography, and elevation and is subject to dramatic changes over the course of seasons and during the passage of weather systems. Because of the variability of water vapor, it is customarily ignored in lists of the fractional composition of the atmosphere's major gases. Thus, while a perfectly dry atmosphere is 78.08% nitrogen, 20.95% oxygen, 0.93% argon,[1] 0.039% carbon dioxide,[2] and an assortment of trace gases, accounting for the presence of water vapor reduces these percentages accordingly. Of the several major gases that comprise the bulk of the atmosphere, only water vapor makes its presence known by condensing into visible clouds of liquid droplets or frozen particles. It is also the most variable of the atmosphere's major gases.

Despite its relatively modest fraction of the atmosphere, water vapor is essential for the hydrologic cycle and is responsible for tropospheric weather and maintaining Earth's average global temperature above the freezing point. The vapor phase of water eventually leads to precipitation that profoundly influences agriculture and the transformation of the landscape by erosion. Water vapor also enhances haze, particularly when it condenses on sulfate and other hygroscopic aerosols and significantly reduces visibility by increasing the aerosol optical thickness of the atmosphere.[3] Because people are cooled by the evaporation of perspiration, water vapor in the ambient air has a significant impact on a person's comfort level on warm days. Water vapor is also the most significant of the greenhouse gases.[4] Calculations made well over a century ago showed that the oceans would be frozen were it not for the warming effect of the atmosphere. Tyndall, who discovered that water vapor and carbon dioxide absorb infrared radiation, expressed it best when he wrote that water vapor "is a blanket more necessary to the vegetable life of England than clothing is to man. Remove for a single summer-night the aqueous vapor from the air… and the sun would rise upon an island held fast in the iron grip of frost."[5]

THE HYDROLOGIC CYCLE

The environmental factors that lead to evaporation and condensation explain and control the hydrologic cycle, the continuous cycling of water between Earth and its atmosphere. Water molecules are constantly being evaporated from the oceans and other surface water unless the adjacent air is saturated with water vapor. Additional vapor arises from the evaporation of small droplets in the spray of waves. The rate of evaporation is increased appreciably when surface water is warmed by sunlight. Some of this heat is transferred to the overlying air, and this gives rise to

Encyclopedia of Natural Resources DOI: 10.1081/E-ENRA-120047621

convection as the warm air expands and its reduced density causes it to rise, along with its water vapor. Cool air can contain less water vapor than warm air, and as the warm air rises, it may cool to a point where it becomes fully saturated with water vapor. The water vapor will then condense into droplets or deposit as ice crystals onto microscopic particles of dust, sea salt, bacteria, and other aerosols, which are collectively described as condensation or ice nuclei, respectively.

Visible clouds are formed when the water vapor in large volumes of moist air is transformed into liquid droplets or ice crystals. While clouds contain considerable amounts of liquid or frozen water, they remain suspended in the air because their density is slightly less than or equal to the volume of dry air. Cloud droplets or ice crystals are initially only a fraction of a micrometer in diameter. They grow in size as they merge into drops a few millimeters in diameter that fall from the cloud when the tug of gravity exceeds the force of convective updrafts or as ice crystals that grow by deposition of water vapor and by the collision and collection of ice crystals to make snowflakes. Falling snowflakes can melt into raindrops if the lower atmosphere is above the freezing point.

The interconnected bodies of water known as oceans cover 71% of Earth's surface. Most water vapor originates from the oceans, where most precipitation falls and where it may be stored for many years before returning to the atmosphere. Precipitation that falls over land may be stored in soil or in aquifers, or it may run off into streams and rivers and be captured in reservoirs or returned to the ocean. Some precipitation is stored in glaciers. The hydrologic cycle is completed when the portion of the atmosphere depleted of its moisture by precipitation is restocked with fresh evaporation from bodies of water, with lesser contributions provided by transpiration from plants, exhalation and evaporated perspiration from animals and insects and sublimation from snow and ice. Volcanic eruptions, natural steam vents, gas wells, steam-powered and water-cooled electricity generators, and other processes also inject small amounts of water vapor into the atmosphere. Some of the water vapor in the upper atmosphere is a byproduct of the oxidation of methane.[6]

TOTAL ABUNDANCE OF GLOBAL WATER VAPOR

The total amount of water stored in the atmosphere can be calculated from the liquid equivalent of the globally averaged water vapor between the surface and the top of the atmosphere, a meteorological parameter with many names, including precipitable water (PW), integrated precipitable water (IPW), and total column water vapor (TCWV). TCWV is expressed as the depth of liquid water that will result if the total column is brought to the surface at

standard temperature and pressure. Although column water vapor generally refers to the total precipitable water over a specific location, it may also apply to specific columns anywhere within the atmosphere.

Although TCWV over a specific site can be remotely inferred by a ground-based instrument to an accuracy of approximately 2–10%, expanding the measurement to the entire globe and providing regular updates is a challenging task that has yet to be satisfactorily completed. Important questions have been raised about some of the methods and differences in the findings of the major projects that have attempted to provide a detailed understanding of the global distribution of TCWV and its regional and global trends.[7] These studies blended data from various satellite instruments and upper air soundings by balloon-borne radiosondes. The National Centers for Environmental Prediction (NCEP)–National Center for Atmospheric Research (NCAR) reanalysis-2 project found a global average TCWV of 24.68 mm from 1988 to 1999.[8] The NASA Water Vapor Project (NVAP) found a global average TCWV of 24.46 mm from 1988 to 2001.[9] Despite problems with both studies, their global TCWV findings were within a few percent of an earlier estimate of 25 mm.[10]

The liquid equivalent of the water stored in the atmosphere can be estimated by subtracting the volume of Earth calculated from its volumetric mean radius of 6371 km[11] from the same radius to which the thickness of the atmosphere's TCWV has been added. The 1988–2001 global TCWV[9] gives an average total volume of 12,476 km³ of water stored in the atmosphere. While this is about five times the volume of all Earth's rivers,[12] it is slightly less than 0.001% of Earth's oceans.[13]

GLOBAL DISTRIBUTION OF WATER VAPOR

The total water vapor over both hemispheres has a distinctive annual cycle, with average TCWV significantly higher in summer than in winter, especially in the northern hemisphere. This is clearly shown in the summer and winter satellite images of the United States in Figs. 1 and 2. Because the hemispherical variations are asymmetrical, the global average TCWV also has an annual cycle, with one study giving 23 mm from November to January and 26.5 mm during July.[14] Spatial variations are substantial, with the polar regions and the Himalaya mountain range having a minimum TCWV less than 1 mm and the equatorial zone having a maximum TCWV of 70 mm or more. Annual TCWV measured by a GPS receiver at Hawaii's Mauna Loa Observatory, an alpine site (3.4 km above the mean sea level) surrounded by ocean, ranges from less than 1 to 15 mm, whereas TCWV at nearby sea level sites ranges from 20 to 50 mm.[15]

The greatest abundance of water vapor is stored in a band that encircles Earth known as the Intertropical

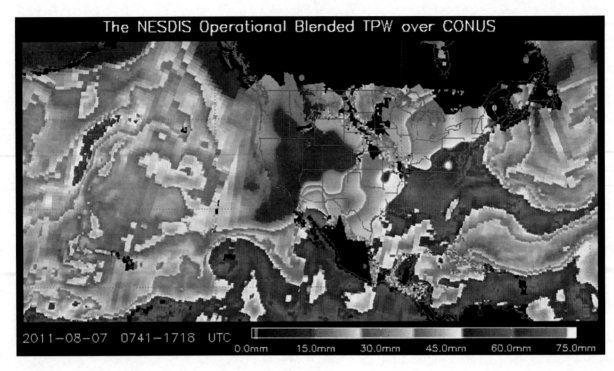

Fig. 1 Enhanced total water vapor over the continental United States on a summer day (7 August 2011). NESDIS Operational Blended TPW Products.

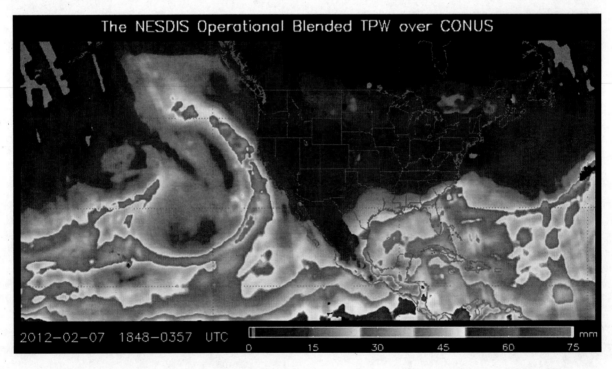

Fig. 2 Reduced total water vapor over the continental United States on a winter day (7 February 2012). NESDIS Operational Blended TPW Products.

Convergence Zone (ITCZ). This especially moist region swings from generally north of the Equator in July to generally south of the Equator in January. The ITCZ is obvious in the satellite image in Fig. 3. Heating from extended hours of direct solar radiation transports considerable moist air in the ITCZ up into the middle and upper troposphere, where its water vapor condenses into heavy downpours. The atmospheric water vapor lost to precipitation is replenished by evaporation the next day.

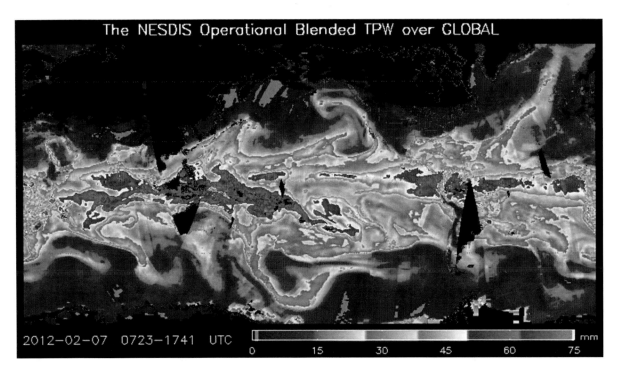

Fig. 3 Expanded view of Fig. 2 showing global total water vapor. Note the significantly enhanced moisture over the tropics in the Intertropical Convergence Zone. Dark areas, mainly over land, indicate an absence of data (7 February 2012). NESDIS Operational Blended TPW Products.

VERTICAL DISTRIBUTION OF WATER VAPOR

While water vapor is present from the surface to the stratosphere, more than half is usually stored in the first few kilometers nearest to the surface. If Earth could be reduced to the size of a basketball, most of the planet's water vapor would be found in a film of air around the ball having an average thickness of several sheets of office paper. The amount of water vapor in a given parcel of air can be expressed as the ratio of the mass of the water vapor to that of an equivalent volume of completely dry air. This mixing ratio declines exponentially with elevation[16] until the water vapor mixing ratio in the stratosphere between 16 and 28 km declines to about 1/1000 of that at sea level.[17]

RESIDENCE TIME IN THE ATMOSPHERE

Because annual precipitation matches the sum of evaporation and the other processes that inject water vapor into the atmosphere, the mean residence time of a water molecule in the atmosphere can be estimated by dividing the average of global TCWV by the average global precipitation of 2.5 mm/day.[18] A global TCWV of 24.46 mm[9] gives a typical residence time of nearly 10 days. Some water vapor may return to a liquid state almost immediately, whereas vapor that reaches the stratosphere may remain there considerably longer than 10 days. Over geological time a given

molecule of water spends minuscule time in the atmosphere. It is much more likely to be found immersed in an ocean, stored in an aquifer or locked in a glacier.

MEASURING ATMOSPHERIC WATER VAPOR

The measurement of water vapor on a global scale is of increasing interest in view of its significance as a greenhouse gas and its role in climate models. TCWV is measured from the surface by various kinds of sun photometers that monitor changes in near-infrared sunlight caused by the absorption of water vapor[19,20] and by microwave radiometers that receive frequencies emitted by water molecules.[21] TCWV can also be measured by an inexpensive infrared thermometer pointed at a cloud-free zenith sky.[22] Water vapor delays the microwave signals from GPS satellites, and this provides a high-quality method for inferring TCWV.[23] The U.S. National Oceanic and Atmospheric Administration's (NOAA) Ground-Based GPS-IPW Network processes data from a vast array of more than 550 GPS receivers across the United States and a number of other countries and islands. The network provides automatic TCWV measurements that are posted online every half hour.[15] This GPS network is arguably the most precise source of site-specific TCWV data, and its expansion around the world would greatly improve the understanding of water vapor and its cycles and trends over land.

Dew—Wind

Since 1930 the vertical distribution of water vapor has been measured by humidity sensors flown from upper air sounding balloons.[24] Integrating the balloon humidity measurements provides the TCWV. Studies of the global distribution of water vapor based only on upper air sounding balloon measurements were biased because most sounding balloons are launched from land, which comprises only 29% of Earth's surface. Additional biases were caused by different humidity sensor designs and inconsistent performance, especially at reduced temperatures.

Earth satellites equipped with water vapor sensing instruments have dramatically improved the understanding of the global distribution of water vapor. Most satellite instruments monitor the infrared or microwave wavelengths emitted by water vapor.[25] Selective monitoring of these wavelengths permits TCWV and the water vapor at various levels in the atmosphere to be measured day and night. Some satellite instruments monitor TCWV by measuring changes in the water vapor absorbing near-infrared sunlight reflected from both oceans and land.[26] Near real-time water vapor imagery from various satellite sensors is available online, including the National Environmental Satellite, Data, and Information Service (NESDIS) Operational Blended TPW Products.[27] As shown in Figs. 1 and 2, NESDIS imagery is particularly informative over the continental United States due to the inclusion of TCWV measurements by NOAA's GPS network.

CONCLUSION

Water vapor, the most variable of the major gases that form Earth's atmosphere, profoundly influences agriculture, weather, climate, and the transformation of the landscape by erosion. Because of its direct and indirect roles in weather and climate, water vapor is a subject of intense scientific scrutiny.[28] Satellite observations of the water vapor stored in the atmosphere have dramatically improved our understanding of its global distribution, abundance and seasonal cycles. Yet the identification and analysis of regional and global trends are made difficult by uncertainties in both the accuracy of instrumental measurements and how best to merge data from ground, upper air, and satellite instruments. A routine understanding of the global distribution of water vapor and its trends will require years of additional observations by a mix of carefully planned and calibrated ground- and space-based instruments. Long-range planning and closer collaboration among the various international institutions and agencies that monitor water vapor and analyze its trends will enhance this process.

ACKNOWLEDGMENTS

The author is grateful to the atmospheric scientists who provided advice about water vapor and its measurement that led to his TCWV measurements since 1990 and this entry, including Seth Gutman, Brent Holben, Yoram Kaufman, Roger Pielke Sr., Robert Roosen, Glenn Shaw, and Frederick Volz.

REFERENCES

1. Lide, D.R., Ed.; Atmospheric composition. In *CRC Handbook of Chemistry and Physics 2006–2007*, 87th Ed.; Taylor and Francis: Boca Raton, FL, 2007; 14–3 pp.
2. http://scrippsco2.ucsd.edu/data/in_situ_co2/monthly_mlo.csv (accessed 6 February 2012).
3. Tang, I.N. Chemical and size effects of hygroscopic aerosols on light scattering coefficients. J. Geophy. Res. **1996**, *101* (D14), 19245–19250.
4. Meckler, S.B. *Special Report: Water Vapor in the Climate System*; American Geophysical Union: Washington, DC., 1995.
5. Tyndall, J. On radiation through the Earth's atmosphere. Philosophical Mag. **1863**, *4* (25), 200–206.
6. McCormick, M.P.; Chou, E.W.; McMaster, L.R.; Chu, W.P.; Larsen, J.C.; Rind D.; Outmans, S. Annual variations of water vapor in the stratosphere and upper troposphere observed by the stratospheric aerosol and gas experiment II. J. Geophy. Res. **1993**, *98* (D3), 4867–4874.
7. Trenberth, K.E.; Fusillo, J.; Smith, L. Trends and variability in column-integrated water vapor. Climate Dynamics **2005**, *24* (7–8), 741–758.
8. Amen, G.G.; Kumar, P. NVAP and Reanalysis-2 global precipitable water products: intercomparison and variability studies. Bull. Amer. Meteor. Soc. **2005**, *86*, 245–256.
9. Vonder Haar, T.H.; Forsythe, J.M.; McKague, D.; Randel, D.L.; Ruston, B.C.; Woo, S. *Continuation of the NVAP Global Water Vapor Data Sets for Pathfinder Science Analysis*, Science and Technology Corporation Technical Report 3333, October 2003, Fig. 24, 26 http://eosweb.larc.nasa.gov/PRODOCS/nvap/sci_tech_report_3333.pdf (accessed at 06 February 2012).
10. Trenberth, K.E.; Guillemot, C.J. The total mass of the atmosphere. J. Geophys. Res. **1994**, *99* (D11), 23,079–23,088.
11. Moritz, H. Geodetic Reference System 1980. J. Geodesy **2000**, *74* (1), 128–162.
12. Claudio Cassardo, C.; Jones, J.A.A. Managing water in a changing world. Water **2011**, *3*, 618–628 http://www.mdpi.com/2073-4441/3/2/618/pdf (accessed 6 Feb. 2012).
13. Trenberth, K.E.; Smith, L.; Qian, T.; Dai, A.; Fusillo, J. Estimates of the global water budget and its annual cycle using observational and model data. J. Hydrometeorol. **2007**, *8* (4), 758–769.
14. Randel, D.L.; Vonder Haar, D.H.; Ringerud, M. A.; Stephens, G.L.; Greenwald, T.J.; Combs, C.L. A new global water vapor dataset. Bull. Amer. Meteor. Soc. **1996**, *77*, 1233–1246.
15. http://gpsmet.noaa.gov (accessed 10 Feb 2012).
16. U.S. Standard Atmosphere, 1976. *National Oceanic and Atmospheric Administration, National Aeronautics and Space Administration, U.S. Air Force*; U.S. Government Printing Office: Washington, DC, 1976; Table 20, 44.
17. Oltmans, S.J.; Vömel, H.; Hofmann, D.J.; Rosenlof, K.H.; Kley, D. The increase in stratospheric water vapor from balloonborne, frostpoint hygrometer measurements at Washington, D.C., and Boulder, Colorado. Geophys. Res. Lett. **2000**, *27* (21), 3453–3456.

Dew—Wind

18. Huffman, G.J.; Adler, R.F.; Arkin, P.; Chang, A.; Ferraro, R.; Gruber, A.; Janowiak, J.; McNab, A.; Rudolf, B.; Schneider, U. The Global Precipitation Climatology Project (GPCP) combined precipitation dataset. Bull. Amer. Meteor. Soc. **1997**, *78* (1), 5–20.

19. Fowle, F.E. The spectroscopic determination of aqueous vapor. Astrophys. J. **1912**, *35*, 149–162.

20. Mims III, F. M. An inexpensive and stable LED Sun photometer for measuring the water vapor column over South Texas from 1990 to 2001. Geophys. Res. Lett. **2002**, *29* (13), 20–21 to 20–24.

21. Liljegren, J.C. Two-channel microwave radiometer for observations of total column precipitable water vapor and cloud liquid water path. In *Fifth Symp. on Global Change Studies*; Nashville, T.N., Ed., American Meteorological Society: Boston, MA, 1994; 262–269 pp.

22. Mims, F.M.; Chambers, L.H.; Brooks, D.R. Measuring total column water vapor by pointing an infrared thermometer at the sky. Bull. Amer. Meteor. Soc. **2011**, *92* (10), 1311–1320.

23. Gutman, S.; Benjamin, S. The role of ground-based GPS meteorological observations in numerical weather modeling. GPS Solutions **2001**, *4* (4), 16–24.

24. Pettifer, R. From observations to forecasts—Part 2. The development of in situ upper air measurements. Weather **2009**, *64* (11), 302–308.

25. Schmit, T.J.; Feltz, W.F.; Menzel, W.P.; Jung, J.; Noel, A.P.; Heil, J.N.; Nelson, J.P.; Wade, G.S. Validation and use of GOES sounder moisture information. Wea. Forecasting **2002**, *17* (1), 139–154.

26. Kaufman, Y.J.; Gao, B.C. Remote sensing of water vapor in the near IR from EOS/MODIS. IEEE Trans. Geosci. Remote Sens. **1992**, *30*, 871–884.

27. http://www.osdpd.noaa.gov/bTPW/index.html (accessed 6 February 2012).

28. Vonder Haar, T.H.; Bytheway, J.L.; Forsythe, J.M. Weather and climate analysis using improved global water vapor observations. Geophys. Res. Letts. **2012**, *39* (15), L15802, doi:10.1029/2012GL052094.

Wind Speed Probability Distribution

Adam H. Monahan

School of Earth and Ocean Sciences, University of Victoria, Victoria, British Columbia, Canada

Abstract

The probability distribution of surface wind speed is a fundamental characterization of the range of speed values taken and their relative frequency of occurrence. This entry first presents an overview of the general features of any wind speed probability distribution, followed by a discussion of empirical models of the wind speed distribution and the relationship of this distribution to that of vector winds. Finally, a brief discussion of physically based models of the wind speed distribution is presented.

INTRODUCTION

The fact that the atmosphere is in ceaseless motion across the Earth's surface is of fundamental importance to a broad range of physical, biological, and chemical processes in the Earth system, as well as being of societal and economic significance. Lower atmospheric winds exercise an important control on the fluxes of mass, momentum, and energy between the atmosphere and the underlying surface. Extreme surface winds represent hazards to both the built and natural environments. Furthermore, the moving air is a reservoir of energy, which has been used by humankind for millennia and is a central element in the portfolio of energy resources alternative to fossil fuels. An elementary fact about surface winds is that they are not steady: the motion of the air changes on timescales from seconds through to days, years, and beyond. This variability is a key aspect of many of the contexts in which winds are of are importance. For example, as the wind power density is proportional to the cube of the wind speed, the mean power density is not that produced by the mean wind speed: higher order statistical moments of the wind speed influence the mean power. It is therefore important to develop mathematical tools to characterize this variability in the wind.

The most natural of such tools is the probability distribution, which describes the range of values a quantity takes and their relative likelihood. As the velocity of the air is a continuous variable (rather than one taking discrete values), we can represent the probability distribution in terms of the associated probability density function (pdf). In particular, denoting the wind speed as w and its pdf as $p_w(w)$, the probability of observing a speed w between the values w_1 and w_2 is

$$\text{prob}(w_1 \leq w \leq w_2) = \int_{w_1}^{w_2} p_w(w)\,dw. \tag{1}$$

The pdf carries all information about the range and relative likelihood of wind speeds. In many contexts, the statistical moments are required. These are defined by integrals over $p_w(w)$, and include the mean, standard deviation, and skewness:

$$\text{mean}(w) = \int_0^{\infty} w p_w(w)\,dw \tag{2}$$

$$\text{std}(w) = \text{mean}\left\{(w - \text{mean}(w))^2\right\} \tag{3}$$

$$\text{skew}(w) = \text{mean}\left\{\left(\frac{w - \text{mean}(w)}{\text{std}(w)}\right)^3\right\}. \tag{4}$$

Different moments carry information about different aspects of wind variability. The mean and standard deviation are measures of magnitude: they represent respectively the overall magnitude of the wind speed and the strength of the variability around it. In contrast, the skewness is a measure of the shape of the distribution. A skewness value of zero indicates that the pdf is symmetric around its mean, so anomalies (relative to the mean) of equal size and opposite sign are equally likely. A positive skewness indicates that positive anomalies of a given size are more likely than negative values of the same magnitude; for negative skewness, the asymmetry favors negative anomalies. Other measures of shape exist, such as kurtosis (a measure of the flatness of the distribution); as these are more sensitive to sampling variability than lower-order moments, they are less often used in practical applications.[1] As noted earlier, the moments of w enter into the calculation of mean wind power density Π: assuming a constant air density ρ,

Encyclopedia of Natural Resources DOI: 10.1081/E-ENRA-120049148

Dew—Wind

$$\text{mean}(\Pi) = \frac{1}{2}\rho \int_0^\infty w^3 p_w(w)\,dw = \frac{1}{2}\rho\,(\text{mean}^3(w)$$
$$+ 3\text{mean}(w)\text{std}^2(w) + \text{skew}(w)\text{std}^3(w)). \tag{5}$$

Computations of the mean power R captured by a turbine of cross-sectional area A,

$$R = \frac{1}{2}A\rho \int_0^\infty P(w)p_w(w)\,dw, \tag{6}$$

account for the physical response of the turbine through the power curve $P(w)$.[2] The integral in Eqn. (6) cannot generally be expressed as a simple function of the moments of w: computation of R requires knowledge of the full distribution of wind speeds.

Physical controls on wind variability involve processes on scales from global-scale differential heating to the millimeter-scale dissipation of turbulence kinetic energy by viscosity. These processes are reviewed in standard treatments of dynamic meteorology[3,4] or boundary layer meteorology.[5,6] While energy is present across all time-scales from seconds through years to beyond, it is not evenly distributed across scales. There are astronomically forced peaks in variability at the daily and annual scales: winds often display marked diurnal and seasonal variations. As well, in many settings it is observed that there are concentrations of energy on timescales less than several minutes and greater than an hour or two. Between these is a "spectral gap" that allows us to distinguish between faster, smaller-scale "turbulent" winds and slower, larger-scale "eddy averaged" winds.[6] It is the second of these, which is generally defined in terms of winds averaged on timescales from 10 minutes to an hour and that is the focus of this review.

There is an extensive literature on the probability distribution of velocities in turbulent flow.[6] For eddy-averaged winds, the vertical component of the wind (relative to the local surface) is generally much smaller than the horizontal components. As such, we will neglect the vertical wind component and write the wind speed in terms of the orthogonal vector wind components u and v as $w = \sqrt{u^2 + v^2}$. Velocity is a dimensional quantity, and can be reported in a range of different units. For concreteness, we will refer to velocities in ms^{-1}. The vector components have their own pdfs, which are closely related to that of wind speed (as will be discussed in the later section, "Relation of Wind Speed and Vector Wind Component pdfs"). Because it is of importance to a broader range of applications, the primary focus of this entry will be the pdf of wind speed. Furthermore, this entry will consider the entire probability distribution of surface winds without specific discussion of extreme events in the tail of the pdf. A general discussion of extreme events is presented in the entry "Climate: Extreme Events" in this encyclopedia.[7]

GENERAL FEATURES OF THE WIND SPEED PDF

There are a number of basic constraints that must be satisfied by any wind speed pdf. These are as follows:

1. $p_w(w) = 0$ for $w < 0$ ms^{-1} (wind speeds must be positive).
2. $p_w(w) \geq 0$ for $w \geq 0$ ms^{-1} (any pdf must be nonnegative).
3. $p_w(w)$ must be normalized to unity (the wind speed must take some value):

$$\int_0^\infty p_w(w)\,dw = 1. \tag{7}$$

4. $p_w(0) = 0$ unless the joint pdf of the vector wind components is infinite at $(u, v) = (0, 0)$ ms^{-1}. This point is addressed in more detail in the section, "Relation of Wind Speed and Vector Wind Component pdfs."

From the normalization constraint (Eqn. 7), it follows that the units of the pdf $p_w(w)$ are the inverse of the units of w. That is, for w measured in ms^{-1}, the units of $p_w(w)$ are m^{-1}s.

Beyond these general requirements of any surface wind speed pdf, there are a number of other features, which are generally observed. Mean wind speeds over the oceans are generally larger than those over land, with the largest values found in the middle latitudes and the subtropics. Variability in w is generally largest in the midlatitude storm tracks.[4] Over land, $p_w(w)$ is generally positively skewed;[8] over the oceans, positive skewness is found in the Northern Hemisphere middle latitudes, near to zero skewness over the Southern Ocean, and negative skewness in much of the tropics and subtropics.[9] Over land, mean(w) and std(w) are generally larger during the day than at night, as a result of the influence of surface stratification on the downward transport of momentum.[5,10] In coastal areas, this diurnal cycle can be reversed over water as a result of changes in stratification due to horizontal temperature advection.[11] Land surface winds tend to be more strongly skewed at night than during the day.[8] Wind speed pdfs generally have a single maximum value (i.e., there is a single most likely speed), although in regions of complex topography multimodal wind speed pdfs are observed.[12,13]

Wind observations often include a nonzero frequency of calms (times when $w = 0$ ms^{-1}) resulting from a finite cut-off speed below which w cannot be measured by the anemometer. To account for this, hybrid pdfs in which a standard parametric distribution is augmented with a delta function at $w = 0$ are sometimes used.[14] As these calms result from limited instrumental precision rather than a true nonzero probability of $w = 0$ ms^{-1}, such hybrid distributions will not be further considered in this review.

EMPIRICAL MODELS OF THE WIND SPEED PROBABILITY DISTRIBUTION

A number of parametric distributions have been suggested for the empirical modelling of the wind speed probability distribution.[14–18] The most common of these in present use is the two-parameter Weibull distribution:

$$p_w^{wbl}(w) = \frac{b}{a}\left(\frac{w}{a}\right)^{b-1}\exp\left[-\left(\frac{w}{a}\right)^b\right] \quad w \geq 0. \quad (8)$$

The parameters a and b denote, respectively, the scale and shape. We can use the notation w ~ $W(a, b)$ to indicate that w is Weibull with parameters a and b. Explicit formulas for the moments of a Weibull distributed variable follow from the general result:

$$\text{mean}\left(w^k\right) = a^k \Gamma\left(1 + \frac{k}{b}\right), \quad (9)$$

where $\Gamma(x)$ is the gamma function. In particular:

$$\text{mean}(w) = a\Gamma\left(1 + \frac{1}{b}\right) \quad (10)$$

$$\text{std}(w) = a\left[\Gamma\left(1 + \frac{2}{b}\right) - \Gamma^2\left(1 + \frac{1}{b}\right)\right]^{1/2} \quad (11)$$

$$\text{skew}(w) = \frac{\Gamma\left(1 + \frac{3}{b}\right) - 3\Gamma\left(1 + \frac{1}{b}\right)\Gamma\left(1 + \frac{2}{b}\right) + 2\Gamma^3\left(1 + \frac{1}{b}\right)}{\left[\Gamma\left(1 + \frac{2}{b}\right) - \Gamma^2\left(1 + \frac{1}{b}\right)\right]^{1/2}}. \quad (12)$$

Both skew(w) and the ratio mean(w)/std(w) are independent of the scale parameter a: for a Weibull distributed variable, the skewness is a unique function of the ratio of the mean to the standard deviation. To a good approximation,

$$b = \left(\frac{\text{mean}(w)}{\text{std}(w)}\right)^{1.091}. \quad (13)$$

A number of methods exist for estimating the parameters of the Weibull distribution from observations.[14,19] Among these, the most straightforward is the method of moments, which makes use of Eqns. (10) and (13).

Observations indicate that $b \simeq 2$ is the most common value of the shape parameter over land.[8] The special case of $p_w^{wbl}(w)$ with $b = 2$ is the Rayleigh distribution:

$$p_w^{ral} = \frac{w}{\sigma^2}\exp\left(-\frac{w}{2\sigma^2}\right). \quad (14)$$

While the Weibull pdf is often a useful model for $p_w(w)$, it is not an exact characterization and deviations of various degrees from Weibull behavior are common.[14,20]

It is worth noting that if w is Weibull, so is w^k: w ~ $W(a, b)$ implies w^k ~ $W(a^k, b/k)$. In particular, if the speed is Weibull, then the wind power density is as well. Furthermore, if w is Rayleigh: w ~ $W(a, 2)$, then $w^{1/2}$ ~ $W(a^{1/2}, 4)$ has a skewness near to zero. This result is consistent with the fact that over land surfaces, $w^{1/2}$ can often be approximated as Gaussian.[21]

It is also possible to model the joint distribution of wind speed and direction, $p_{w\theta}(w, \theta)$, with specified marginal distributions $p_w(w)$ and $p_\theta(\theta)$:

$$p_{w\theta}(w, \theta) = 2\pi g\left(2\pi F_w(w) - 2\pi F_\theta(\theta)\right)p_w(w)p_\theta(\theta). \quad (15)$$

In Eqn. (15), $F_w(w)$ and $F_\theta(\theta)$ are respectively the cumulative distribution functions of w and θ, and g(ζ) is a pdf on $0 \leq \zeta < 2\pi$, which specifies the dependence of w and θ. [22,23]

RELATION OF WIND SPEED WITH VECTOR WIND COMPONENT PDFS

A natural question in the discussion of the wind speed pdf is its relation to the pdf of vector wind components, as was noted in many of the earliest studies of $p_w(w)$.[24–26] Vector winds can be described in terms of either their components u and v (e.g., eastward and northward) or in their speed w and direction θ:

$$(u, v) = (w\cos\theta, w\sin\theta) \quad (16)$$

(where the orientation of $\theta = 0$ is arbitrary and can be chosen for convenience). We will denote the joint pdf of (u, v) by $p_{uv}(u, v)$. While $p_{uv}(u, v)$ and $p_{w\theta}(w, \theta)$ are different functions (with different units), these must be related as the description of vector winds in terms of components is equivalent to that in terms of speed and direction. The probability that the vector winds fall within a given range should be independent of the coordinate system in which we choose to describe them, so

$$p_{uv}(u, v)\,du\,dv = p_{w\theta}(w, \theta)\,dw\,d\theta, \quad (17)$$

where the infinitesimal areas $du\,dv$ and $dw\,d\theta$ are related through the standard change of variables from Cartesian to plane polar coordinates: $du\,dv = w\,dw\,d\theta$. It follows that the relationship between the joint distributions is

$$p_{w\theta}(w, \theta) = wp_{uv}(w\cos\theta, w\sin\theta). \quad (18)$$

To obtain the marginal pdf for w alone, we integrate $p_{w\theta}(w, \theta)$ over θ to obtain:

$$p_w(w) = w\int_{-\pi}^{\pi} p_{uv}(w\cos\theta, w\sin\theta)\, d\theta. \tag{19}$$

From this result, we can relate any expression for the pdf of the vector wind components to that of the wind speed. As well, we see that $p_w(0) \neq 0$ only if $p_{uv}(0, 0)$ is infinite (which is not observed).

If we make the approximation that the vector winds are Gaussian with mean $(\bar{u}, \bar{v})$ and isotropic, uncorrelated variability of standard deviation σ, then

$$p_{uv}^{gau}(u, v) = \frac{1}{2\pi\sigma^2}\exp\left(-\frac{(u-\bar{u})^2}{2\sigma^2} - \frac{(v-\bar{v})^2}{2\sigma^2}\right). \tag{20}$$

Evaluation of the integral in Eqn. (19) yields the Rice distribution

$$p_w^{rice}(w) = \frac{w}{\sigma^2}\exp\left(-\frac{\bar{U}^2 + w^2}{2\sigma^2}\right)I_0\left(\frac{\bar{U}w}{\sigma^2}\right), \tag{21}$$

where $\bar{U}^2 = \bar{u}^2 + \bar{v}^2$ and $I_k(z)$ are the associated Bessel function of the first kind of order k.[27] In this case, the mean and standard deviation of the wind speed are given by

$$\text{mean}(w) = \sqrt{\frac{\pi}{2}}\sigma\exp\left(\frac{-\bar{U}^2}{4\sigma^2}\right)\left[\left(1 + \frac{\bar{U}^2}{2\sigma^2}\right)\right.$$
$$\left. I_0\left(\frac{\bar{U}^2}{4\sigma^2}\right) + \frac{\bar{U}^2}{2\sigma^2}I_1\left(\frac{\bar{U}^2}{4\sigma^2}\right)\right] \tag{22}$$

$$\text{std}(w) = \sqrt{2\sigma^2 + \bar{U}^2 - \text{mean}^2(w)}. \tag{23}$$

It is possible to generalize Eqn. (21) to allow for unequal variances in the vector wind components, but the resulting expressions (involving infinite series of Bessel functions) are awkward.[15,28] Non-Gaussianity in the vector wind components can also be accounted for, although the resulting expressions are either analytically intractable or violate the positivity constraint of the section, "General Features of the Wind Speed pdf." Nevertheless, it has been shown that non-Gaussianity in the vector winds must be accounted for to accurately model wind speed skewness over the oceans.[15]

MECHANISTIC MODELS OF THE WIND SPEED PROBABILITY DISTRIBUTION

While most studies of the wind speed pdf have been empirical, recent efforts have been made to develop physically based models of $p_{uv}(u, v)$ and $p_w(w)$. Through this approach, the pdfs are expressed in terms of dynamically meaningful parameters rather than abstract statistical parameters. For example, Monahan used an idealized slab model of the sea surface boundary layer momentum budget driven by fluctuating large-scale pressure gradients[9] to express $p_w(w)$ in terms of the boundary layer depth, boundary layer top entrainment velocity, surface drag coefficient, and pressure gradient statistics.[9] He et al. used a generalized version of this model[8] to demonstrate that surface buoyancy fluxes and the character of the land surface have important influences on the pdf of wind speed.[8] A subsequent analysis suggested that the long positive tail of the nighttime wind speed pdf is a result of intermittent turbulent mixing at the top of the normally quiescent nocturnal boundary layer.[20] Much more work remains to be done on this problem; the mechanistic study of surface wind pdfs remains in its infancy.

CONCLUSIONS

The wind speed probability distribution is a central ingredient in studies of wind hazards, wind energy assessment, and fluxes between the atmosphere and the underlying surface. While the structure of $p_w(w)$ can be constrained by some fundamental general requirements, it is not known to be fully characterized by any single family of distributions. Characterizations of $p_w(w)$ can be empirical (by parametric distributions such as the Weibull), or derived from representations of the joint distribution of the vector wind components. An emerging area of study seeks to develop physically based models of $p_w(w)$. These approaches offer distinct and complementary insights into the pdf of surface winds.

REFERENCES

1. Wilks, D.S. *Statistical Methods in the Atmospheric Sciences*; Academic Press: 2005.
2. Burton, T.; Jenkins, N.; Sharpe, D.; Bossanyi, E. *Wind Energy Handbook*; Wiley: Chichester, UK, 2011.
3. Holton, J.R. *An Introduction to Dynamic Meteorology*; Academic Press: 2004.
4. Peixoto, J.P.; Oort, A.H. *Physics of Climate*; American Institute of Physics: New York, 1992.
5. Arya, S.P. *Introduction to Micrometeorology*; Academic Press: 2001.
6. Stull, R.B. *An Introduction to Boundary Layer Meteorology*; Kluwer: Dordrecht, 1997.
7. Sura, P. Extreme climate events. In *Encyclopedia of Natural Resources*; Wang, Y., Ed.; Taylor and Francis: 2012.
8. He, Y.; Monahan, A.H.; Jones, C.G.; Dai, A.; Biner, S.; Caya, D.; Winger, K. Land surface wind speed probability distributions in North America: Observations, theory, and regional climate model simulations. J. Geophys. Res. **2010**, *115*, doi: 10.1029/2008JD010708.
9. Monahan, A.H. The probability distribution of sea surface wind speeds. Part I: Theory and sea winds observations. J. Clim. **2006**, *19*, 497–520.

10. Dai, A.; Deser, C. Diurnal and semidurnal variations in global surface wind and divergence fields. J. Geophys. Res. **1999**, *104*, 31109–31125.

11. Barthelmie, R.J.; Grisogono, B.; Pryor, S.C. Observations and simulations of diurnal cycles of near-surface wind speeds over land and sea. J. Geophys. Res. **1996**, *101*, 21327–21337.

12. Romero-Centeno, R.; Zavala-Hidalgo, J.; Gallegos, A.; O'Brien, J.J. Isthmus of Tehuantepec wind climatology and ENSO signal. J. Clim. **2003**, *16*, 2628–2639.

13. Jiménez, P.A.; Dudhia, J.; Navarro, J. On the surface wind probability density function over complex terrain. Geophys. Res. Lett. **2011**, *38*, L22803, doi:10.1029/2011GL049669.

14. Carta, J.A.; Ramírez, P.; Velázquez, S. A review of wind speed probability distributions used in wind energy analysis. Case studies in the Canary Islands. Ren. Sust. Energy Rev. **2009**, *13*, 933–955.

15. Monahan, A.H. Empirical models of the probability distribution of sea surface wind speeds. J. Clim. **2007**, *20*, 5798–5814.

16. Hennessey, J.P. Some aspects of wind power statistics. J. Appl. Meteor. **1977**, *16*, 119–128.

17. Li, M.; Li, X. MEP-type distribution function: a better alternative to Weibull function for wind speed distributions. Ren. Energy **2005**, *30*, 1221–1240.

18. Morrissey, M.L.; Greene, J.S. Tractable analytic expressions for the wind speed probability density functions using expansions of orthogonal polynomials. J. App. Meteor. Clim. **2012**, *51*, 1310–1320.

19. Justus, C.G.; Hargraves, W.R.; Mikhail, A.; Graber, D. Methods for estimating wind speed frequency distributions. J. Appl. Meteor. **1978**, *17*, 350–353.

20. Monahan, A.H.; He, Y.; McFarlane, N.; Dai, A. The probability distribution of land surface wind speeds. J. Clim. **2011**, *24*, 3892–3909.

21. Brown, B.J.; Katz, R.W.; Murphy, A.H. Time series models to simulate and forecast wind speed and wind power. J. Clim. Appl. Meteor. **1984**, *23*, 1184–1195.

22. Johnson, R.A.; Wehrly, T.E. Some angular-linear distributions and related regression models. J. Am. Stat. Assoc. **1978**, *73*, 602–606.

23. Carta, J.A.; Ramírez, P.; Bueno, C. A joint probability density function of wind speed and direction for wind energy analysis. Energy Conversation. Manag. **2008**, *49*, 1309–1320.

24. Brooks, C.E.P.; Durst, C.S.; Carruthers, N. Upper winds over the world. Part I, The frequency distribution of winds at a point in the free air. Q. J. Roy. Met. Soc. **1946**, *72*, 55–73.

25. Crutcher, H.L. On the standard vector-deviation wind rose. J. Meteor. **1957**, *14*, 28–33.

26. Davies, M. Non-circular normal wind distributions. Quart. J. Roy. Meteor. Soc. **1958**, *84*, 277–279.

27. Rice, S.O. Mathematical analysis of random noise (part 2). Bell Syst. Tech. J. **1945**, *24*, 46–156.

28. Weil, H. The distribution of radial error. Ann. Math. Stat. **1954**, *25*, 168–170.

Index

Volume I: Pages 1–588; *Volume II:* Pages 589–1088.
Note: Page references in roman refer to Volume I and page references in italics refer to Volume II.

I-1

Land capability classification (LCC), 296
　capability classes, 300
　capability subclasses, 301
　capability units, 301
　classifications, 299
　historical perspectives, 299–300
Land-change models, 331–332
Land classification, 299
Land conversion, 416
Land cover and land use (LCLU), *621–623*
Land cover change and impervious cover
　environmental impacts, *750*
　extending utility of, *751–752*
　measuring and mapping, *750–751*
Land degradation, 124–125
Land degradation and desertification (LDD)
　causes, 125–127
　drylands, 130–131
　interactions and feedback loops, 128–129
　land users, 127–128
　trends, 129–130
Landfarming, bioremediation, 44
Landform characteristics, coastal zones
　barrier islands, *663*
　beaches, *662–663*
　coral reefs, *664*
　estuaries, *664*
　mangrove forests, *663–664*
　salt marshes, *663*
Landlord ports, *785*
Land plants
　adaptations, 308
　evolutionary relationship, 309–310
　origin, 308–309
Land protection and stewardship
　description, 119–120
　management goals, 120–122
　management plans, 122
　monitoring, 122
　site assessment and baseline inventory, 120
　synthesis, reflection, and adaptive
　　stewardship, 122
Land quality, 295
Landscape
　definition of, 324
　disturbances, 325
　genetics, 243–244
　processes, 325
　succession, 324–325
Landscape connectivity
　conservation, restoration, exotic species
　　management, 320
　definition, 317
　measuring, 318–320
　vs. patch-level connectivity, 317–318
　structural *vs.* functional connectivity, 317
Landscape modeling
　applications, 326
　definition, 326
　functional aspects, 325–326
Landscape-scale ecotone, 189
Land surface temperature (LST)
　definition, 312
　remote sensing
　　physics of, 312–313
　　technical approaches, 313–315
　　validation and evaluation, 315–316

Land systems
　coupled human-environment system, 156
　definition, 156
　ecosystem goods and services, 156–157
　land use change and ecosystem services, 157
LandTrendr, 386
Land-use and land-cover change (LULCC)
　agriculture, 303–304
　biophysical and human-related factors,
　　328–329
　causes and consequences, 328–331
　description, 302, 328
　hydrologic impacts, *892–893*
　hydrologic modeling, *893*
　impacts, *892*
　monitoring and modeling, 331–332
　theoretical foundations, 331
　urban, 304–305
　WWDR4, 302–303
Land use change, 157
Land use controls, surface water, *869*
Land use pattern, 144–145
Lateral dispersion, *721*
LCC, *see* Land capability classification
LCLU, *see* Land cover and land use
LDD, *see* Land degradation and desertification
Leaf transpiration rate, 507–508
Leaves, elevated CO_2 levels
　canopy scale, 340–341
　cellular scale, 338–339
　ecosystem scale, 340–341
　molecular scale, 338–339
　whole leaf scale, 339–340
　whole plant scale, 340
LID, *see* Low-impact development
LiDAR, *see* Light detection and ranging
Light detection and ranging (LiDAR), 101
Livestock management, genetic diversity, 241
Living natural resources, 138
Longitudinal dispersion, *721*
Longshore current, 108
Long-term anomaly trends, *996–998*
Low-impact development (LID)
　characteristics, *763–764*
　definition, *763*
　limitations, *764*
　site design and management objectives
　　conservation measures, *765*
　　distributed integrated management
　　　practices, *765*
　　flow attenuation, *765*
　　goals for, *764–765*
　　minimization techniques, *765*
　　pollution prevention measures, *765*
　stormwater control, *837*
Low-pressure systems, *1037*
LST, *see* Land surface temperature
LULCC, *see* Land-use and land-cover change
Luxury consumption, 461

Macroalgae, *767*
Macrofauna, 183, 455
Madagascar, tropical rain forests, 226
Madden–Julian oscillation (MJO), *1046*
Major biome
　Boreal Forest biome, 91
　definition, 88

Desert biome, 91
　Polar biome, 89, 91
　soil of, 88–90
　Temperate Grassland and Forest biome, 91
　Tropics biome, 91
Major flooding, *1031*
Manatees, marine mammal, *773*
Mangrove ecosystems, 169
Mangrove forests
　description, 343
　ecophysiological challenge
　　hydrologic energy, 346–347
　　hydroperiod, 346
　　integration and spatial scales, 348–349
　　nutrients, 347–348
　　other organisms, 348
　　rainfall, 345
　　redox, 348
　　salinity, 345–346
　　temperature, 344–345
　　wind, 348
　landform characteristics, coastal zones,
　　663–664
Mapping, Riparian wetlands
　ancillary data approach, *828*
　challenges, *826–827*
　mixed method approach, *828*
　on-site mapping, *827*
　quality and standards, *827*
　remote sensing approach, *827–828*
Marine aquaforests, 18
Marine benthic productivity
　abiotic conditions
　　changing conditions, *768–769*
　　resource availability, *768*
　conditions controlling, *767–768*
　loss of habitats, *769*
　macroalgae, *767*
Marine geomorphology, *639*
Marine mammals
　advanced research techniques, *774*
　definition, *772*
　dolphins, *772–773*
　dugongs, *773*
　fur seals, *773*
　manatees, *773*
　phylogenetic discoveries, *773–774*
　sea lions, *773*
　sea otter, *773*
　true seals, *773*
　walruses, *773*
　whales, *772–773*
Marine protected areas (MPAs)
　benefits, *776–777*
　definition, 389, *776*
　fisheries, *777*
　networks, *778*
　reasons for, *776*
　remote sensing, 392
　social benefits, *777*
Marine resources management
　decision processes
　　context, *781–782*
　　instruments, *782–783*
　　means, *782*
　direct use values, *779–780*
　indirect use values, *779–780*